Control Systems

Control Systems

Dr. Veenadevi S.V.
Associate Professor
Dept. of Electronics and Communication Engineering
RV College of Engineering, Bangalore

Mrs. Sujatha Hiremath
Assistant Professor
Dept. of Electronics and Communication Engineering
RV College of Engineering, Bangalore

Mrs. Sowmya Nag K.
Assistant Professor
Dept. of Electronics and Communication Engineering
RV College of Engineering, Bangalore

I.K. International Pvt. Ltd.
NEW DELHI

Published by
I.K. International Pvt. Ltd.
4435-36/7, Ansari Road, Daryaganj
New Delhi-110 002 (India)
E-mail: info@ikinternational.com
Website: www.ikbooks.com

ISBN: 978-93-90620-70-8

Published by Krishan Makhijani for I.K. International Pvt. Ltd., 4435-36/7, Ansari Road, Daryaganj, New Delhi–110 002 and Printed by Rekha Printers Pvt. Ltd., Okhla Industrial Area, Phase II, New Delhi–110 020.

Preface

Control systems is a one semester course of the widely used for the branch of Electrical Sciences. This book is prepared keeping in mind that the students of many branches like undergraduate students (B.E./B.Tech degree) Electrical, Electronics, Instrumentation, Communication, Computer Science, Mechanical, Chemical and Biomedical Engineering of many universities in India. It will also be useful for candidates appearing for AMIE, IETE, GATE, and UPSC Engineering Services entrance examinations. The book provides a comprehensive study of the subject covering basic as well as advanced concepts. It would be equally helpful to practising engineers to understand theoretical aspects of their profession.

The subject matter in each chapter has been developed in a systematic manner with emphasis on basic concepts. The illustrations of concepts done by means of a large number of selected worked examples. Step-by-step procedures for solving problems are provided. Most simplified methods have been provided for solving problems. At the end of each chapter a number of review questions with answers, Gate questions and unsolved problems have been included for practice.

The book begins with a discussion on control systems covering the fundamental concepts: time domain analysis to frequency domain analysis. This book contains the following chapters:

Chapter 1 presents the fundamental and basic concepts of control systems.

Chapter 2 presents the transfer function. It characterizes the input-output relationship of the system. Also deals with mathematical modeling of physical systems including mechanical and electrical systems, force-voltage and force-current analogy systems.

Chapter 3 discusses block diagram representation of systems which simplify the input and output relationship of any complicated system. Also discusses signal flow graph. It represents the system in another pictorial form and is regarded as a simplified version of a block diagram. Mason's gain formula and conversion of block diagram to signal flow graph and vice versa.

Chapter 4 deals with the time response analysis. It is imperative for design and analysis of control systems.

Chapter 5 presents the stability of control systems and discusses Routh-Hurwitz criteria in details to get absolute stability of a system.

Chapter 6 presents Root locus method. The Root locus technique provides knowledge of movement of poles by graphical method in *S*-plane.

Chapter 7 deals with frequency domain analysis which is in analytical method and can be used for analysis and design. Bode plot is the plot of magnitude as well as phase angle versus frequency representing a sinusoidal transfer function. It shows both high frequency and low frequency characteristics of transfer function in a single diagram.

Chapter 8 deals with polar plot. It is possible to plot the magnitude and phase angle of a sinusoidal function on polar co-ordinates varying from 0 to infinity. Also presents Nyquist criteria by which we can find the absolute stability as well as relative stability of any system.

Chapter 9 gives the introduction to compensators and PI, PD, PID controllers.

Chapter 10 is on state variable approach. State variable analysis automatically takes care of initial conditions and it is also possible to analyze time varying or time invariant, linear or nonlinear, single or multiple input-output systems. In conventional control theory, the initial conditions are assumed to be zero and it is applicable to linear time invariant system having single input and output. It also covers the concepts of controllability and absorbability.

Suggestions and comments for further improvement of this book are welcome.

Dr. Veenadevi S.V.
Mrs. Sujatha Hiremath
Mrs. Sowmya Nag K.

Acknowledgements

It is my pleasure to acknowledge the support during the preparation of this book.

We wish to thank the management of Rashtreeya Sikshana Samithi Trust (RSST), Bangalore and the Principal Dr. K.N. Subramanya, RV College of Engineering, Bangalore, for their encouragement while preparing the book.

We also thank our HOD Dr. K.S. Geetha, Prof. & HOD, Electronics & Communication Engineering Department, RV College of Engineering, Bangalore for her constant encouragement and support.

We would like to thank all our colleagues and non teaching staff, ECE dept., RVCE, Bangalore.

Finally, we wish to express our sincere gratitude our family members for their patience and constant inspiration in the preparation of this book.

Dr. Veenadevi S.V.
Mrs. Sujatha Hiremath
Mrs. Sowmya Nag K.

Contents

1 Introduction to Control Systems

1.1 INTRODUCTION TO CONTROL SYSTEMS

Control systems play vital roles in our day-to-day life. The notion of automatic control system is basic and well mixed in every fact of our life. Automatic control systems play important roles in the advancement and improvement of engineering skills. There is a series of implementations of control systems, from a simple bread toaster to a complex power plant, and also in the working of satellites, guided missiles, etc. An automatic control has become an important part of modern manufacturing and industrial processes.

Example: Controlling process, temperature and humidity.

1.1.1 Basic Definitions

System. An arrangement or combination of different physical components that are connected or related together to form an entire unit to achieve a certain objective is called a system. A classroom is an example of a physical system.

Control. The meaning of control is to regulate, direct or command a system so that the desired objective is obtained.

Input. Input is the applied signal or excitation or stimulus or command signal that is applied to a control system to get a specified output or response.

Output. It is the actual response that is obtained from a control system due to the application of the input.

Disturbance. The signal that has some adverse effect on the value of the output of a system is called disturbance. If a disturbance is produced within the system, it is termed an internal disturbance otherwise it is known as an external disturbance.

Control System. It is an arrangement of different physical elements linked or connected in such a manner so as to regulate, direct or command itself to achieve a certain objective. A control system must have input, output and control action. Figure 1.1 shows the relationship between the input and the output.

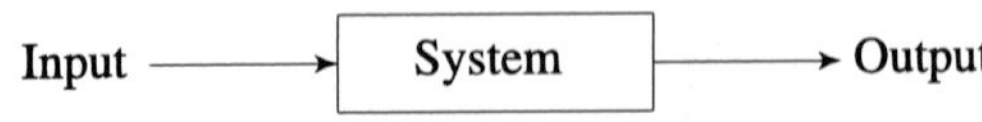

Fig. 1.1

The complexity of the problems of man's existence is growing at an ever increasing rate and it may be expected to continue to do so in future. As such, constantly, human effort and capacity is replaced by machines and control systems. It is indeed not surpising that such replacements perform the job more efficiently than their human counterpart. These machines or processes are not susceptible to so much fatigue or as many blunders and errors as human begins are. Moreover, they may operate at relatively higher speed and precision and can be designed with very little difficulty.

For the past few decades, the trend of the modern civilization has been in the direction of greater control. Its importance has grown tremendously in almost every field of technical endeavour. Thermostats regulate the temperature in air conditioners, refrigerators, ovens and furnaces.

Numerous control arrangements find their ways into industrial and military applications, for the control of quantities such as position, speed, tension, temperature, humidity, pressure flow, etc.

Complex living has demanded automation of large scale problems, thus, for example, transportation problems concerning dispatching, switching and signalling rail roads, reservation system for air passengers, air traffic and landing system, etc., have been partly or fully automated with the limited supply of natural resources and keen competition in allied businesses, one of the burning problems of the day is that of optimization of cost and material. Thus, the electric power generation scheduling is automated to reduce cost. Rocket engineer designs his rocket control arrangement to place the satellite in a desired orbit i.e., send a spacecraft on a mission with a specific payload requiring minimum fuel. Recent efforts in the area of weather prediction, hand written character recognition, medical diagnosis and other types of pattern recognition deal with the most sophisticated type of control systems, known as learning systems.

A system is a coordinated unit of individual elements performing a specific function. It produces an output corresponding to a given input according to some rule. Thus, electrical, mechanical, hydraulic, pneumatic, chemical, analog, digital, etc., or any other elements, devices or processes and/or interacting combinations thereof may be regarded as systems.

1.1.2 Requirement of a Good Control System

- Accuracy: Accuracy should be very high as any error arising should be corrected. Accuracy can be improved by using feedback element.
- Sensitivity: Any control system should be insensitive to environmental changes or internal disturbances or any other parameters.
- Noise: It is an undesired input signal. A good control system should be insensitive to such input signal and able to reduce the effects of undesired inputs.
- Stability: A good control system response is stable for all variations.

- Bandwidth: It defines the operating frequency range in decibels. The bandwidth of any system for frequency response of good control system should be large.
- Speed: A good control system should have high speed. The output of the system should be as fast as possible.

Example 1: Electric switch: controlling the flow of electricity flipping the switch ON or OFF may be considered as the input and output is the flow or non-flow of electricity.

Example 2: Human temperature control system perspiration system: When the temperature of the air exterior to the skin becomes too high the sweat glands secrete heavily, introducing cooling of the skin by evaporation.

Example 3: Person driving an automobile: The driver controls the output by constantly measuring it with his or her eyes and brain, correcting it with his or her hands on the steering wheel. Therefore, the major components are the driver's hands, eyes, brain and the vehicle.

1.2 FEEDBACK

The purpose of feedback is to reduce the error between the reference input and the system output. System without feedback is shown in Fig. 1.2. The error between the system input and output can be reduced by using a feedback system as shown in Fig. 1.5.

The feedback has effects on the system's performance characteristics: stability, bandwidth, overall gain, disturbance and sensitivity.

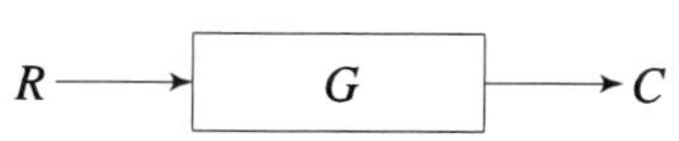

⇒ System without feedback

Fig 1.2

The reduction of system error is merely one of the many important effects that feedback may have upon a system. When feedback is deliberately introduced for the purpose of control, its existence is easily identified. However, there are numerous situations wherein a physical system that normally recognizes as an inherently non-feedback system turns out to have feed back when it is observed in a certain manner.

This viewpoint will inevitably admit feedback in a large number of systems that ordinarily would be identified as a non-feedback systems.

Examples

Thermometer. Consider the action of temperature on a thermometer. We know that the length of the mercury column is directly dependent upon the temperature around the mercury bulb.

We may identify the thermometer as a system whose input $u = \theta$ is the temperature and the output $x = l$ is the length of the mercury thread capillary tube. This is shown in Fig. 1.3.

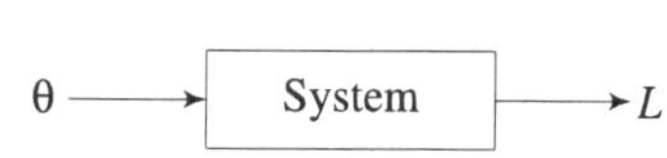

Fig. 1.3

When the input is a step change in temperature, the output length l cannot change immediately to its steady state value.

Human being as a control system. Feedback control system or processes are observed everywhere, in living organisms including human beings, in automatic machines and processes, in organisations, in managements and in society. The most sophisticated and complex feedback control system is that of a human being with certain limitations in speed, accuracy and reliability.

Consider, for instance, the case of a person who is to push a doorbell. We will consider his actions from his decisions to reach the doorbell to the instant he reaches it. The reference input is the coordinates of the position of the doorbell while the position of his particular finger to press the button as the output. His eyes act as the error sensing device which continuously determine the distance of the finger from the doorbell and send via (the controller), i.e., the brain, signals to the power actuator, i.e., the arm which by means of the motor action is lifted in a direction to reduce the error, i.e., the distance between the doorbell and the finger. This action continues till the error is zero. This is shown in the form of a block diagram in Fig. 1.4.

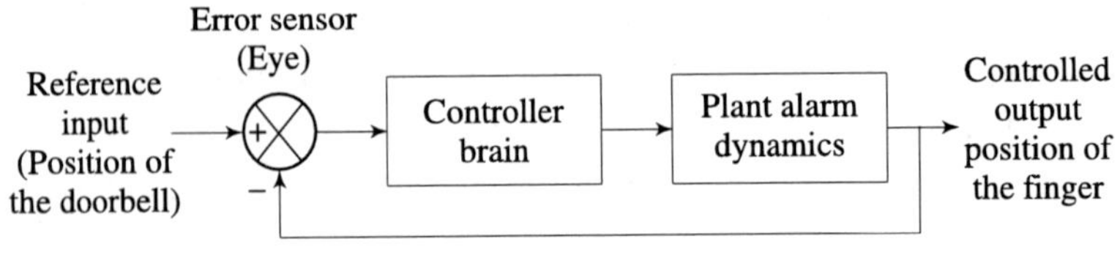

Fig. 1.4

Consider simple feedback system.

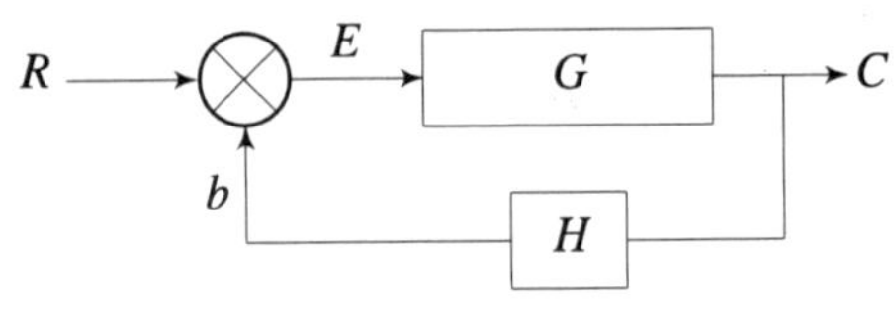

where
R = Input signal
C = Output signal
G & H = Constant gain
b = Feedback signal
E = Error

Fig 1.5

Input and output relation of a system is overall gain (M).

$$M = \frac{C}{R}$$

$$C = GE$$

where $$E = R \pm b$$

where $$b = CH$$

$\therefore$ $$E = R \pm CH$$

$\therefore$ $$C = G(R \pm CH)$$

$$C = RG \pm CGH$$

$$C \pm CGH = RG$$

$$C(1 \pm GH) = RG$$

$$\therefore \quad \frac{C}{R} = \frac{G}{1 + GH} \qquad ...(1.1)$$

Equation (1.1) shows input and output relation of the system with negative feedback.

1.3 FEEDBACK CONTROL EFFECTS OF OVERALL GAIN, STABILITY, SENSITIVITY AND DISTURBANCE

Overall gain. Feedback affects the gain (G) of a system factor ($1 + GH$). In a practical control system, G and H are functions of frequency as in Eq (1.1). So the magnitude of ($1 + GH$) may be greater than one in one frequency range but less than 1 in another. "Feedback could increase the system gain in frequency range but decrease it in another".

Stability. Stability is a notion that describes whether the system will be able to follow the input command. A system is said to be unstable if its output is out of control. To investigate the effect of feedback on stability, the output of the system is infinite for any finite input and the system is said to be unstable. Therefore, we may state that feedback can cause a system that is originally stable to become unstable.

A system is unstable if its output is out of control. To investigate the effect of feedback on stability,

$$M = \frac{G}{1 + GH}, \qquad ...(1.2)$$

In Eq. (1.2), if $GH = -1$, the output of the system is infinite for any finite input and the system is said to be unstable. So that feedback can cause a system that is originally stable to become unstable.

If feedback is used improperly, it can be harmful. If we introduce another feedback loop through a negative feedback gain of F, the input-output relation of the overall system is shown in Eq. (1.3).

$$Y = \frac{G}{1 + GH + GF} \qquad ...(1.3)$$

The properties of G and H are such that the inner loop feedback system is unstable, because $GH = -1$, the overall system can be stable by proper selection of the outer loop feedback gain, F.

Feedback can improve the stability or be harmful to stability if it is not applied properly.

External disturbance or noise. Example of these signals are thermal noise voltage in electronic circuits and brush or commutator noise in electric motors. External disturbances, for example, wind gust acting on an antenna, is also quite common in control system. Therefore, in the design of a control system, considerations should be given so that the system is insensitive to noise and disturbance and sensitive to input commands. The effect of feedback on noise and disturbance depend greatly on these extraneous signals in the system. In many situations feedback can reduce the effect of noise and disturbance on system performance sensitivity.

All physical systems are subject to some types of extraneous signals or noise during operation.

Feedback can reduce the effect of noise and disturbance on system performance.

Sensitivity. Sensitivity considerations often are important in the design of control systems. We cannot always consider the parameter of a control system to be completely stationary over the entire operating life of the system. The winding resistance of an electric motor changes as the temperature of the motor rises during operation.

All physical elements have properties that change with environment and age. Feedback can increase or decrease the sensitivity of a system.

In general, a good control system should be very insensitive to parameter variation but sensitive to input commands. Let us investigate what effect feedback has on the sensitivity to parameter variations. We consider G to be a gain parameter that may vary. The sensitivity of the gain of the overall system, M, to the variation in G is defined as

$$S_G^M = \frac{\partial M/M}{\partial G/G} = \frac{\text{Percentage change in } M}{\text{Percentage change in } G}$$

where ∂M denotes the incremental change in M due to the incremental change in G, ∂G

$$S_G^M = \frac{\partial M}{\partial G}\frac{G}{M} = \frac{1}{1 + GH}$$

This relation shows that if GH is a positive constant, the magnitude of the sensitivity function can be made arbitrarily small by increasing GH, provided that the system remains stable.

It is apparent that in an open-loop system, the gain of the system will respond in a one to one fashion to the variation in G, i.e., $S_G^M = 1$. In practice, GH is a function of frequency. The magnitude of $1 + GH$ may be less than unity over some frequency ranges, so that feedback could be harmful to the sensitivity to parameter variations in certain cases. In general, the sensitivity of the system gain of a feedback system to parameter variations depends on where the parameter is located.

1.4 CLASSIFICATION OF CONTROL SYSTEMS

Control systems may be classified in a number of ways depending on the purpose of classification.

1. Linear and nonlinear control systems.
2. Analog and digital control systems.
3. Open-loop and closed-loop control systems
4. Single input and single output control systems
5. Multiple input and multiple output control systems.

1. Linear and Nonlinear Control Systems

A linear control system satisfies the principle of superposition and homogeneity.

Using Ohm's law, the excitation or input is the current I, and the response or output is the voltage V. When the element is subjected to a current I_1, it provides a response V_1.

Again, when the element is subjected to a current I_2 it provides a response V_2. If the principle of superposition is true, then the excitation $(I_1 + I_2)$ must provide a response $(V_1 + V_2)$ as shown in Eq. (1.4). Also, it is necessary that the magnitude scale factor be preserved for a linear element. If the element is subjected to an excitation βI, where β is a constant multiplier, then if principle of homogeneity is true, the response of the element must be βV as shown in Eq. (1.5).

$$f(x + y) = f(x) + f(y) \quad \text{...(1.4)}$$

$$f(\beta x) = \beta f(x) \quad \text{...(1.5)}$$

If any control system satisfies Eq. 1.4 (principle of superposition) and Eq. (1.5) (homogeneity) such system is known as a linear system. If it does not satisfy, then such system is known as a nonlinear system.

2. Analog and Digital Control Systems

Analog system. In an analog system the input signals that appear in the system are continuous in nature which represents physical measurements.

Digital system. In a digital system the input signals are discrete or discontinuous values to represents information.

3. Open-loop and Closed-loop Control Systems

Open-loop Control System. A system in which the control action is totally independent of the output of the system. Any physical system which does not automatically corrects the variation in its output, in which the output quantity has no effect upon the input quantity. This means that the output is no feedback to the input for correction.

Examples:

- *Electric hand driver.* Control action has nothing to do with the status of the output.
- *Bread toaster.* Toaster simply heats the bread for some time.
- Automatic coffee/milk server.

Advantages of Open-loop System

- They are simple in construction and design.
- They are economical (cheaper).

- Easy for maintenance.
- Not much problems of stability (generally they are stable).

Disadvantages of Open-loop System

- *Inaccurate.* Results are obtained with parameters, variations, internal disturbances.
- Unreliable.
- Recalibration of the controller is required for maintaining quality and accuracy.

Closed-loop Control System. A system in which the control action is dependent on the output or changes in output is called a closed-loop system. The output has an effect upon the input quantity in such a manner as to maintain the desired output value. The open-loop system can be modified as closed-loop system by providing a feedback. The provision of feedback automatically corrects the changes in output due to disturbances. Hence, closed-loop system is also called automatic control system.

Examples:

- *Automatic electric iron.* Regulates the temperature. For a given cloth select the temperature in a specified range, when the temperature falls below the range, heating is started and when it exceeds the range heating is stopped.
- Voltage stabilizer accepts variable voltage input and gives fixed voltage as output.
- *Walking on the road.* Feedback brain.

Advantages

- Accuracy is very high as any error arising is corrected.
- *Insensitivity.* Environmental or parametric changes, internal disturbances.
- Higher bandwidth.

Disadvantages

- Complicated in design and maintenance costlier.
- Stability is a major problem in closed-loop system and more care is needed to design a stable closed-loop system.
- The feedback in closed-loop system may lead to oscillatory response.

A man walking with closed eyes: Consider a man walking with his eyes closed. If, suppose an obstacle is present in his way, he will not be able to detect it. He will not be able to divert his path and he will hit the obstacle, as his eyes as closed, i.e., there is no feedback signal from his eyes to the brain.

Example: Speed control of a turntable by open-loop control system

In the given open-loop system, no feedback signal is given from the turntable to the speed setter.

The speed control of the turntable is done only by varying the input voltage. Any change in the speed setting changes the turntable's speed. The desired speed is not affected by any change in the systems output thereby making the output independent of the input. The input voltage is varied by external power, i.e, manpower, the turntable is made to run at desired speed

1.5 LINEAR VERSUS NONLINEAR CONTROL SYSTEM

A system is said to be linear, if the principle of superposition applies. The principle of superposition states that the response produced by the simultaneous applicable of two different function is equal to the sum of the two individual responses. Hence for the linear system, the response to several inputs, can be calculated by treating one input at a time and adding the result. Linear systems do not exist in practice, since all physical systems are nonlinear to some extent. Linear feedback control systems are idealized models fabricated by the analyst purely for the simplicity of analysis and design. When the magnitudes of signals in a control system are limited to ranges in which system components exhibit linear characteristics, the system is essentially linear, but when the magnitudes of signals are extended beyond the range of the linear operation, depending on the severity of the nonlinearity, the systems should no longer be considered linear. Quite often, nonlinear characteristics are intentionally introduced in a control system to improve its performance, provide more effective control.

For linear systems, there is a wealth of analytical and graphical techniques for design and analysis purposes. Nonlinear system on the other hand are usually difficult to treat mathematically and no general methods are available for solving a wide class of nonlinear control systems.

Time invariant versus time varying control systems when the parameters of a control system are stationary with respect to time during the operation of the system, the system is called a time invariant system. If the parameters of a control system vary with respect to time during the operation of the system, the system is called a time varying system. In practice, most physical systems contain elements that drift or vary with time. Example, the winding resistance of an electric motor will vary when the motor is first being excited and its temperature is rising The analysis and design of linear time varying systems are usually much more complex than that of the linear time invariant systems.

Continuous data versus discrete data control systems. A continuous data control system is one in which the signals at various parts of the system are all functions of the continuous time variable and among all continuous data control systems, the signals may be further classified as AC or DC. An AC control system usually means that the signals in the system are modulated by Some form of modulation scheme A dc control system simply implies that the signal are unmodulated, but they are still signals according to the conventional definition.

A system may incorporate a mixture of ac and dc components using modulators and demodulators to match the signals at various points in the system. Discrete data control system differ from the continuous data control systems in that the signals at one to more points of the system are in the form of either a pulse train or digital code. Usually, discrete data control system are subdivided into sampled data control system and digital control systems in sampled data control system refer to a more general class of discrete data control system, in which the signals are in the form of pulse data. A digital control system refers to the use of a digital computer or controller in the system so that the signals are digitally coded such as in binary code. In general, a sampled data control system receives data or information only intermittently at specific instants of time. A sampled data system can also be classified as an ac system, since the signal of the system is pulse modulated. Because digital computer provide many advantages in size and flexibility, computer control has became increasingly popular in recent years.

Comparison of open-loop and closed-loop control systems.

Open-loop System	Closed-loop System
No feedback	Feedback exists
No error detector	Error detector is present
Inaccurate	Accurate
Highly sensitive to parameter changes	Less sensitive to parameter changes
Small bandwidth	Large bandwidth
Cheap (economical)	Costly

Examples

The electric switch is an open-loop system, because the control action is equal to the input and therefore independent of the output.

Automatic washing machine, human walking on the road and automatic bread toaster are closed-loop systems, the control action is clearly a function of the output.

4. Single Input Single Output (SISO). If the single input variable is applied to the system and the output variable is also single then the system is called SISO.

5. Multi Input Multi Output (MIMO). If the input and output variables are more than one then the system is called MIMO system.

1.6 SERVOMECHANISMS

A servomechanism is a power amplifying feedback control system in which the controlled variable or the output is a mechanical position or its time derivatives such as velocity or acceleration. The motors that are used in automatic control systems are called servomotors. When the objective of the system is to control the position of an object then the system is called servomechanism. Servomotors are used to convert an electrical signal (control voltage) applied to them into an angular displacement of the shaft. They can either operate continuous duty or step duty depending on their construction.

There are varieties of servomotors available for control system applications. The suitability of a motor for a particular application depends on the characteristics of the system, the purpose of the system and its operating conditions. In general, a servomotor should have the following features:

1. Linear relationship between the speed and electric control signal
2. Steady state stability
3. Wide range of speed control
4. Linearity of mechanical characteristics throughout the entire speed range

5. Low mechanical and electrical inertias
6. Fast response

Depending on the supply required to run the motor, they are broadly classified as DC servomotors and AC servomotors. The DC motors are more expensive than AC motors. DC servomotors have linear characteristics and so are easier to control.

The advantages of DC servomotors are the following:

1. Higher output than from a 50 Hz motor of same size.
2. Linearity of characteristics is achieved easily.
3. Easier speed control from zero speed to full speed in both directions.
4. High torque to inertia ratio that gives them quick response to control signals.
5. DC motors are light weight, have low inertia and low inductance armature that can respond quickly to commands for a change in position or speed.
6. Low electrical time constants and low mechanical time constants.
7. DC motors are capable of delivering over 3 times their rated torque for a short time but AC motors start at 2 to 2.5 times their rated torque.

Advantages of AC motors are low cost, higher efficiency and less maintenance since there is no commutator and brush. The disadvantage of AC motor is that characteristics are quite nonlinear and these motors are more difficult to control especially for positioning applications.

The AC motors are best suited for low power applications such as instrument servo and computer related equipment. The three-phase induction motors with pulse width modulated power amplifier are currently gaining popularity in high power control application.

EXERCISE

Short Answer Questions

1. Define the following
   (i) System (ii) Control system (iii) Input (iv) Output
2. Differentiate between linear and nonlinear control systems.
3. Explain classification of control systems.
4. Define open-loop and closed-loop control systems. Also explain advantages and disadvantages with an example.
5. State the effects of feedback on the control system.
6. What do you mean by servomechanism?
7. What are the major types of control systems?
8. What is feedback? What type of feedback is employed in control systems?
9. Define the following terms
   (i) System (ii) Control system (iii) Input (iv) Output
   (v) Disturbance

10. Define open-loop and closed-loop control systems. Mention their merits and demerits.
11. What is feedback? Explain the effects of feedback?
12. Differentiate between
    (i) linear and nonlinear systems
    (ii) continuous and discrete data systems
13. What are the limitations of open-loop systems over closed-loop system? List the advantages of closed-loop system over open-loop system.
14. Explain servomechanism with suitable examples.

Fill in the Blanks

1. The principle of homogeneity and superposition is applied to __________.

Ans: linear time invariant system.

2. Human system is an example of __________.

Ans: multi-variable feedback control system.

3. An automatic iron is an example of __________.

Ans: single feedback control system.

4. The principle of homogeneity and superposition is applied to __________.

Ans. linear time-invariant system

5. In an open-loop control system __________ of the variables have any effect on the input signal.

Ans. none

6. Sprinkler the is an example __________.

Ans. an open-loop control system

7. Human system is __________.

Ans. a multi-variables feedback control system.

True or False

8. Tachometer feedback in a DC position control system enhances stability.

Ans. True

The tachometer acts as a derivative controller, which reduces the type number of the system and hence improves stability.

9. The transfer function (T.F.) of a linear system is the ratio of output, $V_O(t)$, and input, $V_i(t)$.

Ans. False

The T.F. of a linear system is the ratio of Laplace transform of the output and that of the input with all initial conditions zero.

10. A positive feedback signal improves the performance of an automatic control system

Ans. True

Unsolved Questions

11. What is the effect of negative feedback in a closed-loop control system?

Ans. It increases bandwidth, reduces gain and improves distance rejection.

12. Define the following:

(i) System (ii) Plant (iii) Controller (iv) Input
(v) Output (vi) Control system (vii) Disturbance

Ans. (i) *System:* An arrangement or a combination of different physical components which are connected or related together to form an entire unit to achieve a certain objective is called a system.

(ii) *Plant:* It is defined as the portion of a system which is to be controlled or regulated. It is also called a process

(iii) *Controller:* It is the element of the system itself, or may be external to the system. It controls the plant or the process.

(iv) The applied signal or excitation signal that is applied to a control system to get a specified output is called *input.*

(v) The actual response that is obtained from a control system due to the application of input is termed *output.*

(vi) *Control system:* It is an amalgamation of different physical elements linked in a manner as to regulate, direct or command the process itself to obtain a certain objective.

(vii) *Disturbance:* The signal has some adverse effect on the value of the output of a system is called disturbance. If disturbance is produced within the system, it is called internal disturbance otherwise it is known as an external disturbance.

13. What do you mean by servomechanism?

Ans. A servomechanism is power amplifying feedback control system in which the controlled variable or the output is a mechanical position or its time derivatives such as velocity or acceleration.

14. Define the following control systems:

(i) Time-invariant (ii) Time-variant (iii) Continuous (iv) Discrete
(v) Deterministic (vi) Stochastic

Ans. (i) If the parameters of a control system do not vary with time, it is termed a time-invariant control system.

(ii) If the parameters of a control system vary with time, this control system is termed a time-varying control system. Such systems do not depend on the nature of *i/ps* and *o/ps*.

(iii) If all the system variables of a control system are functions of time, it is termed a continuous time control system.

(iv) If one or more system variables of a control system are known certain discrete time, it is termed a discrete–time control systems.

(v) If the response to input and to external disturbances of a control system is predictable and repetitive, the control system is known as deterministic.

(vi) If the response to input, and to external disturbance, of a control system is not predictable and repetitive, the control system is said to be stochastic.

15. Define linear and nonlinear control systems.

Ans. *Linear system:* A control system is called linear if it satisfies the additive property as well as the homogeneous property.

Nonlinear: A control system is called nonlinear if it does not satisfy both the additive property and the homogeneous property.

16. Define open-loop and closed-loop control systems.

Ans. *Open-loop system:* A system in which control action does not depend on output is known as an open-loop system.

Closed-loop system: If controlling actions of a system are somehow dependent on output or changes in output, this system is called a closed-loop system.

17. State the effects of feedback.

Ans. Feedback effects are

(i) There is reduction of parameter variations by a factor $1 + G(s)H(s)$.

(ii) There is improvement in sensitivity.

(iii) There may be reduction in stability.

(iv) Gain is reduced by a factor $\dfrac{1}{G(s)H(s) + 1}$.

18. State the applications of open-loop and closed-loop control systems.

Ans. Applications of open-loop control systems:

- Automatic washing machines
- Bread toaster
- Electric hair drier
- Electric lift
- Traffic signals etc.

Applications of closed-loop systems:

- Water level controller
- Temperature control system
- Voltage stabilizer
- Electric iron
- Railway reservation status.

19. Pick the linear system among the following:

(i) $\dfrac{d^2y(t)}{dt^2} + 8\dfrac{dy(t)}{dt} + y(t) = u(t)$ (ii) $y \cdot \dfrac{dy(t)}{dt} + 7\dfrac{dy(t)}{dt} + 8u(t)$

(iii) $3\dfrac{d^2y(t)}{dt^2} + 2t\dfrac{dy(t)}{dt} + 3t^2\, y(t) = 6$

Ans. (i) $\dfrac{d^2y(t)}{dt^2} + \dfrac{8\, dy(t)}{dt} + y(t) = u(t)$

20. Pick the nonlinear systems from the following:

(i) $\dfrac{dy(t)}{dt} + 8y(t) = 24(t)$ (ii) $\dfrac{d^2y(t)}{dt^2} + 6\dfrac{dy(t)}{dt} + 8\ y(t) = 3u(t)$

(iii) $t\left[\dfrac{d^2y(t)}{dt^2}\right] + 2\dfrac{dy(t)}{dt} + 3y(t) = 2u(t)$

(a) (i) and (ii) (b) (ii) only (c) (i) and (iii) (d) (iii) only

Ans. (c) (i) and (iii)

21. Consider the following statement regardings a linear system $y = f(x)$

(1) $f(x_1 + x_2) = f(x_1) + f(x_2)$ (2) $f[x(t + T)] = f(x(t) + x(T)]$

(3) $f(Kx) = 3f(x)$

Of these statements

(a) 1, 2 and 3 are correct (b) 1 and 2 are correct

(c) 3 is correct (d) 1 and 3 are correct

Ans. (d) 1 and 3 are correct

2 Mathematical Models of Physical Systems

2.1 INTRODUCTION

Control system is a collection of physical objects (components) connected together to serve an object. The input-output relations of various physical components of a system are governed by differential equations. The mathematical model of a control system constitutes a set of differential equations. The response or output of the system can be studied by solving the differential equations for various input conditions.

2.2 MATHEMATICAL MODELS OF PHYSICAL SYSTEMS

The process of obtaining the desired mathematical description of the system is known as modeling. The set of mathematical equations describing the behaviour of a system is called mathematical modeling of the system.

A mathematical model is not unique to a given system. A system may be represented in many different ways and therefore many mathematical models depend on the requirement.

Once mathematical model of a system is obtained, various analytical and computer methods can be used for analysis and synthesis purposes. Most of the control systems contain mechanical or electrical or both types of elements. The mathematical modeling will be linear if the differential equations describing the system are constant coefficients or the coefficients may be functions of independent variables. If the coefficient of the differential equation describing the system is constant then the model is linear time invariant. If the coefficients of the differential equations governing the system are functions of the time then model is linear time varying.

2.3 TRANSFER FUNCTION

The relationship between input and output of a system is given by the transfer function.

It is defined as the ratio of the Laplace transfer of the output (or response) to the Laplace transfer of the input (or excitation) under the assumption of zero initial conditions.

$$\text{Transfer function} = \frac{\text{Laplace transform of output}}{\text{Laplace transform of input}}, \text{ with zero initial conditions.}$$

From the differentiation property of the LT

$$L\left[\frac{di(t)}{dt}\right] = [SI(s) - i(0^-)]$$

$$= [SI(s)] \text{..........Assuming initial condition is zero.}$$

$$L\left[\frac{d^2i}{dt^2}\right] = [S^2I(s) - S\,i(0^-) - \frac{dx}{dt}(0^-)]$$

$$L\frac{d^2i}{dt^2} = [S^2I(s)] \text{................ Assuming initial conditions is zero.}$$

Transfer function can be represented in three forms.

1. Polynomial form

$$T = \frac{bmS^m + b_{m-1}S^{m-1} + b_{m-2}S^{m-2} + \ldots b_0}{a_nS^n + a_{n-1}S^{n-1} + a_{n-2}S^{n-2} + \ldots a_0}$$

$$= \frac{C(S)}{R(S)}$$

2. Pole zero form

$$T = K\frac{(S - b_1)(S - b_2)(S - b_3) \ldots (S - b_n)}{(S - a_1)(S - a_2)(S - a_3) \ldots (S - a_n)}$$

$$= \frac{\text{No of Zero}}{\text{No of Pole}}$$

where K is the gain factor.

3. Time constant form

$$T = \frac{K(TaS + 1)(TbS + 1)(TcS + 1)}{S^N (TaS + 1)(TbS + 1)(TcS + 1)}$$

2.4 MODELING OF MECHANICAL SYSTEM

The motion of a mechanical system may be translational, rotational or combination of both. Translational motion is the motion that takes place along a straight line. Rotational motion is the motion about a fixed axis. For a mechanical system, input may be forces and the output or response may be displacement or velocity.

2.4.1 Translational Motion

In this, the systems are characterized by linear displacement, linear velocity and linear acceleration. The equations of motion are formulated from Newton's 2nd law of motion. It states that the sum of applied forces is equal to the sum of opposing forces on a body. That is, the algebraic sum of the forces acting as a rigid body in a given direction is equal to the product of the mass of the body and its acceleration in the same direction.

It can be expressed as

$$\sum \text{Forces} = ma = M\frac{dv}{dt} = M\frac{d^2x}{d^2t}$$

f = applied force, in newtons

m = mass of the body, in kg

v = linear velocity of the body, in m/sec $= \dfrac{dx}{dt}$

x = linear displacement of the body in metre

a = acceleration of the body in the direction considered $\dfrac{dv}{dt} = \dfrac{d^2x}{dt^2}$ m/sec^2

The model of mechanical translational systems can be obtained by using three basic elements, i.e.,

(i) Mass
(ii) Spring
(iii) Friction, or dashpot (damper).

These three elements represent three essential phenomena which occur in various ways in mechanical systems. Spring and the mass are energy storage elements. Damper is an energy absorbing element.

Mass (M). The weight of the mechanical system is represented by the element mass and it is assumed to be concentrated at the centre of the body.

In the physical model of a mass, it is assumed that its mass is concentrated at its centre. The function of mass in linear motion is to store kinetic energy.

Displacement. Action of moving something from its place or position.

Velocity. The speed of something in a given direction.

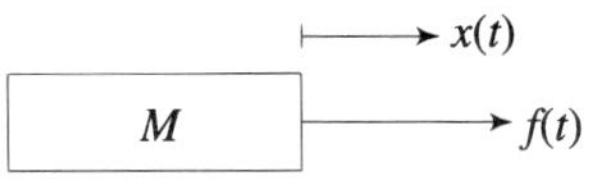

Suppose that a force $f(t)$ is applied to mass. The applied force produces displacement $x(t)$ in the direction of the applied force. The relation between force $f(t)$ and M is given by

$$f = Ma = M\frac{dv}{dt} = M = \frac{d^2x}{dt^2}$$

2.4.2 Spring

The elastic deformation of the body can be represented by a spring. If it is stretched, it tries to contract and if it is compressed it tries to expand to its normal length. It stores potential energy during the variation of its shape due to elastic deformation resulting from the application of a force.

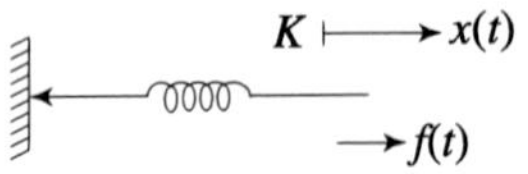

$f \,\alpha$ displacement

$f(t) = K.x\,(t)$

where K = Spring constant

Suppose that the spring has displacement at both ends as shown below.

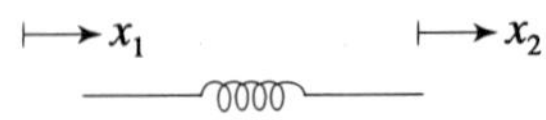

The force is applied at x_1, therefore, net displacement in the spring $(x_1 - x_2)$.

$$f \,\alpha\, (x_1 - x_2)$$

$$f = K(x_1 - x_2)$$

2.4.3 Friction (Damper/Dashpot)

The friction existing in mechanical system can be represented by damper. The damper is a piston moving inside a cylinder filled with viscous fluid. When there is a motion between physical elements frictional forces exist. There are 3 types of friction in practical systems: viscous friction, static friction and Coulombs friction.

Viscous friction. This is the force of friction between moving surfaces separated by viscous fluid. The damping or viscous friction B characterizes the mechanical element that absorbs energy.

It is necessary to introduce damping to improve the dynamics of the system. Damping occurs between a piston and a cylinder.

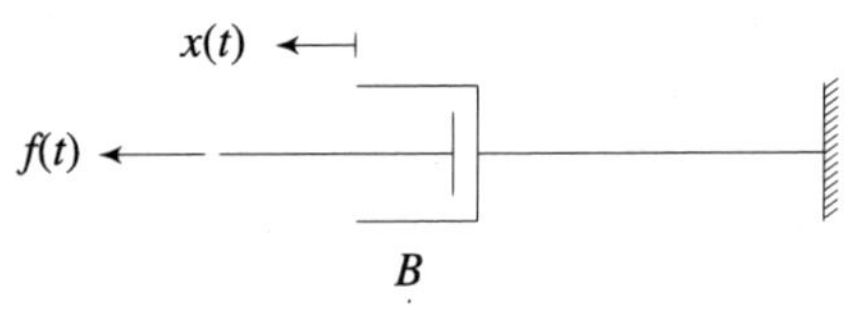

When a force f is applied, damping force which is proportional to the velocity.

$f_B = B\, dx/dt$ $\qquad$ B = damping coefficient

by Newton's law $f = f_B$

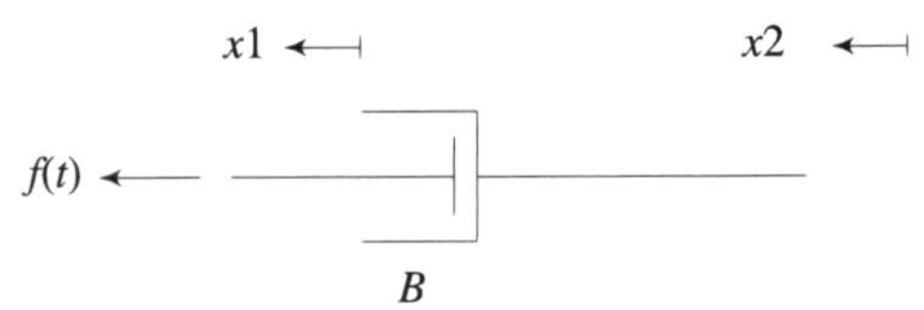

$$f = B\frac{dx}{dt}$$

$$f \,\alpha\, \frac{dx_1}{dt} - \frac{dx_2}{dt}$$

$$f = B\left(\frac{dx_1}{dt} - \frac{dx_2}{dt}\right)$$

$$f = f_B$$

$$f \,\alpha\, \frac{dx_1}{dt} - \frac{dx_2}{dt}$$

The mathematical modeling of a mechanical system is obtained by finding the transfer function. It can be obtained by two methods.

Method I: Free body diagram

In order to obtain the mathematical model of a mechanical system, the 1st step is to draw the free body diagram. The force acting on the mass that causes displacement.

Procedure:

1. Assume the direction of displacement of the mass as positive direction.
2. Find all the forces with their directions acting on the mass. The applied force is in the positive direction of displacement. The forces due to spring, damper and mass are opposite to the direction of displacement.
3. Apply Newton's laws of motion.

Once the free body diagram is obtained, the differential equation can be written easily by equating the sum of applied forces to the sum of opposing forces.

Example: Obtain the differential equations for the mechanical system shown in Fig. 2.1. Find the transfer function for the system.

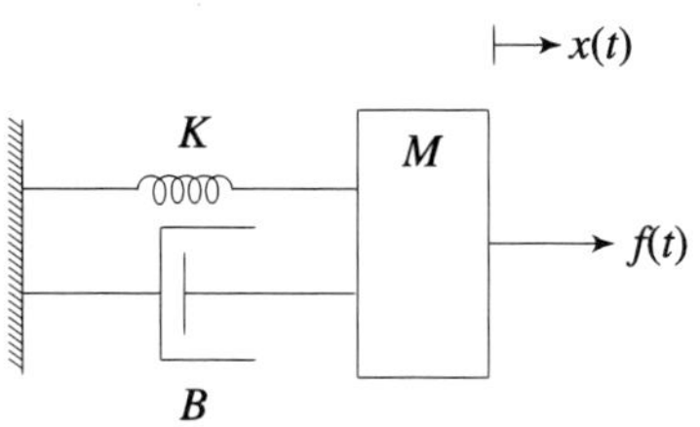

Fig. 2.1

Free body diagram

$$f(t) = M\frac{d^2x(t)}{dt^2} + B\frac{dx(t)}{dt} + Kx(t)$$

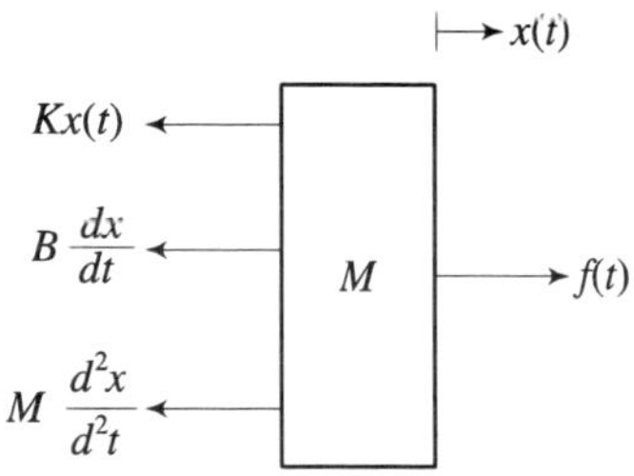

Fig. 2.2

Applying Laplace Transform (LT) on both sides of above equation, assuming zero initial conditions.

$$F(s) = MS^2\, x(s) + BS\, x(s) + K\, x(s)$$

$$F(s) = X(s)\,[MS^2 + BS + K]$$

$$\frac{X(s)}{F(s)} = \frac{1}{MS^2 + BS + K}$$

Method II: Nodal method of developing mechanical equivalent network

1. Identify all the displacements in the mechanical system.
2. Total number of displacements is equal to the total number of nodes.
3. Choose an arbitrary node as a reference node.

4. Connect all mass/inertia elements between the relevant displacement node and the reference node irrespective of its position.
5. Connect spring and damper elements between relevant nodes based on their position.
6. Connect force between the relevant displacement node and reference node.
7. At each of the displacement node apply Newton's laws of motion.
8. Identify the elements which are under the influence of different displacements.
9. Connect all the elements under the respective nodes which are under the influence of respective displacement.

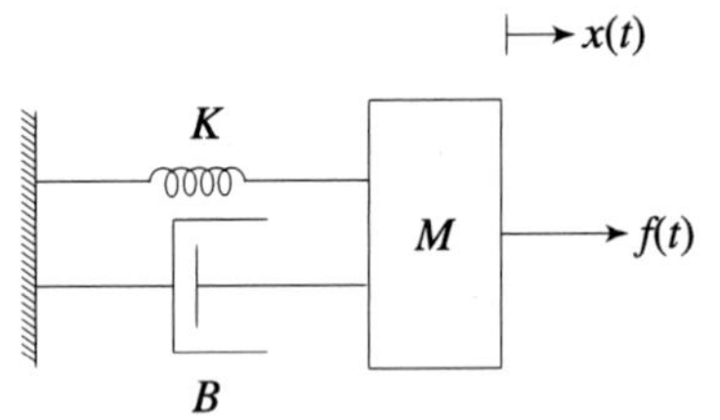

Fig. 2.3

$$f(t) = M\frac{d^2x(t)}{dt^2} + Kx(t) + B\frac{dx(t)}{dt}$$

Applying LT

$$F(s) = MS^2x(s) + Kx(s) + BSx(s)$$

$$\frac{x(s)}{F(s)} = \frac{1}{MS^2 + BS + K}$$

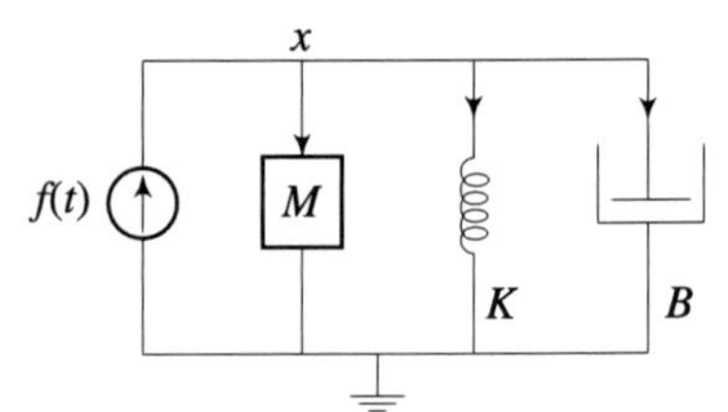

Fig. 2.4

Example 2.1: Draw the free body diagram and write the differential equation describing the dynamics of the system shown in Fig. 2.5 and obtain the transfer function, $\frac{X_2(s)}{F(s)}$.

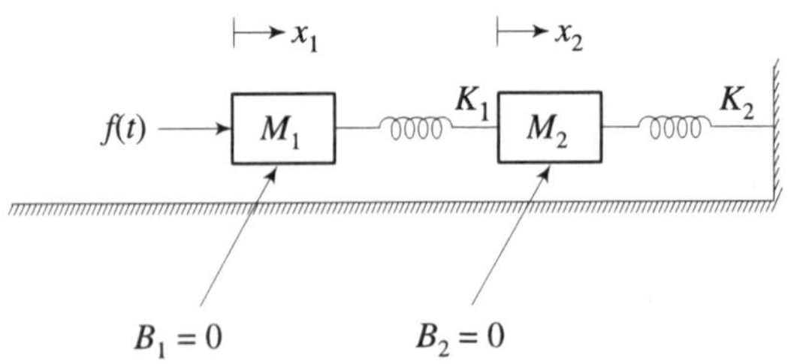

Fig. 2.5

Solution: **Method I** Free body diagram for masses M_1 and M_2

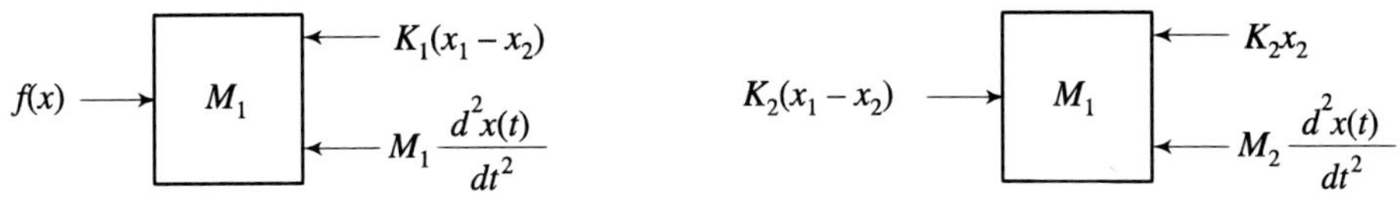

Fig. 2.6

Force equation for mass M_1 is

$$f(t) = M_1 \frac{d^2x(t)}{dt^2} + K_1\ (x_1 - x_2) \qquad ...(2.1)$$

For mass M_2 is

$$K_1(x_1 - x_2) = M \frac{d^2x(t)}{dt^2} + K_2x_2 \qquad ...(2.2)$$

LT of Eqs. (2.1) and (2.2)

$$F(s) = S^2M_1x_1(s) + K_1x_1(s) - K_2x_2(s) \qquad ...(2.3)$$

$$K_1x_1(s) - K_1x_2(s) = S^2M_2x_2(s) + K_2x_2(s) \qquad ...(2.4)$$

From Eq. (2.4)

$$K_1x_1(s) = S^2M_2x_2(s) + K_2x_2(s) + K_1x_2(s)$$

$$X_1(s) = (S^2\ M_2 + K_1 + K_2)\ \frac{x_2(s)}{K_1} \qquad ...(2.5)$$

Substitute Eq. (2.5) in Eq. (2.3),

$$F(s) = (S^2\ M_1 + K_1)\ (S^2\ M_2 + K_1 + K_2)\ \frac{x_2(s)}{K_1} - K_1x_2(s)$$

$$\frac{x_2(s)}{K_1} = \frac{K_1}{(S^2\ M_1 + K_1)(S^2\ M_2 + K_1 + K_2) - K_1^{\ 2}}$$

Method II:

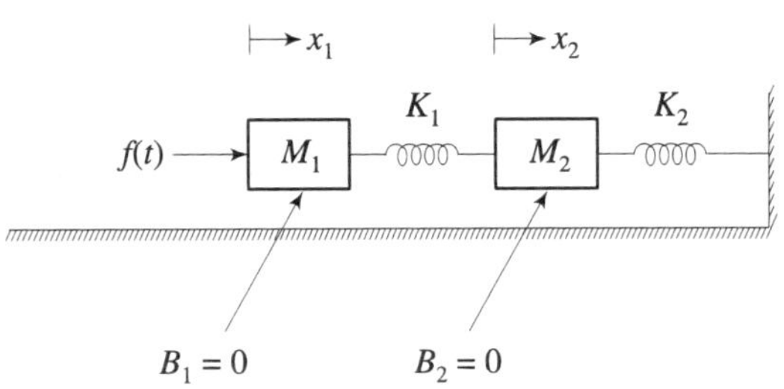

Fig. 2.7

Solution:

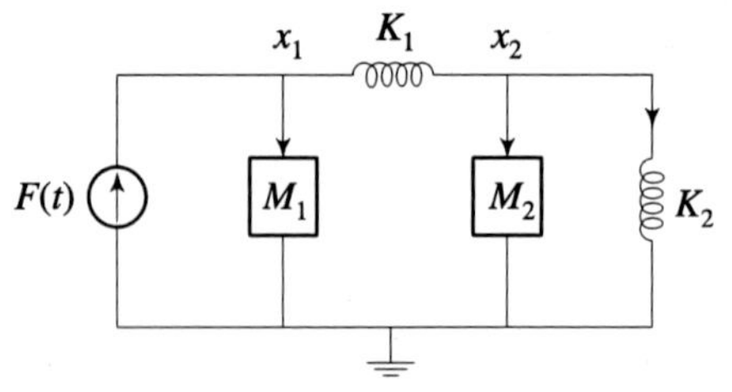

Fig. 2.8

At node x_1

$$f(t) = M_1 \frac{d^2x_1(t)}{dt^2} + K_1(x_1 - x_2)$$

Apply LT for the above equation

$$F(s) = M_1S^2x_1(s) + K_1(x_1(s) - x_2(s))$$

$$F(s) = M_1S^2x_1(s) + K_1x_1(s) - K_1x_2(s) \quad ...(2.6)$$

At node x_2

$$M_2 \frac{d^2x_2}{dt^2} + K_2x_2 = K_1(x_1 - x_2)$$

Apply LT

$$M_2S^2x_2(s) + K_2x_2(s) = K_1(x_1(s) - x_2(s))$$

$$K_1x_1(s) = S^2M_2x_2(s) + K_1x_2(s) + K_1x_1(s)$$

$$x_1(s) = (S^2M_2 + K_1 + K_2)\frac{x_2(s)}{K_1} \quad ...(2.7)$$

Substitute Eq. (2.7) in Eq. (2.6)

$$F(s) = (M_1S^2 + K_1)(S^2M_2 + K_1 + K_2)\frac{x_2(s)}{K_1} - K_1x_2(s)$$

From the above equation,

$$\frac{x_2(s)}{F(s)} = \frac{K_1}{(S^2M_1 + K_1)(S^2M_2 + K_1 + K_2) - K_1^{\,2}}$$

Example 2.2: For the mechanical system shown in Fig. 2.9, write the differential equations.

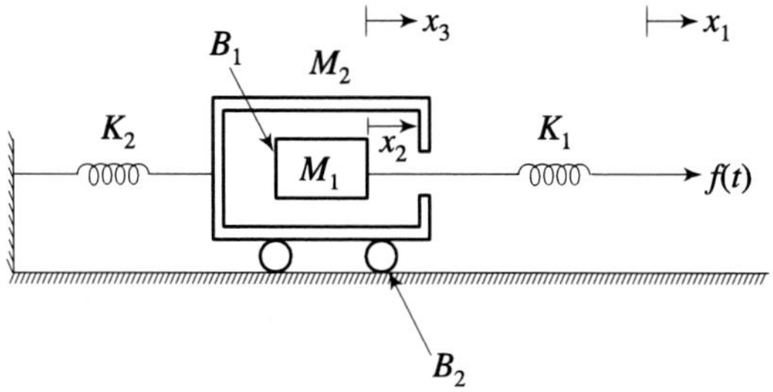

Fig. 2.9

Solution: In this example force is applied to element K_1. There is no mass still assuming x_1 to be the displacement.

Rewrite the Fig. 2.10

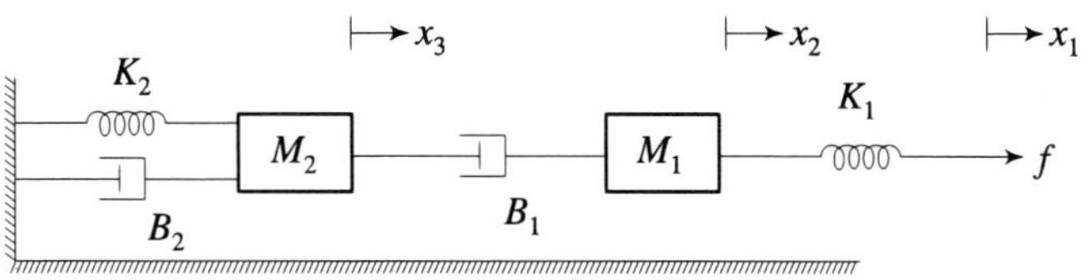

Fig. 2.10(a)

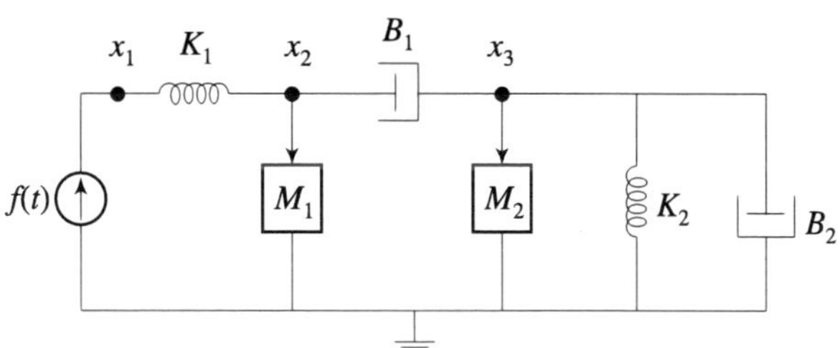

Fig. 2.10(b)

At node x_1

$$f(t) = K_1\,(x_1 - x_2)$$

At node x_2

$$M_1\,\frac{d^2x_2}{dt^2} + B_1\,\frac{d(x_2 - x_3)}{dt} = K_1\,(x_1 - x_2)$$

At node x_3

$$M_2\,\frac{d^2x_3}{dt^2} + B_2\,\frac{dx_3}{dt} + K_2\,x_3 = B_1\,\frac{d(x_2 - x_3)}{dt}$$

Example 2.3: Determine the transfer function $\frac{x_o(s)}{x_i(s)}$ for the system shown in Fig. 2.11.

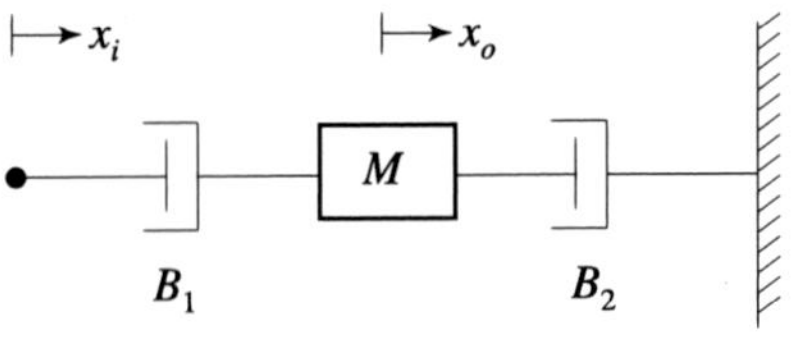

Fig. 2.11

Solution:

$$B_1\left(\frac{dx_i}{dt} - \frac{dx_o}{di}\right) = M\frac{d^2x_o}{dt^2} + B_2\frac{dx_o}{dt}$$

$$B_1\frac{dx_i}{dt} = M\frac{d^2x_0}{dt^2} + (B_1 + B_2)\frac{dx_0}{di}$$

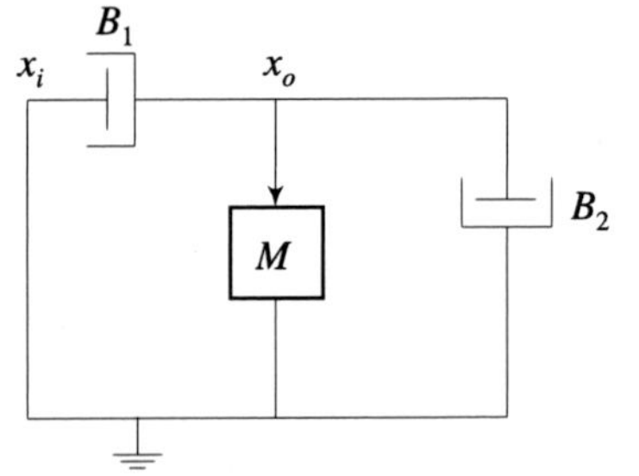

Fig. 2.12

Applying LT in the above equations,

$$B_1Sx_i(s) = MS^2x_o(s) + (B_1 + B_2)\,Sx_o(s)$$

$$B_1Sx_i(s) = x_o(s)[MS^2 + (B_1 + B_2)s]$$

$$\frac{x_o(s)}{x_i(s)} = \frac{B_1S}{MS^2 + (B_1 + B_2)s}$$

$$\frac{x_o(s)}{x_i(s)} = \frac{B_1}{MS^2 + B_1 + B_2}$$

Example 2.4: Write the force equations of the mechanical system shown in Fig. 2.13.

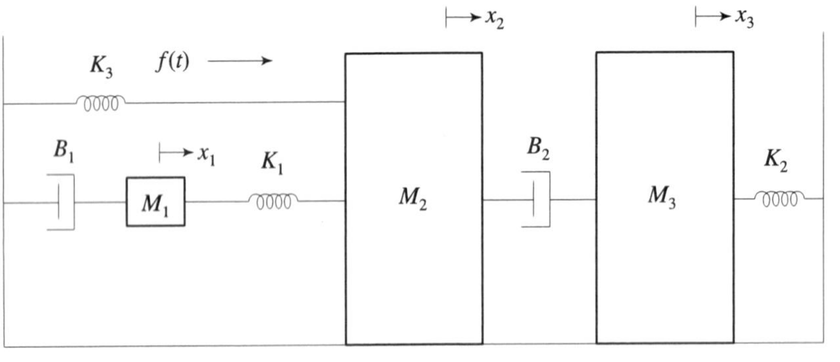

Fig. 2.13

Solution:

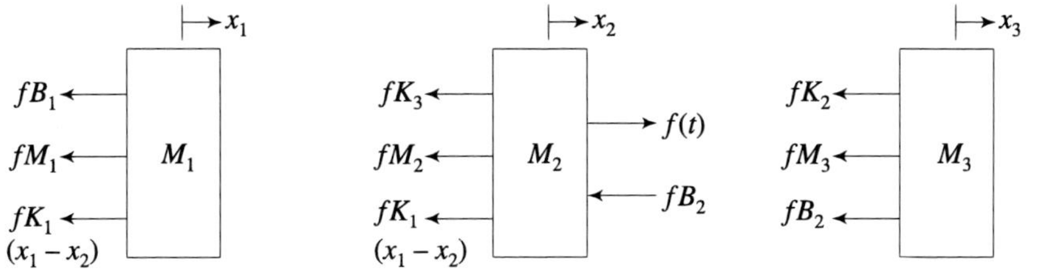

Fig. 2.14

At M_1 or x_1

$$M_1 \frac{d^2x_1}{dt^2} + B_1 \frac{dx_1}{dt} + K_1(x_1 - x_2) = 0$$

At x_2

$$f(t) + K_1 (x_1 - x_2) = M_2 \frac{d^2x_2}{dt^2} + K_3x_2 + B_2 \left(\frac{dx_2}{dt} - \frac{dx_3}{dt}\right)$$

At x_3

$$B_2 \left(\frac{dx_2}{dt} - \frac{dx_3}{dt}\right) = M_3 \frac{d^3x_3}{dt^2} + K_2x_3$$

Example 2.5: Write system equation for the system in shown in Fig. 2.15.

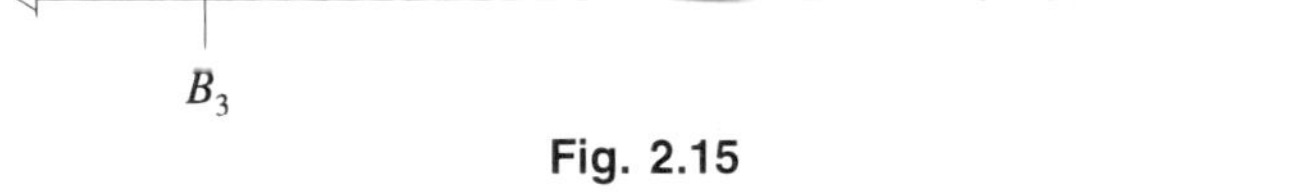

Fig. 2.15

Solution:

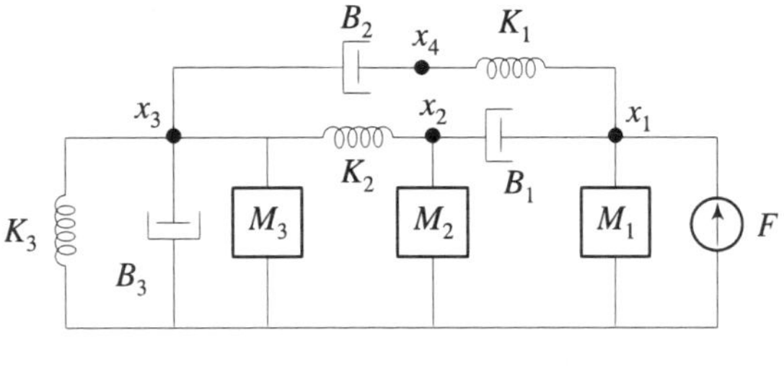

Fig. 2.16

At x_1,

$$F = M_1 \frac{d^2x_1}{dt^2} + B_1 \frac{d(x_1 - x_2)}{dt} + K_1(x_1 - x_4)$$

At x_2,

$$B_1 \frac{d(x_1 - x_2)}{dt} = M_2 \frac{d^2x_2}{dt^2} + K_2 (x_2 - x_3)$$

At x_3,

$$M_3 \frac{d^2x_3}{dt^2} + B_3 \frac{dx_3}{dt} + K_3x_3 + K_2 (x_3 - x_2) + B_2 \frac{d(x_3 - x_4)}{dt} = 0$$

At x_4,

$$K_1 (x_4 - x_1) + B_2 \frac{d(x_4 - x_3)}{dt} = 0$$

Example 2.6: For the mechanical system shown in Fig. 2.17, write the differential equations.

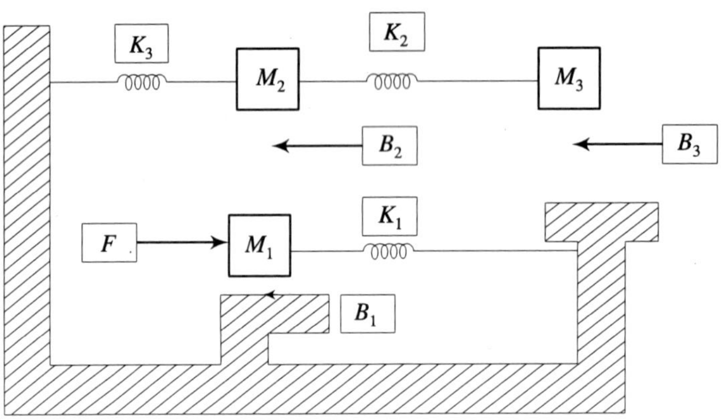

Fig. 2.17

Solution:

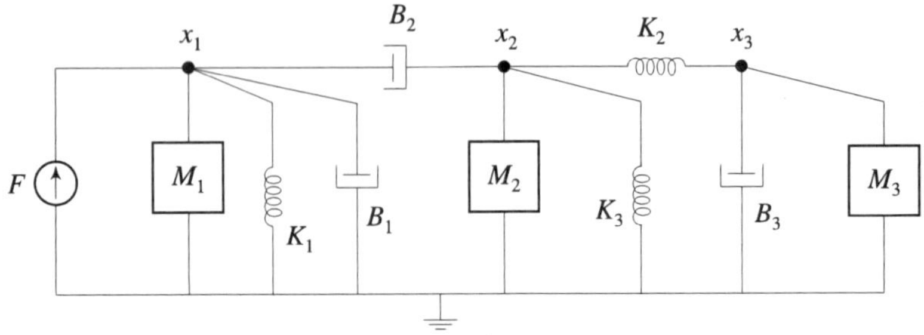

Fig. 2.18

At node x_1,

$$F = M_1 \frac{d^2x_1}{dt^2} + K_1x_1 + B_1 \frac{dx_1}{dt} + B_2 \frac{d(x_1 - x_2)}{dt}$$

At node x_2,

$$M_2 \frac{d^2x_2}{dt^2} + K_3x_2 + K_2(x_2 - x_3) + B_2 \frac{d(x_2 - x_1)}{dt} = 0$$

At node x_3,

$$M_3 \frac{d^2x_3}{dt^2} + B_3 \frac{dx_3}{dt} + K_2 (x_3 - x_2) = 0$$

Example 2.7: For the mechanical network shown in Fig. 2.19, write the system equations.

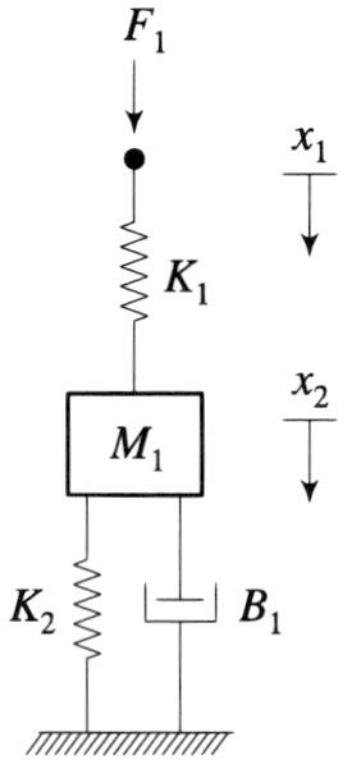

Fig. 2.19

Solution:

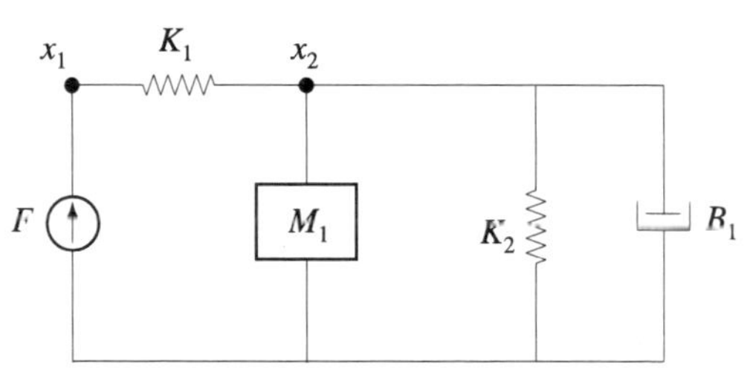

Fig. 2.20

At x_1,

$$F = K_1(x_1 - x_2)$$

At x_2,

$$M_1 \frac{d^2x_2}{dt^2} + K_2x_2 + B_2 \frac{dx_2}{dt} + K_1(x_2 - x_1) = 0$$

$$K_1(x_1 - x_2) = M_1 \frac{d^2x_2}{dt^2} + K_2x_2 + B_2 \frac{dx_2}{dt}$$

$$F = M_1 \frac{d^2x_2}{dt^2} + K_2x_2 + B_2 \frac{dx_2}{dt}$$

2.4.4 Rotational Mechanical System

It is same way as translational mechanical systems except that torque replaces force, angular displacement replaces translational displacement, inertia replaces mass. In rotational system, components undergo rotation instead of translation.

2.4.4.1 Inertia Torque (T_J)

The torque applied to a body having a moment of inertia J produces an angular acceleration. The reaction torque T_J is equal to the product of moment of inertia and acceleration.

The torque equation is

$$T_J = J\,\alpha = J\frac{dw}{dt} = J\frac{d^2\theta}{dt^2}$$

θ = angular displacement

w = angular velocity

$\propto$ = angular acceleration

2.4.4.2 Spring Torque (T_K)

When a torque is applied to a spring, the spring is twisted by an angle θ.

The reaction torque is given by

$$T_K = K(\theta_c - \theta_d)$$

where c and d are the ends of the spring. The angles θ_c and θ_d denote the positions of the two ends of the spring measured from the neutral position.

2.4.4.3 Reaction Damping Torque (T_B)

Damping occurs whenever a body move through a fluid, which may be either a liquid or a gas such as air

$$T_B = B(w_e - w_f) = B\left(\frac{d\theta_e}{dt} - \frac{d\theta_f}{dt}\right)$$

2.4.5 Analogous Quantities in Translational and Rotational Motions

Translational Motion	Rotational Motion
Mass (M)	Inertia (J)
Damper coefficient (B)	Damping constant (B)
Spring constant (K)	Spring constant (K)
Force (F)	Torque (T)
Displacement (x)	Angular displacement (θ)

Velocity $(v) = v = \dfrac{dx}{dt}$ Angular velocity $(w)\ w = \dfrac{d\theta}{dt}$

Acceleration (α) Angular acceleration (α)

Example 2.8: For the mechanical system shown in Fig. 2.21, write equilibrium equations.

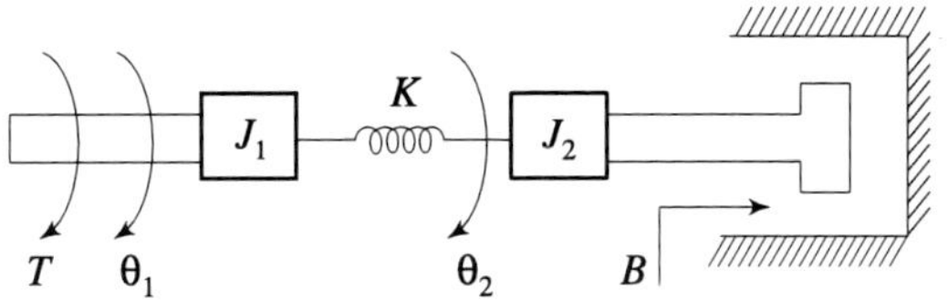

Fig. 2.21

Solution:

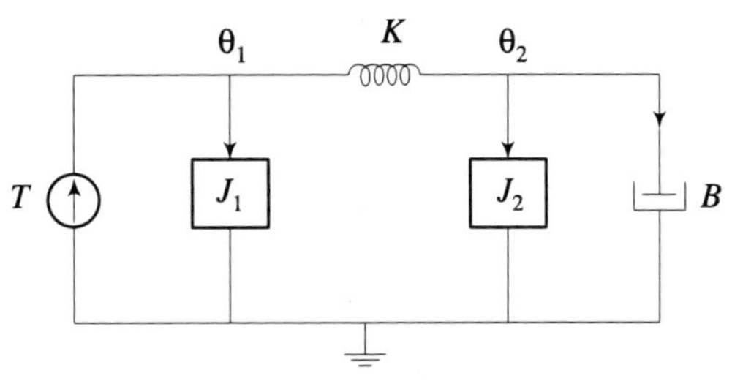

Fig. 2.22

At θ_1,

$$T = J_1 \frac{d^2\theta_1}{dt^2} + K(\theta_1 - \theta_2)$$

At θ_2,

$$K(\theta_1 - \theta_2) = J_2 \frac{d^2\theta_2}{dt^2} + B\frac{d\theta_2}{dt}$$

Example 2.9: The mechanical system consisting of two discs which have damping between them and also between each of them is as shown in Fig. 2.23. For this system write the performance equation.

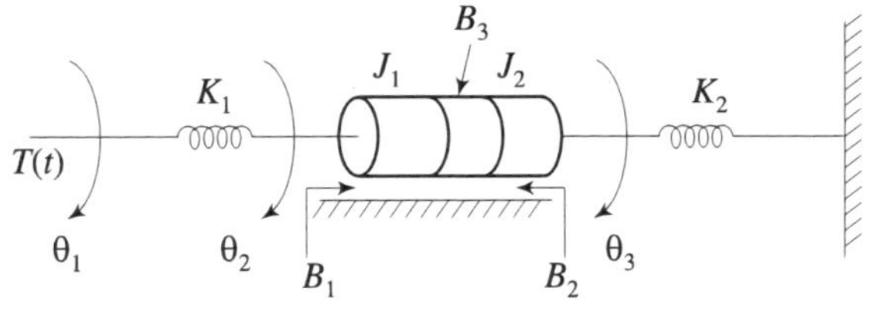

Fig. 2.23

Solution:

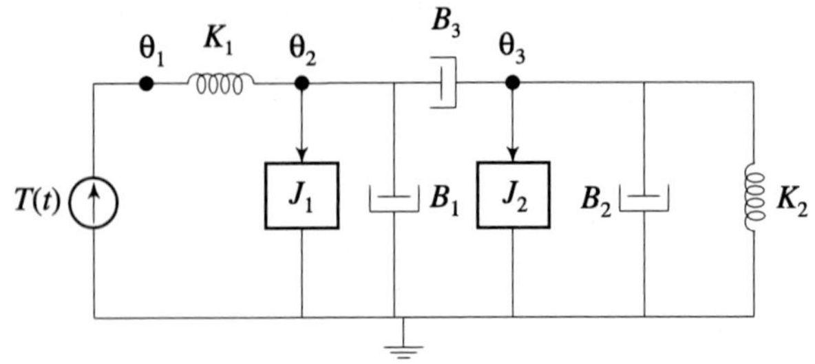

Fig. 2.24

At θ_1,

$$T(t) = K(\theta_1 - \theta_2)$$

At θ_2,

$$K(\theta_1 - \theta_2) = J_1 \frac{d^2\theta_2}{dt^2} + B_1 \frac{d\theta_2}{dt} + B_3 \left(\frac{d\theta_2}{dt} - \frac{d\theta_3}{dt}\right)$$

At θ_3,

$$B_3 \left(\frac{d\theta_2}{dt} - \frac{d\theta_3}{dt}\right) = J_2 \frac{d^2\theta_3}{dt^2} + B_2 \frac{d\theta_3}{dt} + K_2\theta_3$$

2.5 MODELING OF ELECTRICAL SYSTEM

The models of electrical systems can be obtained by using resistor, inductor, capacitor and voltage or current source. The differential equations governing the electrical systems can be formed by writing Kirchhoff's current law equations by choosing various nodes in the network or Kirchhoff's voltage law equations by choosing various closed path in the network. For an electrical system, inputs are in the form of voltage and current sources, output or response is voltage or current at certain points.

The transfer function can be obtained by taking Laplace transform of the differential equations and rearranging them as a ratio of output to input.

Example 2.10: Determine the transfer function if $V_i(t)$ is the input to the system and $V_o(t)$ is the output of the system as shown in Fig. 2.25.

Apply KVL, we can write

$$V_i(t) = R_i(t) + \frac{1}{c}\int i(t)dt \qquad ...(2.8)$$

$$V_i(t) = \frac{1}{c}\int i(t)\, dt \qquad ...(2.9)$$

Applying Laplace Transform of Eqs. (2.8) and (2.9)

$$V_I(s) = R_I\,(s) + \frac{1}{SC}\,I(S)$$

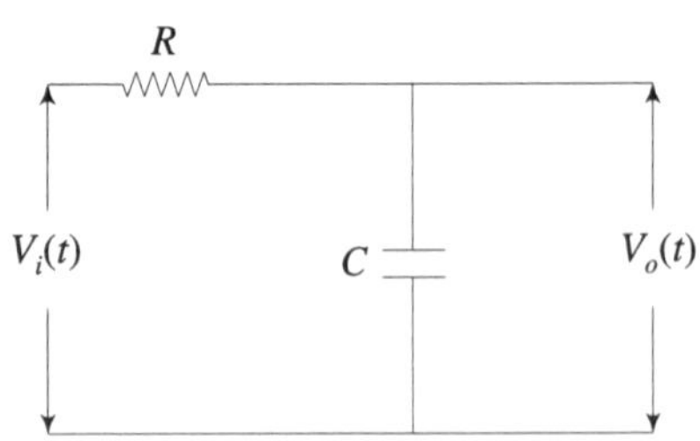

Fig. 2.25

$$V_I(s) = I(s)\left[R + \frac{1}{SC}\right]$$

$$V_I(s) = I(s)\left[\frac{sRc + 1}{SC}\right]$$

$$I(s) = \left[\frac{V_I(s)SC}{1+ sRc}\right] \quad ...(2.10)$$

$$V_o(s) = \frac{1}{SC} I(s) \quad ...(2.11)$$

Substituting Eq. (2.10) in Eq. (2.11)

$$V_o(s) = V_I(S)\left(\left[\frac{SC}{1 + sRc}\right]\right)\left(\frac{1}{SC}\right)$$

$$\frac{V_o(s)}{V_I(s)} = \left(\left[\frac{1}{1 + sRc}\right]\right)$$

Example 2.11: Determine the transfer function of the network shown in Fig. 2.26.

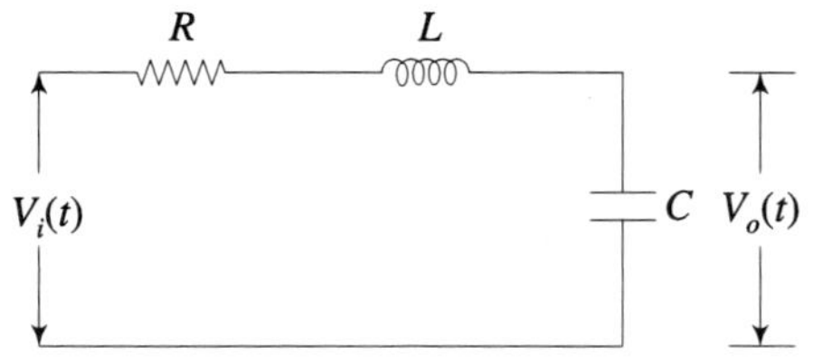

Fig. 2.26

Solution: Using KVL, it can be written as

$$V_i(t) = Ri(t) + L\frac{di(t)}{dt} + \frac{1}{C}\int i(t)dt$$

and
$$V_o(t) = \frac{1}{C}\int i(t)dt$$

Taking Laplace Transform for the above equation, we can write,

$$V_I(s) = RI(s) + SL\, I(s) + \frac{I(s)}{SC}$$

$$V_o(s) = \frac{I(s)}{SC}$$

All the initial conditions have been put to zero

$$I(s) = SC\, V_o\,(s)$$

$$V_I(s) = \left(R + SL + \frac{1}{SC}\right) I(s) = \left(R + SL + \frac{1}{SC}\right) SC\, V_o(s)$$

$$V_I(s) = (S^2LC + SCR + 1)\, V_o(s)$$

$$\frac{V_o(s)}{V_I(s)} = \frac{1}{S^2LC + SCR + 1}$$

Transfer function of the network is given by

$$T(s) = \frac{1}{S^2LC + SCR + 1}$$

Example 2.12: Derive the transfer function of the network shown in Fig. 2.27.

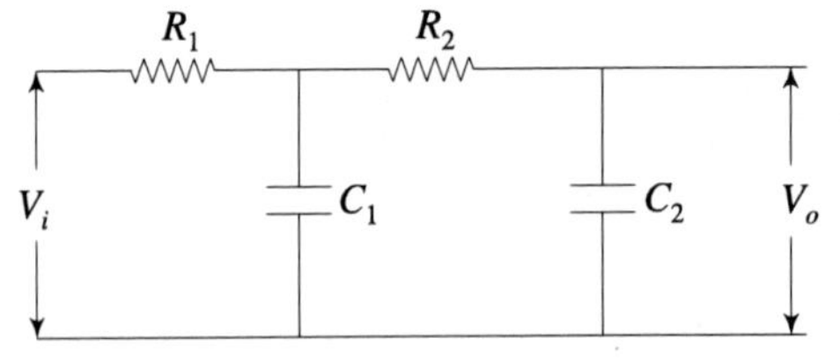

Fig. 2.27

Solution: Redraw the circuit in Laplace domain as shown in Fig. 2.28.

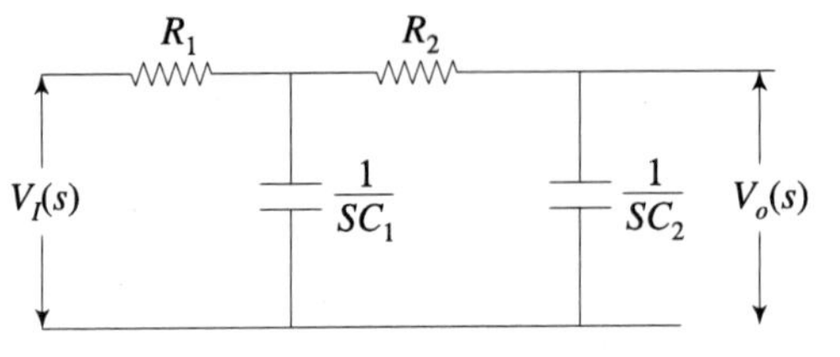

Fig. 2.28

Apply KVL, we get

$$V_I(s) = \left(R_1 + \frac{1}{SC_1}\right) I_1(s) - \frac{1}{SC_1} I_2(s)$$

$$\left(R_2 + \frac{1}{SC_1} + \frac{1}{SC_2}\right) I_2(s) - \frac{1}{SC_1} I_1\,(s) = 0$$

$$\frac{1}{SC_1} I_1(s) = \left(R_2 + \frac{1}{SC_1} + \frac{1}{SC_2}\right) I_2(s)$$

$$I_1(s) = SC_1\left(R_2 + \frac{1}{SC_1} + \frac{1}{SC_2}\right) I_2(s)$$

$$V_I(s) = \left(R_1 + \frac{1}{SC_1}\right) SC_1\left(R_2 + \frac{1}{SC_1} + \frac{1}{SC_2}\right) I_2(s) - \frac{1}{SC_1} I_2(s)$$

$$V_I(s) = \left[\frac{(SC_1R_1 + 1)(SR_2C_1C_2 + C_1 + C_2) - C_2}{SC_1C_2}\right] I_0(s)$$

$$V_o(s) = \frac{I_2(s)}{SC_2}$$

$$V_I(s) = \frac{(SC_1R_1 + 1)(SR_2C_1C_2 + C_1 + C_2) - C_2}{C_1} V_o(s)$$

$$\frac{V_o(s)}{V_I(s)} = \frac{C_1}{(SC_1R_1 + 1)(SR_2C_1C_2 + C_1 + C_2) - C_2}$$

$$= \frac{C_1}{S^2C_1^2R_1R_2C_2 + SC_1^2R_1 + SR_1C_1C_2 + SR_2C_1C_2 + C_1 + C_2 - C_1}$$

$$= \frac{1}{S^2C_1R_1R_2C_2 + SC_1R_1 + SR_1C_2 + SR_2C_2 + 1}$$

$$T(s) = \frac{1}{S^2C_1R_1R_2C_2 + S(R_1C_1 + R_1C_2 + R_2C_2) + 1}$$

Example 2.13: Determine transfer function $\frac{V_o(s)}{V_I(s)}$ of the circuit shown in Fig. 2.29.

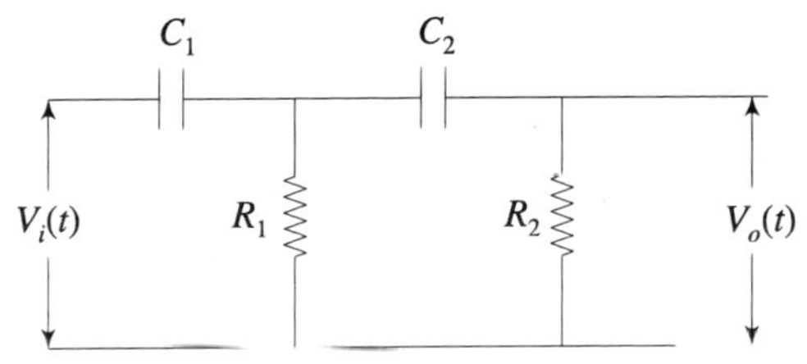

Fig. 2.29

Solution: The circuit shown in Fig. 2.29 when redrawn in Laplace domain is as shown in Fig. 2.30.

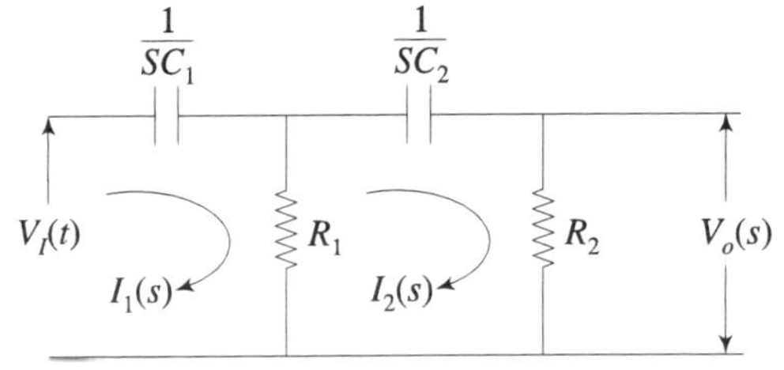

Fig. 2.30

Apply KVL to the loop

$$V_I(s) = \left(R_1 + \frac{1}{SC_1}\right) I_1(s) - R_1 I_2(s)$$

$$\left(R_1 + R_2 + \frac{1}{SC_2}\right) I_2(s) = R_1 I_1(s)$$

$$I_1(s) = \left(R_1 + R_2 + \frac{1}{SC_2}\right) I_2(s)$$

$$V_I(s) = \left(R_1 + \frac{1}{SC_1}\right)\frac{1}{R_1}\left(R_1 + R_2 + \frac{1}{SC_2}\right) I_2(s) - R_1 I_2(s)$$

$$= \frac{(1 + SC_1R_1)\,(1 + SC_2(R_1 + R_2))}{S^2R_1C_1C_2} I_2(s) - R_1 I_2(s)$$

$$= I_2(s)\left[\frac{1 + SC_2(R_1 + R_2) + SC_1R_1 + S^2C_1R_1C_2(R_1 + R_2) - S^2C_1C_2R_1}{S^2R_1C_1C_2}\right]$$

$$= \left[\frac{S^2C_1C_2R_1R_2 + S(R_1C_1 + R_2C_2 + R_1C_2) + 1}{S^2R_1C_1C_2}\right] I_2(s)$$

$$V_0(s) = R_2 I_2(s)$$

$$I_2(s) = \frac{V_o(s)}{R_2}$$

Using these equations

$$V_I(s) = \left[\frac{S^2C_1C_2R_1R_2 + S(R_1C_1 + R_2C_2 + R_1C_2) + 1}{S^2R_1R_2C_1C_2}\right] V_o(s)$$

or

$$\frac{V_o(s)}{V_I(s)} = \left[\frac{S^2R_1R_2C_1C_2}{S^2R_1R_2C_1C_2 + S(R_1C_1 + R_2C_2 + R_1C_2) + 1}\right]$$

Example 2.14: Derive the transfer function of the network shown in Fig. 2.31.

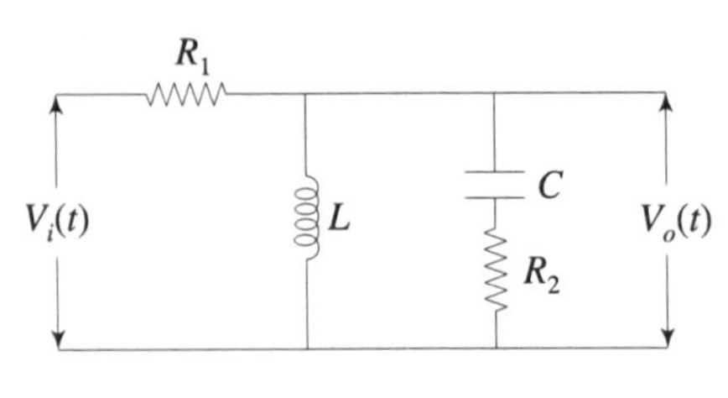

Fig. 2.31

Solution: The circuit shown in Fig. 2.31 redrawn in Laplace domain is as shown in Fig. 2.32.

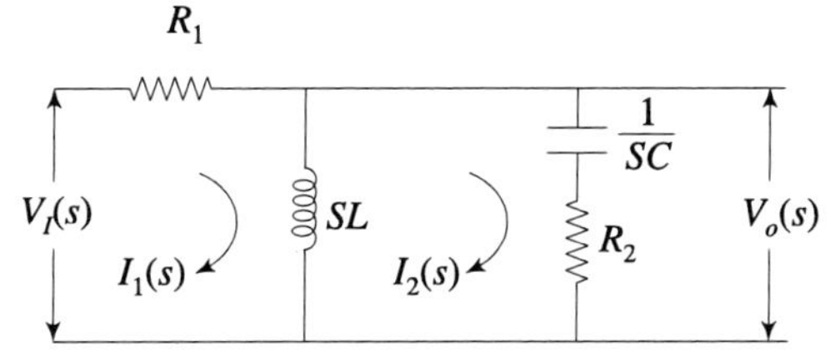

Fig. 2.32

Applying KVL

$$V_I(s) = (R_1 + SL)\, I_1(s) - SLI_2(s)$$

$$\left(SL + \frac{1}{SC} + R_2\right) I_2(s) - SLI_1(s) = 0$$

$$I_1(s) = \frac{1}{SL}\left(SL + \frac{1}{SC} + R_2\right)$$

$$V_I(s) = \left[\frac{(R_1 + SL)(SL + \frac{1}{SC} + R_2)}{SL} - SL\right] I_2(s)$$

$$V_I(s) = \left[\frac{S^2R_1LC + R_1R_2 + R_1 + S^3L^2C + S^2LCR_2 + SL - S^3L^2C}{S^2LC}\right] I_2(s)$$

$$V_I(s) = \left(\frac{S^2R_1LC + S^2LCR_2 + SL + R_1R_2 + R_1}{S^2LC}\right) I_2(s)$$

$$V_o(s) = \left(R_2 + \frac{1}{SC}\right) I_2(s)$$

$$I_2(s) = \left(\frac{SC}{1+SR_2C}\right) V_o(s)$$

$$\therefore \quad V_I(s) = \left(\frac{S^2R_1LC + S^2LCR_2 + SL + R_1R_2 + R_1}{S^2LC}\right)\left(\frac{SC}{1 + SR_2C}\right) V_o(s)$$

$$= \left[\frac{S^2LC(R_1 + R_2) + SL + R_1(R_2 + 1)}{SL(1 + SR_2C)}\right] V_o(s)$$

$$\frac{V_o(s)}{V_I(s)} = \frac{SL(1 + SR_2C)}{S^2LC(R_1 + R_2) + SL + R_1(R_2 + 1)}$$

Example 2.15: Find the transfer function of the electrical network shown in Fig. 2.33.

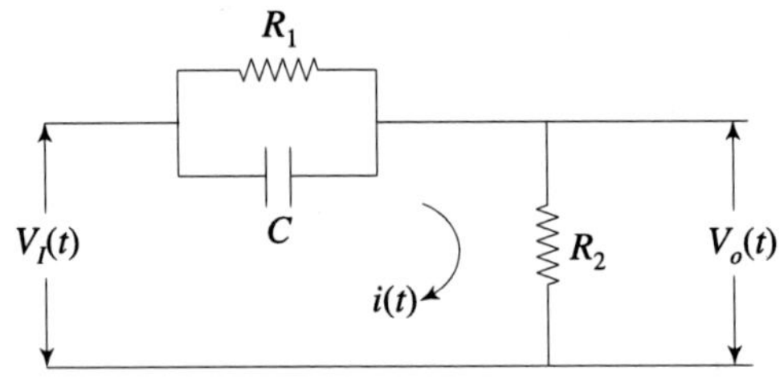

Fig. 2.33

Solution: The circuit shown in Fig. 2.33 is redrawn in Laplace domain is as shown in Fig. 2.34.

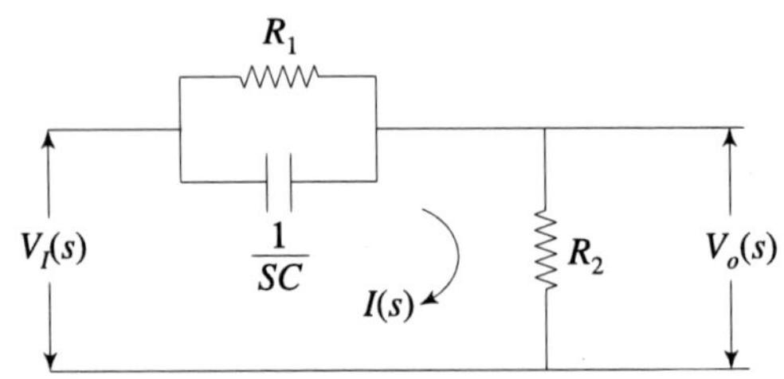

Fig. 2.34

Using nodal analysis, it can be written as

$$\frac{V_I(s) - V_o(s)}{R_1} + \frac{V_I(s) - V_o(s)}{\frac{1}{SC}} = \frac{V_o(s)}{R_2}$$

$$\left(\frac{1}{R_1} + SC\right)V_I(s) = \left(\frac{1}{R_1} + \frac{1}{R_2} + SC\right] V_o(s)$$

$$\frac{V_o(s)}{V_I(s)} = \frac{SC + \left(\frac{1}{R_1}\right)}{SC + \frac{1}{R_1} + \frac{1}{R_2}} = \frac{S + \frac{1}{CR_1}}{S + \frac{1}{R_1C} + \frac{1}{R_2C}}$$

Example 2.16: Determine $\frac{V_2}{V_2}$ for the circuit shown in Fig. 2.35.

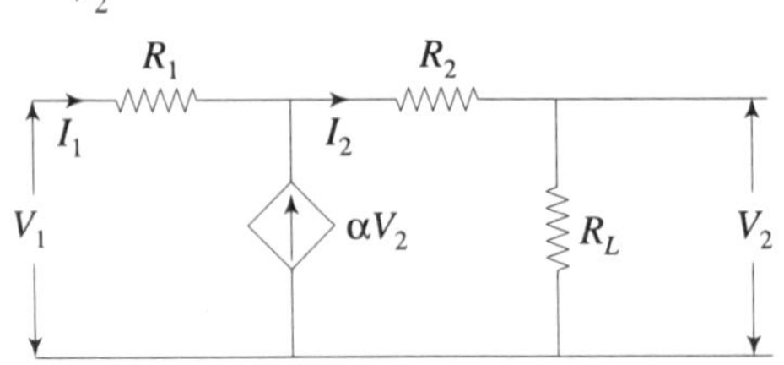

Fig. 2.35

Solution: Using KVL, we get

$$V_I = I_1R_1 + I_2 (R_2 + R_L) \quad ...(2.12)$$

$$V_2 = I_2R_L \quad ...(2.13)$$

Again by KVL, we get

$$I_2 = I_1 + \alpha V_2 \quad ...(2.14)$$

From Eqs. (2.13) and (2.14), we get

$$V_2 = (I_1 + \alpha V_2) R_L$$

or $$(1 - \alpha R_L) V_2 = I_1 R_L$$

$$I_1 = \left(\frac{1 - \alpha R_L}{R_L}\right) V_2 \quad ...(2.15)$$

Using Eqs. (2.15) and (2.12) we get

$$V_1 = \left(\frac{1 - \alpha R_L}{R_L}\right) V_2R_1 + \frac{V_2}{R_L}(R_2 + R_L)$$

$\therefore$ $$I_2 = \frac{V_2}{R_L}$$

$$V_1 = \frac{V_2}{R_L}[(1 - \alpha R_L) R_1 + (R_2 + R_L)]$$

$$\frac{V_2}{V_1} = \frac{R_L}{(1 - \alpha R_L) R_1 + (R_2 + R_L)}$$

This is the transfer function.

Example 2.17: Find $\frac{V_2}{V_1}$ of Fig. 2.36 where opamp is ideal.

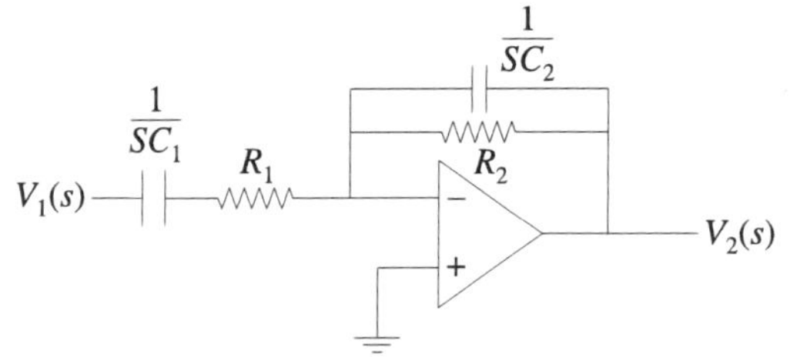

Fig. 2.36

Solution: Figure 2.36 redrawn is as shown in Fig. 2.37.

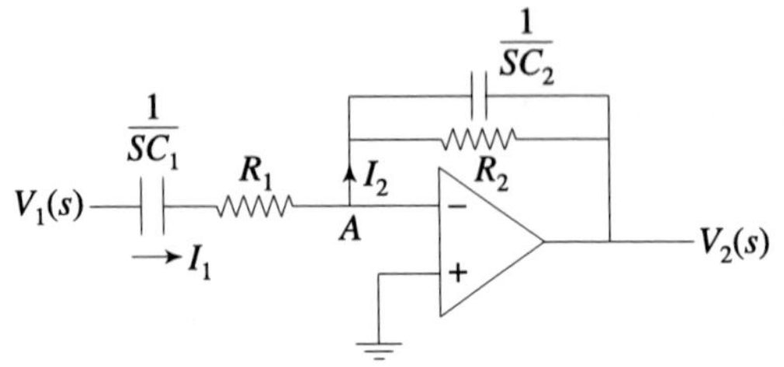

Fig. 2.37

Using KCL at node A, we get

$$I_1 = I_2$$

$$\frac{V_1(s) - O}{R + \frac{1}{C_1 S}} = \frac{O - V_2(s)}{\frac{R_2}{1 + SC_2R_2}}$$

$$\left[\frac{V_1(s)}{\frac{1 + S_1CR_1}{SC_1}}\right]\left(\frac{R_2}{1 + SC_2R_2}\right) = - V_2(s)$$

$$\frac{V_2(s)}{V_1(s)} = -\frac{SC_1R_2}{(1 + SC_2R_2)\,(1 + SR_1C_1)}$$

2.5.1 Procedure for Finding Transfer Function of Electric Networks

The following procedure is used to find transfer function of an electric network.

1. Draw the given network in the S domain with each inductance L replaced by SL and each capacitance by $\frac{1}{SC}$.
2. Replace all sources and time variables with their Laplace transforms. That is, replace $V(t)$ and $i(t)$ by their Laplace transforms $V(s)$ and $I(s)$ respectively.
3. Use KCL, KVL to write network equations.
4. Solve simultaneous equation for the output.
5. Form the transfer function.

Example 2.18: Determine the transfer function $\frac{V_o(s)}{V_I(s)}$ of the RC network shown in Fig. 2.38.

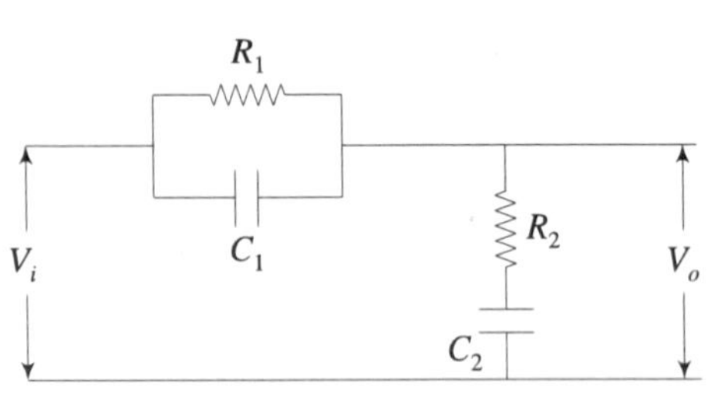

Fig. 2.38

Solution: Redraw the circuit in Laplace domain as show in Fig. 2.39.

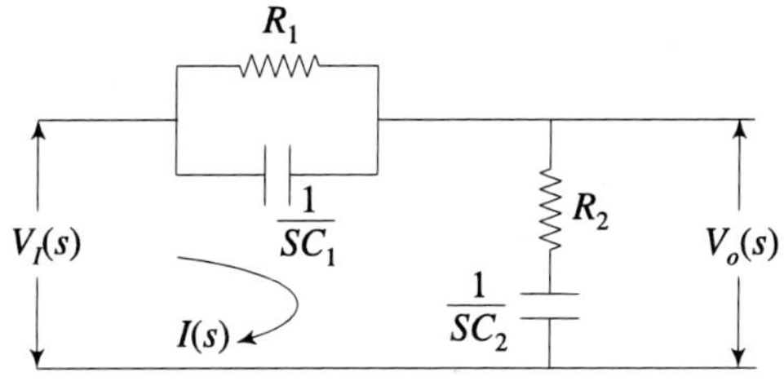

Fig. 2.39

$$V_I(s) = \left[\left(R_1 \parallel^b \frac{1}{SC_1}\right) + \left(R_2 + \frac{1}{SC_2}\right)\right] I(s)$$

$$= \left[\frac{\frac{R_1}{SC_1}}{R_1 + \frac{1}{SC_1}} + \left(R_2 + \frac{1}{SC_2}\right)\right] I(s)$$

$$= \left[\frac{R_1}{R_1C_1S + 1} + R_2 + \frac{1}{SC_2}\right] I(s)$$

$$V_o(s) = \left(R_2 + \frac{1}{SC_2}\right) I(s)$$

$\therefore$ Transfer function

$$\frac{V_o(s)}{V_I(s)} = \frac{\left(R_2 + \frac{1}{SC_2}\right)}{\left(\frac{R_1}{R_1C_1S + 1} + R_2 + \frac{1}{SC_2}\right)}$$

$$\frac{V_o(s)}{V_I(s)} = \frac{\left(\frac{R_2C_2S + 1}{SC_2}\right)}{\frac{R_1C_2S + R_2C_2S + R_1R_2C_1C_2S^2 + 1 + R_1C_1S}{(R_1C_1S + 1)SC_2}}$$

$$\frac{V_o(s)}{V_I(s)} = \frac{(R_1C_1S + 1)\,(R_2C_2S + 1)}{R_1R_2C_1C_2S^2 + (R_1C_1 + R_2C_2 + R_1C_2)S + 1}$$

$$= \frac{R_1R_2C_1C_2\left(S + \frac{1}{R_1C_1}\right)\left(S + \frac{1}{R_2C_2}\right)}{R_1R_2C_1C_2\left[S^2 + \left(\frac{1}{R_1C_1} + \frac{1}{R_2C_2} + \frac{1}{R_2C_1}\right)S + \frac{1}{R_1R_2C_1C_2}\right]}$$

$$\frac{V_o(s)}{V_I(s)} = \frac{\left(S + \frac{1}{R_1C_1}\right)\left(S + \frac{1}{R_2C_2}\right)}{S_2 + \left(\frac{1}{R_1C_1} + \frac{1}{R_2C_2} + \frac{1}{R_2C_1}\right)S + \frac{1}{R_1R_2C_1C_2}}$$

Example 2.19: Draw the mechanical equivalent network for the system and write the nodal equations.

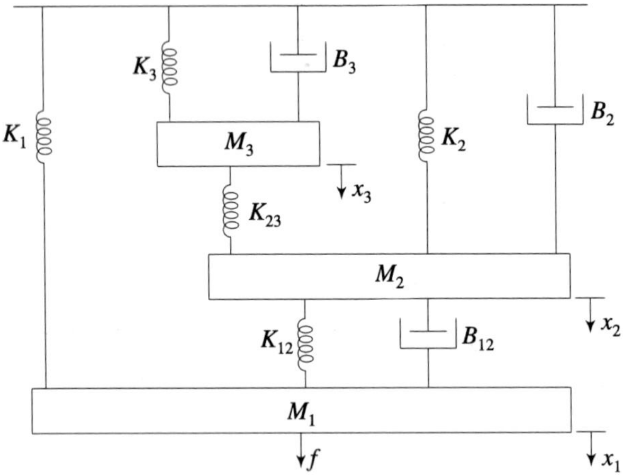

Fig. 2.40

Solution:

Step 1: Since there are three linear displacements x_1, x_2 and x_3, take three nodes x_1, x_2 and x_3 corresponding to these displacements, with an additional reference node.

Step 2: The elements M_1 and K_1 and force f are under the influence of displacement x_1. Therefore, M_1, K_1 and f are connected in parallel between node x_1 and reference node.

Step 3: There is a change in displacement from x_1 to x_2 due to the presence of elements B_{12} and K_{12}. These elements are under the influence of displacement $(x_1 - x_2)$. Therefore, B_{12} and K_{12} are connected in parallel between nodes x_1 and x_2.

Step 4: The elements M_2, K_2 and B_2 are under the influence of displacement x_2. Therefore, M_2, K_2 and B_2 are connected in parallel between node x_2 and reference node.

Step 5: There is a change in displacement from x_2 to x_3 due to the presence of elements k_{23}. The spring element k_{23} is under the influence of displacement $(x_2 - x_3)$. Therefore, K_{23} is connected between nodes x_2 and x_3.

Step 6: The elements M_3, K_3 and B_3 are under the influence of displacement x_3. Therefore, M_3, K_3 and B_3 are connected in parallel between node x_3 and reference node. The mechanical equivalent network of the system is shown in Fig. 2.41.

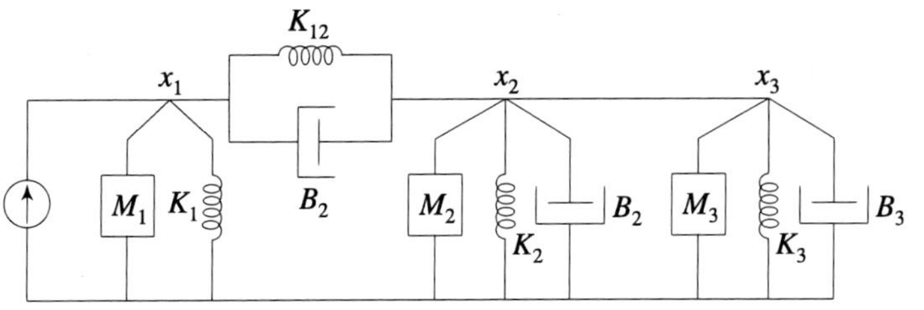

Fig. 2.41

Step 7: Apply force equation at node x_1 to get

$$f(t) = M_1 \frac{d^2x_1}{dt^2} + K_1 x_1 + K_{12}(x_1 - x_2) + B_{12} \frac{d}{dt} (x_1 - x_2)$$

By taking the Laplace transform and assuming zero initial condition, we get

$$F(s) = M_1S^2x_1 (s) + K_1 x_1(s) + K_{12} [x_1(s) - x_2(s)] + B_{12} S [x_1(s) - x_2(s)]$$

Step 8: The force equation at node x_2 is given by

$$K_{12} (x_1 - x_2) + B_{12} \frac{d}{dt} (x_1 - x_2) = M_2 \frac{d^2x_2}{dt^2} + B_2 \frac{dx_2}{dt} + K_2x_2$$

Assuming zero initial conditions and taking Laplace transform of that equation, we get

$$K_{12}[x_1(s) - x_2(s)] + B_{12}S [x_1(s) - x_2(s)] = M_2 S^2x_2(s) + B_2S x_2(s) + K_2x_2(s)$$

Step 9: The force equation at node x_3 is given by

$$K_{23}(x_2 - x_3) = M_3 \frac{d^2x_3}{dt^2} + B_3 \frac{dx_3}{dt} + K_3x_3$$

By taking Laplace transform and assuming zero initial conditions, we get

$$K_{23} [x_2(s) - x_3(s)] = M_3S^2x_3(s) + B_3S x_3(s) + K_3x_3(s)$$

Example 2.20: A mechanical system is shown in Fig. 242. Write down the differential equation at various points.

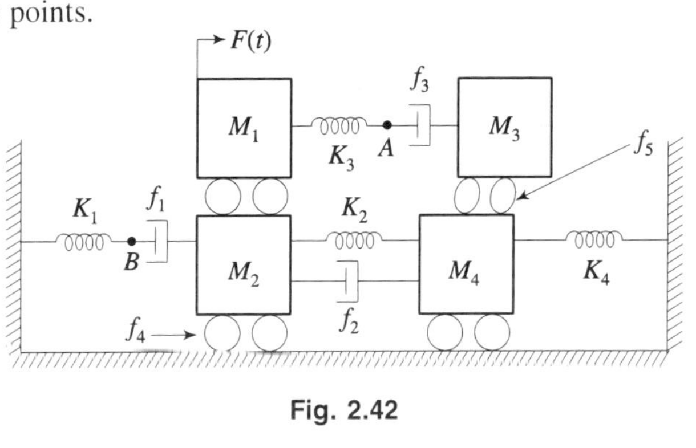

Fig. 2.42

Solution: The output variables for masses M_1, M_2, M_3 and M_4 can be expressed respectively as x_1, x_2, x_3 and x_4. The points *A* and *B* are not connected to any mass, so their displacements can be regarded as x_5 and x_6 respectively.

The equations can be written as,

For mass M_1,

$$M_1 \frac{d^2x_1}{dt^2} + K_3 (x_1 - x_5) = F(t)$$

For mass M_2,

$$M_2 \frac{d^2x_2}{dt^2} + F_1(x_2 - x_6) + f_4 \frac{dx_2}{dt} + f_2(x_2 - x_4) + K_2(x_2 - x_4) = 0$$

For mass M_3,

$$M_3 \frac{d^2x_3}{dt^2} + f_3 \frac{d}{dt} (x_3 - x_5) + f_5 \frac{d}{dt} (x_3 - x_4) = 0$$

For mass M_4,

$$M_4 \frac{d^2x_4}{dt^2} + f_2 \frac{d}{dt} (x_4 - x_2) + K_2(x_4 - x_2) + k_4x_4 = 0$$

For point *A*,

$$f_3 \frac{d}{dt} (x_5 - x_3) + K_3(x_5 - x_1) = 0$$

For point *B*,

$$f_1 \frac{d}{dt} (x_6 - x_1) + K_1x_6 = 0$$

2.6 ANALOGOUS SYSTEMS

If the differential equations for the two systems are of identical form, such systems are called analogous systems. The quantities which occupy corresponding positions in the differential equations are called analogous quantities.

Consider an electrical RLC system,

$$E = L \frac{di}{dt} + Ri + \frac{1}{C}\int idt \quad ...(2.16)$$

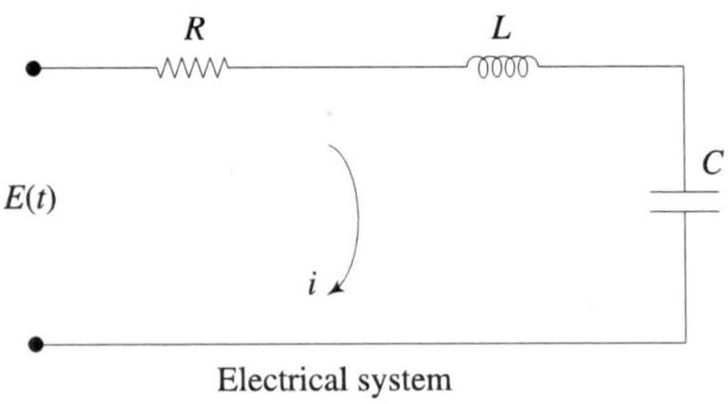

Electrical system

Fig. 2.43

Current is the ratio of flow of electric charge, therefore, $i = \frac{dq}{dt}$ ...(2.17)

Substitute Eq. (2.17) in Eq. (2.16)

$$E = L \frac{d^2q}{dt^2} + R \frac{dq}{dt} + \frac{1}{C} q \quad ...(2.18)$$

$$\text{wkt, } f = M\frac{d^2x}{dt^2} + B\frac{dx}{dt} + Kx \qquad ...(2.19)$$

Comparing Eqs. (2.18) and (2.19), force in mechanical system is analogous to voltage in the electrical system is called force-voltage analogy.

Let us consider a mechanical (both translational and rotational) and an electrical system, as shown Fig. 2.44.

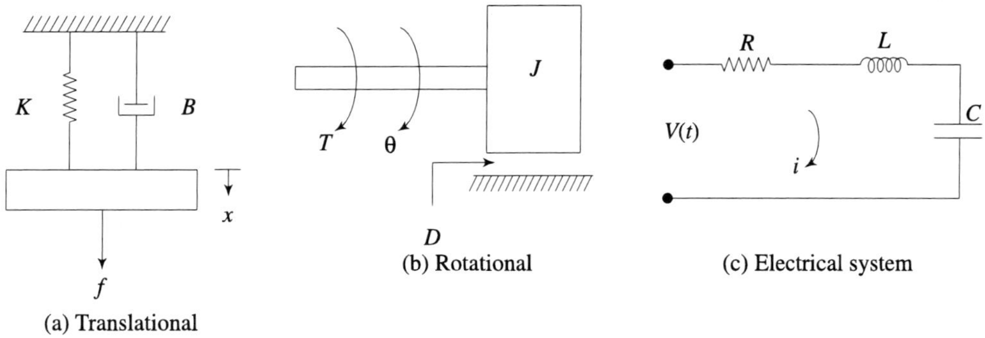

Fig. 2.44

From Fig. 2.44(a) (translational), we have

$$f = M\frac{d^2x}{dt^2} + B\frac{dx}{dt} + kx \qquad ...(2.20)$$

From Fig. 2.44(b) (rotational), we have

$$T = J\frac{d^2\theta}{dt^2} + B\frac{dq}{dt} + k\theta \qquad ...(2.21)$$

From Fig. 2.44(c) (electrical system), we have

$$v(t) - L\frac{d^2\theta}{dt^2} + R\frac{d\theta}{dt} + \frac{1}{c}q \qquad ...(2.22)$$

where $\quad q = \int idt$

The Eqs. (2.20), (2.21) and (2.22) are same. Hence they are analogous systems.

There are two methods to get an analogous system. These are (i) force voltage ($f-v$) analogy and (ii) force current ($f-i$) analogy.

2.6.1 Force–Voltage Analogy

In the case of Force–Voltage analogy, then the following analogies can be written as shown in Table 2.1.

Table 2.1 Force-voltage analogy

Translational	*Electrical*	*Rotational*
Force (*f*)	Voltage (*V*/*E*)	Torque (*T*)
Mass (*M*)	Inductance (*L*)	Inertia (*J*)
Damper (*B*)	Resistance (*R*)	Damper (*D*)
Spring (*K*)	Sprig (1/*C*)	Spring (*K*)
Displacement (*x*)	Charge (*q*)	Displacement (θ)
Velocity (*v*)	Current (*i*)	Velocity (*w*)

2.6.2 Force – Current Analogy

Let us consider electrical network as shown in Fig. 2.45

$$i = \frac{V}{R} + \frac{1}{L}\int V\,dt + C\frac{di}{dt}$$

$$V = \frac{d\varphi}{dt}$$

$$i = \frac{1}{R}\cdot\frac{d\varphi}{dt} + \frac{\varphi}{1} + C\frac{d^2\varphi}{dt^2}$$

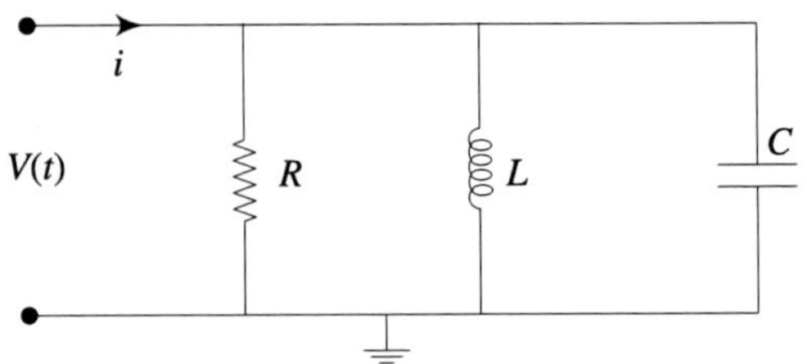

Fig. 2.45

Table 2.2 Force – current analogy

Translational	*Electrical*	*Rotational*
Force (*f*)	Current (*i*)	Torque (*T*)
Mass (*M*)	Capacitance (*C*)	Inertia (*J*)
Spring (*K*)	Reciprocal of inductance (1/*L*)	Damper (*B*)
Damper (*B*)	Conductance (1/*R*)	Spring (*K*)
Displacement (*x*)	Flux linkage (ϕ)	Displacement (θ)
Velocity ($v = dx/dt$)	Voltage (V/E) = $d\phi/dt$	Velocity ($w = d\theta/dt$)

Example 2.21: For the system shown in Fig. 2.46, write mechanical network and obtain the differential equations for the system. Also write the F–V analogous circuits.

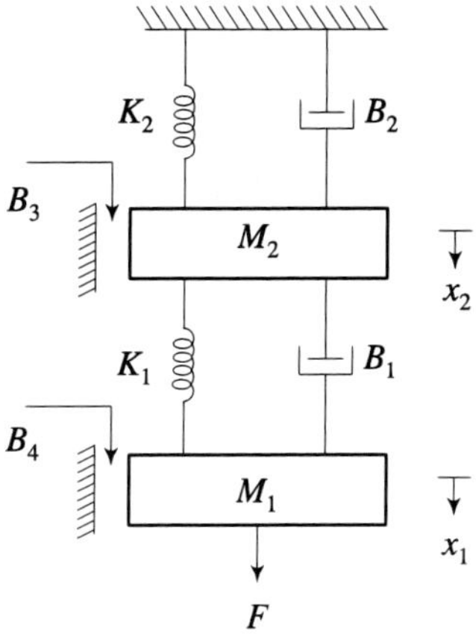

Fig. 2.46

Solution:

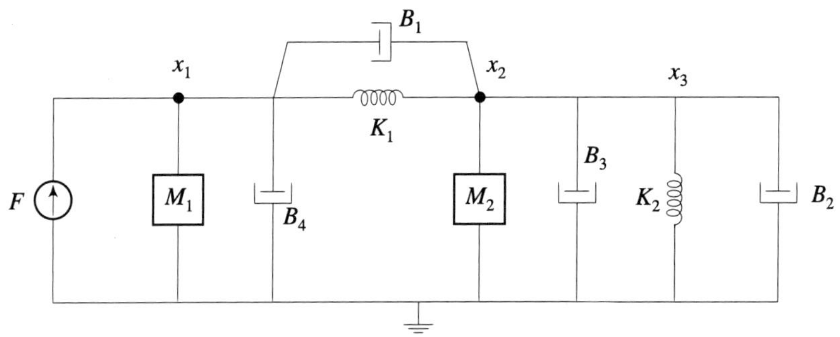

Fig. 2.47

At node x_1,

$$f(t) = M_1 \frac{d^2x_1}{dt^2} + B_4 \frac{dx_1}{dt} + B_1 \left(\frac{dx_1}{dt} - \frac{dx_2}{dt}\right) + K_1 (x_1 - x_2) \quad ...(2.23)$$

At node x_2,

$$B_1 \left(\frac{dx_1}{dt} - \frac{dx_2}{dt}\right) + K_1(x_1 - x_2) = M_2 \frac{d^2x_2}{dt^2} + B_2 \frac{dx_2}{dt} + B_3 \frac{dx_2}{dt} + K_2x_2 \quad ...(2.24)$$

F–V analogy:

$F - E/V$ $\qquad K_1 = \frac{1}{C_1}$

$M_1 - L_1$ $\qquad K_2 = \frac{1}{C_2}$

$$B_1 - R_1 \qquad x_1 = q_1 \text{ or } \int i_1 dt$$

$$B_2 - R_2 \qquad x_2 = q_2 \text{ or } \int i_2 dt$$

$$B_3 - R_3 \qquad dx_1 = i_1$$

$$B_4 - R_4 \qquad dx_2 = i_2$$

$$M_2 - L_2$$

Substituting in Eqs. (2.23) and (2.24)

$$E = L_1 \frac{di_1}{dt} + R_4 i_1 + R_1 (i_1 - i_2) + \frac{1}{C_1}\int (i_1 - i_2)\, dt \qquad ...(2.25)$$

$$R_1 (i_1 - i_2) + \frac{1}{C_1}\int (i_1 - i_2)\, dt = L_2 \frac{di_2}{dt} + R_2 i_2 + R_3 i_2 + \frac{1}{C_2}\int i_2 dt \qquad ...(2.26)$$

Based on Eqs. (2.25) and (2.26), F–V analogous circuit is (Fig. 2.47)

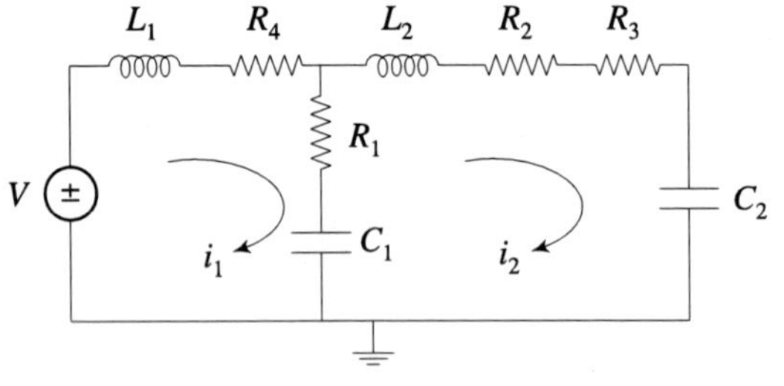

Fig. 2.48

F–I Analogy

$$F = i$$

$$M = C$$

$$B = \frac{1}{R}$$

$$K = \frac{1}{L}$$

$$x = \phi = \int V dt$$

Velocity, V = Voltage $E = \frac{dV}{dt} = E$

$$\frac{d^2V}{dt^2} = \frac{dE}{dt}$$

From Eqs. (2.23) and (2.24)

$$i = C_1 \frac{dE_1}{dt} + \frac{1}{R_4} E_1 + \frac{1}{R_1} (E_1 - E_2) + \frac{1}{L_1} \int (E_1 - E_2) dt$$

or

$$i = C_1 \frac{dE_1}{dt} + \frac{1}{R_4} E_1 + \frac{1}{R_1} (E_1 - E_2) + \frac{1}{L_1} \int (E_1 - E_2)\, dt \qquad ...(2.27)$$

$$\frac{1}{R_1} (E_1 - E_2) + \frac{1}{L_1} (E_1 - E_2)\, dt = C_2 \frac{dE_2}{dt} + \frac{1}{R_2} E_2 + \frac{1}{R_3} E_3 + \frac{1}{L_2} \int E_2 dt \qquad ...(2.28)$$

Based on Eqs. (2.23) and (2.24), the analogous circuit is (Fig. 2.49).

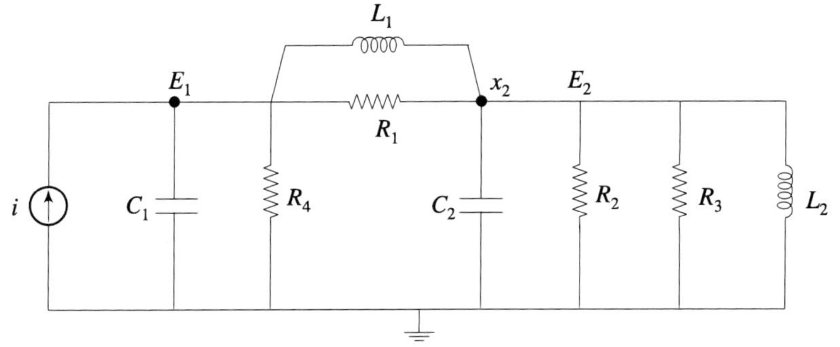

Fig. 2.49

Example 2.22: The transfer function of a system is given by $G(s) = \dfrac{3S + 2}{S^2 + 2S + 3}$. Find the differential equation of the system having input x and output y.

$$G(s) = \frac{y(s)}{x(s)} = \frac{3S + 2}{S^2 + 2S + 3}$$

Solution: $$S^2 y(s) + 2Sy(s) + 3y(s) = 3Sx(s) + 2x(s)$$

Taking inverse LT

$$\frac{d^2y}{dt^2} + \frac{2dy}{dt} + 3y = \frac{3dx}{dt} + 2x$$

Example 2.23: If the transfer function of a system and applied input are e^{-3t} and e^{-4t}. Find the response of the system.

Solution: $h(t) = e^{-3t}$ and $r(t) = e^{-4t}$

Applying LT for above equations,

$$H(s) = \frac{1}{S + 3} \text{ and } R(s) = \frac{1}{S + 4}$$

$$H(s) = \frac{C(s)}{R(s)}$$

$$C(s) = H(s)\,R(s)$$

$$C(s) = \frac{1}{(s+3)(s+4)}$$

$$C(s) = \frac{1}{(s+3)} - \frac{1}{(s+4)}$$

Using partial fraction,

Taking inverse LT

$$C(t) = e^{-3t} - e^{-4t}$$

Example 2.24: Find the system equation of the mechanical systems shown in Fig. 2.50. Also find the *F–V* and *F–I* analogy.

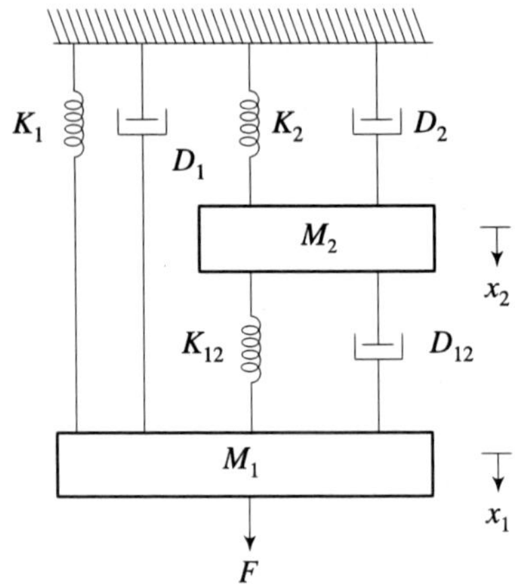

Fig. 2.50

Solution:

At node x_1,

$$f(t) = M_1 \frac{d^2x_1}{dt^2} + K_1x_1 + D_1 \frac{dx_1}{dt} + K_{12}(x_1 - x_2) + D_{12} \frac{d(x_1 - x_2)}{dt}$$

At node x_2,

$$M_2 \frac{d^2x_2}{dt^2} + K_2x_2 + D_2 \frac{dx_2}{dt} = K_{12}(x_1 - x_2) + D_{12} \frac{d(x_1 - x_2)}{dt}$$

Based on the above equations, *F-V* analogy

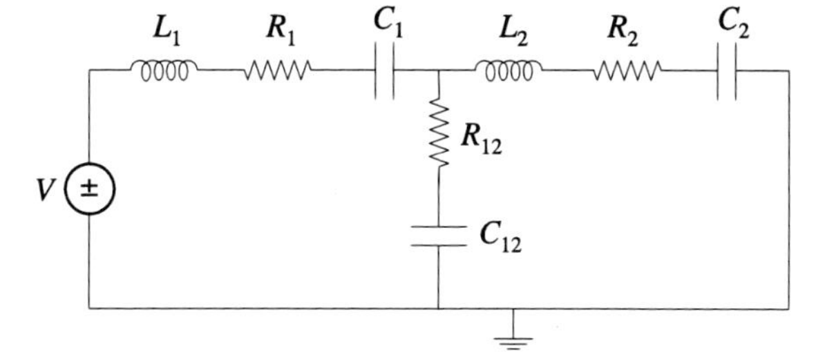

Fig. 2.51

F–I analogy

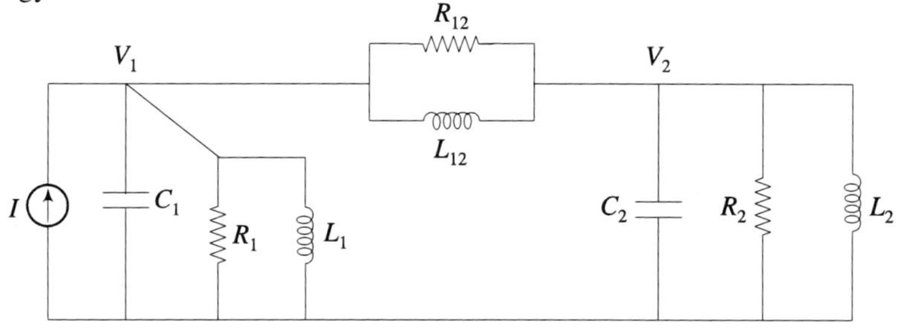

Fig. 2.52

Example 2.25: Determine the transfer function for the network shown in Fig. 2.53. Calculate output voltage for a unit step voltage input when $C_1 = 1\,\mu F$, $R_1 = 1\,M\Omega$, $C_2 = 0.5\,\mu F$ and $R_2 = 1\,M\,\Omega$.

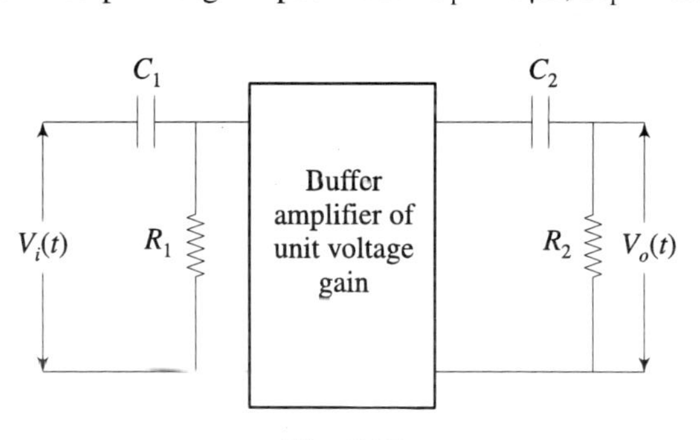

Fig. 2.53

Solution:

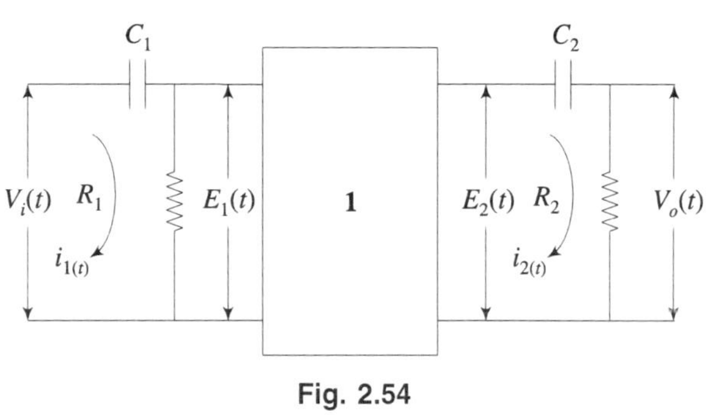

Fig. 2.54

$$V_I(s) = I_1(S)R_1 + \frac{I_1(s)}{SC_1}$$

$$E_2(s) = I_2(S)\ R_2 + \frac{I_2(s)}{SC_2}$$

$$E_1(s) = I_1(S)\ R_1$$

$$V_o(s) = I_2(S)\ R_2$$

$$\text{Transfer function} = \frac{V_o(s)}{V_I(s)} = \frac{I_2(s)R_2}{I_1(s)\left(R_1 + \frac{1}{SC_1}\right)}$$

Example 2.26: For the mechanical system shown in Fig. 2.55.

(a) Draw the mechanical network.

(b) Write the differential equations of the system.

(c) Draw electric circuits based on *F–I* and *F–V* analogy.

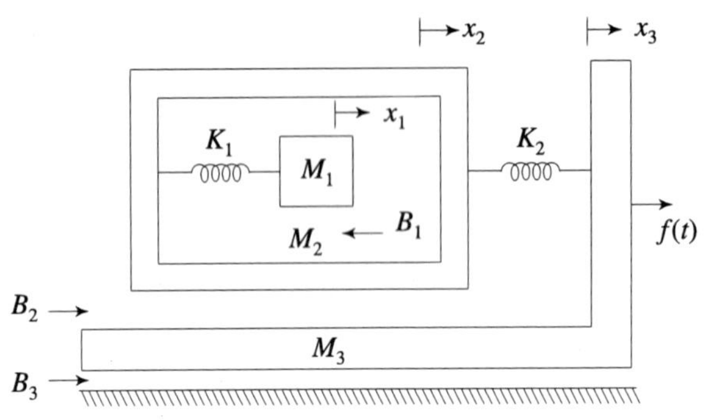

Fig. 2.55

Solution:

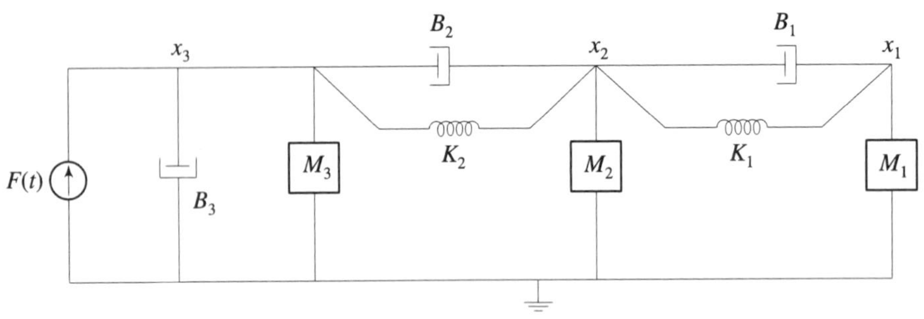

Fig. 2.56

The differential equations for the system are

$$f(t) = M_3 \frac{d^2x_3}{dt^2} + B_3 \frac{dx_3}{dt} + B_2 \frac{d(x_3 - x_2)}{dt} + K_2(x_3 - x_2)$$

$$M_2 \frac{d^2x_2}{dt^2} + B_2 \frac{d(x_2 - x_3)}{dt} + K_2(x_2 - x_3) + B_1 \frac{d(x_2 - x_1)}{dt} + K_1(x_2 - x_1) = 0$$

$$M_1 \frac{dx_1}{dt} + B_1 \frac{d(x_1 - x_2)}{dt} + K_1 (x_1 - x_2)$$

Force-Current (*F–I*) analogy

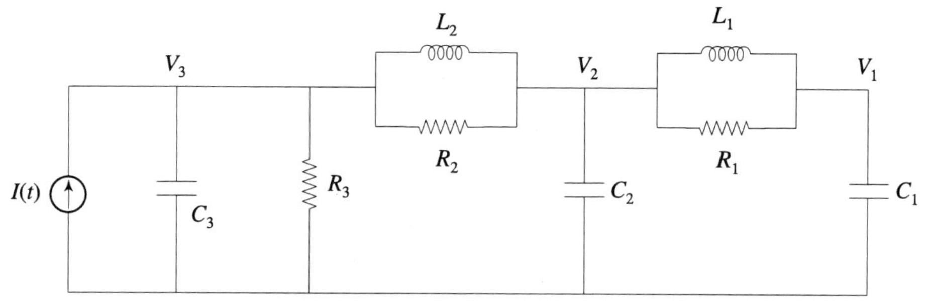

Fig. 2.57

Force-Voltage (*F–V*) analogy

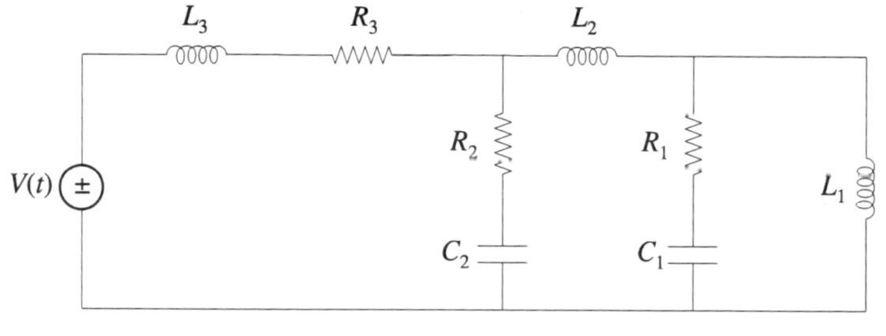

Fig. 2.58

Example 2.27: Write the differential equations governing the mechanical system shown in Fig. 2.59. Draw the force voltage and force current electrical analogous circuit and verify by writing mesh and mode equations.

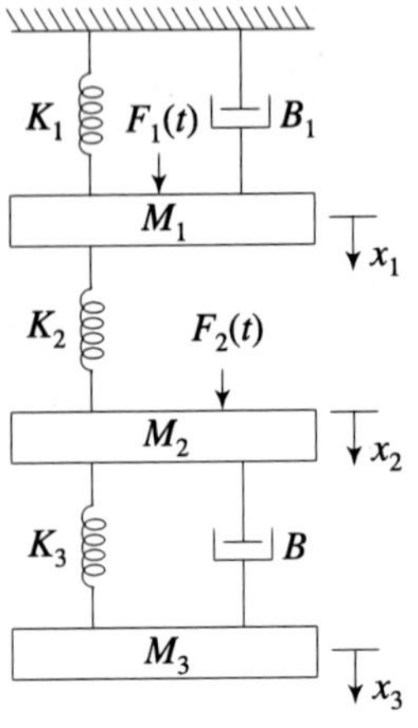

Fig. 2.59

Solution: The given mechanical system has three nodes.

The differential equations governing the mechanical system are given by force balance equations at these nodes. Let the displacement of masses M_1, M_2 and M_3 be x_1, x_2 and x_3 respectively. The corresponding velocities be V_1, V_2 and V_3.

The differential equation at mass M_1

$$M_1 \frac{d^2x_1}{dt^2} + B_1 \frac{dx_1}{dt} + K_1x_1 + K_2(x_1 - x_2) = f_1(t)$$

The differential equation at mass M_2

$$M_2 \frac{d^2x_2}{dt^2} + B_3 \frac{d}{dt}(x_2 - x_3) + K_2 (x_2 - x_1) + K_3 (x_2 - x_3) = f_2(t)$$

The differential equation at mass M_3

$$M_3 \frac{d^2x_3}{dt^2} + B_3 \frac{d}{dt}(x_3 - x_2) + K_3 (x_3 - x_2) = 0$$

On replacing the displacements by velocity in the differential equations governing the mechanical system we get,

i.e., $$\frac{d^2x}{dt^2} = \frac{dv}{dt}, \frac{dx}{dt} = V, \text{ and } x = \int V\,dt$$

$$M_1 \frac{dv_1}{dt} + B_1V_1 + K_1 \int V_1 dt + K_2 \int (V_1 - V_2)dt = f_1(t)$$

$$M_2 \frac{dv_2}{dt} + B_3(V_2 - V_3) + K_2 \int (V_2 - V_1)dt + K_3 \int (V_2 - V_3)dt = f_2(t)$$

$$M_3 \frac{dv_3}{dt} + B_3(V_3 - V_2) + K_3 \int(V_3 - V_2)dt = 0$$

Force voltage analogous circuit

$$\begin{array}{l|l|l} f_1(t) \to e_1(t) & V_1 \to i_1 & M_1 \to L_1 \\ f_2(t) \to e_2(t) & V_2 \to i_2 & M_2 \to L_2 \\ & V_3 \to i_3 & M_3 \to L_3 \end{array}$$

$$\begin{array}{l|l} B_1 \to R_1 & K_1 \to \dfrac{1}{C_1} \\ B_3 \to R_3 & K_2 \to \dfrac{1}{C_2} \\ & K_3 \to \dfrac{1}{C_3} \end{array}$$

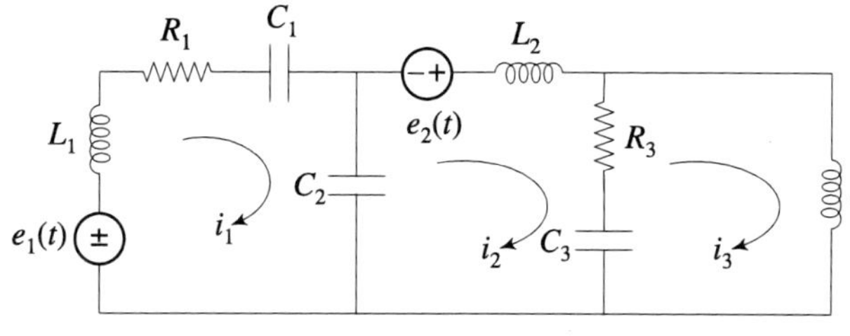

Fig. 2.60

The mesh basis equations using KVL for the circuit

$$L_1 \frac{di_1}{dt} + R_1 i_1 + \frac{1}{C_1} \int i_1 dt + \frac{1}{C_2} \int (i_1 - i_2)\, dt = e_1(t)$$

$$L_2 \frac{di_2}{dt} + R_3(i_2 - i_3) + \frac{1}{C_3} \int (i_2 - i_3)\, dt + \frac{1}{C_2} \int (i_2 - i_1) dt = e_2(t)$$

$$L_3 \frac{di_3}{dt} + R_3\,(i_3 - i_2) + \frac{1}{C_3} \int (i_3 - i_2)\, dt = 0$$

Force current analogous circuit

$$\begin{array}{l|l|l|l|l} f_1(t) \to i_1(t) & V_1 \to V_1 & M_1 \to C_1 & B_1 \to 1/R_1 & K_1 \to 1/L_1 \\ f_2(t) \to i_2(t) & V_2 \to V_2 & M_2 \to C_2 & B_3 \to 1/R_3 & K_2 \to 1/L_2 \\ & V_3 \to V_3 & M_3 \to C_3 & & K_3 \to 1/L_3 \end{array}$$

The node basis equations using KCL for the circuit shown in Fig. 2.61.

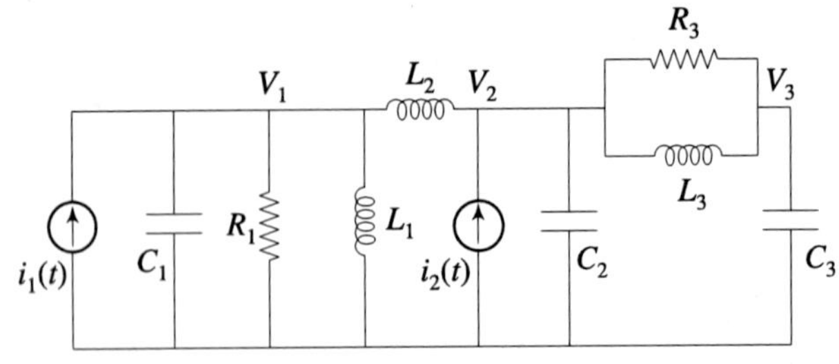

Fig. 2.61

$$C_1 \frac{dV_1}{dt} + \frac{1}{R_1} V_1 + \frac{1}{L_1} \int V_1 dt + \frac{1}{L_2} \int (V_1 - V_2) dt = i_1(t)$$

$$C_2 \frac{dV_2}{dt} + \frac{1}{R_3} (V_2 - V_3) + \frac{1}{L_3} \int (V_2 - V_3) dt + \frac{1}{L_2} \int (V_2 - V_1) dt = i_2(t)$$

$$C_3 \frac{dV_3}{dt} + \frac{1}{R_3} (V_3 - V_2) + \frac{1}{L_3} \int (V_3 - V_2) dt = 0$$

Example 2.28: Write the differential equations governing the mechanical system shown in Fig. 2.62. Draw the force voltage and force current electrical analogous circuits and verify by writing mesh and node equations.

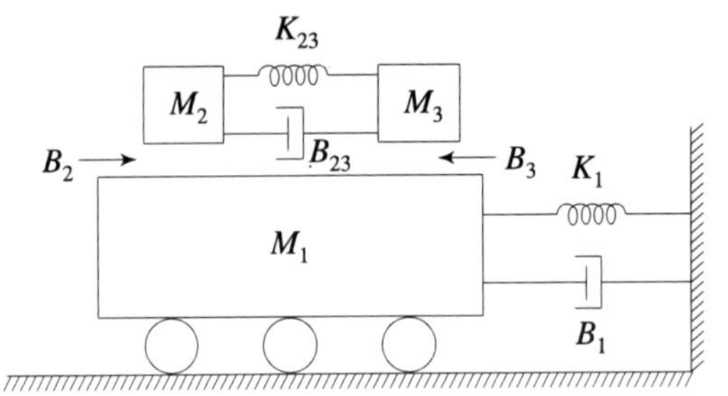

Fig. 2.62

Solution: The given mechanical system has three nodes. The differential equations governing the mechanical system are given by force balanced equations at these nodes.

Let the displacements of masses M_1, M_2 and M_3 be x_1, x_2 and x_3 respectively. The corresponding velocities be V_1, V_2 and V_3.

The differential equation at M_1 is

$$M_1 \frac{d^2x_1}{dt^2} + B_1 \frac{dx_1}{dt} + K_1 x_1 + B_2 \frac{d}{dt}(x_1 - x_2) + B_3 \frac{d}{dt}(x_1 - x_3) = 0$$

The differential equation at M_2 is

$$M_2 \frac{d^2x_2}{dt^2} + B_2 \frac{d}{dt}(x_2 - x_1) + B_{23} \frac{d}{dt}(x_2 - x_3) + K_{23}(x_2 - x_3) = 0$$

The differential equation at M_3 is

$$M_3 \frac{d^2x_3}{dt^2} + B_3 \frac{d}{dt}(x_3 - x_1) + B_{23} \frac{d}{dt}(x_3 - x_2) + K_{23}(x_3 - x_2) = 0$$

On replacing the displacements by velocity in the differential equations

i.e., $$\frac{d^2x}{dt^2} = \frac{dv}{dt}, \frac{dx}{dt} = V, x = \int V dt$$

$\therefore$ $$M_1 \frac{dV_1}{dt} + B_1 V_1 + K_1 \int V_1 \, dt + B_2 (V_1 - V_2) + B_3 (V_1 - V_3) = 0$$

$$M_2 \frac{dV_2}{dt} + B_2 (V_2 - V_1) + B_{23} (V_2 - V_3) + K_{23} \int (V_2 - V_3) \, dt = 0$$

$$M_3 \frac{dV_3}{dt} + B_3 (V_3 - V_1) + B_{23} (V_3 - V_2) + K_{23} \int (V_3 - V_2) \, dt = 0$$

Force voltage analogous circuit

$V_1 \to i_1$	$M_1 \to L_1$	$K_1 \to 1/C_1$	$B_2 \to R_2$
$V_2 \to i_2$	$M_2 \to L_2$	$K_{23} \to 1/C_{23}$	$B_3 \to R_3$
$V_3 \to i_3$	$M_3 \to L_3$	$B_1 \to R_1$	$B_{23} \to R_{23}$

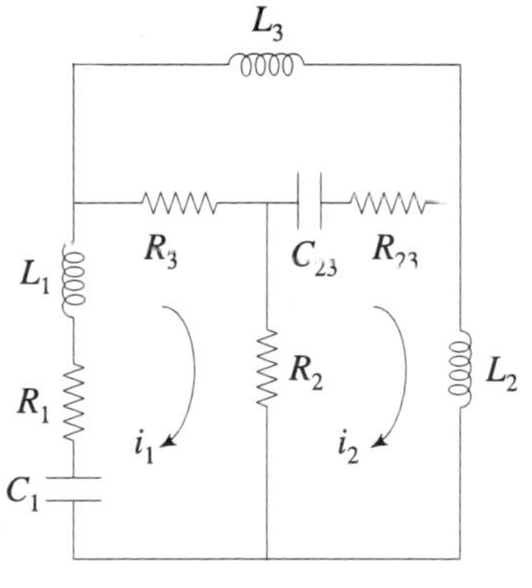

Fig. 2.63

$$L_1 \frac{di_1}{dt} + R_1 i_1 + \frac{1}{C_1} \int i_1 \, dt + R_2 (i_1 - i_2) \, dt + R_3 (i_1 - i_3) = 0$$

$$L_2 \frac{di_2}{dt} + R_2 (i_2 - i_1) + \frac{1}{C_{23}} \int (i_2 - i_3) \, dt + R_{23} (i_2 - i_3) = 0$$

$$L_3 \frac{di_3}{dt} + R_3 (i_3 - i_1) + \frac{1}{C_{23}} \int (i_3 - i_2)\, dt + R_{23} (i_3 - i_2) = 0$$

Force-current analogous circuit

$V_1 \rightarrow V_1$	$M_1 \rightarrow C_1$	$K_1 \rightarrow 1/L_1$	$B_2 \rightarrow 1/R_2$
$V_2 \rightarrow V_2$	$M_2 \rightarrow C_2$	$K_{23} \rightarrow 1/L_{23}$	$B_3 \rightarrow 1/R_3$
$V_3 \rightarrow V_3$	$M_3 \rightarrow C_3$	$B_1 \rightarrow 1/R_1$	$B_{23} \rightarrow 1/R_{23}$

The node basis equations using KCL for the circuit

$$C_1 \frac{dV_1}{dt} + \frac{1}{R_1} V_1 + \frac{1}{L_1} \int V_1 dt + \frac{1}{R_2} (V_1 - V_2) + \frac{1}{R_3} (V_1 - V_3) = 0$$

$$C_2 \frac{dV_2}{dt} + \frac{1}{R_2} (V_2 - V_1) + \frac{1}{L_{23}} \int (V_2 - V_3)\, dt + \frac{1}{R_{23}} (V_2 - V_3) = 0$$

$$C_3 \frac{dV_3}{dt} + \frac{1}{R_3} (V_3 - V_1) + \frac{1}{L_{23}} \int (V_3 - V_2)\, dt + \frac{1}{R_{23}} (V_3 - V_2) = 0$$

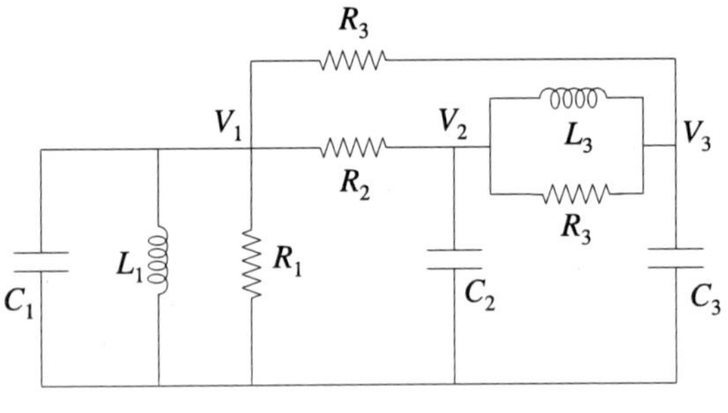

Fig. 2.64

2.6.3 Analogous Mechanical (Rotational) Systems and Electrical Systems

Torque-Voltage (T–V) Analogy

From loop equations,

$$E = L \frac{d^2q}{dt^2} + R \frac{dq}{dt} + \frac{1}{C} q \qquad \text{...(2.29)}$$

$$T = J \frac{d^2\theta}{dt^2} + B \frac{d\theta}{dt} + K\theta \qquad \text{...(2.30)}$$

Comparing Eqs. (2.29) and (2.30)

Mechanical System	***Electrical System***
Torque – T	Voltage – E/V
Inertia – J	Inductance – L
Friction – B	Resistance – R
Spring – K	Capacitance – $1/C$
Angular displacement – θ	Charge – q
Angular velocity – ω	Current – I

Torque – Current (T–I) analogy

From nodal equations.

$$I = C\frac{d^2\phi}{dt^2} + \frac{1}{R}\frac{d\phi}{dt} + \frac{1}{L}\phi \quad ...(2.31)$$

$$T = J\frac{d^2\theta}{dt^2} + B\frac{d\theta}{dt} + K\theta \quad ...(2.32)$$

Comparing Eqs. (2.31) and (2.32)

Mechanical System	***Electrical System***
Torque – T	Current – I
Inertia – J	Capacitance – C
Friction – B	Resistance – $1/R$
Spring – K	Inductance – $1/L$
Angular displacement – θ	Flux – ϕ
Angular velocity – ω	Voltage – E/V

Example 2.29: Obtain the electrical analogous networks based on (i) Torque-Voltage and (ii) Torque-Current analysis of the mechanical system shown in Fig. 2.65.

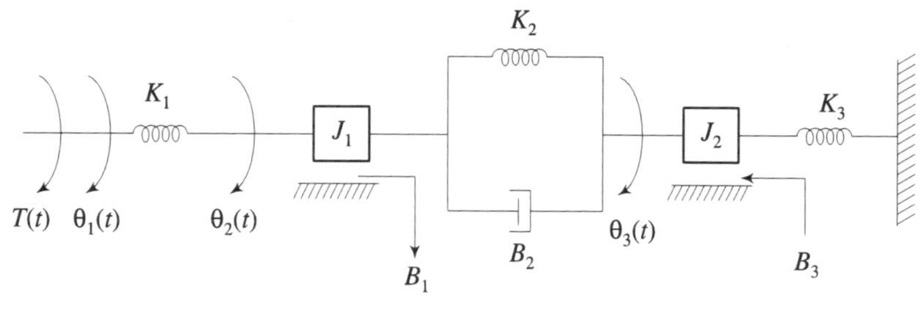

Fig. 2.65

Solution:

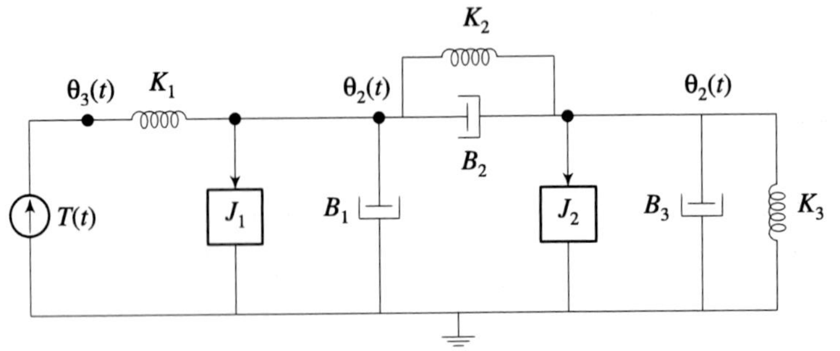

Fig. 2.66

Differential equations

At θ_1,

$$T(t) = K_1(\theta_1(t) - \theta_2(t)) \quad \text{...(2.33)}$$

At θ_2,

$$K_1(\theta_1 - \theta_2) = J_1 \frac{d^2\theta_2}{dt^2} + B_1 \frac{d\theta_2}{dt} + K_2 (\theta_2 - \theta_3) + B_2\left(\frac{d\theta_2}{dt} - \frac{d\theta_3}{dt}\right) \quad \text{...(2.34)}$$

At θ_3,

$$K_2(\theta_2 - \theta_3) + B_2 \left(\frac{d\theta_2}{dt} - \frac{d\theta_3}{dt}\right) = J_2 \frac{d^2\theta_3}{dt^2} + B_3 \frac{d\theta_3}{dt} + K_3\theta_3 \quad \text{...(2.35)}$$

T–V Analogy

From Eqs. (2.33), (2.34) and (2.35)

$$E = \frac{1}{C_1}\int (i_1 - i_2)dt$$

$$\frac{1}{C_1}\int (i_1 - i_2)\, dt = L_1 \frac{di_2}{dt} + R_1 i_2 + \frac{1}{C_2}\int (i_2 - i_3)\, dt + R_2\, (i_2 - i_3)$$

$$\frac{1}{C_2}\int (i_2 - i_3)\, dt + R_2\, (i_2 - i_3) = L_2 \frac{di_2}{dt} + R_3 i_3 + \frac{1}{C_3}\int i_3 c$$

$T = E$
$J = L$
$B = R$
$K = 1/C$
$\theta = q = \int idt$
$\omega = I = d\theta$

T–V Analogy Circuit

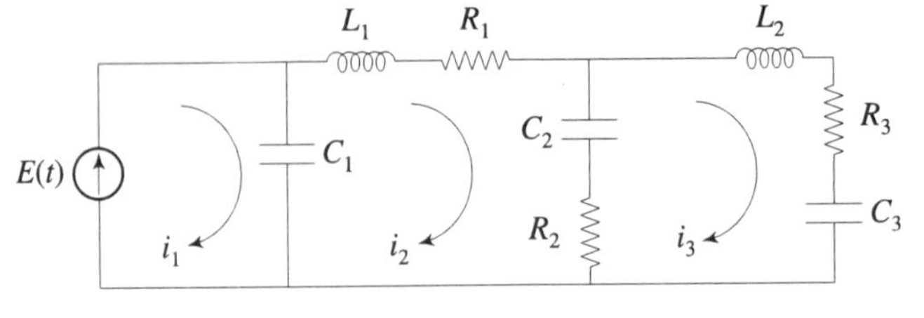

Fig. 2.67

T–I Analogy

From Eqs. (2.33), (2.34) and (2.35)

$T = I$
$J = C$
$B = 1/R$
$K = 1/L$
$\theta = \phi = \int E dt$
$d\theta/dt = E$

$$I = \frac{1}{L_1}\int (E_1 - E_2)dt \qquad \text{...(2.36)}$$

$$\frac{1}{L_1}\int (E_1 - E_2)dt = C_1 \frac{dE_2}{dt} + \frac{1}{R_1} E_2 + \frac{1}{L_2}\int (E_2 - E_3)dt + \frac{1}{R_2}(E_2 - E_3) \qquad \text{...(2.37)}$$

$$\frac{1}{L_2}\int (E_2 - E_3)dt = \frac{1}{R_2}(E_2 - E_3) = C_2 \frac{dE_2}{dt} + \frac{1}{R_3} E_3 + \frac{1}{L_3}\int E_3 dt \qquad \text{...(2.38)}$$

From Eqs. (2.36), (2.37), and (2.38)

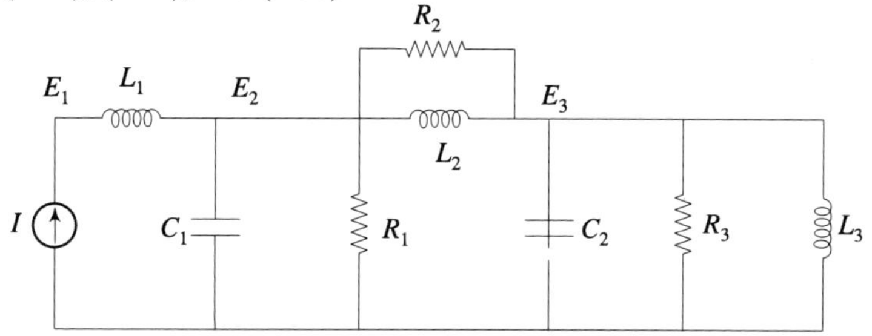

Fig. 2.68

Example 2.30: Write the differential equations governing the mechanical rotational system shown in Fig. 2.69. Draw the torque-voltage and torque-current electrical analogous circuits and verify by writing mesh and node equations.

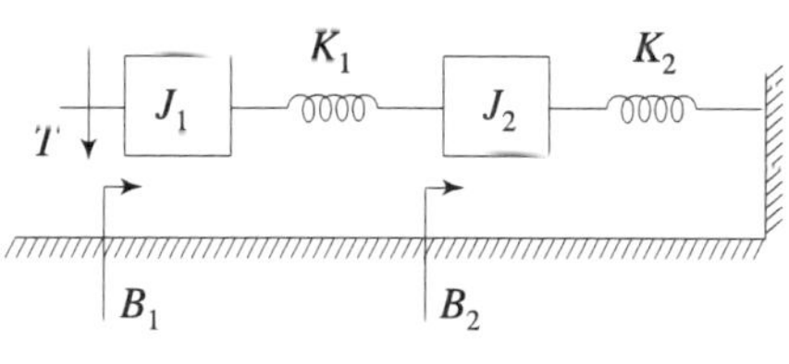

Fig. 2.69

Solution: Let the angular displacements of J_1 and J_2 be θ_1 and θ_2 respectively. The corresponding angular velocities be ω_1 and ω_2

The differential equation at J_1

$$J_1 \frac{d^2\theta_1}{dt^2} + B_1 \frac{d\theta_1}{dt} + K_1 (\theta_1 - \theta_2) = T$$

The differential equation at J_2

$$J_2 \frac{d^2\theta_2}{dt^2} + B_2 \frac{d\theta_2}{dt} + K_2\theta_2 + K_1 (\theta_2 - \theta_1) = 0$$

The angular displacements by angular velocity in the differential equations governing the mechanical rotational system, we get

$$\frac{d^2\theta}{dt^2} = \frac{d\omega}{dt}\frac{d\theta}{dt} = \omega, \; \theta = \int \omega dt$$

$$J_1 \frac{d\omega_1}{dt} + B_1\omega_1 + K_1 \int(\omega_1 - \omega_2)\, dt = T$$

$$J_2 \frac{d\omega_2}{dt} + B_2\omega_2 + K_2 \int \omega_2 dt + K_1 \int(\omega_2 - \omega_1) dt$$

Torque voltage analogous circuit

$T \to e(t)$	$J_1 \to L_1$	$B_1 \to R_1$	$K_1 \to 1/C_1$
$\omega_1 \to i_1$	$J_2 \to L_2$	$B_2 \to R_2$	$K_2 \to 1/C_2$
$\omega_2 \to i_2$			

The mesh basis equations using KVL

$$L_1 \frac{di_1}{dt} + R_1 i_1 + \frac{1}{C_1} \int(i_1 - i_2)\, dt = e(t)$$

$$L_2 \frac{di_2}{dt} + R_2\, i_2 + \frac{1}{C_2} \int i_2\, dt + \frac{1}{C_1} \int(i_2 - i_1)\, dt = 0$$

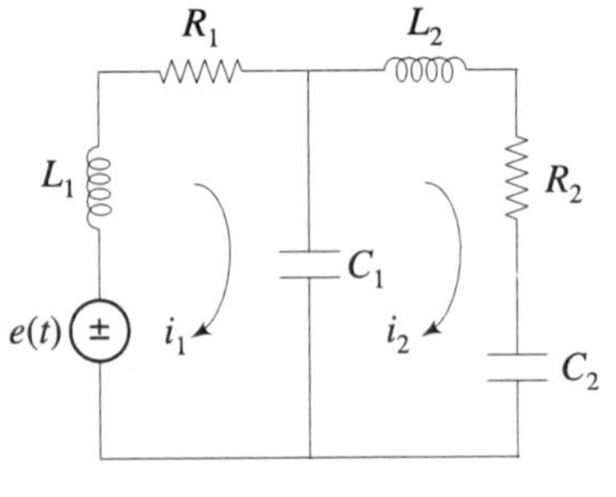

Fig. 2.70

Torque-current analogous circuit

The electrical analogous elements for the elements of mechanical rotational system

$T \rightarrow i(t)$	$B_1 \rightarrow \frac{1}{R_1}$	$\omega_1 \rightarrow V_1$	$J_1 \rightarrow C_1$	$K_1 \rightarrow \frac{1}{L_1}$
	$B_2 \rightarrow \frac{1}{R_2}$	$\omega_2 \rightarrow V_2$	$J_2 \rightarrow C_2$	$K_2 \rightarrow \frac{1}{L_2}$

The node basis equation using KCL

$$C_1 \frac{dV_1}{dt} + \frac{1}{R_1} V_1 + \frac{1}{L_1} \int (V_1 - V_2)dt = i(t)$$

$$C_2 \frac{dV_2}{dt} + \frac{1}{R_2} V_2 + \frac{1}{L_2} \int V_2 dt + \frac{1}{L_1} \int (V_2 - V_1)dt = 0$$

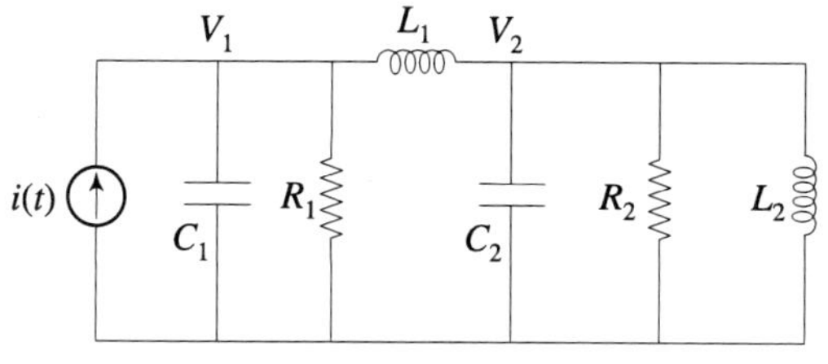

Fig. 2.71

Example 2.31: Write differential equations governing the mechanical rotational system shown in Fig. 2.72. Draw the torque-voltage and torque-current electrical analogous circuits and verify by writing mesh and node equations.

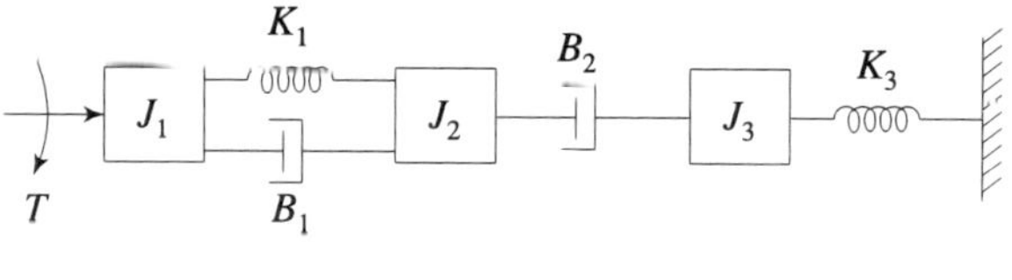

Fig. 2.72

Solution: Let the angular displacements of J_1, J_2 and J_3 be θ_1, θ_2 and θ_3 respectively. The corresponding angular velocities be ω_1, ω_2 and ω_3.

The differential equation at J_1

$$J_1 \frac{d^2\theta_1}{dt^2} + B_1 \frac{d(\theta_1 - \theta_2)}{dt} + K_1(\theta_1 - \theta_2) = T$$

The differential equation at T_2

$$J_2 \frac{d^2\theta_2}{dt^2} + B_2 \frac{d(\theta_2 - \theta_3)}{dt} + B_1 \frac{d(\theta_2 - \theta_1)}{dt} + K_1 (\theta_2 - \theta_1) = 0$$

The differential equation at J_3

$$J_3 \frac{d^2\theta_3}{dt^2} + B_2 \frac{d(\theta_3 - \theta_2)}{dt} + k_1\theta_3 = 0$$

On replacing the angular displacement by angular velocity in the differential equations governing the mechanical rotational system we get,

$$\frac{d^2\theta}{dt^2} = \frac{d\omega}{dt}, \frac{d\theta}{dt} = \omega, \theta = \int \omega dt$$

$$J_1 \frac{d\omega_1}{dt} + B_1(\omega_1 - \omega_2) + K_1 \int (\omega_1 - \omega_2)\, dt = T$$

$$J_2 \frac{d\omega_2}{dt} + B_2(\omega_2 - \omega_3) + B_1(\omega_2 - \omega_1) + K_1 \int (\omega_2 - \omega_1)\, dt = 0$$

$$J_3 \frac{d\omega_3}{dt} + B_2\,(\omega_3 - \omega_2) + K_3 \int \omega_3\, dt = 0$$

Torque-voltage analogous circuit

The electrical analogous elements for the elements of mechanical rotational system are given

$T \to e(t)$	$\omega_1 \to i_1$	$J_1 \to L_1$	$B_1 \to R_1$	$K_1 \to 1/C_1$
	$\omega_2 \to i_2$	$J_2 \to L_2$	$B_2 \to R_2$	$K_3 \to 1/C_3$
	$\omega_3 \to i_3$		$J_3 \to L_3$	

The mesh basis equation using KVL

$$L_1 \frac{di_1}{dt} + R_1\,(i_1 - i_2) + \frac{1}{C_2} \int (i_1 - i_2)\, dt = e(t)$$

$$L_2 \frac{di_2}{dt} + R_2\,(i_2 - i_3) + R_1\,(i_2 - i_1) + \frac{1}{C_1} \int (i_2 - i_1) dt = 0$$

$$L_3 \frac{di_3}{dt} + R_2\,(i_3 - i_2) + \frac{1}{C_3} \int i_3\, dt = 0$$

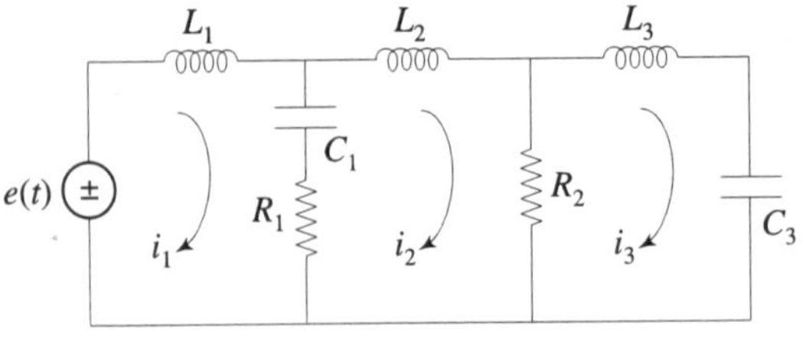

Fig. 2.73

Torque-current analogous circuit

$T \rightarrow i(t)$	$\omega_1 \rightarrow V_1$	$J_1 \rightarrow C_1$	$B_1 \rightarrow \frac{1}{R_1}$	$K_1 \rightarrow \frac{1}{L_1}$
	$\omega_2 \rightarrow V_2$	$J_2 \rightarrow C_2$		
	$\omega_3 \rightarrow V_3$	$J_3 \rightarrow C_3$	$B_2 \rightarrow \frac{1}{R_2}$	$K_3 \rightarrow \frac{1}{L_3}$

The node basis equations using KCL for the circuit

$$C_1 \frac{dV_1}{dt} + \frac{1}{R_1}(V_1 - V_2) + \frac{1}{L_2}\int(V_1 - V_2)\,dt = i(t)$$

$$C_2 \frac{dV_2}{dt} + \frac{1}{R_1}(V_2 - V_1) + \frac{1}{R_2}(V_2 - V_3) + \frac{1}{L_1}\int(V_2 - V_1)\,dt = 0$$

$$C_3 \frac{dV_3}{dt} + \frac{1}{R_1}(V_3 - V_2) + \frac{1}{L_3}\int V_3\,dt = 0$$

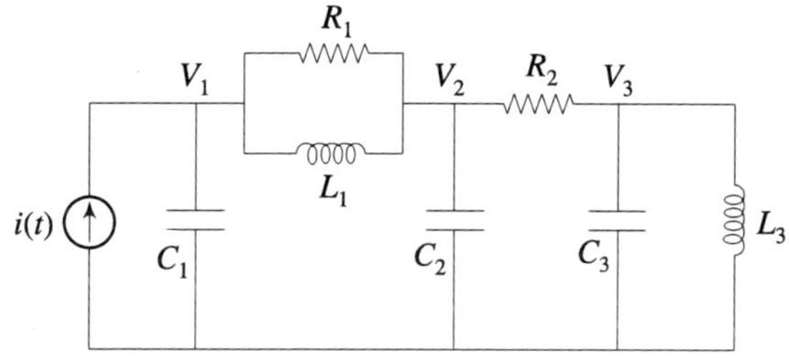

Fig. 2.74

2.7 GEAR TRAINS

It is possible to transmit mechanical energy from one part of a system to another via devices such as gear trains, timing belt over a pulley, or level. Let us consider a gear train with N_1 being the number of teeth on the first gear, N_2 the number of teeth on the second gear, T_1 the input torque on the first gear, θ_1 the angular displacement of the first gear, r_1 the radius of the first gear, T_2 input torque of the second gear θ_2 the angular displacement on the second gear and r_2 radius of the second gear.

Rule 1: The number N of teeth is proportional to the radius r of a gear: $r \propto N$.

Therefore, $r_1 \propto N_1$ and $r_2 \propto N_2$ or $\frac{r_1}{r_2} = \frac{N_1}{N_2}$ or $r_1 N_2 = r_2 N_2$ .....

Rule 2: The distance travelled on a gear is equal.

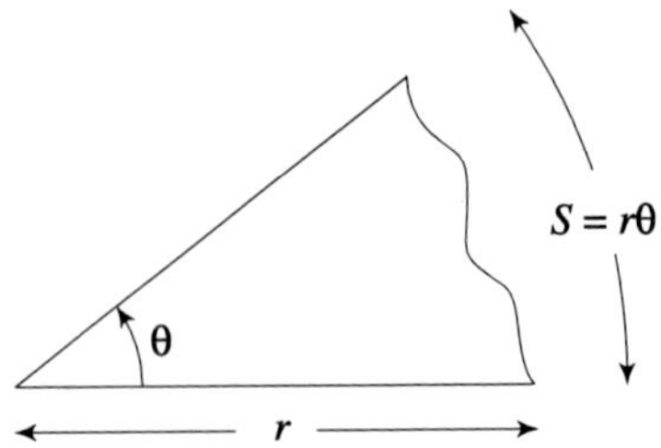

Fig. 2.75 Distance travelled on a gear

Figure 2.76 shows a gear train.

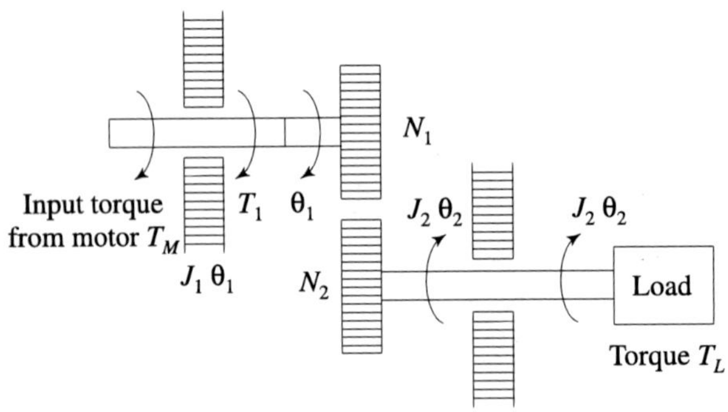

Fig. 2.76 Gear train

From Fig. 2.76 we can write $r_1\,\theta_1 = r_2\,\theta_2$.

Rule 3: Work done $T\theta$ is analogous to $\omega = Fx$. If there is no loss, we can write $T_1\theta_1 = T_2\,\theta_2$.

From equations

$$\frac{T_1}{T_2} = \frac{r_1}{r_2} = \frac{N_1}{N_2} = \frac{\theta_1}{\theta_2}$$

$$T_M = J_1\frac{d^2\theta_1}{dt^2} + D_1\frac{d\theta_1}{dt} + T_1$$

where T_1 is transmitted towards load.

Assuming no loss in the gear, we can write

$$T_2 = J_2\frac{d^2\theta_2}{dt^2} + D_2\frac{d\theta_2}{dt} + T_L$$

where T_2 is the reflected torque on gear secondary and T_L is the load torque.

The relation between T_1 and T_2 is given by

$$\frac{T_1}{T_2} = \frac{N_1}{N_2}$$

$$T_2 = \left(\frac{N_2}{N_1}\right) T_1$$

Using equation

$$\left(\frac{N_2}{N_1}\right) T_1 = J_2 \frac{d^2\theta_2}{dt} + D_2 \frac{d\theta_2}{dt} + T_L$$

$$T_1 = \left(\frac{N_1}{N_2}\right)\left[J_2 \frac{d^2\theta_2}{dt^2} + D_2 \frac{d\theta_2}{dt} + T_L\right]$$

$$T_1 = \left(\frac{N_1}{N_2}\right)\left[J_2 \left(\frac{N_1}{N_2}\right) \frac{d^2\theta_1}{dt^2} + D_2 \left(\frac{N_1}{N_2}\right) \frac{d\theta_2}{dt} + T_L\right]$$

Using
$$T_M = \left[J_1 + J_2 \left(\frac{N_1}{N_2}\right)^2\right] \frac{d^2\theta_1}{dt^2} + \left[D_1 + D_2 \left(\frac{N_1}{N_2}\right)^2\right] \frac{d\theta_1}{dt} + \left(\frac{N_1}{N_2}\right) T_L$$

$$T_M = J_{1\,equ} \frac{d^2\theta_1}{dt^2} + D_{1\,equ} \frac{d\theta_1}{dt} + T_L$$

where
$$J_{1\,equ} = J_1 + \left(\frac{N_1}{N_2}\right)^2 J_2$$

$$D_{1\,equ} = D_1 + \left(\frac{N_1}{N_2}\right)^2 D_2$$

$$T_L = \left(\frac{N_1}{N_2}\right)^2 T_L$$

Equation shows the terms referred to primary. Similarly, we can write

$$T_{2\,equ} = J_{2\,equ} \frac{d^2\theta_2}{dt^2} + D_{2\,equ} \frac{d\theta_2}{dt} + T_2$$

where
$$= J_{2\,equ} = J_2 + \left(\frac{N_1}{N_2}\right)^2 J_1$$

$$D_{2\,equ} = D_2 + \left(\frac{N_1}{N_2}\right)^2 D_1$$

$$T_{M\,equ} = \left(\frac{N_1}{N_2}\right) T_1$$

Equation shows the terms referred to secondary.

Example 2.32: Figure 2.77 shows a gear train system. Find the displacement of the load on the shaft (side 2) if $T_1 = 20\,e^{-2t}$ is applied on side 1.

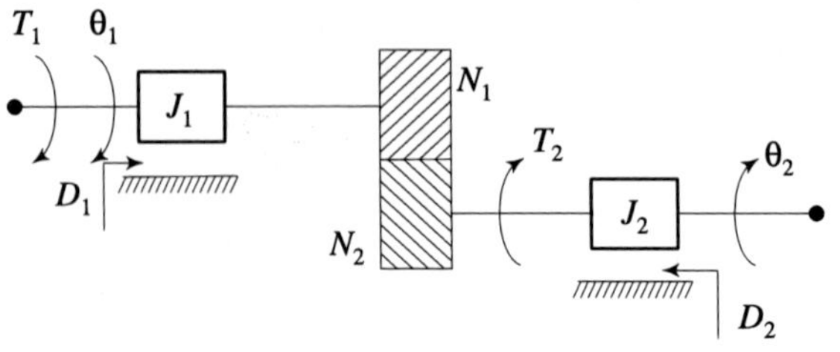

Fig. 2.77

In Fig. 2.67: $J_1 = 6$ kgm^2, $J_2 = 48$ kgm^2,

$D_1 = 3$ Nm/rad/sec, $D_2 = 12$ Nm/rad/sec

$N_1 = 1$ and $N_2 = 4$

Solution: The equivalents *J*, *D* and *T* referring to side – 2 are given by

$$J_{\text{equ}} = J_2 + J_1 \left(\frac{N_2}{N_1}\right)^2$$

$$D_{\text{equ}} = D_2 + D_1 \left(\frac{N_2}{N_1}\right)$$

$$T_2 = T_1 \left(\frac{N_2}{N_1}\right)$$

Again,

$$T_2 = J_{\text{equ}} \frac{d^2\theta_2}{dt^2} + D_{\text{equ}} \frac{d\theta_2}{dt}$$

Taking Laplace transform, we get

$$T(s) = S^2 J_{\text{equ}}\, \theta_2(s) + S\, D_{\text{equ}}\, \theta_2(s)$$

$$J_{\text{equ}} = J_2 + J_1 \left(\frac{N_2}{N_1}\right)^2$$

$$J_{\text{equ}} = 48 + \left(\frac{4}{1}\right)^2 \times 6 = 144 \text{ kgm}^2$$

$$D_{\text{equ}} = D_2 + D_1 \left(\frac{N_2}{N_1}\right)$$

$$D_{\text{equ}} = 12 + 3 \times \left(\frac{4}{1}\right) = 24 \text{ Nm/rad/sec}$$

$$T_2 = T_1 \left(\frac{N_2}{N_1}\right)$$

$$T_2 = \left(\frac{4}{1}\right) 20\, e^{-2t} = 80\, e^{-2t}$$

$$T_2(s) = \frac{80}{S+2}$$

$$\frac{80}{S+2} = S^2 J_{\text{equ}}\, \theta_2(s) + S D_{\text{equ}}\, \theta_2(s)$$

$$= (S^2 J_{\text{equ}} + S D_{\text{equ}})\, \theta_2(s)$$

$$\theta_2(s) = \frac{80}{(s+2)(S^2 + J_{equ} + S D_{equ})} = \frac{80}{(s+2)(144s^2 + 24s)}$$

$$= \frac{80}{(s+2)(144S+24)} = \frac{0.556}{(s+2)(s+0.167)}$$

$$= \frac{A}{S} + \frac{B}{(S+2)} + \frac{C}{(S+0.167)}$$

$$= \frac{1.665}{S} + \frac{0.1516}{(S+2)} + \frac{1.8163}{(S+0.167)}$$

Taking inverse Laplace transform

We get $\theta_2(t) = 1.665 + 0.1516e^{-2t} + 1.8163e^{-0.167t}$

2.7.1 Transfer Function of Armature Controlled DC Motor

The speed of a DC motor is directly proportional to armature voltage and inversely proportional to flux in field winding. In armature controlled DC motor the desired speed is obtained by varying the armature voltage. This speed control system is an electromechanical control system. The electrical system consists of the armature and the field circuit but for analysis purpose, only the armature circuit is considered because the field is excited by a constant voltage. The mechanical system consists of the rotating part of the motor and load connected to the shaft of the motor. The armature controlled DC motor speed control system is shown in Fig. 2.78.

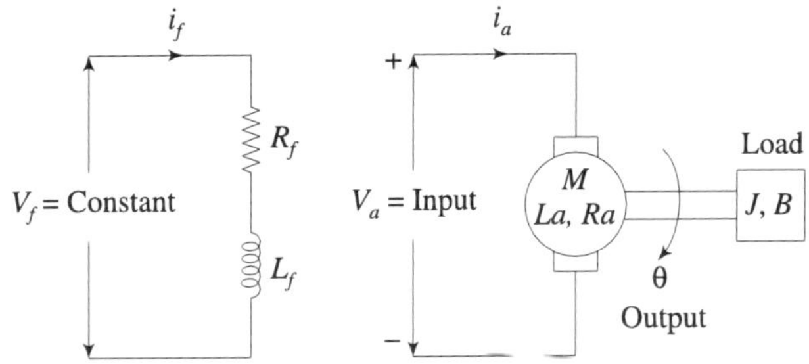

Fig. 2.78

R_a = Armature resistance, Ω

L_a = Armature inductance, H

i_a = Armature current, A

V_a = Armature voltage, V

e_b = Back emf, V

K_t = Torque constant, N-m/A

T = Torque development by motor, N-m

θ = Angular displacement of shaft, radian

J = Moment of inertia of motor and load, kg-m^2/rad

B = Frictional coefficient of motor and load, N-m/(rad/sec)

K_b = Back emf constant, V/(rad/sec)

The equivalent circuit of armature is shown in Fig. 2.79.

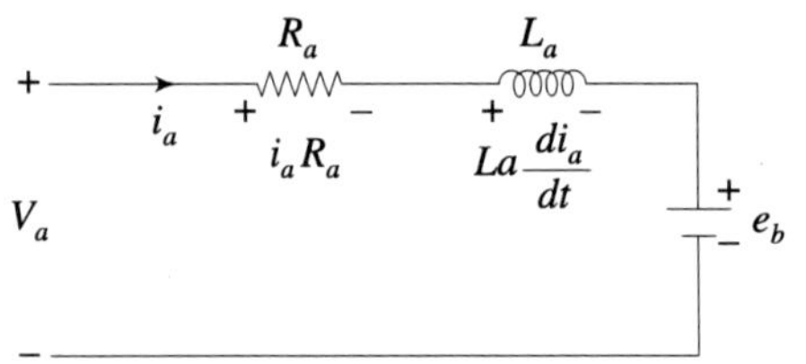

By KVL, we can write $i_aR_a + L_a\frac{di_a}{dt} + e_b = V_a$

Fig. 2.79

Torque of a DC motor is proportional to the product of flux and current. Since flux is constant in this system, the torque is proportional to i_a alone.

$$T \propto i_a$$

$$\therefore \qquad \text{Torque, } T = K_t\, i_a$$

The mechanical system of the motor is shown in Fig. 2.80. The differential equation governing the mechanical system of motor is given by

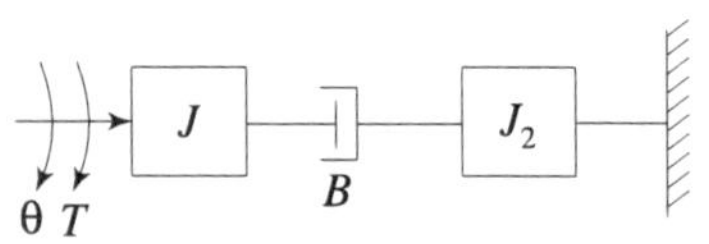

Fig. 2.80

$$J\frac{d^2\theta}{dt^2} + B\frac{d\theta}{dt} = T$$

The back emf of a DC machine is proportional to speed (angular velocity) of shaft

$$\therefore \qquad e_b \propto \frac{d\theta}{dt}, \quad \text{Back emf, } e_b = K_b\frac{d\theta}{dt}$$

The Laplace transform of various time domain signals involved in this system are

$$L[V_a] = V_a(s), \quad L[e_b] = E_b(s), \quad L[T] = T[s]$$

$$L[i_a] = I_a(s), \quad L[\theta] = \theta[s]$$

The differential equations governing the armature controlled DC motor speed control system are

$$i_a R_a + L_a \frac{dia}{dt} + e_b = V_a$$

$$J \frac{d^2\theta}{dt^2} + B \frac{d\theta}{dt} = T$$

$$T = K_t\, i_a, \quad e_b = K_b \frac{d\theta}{dt}$$

Taking Laplace transform of the system differenttial equations with zero initial condition we get

$$I_a(s)\, R_a + L_a\, SI_a(s) + E_b(s) = V_a(s)$$

$$T(s) = K_t \cdot I_a(s)$$

$$JS^2\theta(s) + BS\theta(s) = T(s)$$

$$E_b(s) = K_b S\theta(s)$$

$$K_t I_a(s) = (JS^2 + BS)\theta(s)$$

$$I_a(s) = \left(\frac{JS^2 + BS}{K_t}\right)\theta(s)$$

$$(R_a + SL_a)\, I_a(s) + E_b(s) = V_a(s)$$

$$(R_a + SL_a) \frac{(JS^2 + BS)}{K_t} \theta(s) + K_b\, S\theta(s) = V_a(s)$$

$$\left[\frac{(R_a + SL_a)(JS^2 + BS) + K_b K_t S}{K_t}\right] \theta(s) = V_a(s)$$

$$\theta(s)/V_a(s) = \frac{K_t}{(R_a + SL_a)\,(JS^2 + BS) + K_b\, K_t\, S}$$

$$= \frac{K_t}{R_a JS^2 + R_a BS + LaJS^3 + L_a BS^2 + K_b K_t S}$$

$$= \frac{K_t}{S[JL_a S^2 + (JR_a + BL_a)S + [BR_a + K_b K_t)]}$$

$$= \frac{\dfrac{K_t}{JL_a}}{S\left[S^2 + \left(\dfrac{JR_a + BL_a}{JL_a}\right)S + \left(\dfrac{BRa + K_b K_t}{JL_a}\right)\right]}$$

$$\frac{\theta(s)}{V_a(s)} = \frac{K_t}{(R_a + SL_a)(Js^2 + BS) + K_b K_t S}$$

$$\frac{\theta(s)}{V_a(s)} = \frac{K_t}{R_a\left(\frac{SL_a}{R_a} + 1\right) BS\left(1 + \frac{JS^2}{BS}\right) + K_b K_t S}$$

$$\frac{\theta(s)}{V_a(s)} = \frac{\frac{K_t}{R_a B}}{S\left[(1 + ST_a)(1 + ST_m) + \frac{K_b K_t}{R_a B}\right]}$$

where $\frac{L_a}{R_a} = T_a$ = electrical time constant

$\frac{J}{B} = J_m$ = mechanical time constants

EXERCISE

Objective Type Questions

1. The transfer function $\frac{V_2(s)}{V_1(s)}$ of the circuit shown in Fig. 2.81? (GATE 2013)

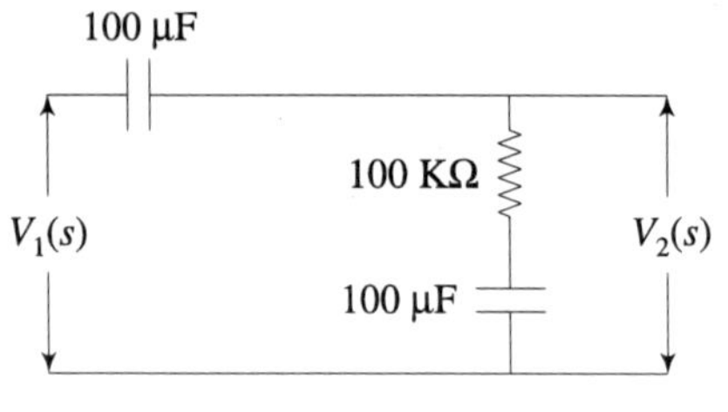

Fig. 2.81

Ans:

$$V_2(s) = V_1(S)\frac{R + \frac{1}{SC}}{R + \frac{1}{SC} + \frac{1}{SC}}$$

$$\frac{V_2(s)}{V_1(s)} = \frac{1 + SCR}{2 + SCR}$$

$$\frac{V_2(s)}{V_1(s)} = \frac{S + 1}{S + 2}$$

$$RC = 100 \times 10^3 \times 100 \times 10^{-6} = 1$$

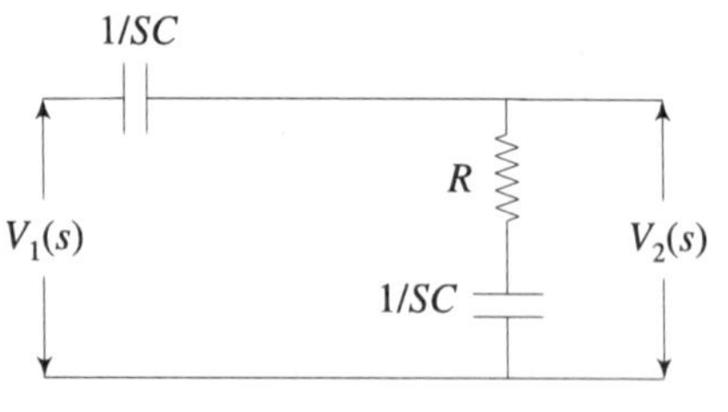

Fig. 2.82

2. A linear time-invariant system has an impulse response $-e^{2t}$, $t > 0$. If the initial conditions are zero and the input is e^{3t}, then output for $t > 0$ is?

Ans:

$$\frac{T}{F} = L$$

$$= \frac{1}{S} - 2$$

Output $\quad C(s) = \frac{T}{F} \times R(s) = \frac{1}{(S-3)(S-2)}$

By partial fraction $\quad = \frac{1}{(S-3)} - \frac{1}{(S-2)}$

Apply the inverse Laplace transfer $\Rightarrow C(t) = e^{3t} - e^{2t}$

3. Let $x(t)$ be the input to a linear, time-invariant system. The required output is $4x(t-2)$. The transfer function of the system is?

Ans: $\quad Y(t) = 4X(t-2)$

Apply $\quad LT = Y(s) = 4 \times (s).e^{-2S}$

$$T/F = \frac{L_{\{\text{Output}\}}}{L_{\{\text{Input}\}}}$$

$$T/F = 4.e^{-2jw} = 4.e^{-2j(2\pi f)} = \mathbf{4.e^{-4j\pi f}}$$

4. The transfer function of mathematical model of the mechanical system given in Fig. is 2.83.

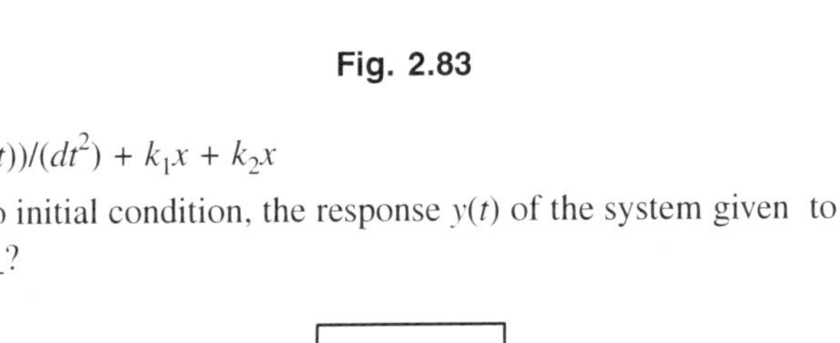

Fig. 2.83

Ans: $u(t) = M\,(d^2x(t))/(dt^2) + k_1x + k_2x$

5. Assuming zero initial condition, the response $y(t)$ of the system given to a unit step input $u(t)$ is __________?

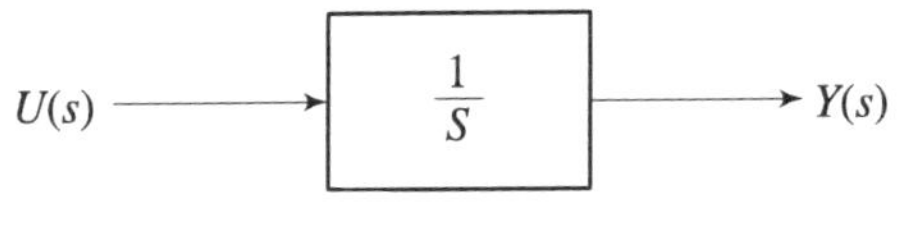

Fig. 2.84

Ans: One

6. The transfer function $V_2(s)/V_1(s)$ of the circuit is __________.

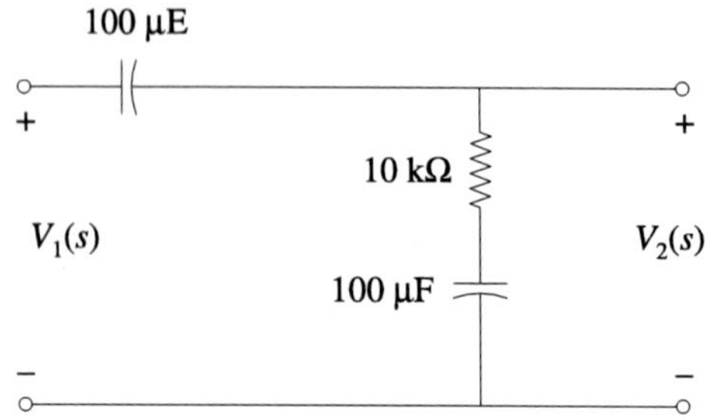

Fig. 2.85

Ans: $V_2(s)/V_1(s) = (S + 1)/(S + 2)$

7. Transfer function $C(s)/R(s)$ having differential equation $(2d^2c(t))/(dt^2) + 2(dc(t))/dt + c(t) = r(t) + 2r(t - 1)$ is __________.

Ans: $C(s)/R(s) = (1 + 2e^{-s})/(2s^2 + 2s + 1)$

8. For the mechanical system shown in Fig. 2.86, derive the transfer function. Also draw the force-voltage and force-current analogous circuits.

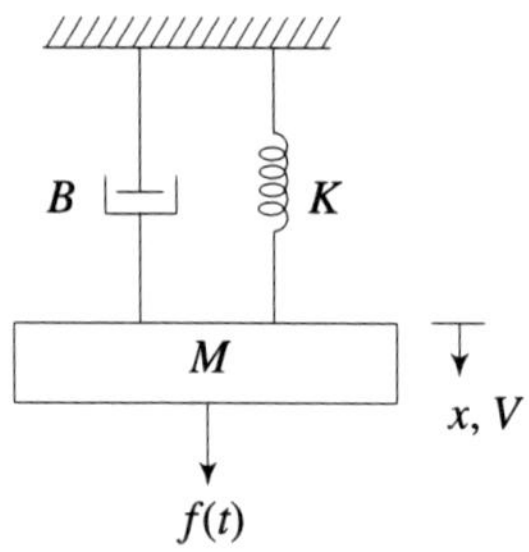

Fig. 2.86

Ans: The transfer function is $\dfrac{X(s)}{F(s)} = \dfrac{1}{(MS^2 + BS + K)}$

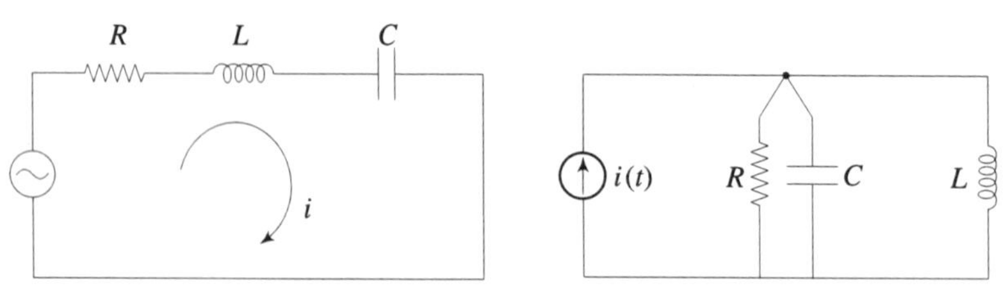

$f(t) \to e(t)$	$M \to L$	$f(t) \to i(t)$	$M \to C$
$V \to i$	$B \to R$	$V \to V$	$B \to 1/R$
	$K \to 1/C$		$K \to 1/L$

Force-voltage analogous circuit   Force-current analogous circuit

Fig. 2.87

9. For the mechanical system shown in Fig. 2.88, draw the force-voltage and force-current analogous circuits.

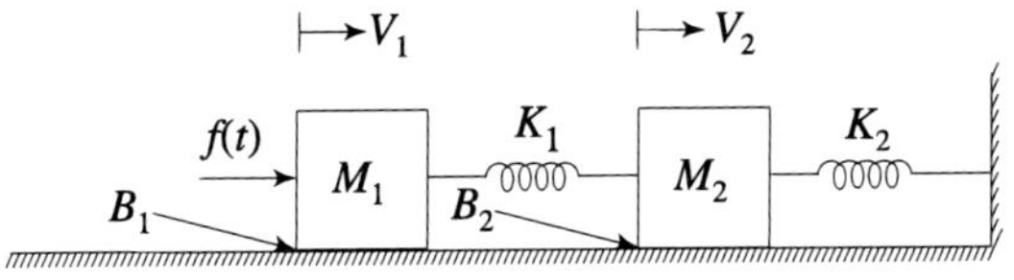

Fig. 2.88

Ans:

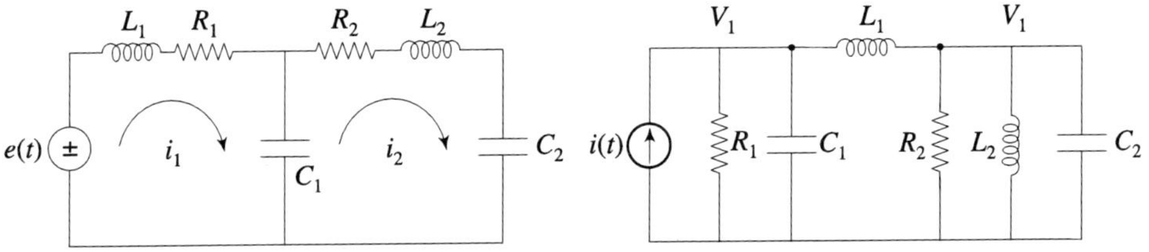

$f(t) \rightarrow e(t)$	$M_1 \rightarrow L_1$	$B_2 \rightarrow R_2$
$V_1 \rightarrow i_1$	$M_2 \rightarrow L_2$	$K_1 \rightarrow 1/C_1$
$V_2 \rightarrow i_2$	$B_1 \rightarrow R_1$	$K_2 \rightarrow 1/C_2$

Force-voltage analogous circuit

$f(t) \rightarrow i(t)$	$M_1 \rightarrow C_1$	$B_2 \rightarrow 1/R_2$
$V_1 \rightarrow V_1$	$M_2 \rightarrow C_2$	$K_1 \rightarrow 1/L_1$
$V_2 \rightarrow V_2$	$B_1 \rightarrow 1/R_1$	$K_2 \rightarrow 1/L_2$

Force-current analogous circuit

Fig. 2.89

10. Write the differential equations governing the mechanical system shown in Fig. 2.90. Also draw the force-voltage and force-current analogous circuit.

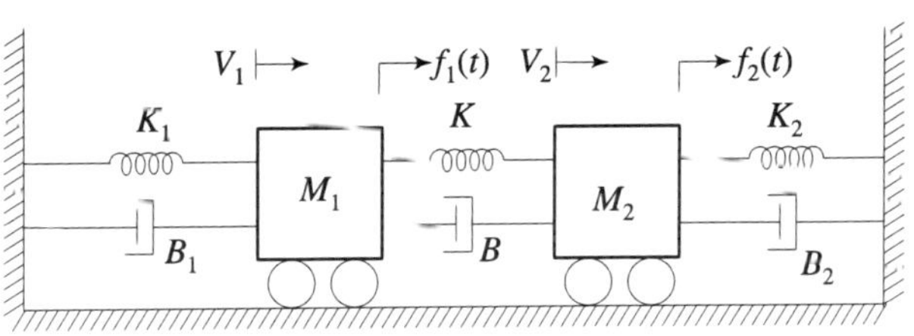

Fig. 2.90

Ans: $$M_1 \frac{dV_1}{dt} + B_1V_1 + B(V_1 - V_2) + K_1 \int V_1 dt + K \int (V_1 - V_2)dt = f_1(t)$$

$$M_2 \frac{dV_2}{dt} + B_2V_2 + B(V_2 - V_1) + K_2 \int V_2 dt + K \int (V_2 - V_1)dt = f_2(t)$$

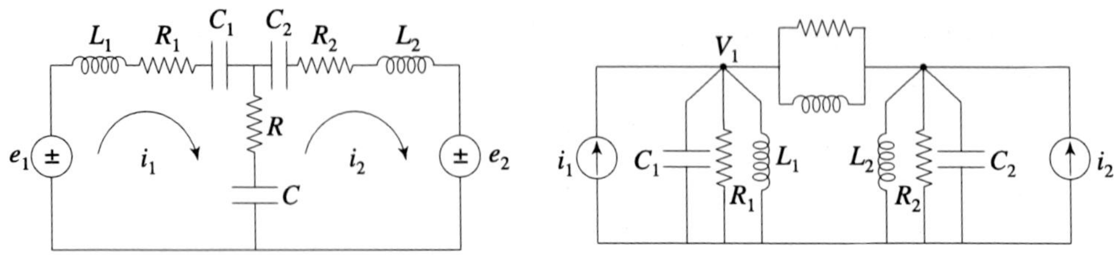

$f_1 \to e_1$	$M_1 \to L_1$	$B \to R$
$f_2 \to e_2$	$M_2 \to L_2$	$K_1 \to 1/C_1$
$V_1 \to i_1$	$B_1 \to R_1$	$K_2 \to 1/C_2$
$V_2 \to i_2$	$B_2 \to R_2$	$K_2 \to 1/C$

Force-voltage analogous circuit

$f_1 \to i_1$	$M_1 \to C_1$	$B \to 1/R$
$f_2 \to i_2$	$M_2 \to C_2$	$K_1 \to 1/L_1$
$V_1 \to V_1$	$B_1 \to 1/R_1$	$K_2 \to 1/L_2$
$V_2 \to V_2$	$B_2 \to 1/R_2$	$K \to 1/L$

Force-current analogous circuit

Fig. 2.91

11. Write the differential equation governing the mechanical system shown in Fig. 2.92. Draw the force-voltage and force-current analogous circuit.

Ans: $M_1 \frac{dV_1}{dt} + K_1 \int V_1 dt + K_2 \int (V_1 - V_3)dt + K_3 \int (V_1 - V_2)\, dt = f(t)$

$$K_3 \int (V_2 - V_1)dt + B\,(V_2 - V_3) = 0$$

$$M_2 \frac{dV_3}{dt} + B(V_3 - V_2) + K_2 \int (V_3 - V_1)\, dt = 0$$

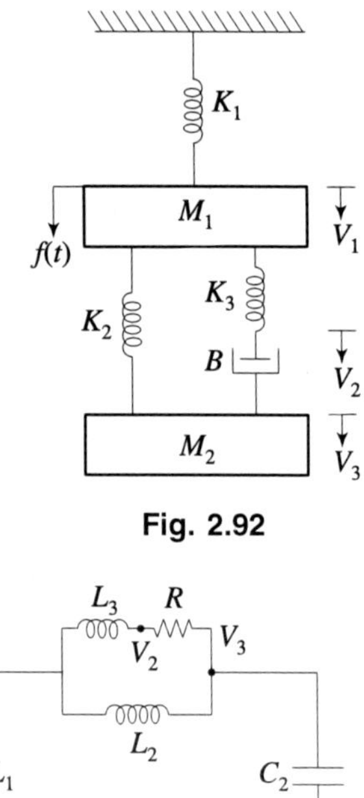

Fig. 2.92

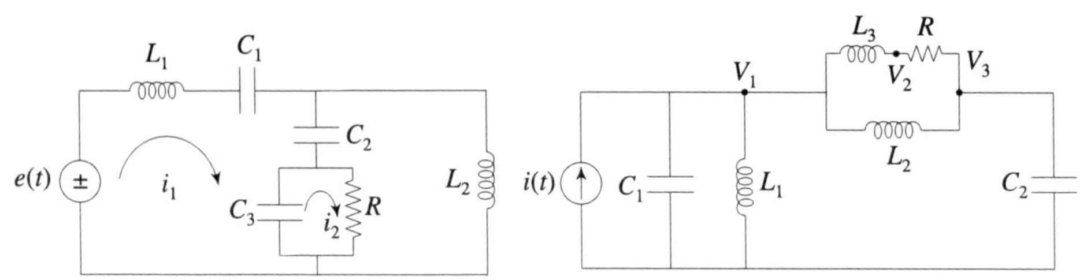

$f(t) \to e(t)$	$M_1 \to L_1$	$K_1 \to 1/C_1$
$V_1 \to i_1$	$M_2 \to L_2$	$K_2 \to 1/C_2$
$V_2 \to i_2$	$B \to R$	$K_3 \to 1/C_3$
$V_3 \to i_3$		

Force-voltage analogous circuit

$f(t) \to i(t)$	$M_1 \to C_1$	$K_1 \to 1/L_1$
$V_1 \to V_1$	$M_2 \to C_2$	$K_2 \to 1/L_2$
$V_2 \to V_2$	$B \to 1/R$	$K_3 \to 1/L_2$
$V_3 \to V_3$		

Force-current analogous circuit

Fig. 2.93

12. Consider the mechanical translational system shown in Fig. 2.94, draw (a) force-voltage and (b) force-current analogous circuits.

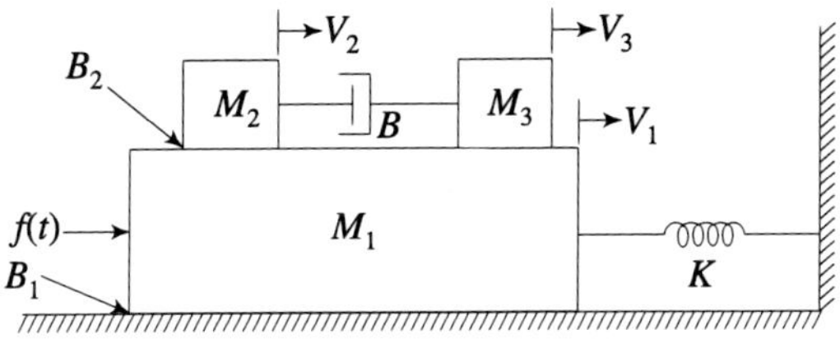

Fig. 2.94

Ans:

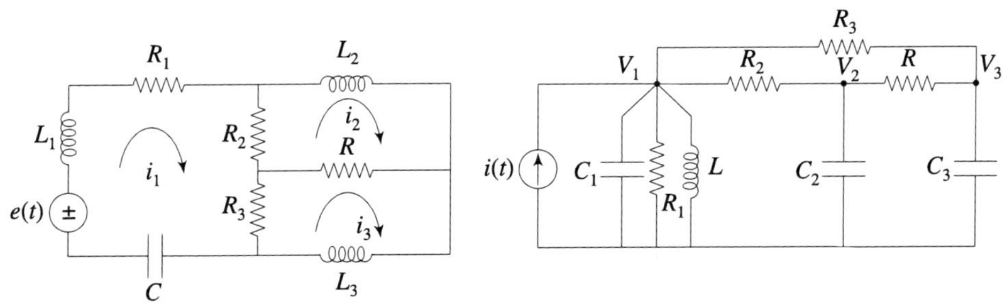

$f(t) \rightarrow e(t)$	$M_1 \rightarrow L_1$	$B_1 \rightarrow R_1$	$f(t) \rightarrow i(t)$	$M_1 \rightarrow C_1$	$B_1 \rightarrow 1/R_1$
$V_1 \rightarrow i_1$	$M_2 \rightarrow L_2$	$B_2 \rightarrow R_2$	$V_1 \rightarrow V_1$	$M_2 \rightarrow C_2$	$B_2 \rightarrow 1/R_2$
$V_2 \rightarrow i_2$	$M_3 \rightarrow L_3$	$B_3 \rightarrow R_3$	$V_2 \rightarrow V_2$	$M_3 \rightarrow C_3$	$B_3 \rightarrow 1/R_3$
$V_3 \rightarrow i_3$	$B \rightarrow R$	$K \rightarrow 1/C$	$V_3 \rightarrow V_3$	$B \rightarrow 1/R$	$K \rightarrow 1/L$

Force-voltage-analogous circuit — Force-current analogous circuit

Fig. 2.95

13. Write the differential equations governing the rotational mechanism system shown in Fig. 2.96. Also draw the torque-voltage and torque-current analogous circuits.

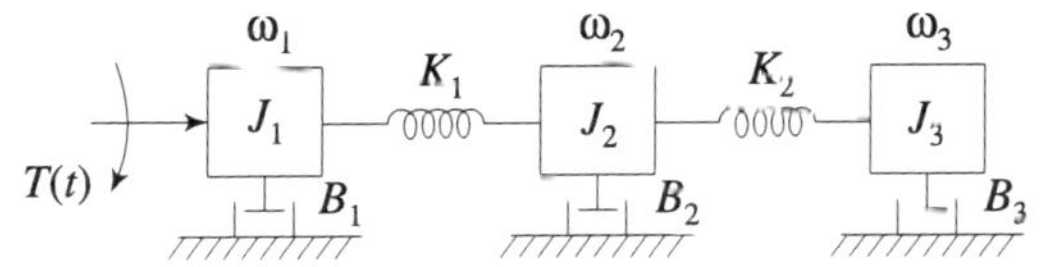

Fig. 2.96

Ans: $J_1 \dfrac{d\omega_1}{dt} + B_1\omega_1 + K_1 \int (\omega_1 - \omega_2)dt = T(t)$

$$J_2 \frac{d\omega_2}{dt} + B_2\omega_2 + K_1 \int (\omega_2 - \omega_1)dt + K_2 \int (\omega_2 - \omega_3)dt = 0$$

$$J_3 \frac{d\omega_2}{dt} + B_3\omega_3 + K_2 \int (\omega_3 - \omega_2)dt = 0$$

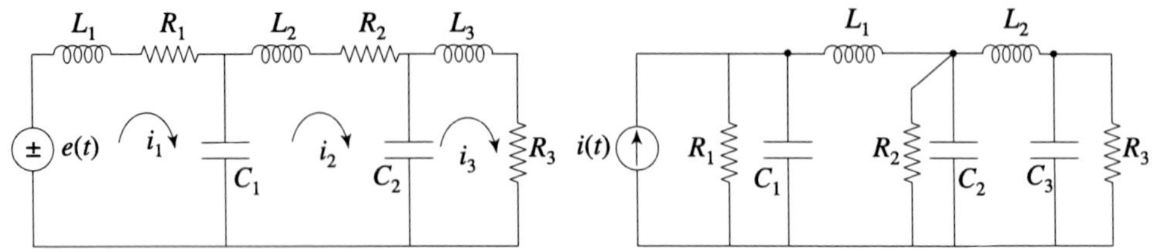

$T(t) \to e(t)$ $J_1 \to L_1$ $B_1 \to R_1$ $K_1 \to 1/C_1$
$w_1 \to i_1$ $J_2 \to L_2$ $B_2 \to R_2$ $K_2 \to 1/C_2$
$w_2 \to i_2$ $J_3 \to L_3$ $B_3 \to R_3$
$w_3 \to i_3$
Torque-voltage analogous circuit

$T(t) \to i(t)$ $J_1 \to C_1$ $B_1 \to 1/R_1$ $K_1 \to 1/L_1$
$w_1 \to V_1$ $J_2 \to C_2$ $B_1 \to 1/R_2$ $K_2 \to 1/L_2$
$w_2 \to V_2$ $J_3 \to C_3$ $B_3 \to 1/R_3$
$w_3 \to V_3$
Torque-current analogous circuit

Fig. 2.97

14. In an electrical circuit the elements, resistance, capacitance and inductance, are connected in parallel action the voltage some E as shown in Fig. 2.98. Draw (a) translational mechanical system, and (b) rotational mechanical analogous system.

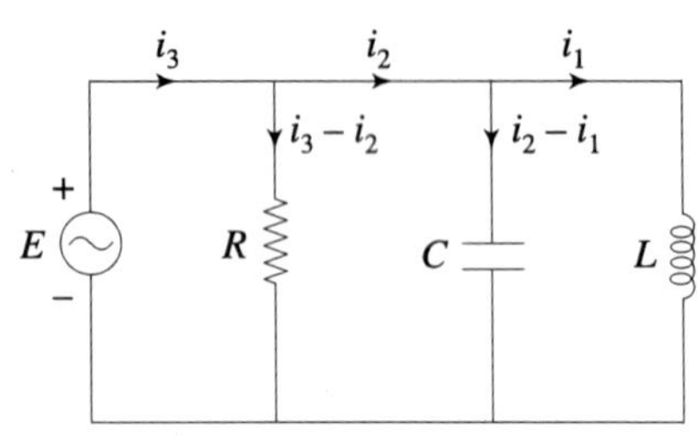

Fig. 2.98

Ans:

$e(t) \to f(t)$ $R \to B$ Analogous
$i_1 \to V_1$ $1/C \to K$ mechanical
$i_2 \to V_2$ $L \to M$ translational system
$i_3 \to V_3$

$e(t) \to T(t)$ $R \to B$ Analogous
$i_1 \to w_1$ $1/C \to K$ mechanical
$i_3 \to w_3$ $L \to j$ rotational system
$i_3 \to w_3$

Fig. 2.99

15. Find the transfer functions of the networks shown in Fig. 2.100

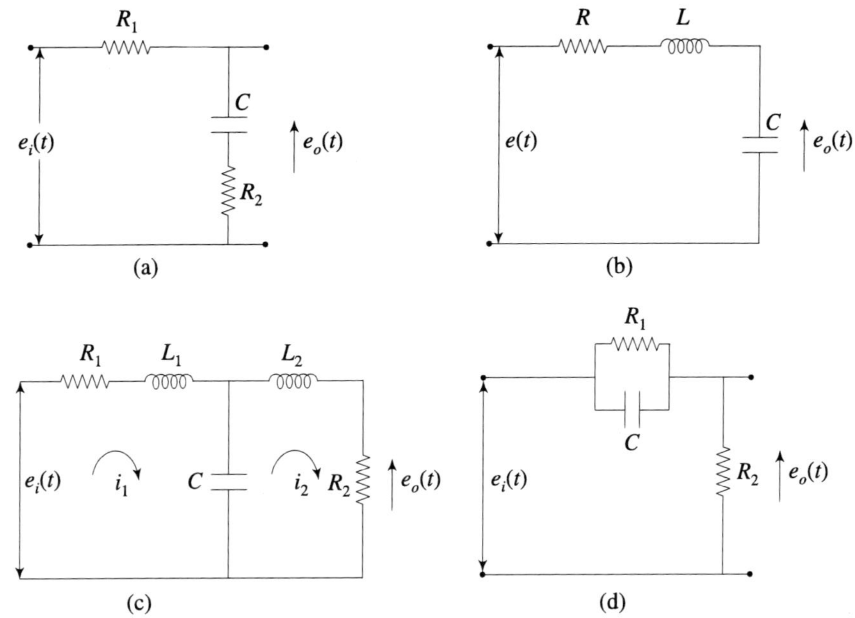

Fig. 2.100

Ans:

(a) $\dfrac{E_0(s)}{E_i(s)} = \dfrac{1 + sR_2C}{1 + S(R_1 + R_2)C}$

(b) $\dfrac{E_0(s)}{E_i(s)} = \dfrac{1}{S^2LC + SRC + 1}$

(c) $\dfrac{E_0(s)}{Ei(s)} = \dfrac{sR_2C}{(S^2L_1C + sR_1C + 1)\,(S^2L_2C + SR_2C + 1) \quad 1}$

(d) $\dfrac{E_0(s)}{E_i(s)} = \dfrac{sR_1R_2C + R_2}{SR_1R_2C + (R_1 + R_2)}$

16. Find the transfer function of the circuit shown in Fig. 2.101

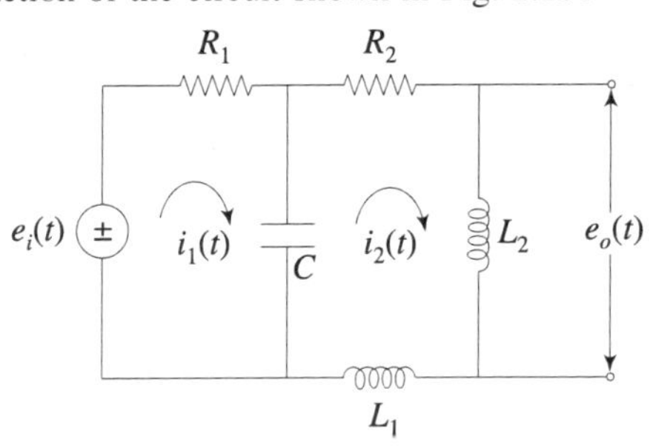

Fig. 2.101

Ans: $$\frac{C(s)}{E(s)} = \frac{s^2LC}{[SR_1C + 1]\,[S^2\,(L_1 + L_2)C + SR_2\,C + 1] - 1}$$

17. What do you mean by transfer function?
18. Find the transfer function of the network shown in Fig. 2.102.

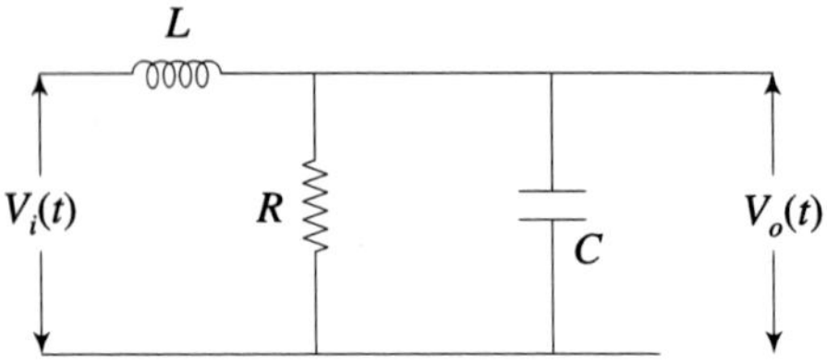

Fig. 2.102

Ans: $$\frac{V_o(s)}{V_i(s)} = \frac{1}{LCS^2 + \frac{L}{R}s + 1}$$

19. Derive the transfer function of the network shown in Fig. 2.103.

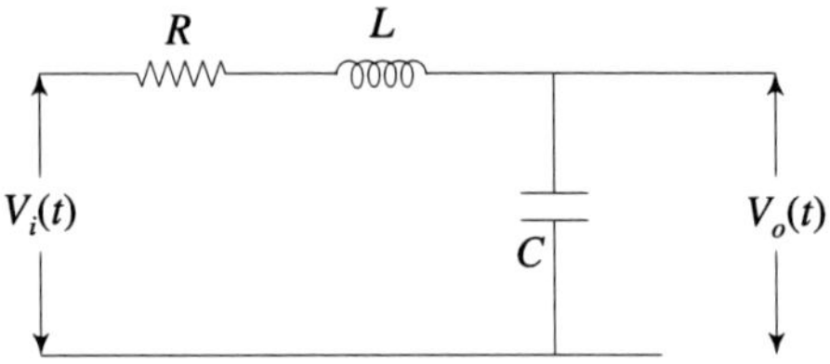

Fig. 2.103

Ans: $$\frac{V_o(s)}{V_i(s)} = \frac{1}{LCS^2 + RCS + 1}$$

20. Figure 2.104 shows a mass M sliding on a horizontal plane with a velocity u. The applied force is f and the coefficient of viscous friction is B. Obtain the equation relating the force and velocity. Draw the electrical analogue of this mass and friction system.

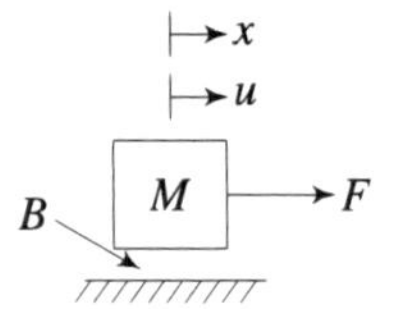

Fig. 2.104

Ans:

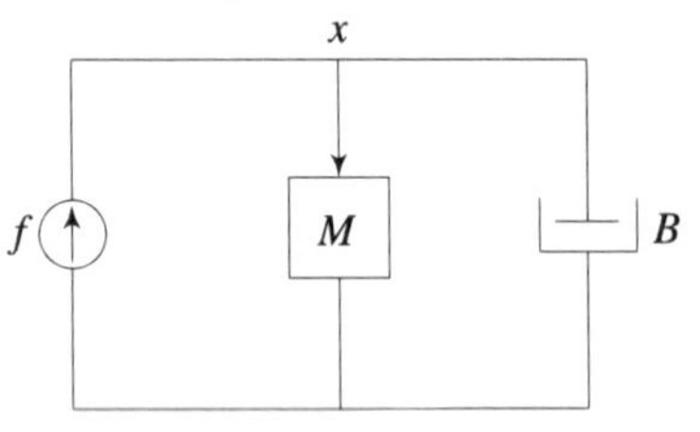

$$f = M\frac{d^2x}{dt^2} + B\frac{dx}{dt}$$

$$f = M\frac{du}{dt} + Bu$$

$$V = L\frac{di}{dt} + Ri$$

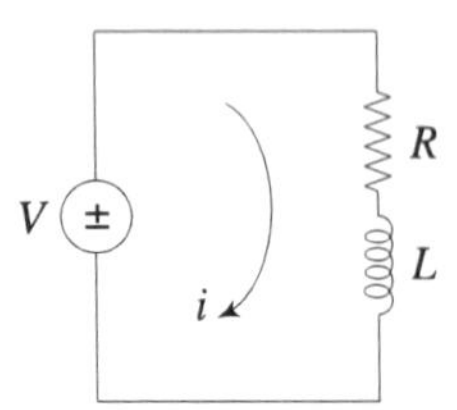

Fig. 2.105

21. Figure 2.106 shows a mass spring friction system. Obtain the equation relating the applied force f and velocity u. Draw the electrical analogue of this system.

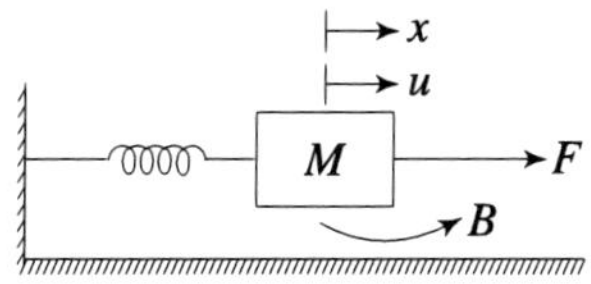

Fig. 2.106

Ans:

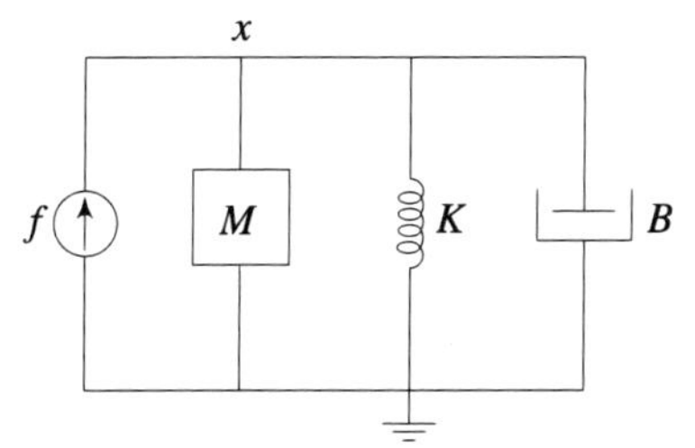

$$f = M\frac{du}{dt} + Bu + \frac{1}{K}\int udt$$

Fig. 2.107(a)

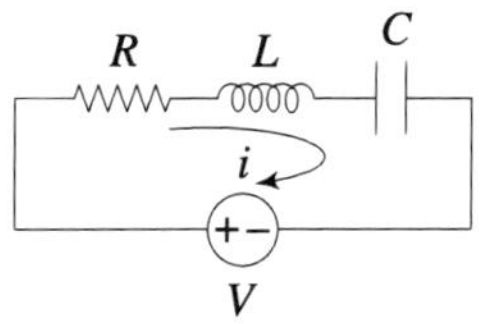

$$V = L\frac{di}{dt} + R_i + \frac{1}{C}\int i\,dt$$

Fig. 2.107(b)

22. Draw the electrical analogue for the mechanical system shown in Fig. 2.108 based on
    (i) force-voltage analogy
    (ii) force-current analogy

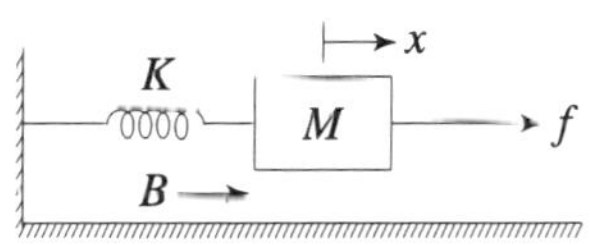

Fig. 2.108

Ans:

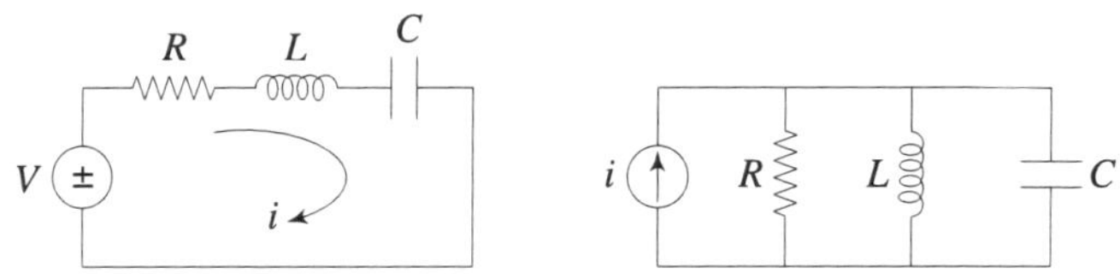

F-V analogue   *F-I* analogue

Fig. 2.109

23. Write the system equations for the mechanical system shown in Fig. 2.110. Determine the transfer functions $\frac{X_2(s)}{F(s)}$ and $\frac{X_1(s)}{F(s)}$.

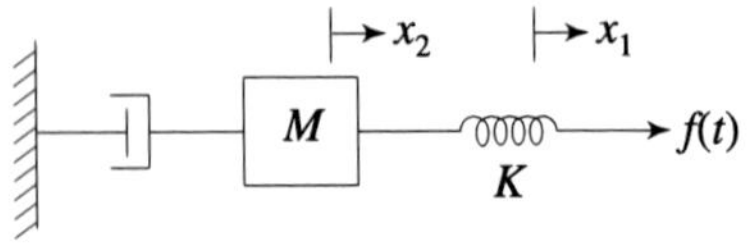

Fig. 2.110

Ans:

$$f(t) = K(x_1 - x_2)$$

$$K(x_1 - x_2) = M\frac{d^2x}{dt^2} + B\frac{dx_2}{dt}$$

$$\frac{X_2(s)}{F(s)} = \frac{1}{S(MS + B)}, \frac{X_1(s)}{F(s)} = \frac{MS^2 + BS + K}{KS(MS + B)}$$

24. Write the differential equations for the mechanical system shown in Fig. 2.111. Also draw the analogous electrical circuit based on force-current analogy.

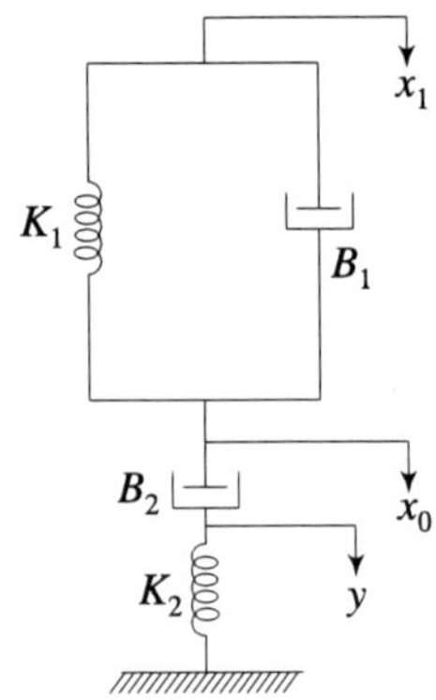

Fig. 2.111

Ans:

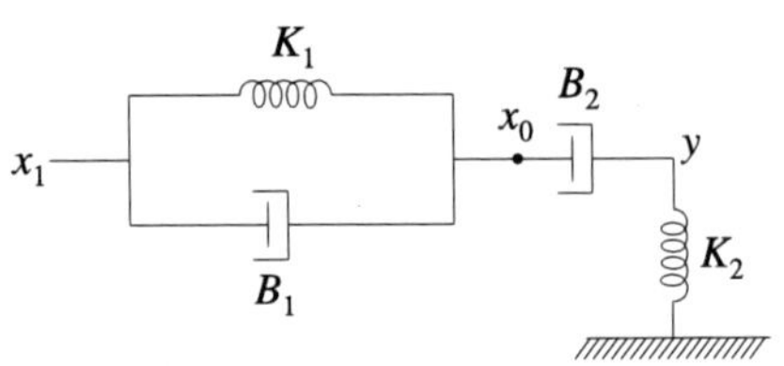

Fig. 2.112

$$K_1(x_0 - x_1) + B_1\left(\frac{dx_0}{dt} - \frac{dx}{dt}\right) + B_2\left(\frac{dx_0}{dt} - \frac{dy}{dt}\right) = 0$$

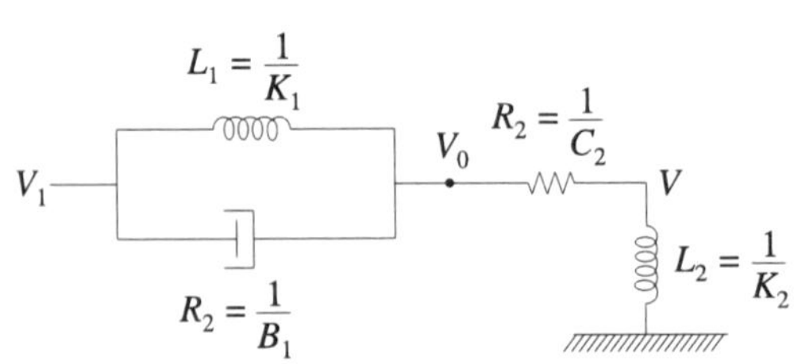

Fig. 2.113

25. Draw an electrical analogue for the mechanical system shown in Fig. 2.114.

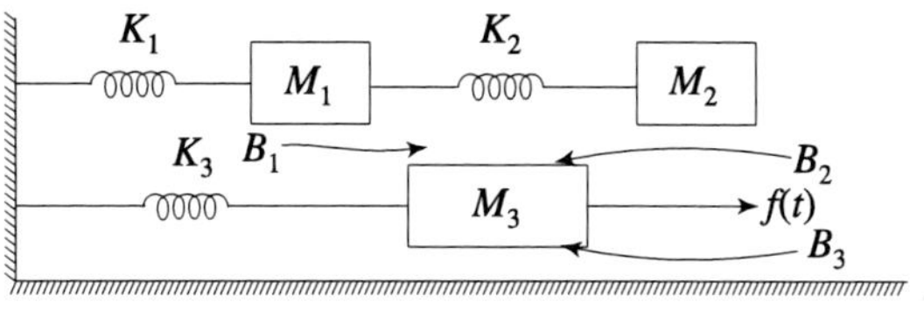

Fig. 2.114

Ans:

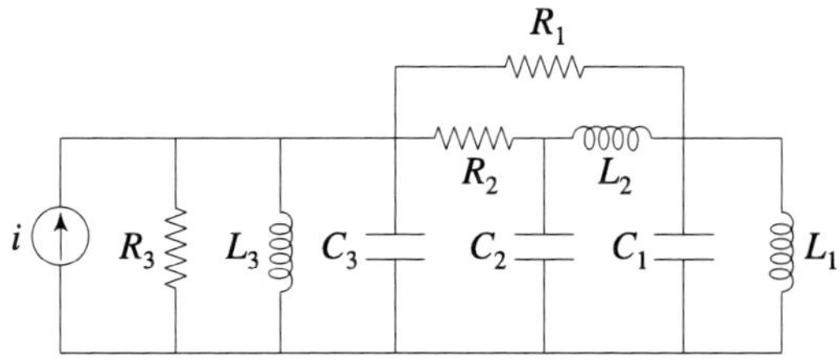

Fig. 2.115

26. Draw the electrical analog for the mechanical system shown in Fig. 2.116.

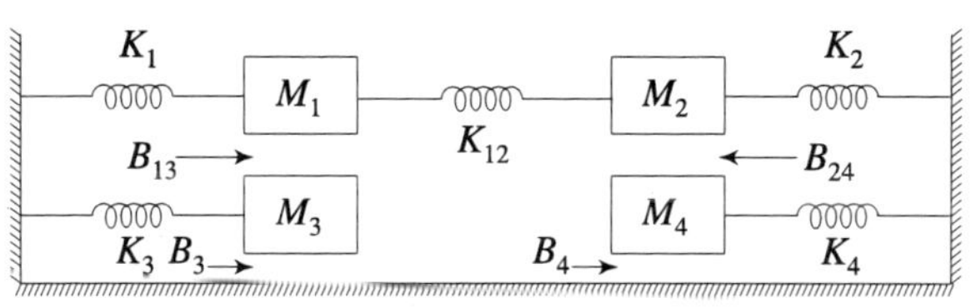

Fig. 2.116

Ans:

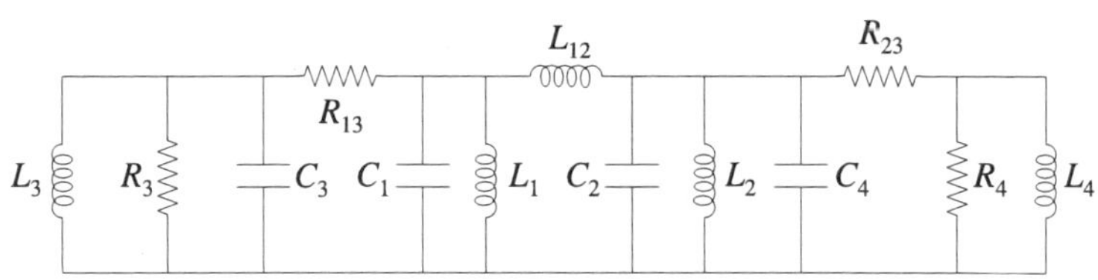

Fig. 2.117

27. Obtain electrical analogy for the mechanical systems shown in Fig. 2.118.

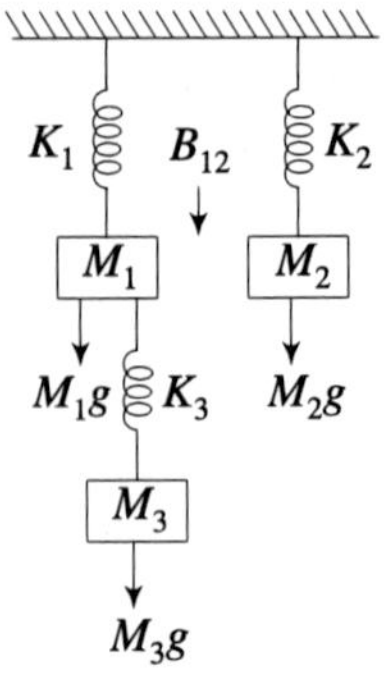

Fig. 2.118

Ans:

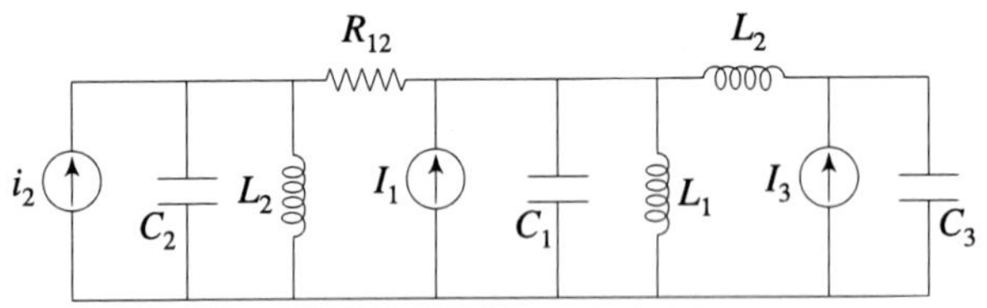

Fig. 2.119

28. Obtain the electrical analogous networks based on

    (i) Torque-voltage and (ii) Torque-current

    Analysis of the mechanical system shown in Fig. 2.120

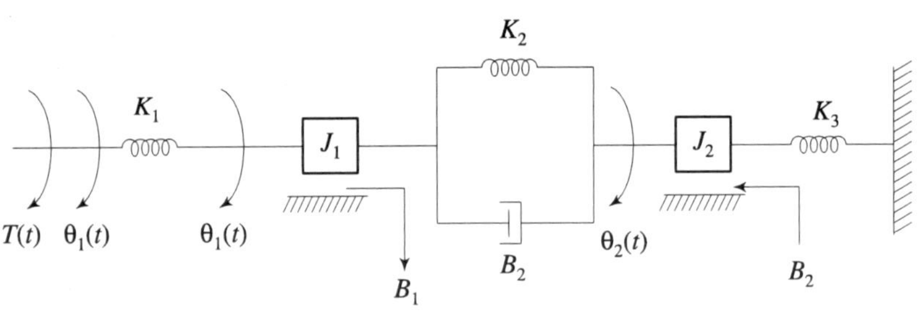

Fig. 2.120

Ans: (i)

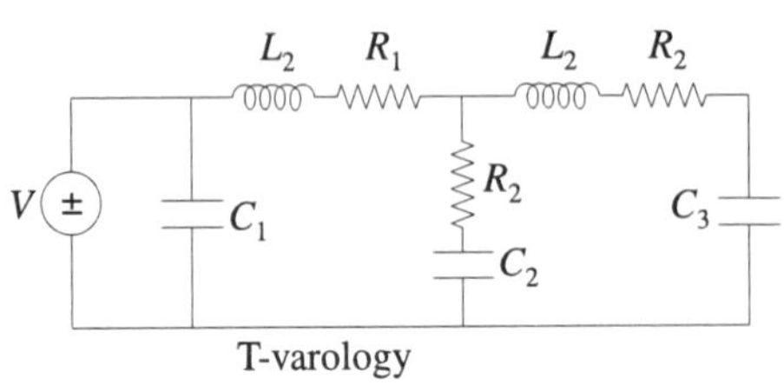

Fig. 2.121(a)

(ii) 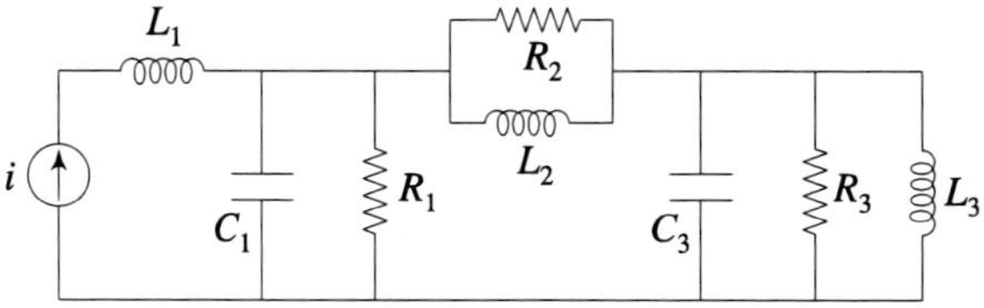

T-currect

Fig. 2.121(b)

29. Derive the system equations in Laplace form for the system shown in Fig. 2.122 and obtain the transfer function $\frac{X_2(s)}{F(s)}$.

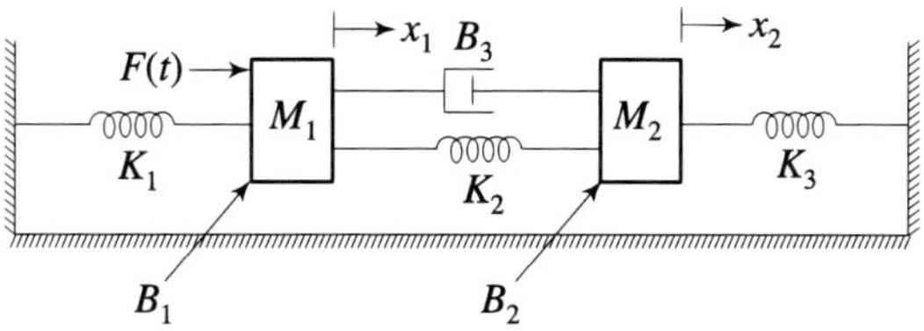

Fig. 2.122

Ans: $[M_1S^2 + (B_1 + B_3)S + (K_1 + K_2)]\, x_1(s) - (B_3S + K_2)\, x_2(s) = F(s)$

$- (B_3S + K_2)\, X_1(s) + [M_2S^2 + (B_2 + B_3)S + (K_2 + K_3)]\, x_2(s) = 0$

$$\frac{X_2(s)}{F(s)} = \frac{B_3S + K_2}{[M_1S^2 + (B_1 + B_2)S + (K_1 + K_2)][M_2S^2 + (B_2 + B_3)S + (K_2 + K_3)] - (B_3S + K_2)\,(B_3S + K_2)}$$

30. Write the equations of motion for the mechanical system shown in Fig. 2.123.

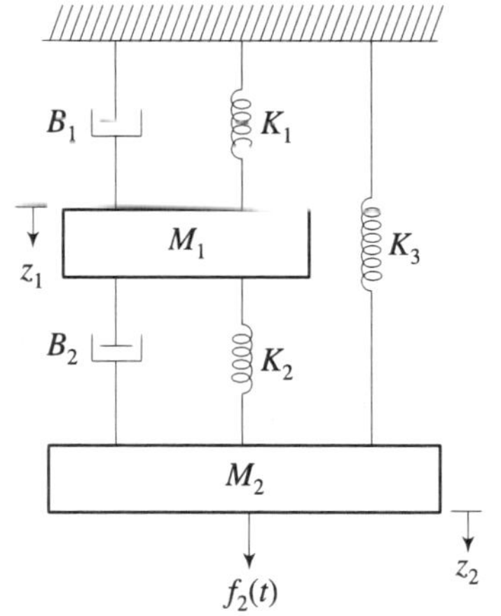

Fig. 2.123

Ans: $M_1\dfrac{d^2z_1}{dt^2} = -B_1\dfrac{dz_1}{dt} - K_1z_1 + B_2\left(\dfrac{dz_2}{dt} - \dfrac{dz_1}{dt}\right) + K_2(z_2 - z_1) + f_1(t)$

$$M_2\frac{d^2z_2}{dt^2} = -K_3z_2 - B_2\left(\frac{dz_2}{dt} - \frac{dz_1}{dt}\right) - K_2(z_2 - z_1) + f_2(t)$$

31. Obtain the force voltage and force current analogies for the mechanical system shown in Fig. 2.124.

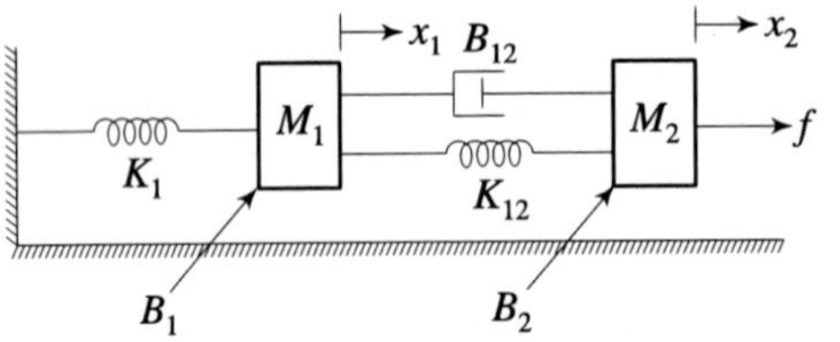

Fig. 2.124

Ans:

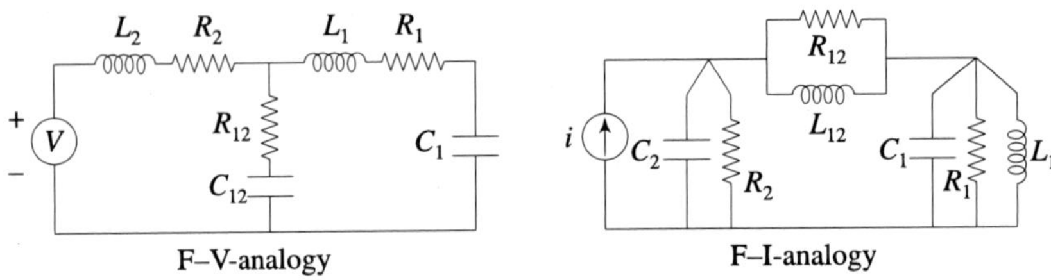

Fig. 2.125

32. Draw the mechanical equivalent network and write the differential equations governing the behaviour of the mechanical system shown in Fig. 2.126.

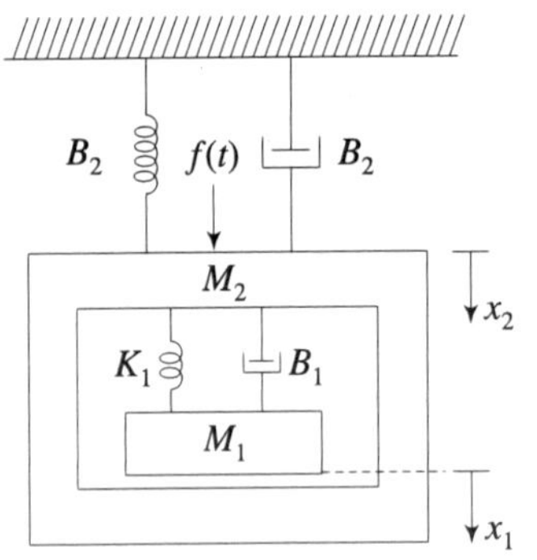

Fig. 2.126

Ans:

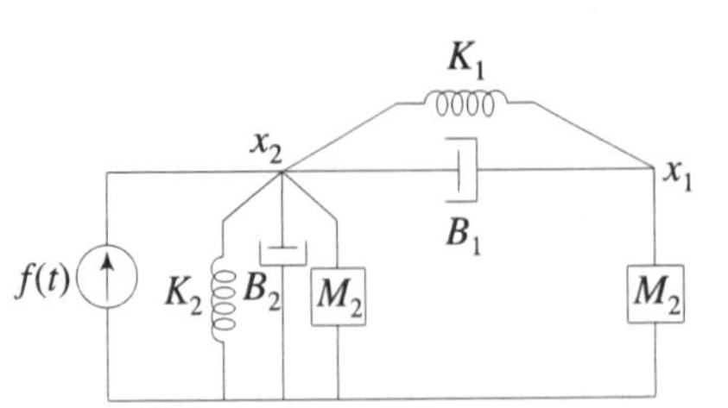

Fig. 2.127

$$M_1 \frac{d^2x_1}{dt^2} + B_1\left(\frac{dx_1}{dt} - \frac{dx_2}{dt}\right) + K_1(x_1 - x_2) = 0$$

$$M_2 \frac{d^2x_2}{dt^2} + B_2 \frac{dx_2}{dt} + K_2\, x_2 + K_1(x_2 - x_1) + B_1\left(\frac{dx_2}{dt} - \frac{dx_1}{dt}\right) = f(t)$$

33. Find the transfer function of the network shown in Fig. 2.128.

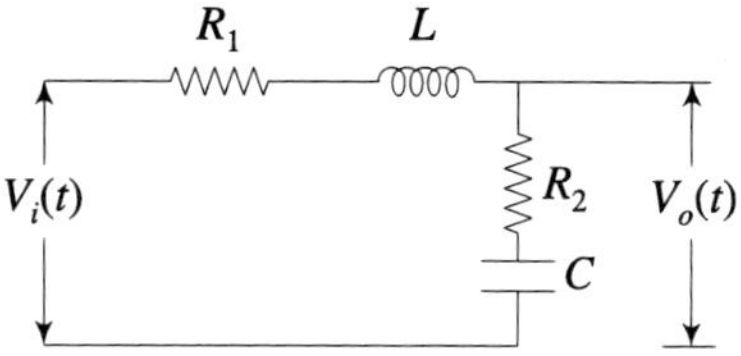

Fig. 2.128

Ans: $\dfrac{sR_2C + 1}{s^2LC + (R_1 + C_2)\, sC + 1}$.

34. Find the transfer function of the network shown in Fig. 2.129.

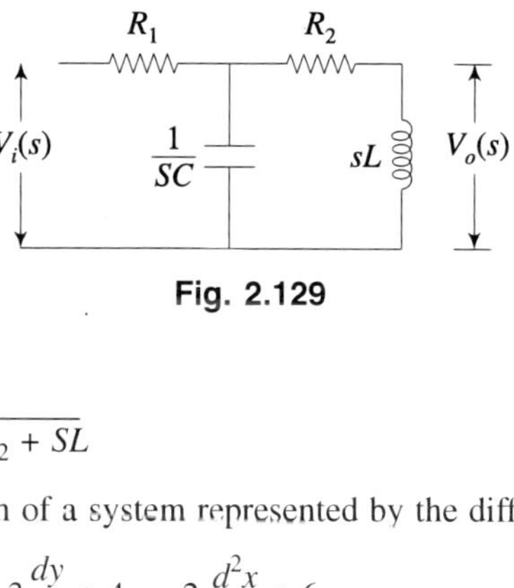

Fig. 2.129

Ans: $\dfrac{SL}{R_1SC(R_2 + SL) + R_1 + R_2 + SL}$

35. Find the transfer function of a system represented by the differential equation

$$\frac{d^2y}{dt^2} + 3\frac{dy}{dt} + 4y = 2\frac{d^2x}{dx^2} + 6x$$

Ans: $\dfrac{C(s)}{R(s)} = \dfrac{2s^2 + 6}{s^2 + 3s + 4}$

36. The dynamic behaviour of a system is described by the equation $\dfrac{dc}{dt} + 12c = 20r$, where r is the input and C is the output. Find the transfer function of the system.

Ans: $\dfrac{C(s)}{R(s)} = \dfrac{20}{s + 12}$

37. The differential equation of a system is $\dfrac{3d^2C(t)}{dt^2} + \dfrac{5dC(t)}{dt} + C(t) = r(t) + 3r\,(t - 2)$. Find the transfer function of the above system.

Ans: $\dfrac{C(s)}{R(s)} = \dfrac{1 + 3e^{-2s}}{3s^2 + 5s + 1}$

38. The impulse response of a system is $e^{-5t}(1 - \cos 3t)$. Determine the transfer function of the system.

Ans: $\dfrac{9}{(s + 1)(s^2 + 25 + 10)}$

39. The impulse response of a system is e^{-5t}. Determine the transfer function.

Ans: $1/s + 5$

40. A system has poles at $S = 4$, $S = -2$ and zero @ $S = -3$. Sketch the pole zero plot and find the transfer function of the system if the value of gain factor is 4.

Ans:

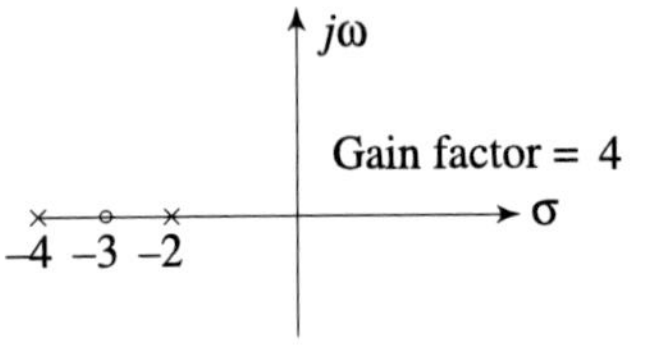

Fig. 2.130

41. Find the pole-zero plot of the following transfer function $\dfrac{20(S + 3)}{(S + 1)(S^2 + 7S + 10)}$

Ans:

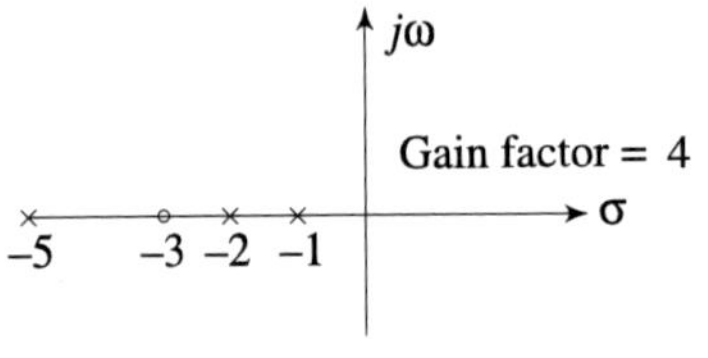

Fig. 2.131

42. The transfer function is defined for __________.

Ans: linear and time-invariant system

43. The transfer function of a system is used to study its __________.

Ans: transient and steady state

44. The transfer function of a system is defined as the ratio of its output to input in __________.

Ans: Laplace transform

45. The steady state output of a unit feedback control system is __________.

Ans: very near to reference input

46. The type of a transfer function denotes __________.

Ans: the number of poles @ origin

47. The transfer function of the network given below is __________.

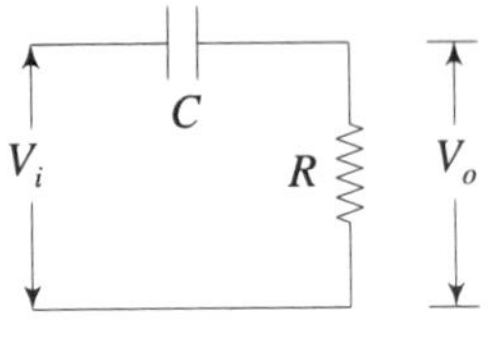

Fig. 2.132

Ans: $\dfrac{sRC}{1 + sRC}$

48. The transfer function of the system shown in Fig. 2.133 is __________.

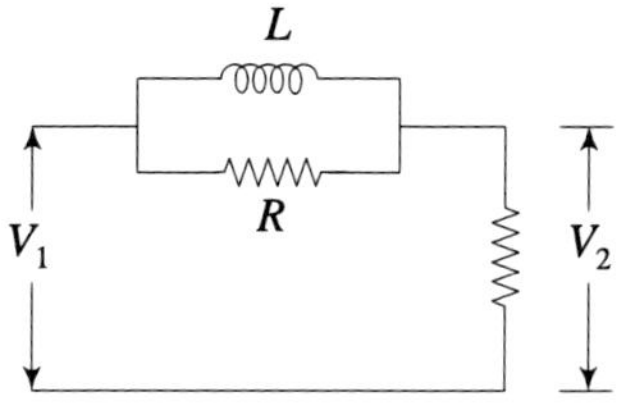

Fig. 2.133

Ans: $\dfrac{R_1 + sL}{sL\left[\dfrac{R_1 + R_2}{R_2}\right] + R_1}$

49. The output of a linear system for a unit step input is given by t^2e^{-1}. The transfer function is given by __________.

Ans: $\dfrac{2s}{(s + 1)^3}$

50. The starting point of the description of a linear system may be systems' __________.

Ans: transfer function or dynamic equations or differential equations.

51. The impulse response of a system $G(s) = 2/(s + 1)(s + 3)$ is given by __________.

Ans: $e^{-t} - e^{-2t}$

52. The input-output relationship of a system is given by __________.

$$r(t) = \frac{d^2c(t)}{dt^2} + \frac{3\,dc(t)}{dt} + 2c(t)$$

Ans: $\dfrac{1}{s^2 + 3s + 2}$

53. Obtain the analogous electrical network of Fig. 2.134 using
    (i) f–v and (ii) f–I analogy

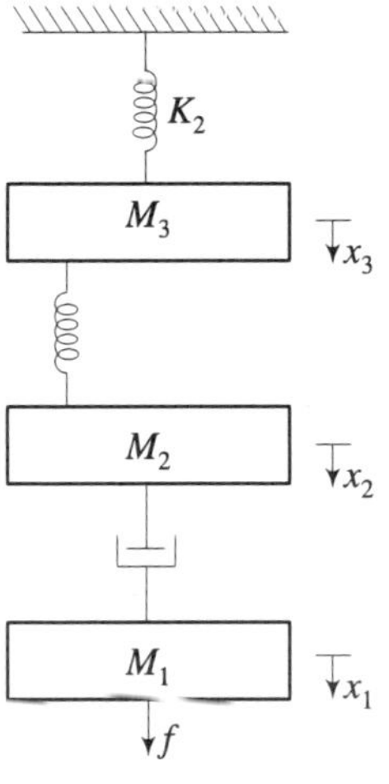

Fig. 2.134

Ans:

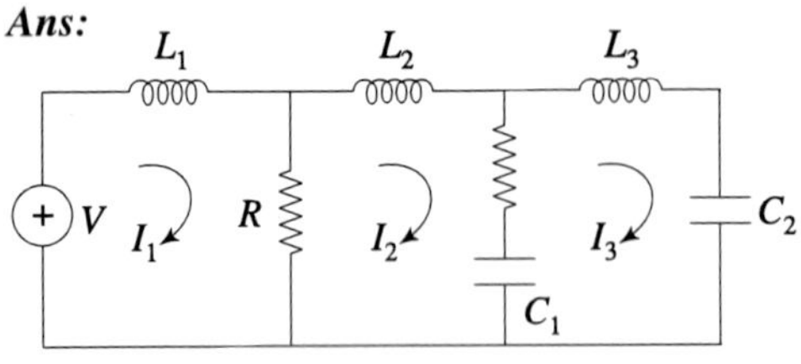

F-V analogy

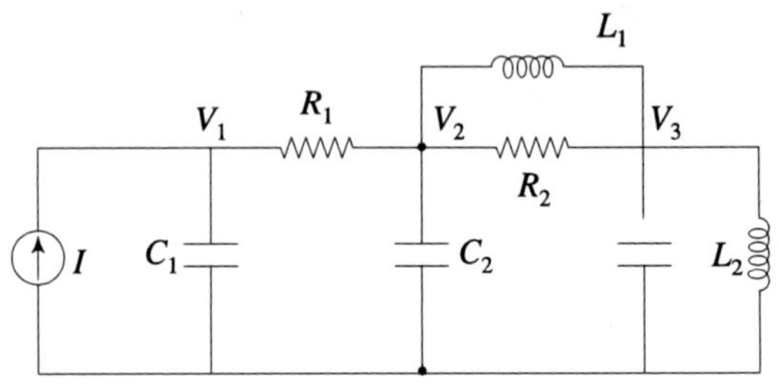

F-I analogy

Fig. 2.135

54. Find the electrical analogous system based on (i) *T-V* and (ii) *T-I* analogy of the system shown in Fig. 2.136.

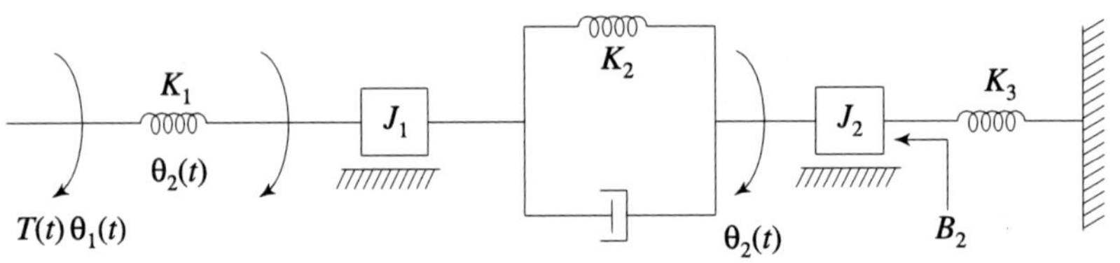

Fig. 2.136

Ans:

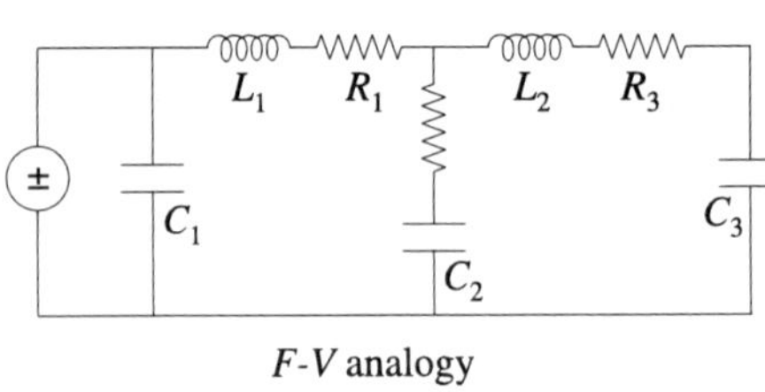

F-V analogy

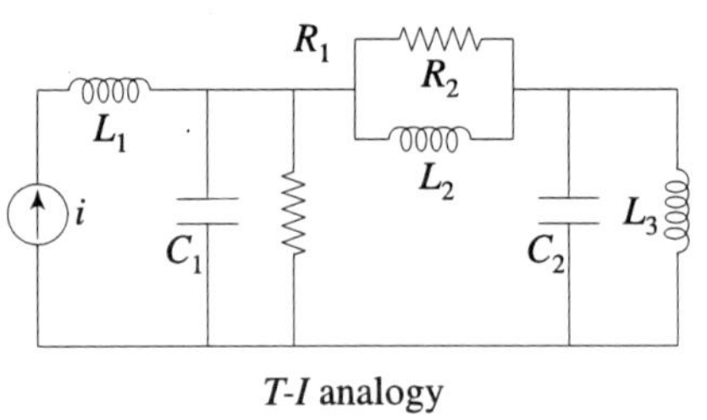

T-I analogy

Fig. 2.137

55. Find *F-V* and *F-I* analogies.

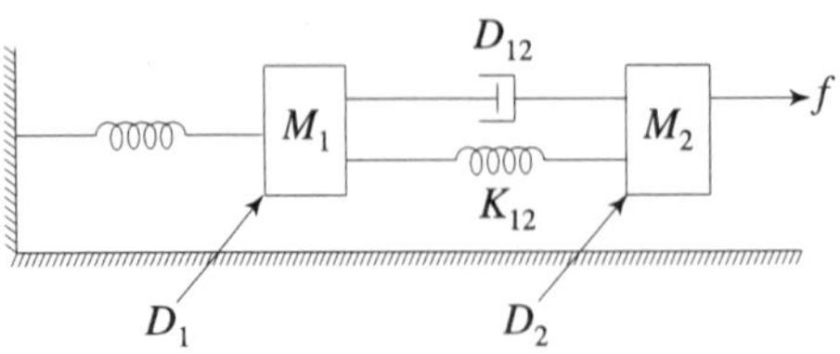

Fig. 2.138(a)

Ans:

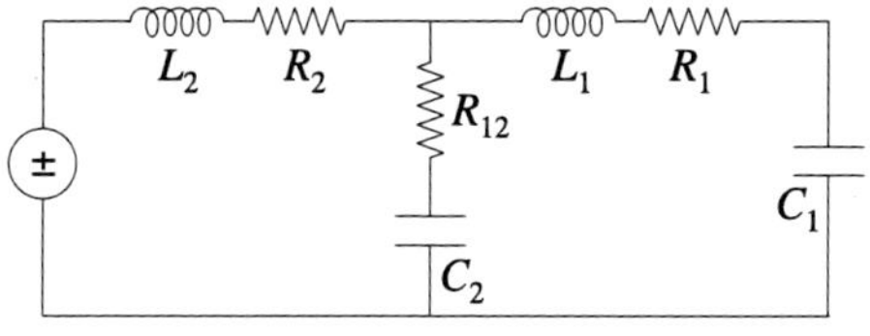

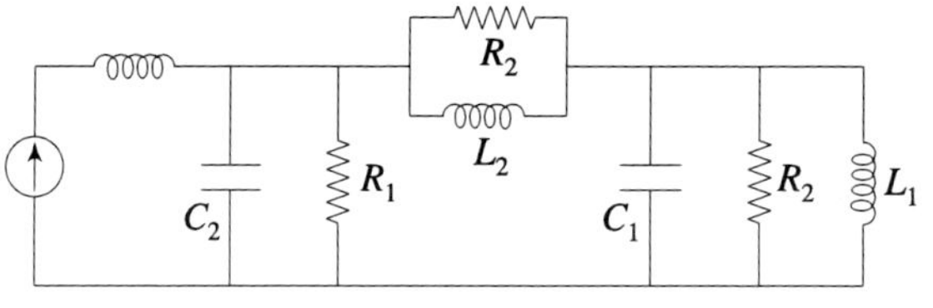

Fig. 2.138(b)

56. The force equation for the system shown in Fig. 2.139 is __________.

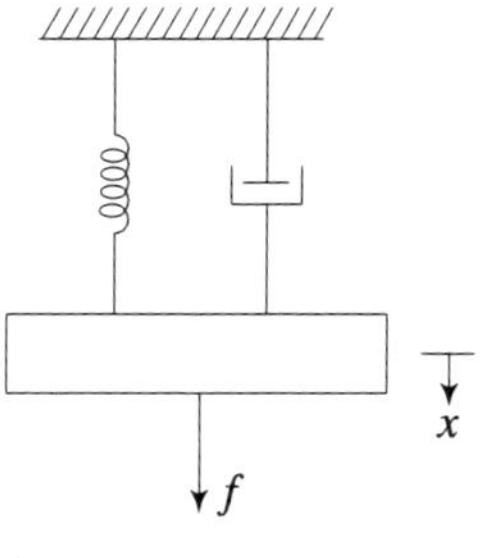

Fig. 2.139

Ans: $F = M\dfrac{d^2x}{dt^2} + D\dfrac{dx}{dt} + Kx$

57. Match the following

(a) Force	(i) Reciprocal of capacitance
(b) Velocity	(ii) Change
(c) Displacement	(iii) Current
(d) Friction coefficient	(iv) Resistance
(e) Spring constant	(v) Inductance
(f) Mass	(vi) Voltage

Ans: (a) → V_i, (b) → (iii), (c) → (iii), (d) → (iv), (e) → (v) (f) → (v)

58. A mass spring system is shown Fig. 2.140 __________ is a order of a system

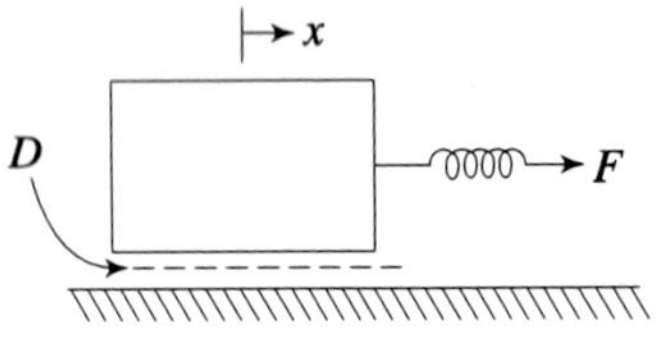

Fig. 2.140

Ans: Second order

59. With a negative feedback, the system gain and stability __________.

Ans: decreases, increases

60. The transfer function of a system in regenerative feedback is given by __________.

Ans: $\dfrac{C(s)}{R(s)} = \dfrac{G(s)}{1 - G(s)H(s)}$.

3 Block Diagram and Signal Flow Graph

3.1 INTRODUCTION

A control system consists of a number of components. In control engineering to show the functions for performing by each component, commonly use a diagram called a block diagram. The block diagram of a system is a pictorial representation of the functions performed by each component and of the flow of signals. The function of each component is represented by line with arrow indicating the flow of signal from output of one block to the input of another. The elements of a block diagram are summing point, take-off point, forward and feedback paths.

3.2 BLOCK DIAGRAM REPRESENTATION OF SYSTEMS (BDR) AND THEIR REDUCTION

A typical block diagram consists of one or more summing junctions, forward and feedback paths take off points.

The basic elements of block diagram:

- Output = gain × input
- Summing point: Summing point is the point at which more than one signal can be added or subtracted.

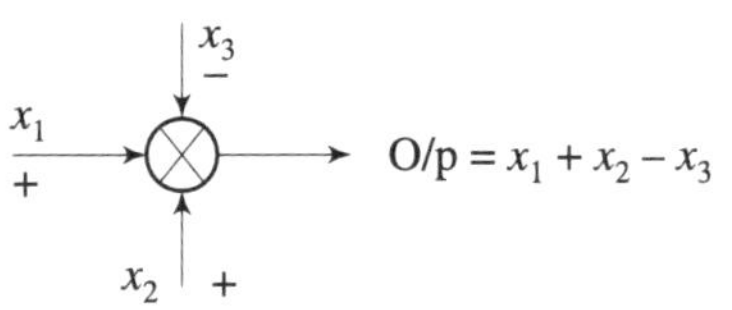

Fig. 3.1

- Take-off Point: The point at which the output signal of any block can be applied to two or more points is known as take-off point.

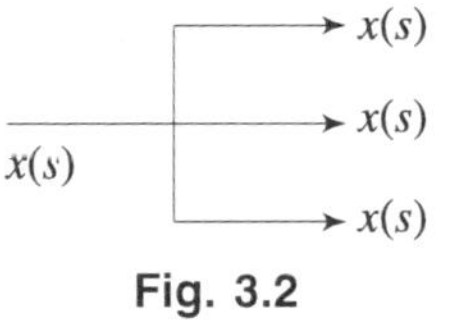

Fig. 3.2

- **Forward path** is the direction of the signal flow from the input to the output.

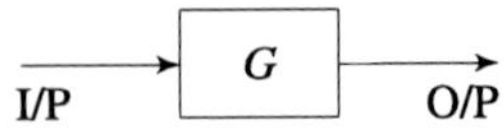

Fig. 3.3

- **Feedback path** is the direction of the signal flow from the output to the input.

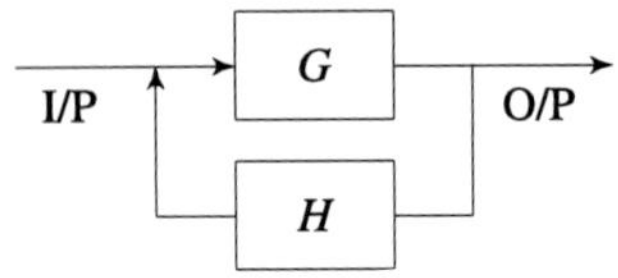

Fig. 3.4

Rules for Block Diagram Reduction

1. Blocks are in series or cascaded.

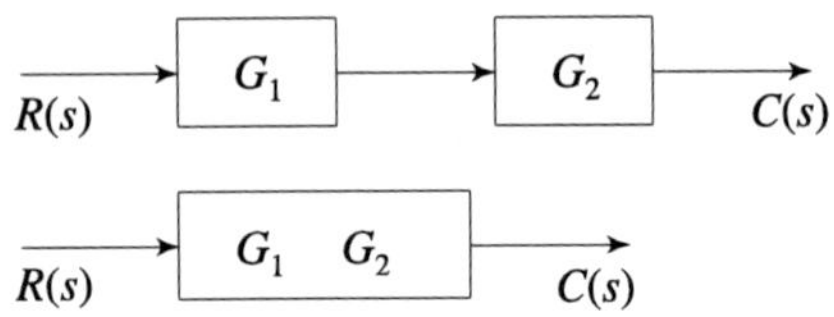

Fig. 3.5

2. Blocks are in parallel.

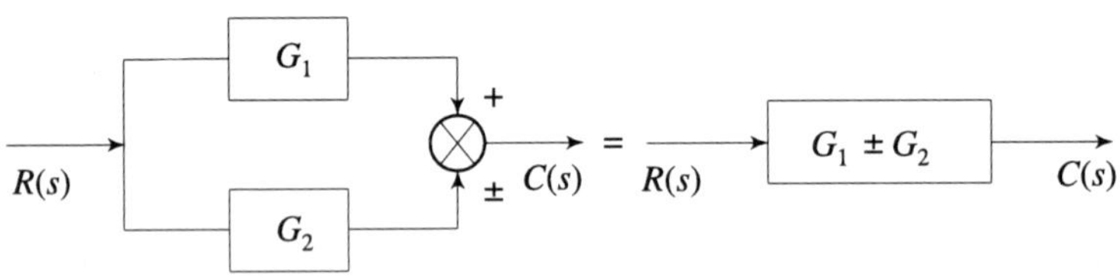

Fig. 3.6

3. Shifting a summing point before the block.

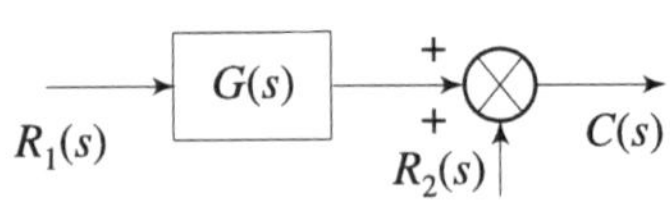

$$C(s) = R_1(s)\,G(s) + R_2(s)$$

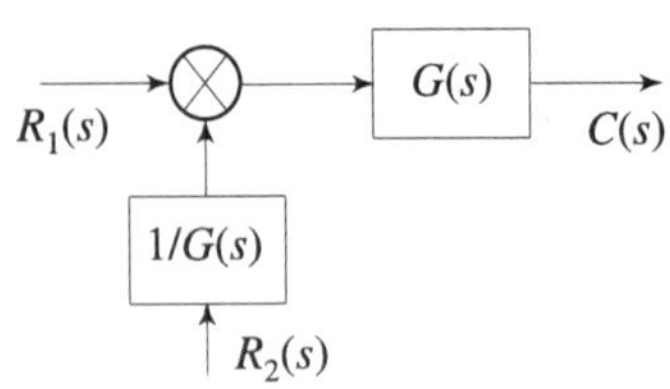

Fig. 3.7

$$C(s) = R_1(s) + R_2(s)\left(\frac{1}{G(s)}\right)$$

$$= \left[R_1(s) + R_2(s)\left(\frac{1}{G(s)}\right)\right] G(s)$$

$$C(s) = R_1(s)G(s) + R_2(s)$$

4. Shifting a summing point after the block.

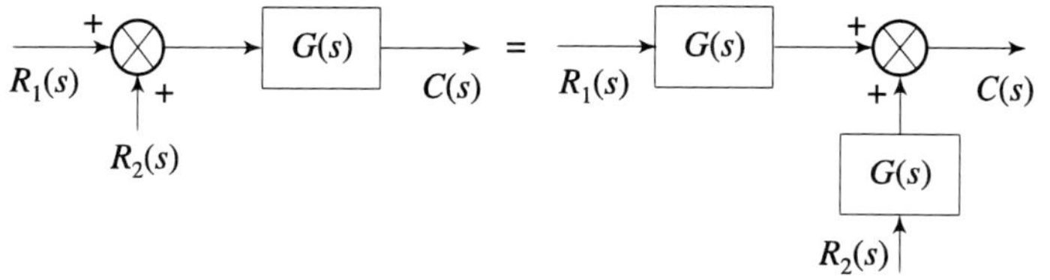

Fig. 3.8

$$C(s) = [R_2(s) + R_2(s)]G(s)$$

$$R_1(s)\ G(s) + R_2(s)\ G(s) = C(s)$$

$$\therefore \quad C(s) = G(s)[R_1(s) + R_2(s)]$$

5. Shifting a take-off point before the block.

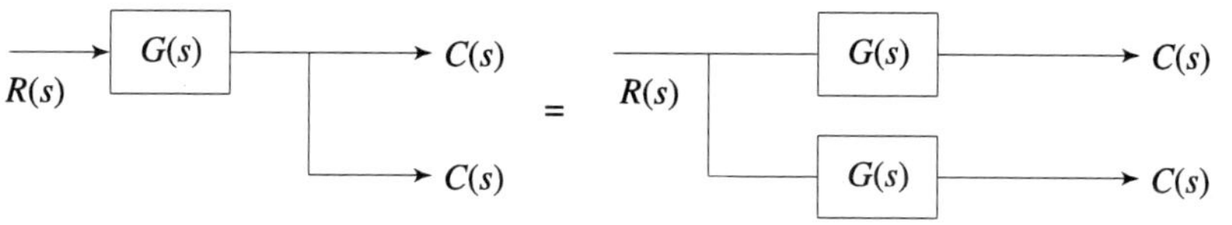

$C(s) = R(s)\ G(s)$

Fig. 3.9

6. Shifting a take-off point after the block.

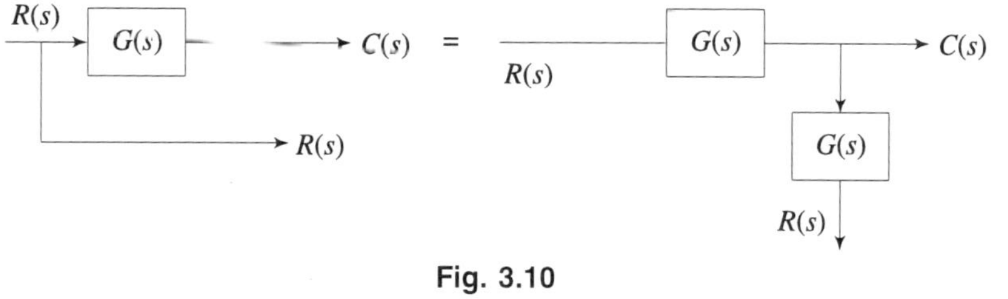

Fig. 3.10

7. Interchanging a summing point.

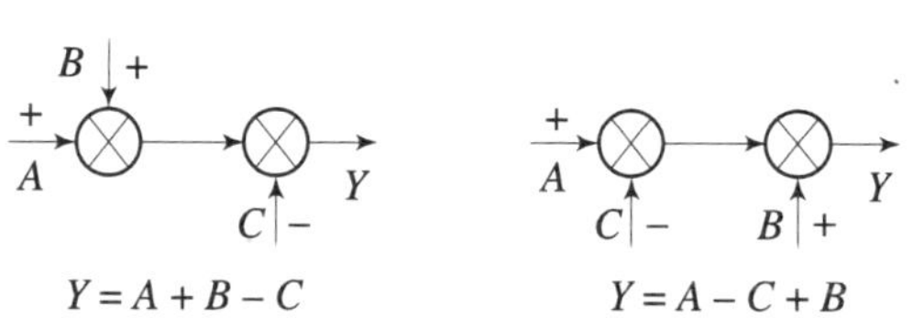

$Y = A + B - C$ $\qquad$ $Y = A - C + B$

Fig. 3.11

8. Splitting a summing point.

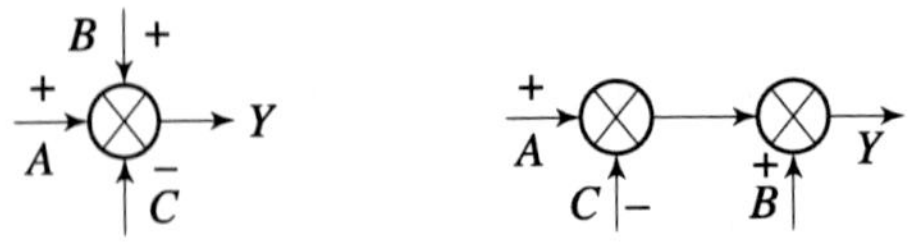

$Y = A + B - C$ $\qquad$ $Y = A - C + B$

Fig. 3.12

9. Combining summing point.

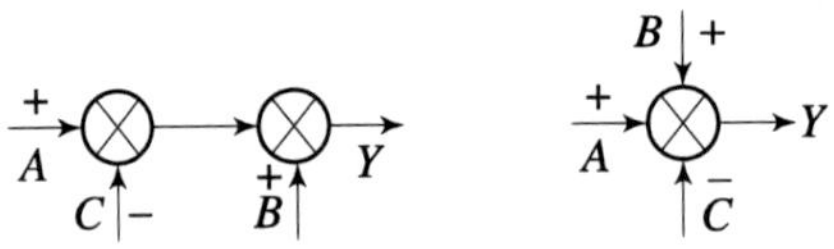

$Y = A - C + B$ $\qquad$ $Y = A + B - C$

Fig. 3.13

If a block diagram consists of many feedback loops, then find the innermost feedback loop.

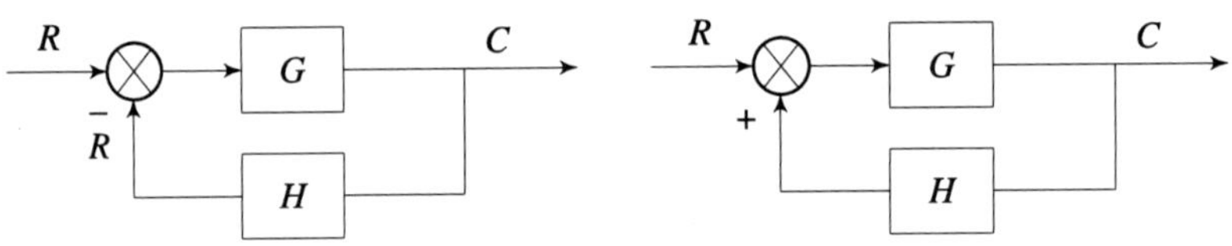

$$\frac{C}{R} = \frac{G}{1+GH} \qquad \frac{C}{R} = \frac{G}{1-GH}$$

Fig. 3.14

Example 3.1: Find the transfer function $\dfrac{C(s)}{R(s)}$ for the block diagram shown in Fig. 3.15.

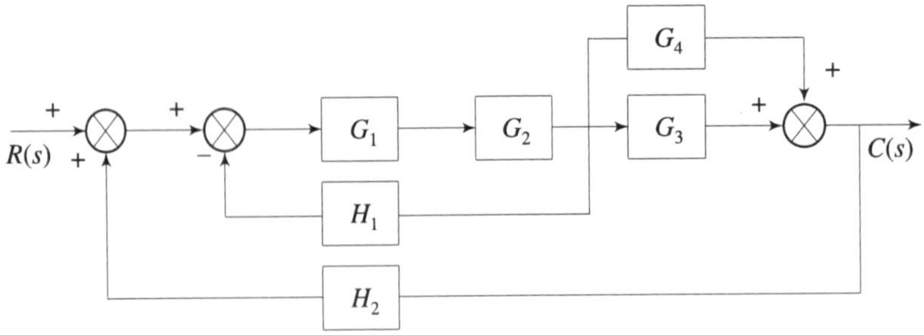

Fig. 3.15

Solution:

Step 1: Reducing cascading $G_1\ G_2$ and parallel G_3 and G_4.

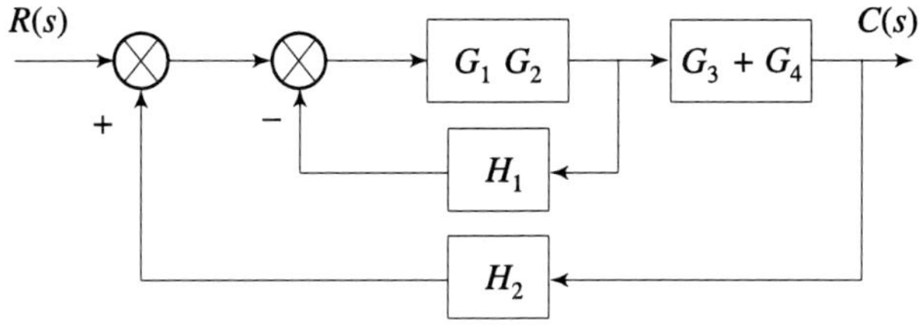

Fig. 3.16

Step 2: Eliminating negative feedback path H_1.

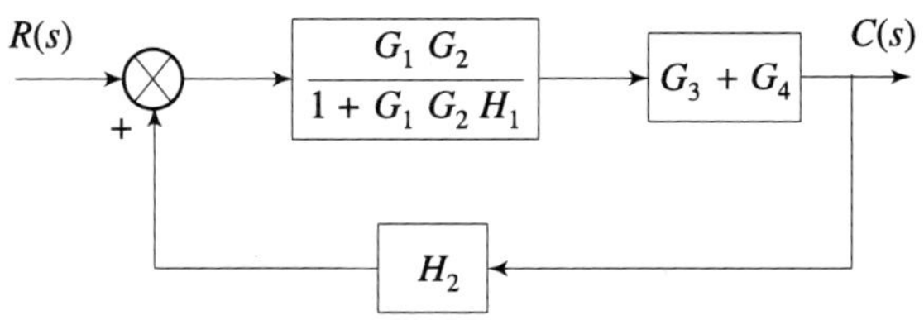

Fig. 3.16(a)

Step 3: Reducing cascading and eliminating positive feedback.

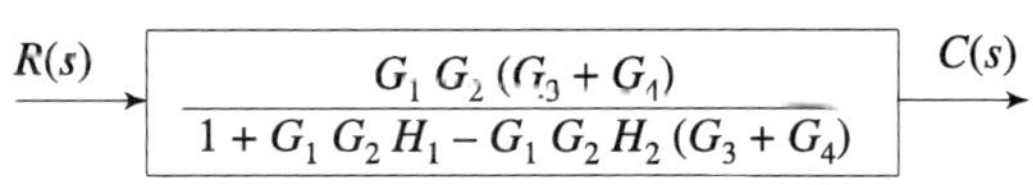

$$= \frac{\left(\dfrac{G_1G_2}{1 + G_1G_2H_1}\right)(G_3 + G_4)}{1 - \left(\dfrac{G_1G_2}{1 + G_1G_2H_1}\right)(G_3 + G_4)\ H_2}$$

$$\frac{C(s)}{R(s)} = \frac{G_1G_2\ (G_3 + G_4)}{1 + G_1G_2H_1 - G_1G_2H_2(G_3 + G_4)}$$

Example 3.2: Find the transfer function for the block diagram shown in Fig. 3.17.

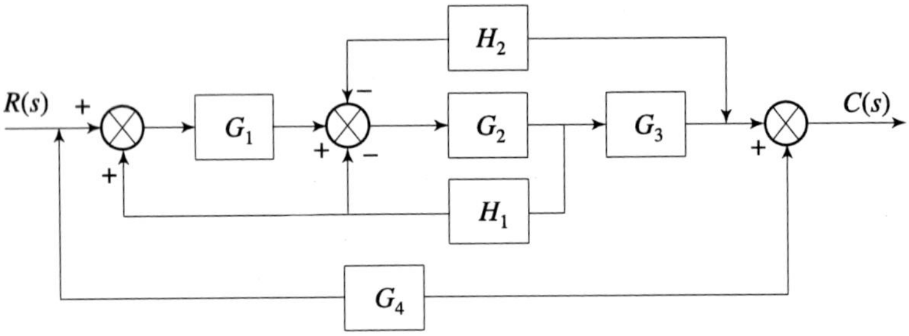

Fig. 3.17

Solution: Rearranging Fig. 3.18,

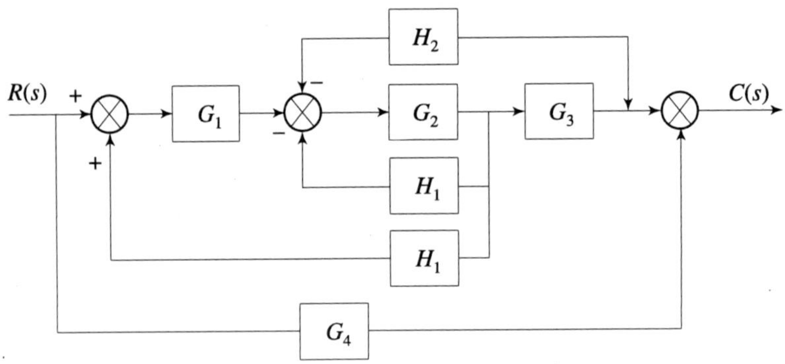

Fig. 3.18

Step 1: Eliminating negative feedback path H_1.

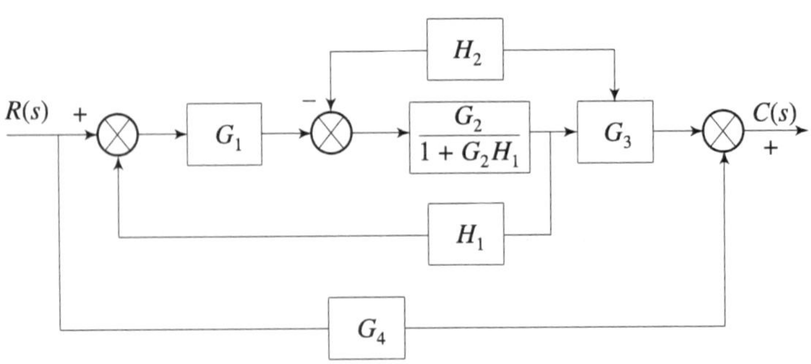

(i)

Step 2: Shifting Take-off point H_1 after G_3.

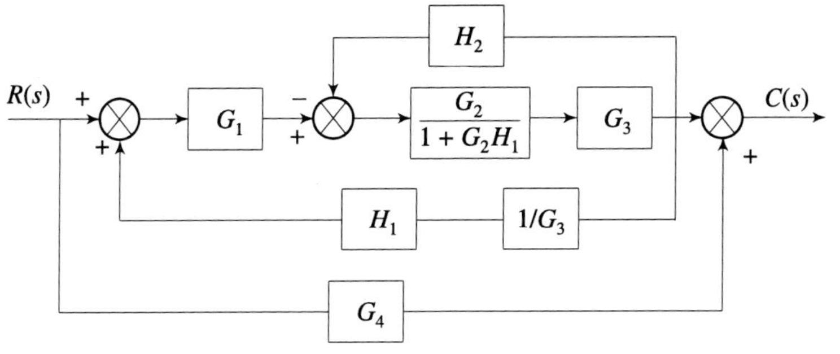

(ii)

Step 3: Cascading and eliminating negative feedback path

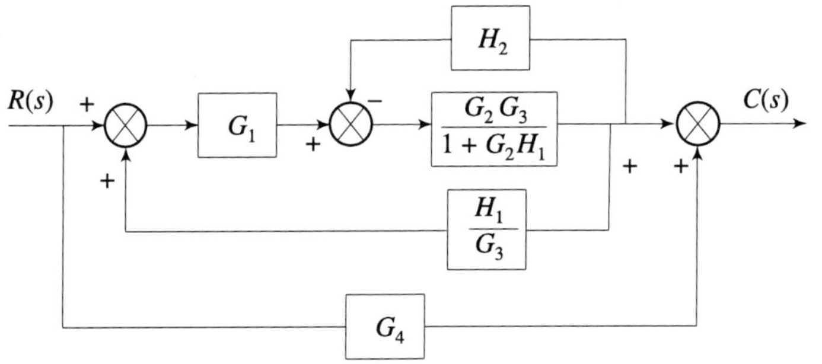

(iii)

Step 4: Reducing cascading

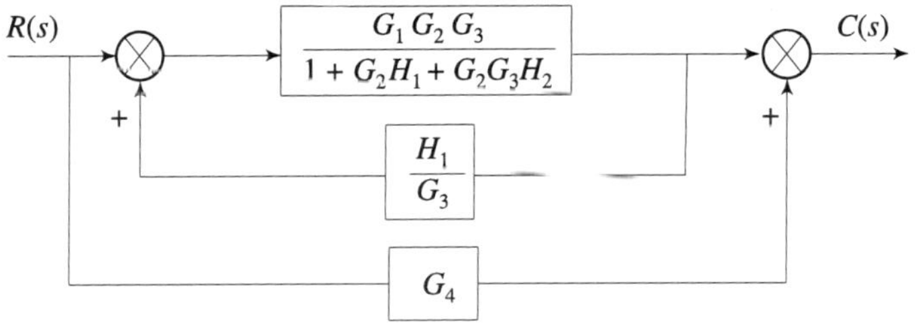

(iv)

Step 5: Eliminating positive feedback path

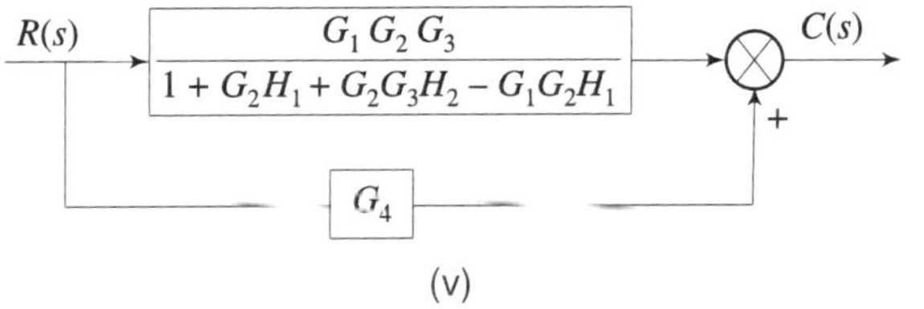

(v)

Step 6: Reducing parallel block G_4

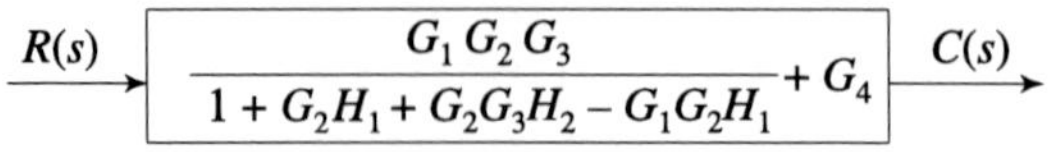

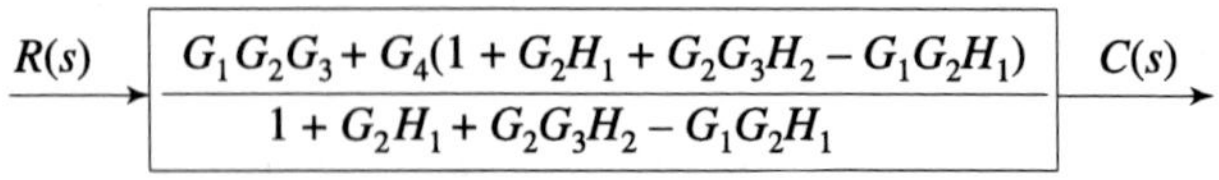

$$\frac{C(s)}{R(s)} = \frac{G_1G_2G_3 + G_4(1 + G_2H_1 + G_2G_3H_2 - G_1G_2H_1)}{1 + G_2H_1 + G_2G_3H_2 - G_1G_2H_1}$$

(vi)

Example 3.3: Find the transfer function for the block diagram shown in Fig. 3.19.

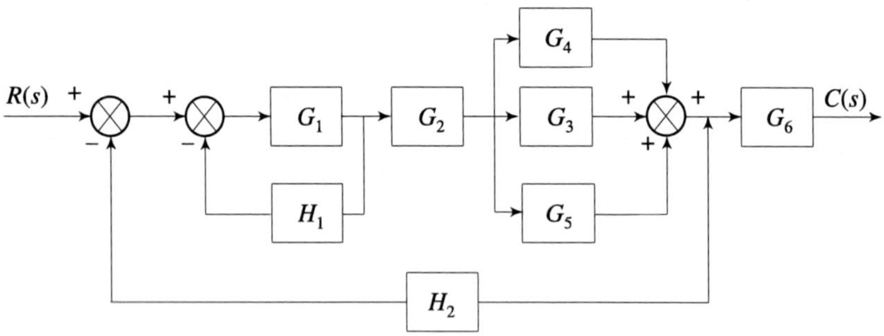

Fig. 3.19

Solution:

Step 1: Reduce parallel blocks and negative feedback path H_1

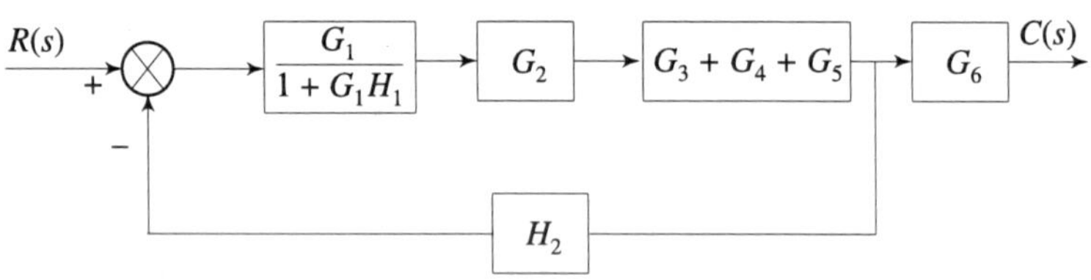

(i)

Step 2: Reducing cascading block

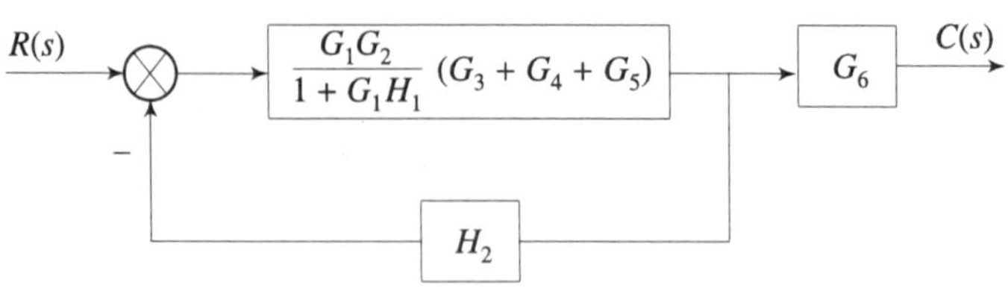

(ii)

Step 3: Eliminating negative feedback path H_2

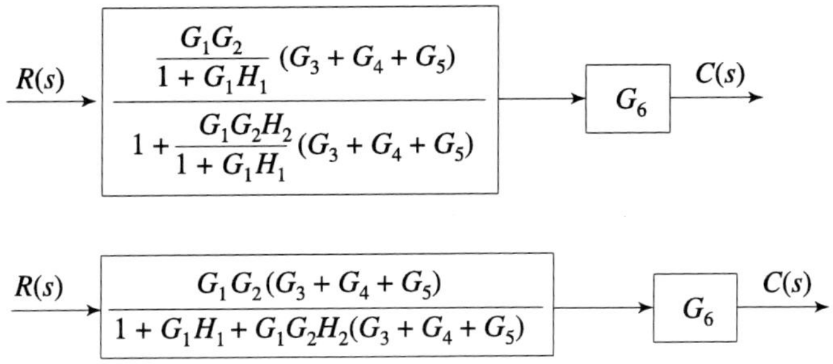

(iii)

Step 4: Eliminating cascading block

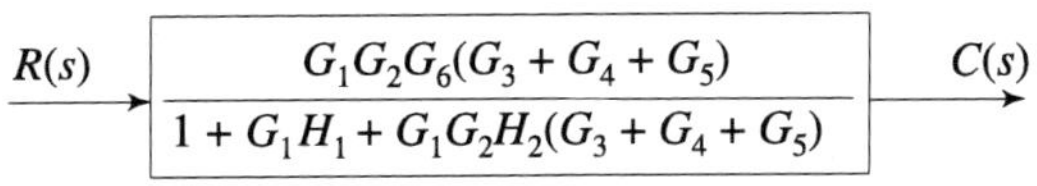

(iv)

Example 3.4: Simplify the block diagram shown in Fig. 3.20 and obtain the transfer function.

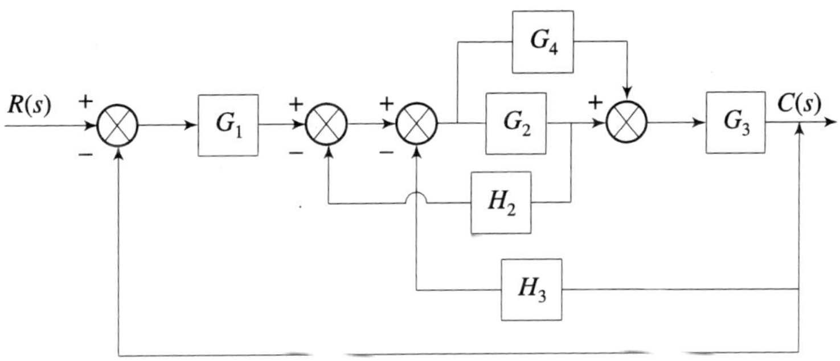

Fig. 3.20

Solution: Shifting take-off point before G_2

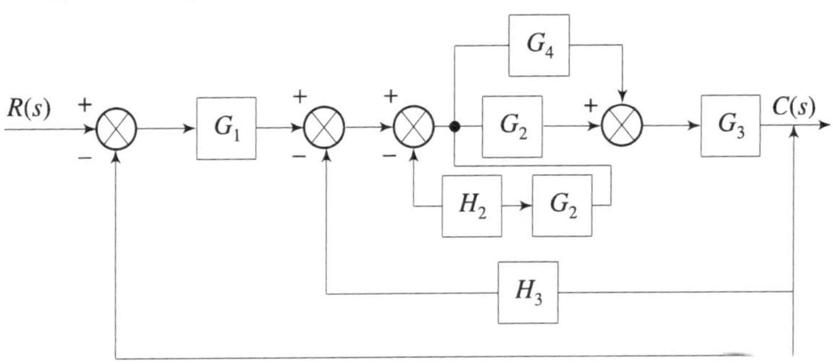

(i)

Reducing parallel block and cascading

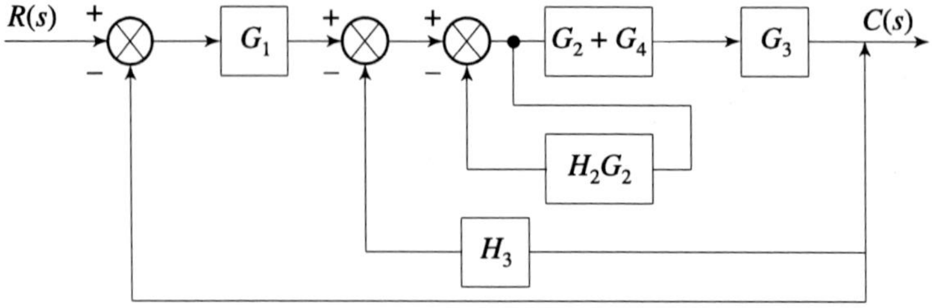

(ii)

Shifting take-off point after $(G_2 + G_4)$

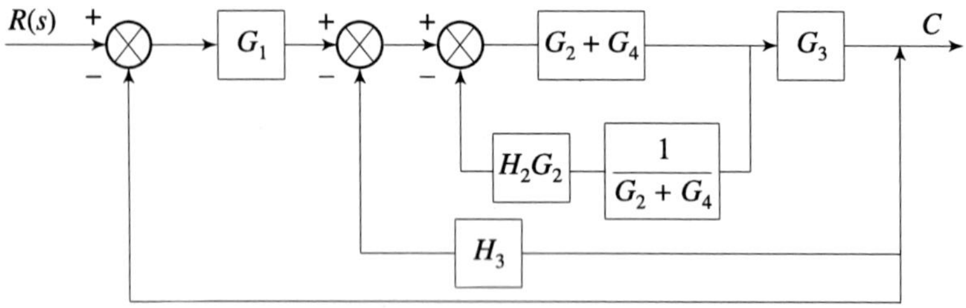

(iii)

Reducing negative feedback and cascading G_3

$$= \frac{(G_2 + G_4)}{1 + G_2H_2 \dfrac{1}{(G_2 + G_4)} (G_2 + G_4)} = \frac{(G_2 + G_4)}{1 + G_2H_2}$$

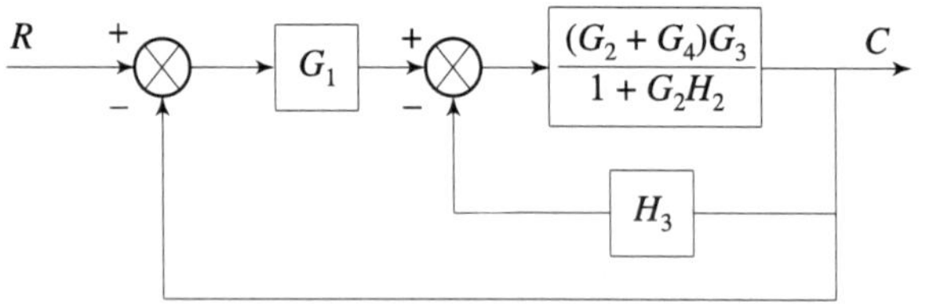

(iv)

Reducing negative feedback

$$= \frac{\dfrac{(G_2 + G_4)G_3}{1 + G_2H_2}}{1 + \dfrac{(G_2 + G_4)G_3}{1 + G_2H_2} (H_3)} = \frac{G_3(G_2 + G_4)}{1 + G_2H_2 + G_3H_3(G_2 + G_4)}$$

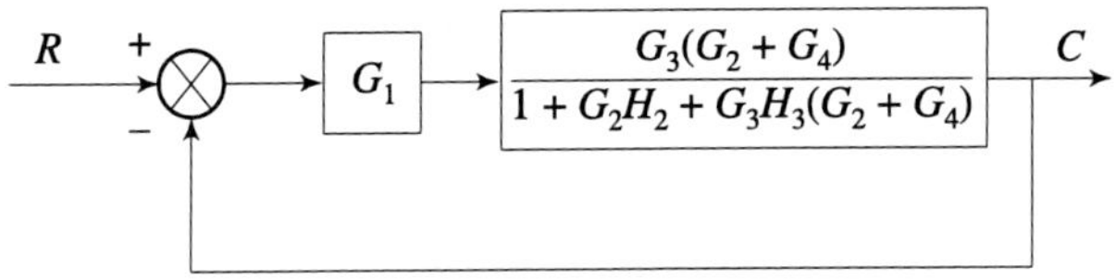

(v)

Cascading and eliminating negative feedback

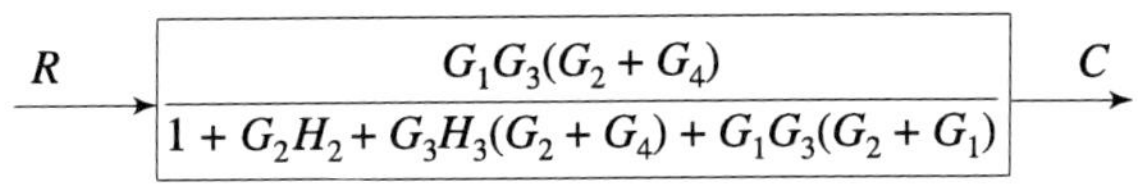

(vi)

Example 3.5: For the system represented by the block diagram shown in Fig. 3.21, evaluate the closed loop transfer function when the input R is

(i) At station I

(ii) At station II

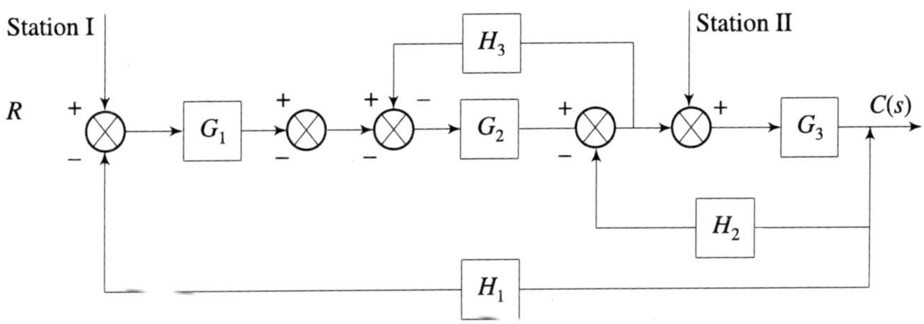

Fig. 3.21

Solution: (i) Consider the input R is at Station I and the input at Station II is made zero.

Let the output be C_1, since there is no input at Station II that summing point can be removed and resulting block diagram is shown in Fig. 3.22.

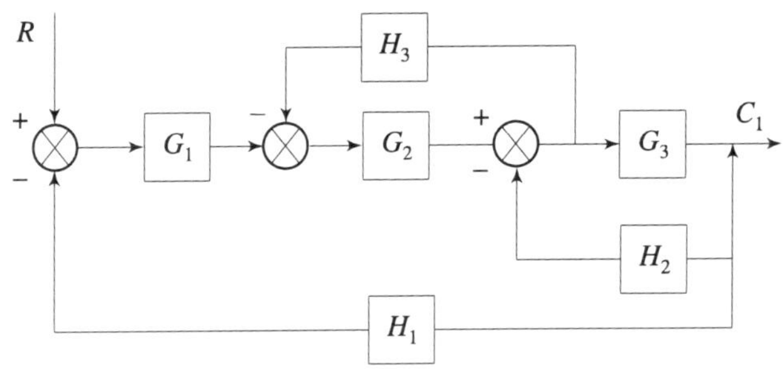

Fig. 3.22

Shift the take-off point of feedback H_3 beyond G_3 and rearrange the branch points.

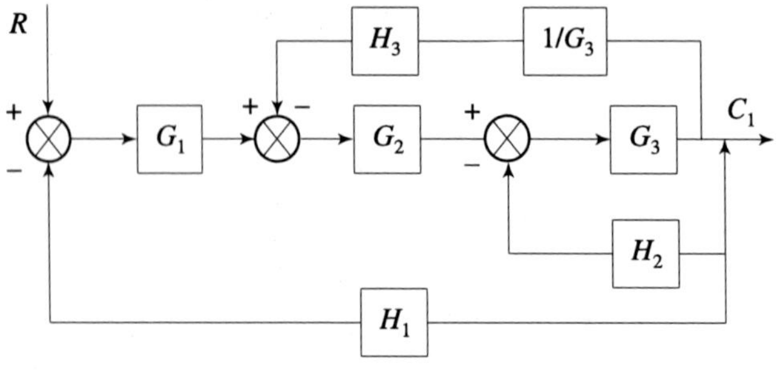

(i)

Eliminating the feedback H_2 and combining blocks in cascade

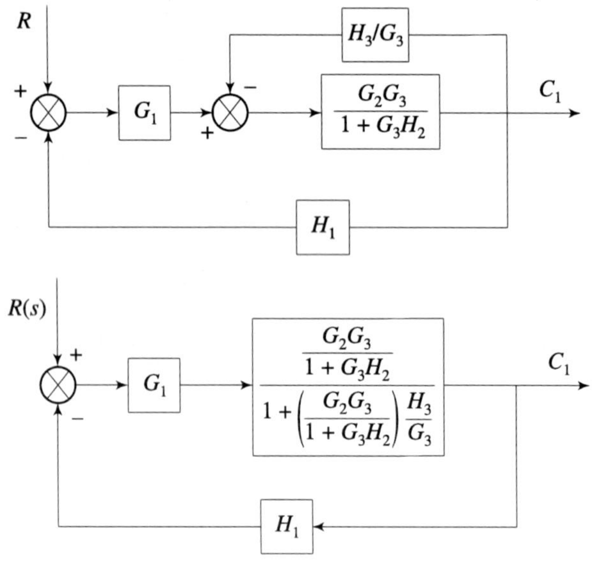

(ii)

Combining the blocks in cascade

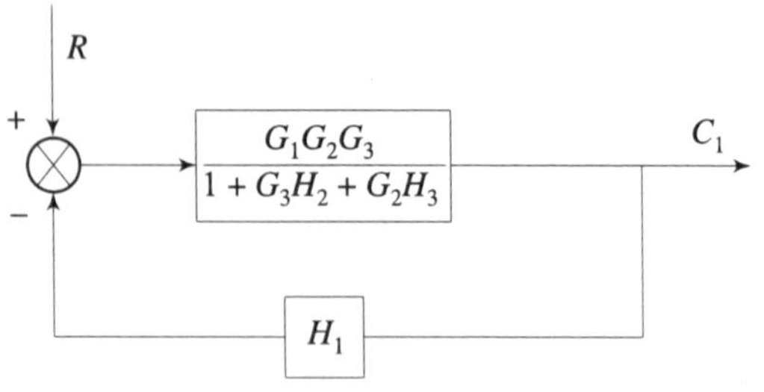

(iii)

Eliminating feedback path H_1

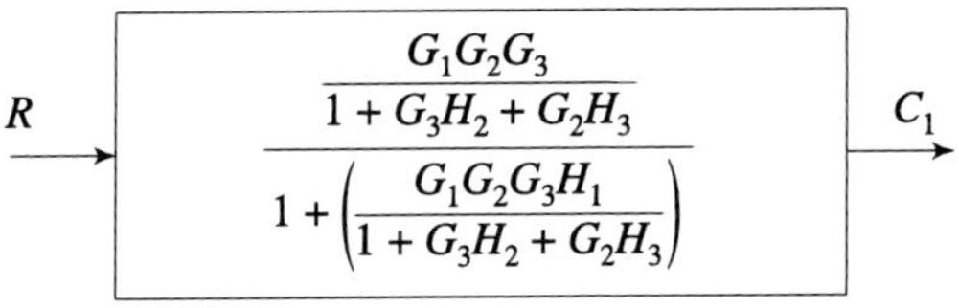

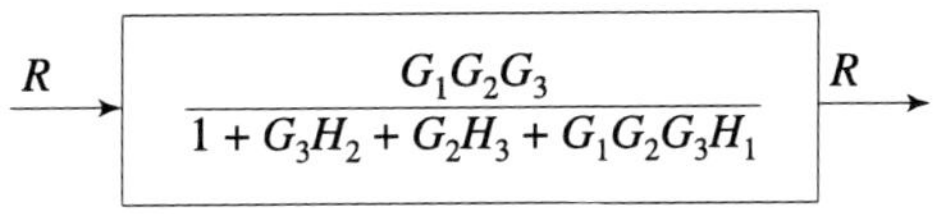

(iv)

$$\frac{C_1(S)}{R(S)} = \frac{G_1G_2G_3}{1 + G_3H_2 + G_2H_3 + G_1G_2G_3H_1}$$

(ii) Consider the input R at Station II, the input at Station I is made zero. Let output be C_2. Since there is no input in Station I that corresponding summing point can be removed and a negative sign can be attached to be feedback path gain H_1.

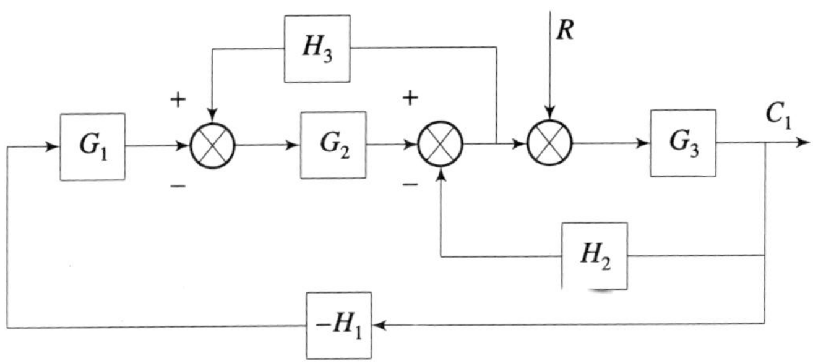

(i)

Combining the blocks in cascade, shifting the summing point of H_2 before G_2 and rearranging the branch points.

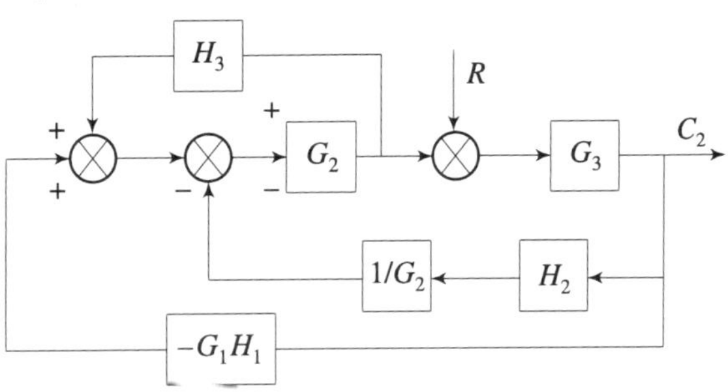

(ii)

Interchanging summing and combining the blocks in cascade.

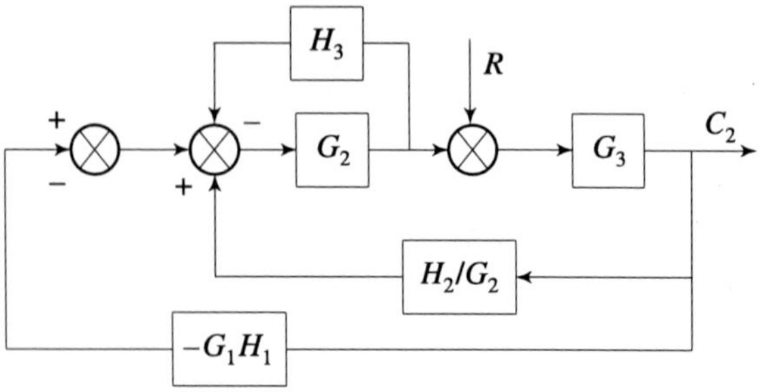

(iii)

Combining parallel blocks and eliminating feedback path.

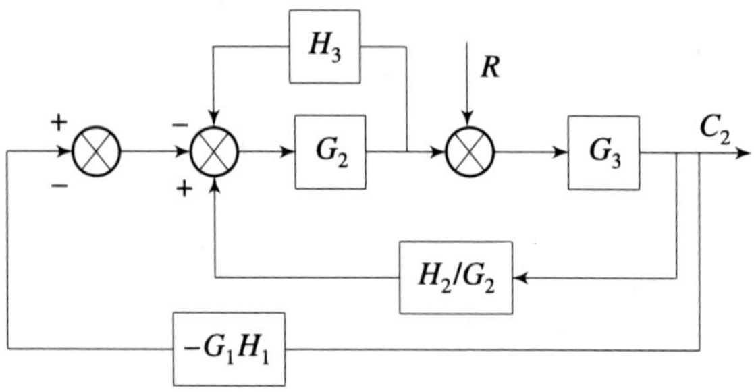

(iv)

Combining the blocks in cascade

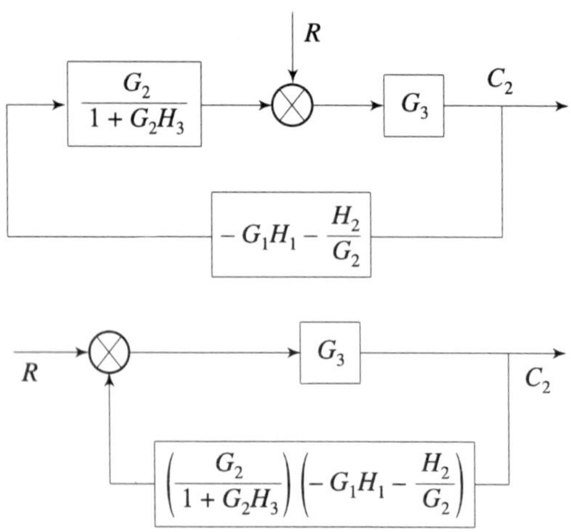

(v)

Eliminating feedback path

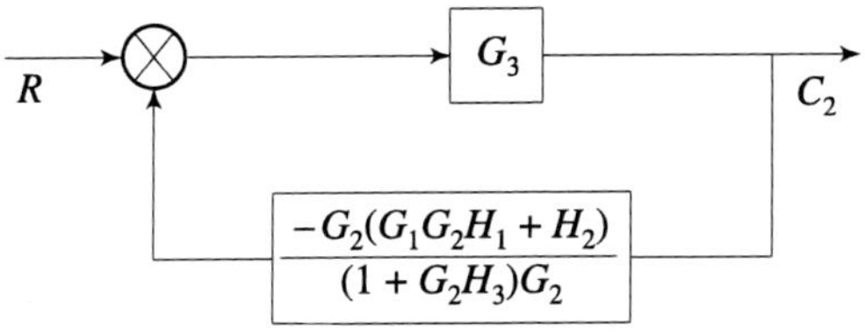

(vi)

$$\frac{C_2}{R} = \frac{G_3}{1 - \left(\frac{-(G_1G_2H_1 + H_2)}{1 + G_2H_3}\right)G_3}$$

$$\frac{C_2}{R} = \frac{G_3}{\left(\frac{1 + G_2H_3 + G_3(G_1G_2H_1 + H_2)}{1 + G_2H_3}\right)}$$

$$\frac{C_2}{R} = \frac{G_3(1 + G_2H_3)}{1 + G_2H_3 + G_3(G_1G_2H_1 + H_2)}$$

The transfer function of the system with input at Station I is

$$\frac{C_1}{R} = \frac{G_1G_2G_3}{1 + G_3H_2 + G_2H_3 + G_1G_2G_3H_1}$$

The transfer function of the system with input at Station II is

$$\frac{C_2}{R} = \frac{G_3(1 + G_2H_3)}{1 + G_2H_3 + G_3(G_1G_2H_1 + H_2)}$$

Example 3.6: For the system represented by the block diagram shown in Fig. 3.23, find $\frac{C_1}{R_1}$ and $\frac{C_2}{R_1}$.

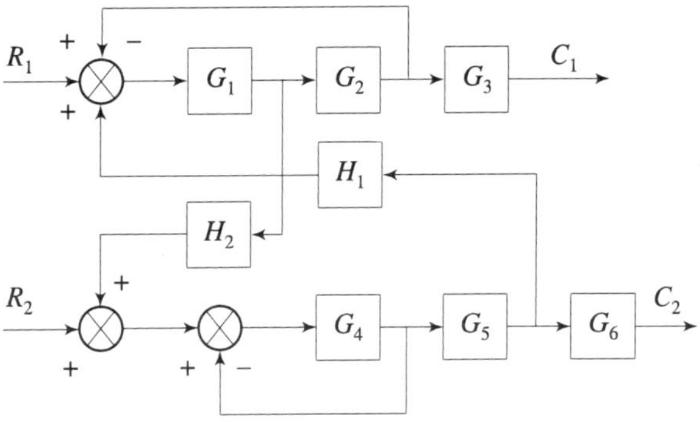

Fig. 3.23

Solution:

Case 1: To find $\frac{C_1}{R_1}$

In this case set $R_2 = 0$ and consider only one output C_1. Hence, we can remove the summing point which adds R_2 and need not consider G_6, since G_6 is on the open path. Eliminating the feedback path.

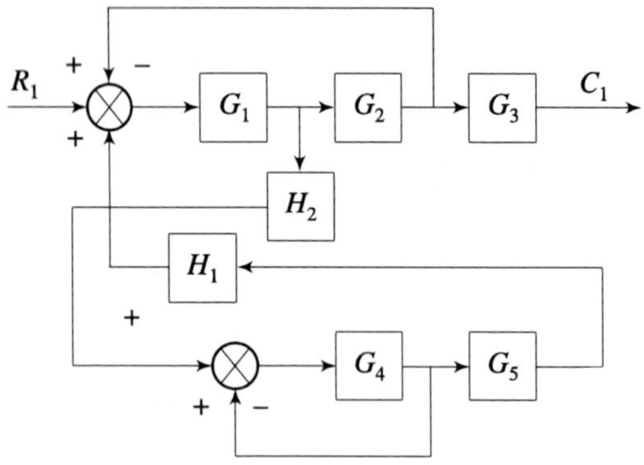

(i)

Combining the blocks in cascade and splitting the summing point.

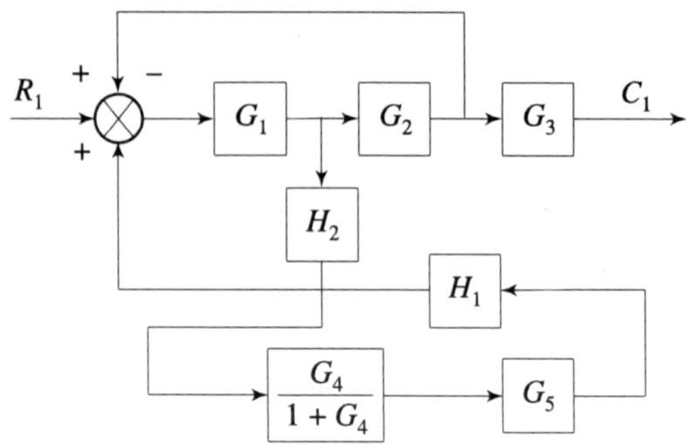

(ii)

Eliminating the feedback path

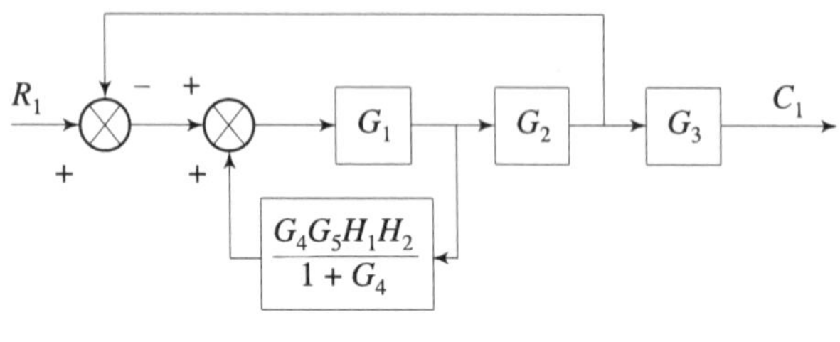

(iii)

Combining the blocks in cascade

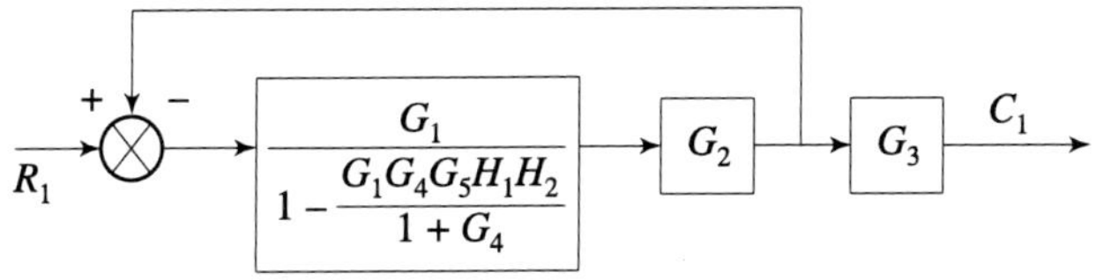

(iv)

Eliminating the feedback path

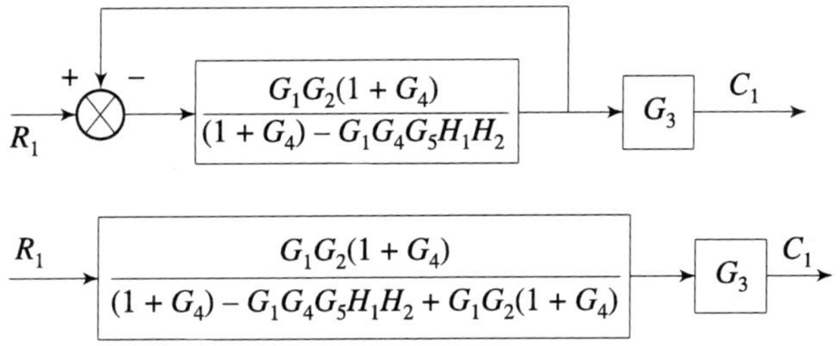

(v)

$$\frac{C_1}{R_1} = \frac{G_1G_2G_3(1 + G_4)}{(1 + G_4)(1 + G_1G_2) - G_1G_4G_5H_1H_2}$$

Case 2: To find $\frac{C_2}{R_1}$

In this case set $R_2 = 0$ and consider only one output C_2. Hence, we can remove the summing point which adds R_2 and need not consider G_3. Since G_3 is on the open path. The resulting block diagram is

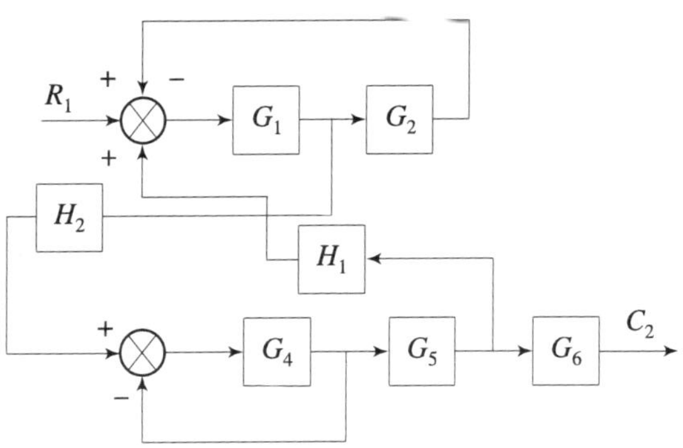

(i)

Eliminating the feedback path

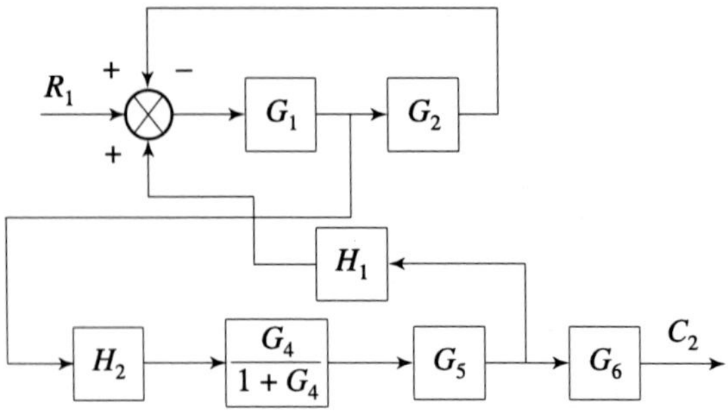

(ii)

Combining blocks in cascade and splitting the summing point

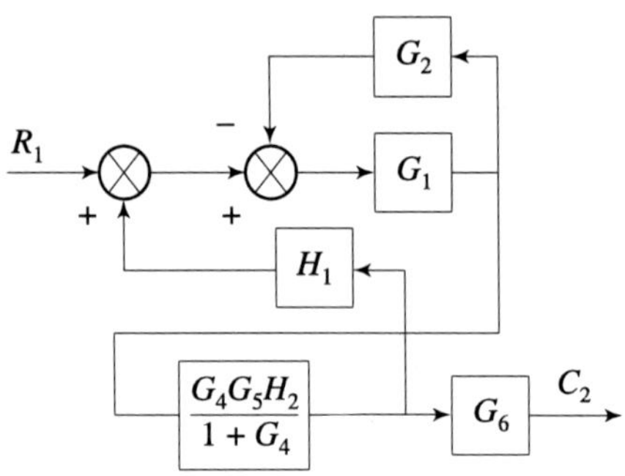

(iii)

Eliminating the feedback path

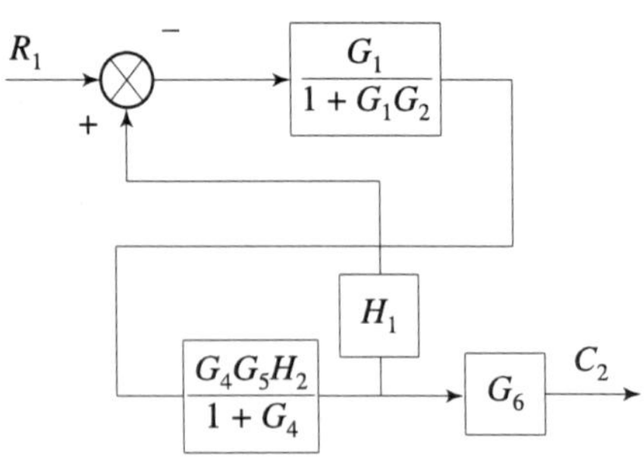

(iv)

Combining the blocks in cascade

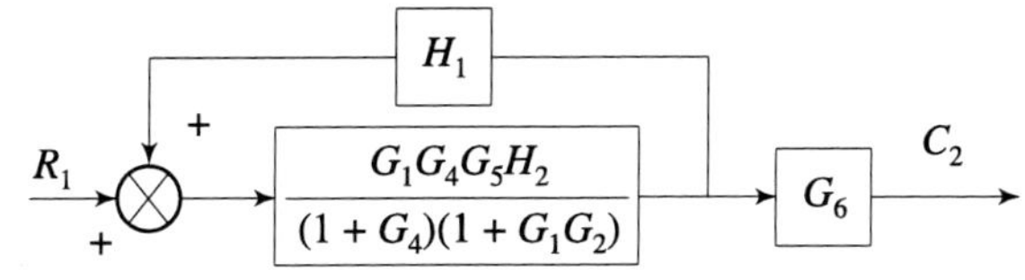

(v)

Combining the blocks in cascade

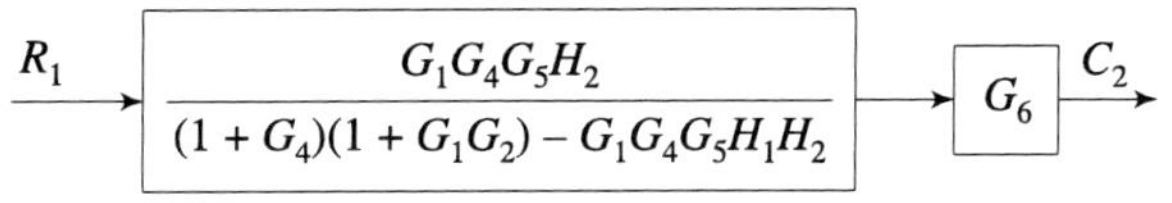

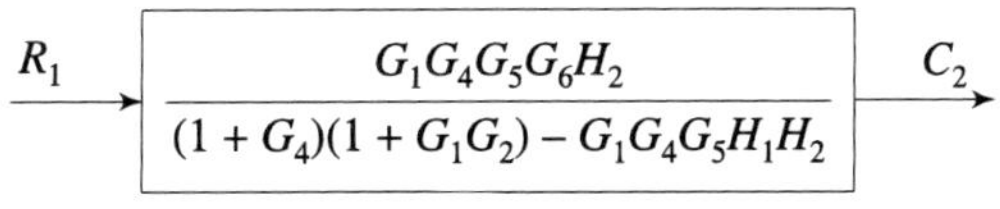

$$\frac{C_2}{R_1} = \frac{G_1G_4G_5G_6H_2}{(1+G_4)(1+G_1G_2) - G_1G_4G_5H_1H_2}$$

Example 3.7: Using block diagram reduction technique, find closed-loop transfer function of the system whose block diagram is shown in Fig. 3.24 (example of interchanging summing point).

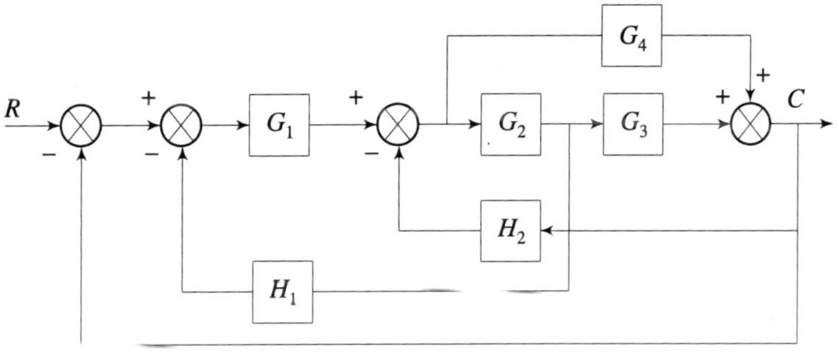

Fig. 3.24

Solution: Moving the take-off point (branch point) before the block G_2

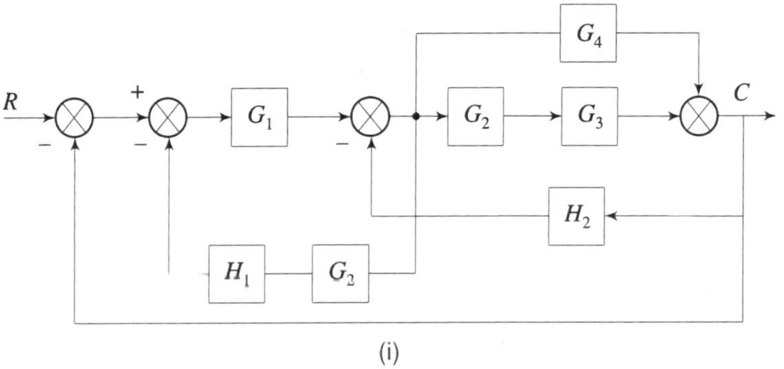

(i)

Combining the blocks in cascade and eliminating parallel blocks

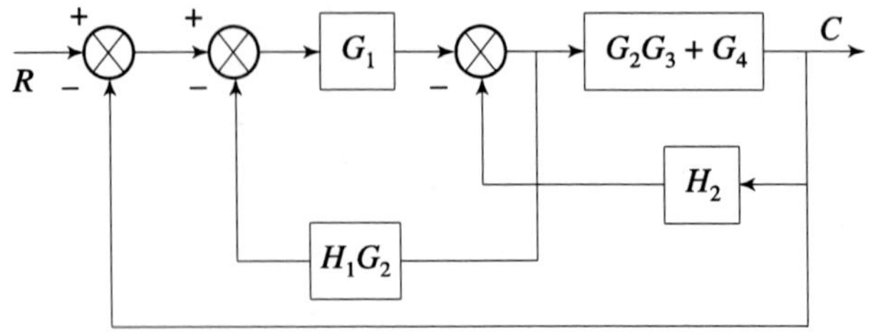

(ii)

Moving summing point before the block.

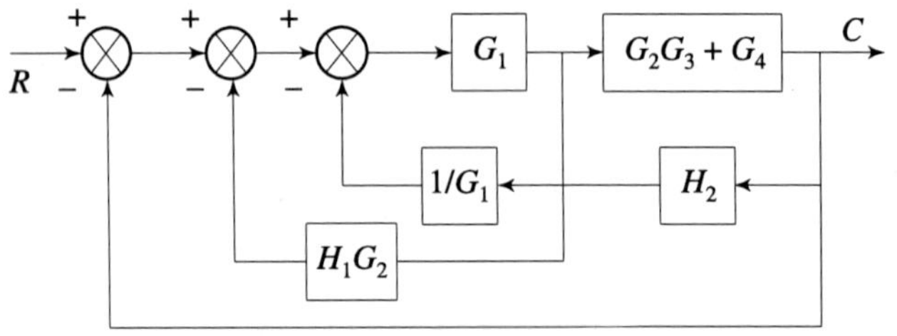

(iii)

Interchanging summing points and modifying branch points.

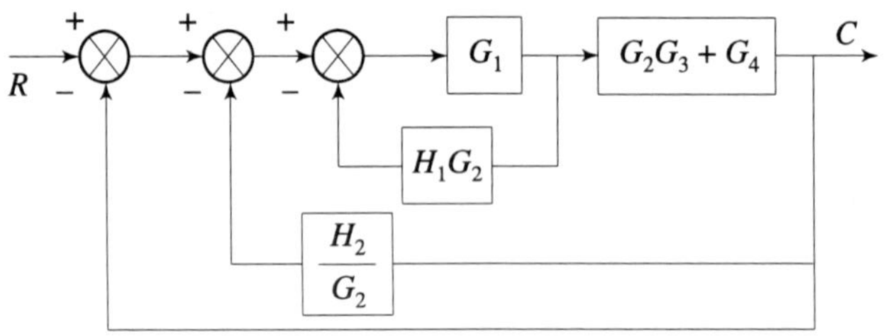

(iv)

Eliminating the feedback path and combining blocks in cascade.

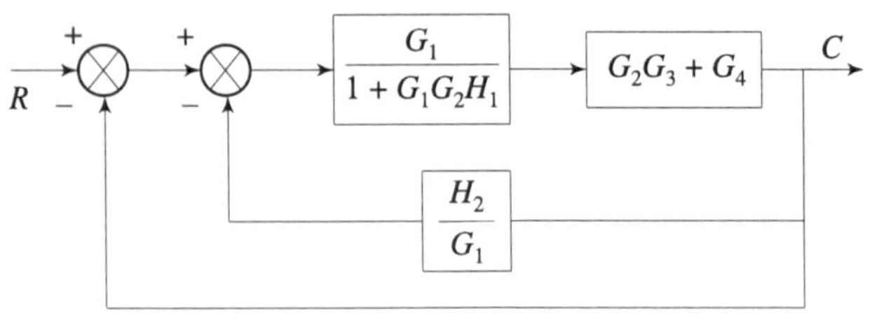

(v)

Eliminating the feedback path

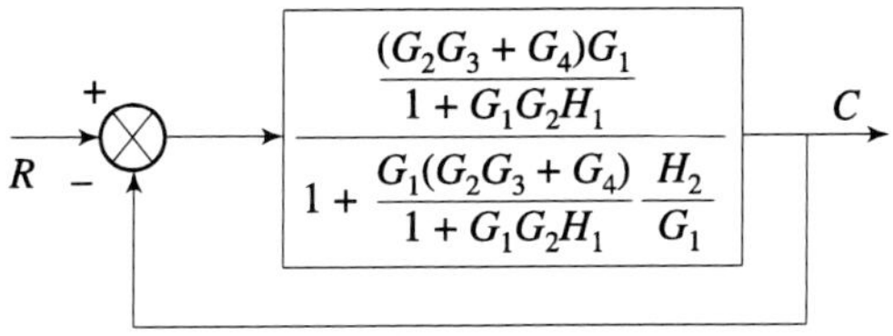

(vi)

Eliminating the unity feedback path

$$\frac{C}{R} = \frac{G_1G_2G_3 + G_1G_4}{1 + G_1G_2H_1 + G_2G_3H_2 + G_4H_2 + G_1G_2G_3 + G_1G_4}.$$

Example 3.8: The block diagram of a closed-loop system shown in Fig. 3.25. Using the block diagram reduction technique, determine the closed-loop transfer function $\frac{C(s)}{R(s)}$ (example of splitting summing point).

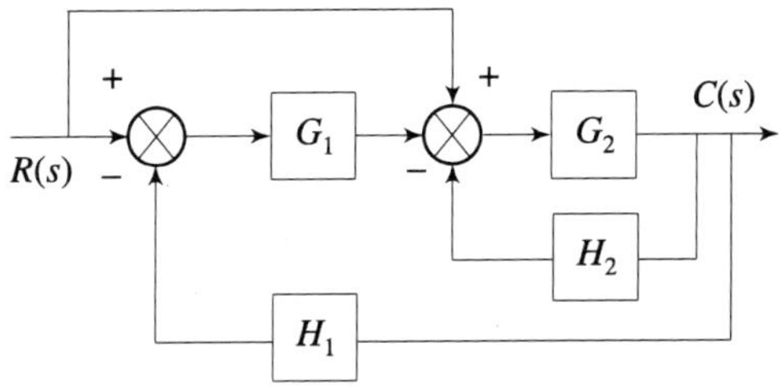

Fig. 3.25

Splitting the summing point

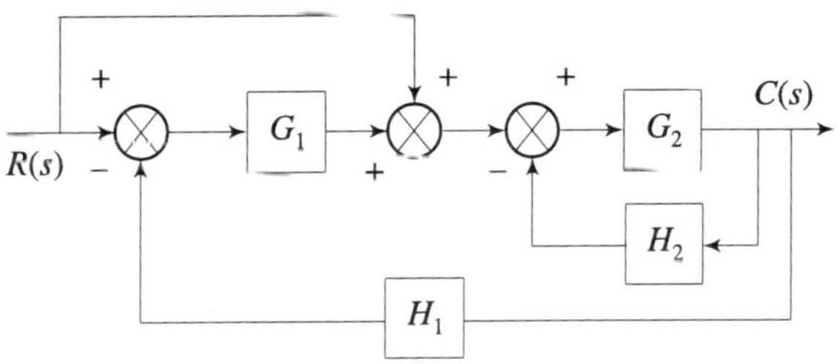

(i)

Eliminating the feedback path

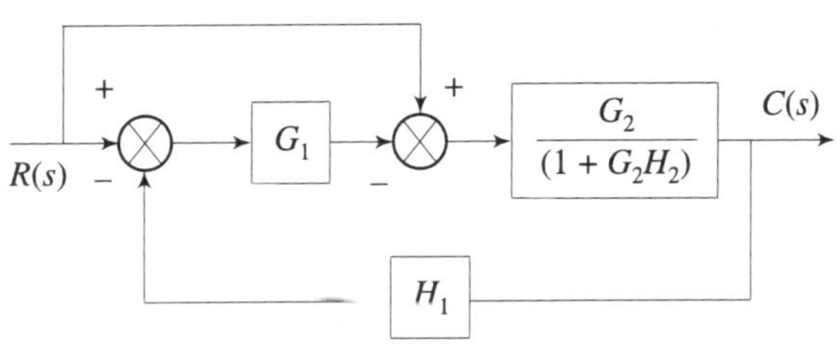

(ii)

Moving the summing point after the block

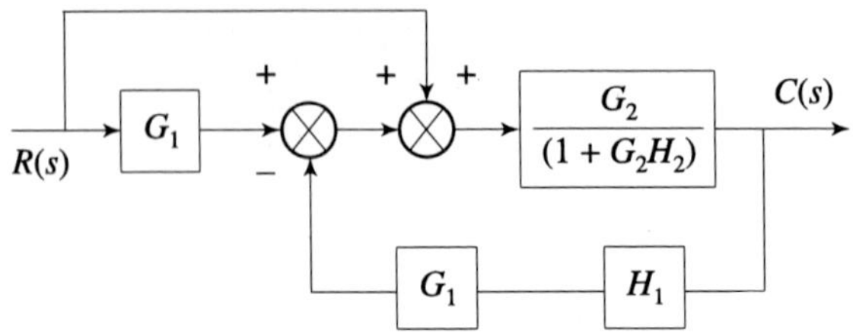

(iii)

Interchanging the summing points and combining the blocks in cascade

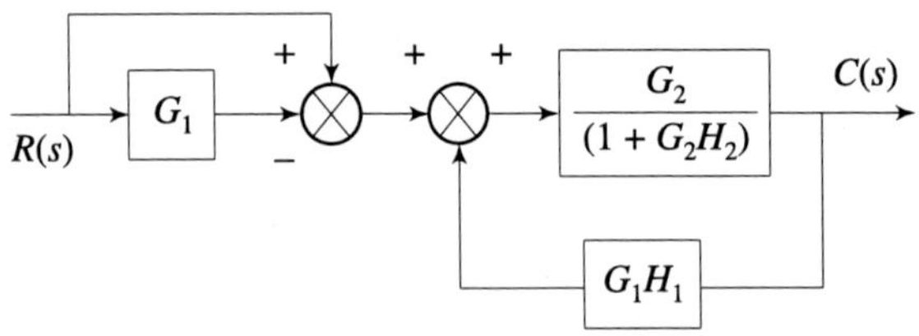

(iv)

Eliminating the feedback path

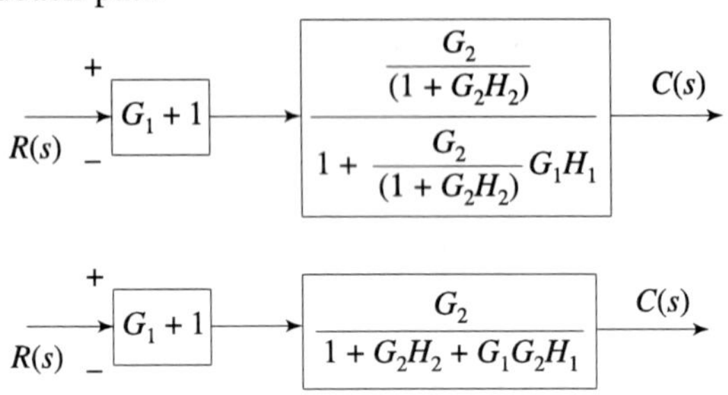

(v)

$$\frac{C(s)}{R(s)} = \frac{G_2(G_1 + 1)}{1 + G_2H_2 + G_1G_2H_1}$$

Example 3.9: For the system shown in Fig. 3.26, determine $\frac{I_o(s)}{V_i(s)}$ by block diagram reduction technique.

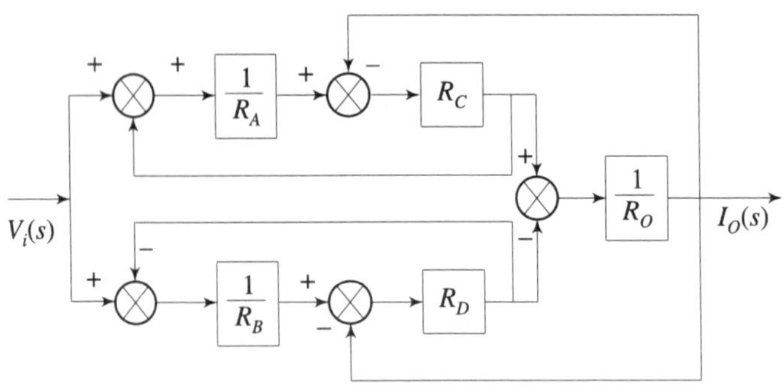

Fig. 3.26

Solution: Shifting summing point before block $\frac{1}{R_A}$ and $\frac{1}{R_B}$.

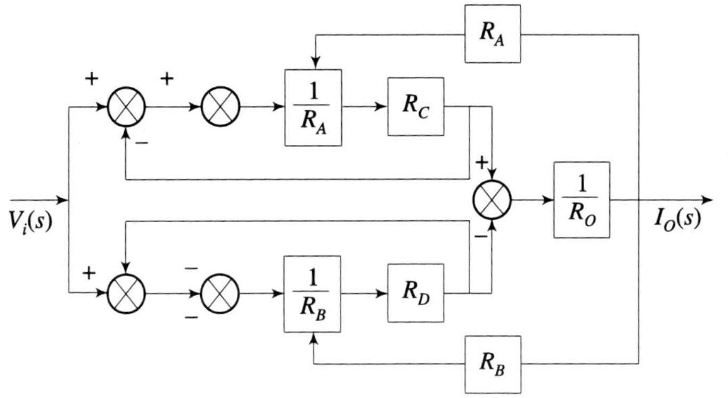

(i)

Combining summing point

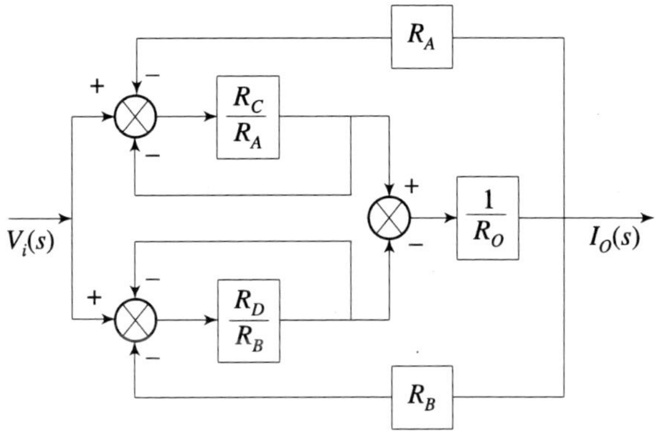

(ii)

Eliminating feedback loop

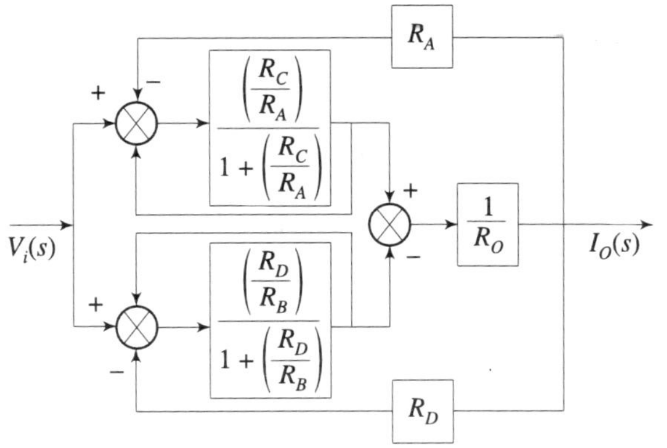

(iii)

Shifting summing point after blocks $\dfrac{R_C}{R_A + R_C}$ and $\dfrac{R_D}{R_B + R_D}$

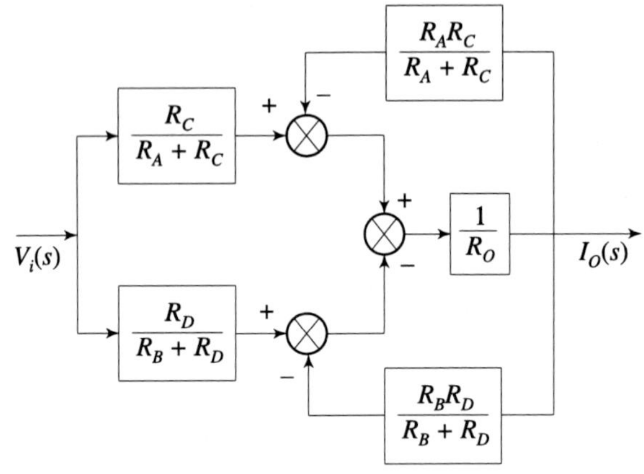

(iv)

Combine 3 summing points and reduce.

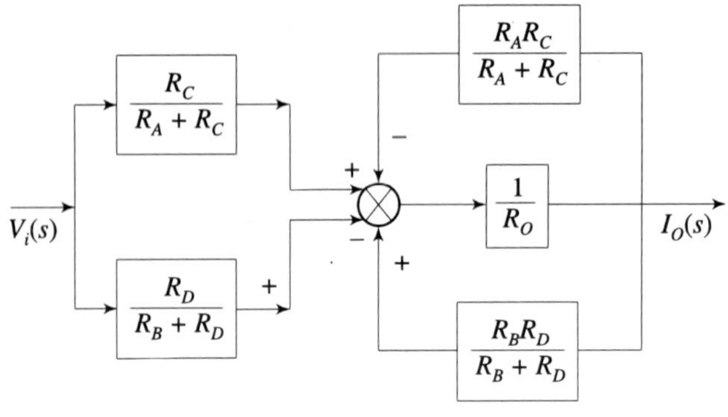

(v)

Reducing parallel combination and feedback.

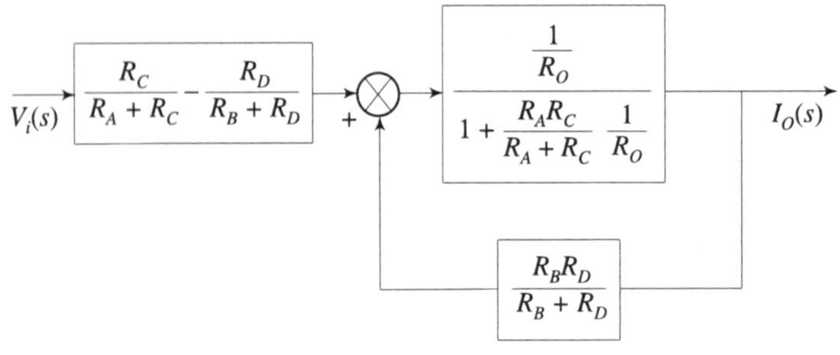

(vi)

Eliminating Feedback

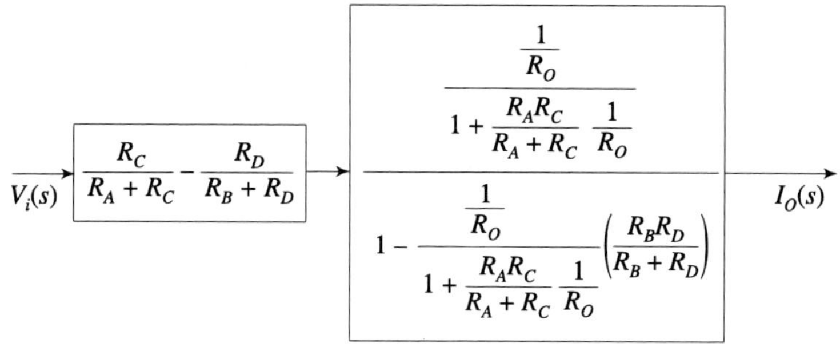

(v)

$$\frac{I_O(S)}{V_i(S)} = \frac{R_C R_B - R_A R_D}{(R_O R_A + R_O R_C + R_A R_C)(R_B + R_D) - R_B R_D (R_A + R_C)}$$

Example 3.10: Obtain $\frac{C(s)}{R(s)}$ of the system shown in Fig. 3.27 by using block diagram reduction rules.

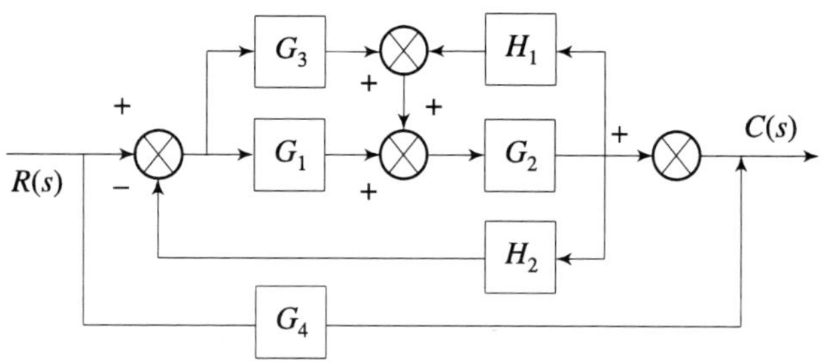

Fig. 3.27

Solution: G_3 and G_1 blocks are in parallel and combining the summing points.

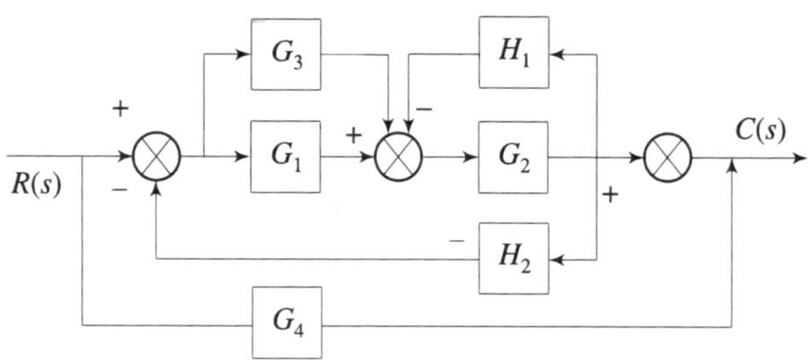

(i)

Eliminate parallel block and feedback path

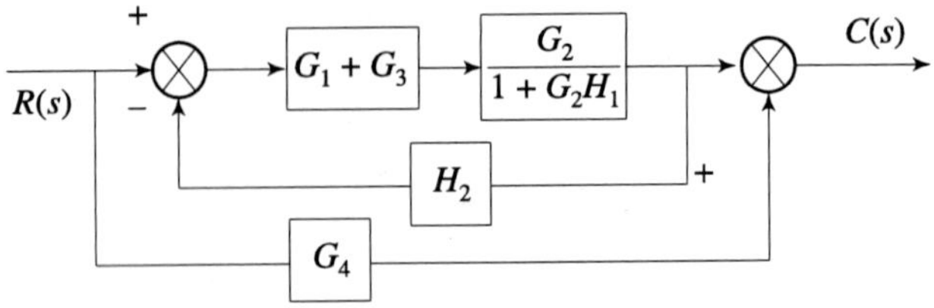

(ii)

Eliminate feedback and cascading

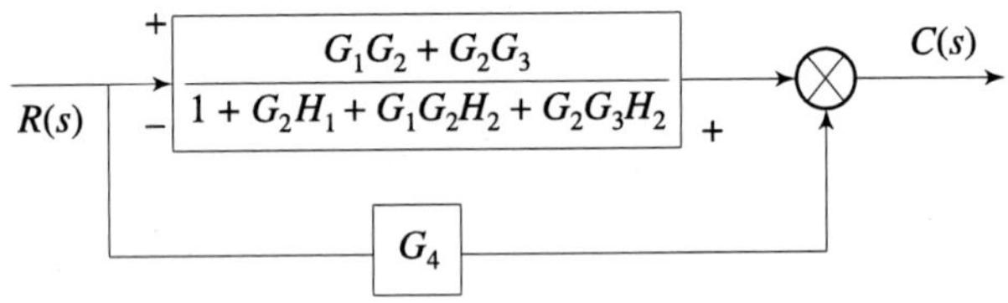

(iii)

Reducing parallel block

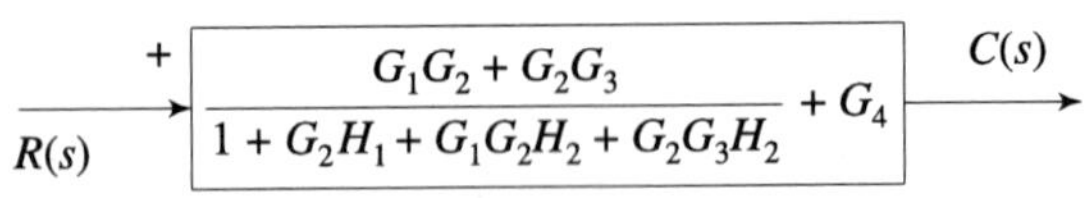

(iv)

Multiple Input Single Output (MISO)

Example 3.11: Find the output of the system shown in Fig. 3.28, $\frac{C}{R}$ and $\frac{C}{x}$

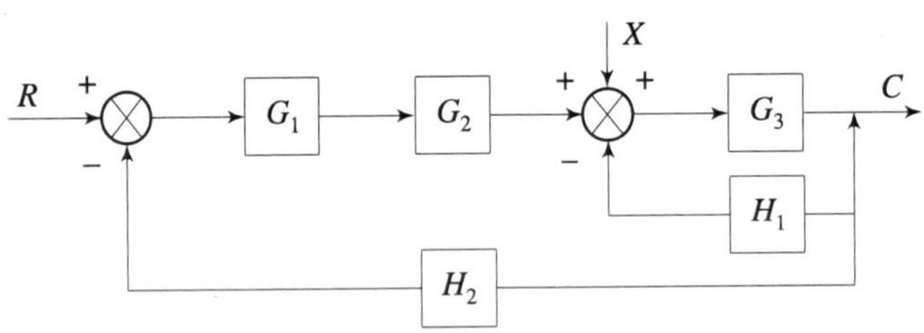

Fig. 3.28

To find $\frac{C}{R}$, putting $x = 0$

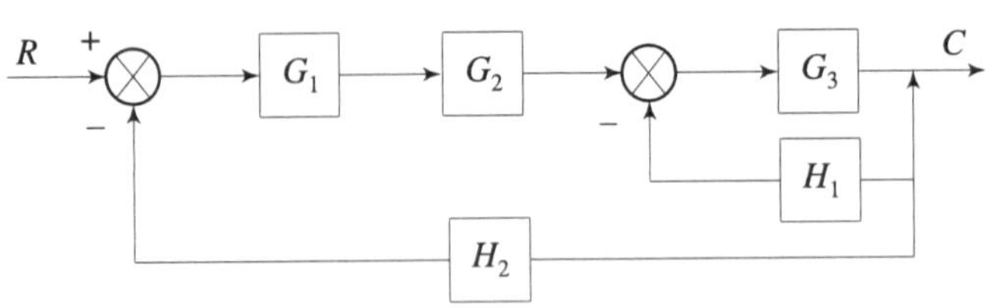

(i)

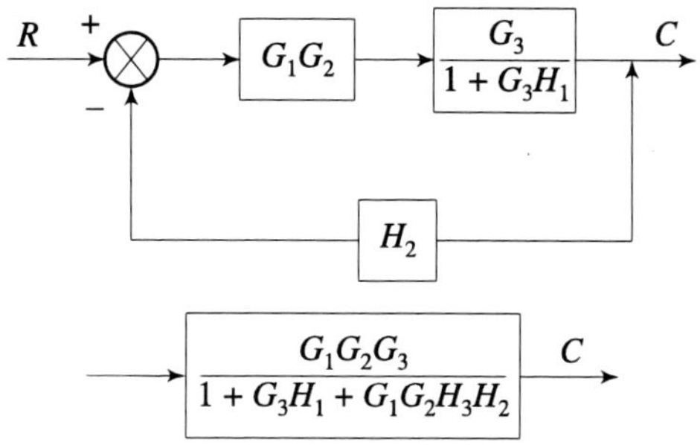

(ii)

$$\frac{C(s)}{R(s)} = \frac{G_1G_2G_3}{1 + H_1G_3 + G_1G_2G_3H_2}$$

To find $\frac{C}{X}$. Putting $R = 0$

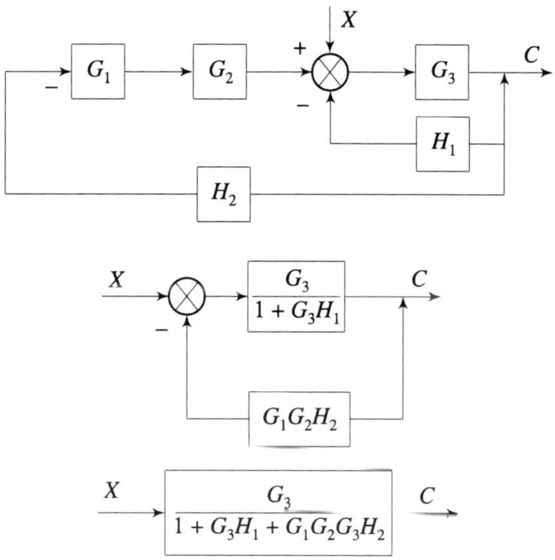

(iv)

$$\frac{C(s)}{X(s)} = \frac{G_3}{1 + H_1G_3 + G_1G_2G_3H_2}$$

Example 3.12: Find the transfer function $\frac{C(s)}{R(s)}$ of the block diagram shown in Fig. 3.29.

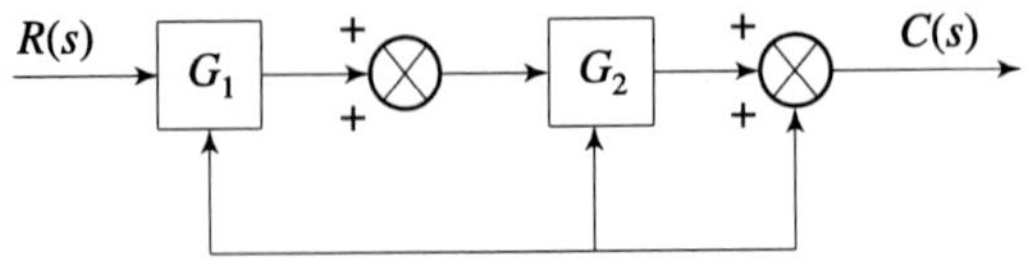

Fig. 3.29

Solution: Modify the given block diagram as shown in Fig. (i)

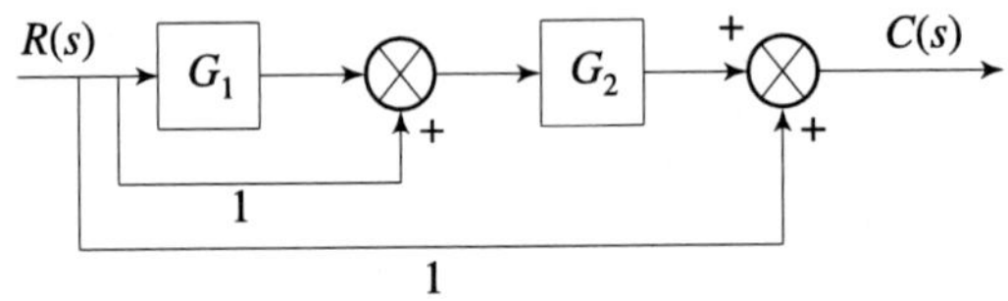

(i)

Eliminate the parallel loop containing block G_1

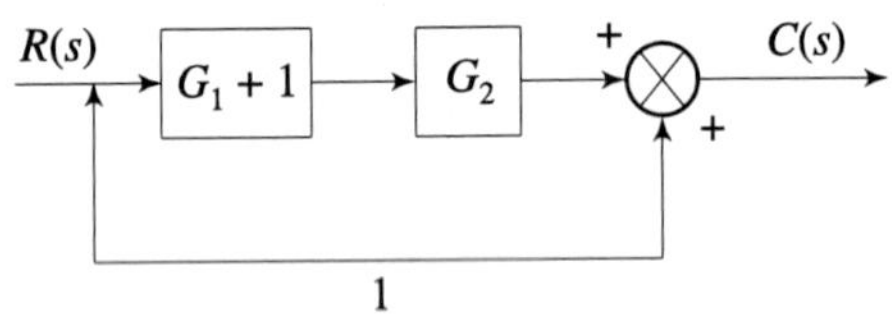

(ii)

Combining cascaded blocks $(G_1 + 1)$ and G_2

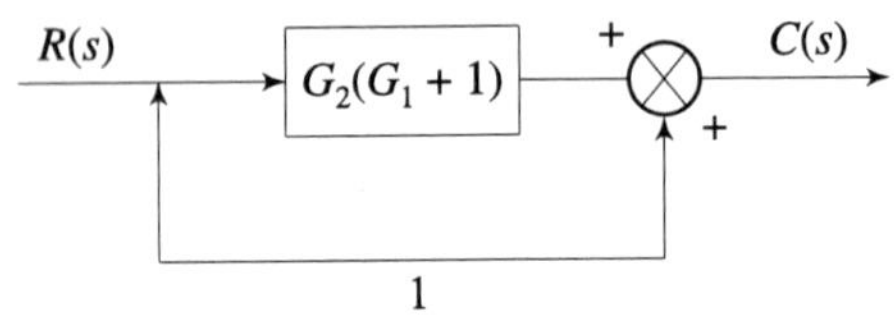

(iii)

Eliminating parallel block

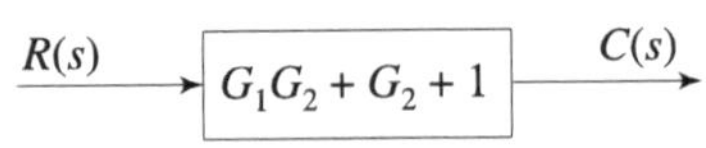

(iv)

$\therefore$ The transfer function $= \frac{C(s)}{R(s)} = G_1 G_2 + G_2 + 1.$

Example 3.13: Find the transfer function of the closed-loop system shown in Fig. 3.30.

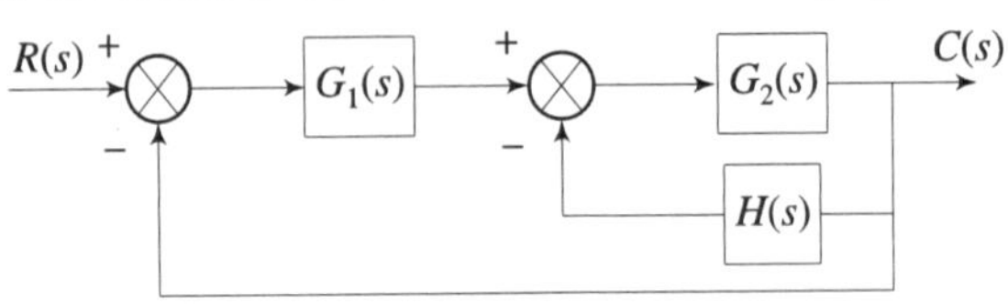

Fig. 3.30

Solution: Eliminating negative feedback of the block $H(S)$

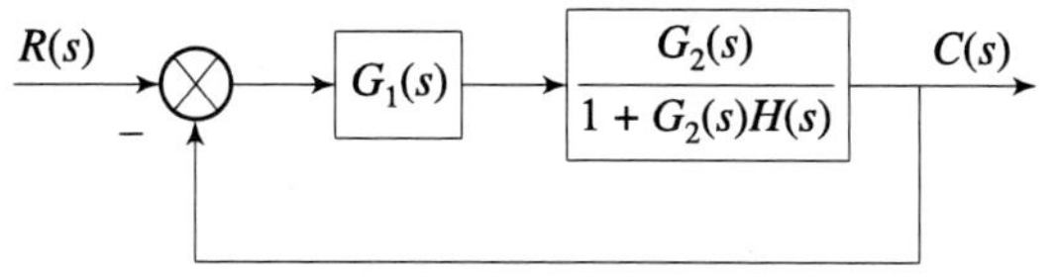

(i)

Combining cascading blocks

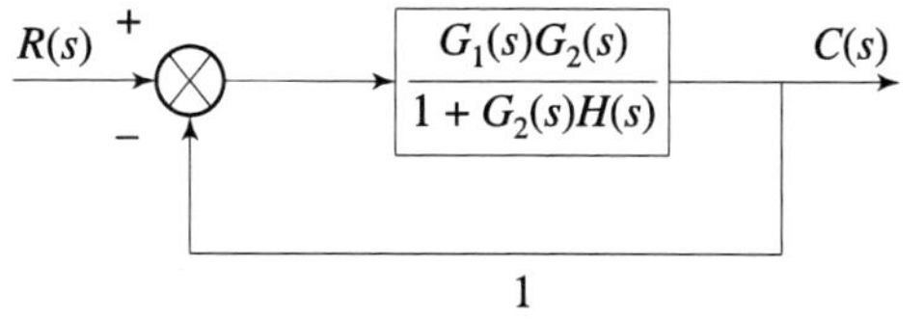

(ii)

Eliminating negative unity feedback

$$\frac{C(s)}{R(s)} = \frac{\dfrac{G_1(s)G_2(s)}{1 + G_2(s)\,H(s)}}{1 + \dfrac{G_1(s)G_2(s)}{1 + G_2(s)\,H(s)}} = \frac{G_1(s)G_2(s)}{1 + G_2(s)H(s) + G_1(s)G_2(s)}$$

∴ The transfer function

$$\frac{C(s)}{R(s)} = \frac{G_1(s)G_2(s)}{1 + G_2(s)\,H(s) + G_1(s)G_2(s)}$$

Example 3.14: Find the transfer function for the block diagram shown in Fig. 3.31.

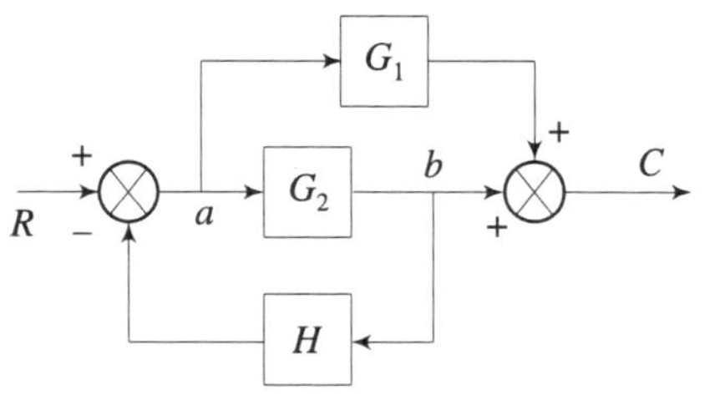

Fig. 3.31

Solution: Move the take-off point from b to a as shown in Fig. (i)

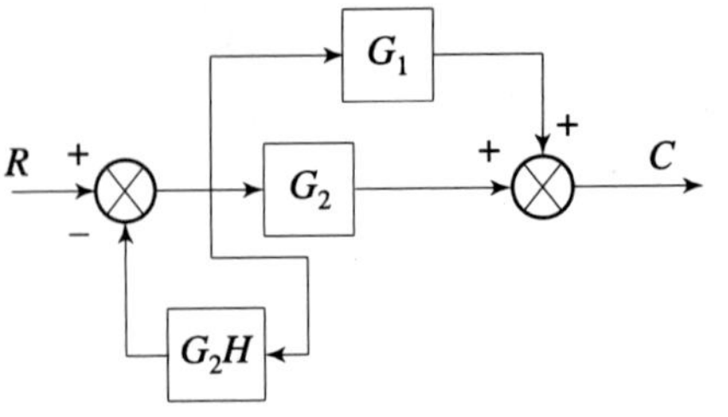

(i)

Combine two blocks G_1 and G_2 in parallel to give a single block $(G_1 + G_2)$ as shown in Fig. (ii)

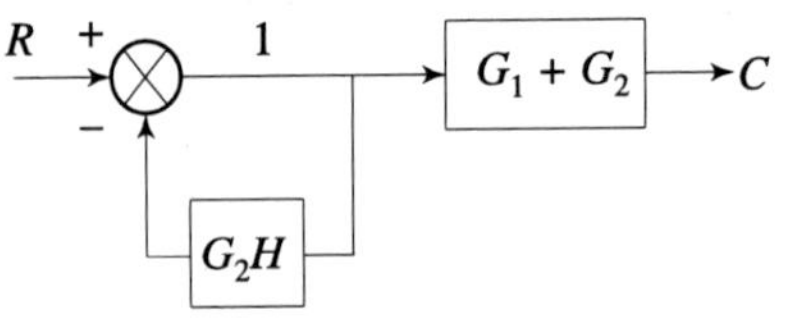

(ii)

Eliminating negative feedback

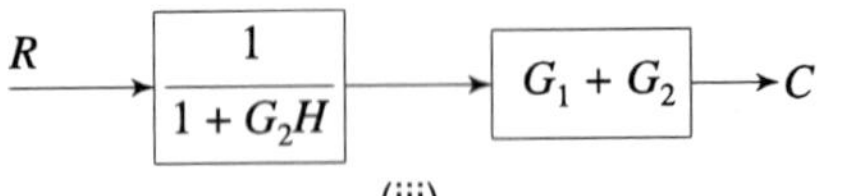

(iii)

Combining cascade blocks

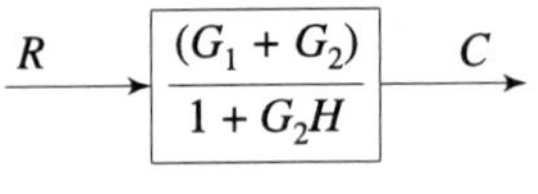

(iv)

Transfer function is given by

$$\boxed{\frac{C}{R} = \frac{G_1 + G_2}{1 + G_2 H}}$$

Example 3.15: Find the transfer function of the system shown in Fig. 3.32.

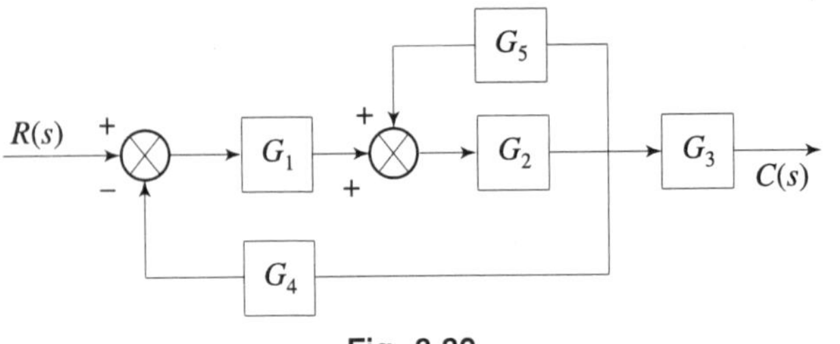

Fig. 3.32

Solution: Eliminating positive feedback of the block G_5.

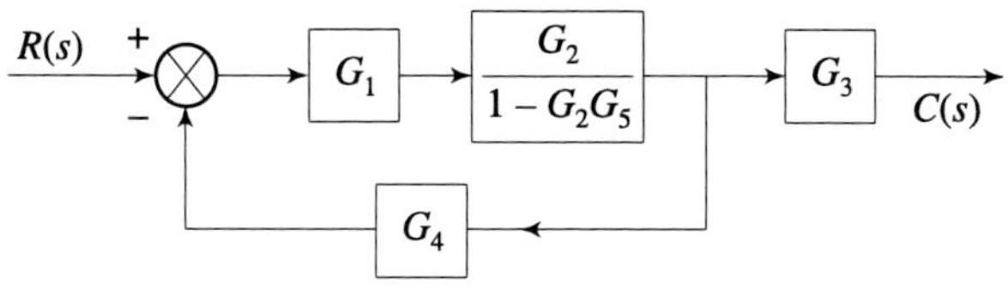

(i)

Shift the takeoff point after the block G_3

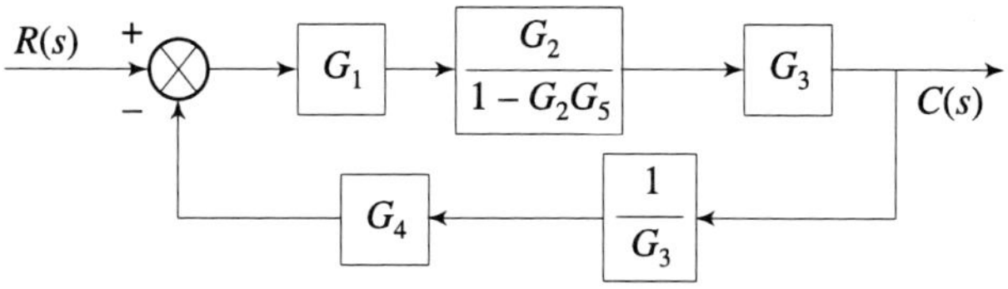

(ii)

Combining cascading blocks

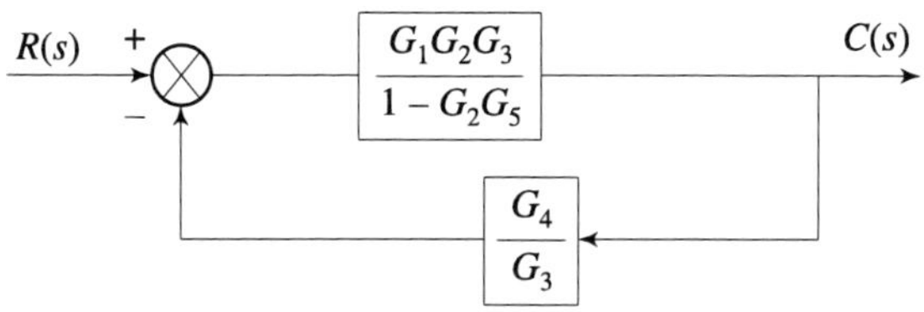

(iii)

Eliminating the negative feedback

$$\frac{C(s)}{R(s)} = \frac{\dfrac{G_1G_2G_3}{1 - G_2G_5}}{1 + \dfrac{G_1G_2G_3}{1 - G_2G_5}\dfrac{G_4}{G_3}} = \frac{G_1G_2G_3}{1 - G_2G_5 + G_1G_2G_4}$$

Example 3.16: From the block diagram shown in Fig. 3.33, determine the overall gain by block diagram reduction technique.

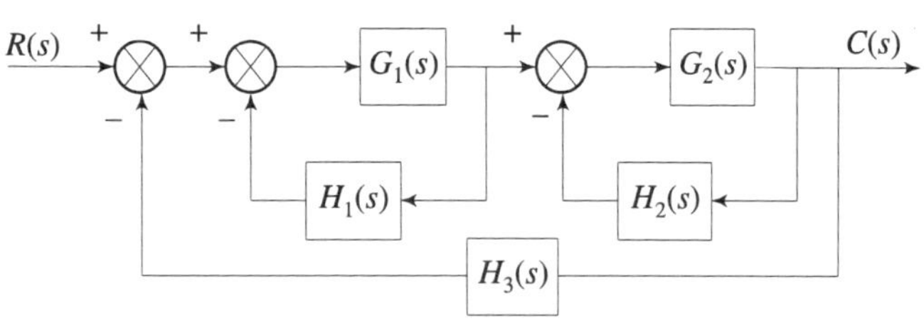

Fig. 3.33

Solution: Eliminating the negative feedback

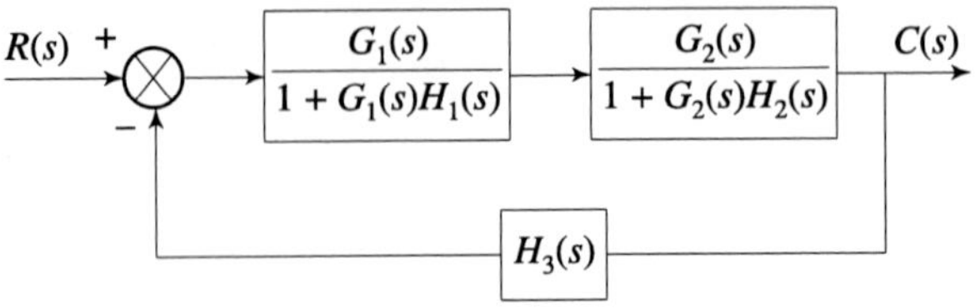

(i)

Combining cascading blocks

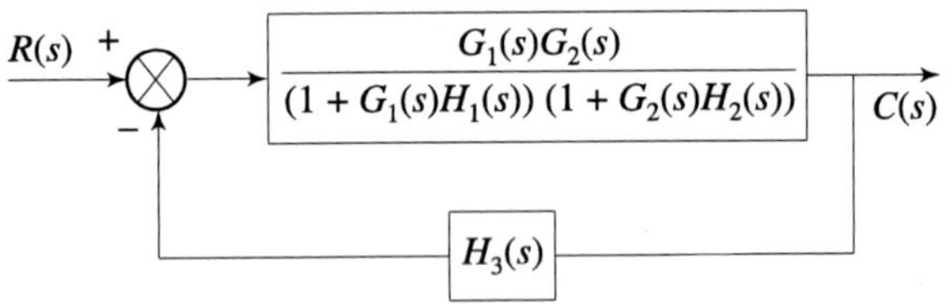

(ii)

Removing the negative feedback path

$$\frac{C(s)}{R(s)} = \frac{\dfrac{G_1(s)G_2(s)}{(1 + G_1(s)H_1(s))\,(1 + G_2(s)\,H_2(s))}}{1 + \dfrac{G_1(s)\,G_2(s)}{(1 + G_1(s)H_1(s))\,(1 + G_2(s)\,H_2(s)}H_3(s)}$$

$$\frac{C(s)}{R(s)} = \frac{G_1(s)\,G_2(s)}{(1 + G_1(s)\,H_1(s))\,(1 + G_2(s)H_2(s)) + G_1(s)G_2(s)\,H_3(s)}$$

Example 3.17: Example of multiple input and single output. From the block diagram shown in Fig. 3.34, find the transfer function $\dfrac{C(s)}{R(s)}$ and $\dfrac{C(s)}{D(s)}$.

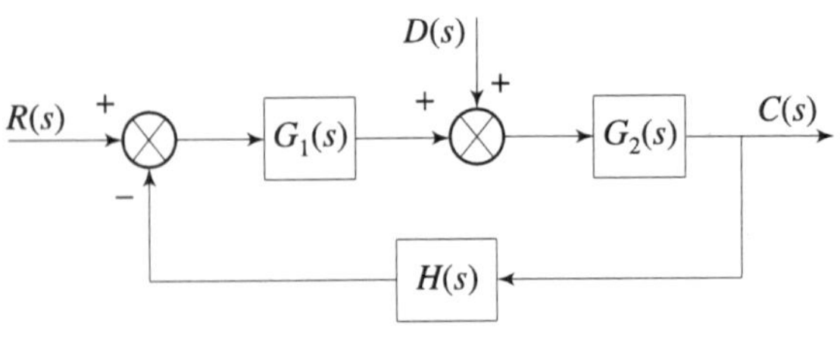

Fig. 3.34

Solution: The given system is a two-input system. Finding the output by considering one input at a time.

I. Considering $R(s)$ and another input $D(s)$ is set to be zero. The summing point at $D(s)$ is removed and resulting block diagram is shown in Fig. (i)

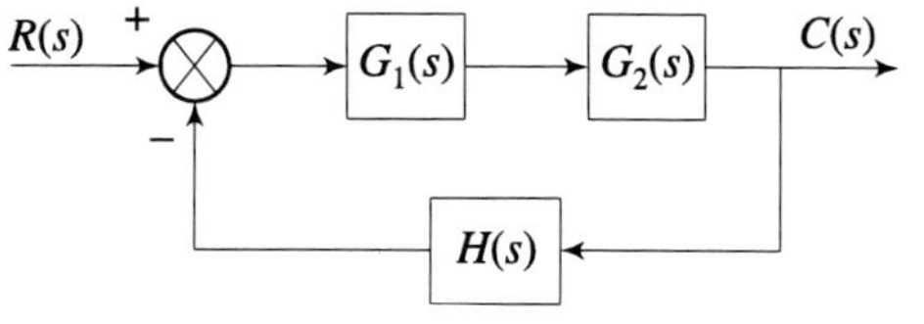

(i)

Combining cascading blocks

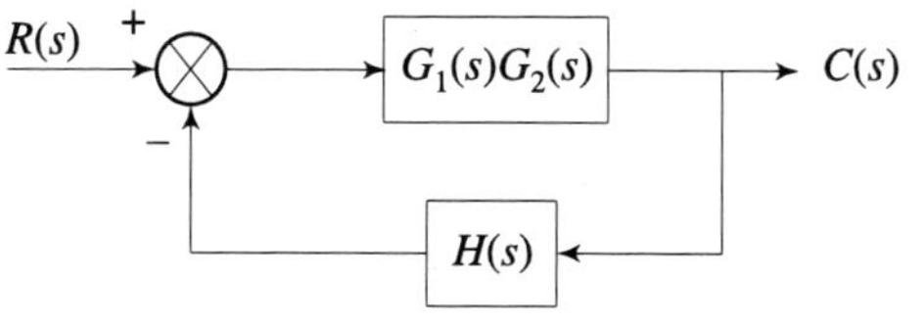

(ii)

Eliminating negative feedback

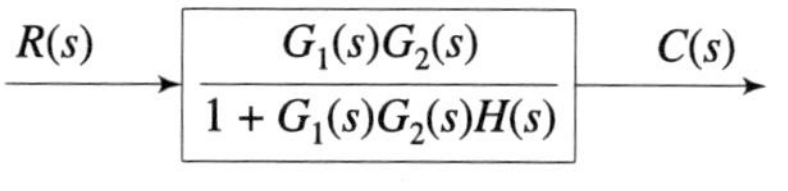

(iii)

The transfer function

$$\frac{C(s)}{R(s)} = \frac{G_1(s)G_2(s)}{1 + G_1(s)G_2(s)H(s)}$$

II. Considering $D(s)$ and $R(s)$ is set to be zero. The summing point at $R(s)$ is removed putting – 1 into a block, representing the negative feedback effect. The resulting block diagram is shown in Fig. (iv)

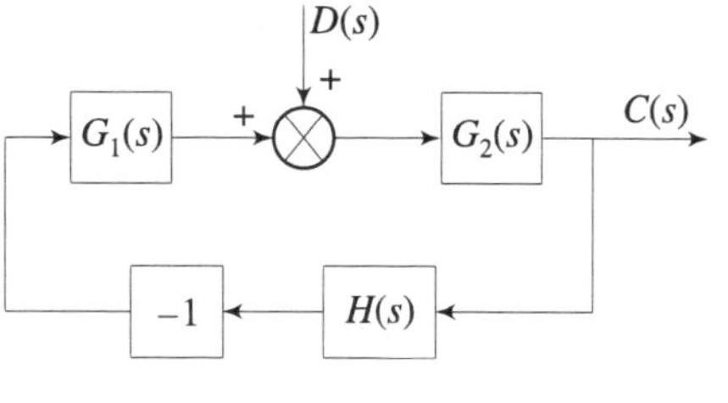

(iv)

Rearranging the block diagram as shown in Fig. (v)

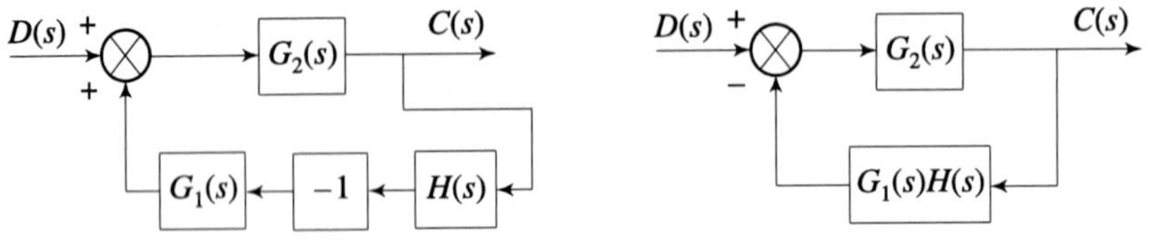

(v)

Eliminating negative feedback

$$\frac{C(s)}{D(s)} = \frac{G_2(s)}{1 + G_1(s)G_2(s)H(s)}.$$

Example 3.18: Using block diagram reduction technique for the system shown in Fig. 3.35, find the transfer function.

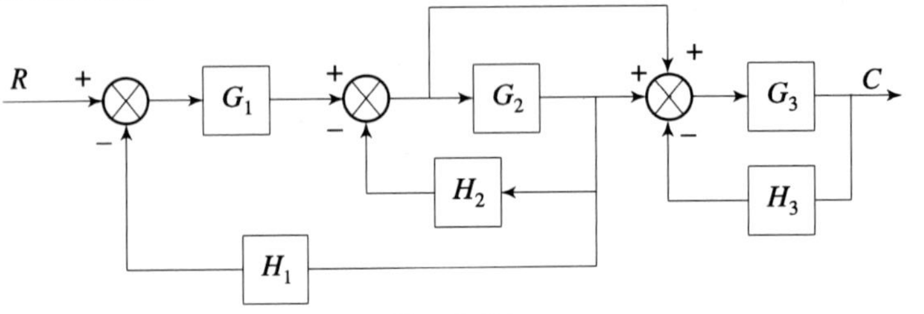

Fig. 3.35

Solution: Move the block G_2 to the left of the take-off point and reduce the negative feedback system consisting of G_3 and H_3.

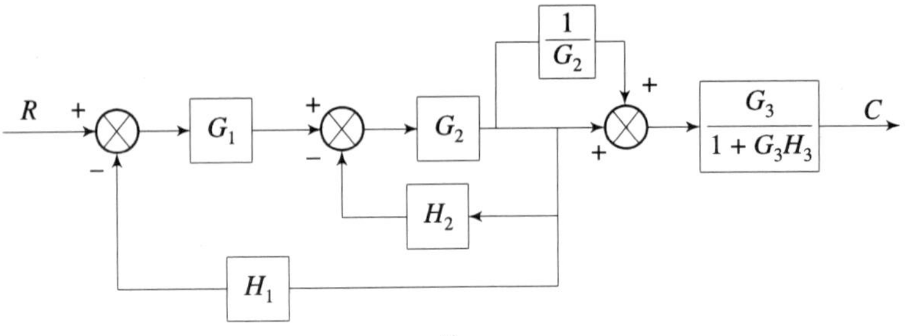

(i)

Add the parallel consisting of $\frac{1}{G_2}$ and unity. Shifting G_1 to the right of the summing point.

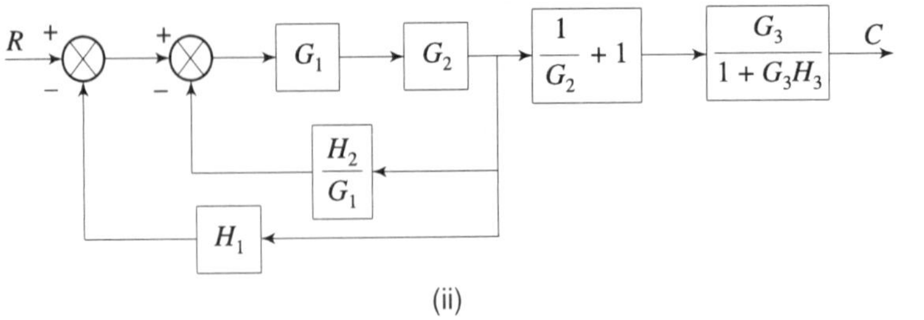

(ii)

Combining in cascade blocks and eliminating negative feedback

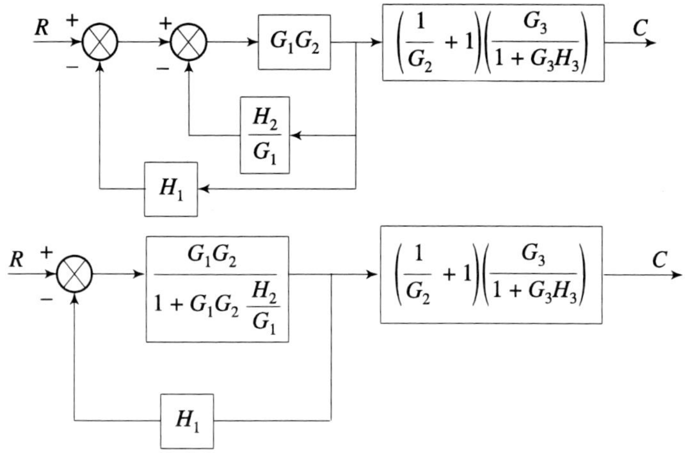

(iii)

Eliminating negative feedback

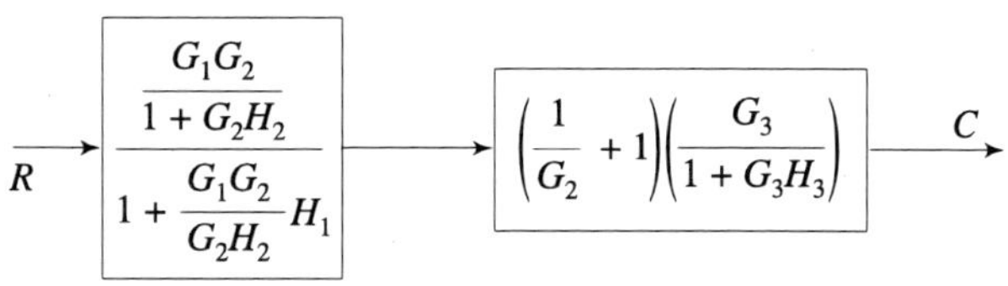

(iv)

Combining cascading blocks

$$\left(\frac{G_1G_2}{1 + G_2H_2 + G_1G_2H_1}\right)\left(\frac{1 + G_2}{G_2}\right)\left(\frac{G_3}{1 + G_3H_3}\right)$$

The transfer function

$$\frac{C}{R} = \frac{G_1G_3(1 + G_2)}{(1 + G_2H_2 + G_1G_2H_1)\,(1 + G_3H_3)}$$

Example 3.19: Obtain the output of the system shown in Fig. 3.36. Find $\frac{C}{R_1}$ and $\frac{C}{R_2}$.

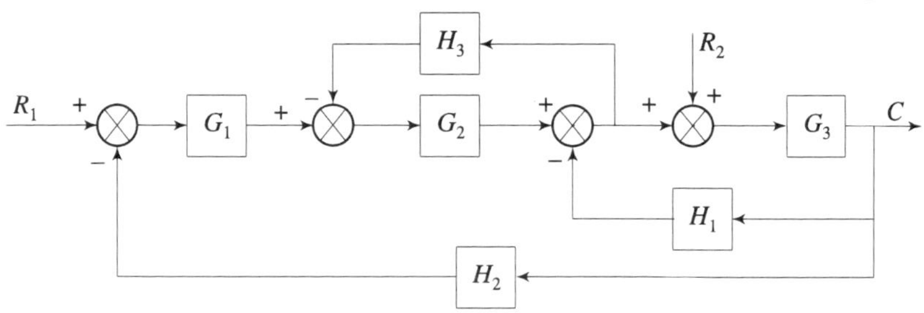

Fig. 3.36

Solution:

I. To find $\frac{C}{R_1}$

First assume that the input R_2 is made zero. The summing point at R_2 is removed and the resulting block diagram is shown in Fig. (i)

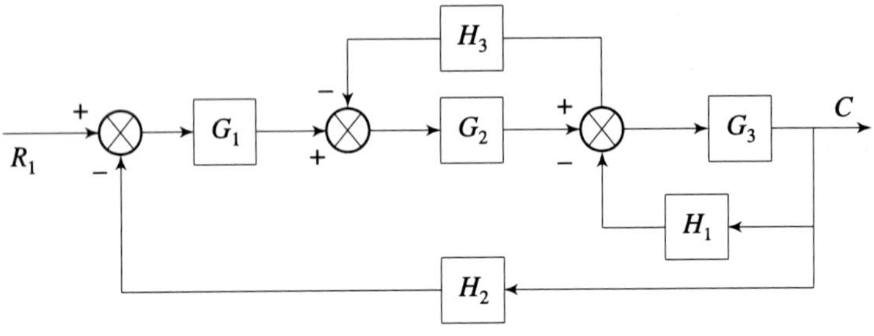

(i)

Move the take-off point after block G_3

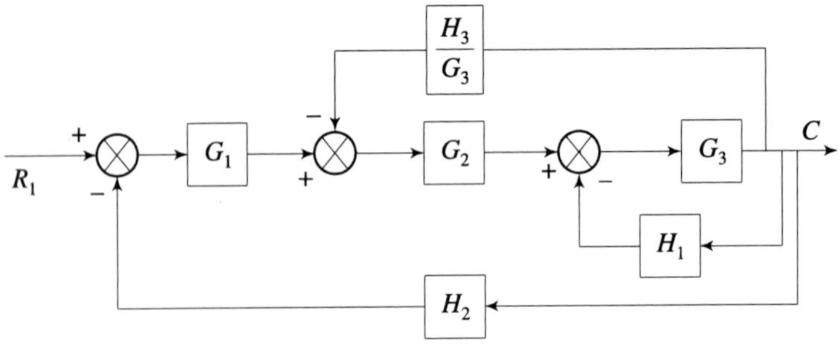

(ii)

Eliminating the negative feedback and combining G_2

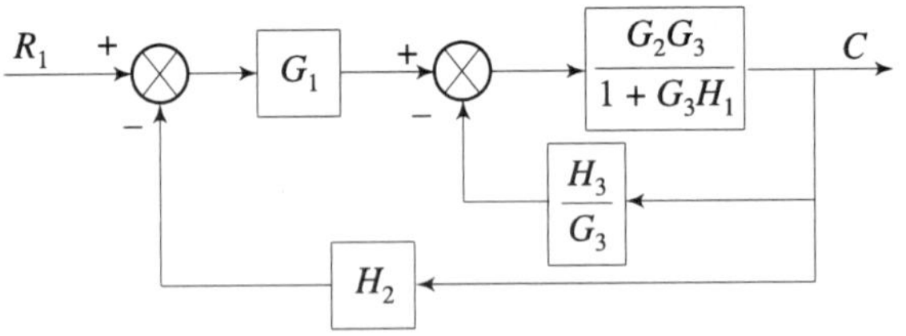

(iii)

Eliminating negative feedback

$$\frac{\frac{G_2G_3}{1 + G_3H_1}}{1 + \frac{G_2G_3}{1 + G_3H_1}\frac{H_3}{G_3}} = \frac{G_2G_3}{1 + G_3H_1 + G_2H_3}$$

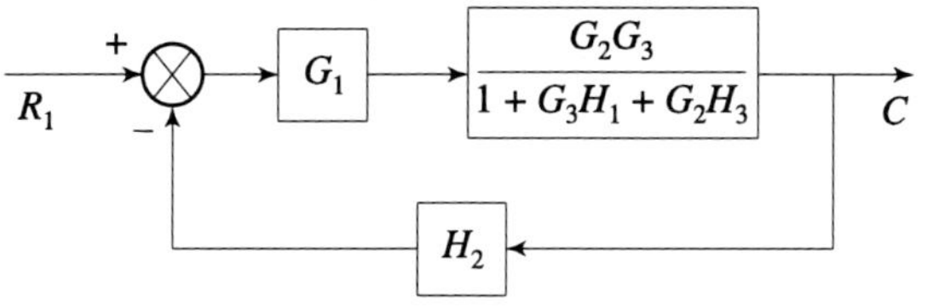

(iv)

Combining cascading block and eliminating negative feedback

$$\frac{C}{R_1} = \frac{\dfrac{G_1G_2G_3}{1 + G_3H_1 + G_2H_3}}{1 + \dfrac{G_1G_2G_3}{1 + G_3H_1 + G_2H_3} \cdot H_2}$$

$$\frac{C}{R_1} = \frac{G_1G_2G_3}{1 + G_3H_1 + G_2H_3 + G_1G_2G_3H_2}$$

II. To find $\dfrac{C}{R_2}$

Assume that the input R_1 is made zero, the summing point at R_1 is removed and a negative sign is attached to the feedback path gain, H_2.

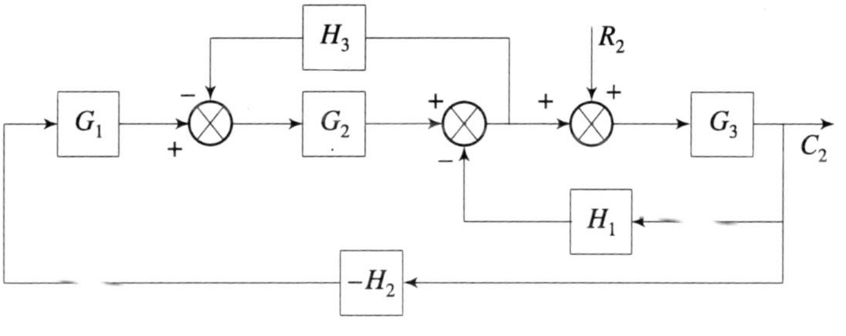

(v)

Combining the block G_1 and $-H_2$ in cascade and rearrange the diagram.

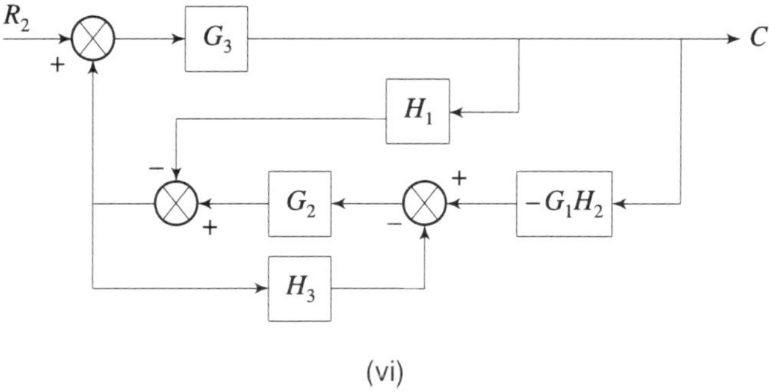

(vi)

Shift the summing point before block G_2

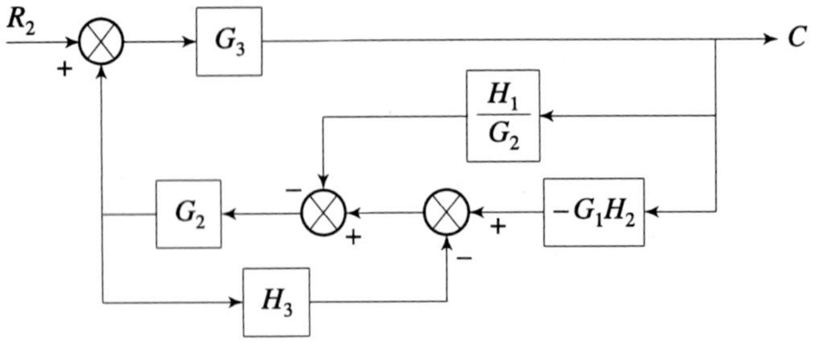

(vii)

Interchanging the summing points

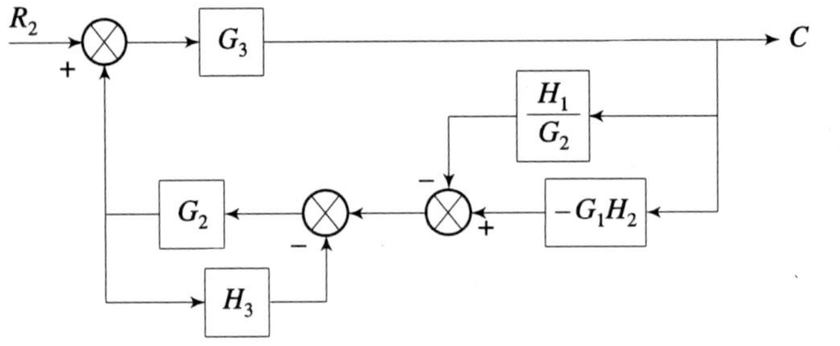

(viii)

Eliminating negative feedback loop and combining parallel shown in Fig. (ix)

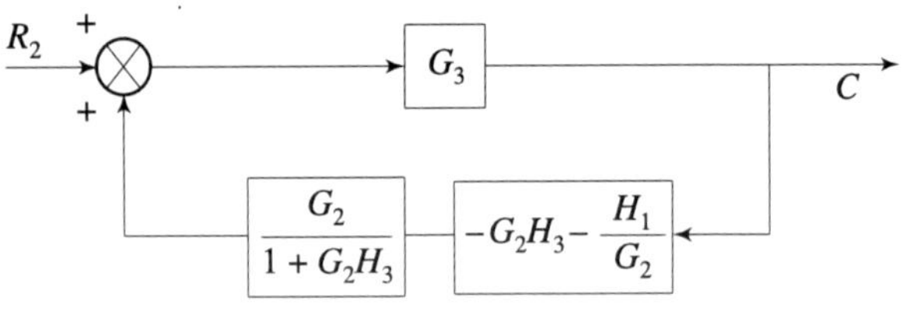

(ix)

Combining cascading block and eliminating positive feedback loop.

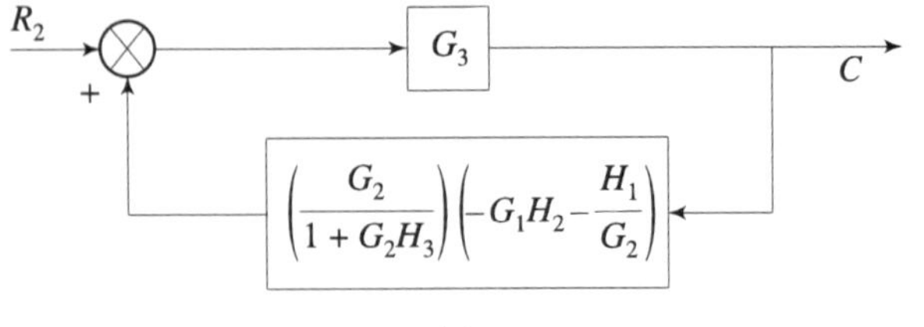

(x)

$$\left(\frac{G_2}{1 + G_2H_3}\right)\left(-G_1H_2 - \frac{H_1}{G_2}\right) = \left(\frac{G_2}{1 + G_2H_3}\right)\left(\frac{-G_1G_2H_2 - H_1}{G_2}\right)$$

$$= \frac{-H_1 - G_1G_2H_2}{1 + G_2H_3}$$

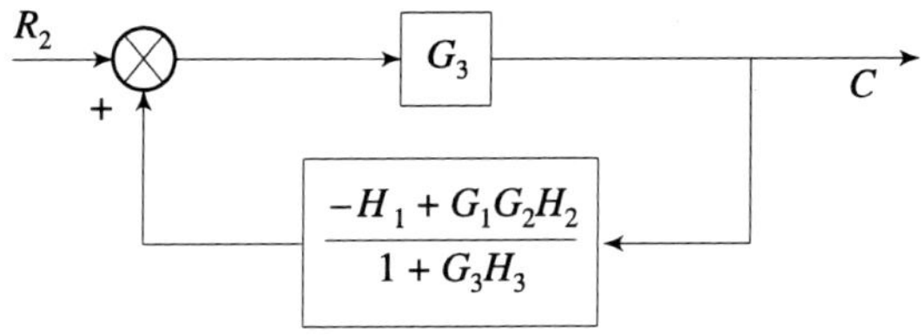

(xi)

$$\frac{C}{R_2} = \frac{G_3}{1 - \left\{\frac{-H_1(G_1G_2H_2)}{1 + G_2H_3} \cdot G_3\right\}}$$

$$= \frac{G_3}{\dfrac{1 + G_2H_3 + H_1G_3 + G_1G_2G_3H_2}{1 + G_2H_3}}$$

$$\frac{C}{R_2} = \frac{G_3(1 + G_2H_3)}{1 + G_2H_3 + G_3H_1 + G_1G_2G_3H_2}$$

Example 3.20: Find the overall transfer function of the system shown in Fig. 3.37.

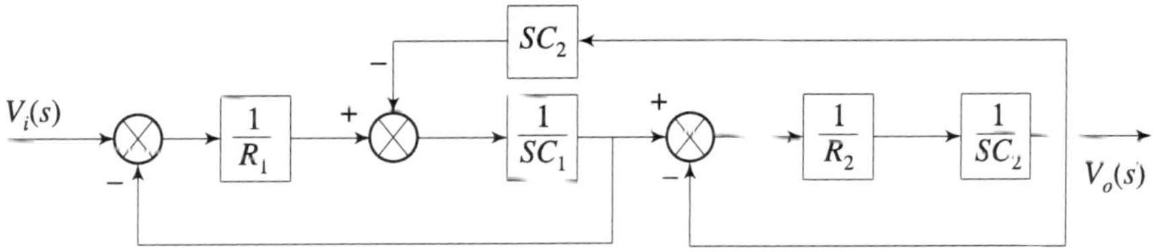

Fig. 3.37

Solution: Combine cascading blocks and eliminate negative unity feedback.

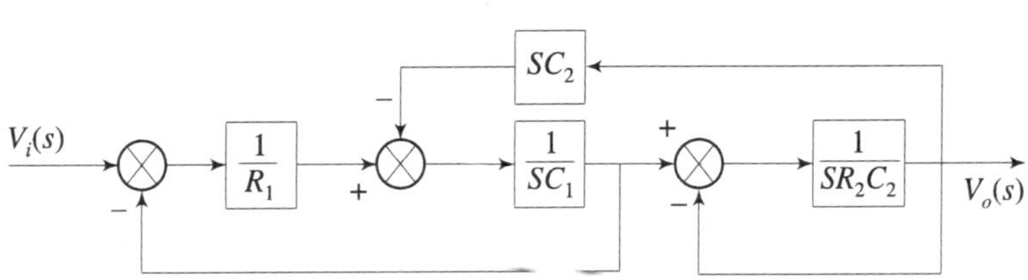

(i)

Eliminating negative unity feedback

$$\frac{\frac{1}{SR_2C_2}}{1+\frac{1}{SR_2C_2}.(1)} = \frac{1}{1+SR_2C_2}$$

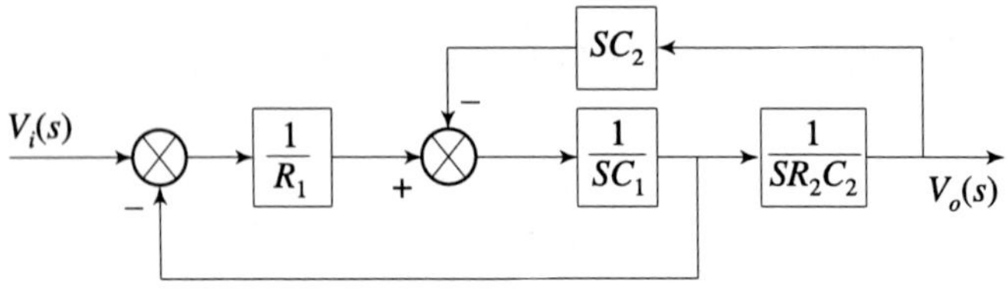

(ii)

Shift the take-off point after the block $\frac{1}{1+SR_2C_2}$

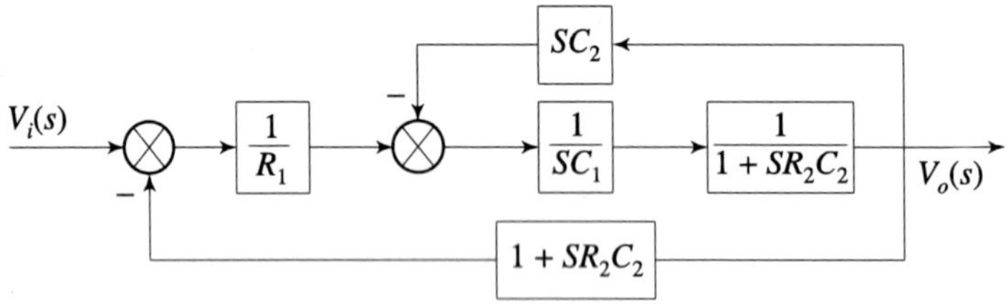

(iii)

Combining cascading block

$$\left(\frac{1}{SC_1}\right)\left(\frac{1}{1+SR_2C_2}\right) = \frac{1}{SC_1+S^2C_1C_2R_2}$$

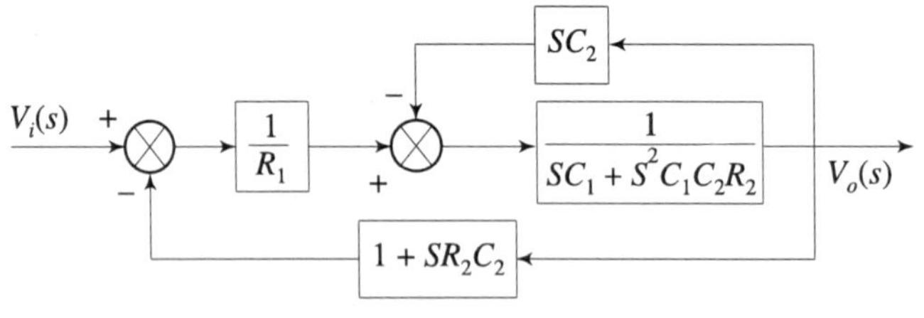

(iv)

Eliminating negative feedback SC_2 combining cascading block $\frac{1}{R_1}$

$$= \frac{\frac{1}{SC_1+S^2C_1\,C_2R_2}}{1+\frac{1}{SC_1+S^2C_1\,C_2R_2}\cdot SC_2}$$

$$= \left(\frac{1}{SC_1 + S^2\,C_1C_2R_2 + SC_2}\right)\frac{1}{R_1}$$

$$= \frac{1}{S^2C_1\,C_2\,R_1R_2 + SC_1\,R_1 + SC_2R_1}$$

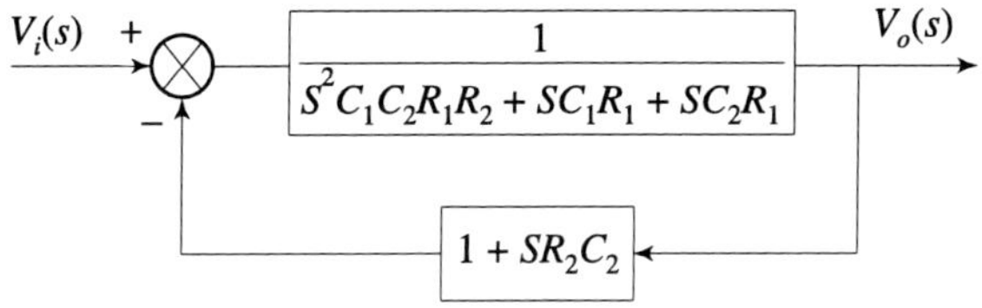

(v)

Eliminating negative feedback

$$= \frac{\dfrac{1}{S^2C_1C_2R_1R_2 + SC_1R_1 + SC_2R_1}}{1 + \dfrac{1}{S^2C_1C_2R_1R_2 + SC_1R_1 + SC_2R_1}\cdot(1 + SR_2C_2)}$$

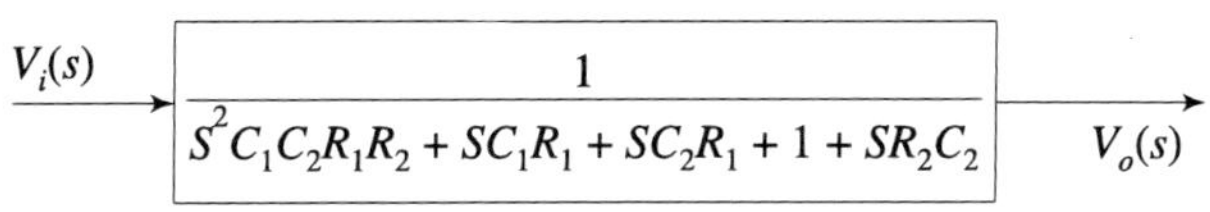

The overall transfer function

$$\frac{V_o(s)}{V_i(s)} = \frac{1}{1 + SC_1R_1 + SC_2R_1 + SR_2C_2 + S^2C_1CR_1R_2}$$

Example 3.21: Determine the overall transfer function $\dfrac{C(s)}{R(s)}$ for the system shown in Fig. 3.38.

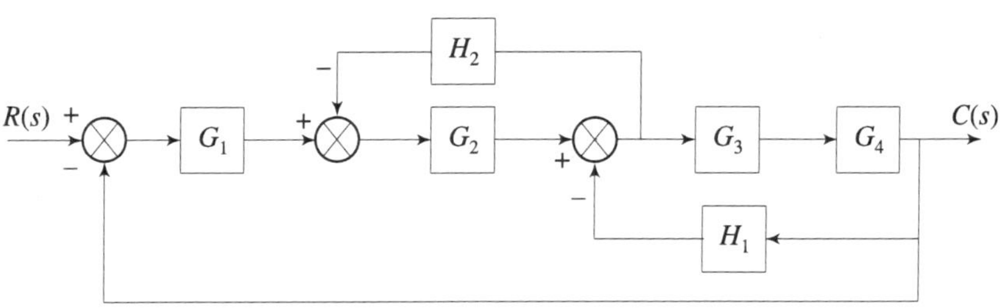

Fig. 3.38

Solution: Moving the take-off point before the block G_4.

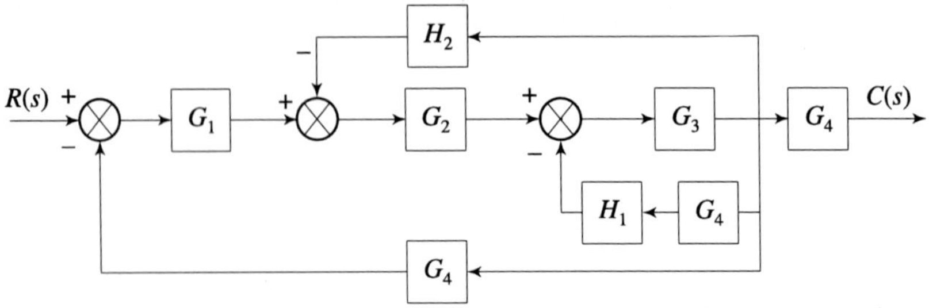

(i)

Eliminating negative feedback and combining cascading block.

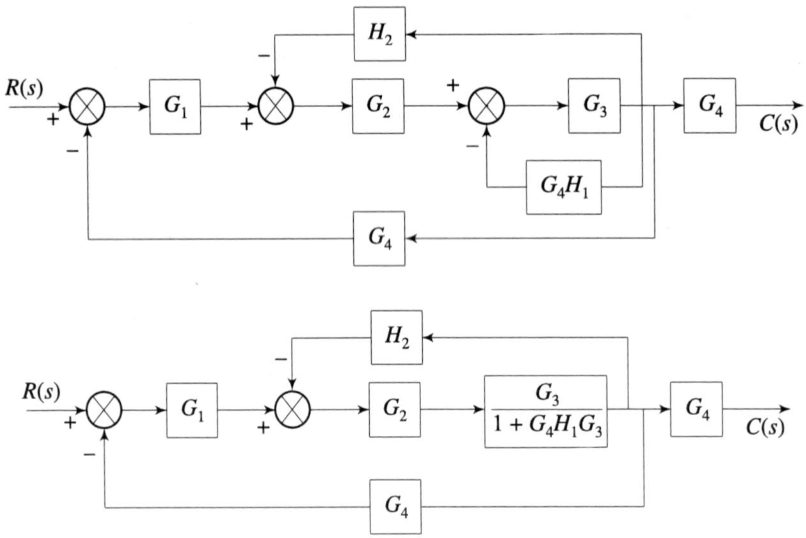

(ii)

Combining the cascading blocks and eliminating negative feedback path

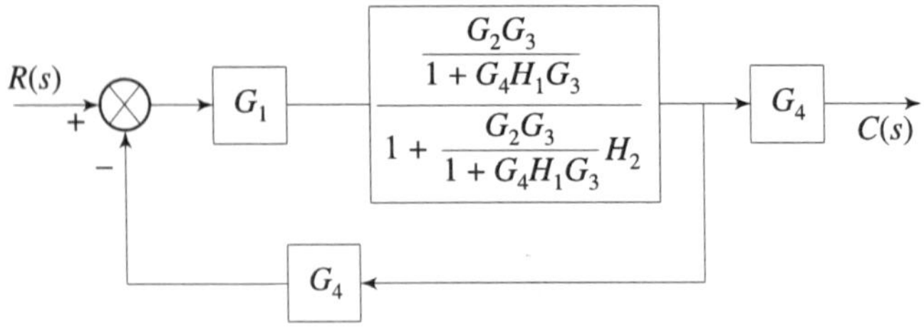

(iii)

Combining the cascading blocks and eliminating negative feedback path

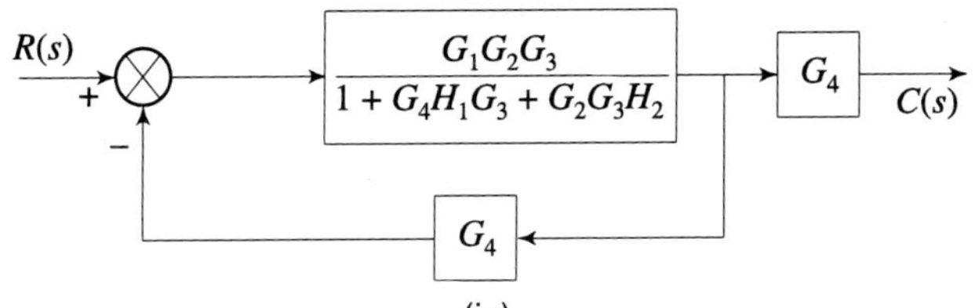

(iv)

$$= \frac{\dfrac{G_1G_2G_3}{1 + G_4H_1G_3 + G_2G_3H_2}}{1 + \dfrac{G_1G_2G_3}{1 + G_4H_1G_3 + G_2G_3H_2} \cdot G_4} = \frac{G_1G_2G_3}{1 + G_4H_1G_3 + G_2G_3H_2 + G_1G_2G_3G_4}$$

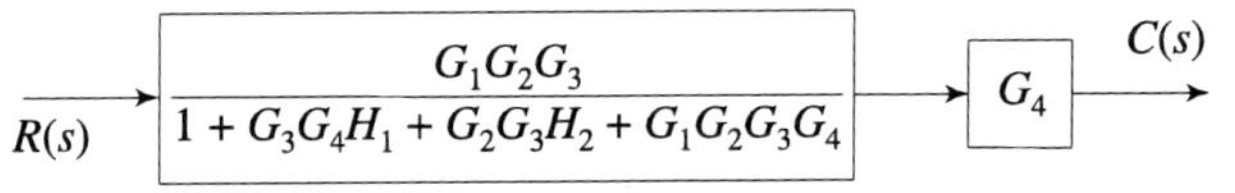

$$\frac{C(s)}{R(s)} = \frac{G_1G_2G_3G_4}{1 + G_3G_4H_1 + G_2G_3H_2 + G_1G_2G_3G_4}$$

Example 3.22: Using block diagram reduction technique find the transfer function $\frac{C(s)}{R(s)}$ for the system shown in Fig. 3.39.

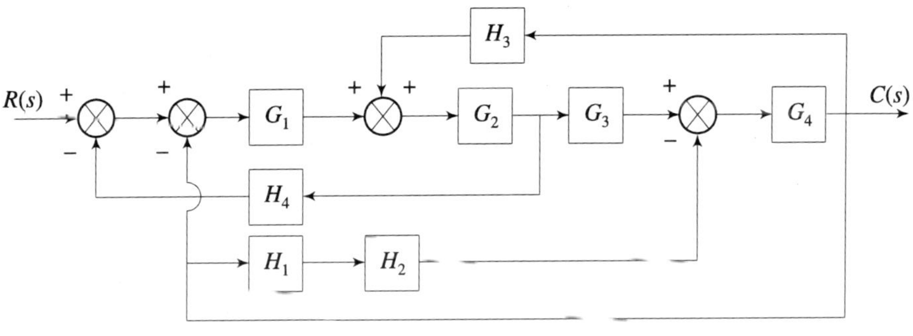

Fig. 3.39

Solution: Rearranging the branch points, combining cascading blocks

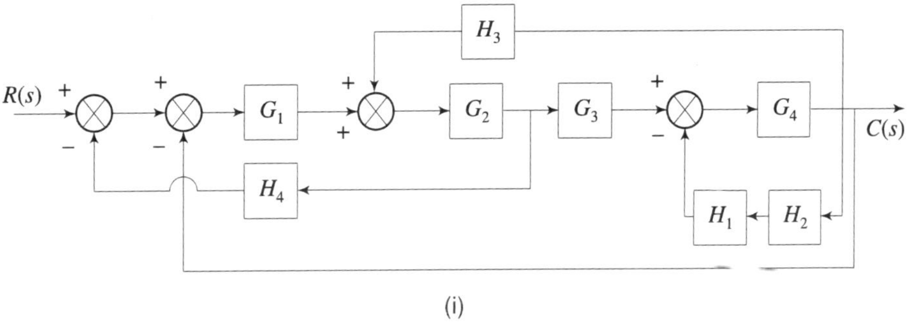

(i)

Shifting the takeoff point after the block G_3

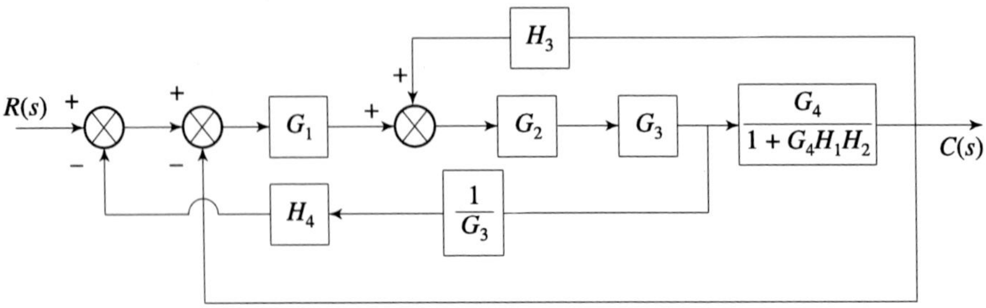

(ii)

Combining cascading block and shifting takeoff point after the block $(G_4/1 + G_4H_1H_2)$

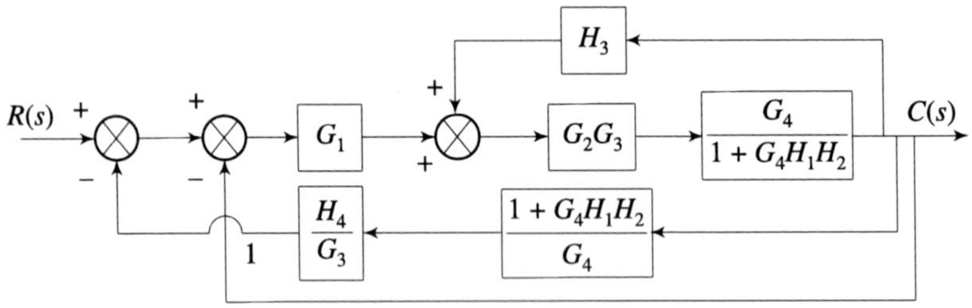

(iii)

Combining cascading blocks and eliminating positive feedback

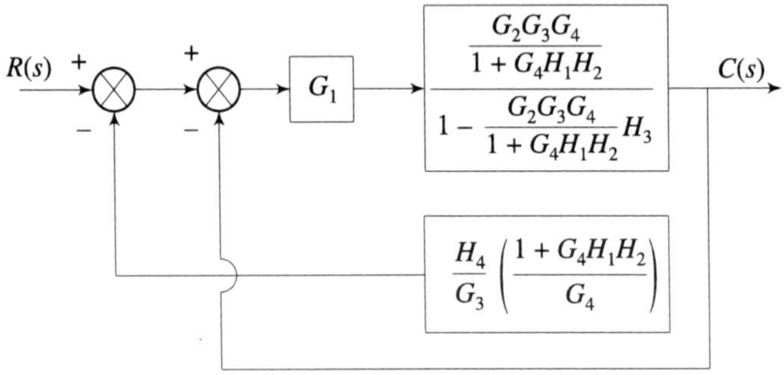

(iv)

Interchanging the summing points

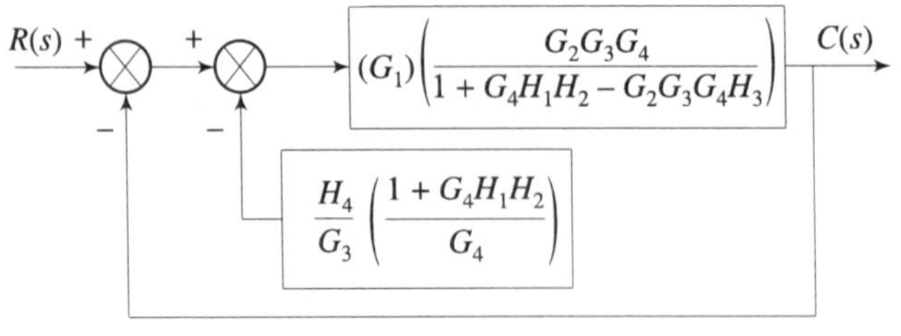

(v)

Eliminating negative feedback

$$= \frac{\dfrac{G_1G_2G_3G_4}{1 + G_4H_1H_2 - G_2G_3G_4H_3}}{1 + \dfrac{G_1G_2G_3G_4}{1 + G_4H_1H_2 - G_2G_3G_4H_3} \cdot \dfrac{H_4(1 + G_4H_1H_2)}{G_3G_4}}$$

$$= \frac{\dfrac{G_1G_2G_3G_4}{1 + G_4H_1\ H_2 - G_2G_3G_4H_3}}{\dfrac{1 + G_4H_1H_2 - G_2G_3G_4H_3 + G_1G_2\ (H_4(1 + G_4H_1H_2))}{1 + G_4H_1H_2 - G_2G_3G_4H_3}}$$

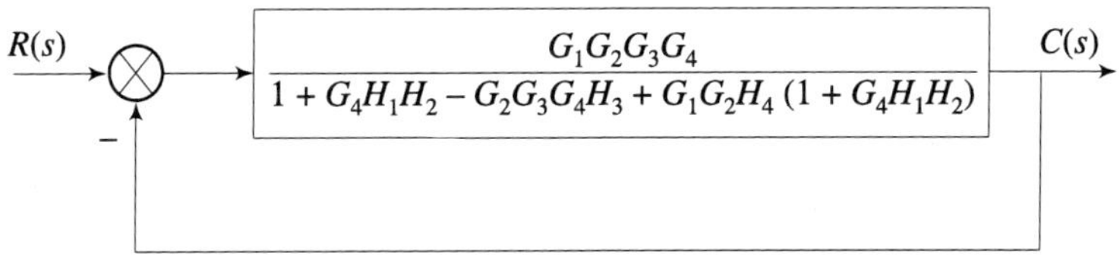

(vi)

Eliminating the negative unity feedback path

$$\frac{C(s)}{R(s)} = \frac{\dfrac{G_1G_2G_3G_4}{1 + G_4H_1H_2 - G_2G_3G_4H_3 + G_1G_2H_4(1 + G_4H_1H_2)}}{1 + \dfrac{G_1G_2G_3G_4}{1 + G_4H_1H_2 - G_2G_3G_4H_3 + G_1G_2H_4\ (1 + G_4H_1H_2)}}$$

$$\frac{C(s)}{R(s)} = \frac{G_1G_2G_3G_4}{1 + G_4H_1H_2 - G_2G_3G_4H_3 + G_1G_2H_4\ (1 + G_4H_1H_2) + G_1G_2G_3G_4}$$

3.3 SIGNAL FLOW GRAPH (SFG)

A signal flow graph is used to represent the control system graphically and it was developed by S.J. Mason. A signal flow graph is a pictorial representation of a system and it displays graphically the transmission of signals in a system. By taking Laplace transform, time domain differential equations governing a control system can be transferred to a set of algebraic equations in S-domain. The signal flow graph of the system can be constructed using these equations. It consists of a network graph in which nodes represent system variable. The nodes are connected by directed branches, each branch connected between two nodes acts as a signal multiplier. Each branch has a gain or a transmittance. When a signal passes through a branch, it gets multiplied by the gain of the branch. Note that the signal flows in only one direction. The direction of signal flow is indicated by an arrow placed on the branch and the multiplication factor is indicated along the branch

It should be noted that signal flow approach and the block diagram approach yield the same information. The advantage in signal flow graph method is that, using Mason's gain formula the overall gain of the system can be computed easily. This method is simpler than the previous block diagram reduction technique.

Basic Terminology

- *Node—(Input node/Source node, Independent node):* It is a node which has only outgoing branches. In the adjascent Fig. 3.40, x_0 is input node.
- *Output node—(Sink node or Dependent node):* The node which has only incoming branches. *Example:* x_4
- *Mixed node:* Mixed node is a node which has both incoming and outgoing branches. *Example:* x_1, x_2 and x_3.

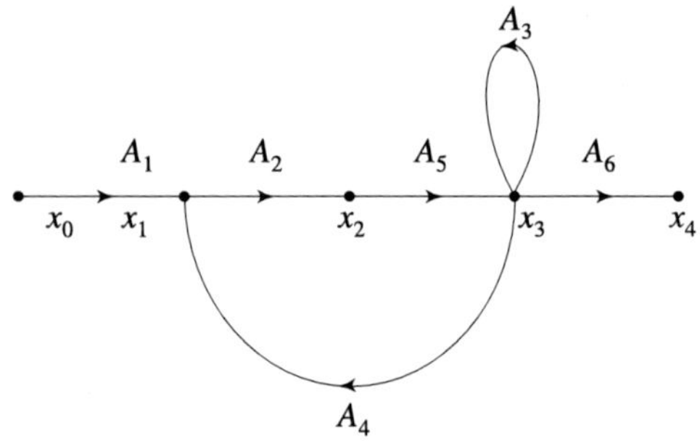

Fig. 3.40

- *Path:* A path is a collection of continuous connected branches in the same direction.
- *Forward path:* It is a path from an input node to an output node and along which no node is encountered more than once. *Example:* $x_0 - x_1 - x_2 - x_3 - x_4$
- *Forward path gain:* It is the product of branches gains which encountered in transferring a forward path. *Example:* $A_1 A_2 A_6$
- *Loop:* It is a path which originates and terminates at the same node and along which no other node is encountered more than once.
- *Loop gain:* It is the product of branch gains which are encountered in transferring loop.

  Example: $A_2 A_5 A_4$
  A_3 is Self loop.
- *Non-touching loops:* Two loops are said to be non-touching loops, if they do not have any common node.

3.3.1 Mason's Gain Formula

It is used to determine the relationship between an input variable and an output variable of the signal flow graph.

According to Mason's gain formula, the overall gain between an input node and an output node is given by

$$M = \frac{\sum_{K=1}^{N} P_K \Delta_K}{\Delta}$$

where P_K = path gain of the Kth forward path.

Δ = Determinant of the graph and is given by

$$\Delta = 1 - \Sigma L_1 + \Sigma L_2 - \Sigma L_3 + \ldots\ldots\ldots$$

ΣL_1 = Sum of the product of loop gains of all individual loops in the graph.

ΣL_2 = Sum of the product of loop gains of all possible combinations of 2 non-touching loops.

ΣL_3 = Sum of the product of loop gains of all possible combinations of 3 non-touching loops.

Δ_K = It is the determinant of that part of the graph which is non-touching loop with the Kth forward path.

N = Total number of forward paths.

Example 3.23: Obtain the transfer function using Mason's gain formula for the signal flow graph shown in Fig. 3.41

There are 2 forward paths.

The forward path gains are

$$P_1 = G_1\ G_2\ G_3\ G_4$$

$$P_2 = G_5\ G_6\ G_7\ G_8$$

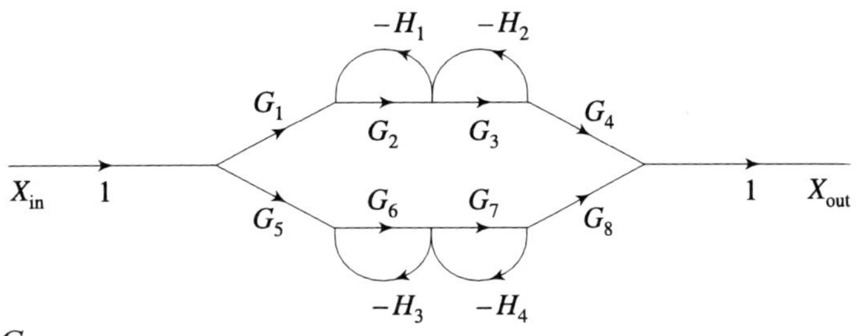

Fig. 3.41

There are 4 individual loops

$$L_{1a} = -H_1G_2$$

$$L_{1b} = -H_2G_3$$

$$L_{1c} = -H_4G_7$$

$$L_{1d} = -H_3G_6$$

$$\Sigma L_1 = -H_1G_2 - H_2G_3 - H_3G_6 - H_4G_7$$

There are 4, two non-touching loops

$$L_{2a} = H_1G_2\ H_3G_6$$

$$L_{2b} = H_1G_2\ H_4G_7$$

$$L_{2c} = H_2G_3\ H_3G_6$$

$$L_{2d} = H_2G_3\ H_4G_7$$

$$\Sigma L_2 = H_1G_2\ H_3G_6 + H_1G_2\ H_4G_7 + H_2G_3\ H_3G_6 + H_2G_3\ H_4G_7$$

There is no possible combination of 3 non-touching loops

$$\Delta = 1 - \Sigma L_1 + \Sigma L_2$$

$$\Delta = 1 - [-H_1G_2 - H_2G_3 - H_3G_6 - H_4G_7] + [H_1G_2\ H_3G_6 + H_1G_2\ H_4G_7 + H_2G_3\ H_3G_6 + H_2G_3\ H_4G_7]$$

To determine Δ_K

To determine Δ_1: Δ_1 is the determinant for that part of graph non-touching 1st forward path.

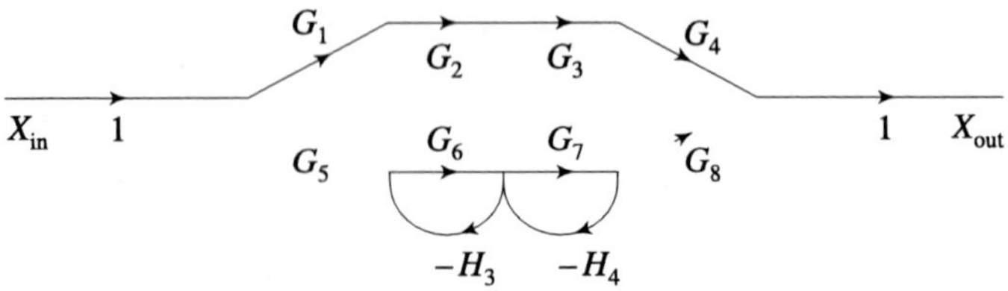

Fig. 3.42

Two individual loops are not touching.

$$\Delta_1 = 1 - (L_{1a} + L_{1b})$$
$$= 1 - (H_3G_6 - H_4G_7)$$
$$\Delta_1 = 1 + H_3G_6 + H_4G_7$$

To determine Δ_2 = part of graph not touching 2nd forward path.

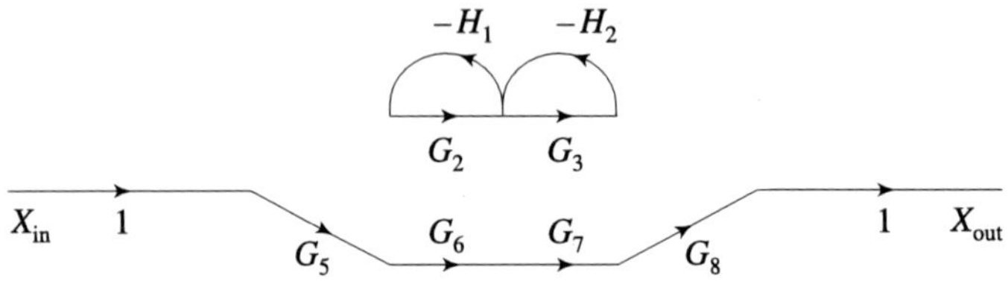

Fig. 3.43

$$\Delta_2 = 1 - (L_x + L_y)$$
$$= 1 - (- G_2H_1 - G_3H_2)$$
$$\Delta_2 = 1 + G_2H_1 + G_3H_2$$

$$M = \frac{\sum_{K=1}^{2} P_K\Delta_K}{\Delta} = \frac{P_1\Delta_1 + P_2\Delta_2}{\Delta}$$

$$M = \frac{G_1G_2G_3G_4(1 + G_6H_3 + G_7H_4) + G_2G_6G_7G_8(1 + G_2H_1 + G_3H_2)}{1 + G_2H_1 + G_3H_2 + G_6H_3 + G_7H_4 + G_2G_6H_1H_3 + G_2G_7H_1H_4 + G_3G_6H_2H_3 + G_3G_7H_2H_4}$$

Example 3.24: Find the transfer function for the Fig. 3.44.

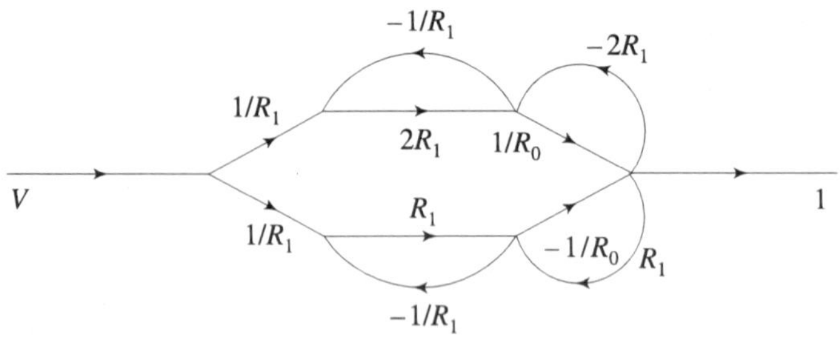

Fig. 3.44

Solution: Number of forward paths = 2

$$M_1 = \frac{1}{R_1} \times 2R_1 \times \frac{1}{R_0} = \frac{2}{R_0}$$

$$M_2 = \frac{1}{R_1} \times R_1 \times -\frac{1}{R_0} = -\frac{1}{R_0}$$

Individual loops = 4

$$L_{1a} = 2R_1 \times -\frac{1}{R_1} = -2$$

$$L_{1b} = R_1 \times -\frac{1}{R_1} = -1$$

$$L_{1c} = \frac{1}{R_0} \times -2R_1 = -\frac{2R_1}{R_0}$$

$$L_{1c} = -\frac{1}{R_0} \times R_1 = -\frac{R_1}{R_0}$$

Number of two non-touching loops = 3

$$L_{2a} = \left(\frac{-1}{R_1} \times 2R_1\right)\left(R_1 \times -\frac{1}{R_1}\right) = 2$$

$$L_{2b} = \left(\frac{-1}{R_1} \times 2R_1\right)\left(\frac{-1}{R_0} \times R_1\right) = \frac{2R_1}{R_0}$$

$$L_{2c} = \left(-2R_1 \times \frac{1}{R_0}\right)\left(R_1 \times \frac{-1}{R_1}\right) = \frac{2R_1}{R_0}$$

$$\Delta = 1 - \left[-2 - 1 \times \frac{2R_1}{R_0} - \frac{R_1}{R_0}\right] + \left[2 + \frac{2R_1}{R_0} + \frac{2R_1}{R_0}\right]$$

To find Δ_K

$$\Delta_1 = 1 - \left(R_1 \times \frac{-1}{R_1}\right) = 2$$

$$\Delta_2 = 1 - \left(2R_1 \times -\frac{1}{R_1}\right) = 1 - (-2) = 3$$

$$M = \frac{M_1\Delta_1 + M_2\Delta_2}{\Delta}$$

$$M = \frac{\frac{2}{R_0} \times 2 + \left(\frac{-1}{R_0}\right) \times 3}{1 - \left(-3 - \frac{3R_1}{R_0}\right) + \left(2 + \frac{4R_1}{R_0}\right)}$$

$$M = \frac{1}{6R_0 + 7R_1}$$

Example 3.25: Using Mason's gain formula, find the gain of the following signal flow graph shown in Fig. 3.45.

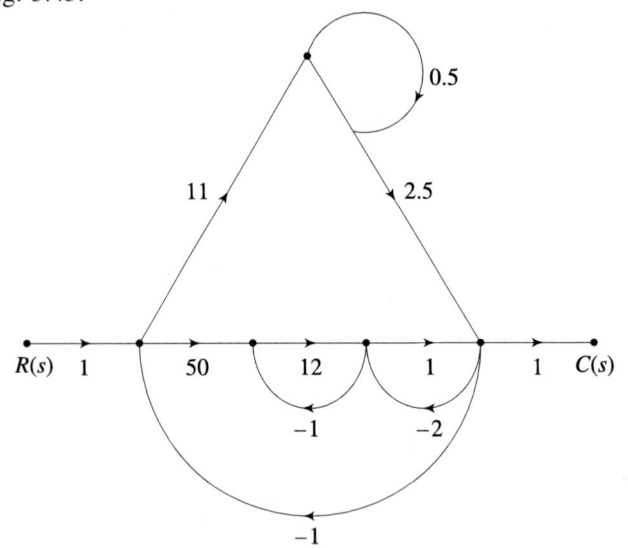

Fig. 3.45

Solution: There are 2 forward paths

$$M_1 = 50 \times 12 = 600$$

$$M_2 = 11 \times 2.5 = 27.5$$

There are 5 individual loops

$$L_{1a} = -12,\ L_{1b} = -2,\ L_{1c} = -600,\ L_{1d} = -27.5,\ L_{1e} = 0.5$$

There are 4, two non-touching loops

$$L_{2a} = -12 \times 1 \times 0.5 = -6$$

$$L_{2b} = -2 \times 0.5 = -1$$

$$L_{2c} = 11 \times 2.5 \times 12 = 330$$

$$L_{2d} = -50 \times 12 \times 0.5 = -300$$

There are no, three non-touching loops

$$\Delta = 1 - (-12 - 2 - 600 - 27.5 + 0.5) + (-6 - 1 + 330 - 300)$$

$$\Delta = 665$$

Finding Δ_1 for M_1

For M_1 path (0.5) loop does not touch it.

$$\Delta_1 = 1 - 0.5 = 0.5$$

Finding Δ_2 for M_2 path –12 loop does not touch it.

$$\Delta_2 = 1 + 12 = 13$$

$$\frac{C(s)}{R(s)} = \frac{600(0.5) + 27.5(13)}{665}$$

$$\frac{C(s)}{R(s)} = 0.988.$$

Example 3.26: Find transfer function for the following SFG shown in Fig. 3.46.

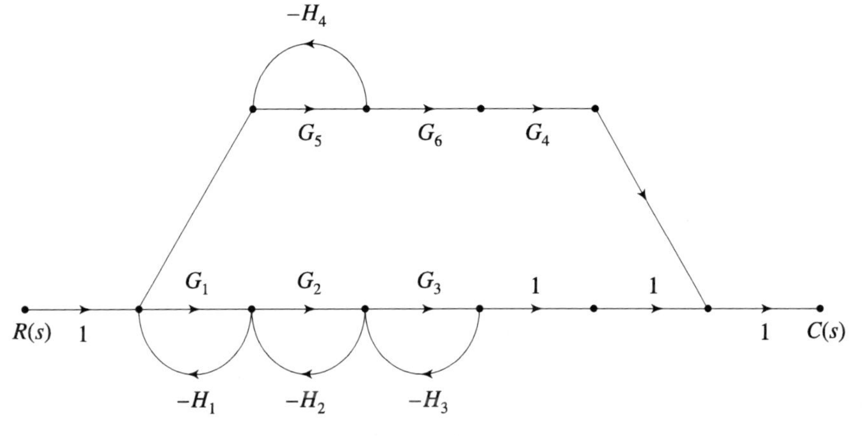

Fig. 3.46

Solution: There are two forward paths

$$M_1 = G_1 G_2 G_3$$

$$M_2 = G_5 G_6 G_4$$

There are 4 individual loops

$$L_{1a} = -G_1H_1, \qquad L_{1b} = -G_2H_2,$$

$$L_{1c} = -G_3H_3, \qquad L_{1d} = -G_5H_4,$$

There are 4, two non-touching loops

$$L_{2a} = G_1G_5H_1H_4,$$

$$L_{2b} = G_2 H_2 G_5 H_4$$

$$L_{2c} = G_3 H_3 G_5 H_4$$

$$L_{2d} = G_1 H_1 G_3 H_3$$

There are 1, three non-touching loops

$$L_{3a} = - G_1H_1G_3H_3\ G_5H_4$$

There are no, four non-touching loops

$$\Delta = (- G_1H_1 - G_2H_2 - G_3H_3 - G_5H_4) + (G_1G_5\ H_1H_4 + G_2\ H_2\ G_5\ H_4 + G_3\ H_3\ G_5\ H_4 + G_1H_1G_3H_3) - (- G_1H_1\ G_3H_3\ G_5H_4)$$

$$\Delta = 1 + G_1H_1 + G_2H_2 + G_3H_3 + G_5H_4 + G_1G_5H_1\ H_4 + G_2H_2\ G_5H_4 + G_3H_3G_5H_4 + G_1H_1G_3H_3 + G_1H_1G_3H_3\ G_5H_4$$

To find Δ_1

For M_1 path, $-G_5H_4$ does not touch

$$\Delta_1 = 1 - (-G_5H_4)$$

$$\Delta_1 = 1 + G_5H_4$$

To find Δ_2

For M_2 path, $-G_2H_2$ and $-\ H_3G_3$ does not touch

$$\therefore \quad \Delta_2 = 1 - (-G_2H_2) - (-G_3H_3)$$

$$\Delta_2 = 1 + G_2H_2 + G_3H_3$$

Transfer function

$$\frac{C}{R} = \frac{M_1\Delta_1 + M_2\Delta_2}{\Delta}$$

$$\frac{C}{R} = \frac{G_1G_2G_3(1 + G_5H_4) + G_4G_5G_6(1 + G_2H_2 + G_3H_3)}{1 + G_1H_1 + G_2H_2 + G_3H_3 + G_5H_4 + G_1G_3H_1H_3 + G_1G_5H_1H_4 + G_2G_5H_2H_4 + G_3G_5H_3H_4 + G_1G_3G_5H_1H_3H_4}$$

Example 3.27: Find transfer function for the following SFG shown in Fig. 3.47.

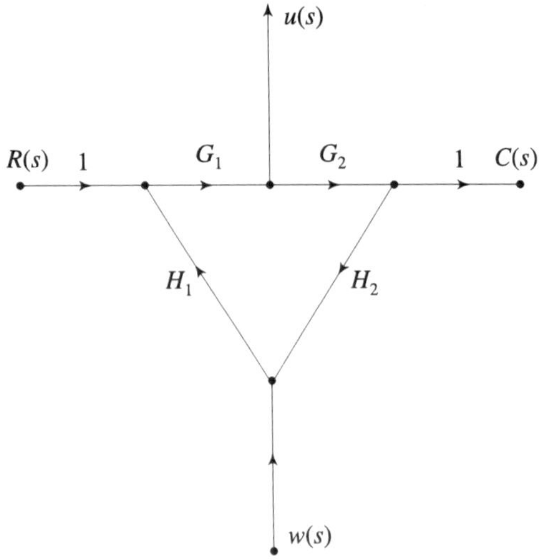

Fig. 3.47

Solution: It is an example of multi-input problem, apply superposition theorem.

Step 1: Keep $R(s)$, set $u(s) = w(s) = 0$

There is 1 forward path

$$M_1 = G_1G_2$$

There is one loop

$$L_{1a} = G_1G_2H_1H_2$$

There are no, 2 or more non-touching loops

$$\Delta = 1 - G_1G_2H_1H_2$$

To find Δ_1 for M_1 all loop touch

$$\Delta_1 = 1 - 0 = 1$$

$$\frac{C(s)}{R(s)} = \frac{G_1G_2}{1 - G_1G_2H_1H_2}$$

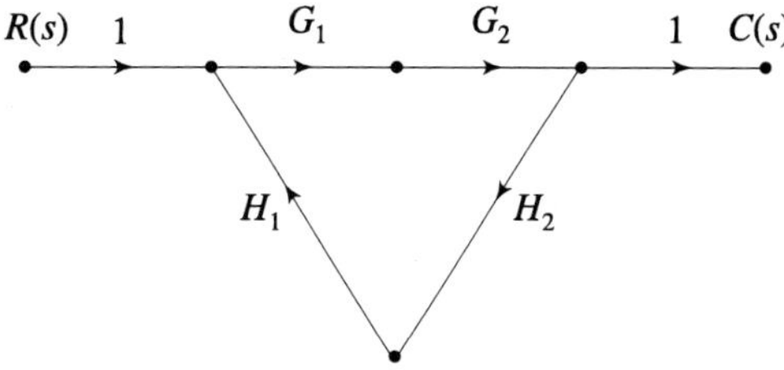

Fig. 3.48

Step 2: Keep $w(s)$ only, set $R(s) = u(s) = 0$

One forward path, $M_1 = H_1G_1G_2$

One individual loop, $L_{1a} = G_1G_2H_1H_2$

There is no non-touching loop

$$\therefore \quad \Delta = 1 - G_1G_2H_1H_2$$

To find Δ_K: For $\Delta_1 = 1 - 0 = 1$

$$M = \frac{G_1G_2H_1}{1 - G_1G_2H_1H_2}$$

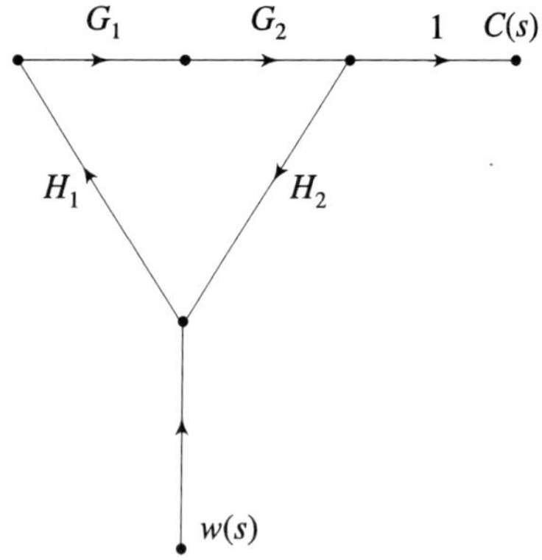

Fig. 3.49

Step 3: Keep $u(s)$ only, set $R(s) = w(s) = 0$

One forward path, $M_1 = G_2$

One individual loop, $L_{1a} = G_2H_1H_2G_1$

There is no non-touching loop

$$\Delta_1 = 1 - 0 = 1$$

$$M = \frac{G_2}{1 - G_1G_2H_1H_2}$$

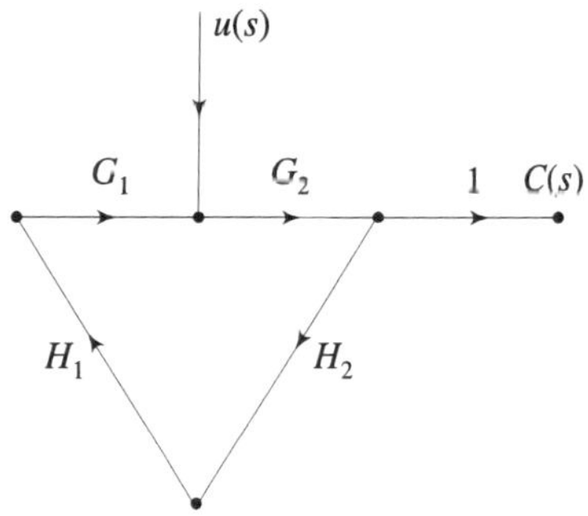

Fig. 3.50

$C(s)$ is total output due to all three inputs.

$$C(s) = \frac{G_1G_2R(s) + G_1G_2H_1w(s) + G_2\,u(s)}{1 - G_1G_2H_1H_2}$$

Rules for signal flow graph

- *Addition rule:* The algebraic sum of all signals entering a node gives the value of the variable of that node. This is known as addition rule.

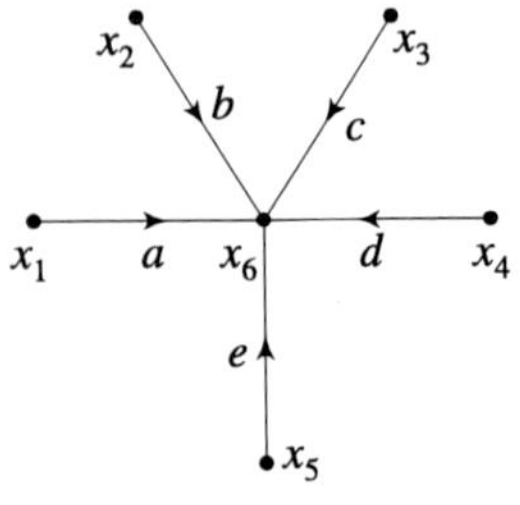

Fig. 3.51

From the Fig. 3.51, $X_6 = ax_1 + bx_2 + cx_3 + dx_4 + ex_5$

- *Transmission rule:* The value of a variable at a node in a signal flow graph is transmitted on every branch leaving that node. This is known as transmission rule.

From Fig. 3.52

$$X_1 = a_1\ X$$

$$X_2 = a_2\ X$$

$$X_n = a_n\ X$$

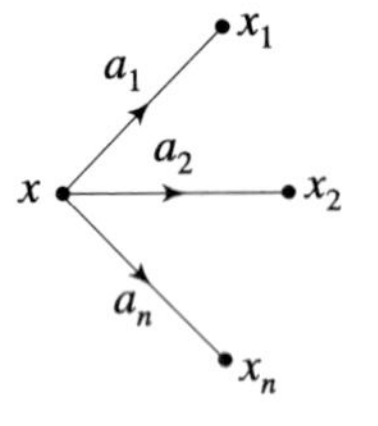

Fig. 3.52

- *Cascade connection:* Cascade signal flow graphs contain only direct series paths from input to towards output

$$x_2 = a_1\ x_1$$

$$x_3 = a_2\ x_2$$

Eliminating x_2

$$x_3 = a_1\ a_2\ X_1$$

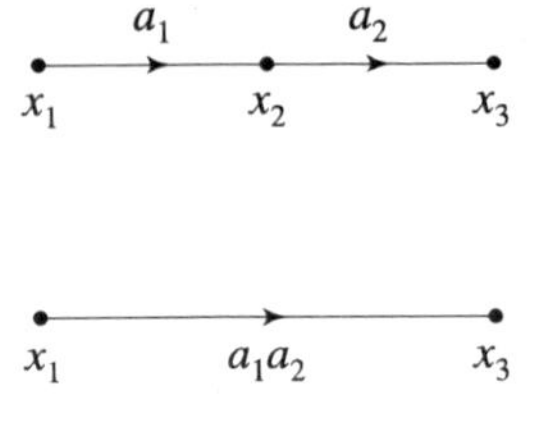

Fig. 3.53

Thus, the gain of a cascade connection is the product of the gains of the branches. It is to be noted that in the resulting graph node x_2 is eliminated.

In general, a series (cascade) connection of branches with transmitted (gains) a_1, a_2, a_3, ... can be replaced by a single branch with an equivalent transmittance (gain) equal to the product of the individual transmittances. This is known as multiplication rule.

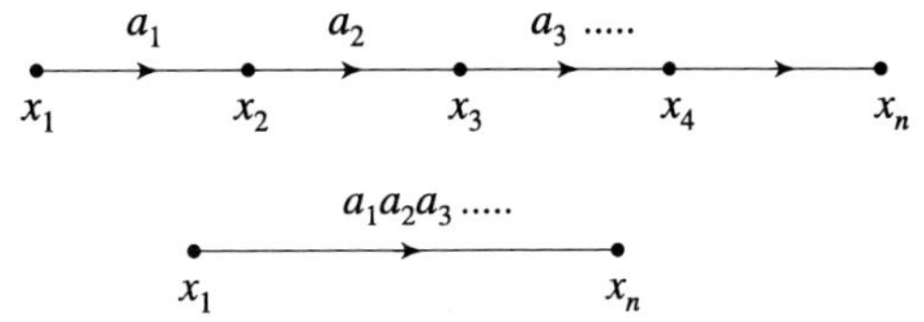

Fig. 3.54

- *Parallel connection:* Consider the signal flow graph shown in Fig. 3.55.

$$X_2 = a_1 X_1 + a_2 X_1$$

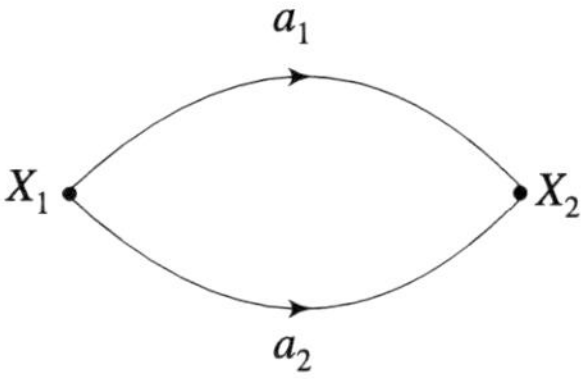

Fig. 3.55

It is seen that the gains of the branches in parallel are added

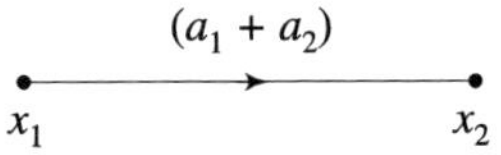

Fig. 3.56

Properties of signal flow graphs

- A branch indicates the functional dependence of one variable on another. Signals travel along branches only in the direction specified by the arrows of the branches.
- A node adds the signals of all incoming branches and transmits this sum to all outgoing branches.
- A mixed node may be treated as an output node by adding an outgoing branch of unity gain, but a mixed node cannot be changed to an input node.
- For a given system, a signal flow graph is not unique. Many different signal flow graphs can be drawn for a given system by writing the system equations differently.

Example 3.28: Find the transfer function $\frac{Y_5}{Y_1}$ for the signal graph as shown in Fig. 3.57.

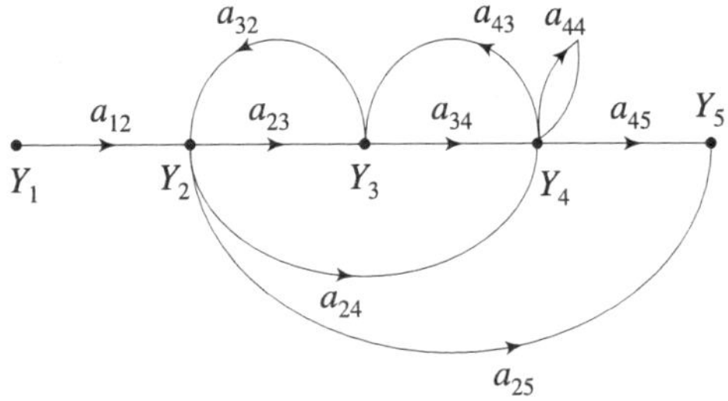

Fig. 3.57

Solution: There are three forward paths between Y_1 and Y_5

Forward path $Y_1 - Y_2 - Y_3 - Y_4 - Y_5$, path gain

$$P_1 = a_{12}\, a_{23}\, a_{34}\, a_{45}$$

Forward path $Y_1 - Y_2 - Y_5$, path gain

$$P_2 = a_{12}\, a_{25}$$

Forward path $Y_1 - Y_2 - Y_4 - Y_5$, path gain

$$P_3 = a_{12}\, a_{24}\, a_{45}$$

There are four individual loops, the loop gains are

$$L_1 = a_{23}\, a_{32}$$

$$L_2 = a_{34}\, a_{43}$$

$$L_3 = a_{24}\, a_{43}\, a_{32}$$

$$L_4 = a_{44}$$

There is only one pair of non-touching loops, the two loops are

$$Y_2 - Y_3 - Y_2 \text{ and } Y_4 - Y_4$$

The product of the gains of the two non-touching loops is

$$L_1 L_4 = a_{23}\, a_{32}\, a_{44}$$

All the loops are in touch with forward paths P_1 and P_3

$$\therefore \quad \Delta_1 = \Delta_3 = 1$$

Two of the loops are in touch with the forward path P_2

These loops are $Y_3 - Y_4 - Y_3$ and $Y_4 - Y_4$

$$\therefore \quad \Delta_2 = 1 - a_{34}\, a_{43} - a_{44}$$

The overall transfer function is given by Mason's rule as

$$\frac{Y_5}{Y_1} = \frac{P_1\Delta_1 + P_2\Delta_2 + P_3\Delta_3}{\Delta}$$

$$\Delta = 1 - (L_1 + L_2 + L_3 + L_4) + L_1\, L_2$$

$$\Delta = 1 - (a_{23}\, a_{32} + a_{34}\, a_{43} + a_{24}\, a_{32}\, a_{43} + a_{44}) + a_{23}\, a_{32}\, a_{44}$$

$\therefore$ Overall transfer function is

$$\frac{Y_5}{Y_1} = \frac{a_{12}\, a_{23}\, a_{34}\, a_{45} + a_{12}\, a_{25}(1 - a_{34}\, a_{43} - a_{44}) + a_{12}\, a_{24}\, a_{45}}{1 - (a_{23}\, a_{32} + a_{34}\, a_{43} + a_{24}\, a_{32}\, a_{43} + a_{44}) + a_{23}\, a_{32}\, a_{44}}$$

Example 3.29: Obtain the transfer function $\frac{Y(s)}{X(s)}$ of the signal flow graph shown in Fig. 3.58.

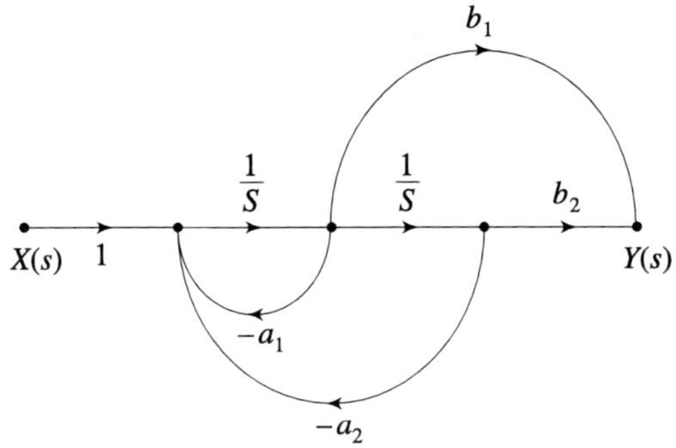

Fig. 3.58

Solution: There are two forward paths with path gains are

$$P_1 = \left(\frac{1}{S}\right)\left(\frac{1}{S}\right) b_2 = \frac{b_2}{S^2}$$

$$P_2 = \left(\frac{1}{S}\right) b_1 = \frac{b_1}{S}$$

There are two individual loops with loop gains

$$L_1 = \frac{1}{S}(-a_1) = \frac{-a_1}{S}$$

$$L_2 = \left(\frac{1}{S}\right)\left(\frac{1}{S}\right)(-a_2) = \frac{-a_2}{S^2}$$

There are no non-touching loops,

$$\Delta = 1 - (L_1 + L_2)$$

$$\Delta = 1 + \frac{a_1}{S} + \frac{a_2}{S_2}$$

Both the forward paths touch the individuals loops

$$\Delta_1 = 1, \Delta_2 = 1$$

The transfer function is given by

$$\frac{Y(s)}{X(s)} = \frac{P_1\Delta_1 + P_2\Delta_2}{\Delta} = \frac{\frac{b_2}{S^2} + \frac{b_1}{S}}{1 + \frac{a_1}{S} + \frac{a_2}{S^2}} = \frac{b_1 S + b_2}{S^2 + a_1 S + a_2}$$

$$\boxed{\frac{Y(s)}{X(s)} = \frac{b_1 S + b_2}{S^2 + a_1 S + a_2}}$$

Example 3.30: Find the overall gain $\frac{C(s)}{R(s)}$ for the signal flow graph shown in Fig. 3.59.

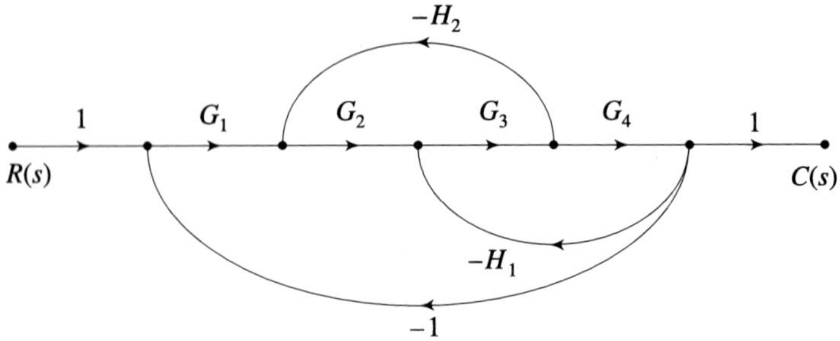

Fig. 3.59

Solution: There is only one forward path and forward path gain

$$P_1 = G_1G_2G_3G_4$$

Individual loop gain, there are three individual loops

$$L_1 = -\ G_3G_4H_1$$
$$L_2 = -\ G_2G_3H_2$$
$$L_3 = -\ G_1G_2G_3G_4$$

There are no possible combinations of two non-touching loops, three non-touching loops

$$\Delta = 1 - (L_1 + L_2 + L_3)$$
$$= 1 - (-\ G_3G_4H_1 - G_2G_3H_2 - G_1G_2G_3G_4)$$

Since no part of the graph is non-touching with forward path $\Delta_1 = 1$

The overall transfer function

$$\frac{C(s)}{R(s)} = \frac{P_1\Delta_1}{\Delta} = \frac{G_1G_2G_3G_4}{1 + G_3G_4H_1 + G_2G_3H_2 + G_1G_2G_3G_4}$$

Example 3.31: The signal flow graph for a feedback control system is shown in Fig. 3.60. Determine the closed-loop transfer function $\frac{C(s)}{R(s)}$.

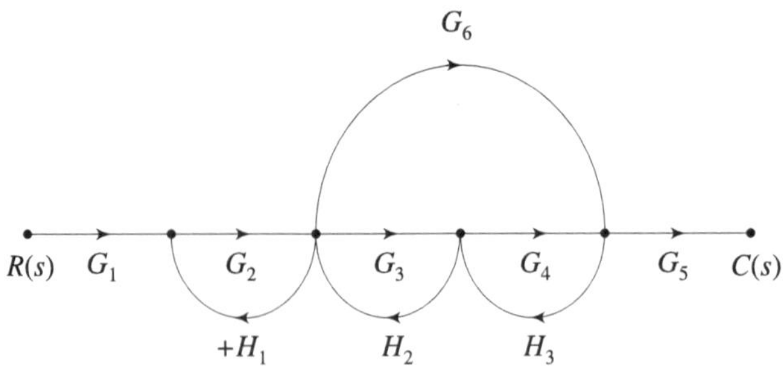

Fig. 3.60

Solution: There are two forward paths and path gains

$$P_1 = G_1G_2G_3G_4G_5$$

$$P_2 = G_1G_2G_6G_5$$

There are four individual loops

$$L_1 = G_2H_1,\ L_2 = G_3H_2,\ L_3 = G_4H_3,\ L_4 = G_6H_2H_3$$

There is only one combination of two non-touching loops. Let the gain product of two non-touching loops be

$$L_{11} = (G_2H_1)\ (G_4H_3)$$

$$\therefore \quad \Delta = 1 - (L_1 + L_2 + L_3 + L_4) + L_{11}$$

$$\Delta = 1 - (G_2H_1 + G_3H_2 + G_4H_3 + G_6H_2\ H_3) + G_2G_4H_1H_3$$

Since there is no part of graph which is non-touching with forward paths 1 and 2.

$$\Delta_1 = \Delta_2 = 1$$

The overall transfer function

$$\frac{C(s)}{R(s)} = \frac{P_1\Delta_1 + P_2\Delta_2}{\Delta}$$

$$\frac{C(s)}{R(s)} = \frac{G_1G_2G_3G_4G_5 + G_1G_2G_6G_5}{1 - G_2H_1 - G_3H_2 - G_4H_3 - G_6H_2H_3 + G_2G_4H_1H_3}$$

Example 3.32: Find the transfer function of the system shown in Fig. 3.61 using Mason's gain formula.

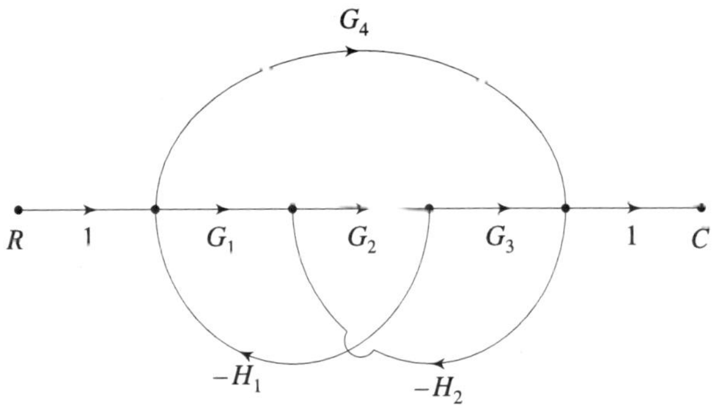

Fig. 3.61

Solution: There are two forward paths and their path gain

$$P_1 = G_1G_2G_3 \quad \text{and} \quad P_2 = G_4$$

There are two individual loops

$$L_1 = -G_1G_2H_1, \quad L_2 = -G_2G_3H_2$$

There are non-touching loops

$$\therefore \qquad \Delta = 1 - (L_1 + L_2) = 1 - (- G_1G_2H_1 - G_2G_3H_2)$$

Both the forward paths touch the individual loops

$$\Delta_1 = \Delta_2 = 1$$

$$\text{The transfer function} = \frac{C}{R} = \frac{P_1\Delta_1 + P_2\Delta_2}{\Delta}$$

$$\frac{C}{R} = \frac{G_1G_2G_3 + G_4}{1 + G_1G_2H_1 + G_2G_3H_2}$$

Example 3.33: Find transfer function of the system shown in Fig. 3.62 using Mason's gain formula.

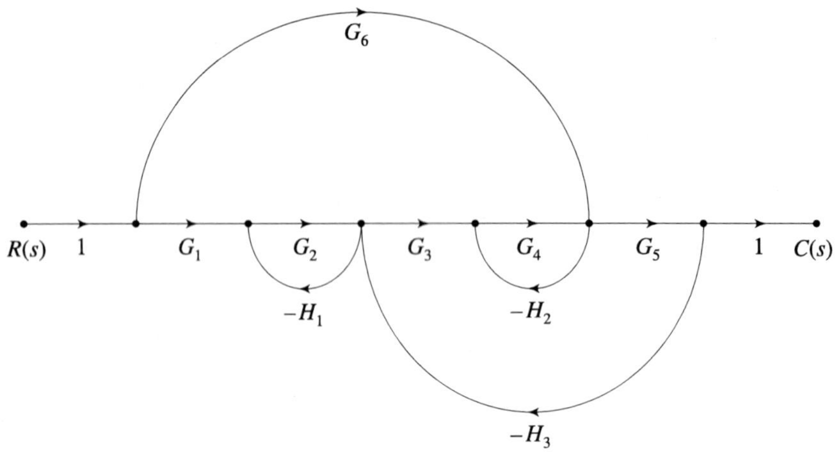

Fig. 3.62

Solution: There are two forward paths and their path gain

$$P_1 = G_1G_2G_3G_4G_5, \quad P_2 = G_6G_5$$

There are three individual loops and their loop gain

$$L_1 = - G_2H_1, \quad L_2 = - G_4H_2, \quad L_3 = - G_3G_4G_5H_3$$

Loop 1 and loop 2 are non-touching and loop gains

$$L_{11} = G_2G_4H_1H_2$$

$$\Delta = 1 - (L_1 + L_2 + L_3) + L_{11}$$

$$\Delta = 1 - (-G_2H_1 - G_4H_2 - G_3G_4G_5H_3) + G_2G_4H_1H_2$$

$$\Delta = 1 + G_2H_1 + G_4H_2 + G_3G_4G_5H_3 + G_2G_4H_1H_2$$

The forward path P_1 touches the individual loops

$$\Delta_1 = 1$$

The forward path P_2, the individual loop G_2H_1 will not touch

$$\Delta_2 = 1 - (-G_2H_1) = 1 + G_2H_1$$

The transfer function

$$\frac{C(s)}{R(s)} = \frac{P_1\Delta_1 + P_2\Delta_2}{\Delta} = \frac{G_1G_2G_3G_4G_5 + G_5G_6(1 + G_2H_1)}{1 + G_2H_1 + G_4H_2 + G_3G_4G_5H_3 + G_2G_4H_1H_2}$$

$$\frac{C(s)}{R(s)} = \frac{G_1G_2G_3G_4G_5 + G_5G_6\,(1 + G_2H_1)}{1 + G_2H_1 + G_4H_2 + G_3G_4G_5H_3 + G_2G_4H_1H_2}$$

Example 3.34: Find transfer function of the system shown in Fig. 3.63 using Mason's gain formula.

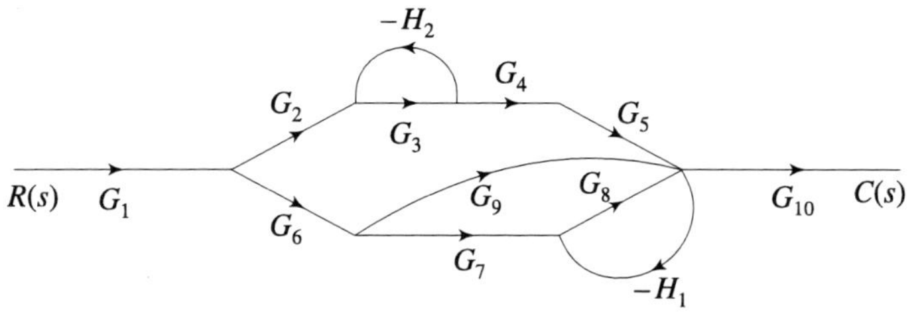

Fig. 3.63

Solution: There are three forward paths

$$P_1 = G_1G_2G_3G_4G_5G_{10}$$

$$P_2 = G_1G_6G_9G_{10}$$

$$P_3 = G_1G_6G_7G_8G_{10}$$

These are two individual loops

$$L_1 = -G_3H_2$$

$$L_2 = -G_8H_1$$

There are one, two non-touching loops

$$L_{11} = (-G_3H_2)\,(-G_8H_1) = G_3G_8H_1H_2$$

$$\therefore \quad \Delta = 1 - (L_1 + L_2) + L_{11} = 1 + G_3H_2 + G_8H_1 + G_3G_8H_1H_2$$

For forward path P_1, $\Delta_1 = 1$

For P_2, $\quad \Delta_2 = 1 - (-G_3H_2) = 1 + G_3H_2$

For P_3, $\quad \Delta_3 = 1 - (-G_3H_2) = 1 + G_3H_2$

The transfer function

$$\frac{C(s)}{R(s)} = \frac{P_1\Delta_1 + P_2\Delta_2 + P_3\Delta_3}{\Delta}$$

$$\frac{C(s)}{R(s)} = \frac{G_1G_2G_3G_4G_5G_{10} + G_1G_6G_9G_{10}\,(1 + G_3H_2) + G_1G_6G_7G_8G_{10}\,(1 + G_3H_2)}{1 + G_3H_2 + G_8H_1 + G_3G_8H_1H_2}$$

Example 3.35: Find the transfer function of the system whose signal flow graph is shown in Fig. 3.64.

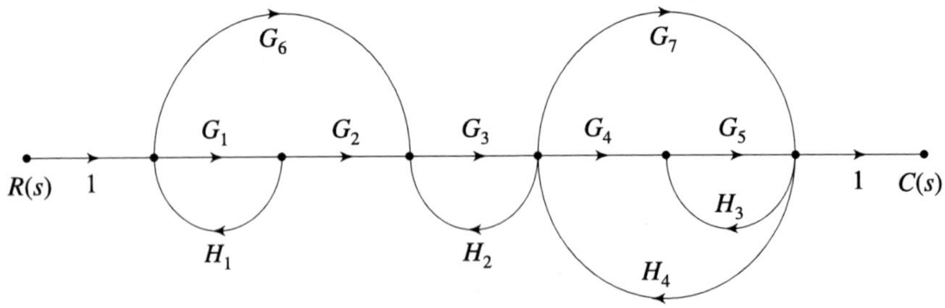

Fig. 3.64

Solution: There are four forward paths and their path gain

$$P_1 = G_1G_2G_4G_5G_3$$

$$P_2 = G_6G_3G_4G_5$$

$$P_3 = G_1G_2G_3G_7$$

$$P_4 = G_6G_3G_7$$

There are five individual loops and loop gain are

$$L_1 = G_1H_1,\quad L_2 = G_3H_2,\quad L_3 = G_5H_3,\quad L_4 = G_4G_5H_4,\quad L_5 = G_7H_4$$

There are 5, two non-touching loops and loop gains are

$$L_{11} = (G_1H_1)\,(G_3H_2) = G_1H_1G_3H_2$$

$$L_{12} = (G_1H_1)\,(G_5H_3) = G_1H_1G_5H_3$$

$$L_{13} = (G_3H_2)\,(G_5H_3) = G_3G_5H_2H_3$$

$$L_{14} = (G_1H_1)\,(G_4G_5H_4) = G_1H_1G_4G_5H_4$$

$$L_{15} = (G_1H_2)\,(G_7H_4) = G_1G_7H_1H_4$$

There are one, three non-touching loops and loop gains

$$L_{22} = (G_1H_1)\,(G_3H_2)\,(G_5H_3) = G_1H_1G_3H_2H_5H_3$$

Since all the loops are touching all the forward paths

$$\Delta_1 = 1,\quad \Delta_2 = 1,\quad \Delta_3 = 1,\quad \Delta_4 = 1$$

$$\Delta = 1 - (L_1 + L_2 + L_3 + L_4 + L_5) + (L_{11} + L_{12} + L_{13} + L_{14} + L_{15} - (L_{22})$$

The closed-loop transfer function

$$\frac{C}{R} = \frac{P_1\Delta_1 + P_2\Delta_2 + P_3\Delta_3 + P_4\Delta_4}{\Delta}$$

$$\frac{C}{R} = \frac{G_1G_2G_3G_4G_5 + G_6G_3G_4G_5 + G_1G_2G_3G_7 + G_6G_3G_7}{\begin{array}{c}1 - (G_1H_1 + G_3H_2 + G_5H_3 + G_4G_5H_4 + G_7H_4) \\ + (G_1H_1G_3H_2 + G_1H_1G_5H_3 + G_3G_5H_2H_3 \\ + G_1H_1G_4G_5H_4 + G_1G_7H_1H_2) - G_1H_1G_3H_2G_5H_3\end{array}}$$

Example 3.36: Obtain the transfer function of the system whose signal flow graph is shown in Fig. 3.65.

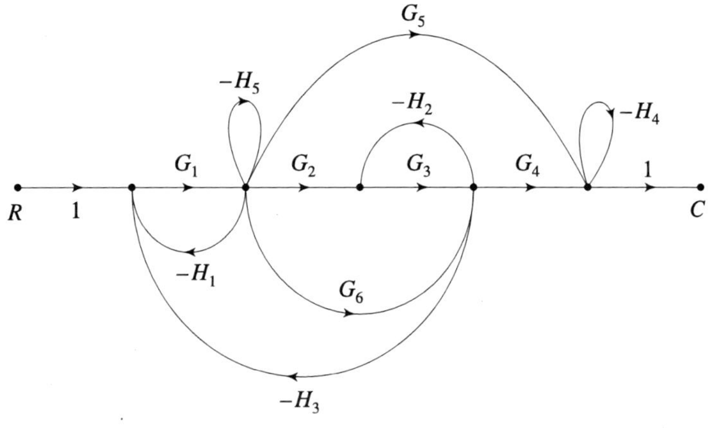

Fig. 3.65

Solution: There are 3 forward paths and the path gain

$$P_1 = G_1G_2G_3G_4$$

$$P_2 = G_1G_5$$

$$P_3 = G_1G_6G_4$$

These are 6 individual loops and loop gain

$$L_1 = -G_1H_1$$

$$L_2 = -H_5$$

$$L_3 = -G_3H_2$$

$$L_4 = -H_4$$

$$L_5 = -G_1G_2G_3H_3$$

$$L_6 = -G_1G_6H_3$$

There are 7, two non-touching loops and loop gains

$$L_{11} = G_1H_1G_3H_2$$

$$L_{12} = G_1H_1H_4$$

$$L_{13} = G_3H_2H_5$$

$$L_{14} = H_5H_4$$

$$L_{15} = G_3H_2H_4$$

$$L_{16} = G_1G_2G_3H_3H_4$$

$$L_{17} = G_1G_6H_3H_4$$

There are 2, three non-touching loops and loop gains

$$L_{21} = -\ G_3H_2H_4H_5$$

$$L_{22} = -\ G_1H_1G_3H_2H_4$$

For P_1, $\Delta_1 = 1$, because all loops are touching the P_1 forward path

For P_2, $\Delta_2 = 1 + G_3H_2$, because L_3 is not touching the P_2

For P_3, $\Delta_3 = 1$, because all loops are touching the P_3 forward path.

$$\Delta = 1 - (L_1 + L_2 + L_3 + L_4 + L_5 + L_6) + (L_{11} + L_{12} + L_{13} + L_{14} + L_{15} + L_{16} + L_{17}) - (L_{21} + L_{22})$$

$$\Delta = 1 - (-G_1H_1 - H_5 - G_3H_2 - H_4 - G_1G_2G_3H_3 - G_1G_6H_3) + (G_1G_3H_1H_2 + G_1H_1H_4 + H_4H_5 + G_3H_2H_5 + G_3H_2H_4 + G_1G_2G_3H_3H_4 + G_1G_6H_3H_4) - (-G_1G_3H_1H_2H_4 - G_3H_2H_4H_5)$$

The overall transfer function

$$\frac{C}{R} = \frac{P_1\Delta_1 + P_2\Delta_2 + P_3\Delta_3}{\Delta}$$

$$\frac{C}{R} = \frac{G_1G_2G_3G_4 + G_1G_5\,(1 + G_3H_2) + G_1G_4G_6}{1 + G_1H_1 + H_5 + G_3H_2 + H_4 + G_1G_2G_3H_3 + G_1G_6H_3 + G_1G_3H_1H_2 + G_1H_1H_4 + H_4H_5 + G_3H_2H_4 + G_1G_2G_3H_3H_4 + G_1G_6\,H_3H_4 + G_1G_3H_1H_2H_4 + G_3H_2H_4H_5}$$

Example 3.37: Find the transfer function of the system having signal flow graph shown in Fig. 3.66.

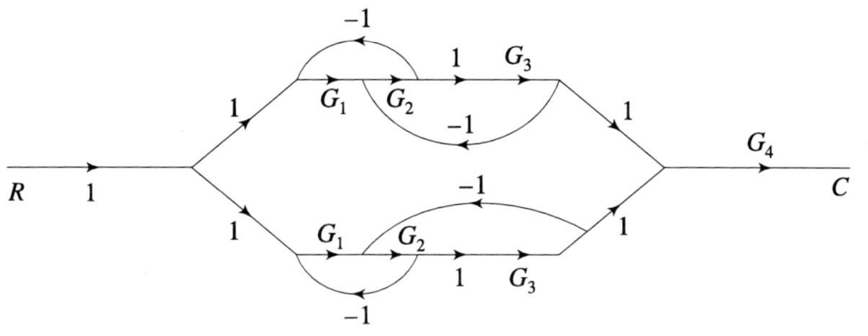

Fig. 3.66

Solution: There are two forward paths with path gain as

$$P_1 = G_1G_2G_3G_4$$
$$P_2 = G_1G_2G_3G_4$$

There are 4 individual loops with gains

$$L_1 = -G_1G_2, \quad L_2 = -G_2G_3, \quad L_3 = -G_1G_2, \quad L_4 = -G_2G_3$$

There are 4, two non-touching loops

$$L_{11} = (-G_1G_2)(-G_1G_2) = G_1^2G_2^2$$
$$L_{12} = (-G_1G_2)(-G_2G_3) = G_1G_2^2G_3$$
$$L_{13} = (-G_2G_3)(-G_1G_2) = G_1G_2^2G_3$$
$$L_{14} = (-G_2G_3)(-G_2G_3) = G_2^2G_3^2$$

$\therefore$ $$\Delta = 1 - (L_1 + L_2 + L_3 + L_4) + (L_{11} + L_{12} + L_{13} + L_{14})$$

$$\Delta = 1 - (-G_1G_2 - G_2G_3 - G_1G_2 - G_2G_3) + G_1^2G_2^2 + G_1G_2^2G_3 + G_1G_2^2G_3 + G_2^2G_3^2$$

$$\Delta = 1 + 2(G_1G_2 + G_2G_3) + G_1^2G_2^2 + 2G_1G_2^2G_3 + G_2^2G_3^2$$

For P_1, $$\Delta_1 = 1 - (-G_1G_2) - (-G_2G_3)$$
$$\Delta_1 = 1 + G_1G_2 + G_2G_3$$

For P_2, $$\Delta_2 = 1 - (-G_1G_2) - (G_2G_3)$$
$$\Delta_2 = 1 + G_1G_2 + G_2G_3$$

The overall transfer function is

$$\frac{C}{R} = \frac{P_1\Delta_1 + P_2\Delta_2}{\Delta}$$

$$\frac{C}{R} = \frac{G_1G_2G_3G_4(1 + G_1G_2 + G_2G_3) + G_1G_2G_3G_4(1 + G_1G_2 + G_2G_3)}{1 + 2(G_1G_2 + G_2G_3) + G_1^2G_2^2 + 2G_1G_2^2G_3 + G_2^2G_3^2}$$

$$\frac{C}{R} = \frac{2G_1G_2G_3G_4\,(1 + G_1G_2 + G_2G_3)}{1 + 2(G_1G_2 + G_2G_3) + G_1^2G_2^2 + 2G_1\,G_2^2G_3 + G_2^2\,G_3^2}$$

$$\frac{C}{R} = \frac{2G_1G_2G_3G_4\,(1 + G_1G_2 + G_2G_3)}{(1 + G_1G_2 + G_2G_3)^2}$$

$$\therefore \qquad \frac{C}{R} = \frac{2G_1G_2G_3G_4}{1 + G_1G_2 + G_2G_3}$$

3.4 CONVERSION FROM BLOCK DIAGRAM TO SFG

Procedure to Draw Signal Glow Graph from Block Diagram

1. Replace the input signal and output signal by nodes.
2. Replace all the summing points by nodes.
3. Replace all the take-off points by nodes.
4. Branch connecting a summing point and take-off point has unity gain, then summing point and take-off point can be combined and represented by a single node.

Example 3.38: Obtain the transfer function of the given block diagram shown in Fig. 3.67 by drawing signal flow graph using Mason's gain formula.

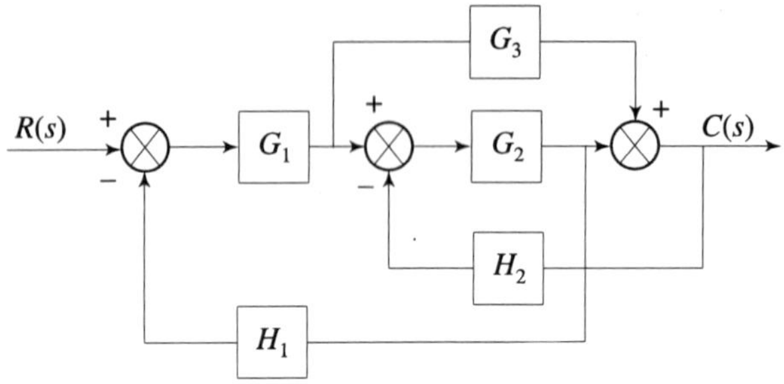

Fig. 3.67

Solution: Mark nodes corresponding to input and output signals. Assign a signal after every summing point and at the take-off point.

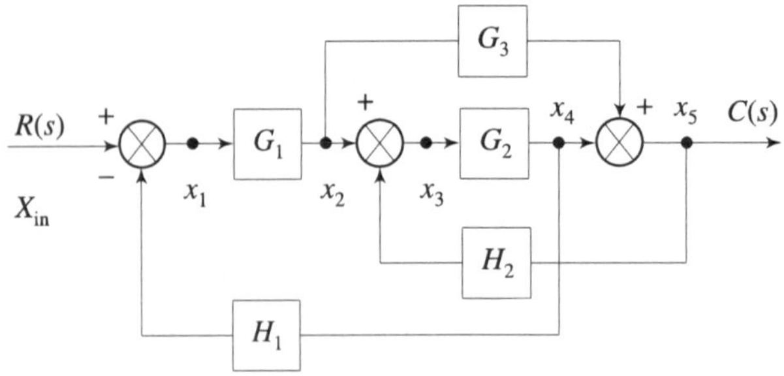

Fig. 3.68

$x_1 = x_{in} - x_4H_1$

$x_2 = x_1G_1$

$x_3 = x_2 - H_2x_5$

$x_4 = x_2G_2$

$x_5 = x_4 + x_2G_3$

2-forward paths

3-individual loops

No non-touching loops

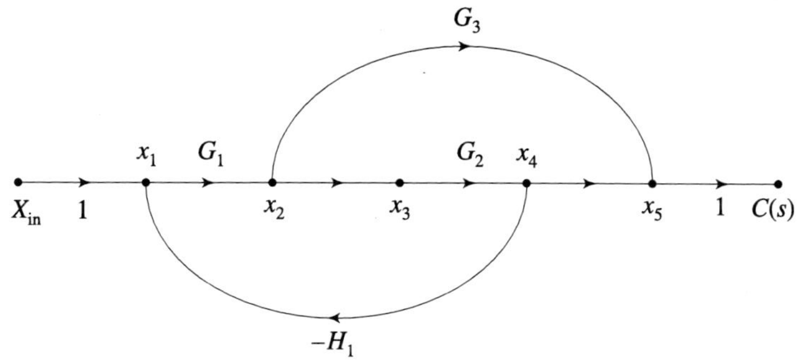

Fig. 3.69

$$\Delta_K = 1$$

$$M = \frac{G_1G_2 + G_1G_3}{1 + G_1G_2H_1 + G_2H_2 - G_1G_3H_2G_2H_1}$$

Example 3.39: The performance equations of a controlled system are given by the following set of linear algebraic equations. Draw the block diagram and determine $\frac{C(s)}{R(s)}$ by reducing the block diagram in steps.

$$E_1(s) = R(s) - H_3(s)C(s)$$

$$E_2(s) = E_1(s) - H_1(s)\,E_4(s)$$

$$E_3(s) = G_1(s)\,E_2(s) - H_2(s)\,C(s)$$

$$E_4(s) = G_2(s)E_3(s)$$

$$C(s) = G_3(s)\,E_4(s)$$

Solution: The block diagram for the various equations are

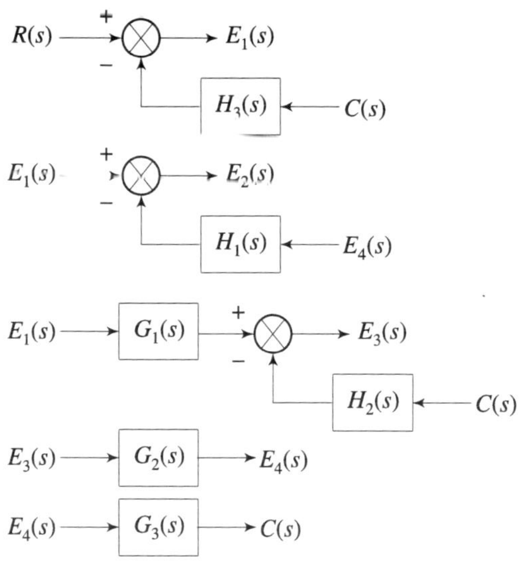

Fig. 3.70

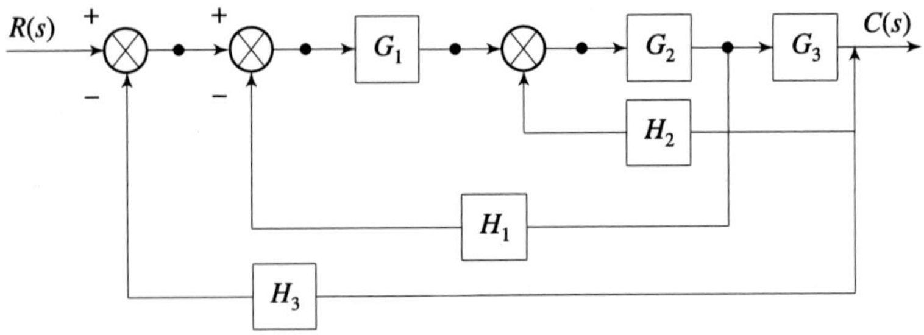

Fig. 3.71

$$\frac{C(s)}{R(s)} = \frac{G_1G_2G_3}{1 + G_1G_2H_1 + G_1G_2G_3H_3 + G_2G_3H_2}$$

Example 3.40: Draw an equivalent signal flow graph and determine $\frac{C(s)}{R(s)}$.

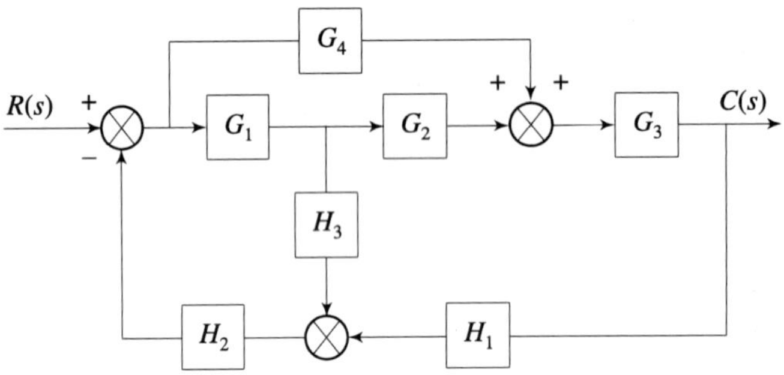

Fig. 3.72

Solution: Number of forward paths = 2

$$M_1 = G_1G_2G_3$$

$$M_2 = G_3G_4$$

There are 3 individual loops

$$L_{1a} = G_3G_4H_1H_2$$

$$L_{1b} = -G_1H_2H_3$$

$$L_{1c} = G_1G_2G_3H_1H_2$$

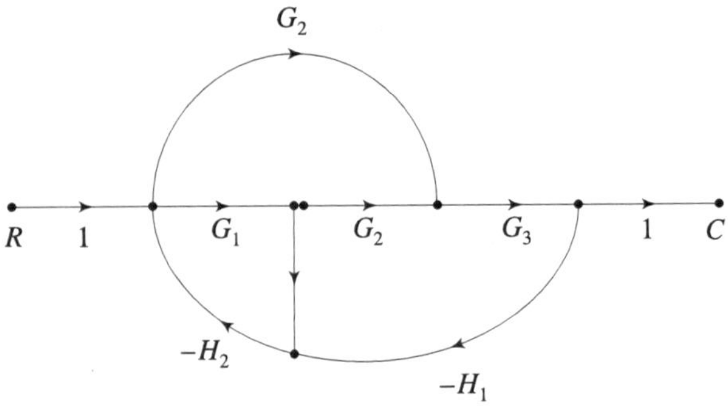

Fig. 3.73

There is no non-touching loops

To find Δ_K

For Δ_1 all the loops touches M_1

$$\Delta_1 = 1, \text{ similarly } \Delta_2 = 1$$

$$\frac{C}{R} = \frac{G_1G_2G_3 + G_3G_4}{1 - G_3G_4H_1H_2 + G_1H_2H_3 - G_1G_2G_3H_1H_2}$$

Example 3.41: Find the transfer function for the system shown in Fig. 3.74 by Mason's gain formula.

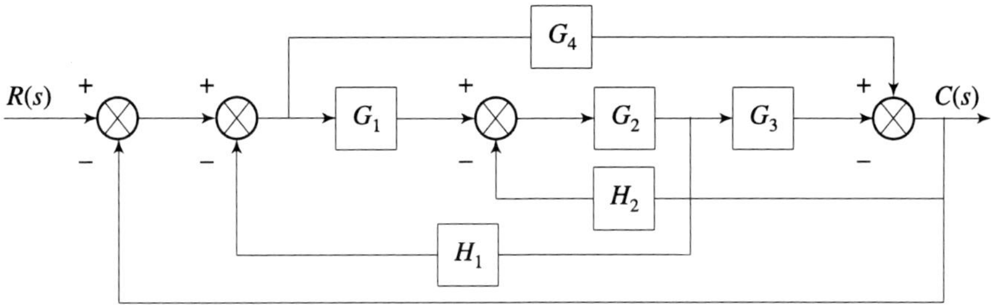

Fig. 3.74

Solution:

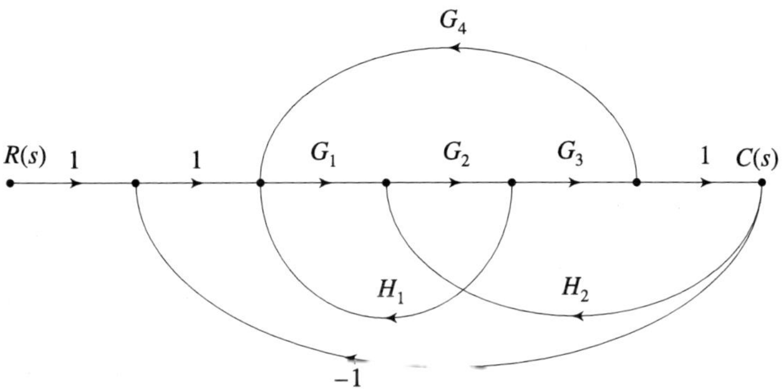

Fig. 3.75

$$\frac{C(s)}{R(s)} = \frac{G_4 + G_1G_2G_3}{1 + G_1G_2H_1 + G_1G_2G_3 + G_2G_3H_2 + G_4 - G_2G_4H_1H_2}$$

Example 3.42: Construct the signal flow graph for the following set of equations.

$$Y_2 = G_1Y_1 - G_2Y_4$$

$$Y_3 = G_3Y_2 + G_4Y_3$$

$$Y_4 = G_5Y_1 + G_6Y_3$$

When Y_4 is the output using Mason's gain formula, find the transfer function of the system.

Solution:

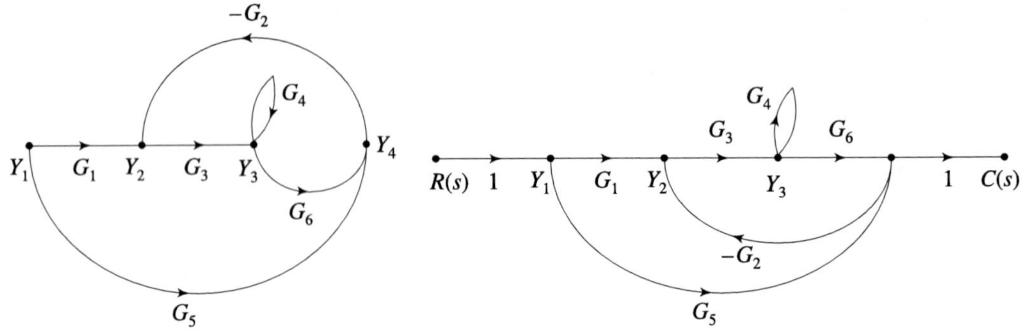

Fig. 3.76

$$M = \frac{G_1G_3G_6 + G_5(1 - G_4)}{1 + G_2G_3G_6 - G_4}$$

3.5 SIGNAL FLOW GRAPHS FOR ELECTRIC NETWORKS

1. Draw the given network in the *S*-domain.
2. Write the equation for different branch currents and node voltages. Write separate equation for separate branch and each element must be considered at least once.
3. Identify different variables and represent each variable by a separate node. The input and output nodes are to be shown separately.
4. Draw the signal flow graph for each equation in step 2 combine all signal flow graphs to get total signal flow graph.
5. Use Manson's gain formula to find the transfer function of the given network.

Example 3.43: Find the transfer function of the network shown in Fig. 3.77 using Mason's gain formula.

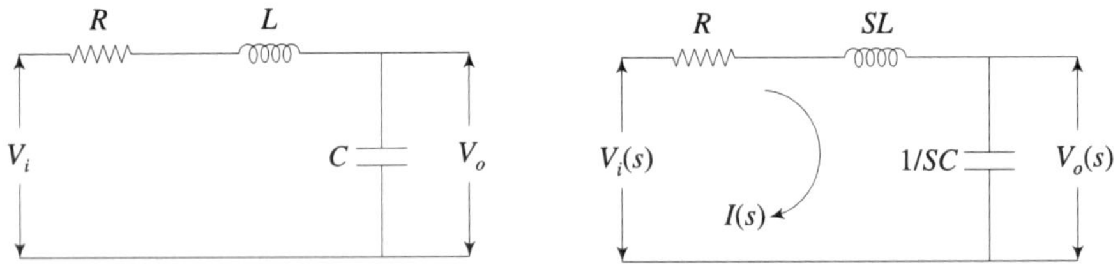

Fig. 3.77(a)

$$I(s) = \frac{V_i(s) - V_o(s)}{R + SL}$$

$$I(s) = \frac{V_i(s)}{R + SL} + V_2(s)\left(-\frac{1}{R + SL}\right)$$

$$V_o(s) = I(s)\,\frac{1}{SL}$$

$$\frac{V_o(s)}{V_i(s)} = \frac{1}{S^2LC + SRC + 1}$$

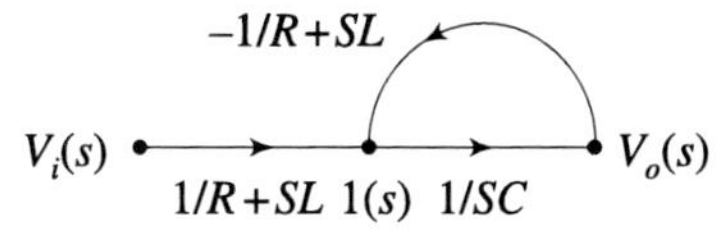

Fig. 3.77(b)

Example 3.44: For the given system equation to form the signal flow graph and solve for $\frac{V_o}{V_1}$ by SFG.

$$I(s) = \frac{V_i - V_o}{R + SL}$$

$$V_o(s) = \frac{1}{SC}\,I(s)$$

Solution: $I(s)$, $V_o(s)$ and $V_i(s)$ are variables

$$I(s) = \frac{V_i - V_o}{R + SL}$$

$$I(s) = \frac{V_i}{R + SL} - \frac{V_o}{R + SL}$$

$$V_o(s) = \frac{1}{SC} I(s)$$

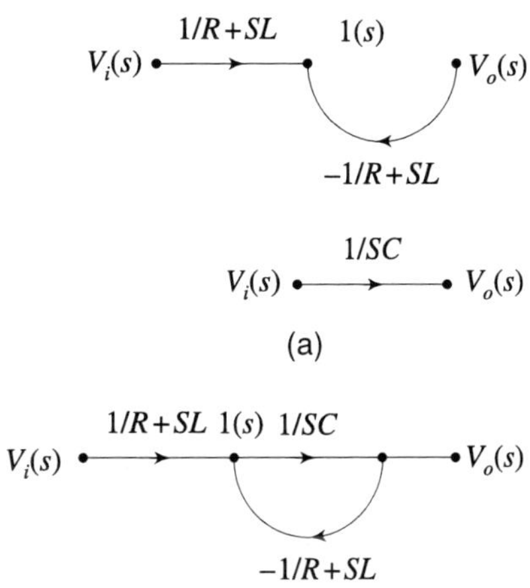

(b)

Fig. 3.78

The given SFG is (Fig. 3.78b)

There is one forward path

$$M_1 = \frac{1}{SC}\left(\frac{1}{R + SL}\right)$$

There is one individual loop

$$L_{1a} = -\frac{1}{SC}\left(\frac{1}{R + SL}\right)$$

To find Δ_K

$$\Delta_1 = 1$$

$$\Delta = 1 + \frac{1}{SC}\left(\frac{1}{R + SL}\right)$$

$$\text{Transfer function} = \frac{V_o(s)}{V_i(s)}$$

$$T = \frac{1}{S^2LC + SCR + 1}$$

Example 3.45: Find the transfer function for the electrical network shown in Fig. 3.79 by drawing the signal flow graph.

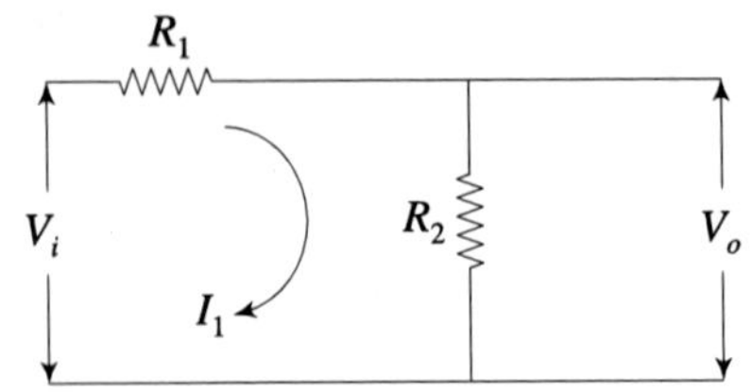

Fig. 3.79

Step 1: Find the Laplace domain representation

Step 2: Find equations for currents and voltages

Step 3: Draw signal flow graph from equations

Solution: **Step 1: Laplace domain representation**

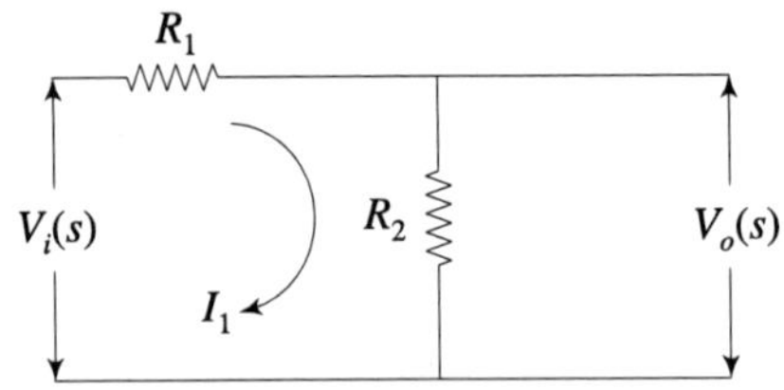

Fig. 3.80

Step 2: Find the equations for currents and voltages.

$$I_1(s) = \frac{V_i(s)}{R_1 + R_2} \qquad ...(3.1)$$

$$V_o(s) = R_2 I_1(s) \qquad ...(3.2)$$

The variables are $V_o(s)$, $V_i(s)$, $I_1(s)$

Step 3: Draw SFG from Eqs. (3.1) and (3.2)

From Eq. (3.1)

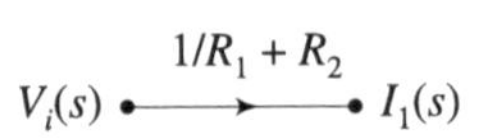

From Eq. (3.2)

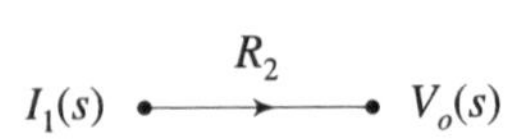

Combining these two SFGs

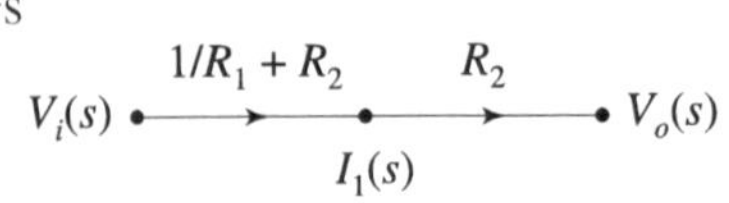

$$\text{Transfer function} = \frac{V_o(s)}{V_i(s)} = \left(\frac{1}{R_1 + R_2}\right) R_2$$

Example 3.46: Find transfer function and draw signal flow graph shown in Fig. 3.81.

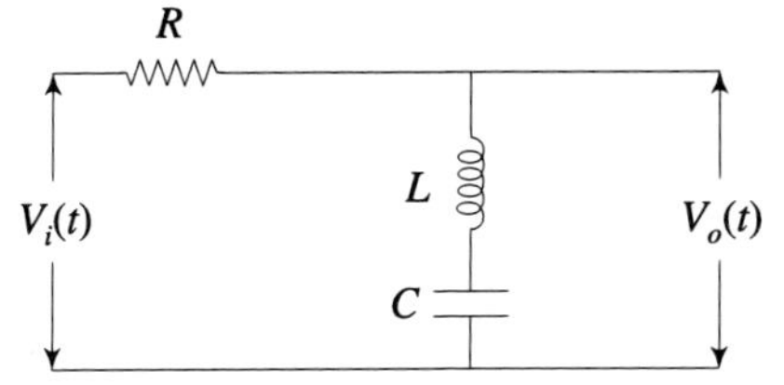

Fig. 3.81

Solution:

Step 1: Laplace domain representation

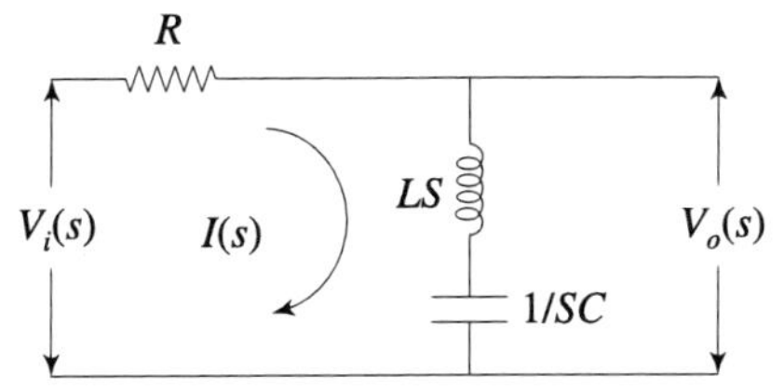

Fig. 3.82

Step 2:

$$I(s) = \frac{V_i(s)}{R + SL + \frac{1}{SC}}$$

$$V_o(s) = \left(SL + \frac{1}{SC}\right) I_1(s)$$

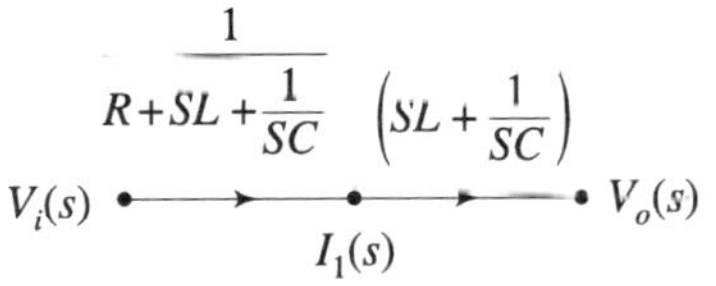

Transfer function

$$= \frac{V_o(s)}{V_i(s)} = \frac{\left(SL + \frac{1}{SC}\right)}{\left(R + SL + \frac{1}{SC}\right)}$$

$$\frac{V_o(s)}{V_i(s)} = \frac{1 + S^2LC}{S^2LC + SRC + 1}$$

Example 3.47: Find transfer function and signal flow graph shown in Fig. 3.83.

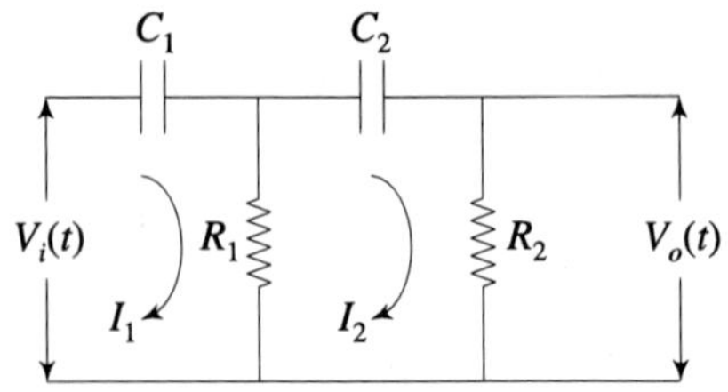

Fig. 3.83

Solution:

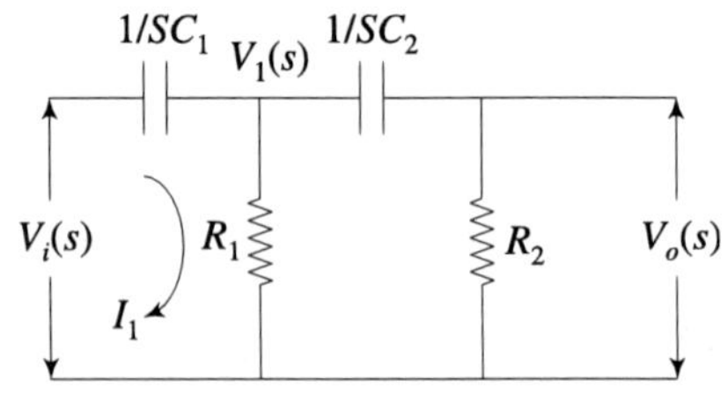

Fig. 3.84

$$I_1(s) = \frac{V_i(s) - V_1(s)}{\dfrac{1}{SC_1}}$$

$$I_1(s) = SC_1(V_i(s) - V_1(s)) \qquad ...(3.3)$$

$$V_1(s) = (I_1 - I_2)\, R_1 \qquad ...(3.4)$$

$$I_2(s) = \frac{V_1(s) - V_o(s)}{\left(\dfrac{1}{SC_2}\right)} = SC_2\,(V_1(s) - V_o(s)) \qquad ...(3.5)$$

$$V_o(s) = I_2 R_2 \qquad ...(3.6)$$

Based on Eqs (3.3), (3.4), (3.5) and (3.6)

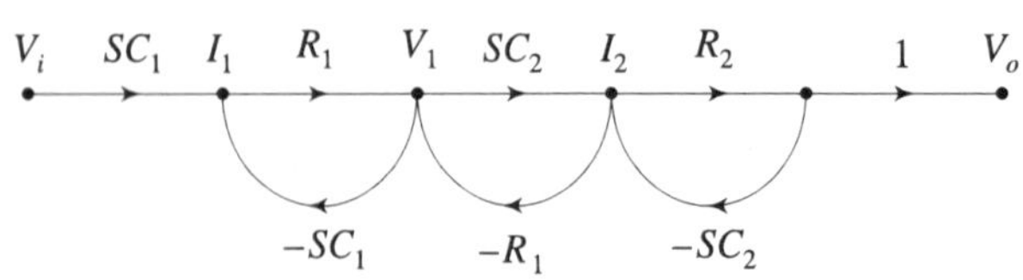

Fig. 3.85

Number of forward path

$$M_1 = SC_1R_1SC_2R_2 = S^2R_1R_2C_1C_2$$

There are 3 individual loops

$$L_{1a} = -SC_1R_1$$

$$L1_b = -SC_2R_1$$

$$L_{1c} = -SC_2R_2$$

There is one, 2 non-touching loops

$$L_{1a} = -S^2C_1C_2R_1R_2$$

$$\Delta = 1 + SC_1R_1 + SC_2R_1 + SC_2R_2 + S^2R_1R_2C_1C_2$$

To find Δ_K

$$\Delta_1 = 1$$

Transfer function

$$\frac{V_o(s)}{V_i(s)} = \frac{S^2R_1R_2C_1C_2}{1 + SC_1R_1 + SC_2R_1 + SC_2R_2 + S^2R_1R_2C_1C_2}$$

Example 3.48: Find transfer function by drawing SFG for the electrical network shown in Fig. 3.86.

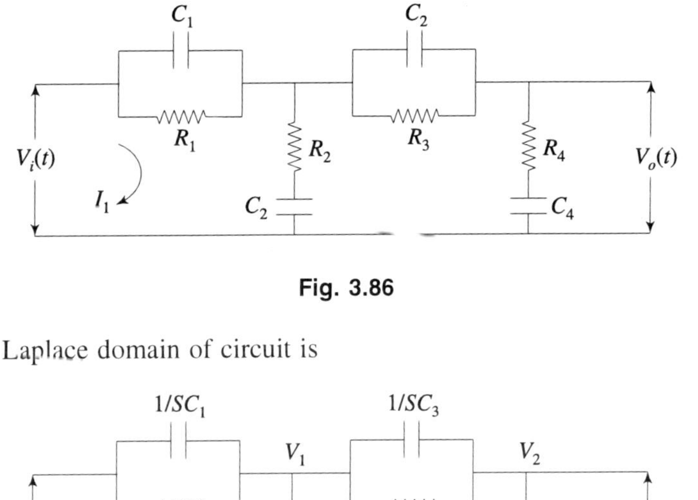

Fig. 3.86

Solution: The Laplace domain of circuit is

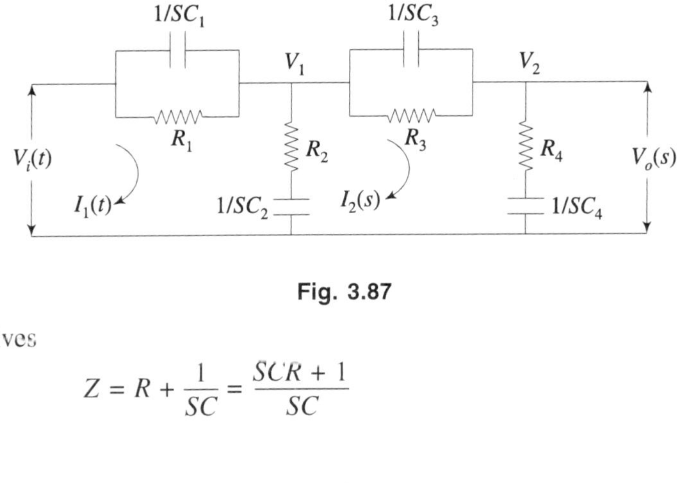

Fig. 3.87

Series RC gives

$$Z = R + \frac{1}{SC} = \frac{SCR + 1}{SC}$$

Parallel RC gives

$$Y = SC + \frac{1}{R} = \frac{SCR + 1}{R}$$

$$Z = \frac{1}{Y} = \frac{R}{SCR + 1}$$

$$I_1(s) = \frac{V_i(s) - V_1(s)}{Z_1} = \frac{V_i(s) - V_1(s)}{\dfrac{R_1}{SC_1R_1 + 1}}$$

$$I_1(s) = \frac{(SC_1R_1 + 1)}{R_1}(V_i(s) - V_1(s))$$

$$V_1(s) = (I_1(s) - I_2(s))Z_2$$

$$V_1(s) = (I_1(s) - I_2(s))\left(\frac{SC_2R_2 + 1}{SC_2}\right)$$

$$I_2(s) = (V_1(s) - V_o(s))\,Z_3$$

$$I_2(s) = (V_1(s) - V_o(s))\left(\frac{SC_3R_3 + 1}{R_3}\right)$$

$$V_o(s) = (I_2(s))Z_4$$

$$V_o(s) = (I_2(s))\left(\frac{SC_4R_4 + 1}{SC_4}\right)$$

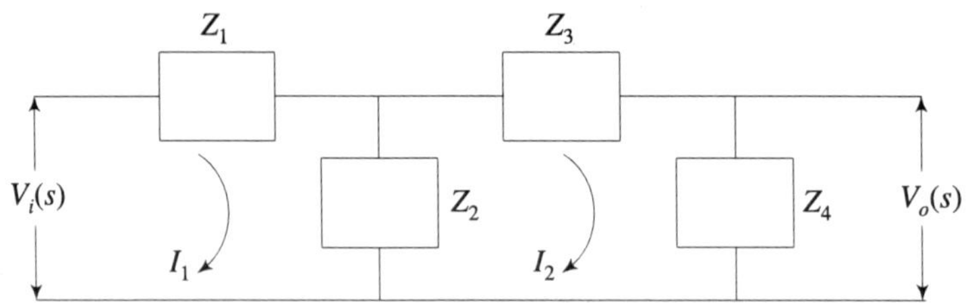

Fig. 3.88

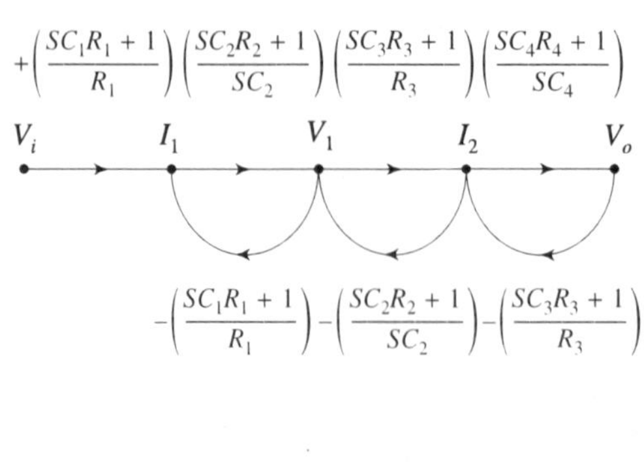

From the given SFG there is only one forward path.

$$M_1 = \left(\frac{SC_1R_1 + 1}{R_1}\right)\left(\frac{SC_2R_2 + 1}{SC_2}\right)\left(\frac{SC_3R_3 + 1}{R_3}\right)\left(\frac{SC_4R_4 + 1}{SC_4}\right)$$

$$M_1 = \frac{(SC_1R_1 + 1)(SC_2R_2 + 1)(SC_3R_3 + 1)(SC_4R_4 + 1)}{S^2R_1C_2R_3C_4}$$

There are three individual loops

$$L_{1a} = -\left(\frac{SC_1R_1 + 1}{R_1}\right)\left(\frac{SC_2R_2 + 1}{SC_2}\right)$$

$$L_{1b} = -\left(\frac{SC_2R_2 + 1}{SC_2}\right)\left(\frac{SC_3R_3 + 1}{R_3}\right)$$

$$L_{1c} = -\left(\frac{SC_3R_3 + 1}{R_3}\right)\left(\frac{SC_4R_4 + 1}{SC_4}\right)$$

There is one, 2 non-touching loops

$$L_{2a} = \left(\frac{SC_2R_2 + 1}{SC_2}\right)\left(\frac{SC_4R_4 + 1}{SC_4}\right)\left(\frac{SC_1R_1 + 1}{R_1}\right)\left(\frac{SC_3R_3 + 1}{R_3}\right)$$

To find Δ_K

$$\Delta_1 = 1$$

$$\Delta = 1 - (L_{1a} + L_{1b} + L_{1c}) + L_{2a}$$

Transfer function $= \dfrac{V_o(s)}{V_i(s)} = \dfrac{M_1\Delta_1}{\Delta}$

Example 3.49: Find the transfer function of the network shown in Fig. 3.89 using Mason's gain formula.

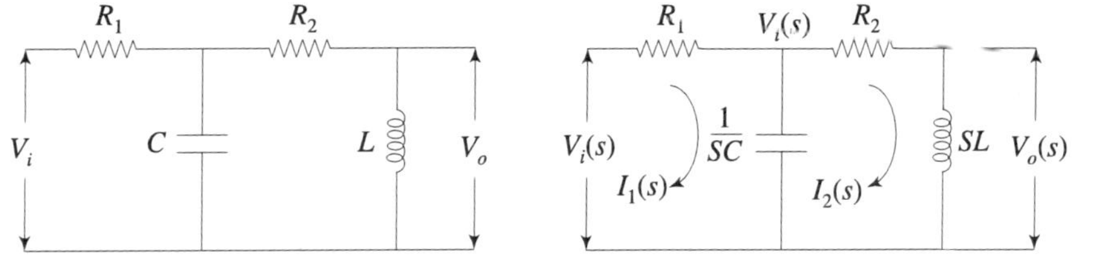

Fig. 3.89

$$I_1(s) = \frac{V_i(s) - V_1(s)}{R_1}$$

$$V_1(s) = (I_1(s) - I_2(s))\frac{1}{SC}$$

$$I_2(s) = \frac{(V_1(s) - V_0(s))}{R_2}$$

$$V_1(s) = (I_2(s))SL$$

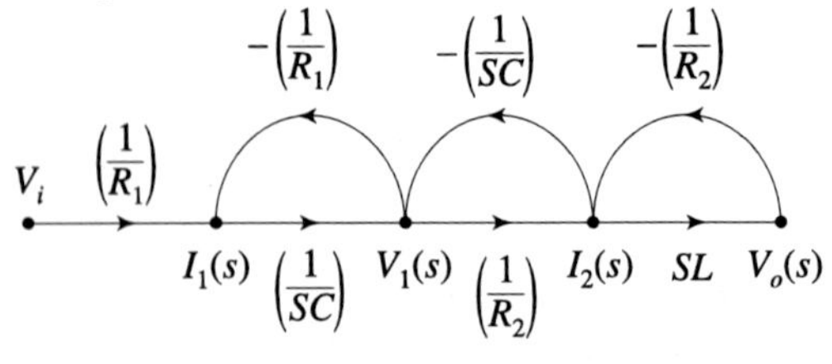

Fig. 3.90

$$\frac{V_o(s)}{V_i(s)} = \frac{SL}{S^2LR_1\,C + S(L + R_1R_2C) + (R_1 + R_2)}$$

Example 3.50: Convert the given block diagram to signal flow graph and determine $\frac{C(s)}{R(s)}$.

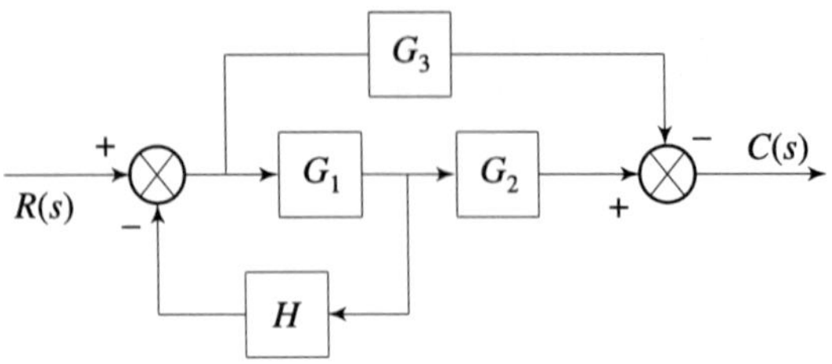

Fig. 3.91

Solution: The nodes are assigned at input, output, at every summing point and takeoff point.

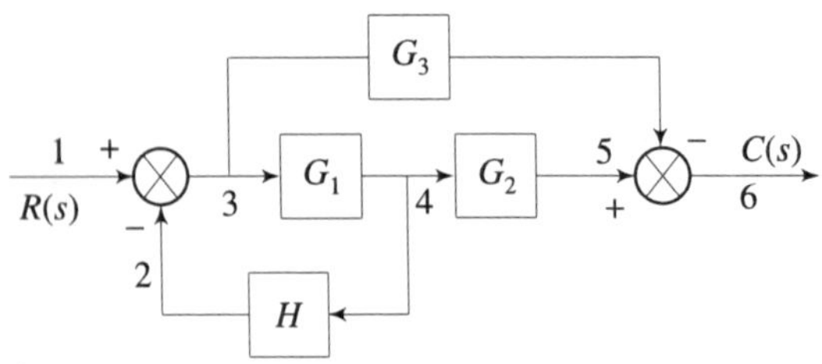

Fig. 3.92

The signal flow graph of the above system is shown in Fig. 3.93.

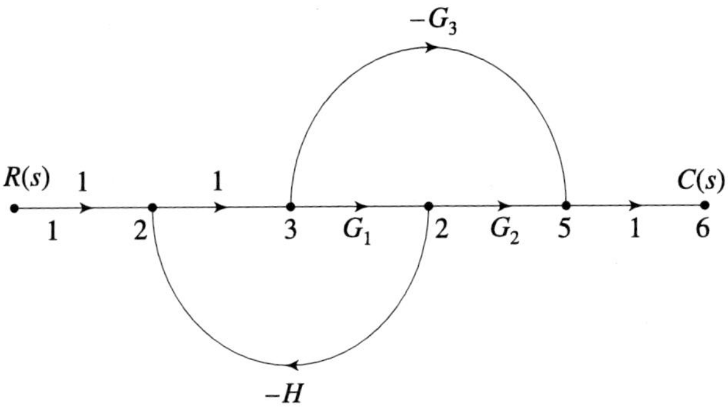

Fig. 3.93

There are two forward paths

$$P_1 = G_1G_2, \quad P_2 = -G_3$$

There is only one individual loop

$$L_1 = -G_1H$$

There is no combination of non-touching loops

$\therefore$
$$\Delta = 1 - (L_1) = 1 - (-G_1H) = 1 + G_1H$$

For P_1 and P_2, all the parts of the graph touched P_1 and P_2

$\therefore$
$$\Delta_1 = \Delta_2 = 1$$

Transfer function,

$$\frac{C(s)}{R(s)} = \frac{P_1\Delta_1 + P_2\Delta_2}{\Delta}$$

$$\frac{C(s)}{R(s)} = \frac{G_1G_2 - G_3}{1 + G_1H}$$

Example 3.51: Convert the block diagram to signal flow graph and determine the transfer function using Mason's gain formula.

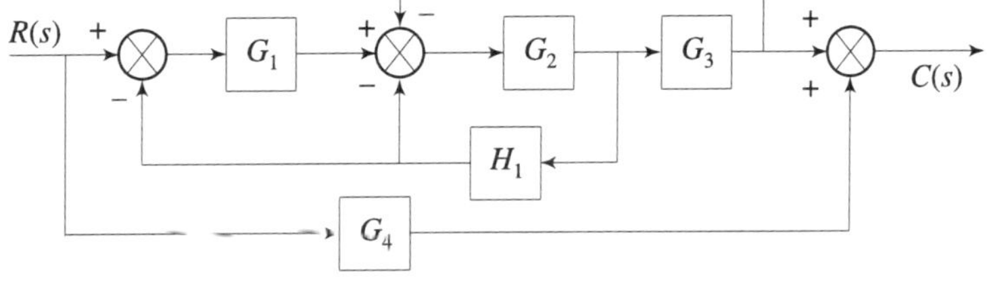

Fig. 3.94

Solution: The nodes are assigned at input, output, at summing point and take-off point.

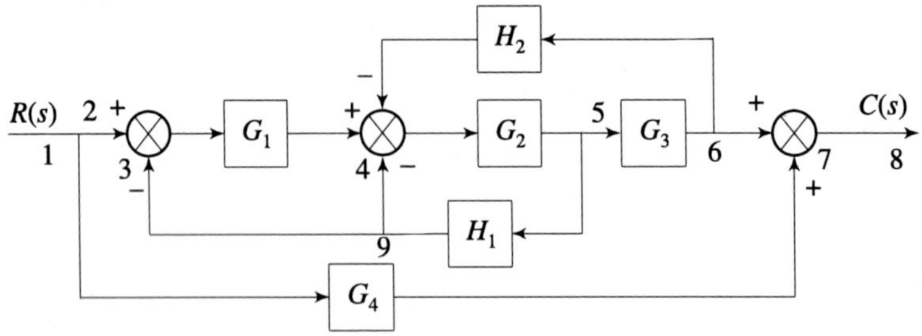

(a)

The signal flow graph of the block diagram

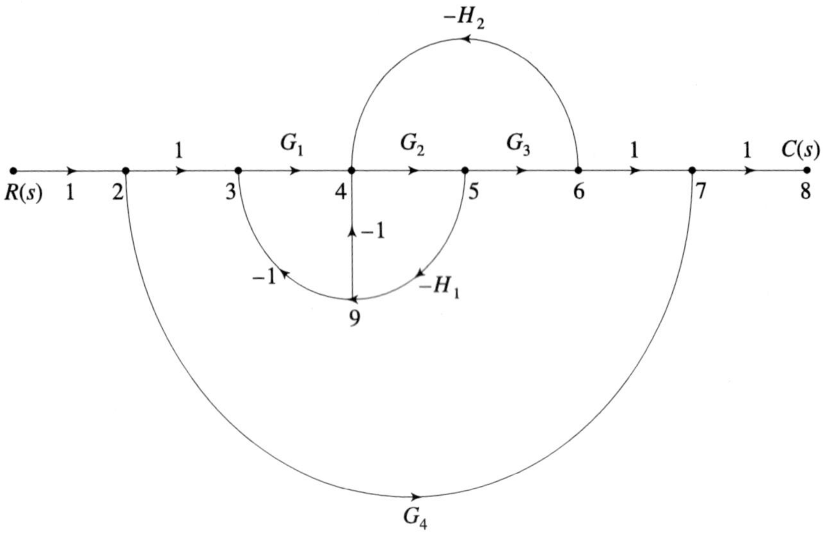

(b)

Fig. 3.95

Forward path gain

$$P_1 = G_1G_2G_3$$

$$P_2 = G_4$$

There are 3 individual loops

$$L_1 = -\, G_1G_2H_1$$

$$L_2 = -\, G_2G_3H_2$$

$$L_3 = -\, G_2H_1$$

There are no possible combinations of two non-touching loops, three non-touching loops so on.

$$\therefore \quad \Delta = 1 - (L_1 + L_2 + L_3)$$

$$\Delta = 1 - (-G_1G_2H_1 - G_2G_3H_2 - G_2H_1)$$

$$\Delta = 1 + G_1G_2H_1 + G_2G_3H_2 + G_2H_1$$

For P_1, $\Delta_1 = 1$, all parts of the graph touch P_1

For P_2, $\Delta_2 = 1 - (-G_1\, G_2H_1 - G_2G_3H_2 - G_2H_1)$

$$\Delta_2 = 1 + G_1G_2H_1 + G_2G_3H_2 + G_2H_1$$

Transfer function

$$\frac{C(s)}{R(s)} = \frac{P_1\Delta_1 + P_2\Delta_2}{\Delta}$$

$$\frac{C(s)}{R(s)} = \frac{G_1G_2G_3 + G_4(1 + G_1G_2H_1 + G_2G_3H_2 + G_2H_1)}{1 + G_1G_2H_1 + G_2G_3H_2 + G_2H_1}$$

Example 3.52: Draw a signal flow graph and evaluate the closed-loop transfer function of a system whose block diagram is shown in Fig. 3.96.

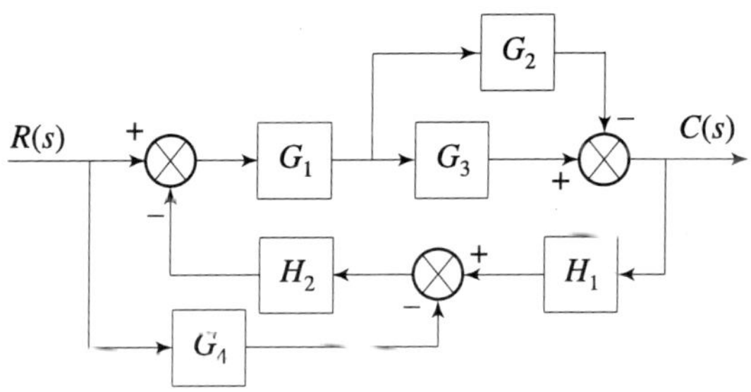

Fig. 3.96

Solution: Assign the nodes

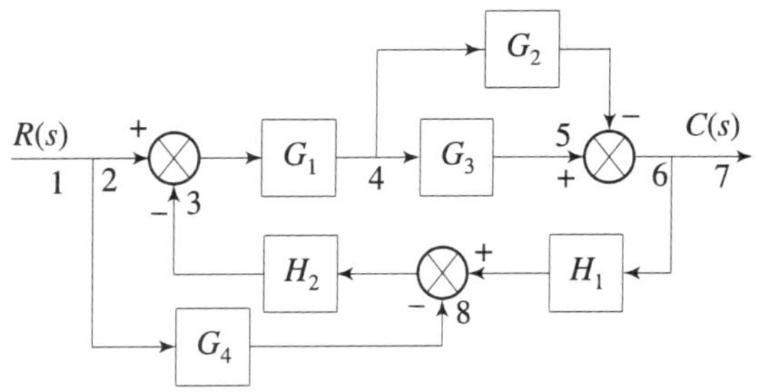

Fig. 3.97

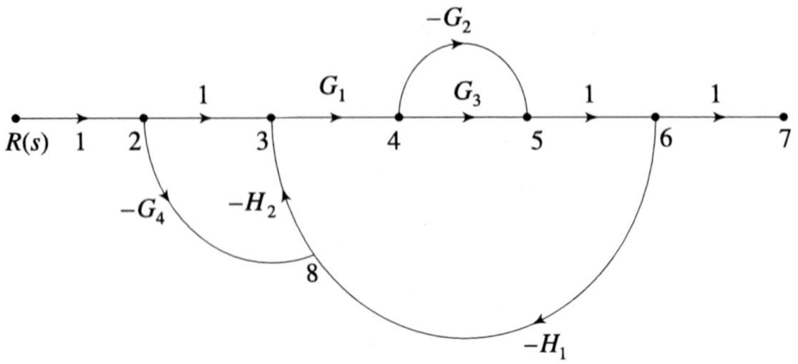

Fig. 3.98

There are 4 forward paths and path gains

$$P_1 = G_1G_3$$
$$P_2 = -G_1G_2$$
$$P_3 = G_1G_3G_4H_2$$
$$P_4 = -G_1G_2G_4H_2$$

There are two individual loops and loop gain

$$L_1 = -G_1G_3H_1H_2$$
$$L_2 = G_1G_2H_1H_2$$

There is no possible combinations of two, three non-touching loops

$$\therefore \qquad \Delta = 1 - (L_1 + L_2)$$
$$\Delta = 1 - (-G_1G_3\ H_1H_2 + G_1G_2H_1H_2)$$
$$\Delta = 1 + G_1G_3\ H_1H_2 - G_1G_2H_1H_2)$$

Since no part of graph is non-touching with the forward paths

$$\Delta_1 = \Delta_2 = \Delta_3 = \Delta_4 = 1$$

The transfer function

$$\frac{C(s)}{R(s)} = \frac{P_1\Delta_1 + P_2\Delta_2 + P_3\Delta_3 + P_4\Delta_4}{\Delta}$$

$$\frac{C(s)}{R(s)} = \frac{(G_1G_3) + (-\ G_1G_2) + G_1G_3G_4H_2 + (-\ G_1G_2G_4H_2)}{1 + G_1G_3H_1H_2 - G_1G_2H_1H_2}$$

$$\frac{C(s)}{R(s)} = \frac{G_1G_3 - G_1G_2 + G_1G_3G_4H_2 - G_1G_2G_4H_2}{1 + G_1G_3H_1H_2 - G_1G_2H_1H_2}$$

Example 3.53: Convert the block diagram shown in Fig. 3.99 to a signal flow graph.

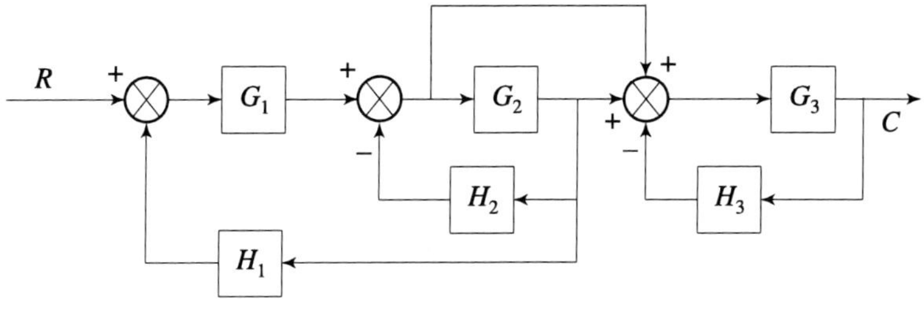

Fig. 3.99

Solution: Assign nodes

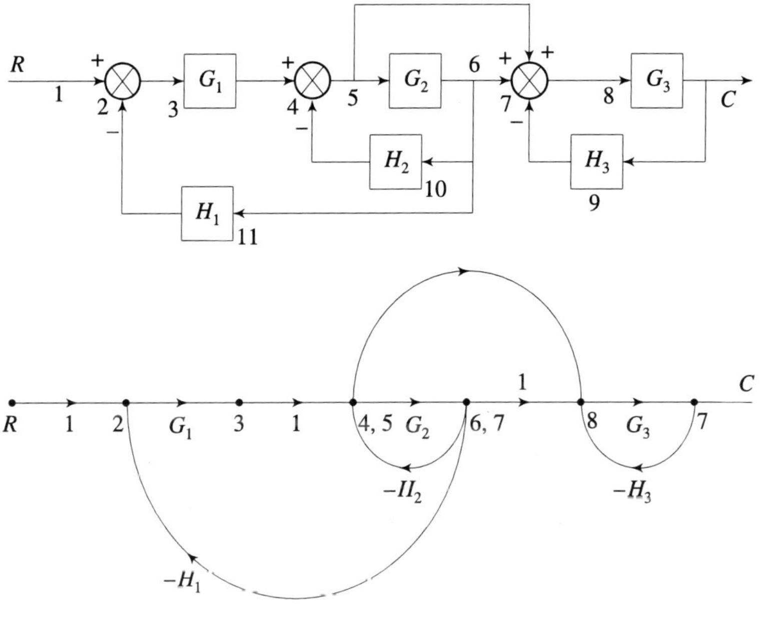

Fig. 3.100

Example 3.54: A system is described by the following set of linear algebraic equations

$$x_2 = a_{12}\, x_1 + a_{22}\, x_2 + a_{32}\, x_3$$

$$x_3 = a_{23}\, x_2 + a_{43}\, x_4$$

$$x_4 = a_{24}\, x_2 + a_{34}\, x_3 + a_{44}\, x_4$$

$$x_5 = a_{25}\, x_2 + a_{45}\, x_4$$

Draw the signal flow graph and obtain the transfer function of the system using Mason's gain formula.

Solution:

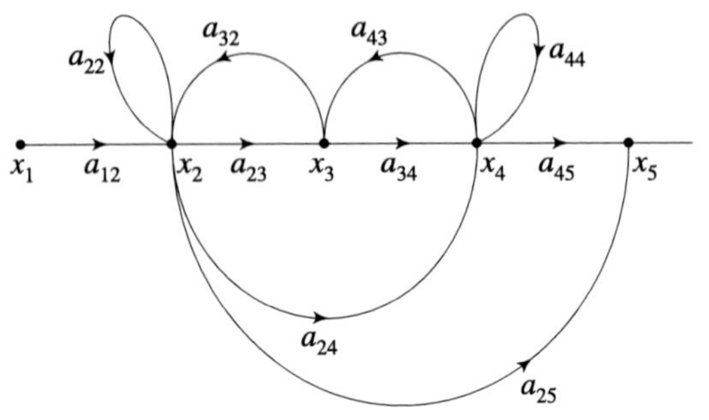

Fig. 3.101

There are 3 forward paths and path gains

$$P_1 = a_{12}\, a_{23}\, a_{34}\, a_{45}$$

$$P_2 = a_{12}\, a_{24}\, a_{45}$$

$$P_3 = a_{12}\, a_{25}$$

There are 5 individual loops

$$L_1 = a_{22},\ L_2 = a_{23}\, a_{32},\ L_3 = a_{34}\, a_{43}$$

$$L_4 = a_{44},\ L_5 = a_{24}\, a_{43}\, a_{32}$$

There are 3, two non-touching loops and the loop gain

$$L_{11} = (a_{22})\,(a_{34}\, a_{43}) = a_{22}\, a_{34}\, a_{43}$$

$$L_{12} = (a_{22})\,(a_{44}) = a_{22}\, a_{44}$$

$$L_{13} = (a_{23}\, a_{32})\,(a_{44}) = a_{23}\, a_{32}\, a_{44}$$

$$\therefore \quad \Delta = 1 - (L_1 + L_2 + L_3 + L_4 + L_5) + (L_{11} + L_{12} + L_{13})$$

$$\Delta = 1 - (a_{22} + a_{23}\, a_{32} + a_{34}\, a_{43} + a_{44} + a_{24}\, a_{43}\, a_{32}) + (a_{22}\, a_{34}\, a_{43} + a_{22}\, a_{44} + a_{23}\, a_{32}\, a_{44})$$

To find Δ_k

For P_1, $\quad \Delta_1 = 1$

For P_2, $\quad \Delta_2 = 1$

For P_3, $\quad \Delta_3 = 1 - (a_{34}\, a_{43} + a_{44})$

The transfer function is

$$\frac{x_5}{x_1} = \frac{P_1\Delta_1 + P_2\Delta_2 + P_3\Delta_3}{\Delta}$$

$$\frac{x_5}{x_1} = \frac{a_{12}\, a_{23}\, a_{34}\, a_{45} + a_{12}\, a_{24}\, a_{45} + a_{12}\, a_{25}\,(1 - a_{34}\, a_{43} - a_{44})}{1 - a_{22} - a_{23}\, a_{32} - a_{34}\, a_{43} - a_{44} - a_{24}\, a_{43}\, a_{32} + a_{22}\, a_{34}\, a_{43} + a_{22}\, a_{44} + a_{23}\, a_{32}\, a_{44}}$$

EXERCISE

Fill in the Blanks

1. The position control of a DC servomoter is given in Fig. 3.102. The values of the parameters are $K_T = 1\text{N} - mt$, $Ra = 1\Omega$, $La = 0.1\text{H}$, J = 5 kg-m^2, $B = 1\text{N-m(rad/sec)}$ and $Kb = 1\text{V (rad/sec)}$. The steady state position response (in radius) due to unit impulse disturbance torque Td is given by __________? (GATE 2015)

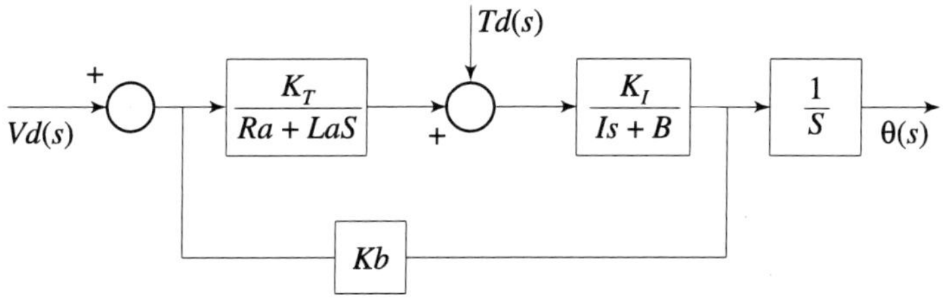

Fig. 3.102

Ans:

$$T \cdot F \frac{\theta(s)}{Td(s)} = \frac{\frac{1}{SCjs + B}}{\left(1 + \frac{K_T}{Ra + 5La}\right)} \left(\frac{1}{Js + B}\right) Kb$$

$$= \frac{-Ra + 5La}{5[(Ra + 5La)(Js + B) + K_T\,Kb]}$$

Final value $\theta(\infty) = \underset{s \to 0}{\text{Lt}}$

or

Steady state $\theta(\infty) = \frac{-Ra}{Ra(B) + K_T Kb} = \frac{-1}{1 + 1} = \mathbf{0.5}$

2. Write an equivalent block diagram of Fig. 3.103.

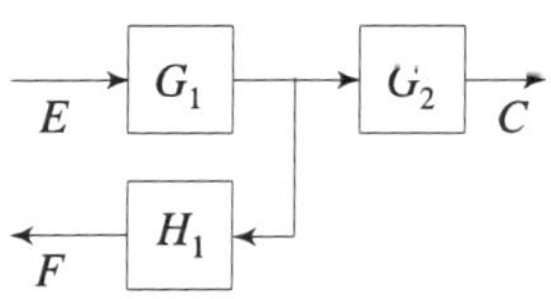

Fig. 3.103

Ans:

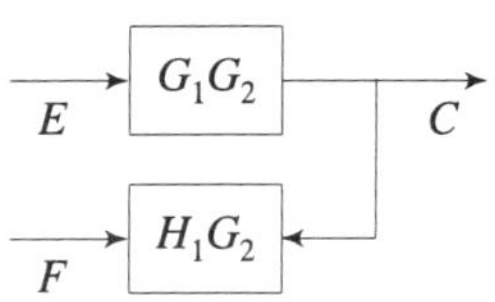

Fig. 3.104

3. Find the $T.F \frac{Y(s)}{R(s)}$ of the system shown in Fig. 3.105.

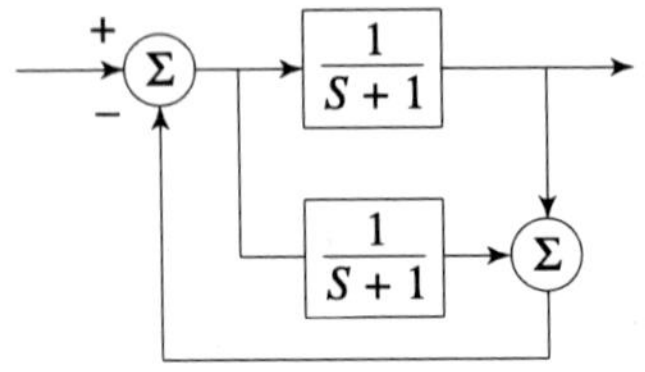

Fig. 3.105

Ans:

$$\frac{Y(s)}{R(s)} = \frac{\frac{1}{S+1}}{1+\frac{1}{S+1}-\frac{1}{S+1}} = \frac{1}{S+1}$$

4. Find the $T.F \frac{Y(s)}{XZ(s)}$, when $X_1(s) = 0$ from the Fig. 3.106. (GATE 2014)

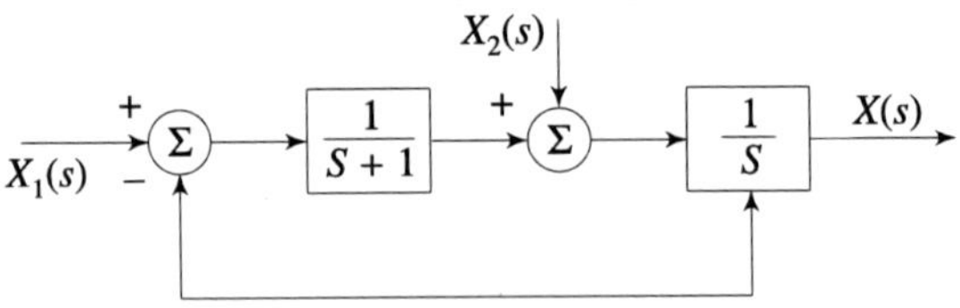

Fig. 3.106

Ans: If $X_1(s) = 0$

$$\frac{Y(s)}{X_2(s)}$$

The block diagram becomes

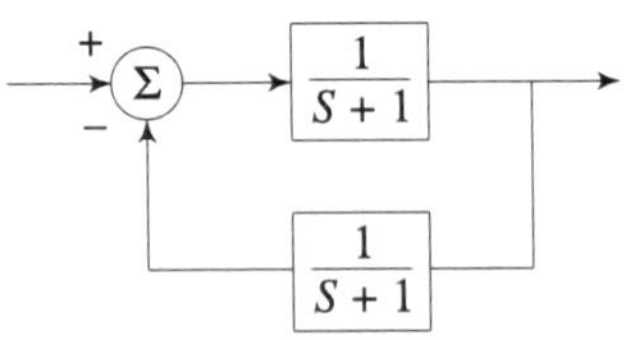

Fig. 3.107

$$\frac{Y(s)}{X_2(s)} = \frac{\frac{1}{S}}{1+\frac{1}{S}\frac{S}{(S+1)}} = \frac{\frac{1}{S}}{\frac{(S+2)}{(S+1)}} = \frac{S+1}{S(S+2)}$$

5. From Fig. 3.108, find the average value of the voltage V_{ab} in steady state condition. (GATE 2015)

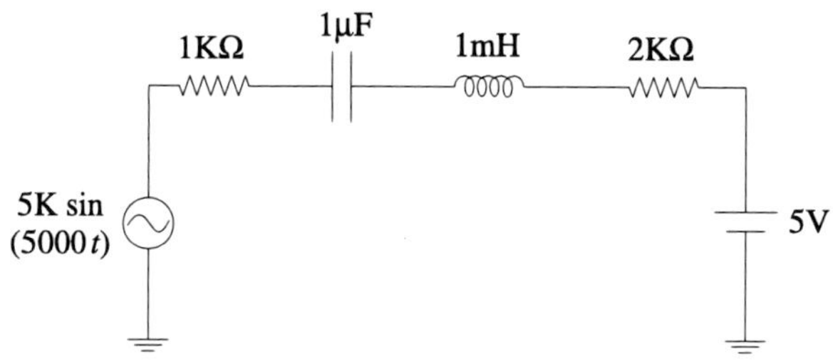

Fig. 3.108

Ans: **Case 1:** 5V source in steady state V_{ab} = 5V

Case 2: 5V sin (5000 t) in steady state,

$$V_{ab} = K \sin(5000t + d)V$$

V_{adb} at steady state

$$V_{ad} = V_{ab} + V_{ab}$$
$$= 5 + K \sin(5000t + a)$$

Average value

$$V_{ab} = \int_0^T V_{ab} = 5\text{V}$$

6. For the signal flow graph shown in Fig. 3.109 the value of $\frac{C(s)}{R(s)}$ is_____. (GATE 2015)

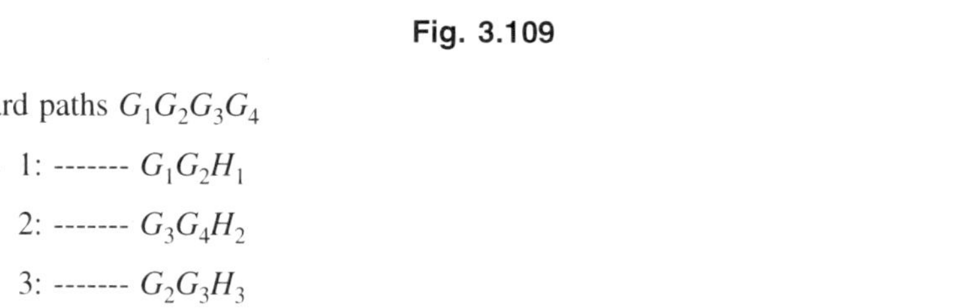

Fig. 3.109

Ans: Forward paths $G_1G_2G_3G_4$

Loops 1: ------- $G_1G_2H_1$

2: ------- $G_3G_4H_2$

3: ------- $G_2G_3H_3$

Non-touching loops $G_1G_2H_1\ G_3G_4H_2$

$$\frac{C(s)}{R(s)} = \frac{G_1G_2G_3G_4}{1 + G_1G_2H_1 + G_3G_4H_2 + G_2G_3H_3 + G_1G_2G_3G_4H_1H_2}$$

7. For the system shown in Fig. 3.110 $T \cdot F \dfrac{C(s)}{R(s)}$.

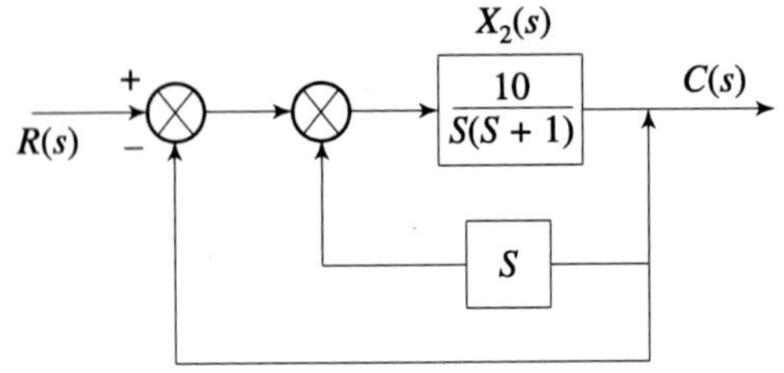

Fig. 3.110

Ans:
$$G_1 = \frac{\dfrac{10}{S(S+1)}}{1+\dfrac{10S}{S(S+1)}} = \frac{10}{S^2+11S}$$

Now close the outer loop

$$\frac{Y(s)}{R(s)} = \frac{\dfrac{10}{S^2+11S}}{1+\dfrac{10}{S^2+11S}} = \frac{10}{S^2+11S+10}$$

8. In the signal flow graph of Fig. 3.111. Find the gain C/r.

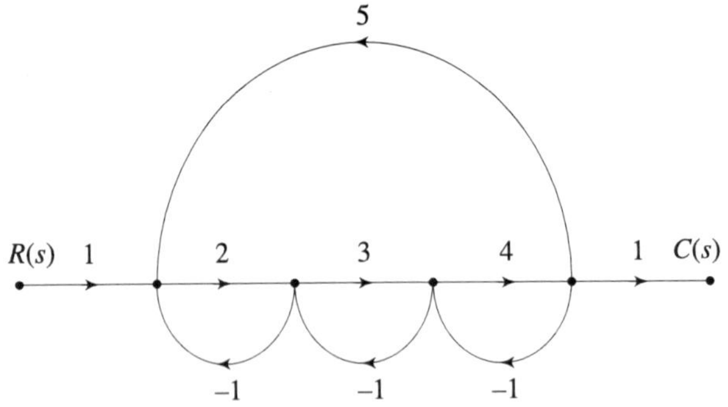

Fig. 3.111

Ans:
$$\frac{C}{r} = \frac{(2 \times 3 \times 4) + (1 \times 5\ (1+3))}{(1+2+3+4+5+(2 \times 4))} = \frac{44}{23}$$

9. An electrical system and its signal flow graph representation are shown in Fig. 3.112 (a) and (b) respectively find the values of G_2 and H.

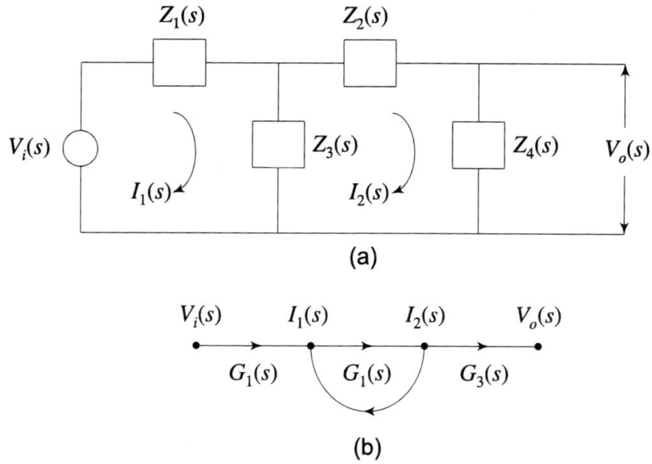

Fig. 3.112

Ans: Applying KVL in 1st loop

$$V_i(s) = I_1(s)\,Z_1(s) + [I_1(s) - I_2(s)]\,Z_3(s)$$

$$V_i(s) = I_1(s)\,[Z_1(s) + Z_3(s)] - I_2(s)\,Z_3(s)$$

$$\frac{V_i(s)}{Z_1(s) + Z_3(s)} = I_1(s) - \frac{I_2(s)\,Z_3(s)}{Z_1(s) + Z_3(S)}$$

Apply KVL in 2nd loop

$$[I_2\,(S) - I_1\,(s)]\,Z_3(s) - I_2(s)\,Z_2(s) + J_2(s)\,Z(s) = 0$$

$$J_2[Z_2(s) + Z_3(s) + Z_4(s)] = 0$$

$$\frac{I_2(s)}{I_1(s)} = \frac{Z_3(s)}{Z_2(s) + Z_3(s) + Z_4(s)} \quad \ldots(1)$$

$$G_2 = \frac{I_2(s)}{I_1(s)} = \frac{Z_3(s)}{Z_2(s) + Z_3(s) + Z_4(s)}$$

From signal flow graph

$$I_1(s) = V_i G_1(s) + I_2(s)\,H(s)$$

From Eq. (1)

$$I_1(s) = V_i\,\frac{1}{Z_1(s) + Z_3(s)} + I_2(s)\,\frac{Z_3(s)}{Z_1(s) + Z_3(S)}$$

$$H(s) = \frac{Z_3(s)}{Z_1(s) + Z_3(s)}$$

10. Find the $T.F \dfrac{C(s)}{R(s)}$ of the system, whose signal is flow graph is shown in Fig. 3.113.

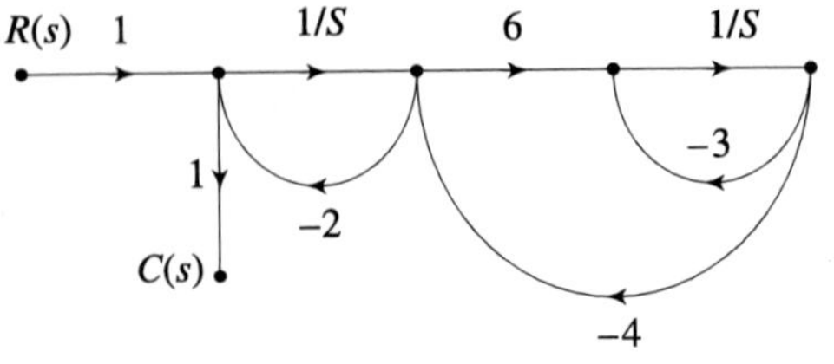

Fig. 3.113

Ans: SFG has 1 forward path (1)

3 individual loops $\left(-\dfrac{2}{S}, \dfrac{3}{S}, -\dfrac{24}{S}\right)$ and one combination of two non-touching loops $\left(\dfrac{6}{S}\right)$.

$$\frac{T}{F} = \frac{C(s)}{R(s)} = \frac{1\left[1 + \dfrac{3}{S} + \dfrac{24}{S}\right]}{1 + \dfrac{2}{S} + \dfrac{3}{S} + \dfrac{24}{S} + \dfrac{6}{S}} = \frac{S(S + 27)}{S^2 + 295 + 6}$$

11. For the discrete-time system shown in Fig. 3.114, find the locations of system T.F. (GATE 2015)

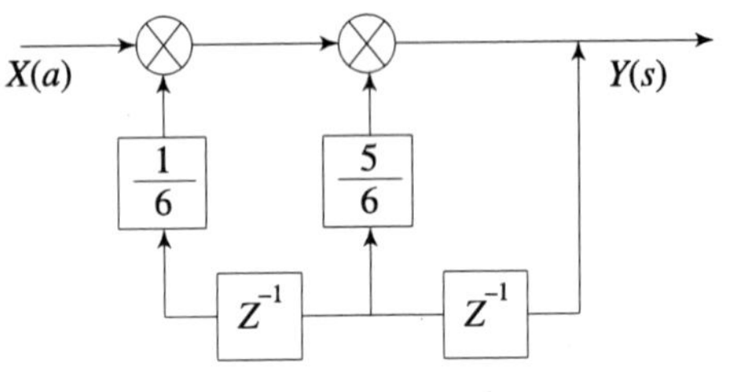

Fig. 3.114

Ans:

$$\frac{Y(Z)}{X(Z)} = H(Z) = \frac{1}{1 - \dfrac{5}{6}Z^{-1} + \dfrac{Z^2}{6}} = \frac{Z^2}{Z^2 - \dfrac{5}{6}Z + \dfrac{1}{6}}$$

$$= \frac{Z^2}{\left(Z - \dfrac{1}{2}\right)\left(Z - \dfrac{1}{3}\right)} = Z = \frac{1}{2}, \frac{1}{3}$$

Poles @ $Z = \dfrac{1}{2}, \dfrac{1}{3}$

12. Draw a signal flow graph for the following set of algebraic equations.

$$y_2 = ay_1 - gy_3$$
$$y_3 = ey_2 - cy_4$$
$$y_3 = hy_2 - dy_4$$

Find the gains $\dfrac{y_2}{y_1}$ and $\dfrac{y_3}{y_1}$

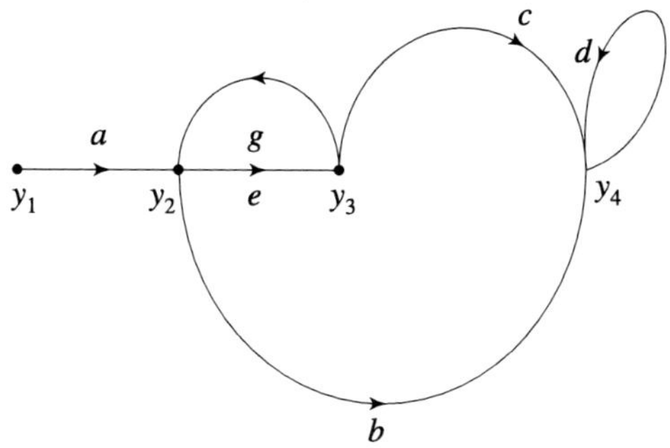

Fig. 3.115

Ans: The SFG has 3 individual loops ($-ge$, bcg, $-d$) and one combination of two non-touching loops (ged), Δ remains same for $\frac{y_2}{y_1}$ and $\frac{y_3}{y_1}$.

$$\frac{y_2}{y_1} = \frac{a(1 + d) - abcg}{1 + d + eg + bcg + edg}$$

$$\frac{y_3}{y_1} = \frac{ae(1 + d) + abc}{1 + d + eg + bcg + edg}$$

13. Consider the signal flow graph shown in Fig. 3.116. The gain C/R is __________ .

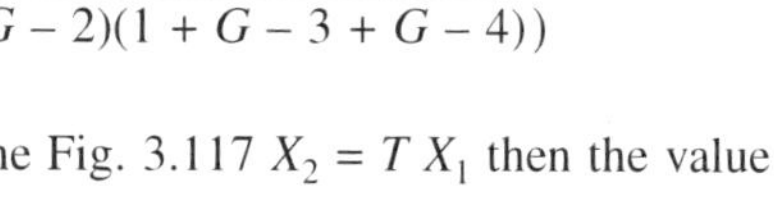

Fig. 3.116

Ans: $C/R = \dfrac{(G - 1\ G - 2\ G - 3\ G - 4)}{((1 + G - 1 + G - 2)(1 + G - 3 + G - 4))}$

14. In the SFG shown in the Fig. 3.117 $X_2 = T\ X_1$ then the value of T is __________ .

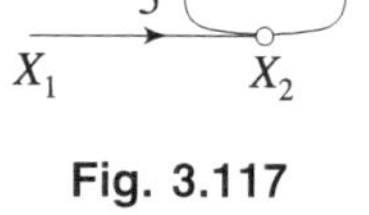

Fig. 3.117

Ans: 10

15. In the SFG the gain C/R will be __________ .

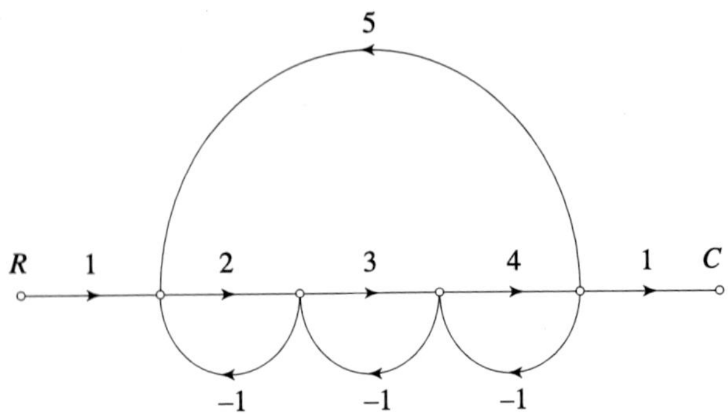

Fig. 3.118

Ans: 1.91

16. Consider the block diagram shown in Fig. 3.119 using the block diagram reduction technique, find *C*/*R*.

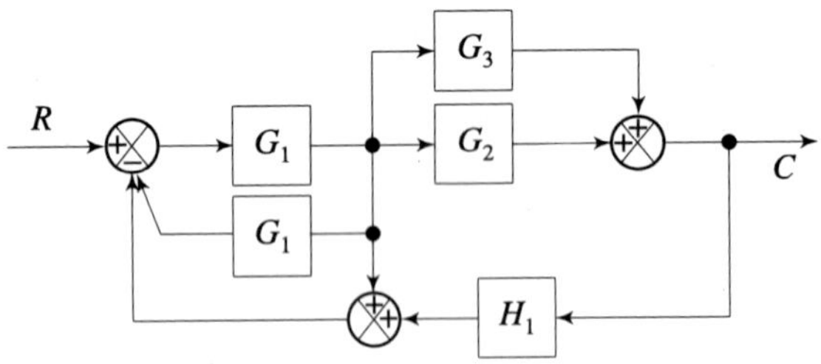

Fig. 3.119

Ans: $\dfrac{C}{R} = \dfrac{G_1G_2 + G_1G_3}{1 + G_1H_2 + G_1 + G_1G_2H_1 + G_1G_3H_1}$

17. Consider the block diagram shown in Fig. 3.120. Using the block diagram reduction technique, obtain *C*/*R*.

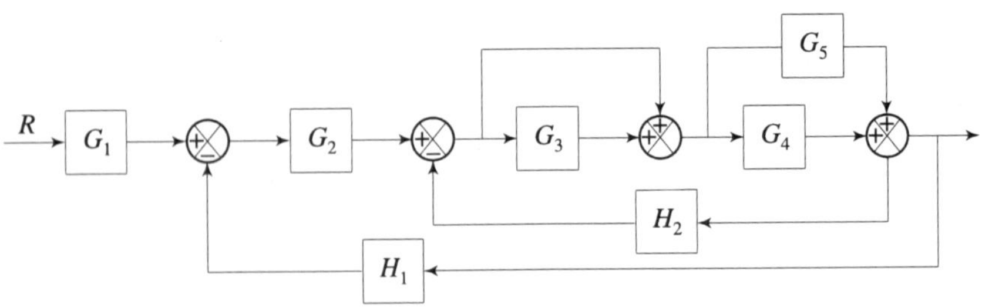

Fig. 3.120

Ans: $\dfrac{C}{R} = \dfrac{G_1G_2\,(1 + G_3)(G_4 + G_5)}{[1 + (1 + G_3)(G_4 + G_5)H_2 + (1 + G_3)(G_4 + G_5)\,G_2H_1]}$

18. For the block diagram shown in Fig. 3.121 obtain C/R.

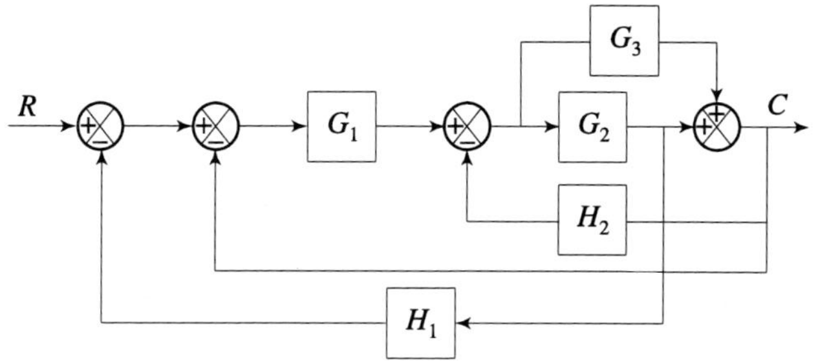

Fig. 3.121

$$\textit{Ans:}\ \frac{C}{R} = \frac{G_1G_2 + G_1G_3}{[1 + G_1G_2H_1 + G_1G_2 + G_1G_3 + G_2H_2 + G_3H_2]}$$

19. For the block diagram shown in Fig. 3.122 obtain C/R.

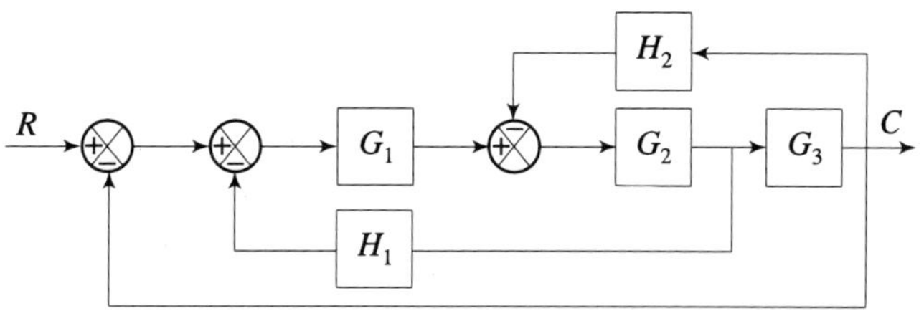

Fig. 3.122

$$\textit{Ans:}\ \frac{C}{R} = \frac{G_1G_2G_3}{1 + G_2G_3H_2 + G_1G_2H_1 + G_1G_2G_3}$$

20. Consider the system shown in Fig. 3.123. Obtain the transfer function by using Mason's gain formula.

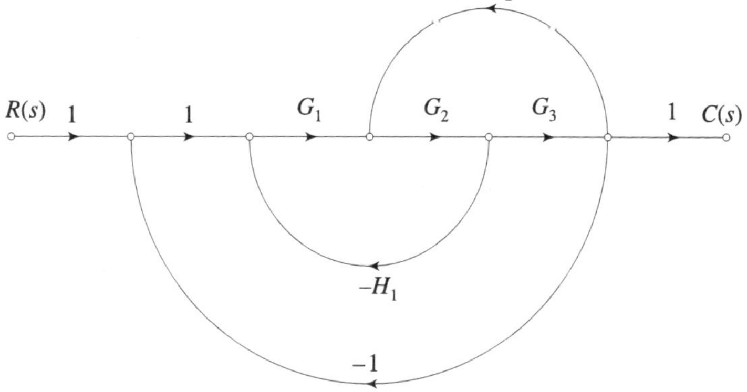

Fig. 3.123

Ans: $\dfrac{C(s)}{R(s)} = \dfrac{G_1G_2G_3}{1 + G_1G_2H_1 + G_2G_3H_2 + G_1G_2G_3}$

21. Convert the block diagram shown in Fig. 3.124 to signal flow graphs and find the transfer function of the system.

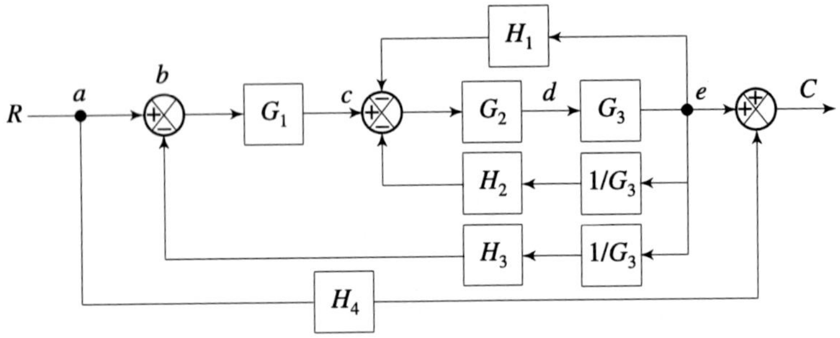

Fig. 3.124

Ans: $\dfrac{G_1G_2G_3 + G_4 + G_2G_3G_4H_1 + G_2G_4H_2 + G_1G_2G_4H_3}{1 + G_2G_3H_1 + G_2H_2 + G_1G_2H_3}$

22. Find the transfer function of the system whose signal flow graph is shown in Fig. 3.125.

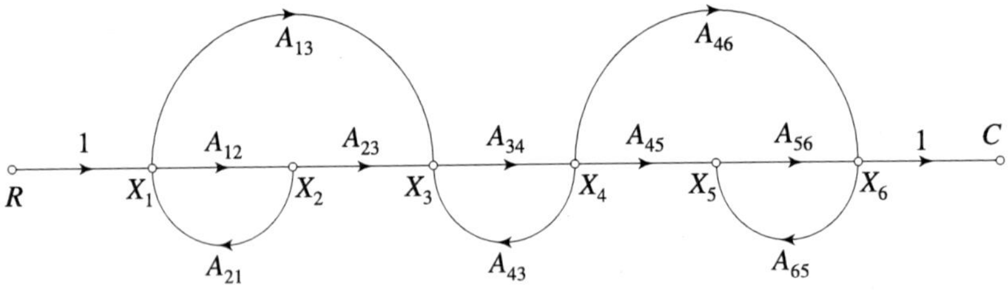

Fig. 3.125

Ans: $\dfrac{C}{R} = \dfrac{A_{12}A_{23}A_{34}A_{45}A_{56} + A_{13}A_{34}A_{45}A_{56} + A_{12}A_{23}A_{34}A_{46} + A_{13}A_{34}A_{46}}{\begin{array}{c}1 - (A_{12}A_{21} + A_{34}A_{43} + A_{56}A_{65}) + (A_{12}A_{21}A_{34}A_{43} \\ + A_{12}A_{21}A_{56}A_{65} + A_{34}A_{43}A_{56}A_{65}) - (A_{12}A_{21}A_{34}A_{43}A_{56}A_{65})\end{array}}$

23. Find the transfer function of the system having signal flow graph as shown in Fig. 3.126.

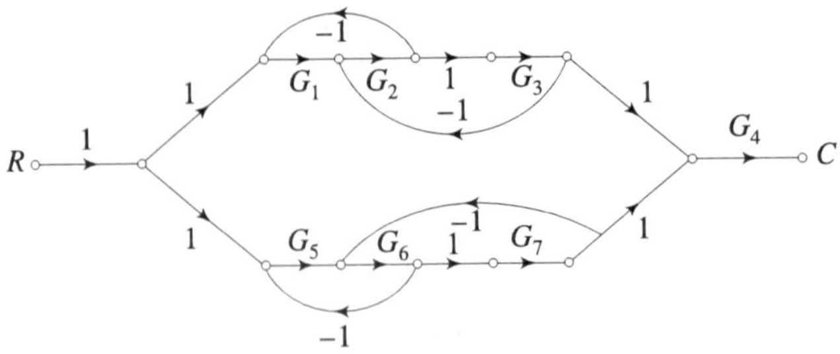

Fig. 3.126

Ans: $$\frac{C}{R} = \frac{G_1G_2G_3G_4\,(1 + G_5G_6 + G_6G_7) + G_4G_5G_6G_7\,(1 + G_1G_2 + G_2G_3)}{1 + G_1G_2 + G_2G_3 + G_5G_6 + G_6G_7 + G_1G_2G_5G_6 + G_5G_6G_2G_3 + G_1G_2G_6G_7 + G_2G_3G_6G_7}$$

24. Consider the signal flow graph shown in Fig. 3.127. Find $\frac{X_8}{X_1}$.

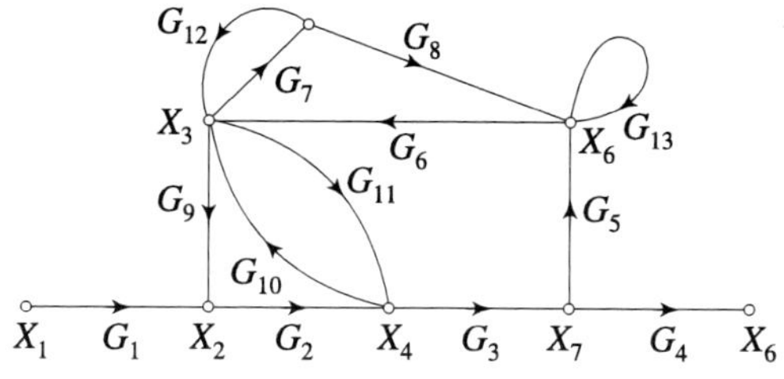

Fig. 3.127

Ans: $$\frac{X_8}{X_1} = \frac{[G_1G_2G_3G_4]\,[1 - (G_7G_{12} + G_6G_7G_8 + G_{13}) + G_7G_{12}G_{13}]}{1 - \{G_2G_9G_{10} + G_{10}G_{11} + G_2G_3G_5G_6G_9 + G_3G_5G_6G_{11} + G_7G_{12} + G_6G_7G_8 + G_{13}\} + G_2G_9G_{10}G_{13} + G_{10}G_{11}G_{13} + G_7G_{12}G_{13}}$$

25. Consider the signal flow graph shown in Fig. 3.128 obtain $\frac{X_8}{X_1}$ and $\frac{X_8}{X_2}$.

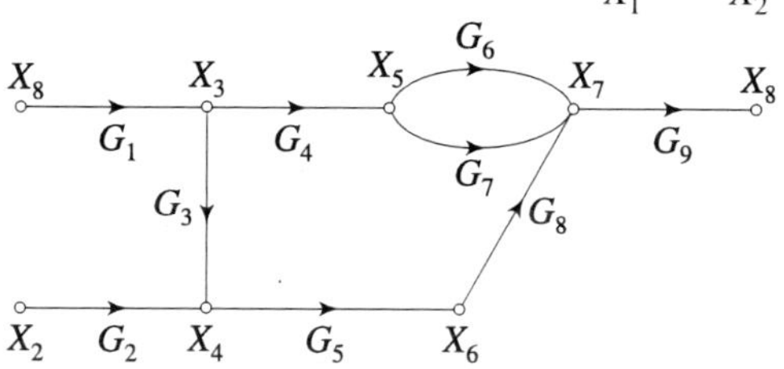

Fig. 3.128

Ans: $\frac{X_8}{X_1} = G_1G_4G_6G_9 + G_1G_4G_7G_9 + G_1G_3G_5G_8G_{10}$

$\frac{X_8}{X_2} = G_2G_5G_8G_9$

26. For the given signal flow graph, identify the number of forward paths and number of individual loops.

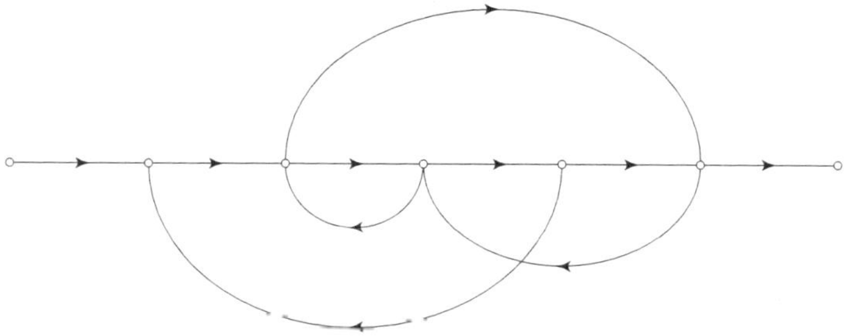

Fig. 3.129

Ans: Number of forward paths = 2

Number of individual loops = 4

27. Determine the transfer function $\frac{C(s)}{R(s)}$ for the block diagram as shown in Fig. 3.130.

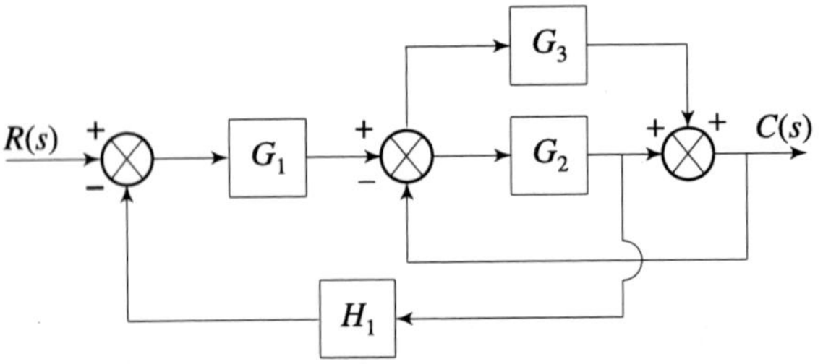

Fig. 3.130

Ans: $$\frac{C(s)}{R(s)} = \frac{G_1(G_2 + G_3)}{1 + G_2 + G_3 + H_1G_1G_2}$$

28. What do you mean by a block diagram? What is meant by summing point and take-off point? State the advantages of block diagram representation?

29. Reduce the block diagram shown in Fig. 3.131 and find the input-output relationship ratio, $\frac{C}{R}$.

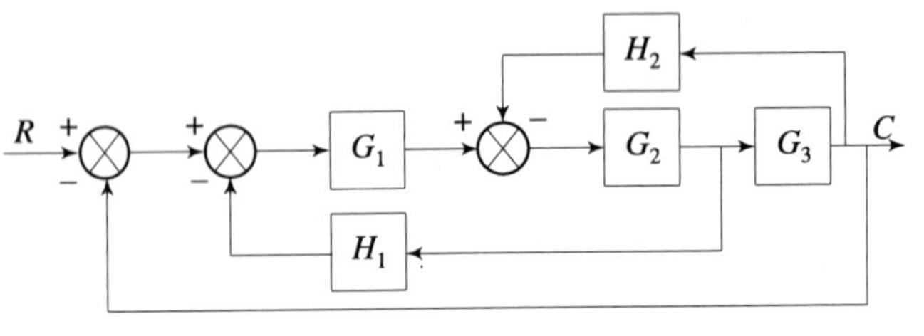

Fig. 3.131

Ans: $$\frac{C}{R} = \frac{G_1G_2G_3}{1 - G_1G_2H_1 + G_1G_3H_2 + G_1G_2G_3}$$

30. Find the equivalent transfer function $\frac{C(s)}{R(s)}$ for the system shown in Fig. 3.132.

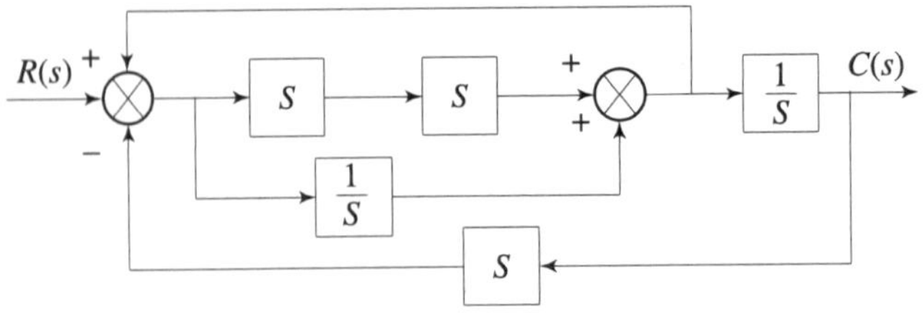

Fig. 3.132

Ans: $\dfrac{C(s)}{R(s)} = \dfrac{s^3 + 1}{2s^4 + s^2 + 2s}$

31. A closed-loop control system is subjected to a disturbance $D(s)$ as shown in Fig, 3.133. Determine the value of $C(s)$.

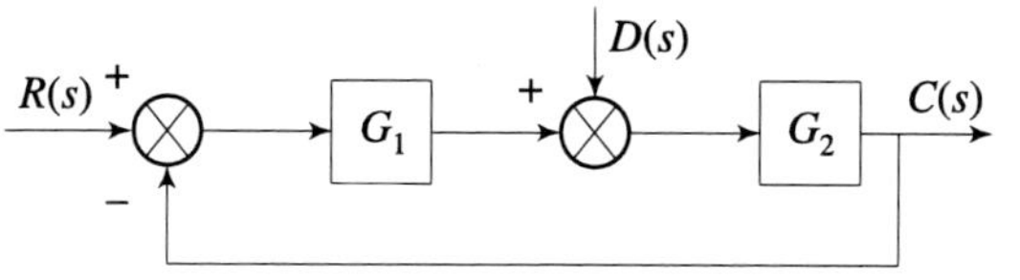

Fig. 3.133

Ans: $C(s) = \left(\dfrac{G_2}{1 + G_1 + G_2}\right)(G_1R(s) + D(s))$

32. A closed-loop control system is subjected to a disturbance $D(s)$ as shown in Fig. 3.134. Determine the output of the system.

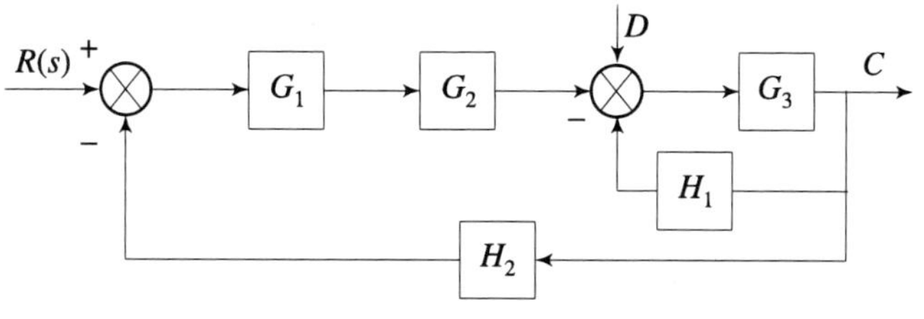

Fig. 3.134

Ans: $C = \dfrac{G_1G_2G_3R}{1 + G_3H_1 + G_1G_2G_3H_2} + \dfrac{G_3D}{1 + G_3H_1 + G_1G_2G_3H_2}$

33. Determine the output θ_0 for the system shown in Fig. 3.135.

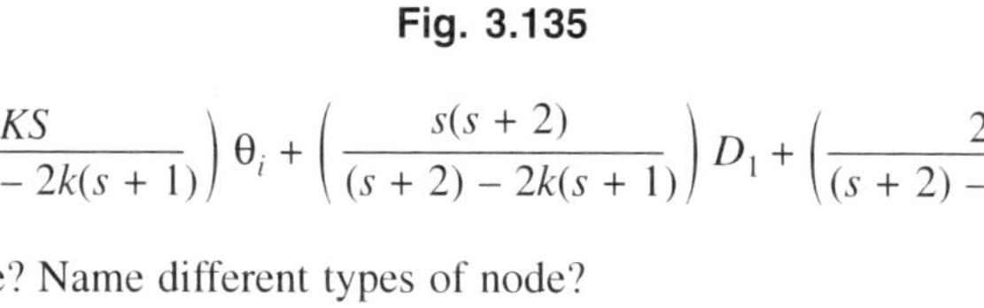

Fig. 3.135

Ans: $\theta_0 = \left[\left(\dfrac{KS}{(s + 2) - 2k(s + 1)}\right)\theta_i + \left(\dfrac{s(s + 2)}{(s + 2) - 2k(s + 1)}\right)D_1 + \left(\dfrac{2k}{(s + 2) - 2k(s + 1)}\right)D_2\right]$

34. What is a node? Name different types of node?
35. Define loop. Name different types of loops. What is meant by the term loop gain?
36. Explain the following terms

    (i) Path (ii) Forward path (iii) Forward path gain (iv) Feedback path

    (v) Loop gain

37. State the Mason's gain formula

38. A system having input and output represented by u and x is described by the following equations. Draw the signal flow diagram.

$$X = x_1 + a_3 u$$
$$X_1 = b_1 x_1 + x_2 + a_2 u$$
$$X_2 = -b_2 x_1 + a_1 u$$

Ans:

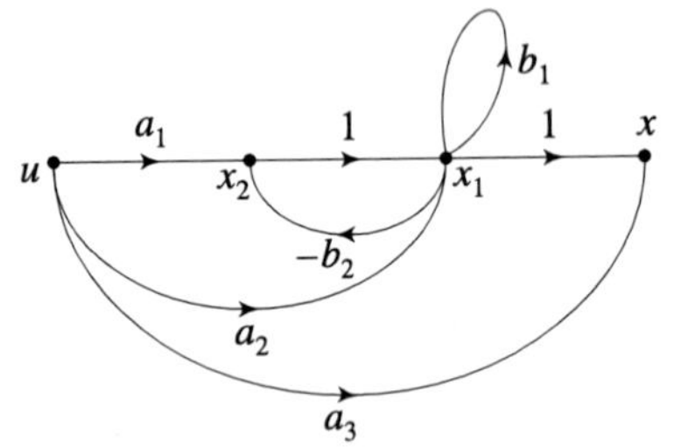

Fig. 3.136

39. Obtain the transfer function $\frac{C}{R}$ of signal flow graph shown in Fig. 3.137. Using Mason's gain formula.

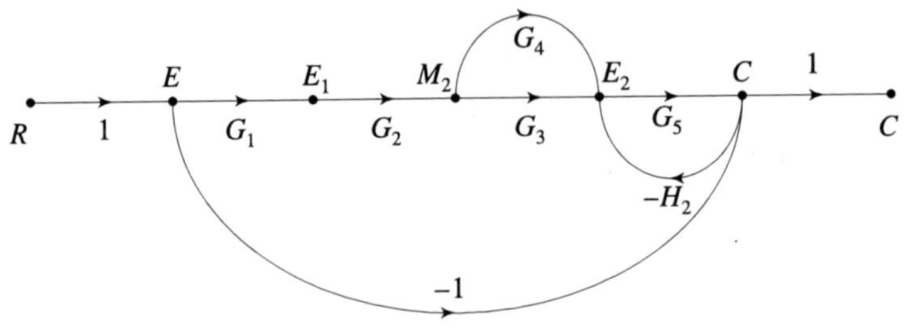

Fig. 3.137

Ans: $$\frac{C}{R} = \frac{G_1G_2G_3G_5 + G_1G_2G_4G_5}{1 + G_2H_1 + G_5H_2 + G_1G_2G_3G_5 + G_1G_2G_4G_5 + G_2G_5H_1H_2}$$

40. Obtain the transfer function $\frac{C}{R}$ of the signal flow graph shown in Fig. 3.138, using Mason's gain formula.

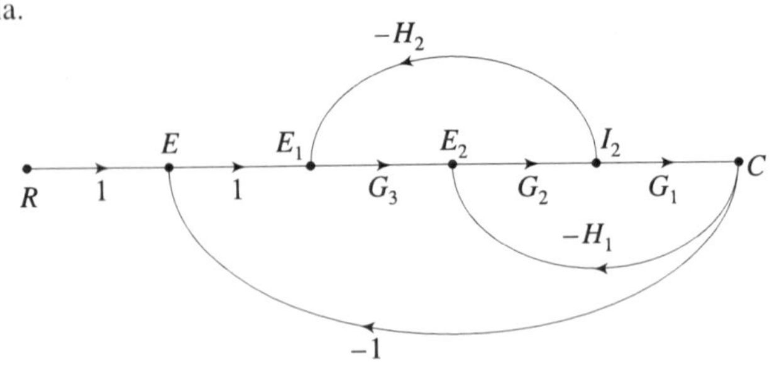

Fig. 3.138

Ans: $\frac{C}{R} = \frac{G_1G_2G_3}{1 + G_2G_3H_2 + G_1G_2H_1 + G_1G_2G_3}$

41. Convert the block diagram of Fig. 3.139 shown to a signal flow graph and determine the closed loop transfer function $\frac{C(s)}{R(s)}$ using Mason's gain formula.

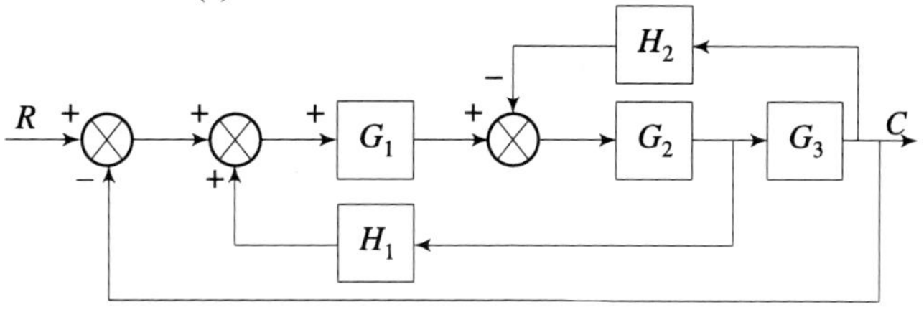

Fig. 3.139

Ans:

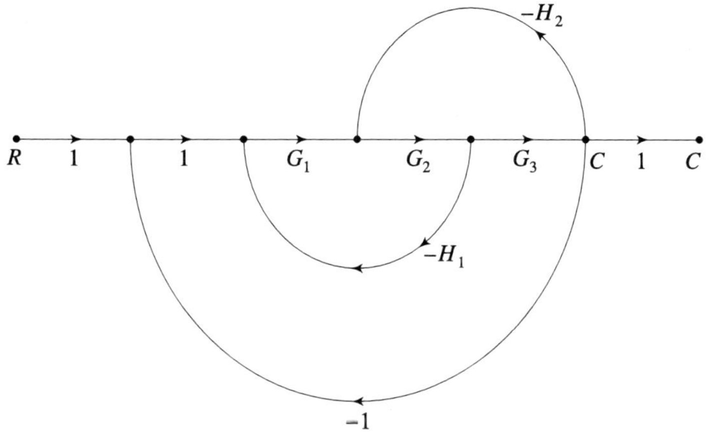

Fig. 3.140

$$\frac{C(s)}{R(s)} = \frac{G_1G_2G_3}{1 - G_1G_2H_1 + G_2G_3H_2 + G_1G_2G_3}$$

42. Determine the closed loop transfer function $C(s)/R(s)$ in the system shown in Fig. 3.141.

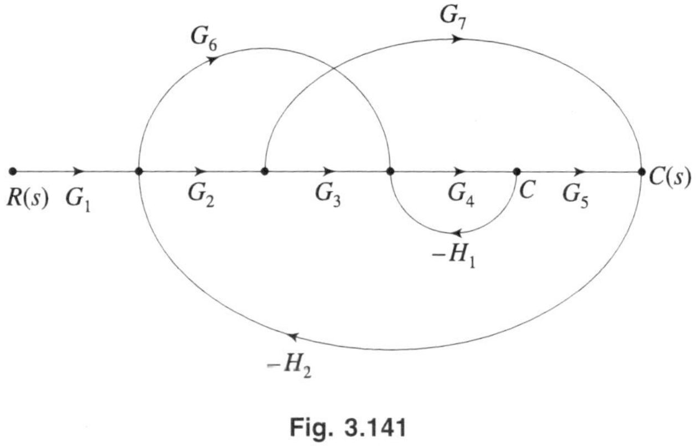

Fig. 3.141

Ans: $$\frac{C(s)}{R(s)} = \frac{G_1G_2G_3G_4G_5 + G_1G_6G_4G_5 + G_1G_2G_7(1 + G_4H_1)}{1 + G_4H_1 + G_2G_7H_2 + G_4G_5G_6H_2 + G_2G_3G_4G_5H_2 + G_4H_1G_2G_7H_2}$$

43. Determine the transfer function $\frac{Y(s)}{R(s)}$ from the signal flow graph shown in Fig. 3.142.

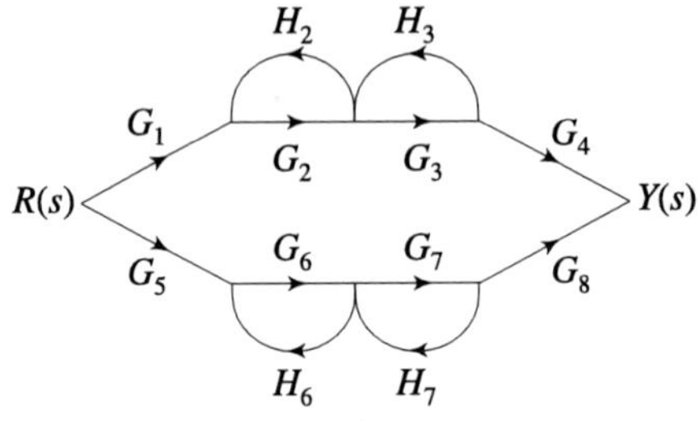

Fig. 3.142

Ans: $$\frac{Y(s)}{R(s)} = \frac{G_1G_2G_3G_4(1 - G_6H_6 - G_7H_7) + G_5G_6G_7G_8(1 - G_2H_2 - G_3H_3)}{1 - G_2H_2 - H_3G_3 - G_6H_6 - G_7H_7 + G_2H_2G_6H_6 + G_2H_2G_7H_7 + G_3H_3G_6H_6 + G_3H_3G_7H_7}$$

44. Draw the signal flow graph and evaluate the closed loop transfer function of a system whose block diagram is shown in Fig. 3.143.

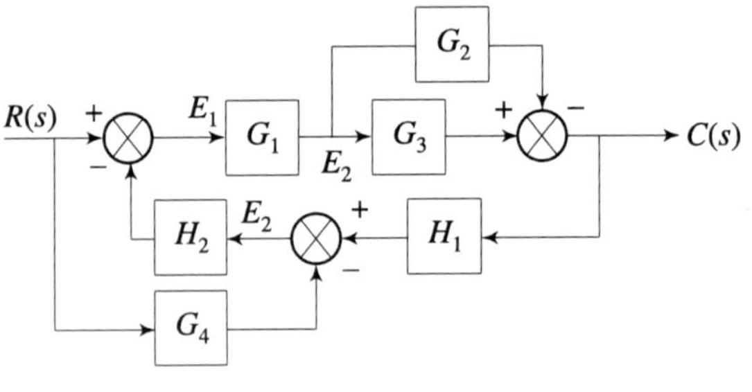

Fig. 3.143

Ans:

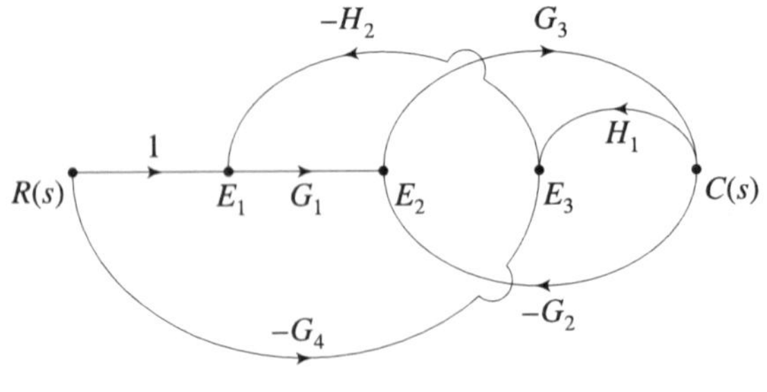

Fig. 3.144

$$\frac{C(s)}{R(s)} = \frac{-G_1G_2 + G_1G_3 - G_1G_2G_4H_2 + G_1G_3G_4H_2}{1 + G_1G_3H_1H_2 - G_1G_2H_1H_2}$$

45. Define the following:

(i) Block diagram (ii) Summing point (iii) Take-off paint

Ans: (i) The shorthand pictorial representation of the cause-and-effect relationship between the input and o/p of a physical system is known as a block diagram.

(ii) The point at which two or more signals can be added or subtracted is known as summing point.

(iii) The point at which output signal of any block can be applied to two or more points is known as take-off point.

46. Find the transfer function of the following

(i)

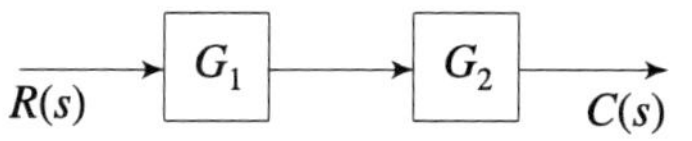

Fig. 3.145

Ans: G_1G_2

(ii)

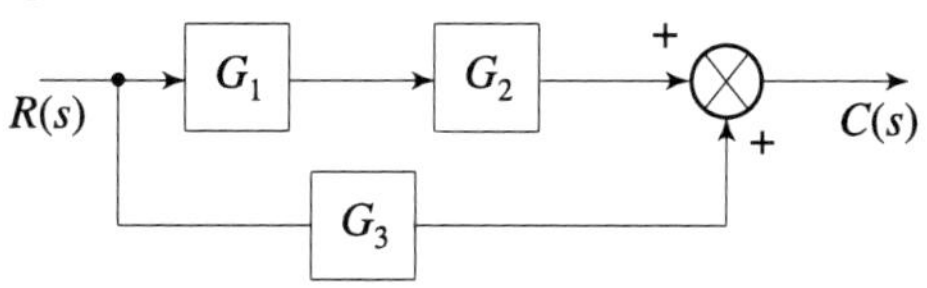

Fig. 3.146

Ans: $G_1G_2 + G_3$

(iii)

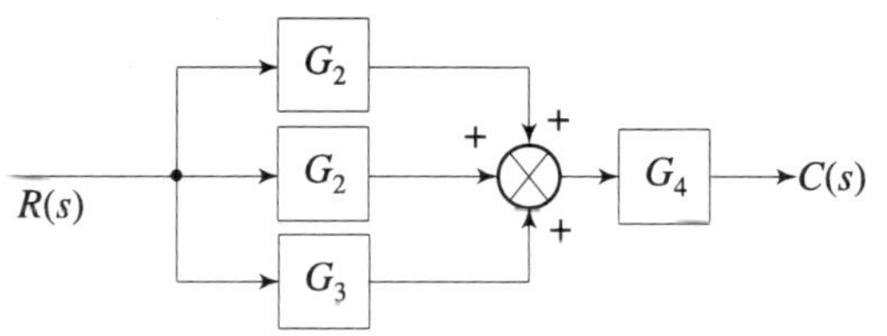

Fig. 3.147

Ans: $(G_1 + G_2 + G_3)\ G_4$

(iv)

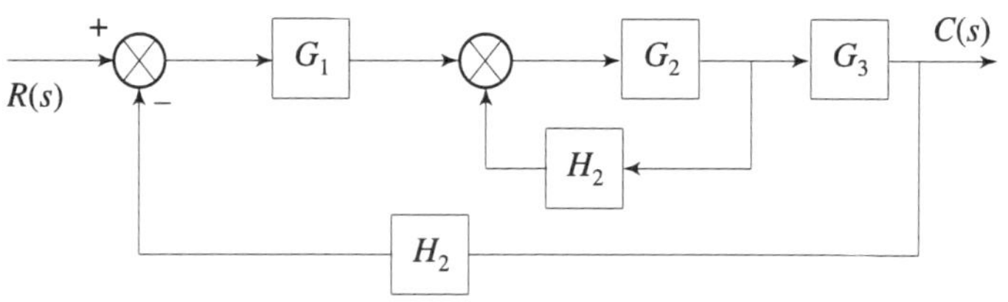

Fig. 3.148

Ans: $\dfrac{G_1G_2G_3}{1 + G_2H_1 + G_1G_2G_3H_2}$

47. Find the transfer function of the following block diagram.

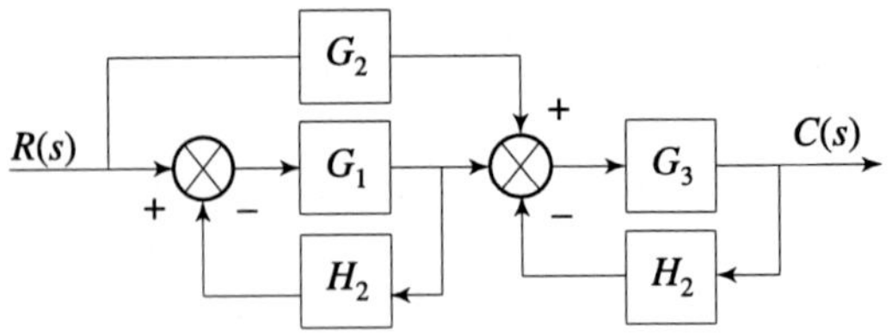

Fig. 3.149

Ans: $\dfrac{(G_1 + G_2 + G_1G_2H_1)G_3}{(1 + G_1H_1)(1 + G_3H_2)}$

48. Find the transfer function of the following block diagram.

Ans:

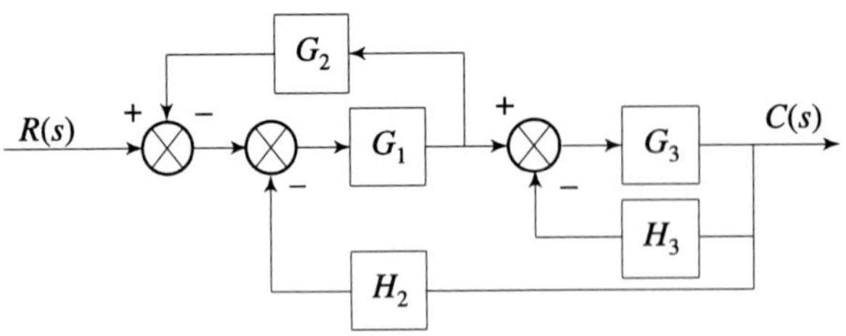

Fig. 3.150

Ans: $\dfrac{G_1G_3}{(1 + G_1H_1)(1 + G_3H_3) + G_1G_3H_2}$

49. Find the transfer function of the following diagram.

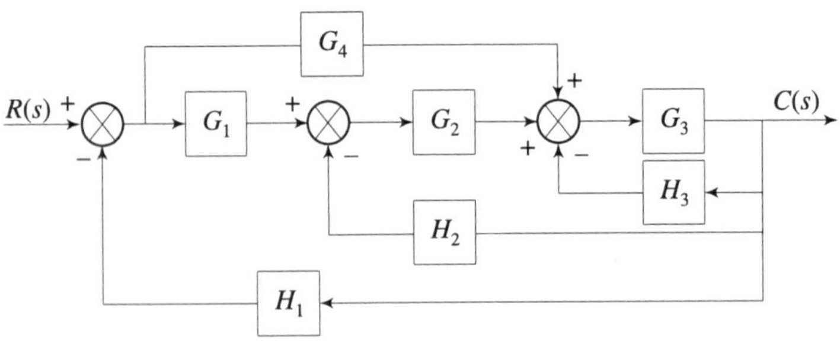

Fig. 3.151

Ans: $\dfrac{(G_1G_2 + G_4)G_3}{1 + G_1G_2G_3H_1 + G_4G_3H_1 + G_2G_3H_2 + G_3H_3}$

50. Determine the transfer function of the following block diagram.

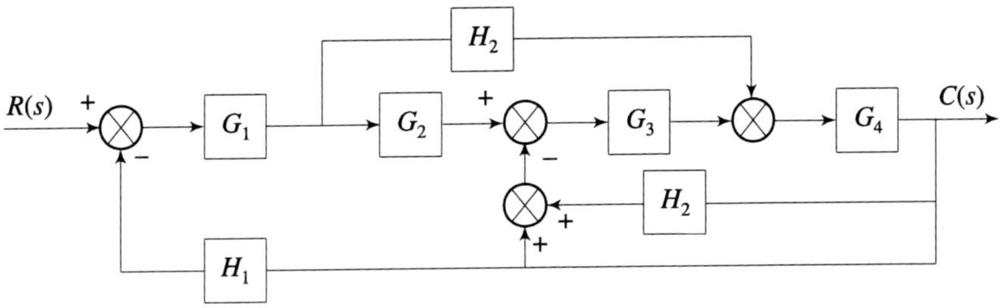

Fig. 3.152

Ans: $\dfrac{G_1G_2G_3G_2 + G_1G_4H_2}{1 + G_3G_4 + G_3G_4H_2 + G_1G_4H_1H_3 + G_1G_2G_3G_4H_1}$

51. The transfer function $\dfrac{C(s)}{R(s)}$ of the given system is equal to

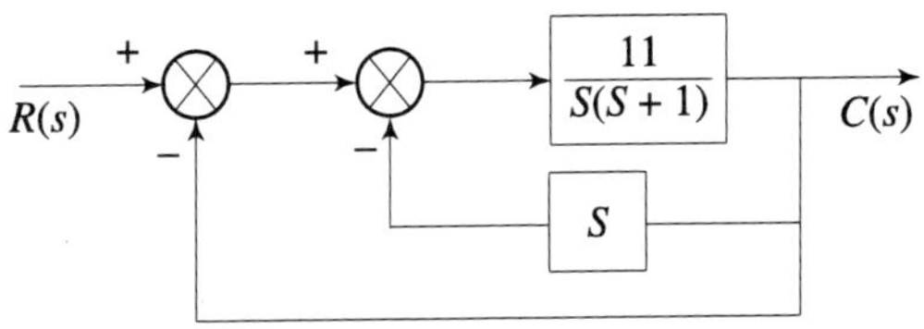

Fig. 3.153

Ans: $\dfrac{11}{s^2 + 11s + 11}$

52. The total gain of the system shown in Fig. 3.154 is __________ .

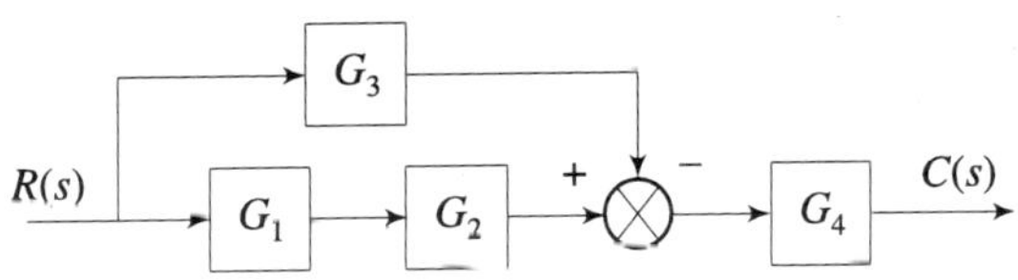

Fig. 3.154

Ans: $(G_1G_2 - G_3)G_4$

53. The given block diagram is equivalent to __________ .

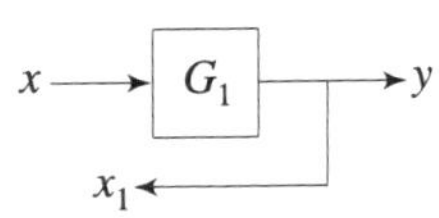

Fig. 3.155

Ans:

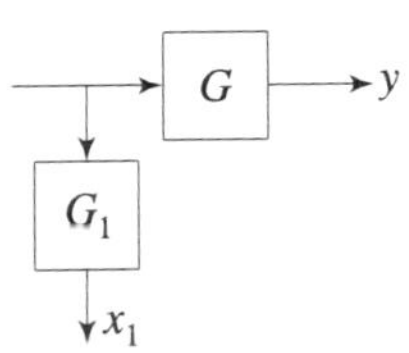

Fig. 3.156

54. Two equivalent block diagrams are shown below G_1 is equal to ________ .

Fig. 3.157

Ans: $\frac{1}{G}$

55. The transfer function of this system is ________ .

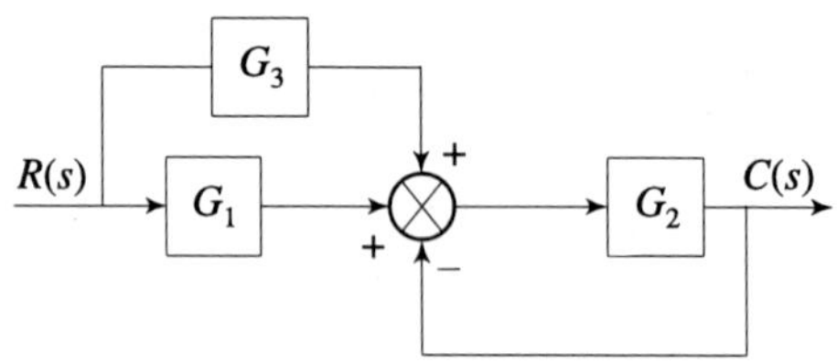

Fig. 3.158

Ans: $(G_1 + G_3)\left(\frac{G_2}{1 + G_2}\right)$

56. For a given block diagram, the output is equal to ________ .

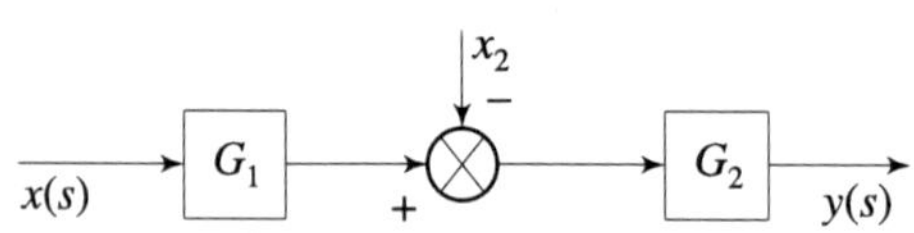

Fig. 3.159

Ans: $(G_1 x_1 - x_2)\, g_2$

57. The block diagram of Fig. 3.160(a) are equivalent if x in Fig. 3.160(b) is equal to ________ .

Ans:

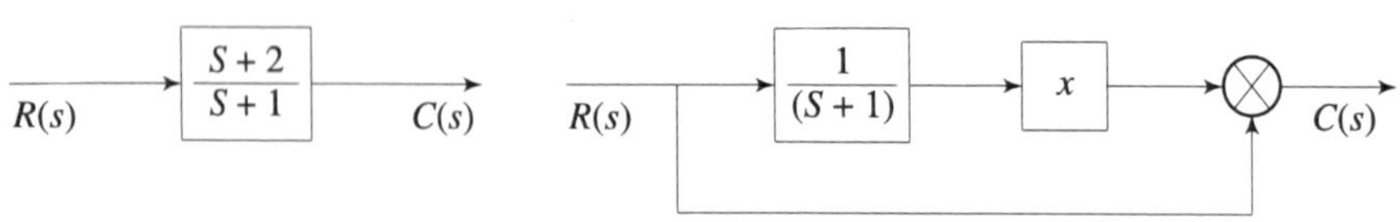

(a) (b)

Fig. 3.160

58. Find the transfer function of the following signal flow graphs.

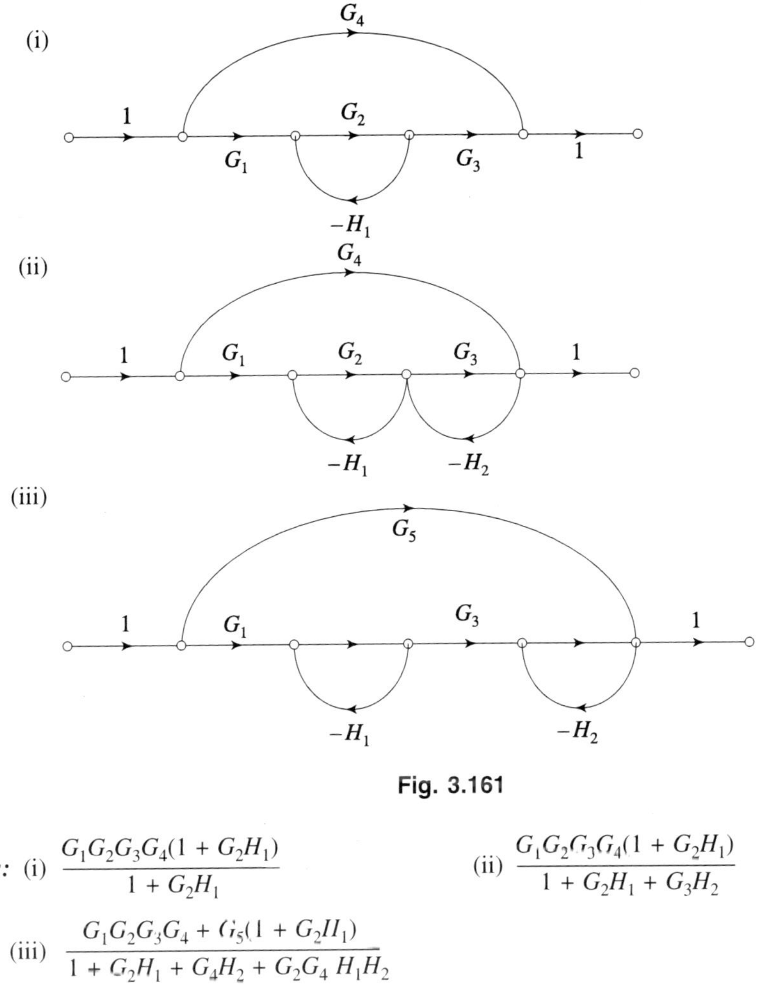

Fig. 3.161

Ans: (i) $\dfrac{G_1G_2G_3G_4(1 + G_2H_1)}{1 + G_2H_1}$ (ii) $\dfrac{G_1G_2G_3G_4(1 + G_2H_1)}{1 + G_2H_1 + G_3H_2}$

(iii) $\dfrac{G_1G_2G_3G_4 + G_5(1 + G_2H_1)}{1 + G_2H_1 + G_4H_2 + G_2G_4\,H_1H_2}$

59. Find the transfer function of signal-flow graph.

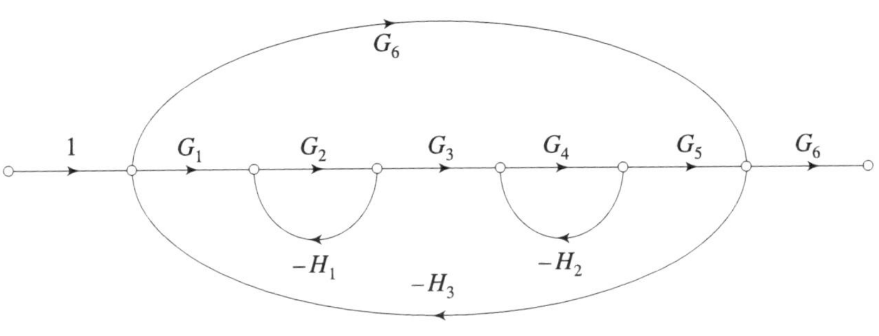

Fig. 3.162

Ans: $$\frac{G_1G_2G_3G_4G_5 + G_6(1 + G_1H_1 + G_4H_2 + G_2G_4\,H_1H_2)}{1 + G_1H_1 + G_4H_2 + G_1G_2G_3G_4\,G_5H_3 + G_6H_3 + G_2G_4H_1H_2 + G_1G_6H_1H_3 + G_4G_6H_2H_3}$$

60. Determine the transfer function.

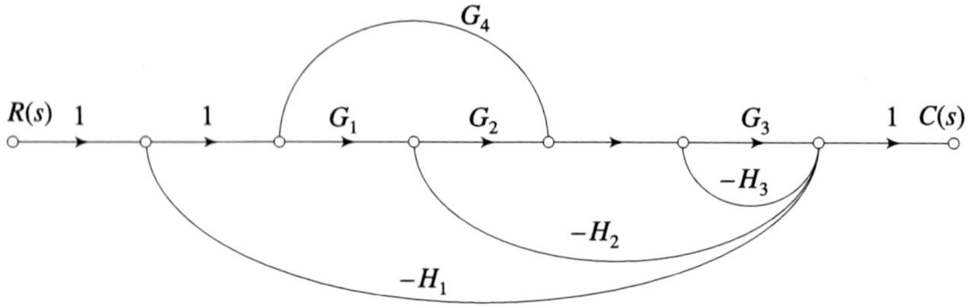

Fig. 3.163

Ans: $$\frac{(G_1G_2 + G_4)G_3}{1 + G_1G_2G_3H_1 + G_4G_3H_1 + G_2G_3H_2 + G_3H_3}$$

61. Determine the transfer function of system.

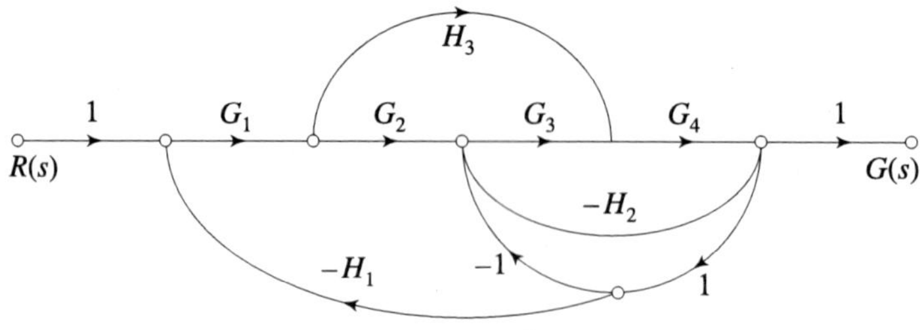

Fig. 3.164

Ans: $$\frac{G_1G_2G_3G_4 + G_1G_4H_3}{1 + G_3G_4 + G_3G_4H_2 + G_1G_4H_1H_2 + G_1G_2G_3G_2H_1}$$

62. 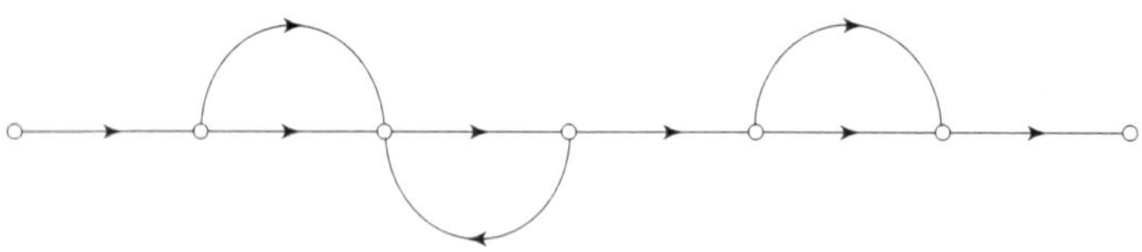

Fig. 3.165

The number of forward path of the given SFG is __________ .

Ans: 7

63. The number of individual loops in the signal-flow graph is __________ .

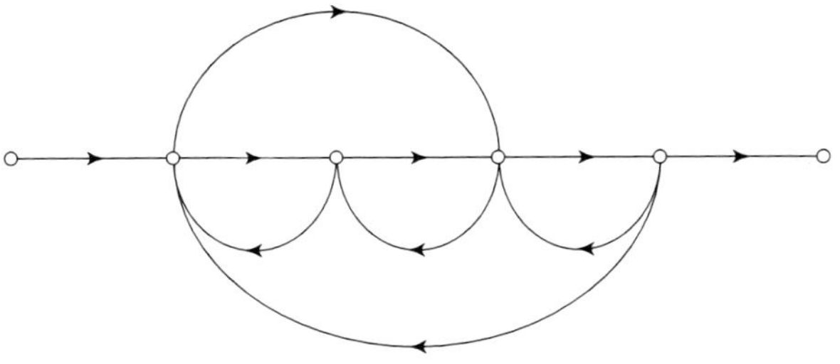

Fig. 3.166

Ans: 6

64. The number of forward paths given by

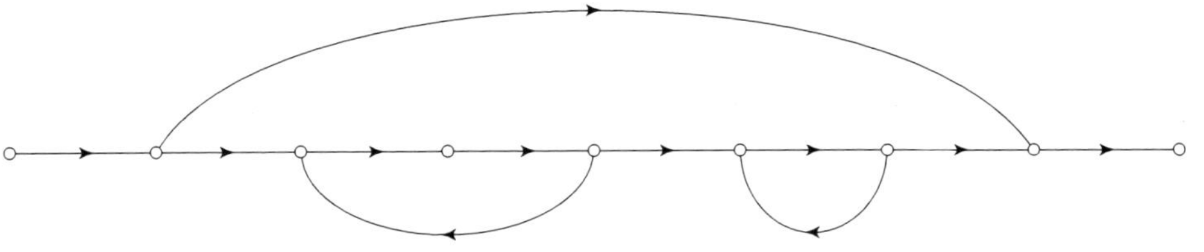

Fig. 3.167

Ans: 2

65. The transfer function of this signal How graph is __________ .

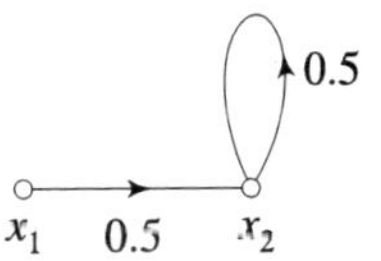

Fig. 3.168

Ans: 12

66. The value of x_3/x_1 is _______ .

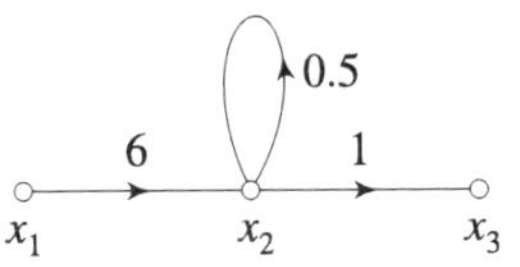

Fig. 3.169

Ans: 12

67. The representation of $\frac{C(s)}{R(s)} = \frac{a_0}{b_0} \frac{(s + c_i)}{(s + d_i)}$ in static diagram will be __________ .

Ans:

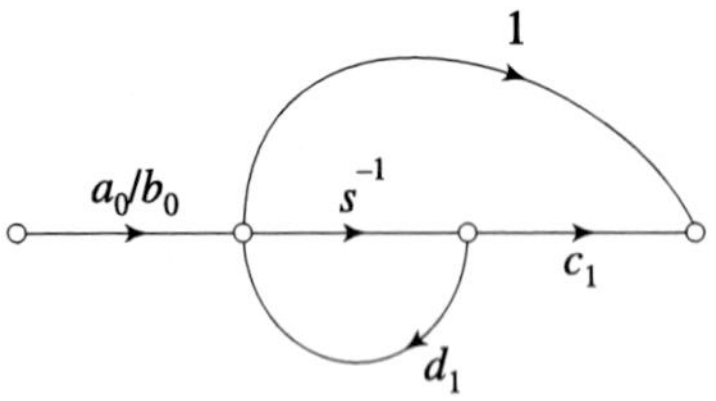

Fig. 3.170

68. The signal flow graph has __________ forward paths and __________ non-touching loops.

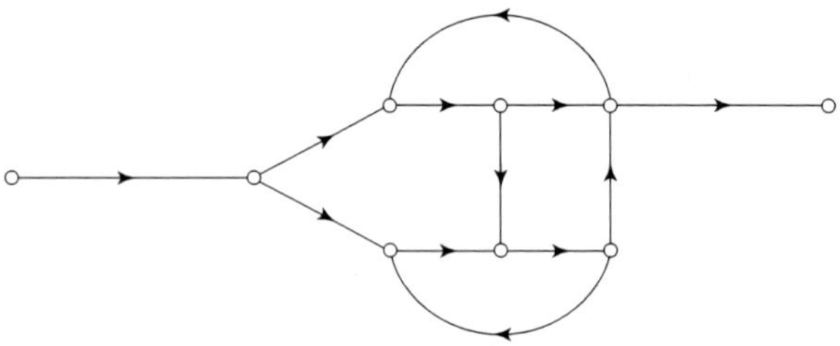

Fig. 3.171

Ans: 3 forward paths and 2 non-touching loops

69. The overall transfer function of the system will be __________ .

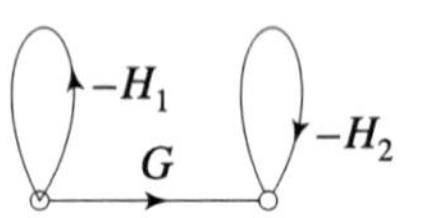

Fig. 3.172

Ans: $\dfrac{C}{R} = \dfrac{G}{(1 + H_1)(1 + H_2)}$

4 Time Response of Feedback Control Systems

4.1 INTRODUCTION

In time domain analysis, time is an independent variable when a system is given an excitation, there is a response (output). This response varies with time and is called the time response. Time domain analysis means the study of output or response of the system as function of time for a given input signal. The time domain analysis can be classified into two parts, that is, transient response and steady state response.

Transient response is defined as that part of the response which goes to zero as time t reaches infinity. It remains only for some time from initial state to final state.

Steady state response is defined as the response of the system as time tends to infinity.

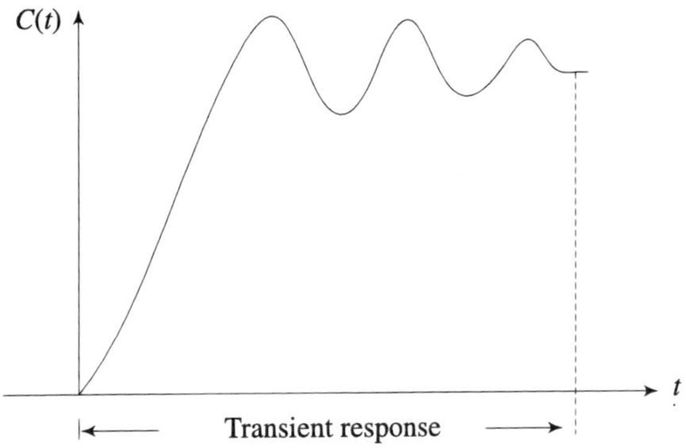

$$\lim_{t \to \infty} C_t(t) = 0$$

Fig. 4.1

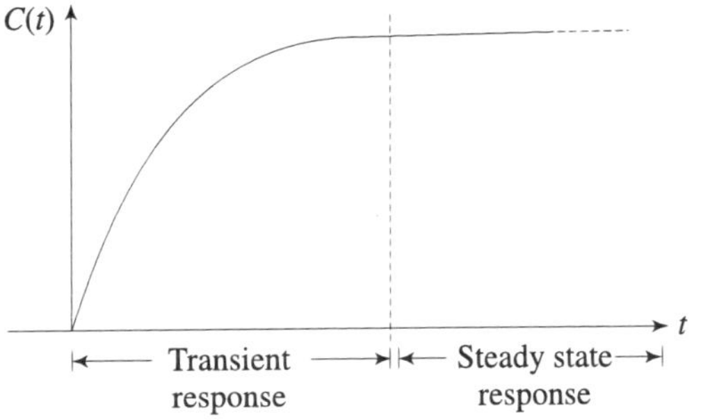

Fig. 4.2

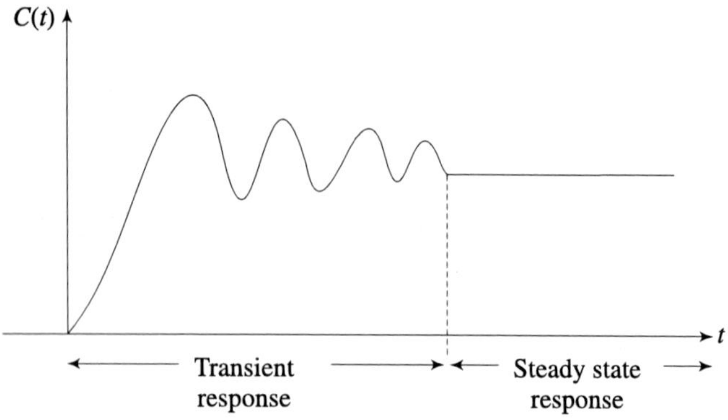

Fig. 4.3

$$Css(t) = \lim_{t \to \infty} C(t)$$

The system shows transient response because of the presence of energy storing elements such as mass, capacitance or inductance in the system. In the absence of energy storing elements the system response consists only of steady state portion.

In many control systems the input to the system is not known in advance. For example, in radar tracking systems the position of the target is one of the inputs to the system. But this input signal changes in random manner. Therefore, there is a need for some standard test input signals.

The test signal to be used for the purpose depends on the nature of input to which the system is subjected to during its operation in the field.

4.2 STANDARD TEST INPUT SIGNALS

4.2.1 Unit Step Input (Step Displacement Input)

This is the sudden change in the reference input signal. Mathematically represented as

$$u(t) = 1, \ t > 0$$

$$= 0, \ t < 0$$

If $r(t) = A\ u(t)$, where A is a constant.

i.e. $$u(t) = A, \ t > 0$$

$$= 0, \ t < 0$$

Laplace transform of $u(t)$, i.e., $L[u(t)] = A \cdot \frac{1}{S}$.

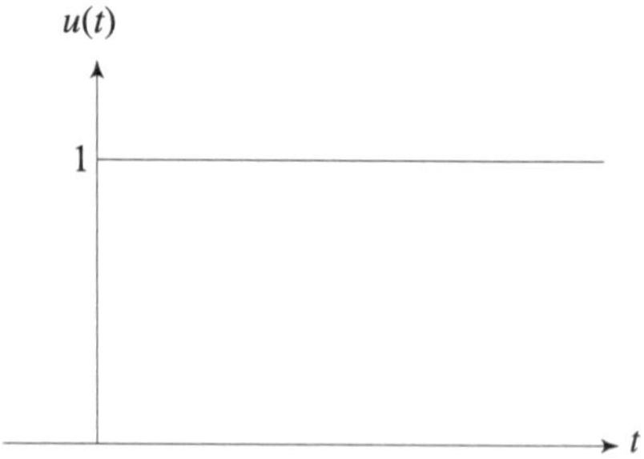

Fig. 4.4

4.2.2 Unit Ramp Function (Step Velocity Input)

If the reference input signal is changing gradually at a constant rate then such a signal is represented by a ramp function.

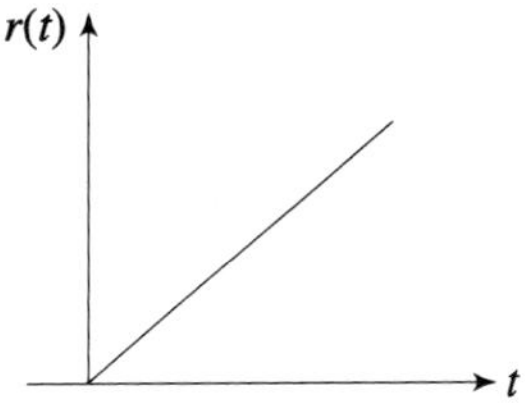

Fig. 4.5

Mathematically represented by

$$r(t) = At, \; t > 0$$
$$= 0, \quad t < 0$$

or
$$r(t) = A \cdot tu(t)$$

Laplace transform $L[r(t)] = A\,\dfrac{1}{S^2}.$

4.2.3 Unit Parabolic Function (Acceleration Input)

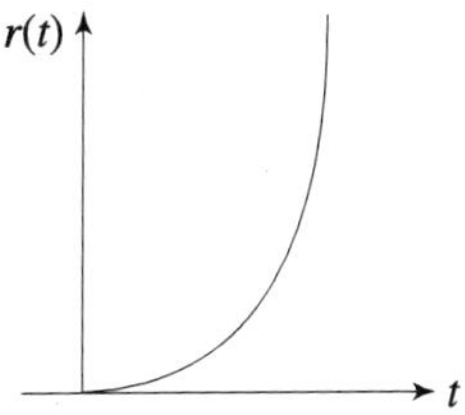

Fig. 4.6

$$r(t) = \frac{A}{2}\,t^2, \; t > 0$$
$$= 0, \qquad t < 0$$

or
$$r(t) = \frac{A}{2}\,t^2\,u(t)$$

$$L[r(t)] = L\left[\frac{t^2}{2}\right] = \frac{1}{S^3}.$$

Impulse input: If the reference input signal of a large amplitude is applied for very short duration, then such an input is represented by an impulse function.

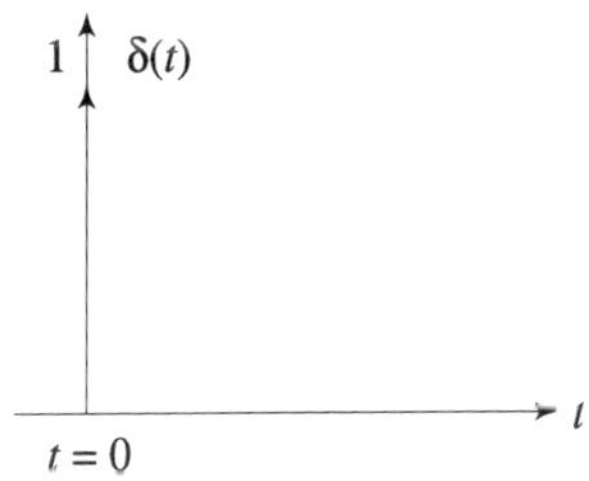

Fig. 4.7

$$\delta(t) = 1, \quad t = 0$$
$$= 0, \quad t \neq 0$$

and
$$\int_{-\infty}^{\infty} \delta(t)\cdot dt = 1$$

Then $\quad L.T.\,[\delta(t)] = 1.$

From the transient response, we can know

(i) When the system begins to respond after an input is given.

(ii) How much time it takes to reach the output for the first time.

(iii) Whether the output oscillates about its final value.

(iv) When does it settle to the final value.

The above information is translated to 'performance index' and will form an important requirement in design.

Steady State Response

(i) How long it took before steady state was reached.

(ii) Whether there is any error between the desired and actual values.

(iii) Whether this error is constant, zero or infinite.
(iv) It judges the accuracy of the system output.

Derivation of Steady State Error

Steady state response is important to judge the accuracy of the output.

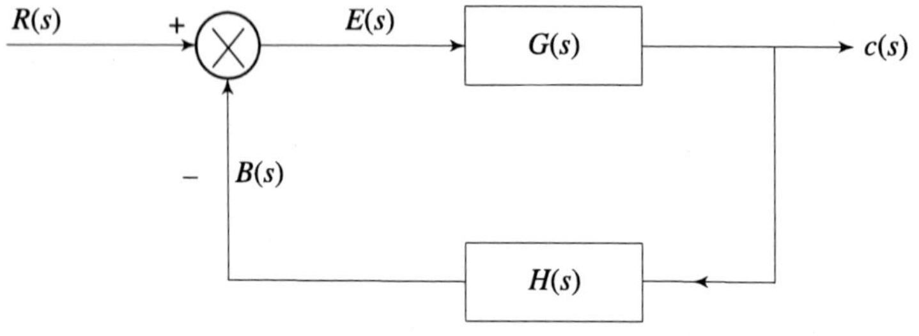

Fig. 4.8

From Fig. 4.8

Steady state error

$$E(s) = R(s) - B(s)$$

$$B(s) = C(s) \cdot H(s)$$

$\therefore$ $$E(s) = R(s) - C(s)\,H(s)$$

Also $$C(s) = G(s)\,E(s)$$

$\therefore$ $$E(s) = R(s) - G(s)\,E(s)\,H(s)$$

$\therefore$ $$E(s)\,[1 + G(s)\,H(s)] = R(s)$$

$$E(s) = \frac{R(s)}{1 + G(s)\,H(s)}$$

If positive feedback is used then,

$$E(s) = \frac{R(s)}{1 - G(s) \cdot H(s)}$$

$\therefore$ $$E(s) = \frac{R(s)}{1 \pm G(s) \cdot H(s)}$$

For a unity feedback system $H(s) = 1$

$\therefore$ $$E(s) = \frac{R(s)}{1 \pm G(s)}.$$ In time domain $e(t) = L^{-1}[E(s)]$

$\therefore$ This is the expression of errors valid for all time. Steady state error is defined as

$$e_{ss} = \underset{t \to \infty}{\text{Lt}}\; e(t)$$

$$= \underset{s \to 0}{\text{Lt}}\; S.E(s)$$ $\because$ From the final value theorem in *L.T.*

$$e_{ss} = \underset{s\to 0}{\text{Lt}}\ S.\ \frac{R(s)}{1 + G(s)\cdot H(s)}$$

Unity gain system, $e(t) = r(t) - c(t)$.

Type of a system: The number of poles of a system having open-loop transfer function $G(s)$ $H(s)$ specifies the type of the system.

4.3 UNIT STEP RESPONSE OF FIRST ORDER SYSTEM

A system is said to be of first order if the highest power of S in the denominator polynomial of the transfer function is unity.

Consider the I (first) order system shown in Fig. 4.9.

$$\text{R(S)} \rightarrow \boxed{\frac{1}{1+\text{TS}}} \rightarrow \text{C(s)}$$

Fig. 4.9

The transfer function of the system is $G(s) = \dfrac{C(s)}{R(s)} = \dfrac{1}{Ts + 1}$

The transfer function of the system is $M(s) = \dfrac{C(s)}{R(s)} = \dfrac{1}{1 + TS}$

For the unit step input $r(t) = u(t)$

$$R(s) = L.T.[r(t)] = \frac{1}{S}$$

The output response is given by $C(s) = R(s)\cdot M(s) = \dfrac{1}{S(Ts + 1)}$

On partial fraction expansion,

$$C(s) = \frac{1}{S(Ts + 1)} = \frac{A}{S} + \frac{B}{Ts + 1}$$

$$1 = A\ (Ts + 1) + BS$$

$$S = 0,\ 1 = A$$

$$S = \frac{1}{T},\ 1 = A\left(T\left(-\frac{1}{T}\right) + 1\right) + B\left(-\frac{1}{T}\right)$$

$$1 = -\frac{1}{T}\cdot B$$

$$B = -T$$

$$C(s) = \frac{1}{S} - \frac{T}{Ts + 1}$$

Taking inverse $L.T. \Rightarrow C(t) = 1 - e^{-t/T},\ t \geq 0$

Let us plot some values of $C(t)$

t	$C(t)$
T	$C(t) = 1 - e^{-1} = 0.632$
$2T$	0.86
$3T$	0.95
$4T$	0.982
$5T$	0.993
∞	1

Here the output $C(t)$ is initially zero and exponentially reaches a final value of unity as shown in Fig. 4.10.

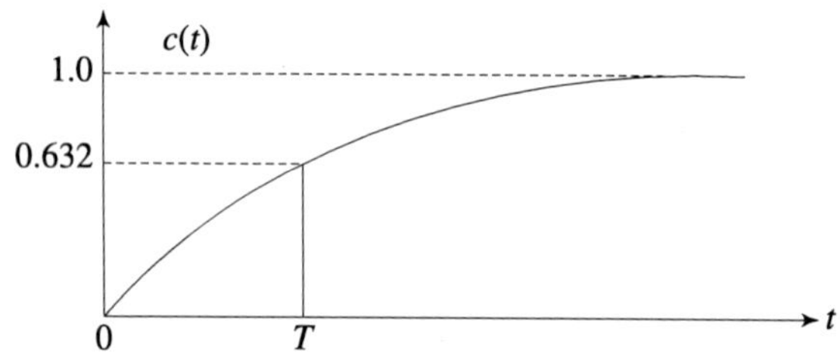

Fig. 4.10

An important characteristic of the response is that at $t = T$, $C(t) = 0.632$ or the response reaches 63.2% of the final value. Here time constant T is the time required for the output to reach 63.2% of its final value.

If a system has a small time constant then it has fast response.

The initial slope of the curve at $t = 0$ is given by

$$\frac{dc(t)}{dt} = \frac{1}{T} e^{-t/T}\Big|_{t=0} = \frac{1}{T}$$

Steady state error

$$e(t) = r(t) - C(t)$$
$$= 1 - (1 - e^{-t/T})$$
$$= 1 - 1 + e^{-t/T}$$
$$e(t) = e^{-t/T}$$
$$e_{ss} = \underset{t\to\infty}{\text{Lt}}\ e(t) = \underset{t\to\infty}{\text{Lt}}\ e^{-t/T} = 0$$

Note:

(i) Steady state error is zero.

(ii) First order system is capable of responding to a unit step unit without any steady state error.

(iii) The speed of response depends on constant [T], smaller the time constant, larger the response.

Time response of first order system for unit impulse input

$$C(s) = \frac{1}{1 + Ts} R(s)$$

$$r(t) = \delta(t)$$

$$R(s) = 1$$

$$C(s) = \frac{1}{Ts + 1} \cdot 1 = \frac{\left(\frac{1}{T}\right)}{(S + 1/T)} \quad \text{...(4.1)}$$

Taking Laplace inverse $C(t) = \frac{1}{T} \cdot e^{-1/T}$

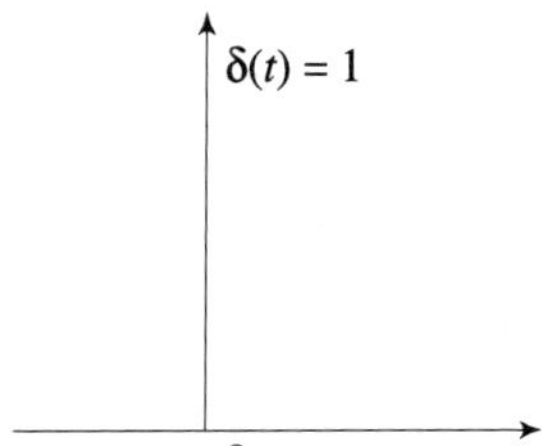

Fig. 4.11

$$e(t) = r(t) - C(t)$$

$$= \delta(t) - \frac{1}{T} e^{-t/T}$$

$$e_{ss} = \underset{t \to \infty}{\text{Lt}} \left[\delta(t) - \frac{1}{T} e^{-t/T}\right)$$

$$= \delta(\infty) - \frac{1}{T} e^{\infty}$$

$$= 0 - 0$$

$$e_{ss} = 0$$

Time response of first order system for unit ramp input

$$C(s) = \frac{1}{1 + Ts} \cdot R(s)$$

$$r(t) = t \cdot u(t)$$

$$R(s) = \frac{1}{S^2}$$

$$C(s) = \frac{1}{1 + Ts} \cdot \frac{1}{S^2} = \left\{\frac{\frac{1}{T}}{S + \frac{1}{T}}\right\} \cdot \frac{1}{S^2}$$

$$= \frac{A}{S} + \frac{B}{S^2} + \frac{C}{S + \frac{1}{T}}$$

$\therefore$ $$B = C(S) \cdot S^2 \Big|_{S^2 = 0} = \frac{\left(\frac{1}{T}\right)}{\left(S + \frac{1}{T}\right)} \cdot \frac{S^2}{S^2}\Big|_{S=0} = 1$$

$\therefore$ $$B = 1$$

$$C = C(S)\cdot\left(S+\frac{1}{T}\right)\Big|_S = -\frac{1}{T}$$

$$= \frac{\frac{1}{T}}{\left(S+\frac{1}{T}\right)}\cdot\left(S+\frac{1}{T}\right)\Big|_S = -\frac{1}{T}$$

$$C = T$$

$$A = -T,\ B = 1,\ C = T$$

$$\therefore \qquad C(S) = \frac{-T}{S} + \frac{1}{S^2} + \frac{T}{S+\frac{1}{T}}$$

Taking inverse L.T.

$$C(t) = -T + t\cdot u(t) + T\cdot e^{-t/T}$$

$$= T(e^{-t/T} - 1) + t$$

$$= t - T(1 - e^{-t/T})$$

The error signal

$$e(t) = r(t) - C(t)$$

$$= t - \left\{t - T(1 - e^{-t/T})\right\}$$

$$= T(1 - e^{-t/T})$$

Steady state error

$$e_{ss} = \underset{t\to\infty}{\text{Lt}}\ e(t) = \underset{t\to\infty}{\text{Lt}}\ T(1 - e^{-t/T})$$

$$= T - e^{\infty} = T - 0 = T$$

$$\therefore \qquad e_{ss} = T$$

The first order system has a steady state error of T, when it is subjected to unit ramp input.

$$C(t) = t - T(1 - e^{-t/T})$$

If T is small, $-t/T$ increases, $e^{-t/T}$ increases

e^{-St} is small, e^{-10t} is large

Smaller the time constant speed of response is more.

Note:

Unit step $C(t) = 1 - e^{-t/T}$, $ess = 0$

Unit impulse $C(t) = \frac{1}{T}e^{-t/T}$, $ess = 0$

Unit ramp $C(t) = t - T(1 - e^{-t/T})$, $ess = T$.

4.4 TIME RESPONSE OF SECOND ORDER SYSTEM

Consider the second order system as shown in Fig. 4.12.

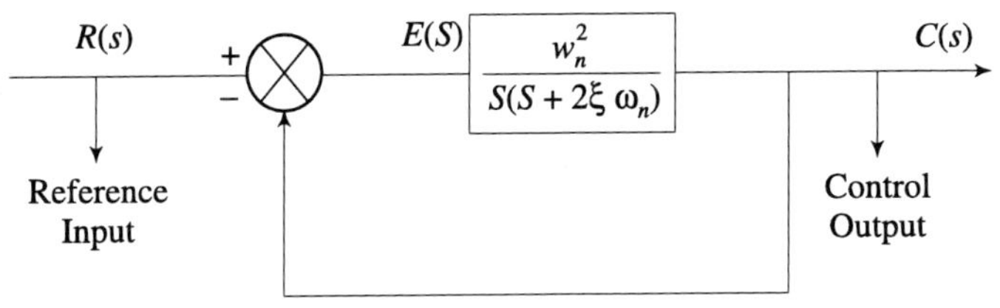

Fig. 4.12

A system is said to be of second order if the highest power of S in the denominator polynomial of the transfer function is two.

The overall transfer function of the system is

$$\frac{C(s)}{R(s)} = \frac{\dfrac{\omega_n^2}{S(S + 2\zeta\omega_n)}}{1 + \dfrac{\omega_n^2}{S(S + 2\zeta\omega_n)}} = \frac{G(s)}{1 + G(s)H(s)}$$

$$= \frac{\omega_n^2}{S(S + 2\zeta\omega_n + \omega_n^2)}$$

$$\frac{C(s)}{R(s)} = \frac{\omega_n^2}{S^2 + 2\zeta\omega_n S + \omega_n^2} \text{ is known as standard form}$$

where ω_n = natural undamped frequency.

ζ = damping ratio of the system.

$\sigma = \zeta\omega_n$ = damping factor.

$\omega_d = \omega_n\sqrt{1 - \zeta^2}$ = damped frequency.

Characteristic equation:

$$1 + G(s)\,H(s) = 0$$

$$\therefore \quad S^2 + 2\,\zeta\omega_n s + \omega_n^2 = 0$$

Poles of transfer function are same as the roots of characteristic equation.

$\zeta = 0$ undamped condition

$0 < \zeta < 1$ underdamped condition

$\zeta > 1$ overdamped condition

$\zeta = 1$ critically damped.

- For critically damped and overdamped systems the transient response of the system does not have oscillations.
- If $0 < \zeta < 1$, then the transient response due to unit step input is oscillatory and magnitude of oscillations becomes zero after some time. The system is said to be underdamped.
- In undamped condition, the magnitude of oscillations in the transient response of the system does not become zero.

Time response of underdamped system to unit step input

$$\frac{C(s)}{R(s)} = \frac{\omega_n^2}{S^2 + 2\omega_n \zeta s + \omega_n^2}$$

$$\therefore \qquad C(s) = \frac{\omega_n^2}{S^2 + 2\zeta\omega_n S + \omega_n^2} R(s)$$

For unit step input, $r(t) = u(t)$

$$R(s) = \frac{1}{S}$$

$$\therefore \qquad C(s) = \frac{\omega_n^2}{S^2 + 2\zeta\omega_n S + \omega_n^2}\left(\frac{1}{S}\right)$$

$$= \frac{A}{S} + \frac{BS + D}{S^2 + 2\zeta\omega_n S + \omega_n^2}$$

Here, $A = 1$, $B = -1$, $D = -2\zeta\omega_n$

$$C(s) = \frac{1}{S} - \left(\frac{S + 2\zeta\omega_n}{S^2 + 2\zeta\omega_n S + \omega_n^2}\right)$$

$$= \frac{1}{S} - \frac{S + \zeta\omega_n}{(S^2 + 2\zeta\omega_n S + \omega_n^2)} - \frac{\zeta\omega_n}{S^2 + 2\zeta\omega_n S + \omega_n^2}$$

$$= \frac{1}{S} - \frac{S + \zeta\omega_n}{(S + \zeta\omega_n)^2 + \omega_d^2} + \frac{\zeta\omega_n}{(S + \zeta\omega_n)^2 + \omega_d^2}$$

where $\omega_d = \omega_n\sqrt{1 - \zeta^2}$

$$S^2 + 2\zeta\omega_n S + \omega_n^2 = S^2 + 2\zeta\omega_n S + \omega_n^2 + \zeta^2\omega_n^2 - \zeta^2\omega_n^2$$

$$= (S + \zeta\omega_n)^2 + \omega_n^2(1 - \zeta^2)$$

$$= (S + \zeta\omega_n)^2 + \omega_d^2$$

where $\omega_d = \omega_n \sqrt{1 - \zeta^2}$

$$C(s) = \frac{1}{S} - \frac{S + \zeta\omega_n}{(S + \zeta\omega_n)^2 + \omega_d^2} - \frac{\zeta\omega_n}{(S + \zeta\omega_n)^2 + \omega_d^2}$$

Taking inverse Laplace transform

$$C(t) = L^{-1}\,(C(s)) = 1 - e^{-\zeta\omega_n t} \cos \omega_d t - \frac{e^{-\zeta\omega_n t}}{\omega_d} \zeta\omega_n \sin \omega_d t$$

$$C(t) = 1 - e^{-\zeta\omega_n t} \cos \omega_d t - \frac{e^{-\zeta\omega_n t}}{W_n\sqrt{1 - \zeta^2}} . \zeta\omega_n \sin \omega_d t$$

$$= 1 - \frac{e^{-\zeta\omega_n t}}{\sqrt{1 - \zeta^2}} (\sqrt{1 - \zeta^2} \cos \omega_d t + \zeta \sin \omega_d t)$$

$$\text{WKT, } L^{-1}\left[\frac{W}{(S + a)^2 + \omega^2}\right] = e^{-at} \sin \omega t$$

$$L^{-1}\left[\frac{S + a}{(S + a)^2 + W^2}\right] = e^{-at} \cos \omega t$$

$$C(t) = 1 - \frac{e^{-\zeta\omega_n t}}{\sqrt{1 - \zeta^2}} \sin (\omega_d t + \varphi)$$

$$C(t) = 1 - \frac{e^{-\zeta\omega_n t}}{\sqrt{1 - \zeta^2}} (\sin \varphi \cos \omega_d t + \cos \varphi \sin \omega_d t)$$

$$C(t) = 1 = \frac{e^{-\zeta\omega_n t}}{\sqrt{1 - \zeta^2}} \sin (\omega_d t + \varphi)$$

When

$$\varphi = \tan^{-1} \frac{\sqrt{1 - \zeta^2}}{\zeta}$$

$$C(t) = 1 \frac{e^{-\zeta\omega_n t}}{\sqrt{1 - \zeta^2}} \sin(\omega_d t + \tan^{-1} \frac{\sqrt{1 - \zeta^2}}{\xi}$$

The error signal

$$e(t) = C(t) - r(t) = -\frac{e^{-\zeta\omega_n t}}{\sqrt{1 - \zeta^2}} \sin (\omega_d t + \varphi)$$

The error exhibits a damped oscillatory behaviour as shown in Fig. 4.13.

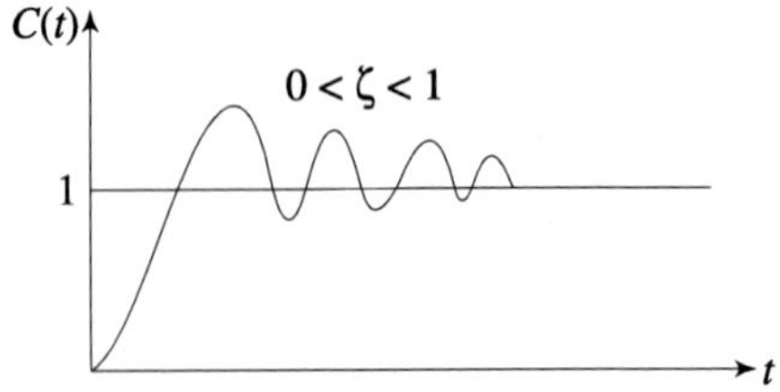

Fig. 4.13

The steady state errors, $e_{ss} = \lim_{t \to \infty} e(t) = 0$

If $\zeta = 0$, the system transient response has undamped oscillations and oscillations continue indefinitely.

$$\text{If } \zeta = 0 \quad \Rightarrow \quad C(t) = 1 - \frac{e^{-0(\omega_n t)}}{\sqrt{1-0^2}} \sin(\omega_d t + \varphi)$$

$$C(t) = 1 - \sin(\omega_n t + 90^\circ)$$

$$C(t) = 1 - \cos \omega_n t$$

The system response to unit step input when $\zeta = 0$ is shown in Fig. 4.14

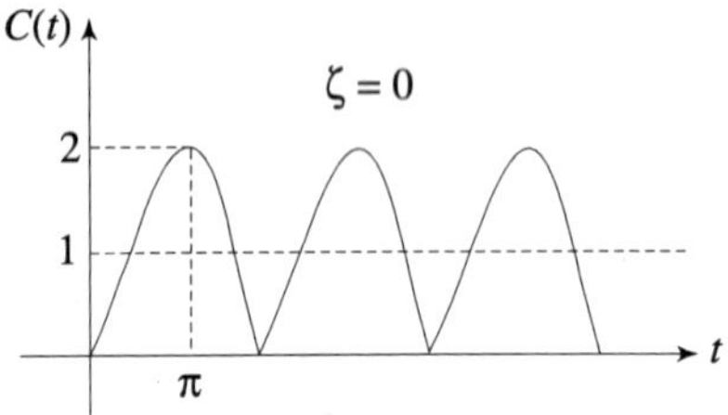

Response of undamped system

Fig. 4.14

Response of critically damped system to unit step input ($\zeta = 1$):

$$\text{For this system, } C(s) = \frac{\omega_n^2}{S(S^2 + 2\omega_n S + \omega_n^2)}$$

$$= \frac{A}{s} + \frac{BS + D}{S^2 + 2\omega_n S + \omega_n^2}$$

where $A = 1$, $B = -1$ and $D = -2\omega_n$

$$\therefore \qquad C(s) = \frac{1}{S} - \frac{S + \omega_n}{S^2 + 2\omega_n S + \omega_n^2} - \frac{\omega_n}{S^2 + 2\omega_n S + \omega_n^2}$$

$$C(s) = \frac{1}{S} - \frac{1}{S+\omega_n} - \frac{\omega_n}{(S+\omega_n)^2}$$

$$C(t) = 1 - \omega_n e^{-\zeta\omega_n t} \cdot t - e^{-\zeta\omega_n t}$$

$$L^{-1}(C(s) = 1 - e^{-\omega_n t} - \omega_d t \cdot e^{-\omega_n t}$$

$$\therefore \qquad C(t) = 1 - e^{-\omega_n t}(1 + \omega_n \cdot t)$$

The response of the system reaches its final value without any oscillations as shown in Fig. 4.15.

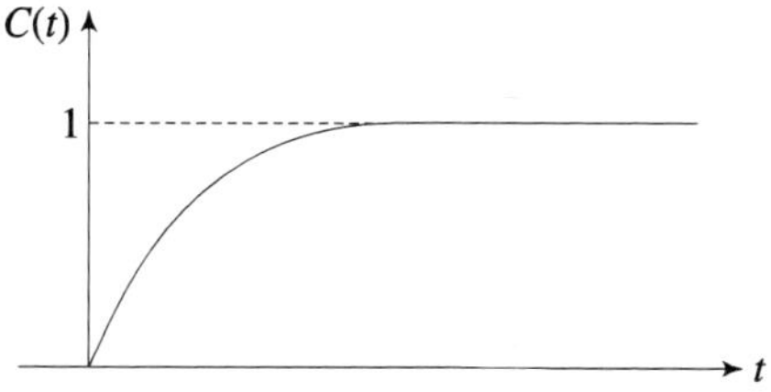

Fig. 4.15

Response of overdamped system to unit step input ($\zeta > 1$)

Here, $$\frac{C(s)}{R(s)} = \frac{\omega_n^2}{S^2 + 2\zeta\omega_n S + \omega_n^2}$$

For unit step input $R(s) = \frac{1}{S}$

$$\therefore \qquad C(s) = \frac{\omega_n^2}{S(S^2 + 2\zeta\omega_n S + \omega_n^2)}$$

$$= \frac{\omega_n^2}{S(S + \zeta\omega_n + \omega_n\sqrt{\zeta^2 - 1})(S + \zeta\omega_n - \omega_n\sqrt{\zeta^2 - 1})}$$

After partial fraction expansion,

$$C(s) = \frac{1}{S} + \frac{1}{2\sqrt{\zeta^2 - 1}\,(\zeta + \sqrt{\zeta^2 - 1})(S + \zeta\omega_n + \omega_n\sqrt{\zeta^2 - 1})}$$

$$= \frac{1}{2\sqrt{\zeta^2 - 1}\,(\zeta - \sqrt{\zeta^2 - 1})(S + \zeta\omega_n - \omega_n\sqrt{\zeta^2 - 1})}$$

Taking inverse L.T.

$$C(t) = 1 + \frac{1}{2\sqrt{\zeta^2 - 1}}\left[\frac{e^{-(\zeta + \sqrt{\zeta^2 - 1})\,\omega_n t}}{\zeta + \sqrt{\zeta^2 - 1}} - \frac{e^{-(\zeta + \sqrt{\zeta^2 - 1}\,\omega_n t}}{(\zeta - \sqrt{\zeta^2 - 1})}\right]$$

From the above equation, we see transient response of the overdamped system to unit step input consists of 2 exponentially decaying curves.

The response of the system reaches its final value without any oscillations as in the case of critical damping but the response is very slow. As shown in Fig. 4.16.

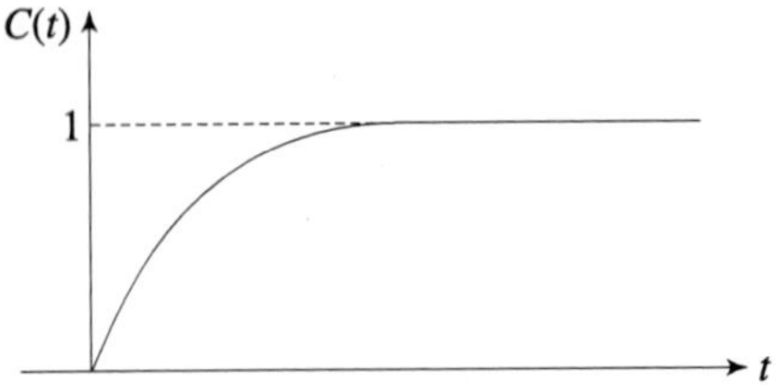

Fig. 4.16

In a control system if a small amount of overshoot is acceptable, it is preferable to select values of ζ between 0.4 and 0.8. This ensures that the system is fast and excessive overshoot is avoided.

If ζ is low, then the overshoot will be excessive. Larger value of ζ results in reduced overshoot but the system response will be slow.

4.5 TRANSIENT SPECIFICATIONS OF SECOND ORDER SYSTEMS

The performance characteristics of a control system are specified in terms of the transient response to unit step input. The transient response of a practical control system exhibits damped oscillations before attaining the steady state. The response of the system is shown in Fig. 4.17.

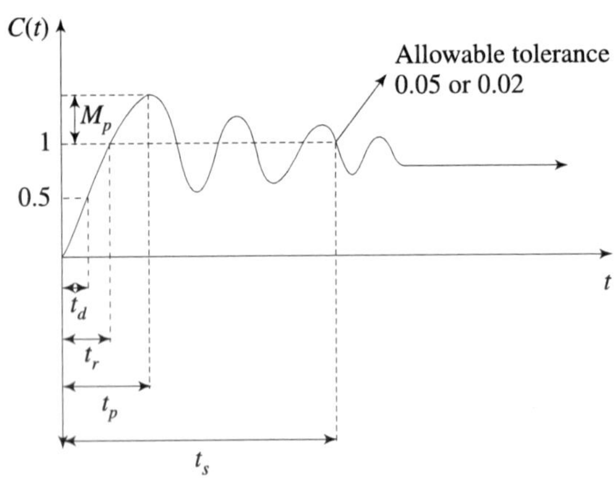

Fig. 4.17

The following quantities are used to specify transient response:

1. **Delay time, (t_d):** Defined as the time required for the response of the system to reach 50% of the final value for the first time.

2. **Rise time (t_r):** Defined as the time required for the response of the system to rise from 10% to 90%, 5% to 95% or 0% to 100% of its final value.

   For underdamaped system 0% to 100% rise time is used whereas for overdamped systems normally 10% to 90% rise time is used.
3. **Peak time (t_p):** The peak time is the time required for the response to reach the first peak of the overshoot.
4. **Maximum overshoot, (M_p):** It is the maximum peak value of the response measured from unity. The overshoot is an indication of the largest error between input and output during transient state. The amount of maximum overshoot indicates the relative stability of the system. If the final steady state value of the system $C(\infty)$ is different from unity.

   $$\text{Maximum percent overshoot} = \frac{C(t_p) - C(\infty)}{C(\infty)} \times 100\%$$
5. **Settling time (t_s):** Defined as the time required for the response to reach and stay within a specified percentage of its final value. Usually, 2% or 5% criterion is used.

Derivation of transient response specifications of second order system

I. Rise time (t_r): The rise time, (t_r) is obtained when the response of the system $C(t)$ reaches unity for the first time.

At $\quad t = t_r,\ C(t) = 1$

$$C(t) = 1 - \frac{e^{-\zeta\omega_n t}}{\sqrt{1-\zeta^2}} \sin\left(\omega_d \cdot t_r + \tan^{-1}\frac{\sqrt{1-\zeta^2}}{\zeta}\right) = 1$$

Since $e^{-\zeta\omega_n tr} \neq 0$, from the above equation.

$$\sin\left(\omega_d \cdot t_r + \tan^{-1}\frac{\sqrt{1-\zeta^2}}{\zeta}\right) = 0$$

$$\omega_d \cdot t_r + \tan^{\ 1}\frac{\sqrt{1-\zeta^2}}{\zeta} - \pi$$

$$t_r = \frac{\pi - \tan^{-1}\dfrac{\sqrt{1-\zeta^2}}{\zeta}}{\omega_d} = \frac{\pi - \theta}{\omega_d}$$

II. Peak time (t_p): At, $t = t_p$, the output $C(t)$ has a peak value $\dfrac{d}{dt}c(t) = \dfrac{e^{-\zeta\omega_n t}}{\sqrt{1-\zeta^2}}$

$$[\zeta\omega_n \sin(\omega_d t + \theta) - \omega_d \cos(\omega_d t + \theta)]$$

At $\quad t = t_p,\ \dfrac{dc(t)}{dt} = 0$

$$\frac{dc(t_p)}{dt} = \frac{e^{-\zeta\omega_n t_p}}{\sqrt{1-\zeta^2}} [\zeta\omega_n \sin(\omega_d t_p + \theta) - \omega_d \cos(\omega_d t + \theta)] = 0$$

Since $\quad e^{-\zeta\omega_n t_p} \neq 0$

$$\zeta\omega_n \sin(\omega_d t_p + \theta) - \omega_d \cos(\omega_d t_p + \theta) = 0$$

$$\zeta\omega_n \sin(\omega_d t_p + \theta) - \omega_n\sqrt{1-\zeta^2} \cos(\omega_d t_p + \theta) = 0$$

$$\sin(\omega_d t_p + \theta - \theta) = 0$$

$$\sin(\omega_d t_p) = 0$$

$$\omega_d \cdot t_p = \pi \quad \text{or} \quad t_p = \frac{\pi}{\omega_d}$$

Note: In general, the condition $\omega_d \cdot t = n\pi$ gives the time at which $\frac{d}{dt} C(t) = 0$.

III. Maximum overshoot (M_p): The maximum overshoot occurs at the peak time,

$$t_p = \frac{\pi}{\omega_d}$$

$$M_p = C(t_p) - 1$$

$$= -\frac{e^{-\zeta\omega_n t_p}}{\sqrt{1-\zeta^2}} \sin(\omega_d t_p + \theta)$$

But $\quad \omega_d t_p = \pi$

$$M_p = -\frac{e^{-\zeta\omega_n t_p}}{\sqrt{1-\zeta^2}} \sin(\pi - \theta)$$

$$= \frac{e^{-\zeta\omega_n t_p}}{\sqrt{1-\zeta^2}} \sin\theta$$

But $\quad \sin\theta = \sqrt{1-\zeta^2} \quad \therefore \quad M_p = e^{-\frac{\zeta\omega_n\pi}{\omega_d}} = e^{-\frac{\zeta\omega_n\pi}{\omega n\sqrt{1-\zeta^2}}} = e^{-\frac{\pi\zeta}{\sqrt{1-\zeta^2}}}$

Maximum per cent overshoot $= e^{-\frac{\pi\zeta}{\sqrt{1-\zeta^2}}} \times 100\%$

IV. Settling time (t_s): It gives an indication of the time required for the system response to reach and stay within a specified percentage of its steady state value (usually 2% or 5%).

The output of underdamped system to a unit step input is

$$C(t) = 1 \pm \frac{e^{-\zeta\omega_n t}}{\sqrt{1-\zeta^2}} \sin(\omega_d t + \theta)$$

It can be observed from the above equation, that the output is always bounded by two exponentially decaying curves.

$$\left(1 + \frac{e^{-\zeta\omega_n t}}{\sqrt{1-\zeta^2}}\right) \text{ and } \left(1 - \frac{e^{-\zeta\omega_n t}}{\sqrt{1-\zeta^2}}\right),$$

Time constant of these curves, $T = \dfrac{1}{\zeta\omega_n}$

$$\omega_n t_s = \frac{4}{\zeta} \qquad \therefore \quad t_s = \frac{4}{\zeta\omega_n} = 4T.$$

For 2% criterion, setting time, $t_s = 4T$

Steady state response: When a certain input is given to a control system, the system response usually has transient portion. The response of the system when the transients have become negligible is known as the steady state response. It is desirable to have steady state response equal to the reference input signal. But in practice due to system imperfections, the steady state output may not match with the reference input. In such a case the system is said to have a steady state error.

Figure 4.18 shows the steady state error of a system subjected to step input.

The upper envelop curve reaches to 5% of the final value of the output at $t = t_s$ if 5% criterion is used. The value of the upper envelop at this instant is 1.05.

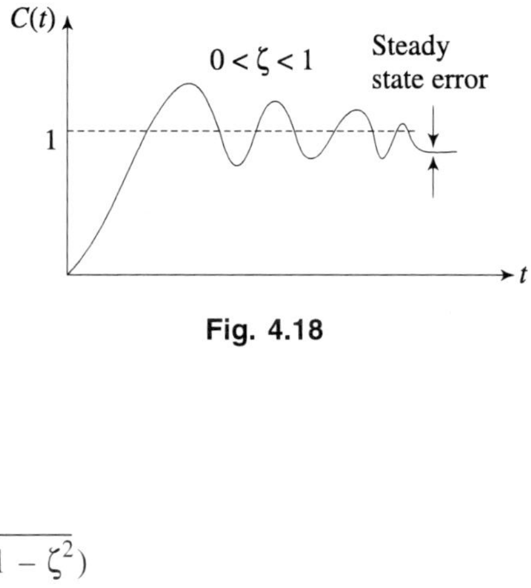

Fig. 4.18

$$1 + \frac{e^{-\zeta\omega_n t_s}}{\sqrt{1-\zeta^2}} = 1.05$$

$$\frac{e^{-\zeta\omega_n t_s}}{\sqrt{1-\zeta^2}} = 0.05$$

From the above equation, $\omega_n \cdot t_s = -\dfrac{1}{\zeta}\log_e\left(0.05\sqrt{1-\zeta^2}\right)$

For small value of ζ in the above equation is approximated to $\omega_n \cdot t_s = \dfrac{3}{\zeta}$

$$\therefore \qquad t_s = \frac{3}{\zeta\omega_n} = 3T$$

Same result is obtained if the lower envelop is used. For 5% criterion, settling time $t_s = 3T$. For 2% criterion, the upper envelop has a value of 1.02 at $t = t_s$.

$$1 + \frac{e^{-\zeta\omega_n t_s}}{\sqrt{1-\zeta^2}} = 1.02$$

From the above equation, $e^{-\zeta\omega_n t_s} = 0.02\sqrt{1-\zeta^2}$

In Fig. 4.18, it is seen that steady state response is not equal to the reference input.

In case of unity feedback system $H(s) = 1$, the error signal is $e(t) = r(t) - c(t)$. In case of non-unity feedback system, the error signal is $e(t) = r(t) - b(t)$.

Where $b(t)$ is the feedback signal.

Classification of a control systems: A simple closed-loop feedback system is shown in Fig. 4.19. The open-loop transfer function $G(s)H(s)$ depends on the components of the control system.

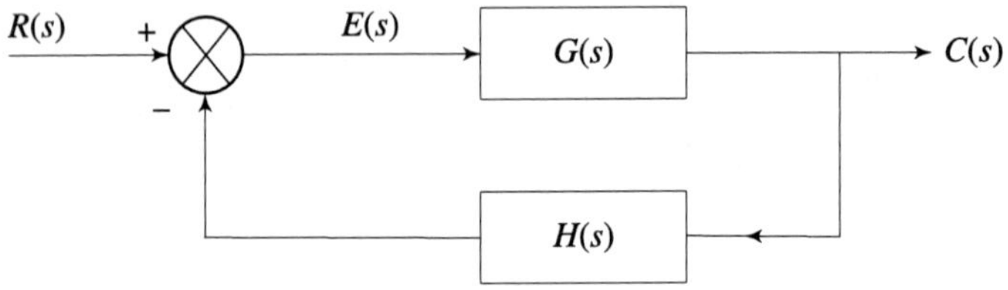

Fig. 4.19

Depending on the system the open-loop transfer function has one of the following forms.

$$G(s)\,H(s) = \frac{K(T_aS + 1)\,(T_bS + 1)\,\ldots\ldots}{S^0(T_1S + 1)\,(T_2S + 1)\,\ldots\ldots} \qquad \ldots(4.2)$$

$$G(s)\,H(s) = \frac{K(T_aS + 1)\,(T_bS + 1)\,\ldots\ldots}{S(T_1S + 1)\,(T_2S + 1)\,\ldots\ldots} \qquad \ldots(4.3)$$

$$G(s)\,H(s) = \frac{K(T_aS + 1)\,(T_bS + 1)\,\ldots\ldots}{S^2(T_1S + 1)\,(T_2S + 1)\,\ldots\ldots} \qquad \ldots(4.4)$$

In general,

$$G(s)\,H(s) = \frac{K(T_aS + 1)\,(T_bS + 1)\,\ldots\ldots}{S^N(T_1S + 1)\,(T_2S + 1)\,\ldots\ldots} \qquad \ldots(4.5)$$

Here K is the gain and N is known as the type of the system. The term S^N represents the number of poles of $G(s)\,H(s)$ at the origin.

A system is called type 0, type 1, type 2, if $N = 0, 1, 2$ respectively. The system of Eq. (4.2) is a type 0 system, the system of Eq. (4.3) is a type 1 system and the system of Eq. (4.4) is a type 2 system.

It is interesting to note that steady state error is not dependent on the order of the system. It is determined by the number of poles at the origin of the open-loop transfer function.

It is found that increase in the type of the system, results in reduction of steady state error but stability of the system is affected.

4.6 STEADY STATE ERRORS

The closed-loop transfer function is shown in Fig. 4.19.

$$\frac{C(s)}{R(s)} = \frac{G(s)}{1 + G(s)\,H(s)}$$

Here, the error signal $E(s) = R(s) - B(s)$

$$E(s) = R(s) - H(s)\,C(s)$$

$$C(s) - G(s)\,E(s)$$

$$\therefore \qquad E(s) = R(s) - H(s)\,G(s)\,E(s)$$

$$E(s)\,(1 + G(s)\,H(s)) = R(s)$$

$$E(s) = \frac{R(s)}{1 + G(s)\,H(s)}$$

Substituting for $C(s)$ in the above equation.

$$E(s) = R(s) - \frac{H(s)\,G(s)\,R(s)}{1 + G(s)\,H(s)}$$

$$= \frac{R(s)\,[1 + G(s)\,H(s) - G(s)\,H(s)]}{1 + G(s)\,H(s)}$$

$$E(s) = \frac{1}{1 + G(s)\,H(s)}\,R(s)$$

By using final value theorem, the steady state error of the system e_{ss} is determined.

Types of steady state error

(1) Static error constants (given error for the standard inputs).

(2) Dynamic error constants (given error for any inputs).

Steady state error due to step input

When a system is subjected to step input $r(t) = A \cdot u(t)$

$$e_{ss} = \lim_{s \to 0} \frac{SR(s)}{1 + G(s) \cdot H(s)}$$

$$R(s) = \frac{A}{S}$$

$$\therefore \qquad e_{ss} = \lim_{s \to 0} \frac{S \cdot (A/s)}{1 + G(s) \cdot H(s)} = \frac{A}{1 + \lim_{s \to 0} G(s) \cdot H(s)}$$

$$e_{ss} = \frac{A}{1 + K_p} \qquad \text{...(4.6)}$$

where K_p = Static position error constant

From equation (4.6) it can be concluded that a system subjected to step input has zero steady state error if K_p is ∞. Otherwise, system has steady state error.

For type 0 system, $G(s)\,H(s) = \dfrac{K(T_aS+1)\,(T_bS+1)\,.....}{(T_1S+1)\,(T_2S+1)\,.....}$

$$\therefore \quad K_p = \lim_{s\to 0} = G(s)\cdot H(s) = \lim_{s\to 0} \frac{K(T_aS+1)\,(T_bS+1)\;.....}{T_1S+1)\,(T_2S+1)\;.....}$$

$$K_p = K$$

For a type 1 or higher system,

$$K_p = \lim_{s\to 0} = \frac{K(T_aS+1)\,(T_bS+1)\;.....}{(T_1S+1)\,(T_2S+1)\;.....} \qquad = \infty\;(N \geq 1)$$

Hence,

For type 0 system, $e_{ss} = \dfrac{A}{1+K_p} = \dfrac{A}{1+K}$

For type 1 and higher systems, $e_{ss} = \dfrac{A}{1+K_p} = 0$

Static portion error constant, K_p

$K_p = \lim_{s\to 0} G(s)\,H(s)$ for unit step input

$$G(s)\,H(s) = \frac{K(1+ST_1)\,(1+ST_2)\;.....}{S^N(1+STa)\,(1+STb)\;.....}$$

for $N = 0$, type 0 system

$N = 1$ type 1 system

But $\quad E(s) = \left\{\dfrac{1}{1+G(s)\,H(s)}\right\} R(s)$

But $\quad R(s) = \dfrac{1}{S}$ unit step input

$$\therefore \quad E(s) = \left\{\frac{1}{1+G(s)\,H(s)}\right\}\frac{1}{S}$$

Steady state error, $e_{ss} = \lim_{s\to 0} S\cdot E(s)$

$$e_{ss} = \lim_{s\to 0}\left\{\frac{1}{1+G(s)\,H(s)}\right\}\frac{1}{S}$$

$$= \frac{1}{1+\lim_{s\to 0} G(s)\,H(s)}$$

$$\therefore \quad e_{ss} = \frac{1}{1+K_p}$$

Type 0 system:

$$G(s)\ H(s) = \frac{K(1 + ST_1)\ (1 + ST_2)\ ...}{S^0(1 + STa)\ ...}$$

$$K_p = \lim_{s \to 0} G(s)\ H(s) = K$$

$$e_{ss} = \frac{1}{1 + K_p} = \frac{1}{1 + K} \text{ For type 0 system}$$

Type 1 system:

$$G(s)\ H(s) = \frac{K(1 + ST_1)\ ...}{S^1(1 + ST_2)\ ...}$$

$$K_p = \lim_{s \to 0}\ G(s)\ H(s) = \infty$$

$$\therefore \qquad e_{ss} = \frac{1}{1 + K_p} = \frac{1}{1 + \infty} = 0$$

Type 1 and above, $e_{ss} = 0$

Steady state error due to ramp input (state velocity error constant, K_v)

Ramp input, $r(t)$ = A.t. $u(t)$, $R(s) = \frac{A}{s^2}$

$$e_{ss} = \frac{S \cdot R(s)}{1 + G(s)\ H(s)} = \lim_{s \to 0} \frac{S\left(\frac{A}{S^2}\right)}{1 + G(s)\ H(s)}$$

$$\lim_{s \to 0} = \frac{A}{S + S \cdot G(s)\ H(s)} = \frac{A}{0 + \lim_{s \to 0} S \cdot G(s)\ H(s)}$$

$$e_{ss} = \frac{A}{K_v}$$

where K_v is known as velocity error constant and is given by

$$K_v = \lim_{s \to 0} S \cdot G(s)\ H(s)$$

For a type 0 system,

$$K_v = \lim_{s \to 0}\ S \cdot G(s)\ H(s)$$

$$= \lim_{s \to 0} \frac{SK(1 + ST_1)...}{S^0(1 + STa)...}$$

$$K_v = 0$$

For type 1 system

$$K_v = \lim_{s \to 0} \frac{SK(TaS + 1)\ (TbS + 1)\ ...}{S(T_1S + 1)\ (T_2S + 1)\ ...} = K$$

$$K_v = K$$

For type 2 system and higher systems,

$$K_v = \lim_{s\to 0} \frac{SK(TaS + 1)\ (TbS + 1)\ ...}{S^N(T_1S + 1)\ (T_2S + 1)\ ...} = \infty\ N \geq 2$$

For type 0 system, $e_{ss} = A/K_v = \infty$

For type 1 system, $e_{ss} = A/K_v = A/K$

For type 2 and higher systems, $e_{ss} = A/K_v = 0$

Steady state error due to parabolic input (K_a)

Here $$r(t) = \frac{A}{2} t^2\ u(t)$$

$$R(s) = \frac{A}{2} \cdot \frac{2}{S^3} = \frac{A}{S^3}$$

$$e_{ss} = \lim_{s\to 0} \frac{s \cdot R(s)}{1 + G(s)\ H(s)} = \lim_{s\to 0} \frac{S(A/S^3)}{1 + G(s)\ H(s)}$$

$$= \lim_{s\to 0} \frac{A}{S^2 + S^2\ G(s)\ H(s)} = \frac{A}{0 + \lim\limits_{s\to 0} S^2\ G(s)\ H(s)} = \frac{A}{K_a}$$

where $K_a = \lim\limits_{s\to 0} S^2 G(s)\ H(s)$ and is known as acceleration error constant.

For type 0 system, $K_a = \lim\limits_{s\to 0} S^2\ G(s)\ H(s)$

$$K_a = \lim_{s\to 0} \frac{S^2K(TaS + 1)(TbS + 1)}{(T_1S + 1)(T_2S + 1)} = 0$$

For type 1 system,

$$K_a = \lim_{s\to 0} S^2\ G(s).\ H(s) = \lim_{s\to 0} \frac{S^2K\ (TaS + 1)\ (TbS + 1)}{(T_1S + 1)\ (T_2S + 1)} = 0$$

For type 2 system,

$$K_v = \lim_{s\to 0}\ S^2G(s)\ H(s)$$

$$= \lim_{s\to 0} \frac{S^2K(TaS + 1)\ (TbS + 1)}{S^2(T_1S + 1)\ (T_2S + 1)}\ .....$$

$$K_v = K$$

Example 4.1: The closed-loop transfer functions of certain second order unity feedback control systems are given below. Determine the type of damping in the systems

(a) $\dfrac{C(s)}{R(s)} = \dfrac{8}{S^2 + 3S + 8}$ (b) $\dfrac{C(s)}{R(s)} = \dfrac{2}{S^2 + 4S + 2}$

(c) $\dfrac{C(s)}{R(s)} = \dfrac{2}{S^2 + 2S + 1}$ (d) $\dfrac{C(s)}{R(s)} = \dfrac{4}{S^2 + 16}$

Solution: The standard form of the transfer function of a second order system is

$$\frac{C(s)}{R(s)} = \frac{\omega_n^2}{S^2 + 2\,\xi\omega_n S + \omega_n^2}$$

We shall compare the given transfer functions with this standard form.

(a) $$\frac{C(s)}{R(s)} = \frac{8}{S^2 + 3S + 8} = \frac{\omega_n^2}{S^2 + 2\,\xi\omega_n S + \omega_n^2}$$

$$= \frac{\omega_n^2}{S^2 + 2\,\xi\,\omega_n\,S + 8}$$

$$\therefore\quad \omega_n^2 = 8,\ \omega_n = \sqrt{8} = 2.58 \text{ rad/s}$$

$$2\,\xi\,\omega_n = 3$$

$$\xi = \frac{3}{2\omega_n} = \frac{3}{2 \times 2.82} = 0.53$$

Since $\xi < 1$, the given system is undamped.

(b) $$\frac{C(s)}{R(s)} = \frac{2}{S^2 + 4S + 2} = \frac{\omega_n^2}{S^2 + 2\,\xi\omega_n\,S + \omega_n^2}$$

$$\omega_n^2 = 2,\ \omega_n = \sqrt{2} = 1.414$$

$$2\xi\omega_n = 4$$

$$\xi = \frac{4}{2\omega_n} = \frac{4}{2 \times 1.44} = 1.41$$

Since $\xi > 1$, the system is overdamped.

(c) $$\frac{C(s)}{R(s)} = \frac{2}{S^2 + 2S + 1} = \frac{\omega_n^2}{S^2 + 2\xi\omega_n s + \omega_n^2}$$

$$\omega_n^2 = 1,\ \omega_n = 1$$

$$2\xi\,\omega_n = 2$$

$$\xi = \frac{2}{2\omega_n} = 1$$

Since $\xi = 1$, the given system is critically damped.

(d) $$\frac{C(s)}{R(s)} = \frac{4}{S^2 + 16} = \frac{\omega_n^2}{S^2 + 2\xi\,\omega_n S + \omega_n^2}$$

$$\omega_n^2 = 16,\ \omega_n = 4$$

$$2\xi\,\omega_n = 0$$

$$\xi = 0$$

Since $\xi = 0$, the given system is undamped.

Example 4.2: The open-loop transfer function of a unity feedback system is $G(s) = \dfrac{4}{S(S+1)}$.

Determine the nature of response of the closed-loop system for a unit step input. Also determine the rise time, peak time, peak overshoot and settling time.

Solution: $G(s) = \dfrac{4}{S(S+1)},\ H(s) = 1$

$$G(S)\,H(s) = \frac{4}{S(S+1)}$$

The closed-loop transfer function is

$$\frac{C(s)}{R(s)} = \frac{G(s)}{1 + G(s)\,H(s)} = \frac{\dfrac{4}{S(S+1)}}{1 + \dfrac{4}{S(S+1)}}$$

$$\frac{C(s)}{R(s)} = \frac{4}{S^2 + S + 4} = \frac{\omega_n^2}{S^2 + 2\xi\,\omega_n S + \omega_n^2}$$

$$\omega_n^2 = 4,\ \omega_n = 2$$

$$2\xi\,\omega_n = 1$$

$$\xi = \frac{1}{2\omega_n} = \frac{1}{2 \times 2} = 0.25$$

Since $\xi < 1$, the given system is underdamped.

$$\omega_d = \omega_n\sqrt{1 - \xi^2} = 2\sqrt{1 - (0.25)^2} = 1.936 \text{ rad/sec}$$

$$\phi = \tan^{-1}\frac{\sqrt{1 - \xi^2}}{\xi} = \tan^{-1}\frac{\sqrt{1 - (0.25)^2}}{1.936} = 1.310 \text{ rad}$$

Rise time, $t_r = \dfrac{\pi - \phi}{\omega_d}$

$$= \frac{3.141 - 1.310}{1.936} = 0.945\,S$$

Peak time, $t_p = \dfrac{\pi}{\omega_d} = \dfrac{3.141}{1.936} = 1.622\,S$

Percentage peak overshoot,

$$M_p = e^{-\frac{\xi\pi}{\sqrt{1-\xi^2}}} \times 100\%$$

$$= e^{\frac{-\pi \times 0.25}{\sqrt{1-(0.25)^2}}} \times 100\%$$

$$M_p = 43.26\%$$

The settling time for 5% error

$$t_s = \frac{3}{\xi\omega_n} = \frac{3}{0.25 \times 2} = 6S$$

The settling time for 2% error

$$t_s = \frac{4}{\xi\omega_n} = \frac{4}{0.25 \times 2} = 8S.$$

Example 4.3: For unity feedback system having an open-loop function, $G(s) = \dfrac{K(S+2)}{S(S^3 + 7S^2 + 12S)}$.

Determine:

(a) Type of system (b) Error constants K_p, K_v, K_a

(c) Steady state error for unity parabolic input.

Solution: For a unity feedback system, $H(s) = 1$

$$G(s)\,H(s) = \frac{K(S+2)}{S^2(S^2 + 7S + 12)}$$

(a) Since $G(s)\,H(s)$ has two poles at the origin of the S-plane, so it is a type 2 system.

(b) $K_p = \lim\limits_{s\to 0} G(s)\,H(s) = \lim\limits_{s\to 0} \dfrac{K(S+2)}{S^2(S^2 + 7S + 12)} = \infty$

$$K_v = \lim_{s\to 0} SG(s)\,H(s) = \lim_{s\to 0} \frac{K(S+2)}{S(S^2 + 7S + 12)}$$

$$K_v = \infty$$

$$K_a = \lim_{s\to 0} S^2G(s)\,H(s) = \lim_{s\to 0} \frac{K(S+2)}{S^2+7S+12} = \frac{K}{6}$$

(c) Steady state error, $e_{ss} = \lim_{s\to 0} \dfrac{SR(s)}{1+G(s)\,H(s)}$

For unit parabolic input, $R(s) = \dfrac{1}{S^3}$

$$\therefore\ e_{ss} = \lim_{s\to 0} \frac{\left(\frac{1}{S^2}\right)}{1+\dfrac{K(S+2)}{S^2(S+4)(S+3)}}$$

$$e_{ss} = \lim_{s\to 0} \frac{(S+4)\,(S+3)}{S^2(S+4)(S+3)+K(S+2)} = \frac{6}{K}.$$

Example 4.4: The block diagram of the system is shown in Fig. 4.20. Assuming unity feedback and $K = 400$, calculate

(a) The output $C(t)$ for unit step input
(b) Steady state error for unit ramp input
(c) Determine K if the error to a ramp input is 0.02

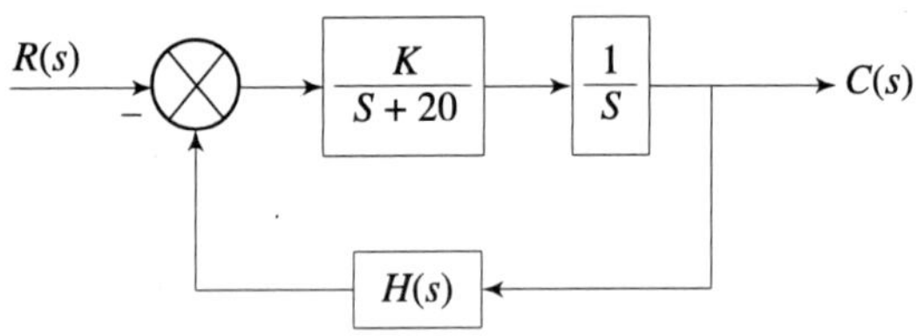

Fig. 4.20

Solution: $G(s) = \left(\dfrac{K}{S+20}\right)\left(\dfrac{1}{S}\right)$, $H(s) = 1$, $K = 400$

(a) $$\frac{C(s)}{R(s)} = \frac{G(s)}{1+G(s)H(s)} = \frac{\dfrac{K}{S(S+20)}}{1+\dfrac{K}{S(S+20)}}$$

$$\frac{C(s)}{R(s)} = \frac{K}{S^2+20S+K} = \frac{400}{S^2+20S+400}$$

The standard form of the transfer function of a second order system is

$$\frac{C(s)}{R(s)} = \frac{400}{S^2+20S+400} = \frac{\omega_n^2}{S^2+2\xi\omega_n S+\omega_n^2}$$

$$\omega_n^2 = 400,\ \omega_n = \sqrt{400} = 20$$

$$2\xi\omega_n = 20$$

$$\xi = \frac{20}{2\omega_n} = \frac{20}{2 \times 20} = 0.5$$

$$\omega_d = \omega_n \sqrt{1 - \xi^2} = 20\sqrt{1 - (0.5)^2} = 17.32 \text{ rad/sec}$$

$$\phi = \tan^{-1}\left(\frac{\sqrt{1 - \xi^2}}{\xi}\right) = \tan^{-1}\frac{\sqrt{1 - (0.5)^2}}{0.5} = 1.047 \text{ rad}$$

The response $C(t)$ is given by

$$C(t) = 1 - \frac{e^{-\xi\omega_n t}}{\sqrt{1 - \xi^2}} \sin(\omega_d t + \phi)$$

$$C(t) = 1 - \frac{e^{-0.5 \times 20t}}{\sqrt{1 - (0.5)^2}} \sin(17.32t + 1.047)$$

$$C(t) = 1 - 1.154\, e^{-10t} \sin(17.32t + 1.047)$$

(b) $K_v = \lim_{s \to 0} SG(s)\, H(s)$

$$K_v = \lim_{s \to 0} S \frac{K}{S(S + 20)} = \frac{K}{20} = \frac{400}{20} = 20$$

Steady state error for unit ramp input is

$$e_{ss} = \frac{1}{K_v} = \frac{1}{20} = 0.05$$

(c) $e_{ss} = \dfrac{1}{K_v}$

$$e_{ss} = 0.02,\ K_v = \frac{K}{20}$$

$$0.02 = \frac{1}{\left(\frac{K}{20}\right)}$$

$$K = \frac{20}{0.02} = 1000.$$

Example 4.5: Determine static error coefficients for a unity feedback system given by

$$G(s) = \frac{K}{S^2(S + 20)(S + 30)}$$

Solution: $H(S) = 1$

$$G(s)\,H(s) = \frac{K}{S^2(S+20)(S+30)}$$

Arranging $G(s)H(s)$ in the time constant form

$$G(s)H(s) = \frac{K}{S^2.20\left(1+\frac{S}{20}\right)30\left(1+\frac{S}{30}\right)}$$

$$= \frac{\frac{K}{600}}{S^2\left(1+\frac{S}{20}\right)\left(1+\frac{S}{30}\right)}$$

$$K_p = \lim_{s\to 0} G(s)\,H(s) = \lim_{s\to 0} \frac{\left(\frac{K}{600}\right)}{S^2\left(1+\frac{S}{20}\right)\left(1+\frac{S}{30}\right)}$$

$$K_p = \infty$$

$$K_v = \lim_{s\to 0} SG(s)\,H(s) = \lim_{s\to 0} \frac{S\left(\frac{K}{600}\right)}{S^2\left(1+\frac{S}{20}\right)\left(1+\frac{S}{30}\right)}$$

$$K_v = \infty$$

$$K_a = \lim_{s\to 0} S^2G(s)\,H(s) = \lim_{s\to 0} \frac{S^2\left(\frac{K}{600}\right)}{S^2\left(1+\frac{S}{20}\right)\left(1+\frac{S}{30}\right)}$$

$$K_a = \frac{K}{600}$$

Example 4.6: Determine the type and order of the unity feedback control systems whose open-loop transfer functions are:

(a) $G(s) = \dfrac{K}{S(S^2+4S+200)}$ (b) $G(s) = \dfrac{K(1+2S)(1+4S)}{S^2(S^2+2S+10)}$

Find also the static error coefficients and the errors for unit step and unit ramp inputs.

Solution: $G(s) = \dfrac{K}{S(S^2+4S+200)}$, $H(s) = 1$

$$G(s)H(s) = \frac{K}{S(S^2+4S+200)}.$$

The transfer function has one pole at the origin of the S-plane, so it is a type 1 system. The highest power of S present in the denominator is 3, so it is a third order system.

$$G(s)\ H(s) = \frac{\left(\frac{K}{200}\right)}{S\left(1 + \frac{4}{200}S + \frac{S^2}{200}\right)}.$$

Static error coefficients,

$$K_p = \lim_{s\to 0} G(s)\ H(s) = \lim_{s\to 0} \frac{\left(\frac{K}{200}\right)}{S\left(1 + \frac{4}{200}S + \frac{S^2}{200}\right)}$$

$$K_p = \infty$$

$$K_v = \lim_{s\to 0} SG(s)H(s) = \lim_{s\to 0} \frac{S\left(\frac{K}{200}\right)}{S\left(1 + \frac{4}{200}S + \frac{S^2}{200}\right)}$$

$$K_v = \frac{K}{200}$$

$$K_a = \lim_{s\to 0} S^2G(s)H(s)$$

$$K_a = \lim_{s\to 0} \frac{S^2\left(\frac{K}{200}\right)}{S\left(1 + \frac{4}{200}S + \frac{S^2}{200}\right)} = 0$$

Steady state errors

For unit step input, $A = 1$

Steady state error,

(a) $$e_{ss} = \frac{A}{1 + K_p} = \frac{1}{1 + \infty} = 0$$

For unit ramp input, $A = 1$

$$e_{ss} = \frac{A}{K_v} = \frac{1}{\left(\frac{K}{200}\right)} = \frac{200}{K}$$

(b) $$G(s) = \frac{K(1 + 2S)\ (1 + 4S)}{S^2(S^2 + 2S + 10)},\ H(s) = 1$$

$\therefore$ $$G(s)H(s) = \frac{K(1 + 2S)\ (1 + 4S)}{S^2(S^2 + 2S + 10)}$$

$$= \left(\frac{K}{10}\right) \frac{(1 + 2S)(1 + 45)}{S^2\left(1 + \frac{2}{10}S + \frac{S^2}{20}\right)}$$

The transfer function, $G(s)\ H(s)$, has two poles at the origin of the S-plane. Hence, it is a type 2 system The highest power of S present in the denominator of $G(s)\ H(s)$ is 4. Hence, it is a fourth order system

Static error coefficients

$$K_p = \lim_{s\to 0} G(s)H(s) = \lim_{s\to 0} \frac{\left(\frac{K}{10}\right)(1+2S)(1+4S)}{S^2\left(1+\frac{2}{10}S+\frac{S^2}{10}\right)}$$

$$K_p = \infty$$

$$K_v = \lim_{s\to 0} SG(s)H(s) = \lim_{s\to 0} \frac{S\left(\frac{K}{10}\right)(1+2S)(1+4S)}{S^2\left(1+\frac{2}{10}S+\frac{S^2}{10}\right)}$$

$$K_v = \infty$$

$$K_a = \lim_{s\to 0} S^2G(s)H(s) = \lim_{s\to 0} \frac{S^2\left(\frac{K}{10}\right)(1+2S)(1+4S)}{S^2\left(1+\frac{2}{10}S+\frac{S^2}{10}\right)}$$

$$K_a = \frac{K}{10}$$

Steady state errors

For unit step input, $A = 1$

Steady state errors, $e_{ss} = \dfrac{A}{1+K_p} = \dfrac{1}{1+\infty} = 0$

For unit ramp input, $A = 1$

$$e_{ss} = \frac{A}{K_v} = \frac{1}{\infty} = 0.$$

Example 4.7: Consider a unity feedback system with a closed-loop transfer function

$$\frac{C(s)}{R(s)} = \frac{KS+b}{S^2+aS+b}$$

Determine the open-loop transfer function $G(s)$.

Show that the steady state error with unit ramp input is given by $\left(\frac{a-K}{b}\right)$.

Solution: For a unity feedback system, $H(s) = 1$

Closed-loop transfer function

$$\frac{C(s)}{R(s)} = \frac{G(s)}{1+G(s)H(s)} = \frac{G(s)}{1+G(s)(1)}$$

$$\frac{C(s)}{R(s)} = \frac{G(s)}{1 + G(s)} = \frac{KS + b}{S^2 + aS + b}$$

By cross-multiplication, we get

$$(S^2 + aS + b)\,(G(s)) = (KS + b)\,(1 + G(s))$$

$$G(s)\,[(S^2 + aS + b) - (KS + b)] = KS + b$$

$$G(s) = \frac{KS + b}{S(S + (a - K)]}$$

The velocity error constant

$$K_v = \lim_{s \to 0} SG(S) = \lim_{s \to 0} \frac{S(KS + b)}{S(S + (a - K))}$$

$$K_v = \frac{b}{a - K}$$

The steady state error for unit ramp input,

$$e_{ss}(t) = \frac{1}{K_v} = \frac{1}{\dfrac{b}{a - K}} = \frac{a - K}{b}.$$

Example 4.8: For a unity feedback system having open-loop transfer function as

$$G(s) = \frac{K(S + 2)}{S^2(S^2 + 7S + 12)}$$

Determine (i) type and order of the system, (ii) error constants, and (iii) steady state error for parabolic input.

Solution: (i) Type 2 and order 4 system

(ii) For unity feedback system, $H(s) = 1$

$$G(s)\,H(s) = \frac{K(S + 2)}{S^2\,(S^2 + 7S + 12)}$$

The position error constant

$$K_p = \lim_{s \to 0} G(s)H(s)$$

$$K_p = \lim_{s \to 0} \frac{K(S + 2)}{S^2(S^2 + 7S + 12)}$$

$$K_p = \infty$$

The velocity error constant

$$K_v = \lim_{s \to 0} SG(s)H(s)$$

$$= \lim_{s\to 0} \frac{S(S+2)K}{S^2(S^2+7S+12)} = \infty$$

$$K_v = \infty$$

The acceleration error constant

$$K_a = \lim_{s\to 0} S^2G(s)H(s)$$

$$K_a = \lim_{s\to 0} \frac{S^2K(S+2)}{S^2(S^2+7S+12)} = \frac{2K}{12} = \frac{K}{6}$$

(iii) For parabolic input

$$R(s) = \frac{1}{S^2}$$

$$e_{ss} = \lim_{s\to 0} \frac{SR(s)}{1+G(s)H(s)}$$

$$e_{ss} = \lim_{s\to 0} \frac{S\left(\frac{1}{S^2}\right)}{1+\frac{K(S+2)}{S^2(S^2+7S+12)}}$$

$$e_{ss} = \lim_{s\to 0} \frac{\left(S\frac{1}{S^2}\right)}{\frac{S^2}{S^2}+\frac{K(S+2)}{S^2+7S+12}}$$

$$= \lim_{s\to 0} \frac{S}{S^2+\frac{K(S+2)}{S^2+7S+12}}$$

$$e_{ss} = 0.$$

Example 4.9: The unity feedback system is characterised by an open-loop transfer function, $G(s) = \frac{K}{S}(S+10)$.

Determine the gain K, so that the system will have a damping ratio of 0.5 for this value of K. Determine settling time, peak overshoot and time to peak overshoot for a unit step input.

Solution: The closed-loop transfer function

$$\frac{C(s)}{R(s)} = \frac{G(s)}{1+G(s)}$$

$$G(s) = \frac{K}{S}(S+10)$$

$$\frac{C(s)}{R(s)} = \frac{\frac{K}{S}(S+10)}{1+\frac{K}{S}(S+10)} = \frac{K}{S(S+10)+K} = \frac{K}{S^2+10S+K}$$

The value of K can be evaluated by comparing the system transfer function with standard form of second order transfer function.

$$\frac{C(s)}{R(s)} = \frac{\omega_n^2}{S^2+2\xi\,\omega_n S+\omega_n^2} = \frac{K}{S^2+10S+K}$$

On comparing, we get

$$\omega_n^2 = K, \qquad 2\xi\,\omega_n = 10$$

$$\omega = \sqrt{K} \qquad \xi = 0.5,\ \omega_n = \sqrt{K}$$

$$2 \times 0.5 \times \sqrt{K} = 10$$

$$\sqrt{K} = 10$$

$$\therefore \qquad k = 100,\ \omega_n = 10 \text{ rad/sec}$$

The value of gain $K = 100$

% peak overshoot, $\%\ M_p = e^{-\xi\pi/\sqrt{1-\xi^2}} \times 100$

$$M_p = e^{-0.5\pi/\sqrt{1-(0.5)^2}} \times 100$$

$$M_p = 0.163 \times 100$$

$$M_p = 16.3\%$$

Peak time,
$$t_p = \frac{\pi}{\omega_d} = \frac{\pi}{\omega_n\sqrt{1-\xi^2}} = \frac{\pi}{10\sqrt{1-(0.5)^2}}$$

$$t_p = 0.363 \text{ sec.}$$

Example 4.10: A unity feedback control system is characterized by the following open-loop transfer function, $G(s) = \frac{(0.4S+1)}{S(S+0.6)}$. Determine its transient response for unit step input and sketch the response. Evaluate the maximum overshoot and the corresponding peak time.

Solution: The closed-loop transfer function

$$\frac{C(s)}{R(s)} = \frac{G(s)}{1+G(s)\,H(s)}$$

$$\frac{C(s)}{R(s)} = \frac{\frac{(0.4S+1)}{S(S+0.6)}}{1+\frac{(0.4S+1)}{S(S+0.6)}} = \frac{0.4\,S+1}{S(S+0.6)+0.4S+1}$$

$$\frac{C(s)}{R(s)} = \frac{0.4S + 1}{S^2 + 0.6S + 0.4S + 1} = \frac{0.4S + 1}{S^2 + S + 1}$$

$$C(s) = \frac{(0.4\,S + 1)}{S^2 + S + 1} \,.\, R(s)$$

For step input, $R(s) = \frac{1}{S}$

$$\therefore \qquad C(s) = \frac{1}{S}\,\frac{(0.4\,S + 1)}{(S^2 + S + 1)} = \frac{0.4\,S + 1}{S(S^2 + S + 1)}$$

By partial fraction expansion $C(s)$ can be expressed as

$$C(s) = \frac{0.4S + 1}{S(S^2 + S + 1)} = \frac{A}{S} + \frac{BS + C}{S^2 + S + 1}$$

By solving $A = C(s)\cdot S\Big|_{S=0} = \frac{0.4S + 1}{S^2 + S + 1}\Big|_{S=0} = 1$

$$\frac{0.4\,S + 1}{S(S^2 + S + 1)} = \frac{A}{S} + \frac{BS + C}{S^2 + S + 1}$$

Cross multiplication, we get

$$0.4S + 1 = A\,(S^2 + S + 1) + S\,(BS + C)$$

$$0.4S + 1 = AS^2 + AS + A + BS^2 + CS$$

On equating coefficients of S^2, we get

$$O = A + B \quad \Rightarrow B = -A = -1$$

On equating coefficients of S, we get

$$0.4 = A + C$$

$$\therefore \qquad C = 0.4 - A = -0.6$$

$$\therefore \qquad C(s) = \frac{1}{S} + \frac{-S - 0.6}{S^2 + S + 1} = \frac{1}{S} - \frac{S + 0.6}{S^2 + S + 0.25 + 0.75}$$

$$= \frac{1}{S} - \frac{S + 0.5 + 0.1}{(S + 0.5)^2 + 0.75} = \frac{1}{S} - \frac{S + 0.5}{(S + 0.5)^2 + 0.75} - \frac{0.1}{\sqrt{0.75}}\,\frac{\sqrt{0.75}}{(S + 0.5)^2 + 0.75}$$

The time domain response is obtained by taking inverse Laplace transform of $C(s)$

$\therefore$ The response in time domain,

$$C(t) = L^{-1}[C(s)] = L^{-1}\left[\frac{1}{S} - \frac{S + 0.5}{(S + 0.5)^2 + 0.75} - \frac{0.1}{\sqrt{0.75}}\frac{\sqrt{0.75}}{(S + 0.5)^2 + 0.75}\right]$$

$$C(t) = 1 - e^{-0.5t}\cos\sqrt{0.75t} - \frac{0.1}{\sqrt{0.75}}e^{-0.5t}\sin\sqrt{0.75}t$$

$$C(t) = 1 - e^{-0.5t}\left[0.1155\sin(\sqrt{0.75}t) + \cos\sqrt{0.75}t\right]$$

The transient response is the part of the output which vanishes as t tends to infinity. Here as t tends to infinity the exponential component $e^{-0.5t}$ tends to zero. Hence, the transient response is given by the damped sinusoidal component.

The transient response,

$$C(t) = e^{-0.5t}\,[0.1155\sin(\sqrt{0.75}\,t) + \cos(\sqrt{0.75}t]$$

The value of ξ and ω_n can be estimated by comparing the characteristic equation of the system transfer function with standard form of second order characteristic equation

$$\therefore \qquad S^2 + 2\xi\,\omega_n\,S + \omega_n^2 = S^2 + S + 1$$

On comparing, we get

$$\omega_n^2 = 1 \qquad\qquad 2\xi\omega_n = 1$$

$$\omega_n = 1 \text{ rad/sec} \qquad \therefore \qquad \xi = \frac{1}{2\omega_n} = \frac{1}{2} = 0.5$$

Maximum overshoot, $M_p = e^{-\xi\pi/\sqrt{1-\xi^2}}$

$$M_p = e^{-\xi\pi/\sqrt{1-(0.5)^2}} = 0.163$$

$$\% M_p = M_p \times 100 = 0.163 \times 100 = 16.3\%$$

Peak time, $$bp = \frac{\pi}{\omega_d} = \frac{\pi}{\omega_n\sqrt{1-\xi^2}} = \frac{\pi}{\sqrt{1-(0.5)^2}} = 3.628 \text{ sec.}$$

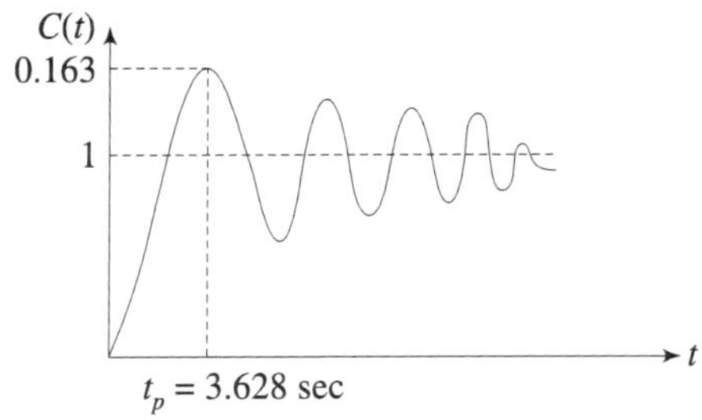

Fig. 4.21

Example 4.11: A unity feedback control system has an amplifier with gain, $K_A = 10$ and gain ratio, $G(s) = \frac{1}{S(S+2)}$. In the feedforward path. A derivative feedback, $H(s) = SK_0$, is introduced as a minor loop around $G(s)$. Determine the derivative feedback constant, K_0, so that the system damping factor is 0.6.

Solution: The given system can be represented by the block diagram shown in Fig. 4.22.

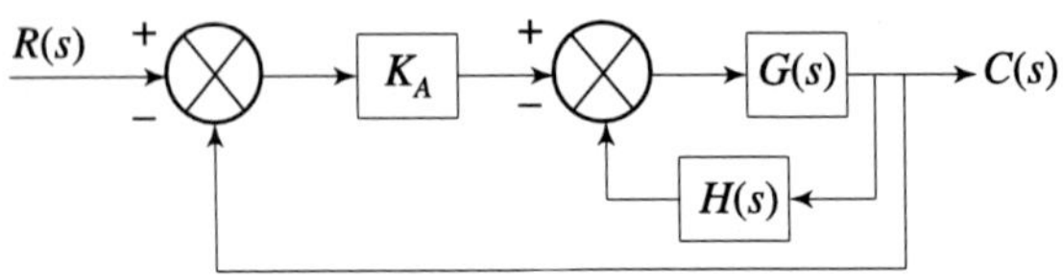

Fig. 4.22

$$K_A = 10,\ G(s) = \frac{1}{S(S+2)},\ H(s) = SK_0$$

Reducing negative feedback loop

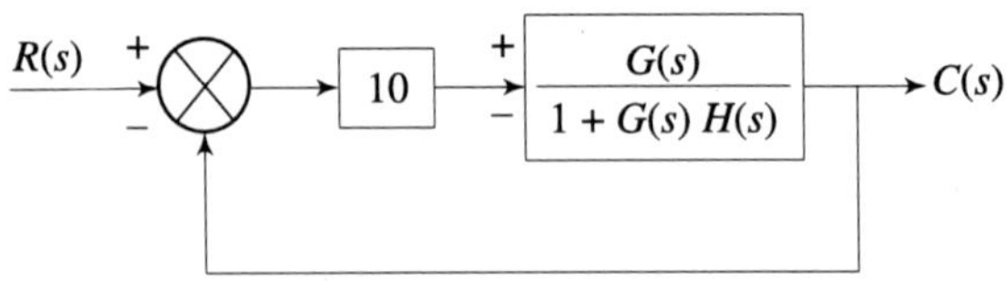

Fig. 4.23

$$\frac{G(s)}{1 + G(s)\,H(s)} = \frac{\frac{1}{S(S+2)}}{1 + \frac{1}{S(S+2)}SK_0} = \frac{1}{S(S+2) + 2K_0}$$

$$= \frac{1}{S^2 + 2S + SK_0} = \frac{1}{S^2 + (2 + K_0)S}$$

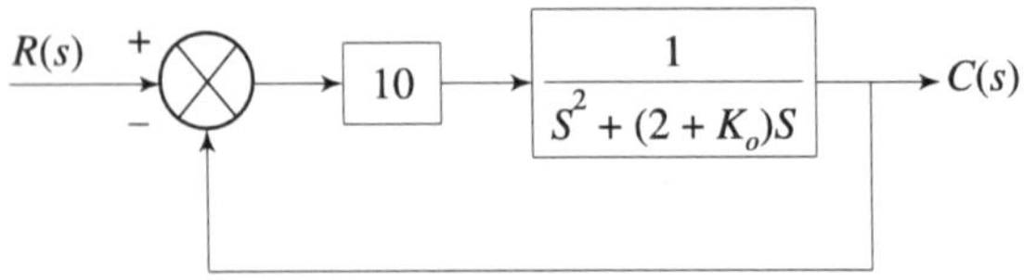

Fig. 4.24

Combining blocks in cascade.

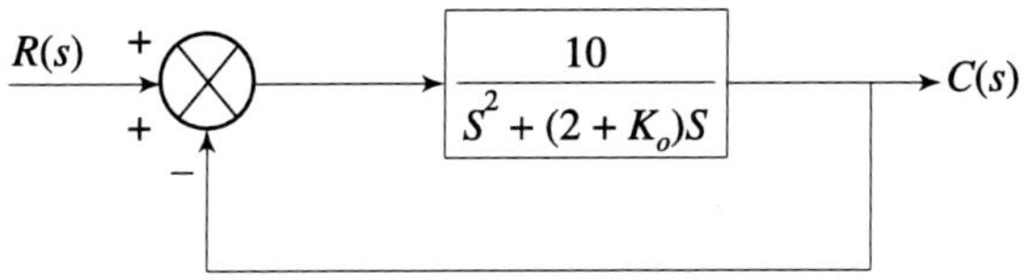

Fig. 4.25

Reducing the unity feedback path,

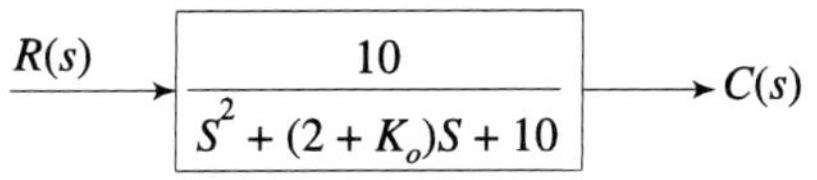

Fig. 4.26

The closed-loop transfer function,

$$\frac{C(s)}{R(s)} = \frac{10}{S^2 + (2 + K_0)\,S + 10}$$

The given system is a second order system. The value of K_0 can be determined by comparing the system function with standard form of second order transfer function

$$\frac{C(s)}{R(s)} = \frac{\omega_n^2}{S^2 + 2\xi\,\omega_n S + \omega_n^2} = \frac{10}{S^2 + (2 + K_0)\,S + 10}$$

$$\therefore \quad \omega_n^2 = 10, \qquad (2 + K_0) = 2\xi\,\omega_n$$

$$\omega_n = \sqrt{10} \qquad K_0 = 2\xi\,\omega_n - 2$$

$$K_0 = 2 \times 0.6 \times 3.162 - 2$$

$$\therefore \quad K_0 = 1.7944.$$

Example 4.12: A unity feedback control system has an open-loop transfer function, $G(s) = \frac{10}{S(S + 2)}$. Find the rise time, percentage overshoot, peak time and settling time for a step input of 12 units.

Solution: The closed-loop transfer function,

$$\frac{C(s)}{R(s)} = \frac{G(s)}{1 + G(s)} = \frac{\frac{10}{S(S+2)}}{1 + \frac{10}{S(S+2)}}$$

$$\frac{C(s)}{R(s)} = \frac{10}{S^2 + 2S + 10} = \frac{\omega_n^2}{S^2 + 2\xi\,\omega_n S + \omega_n^2}$$

On comparing, we get

$$\omega_n^2 = 10 \qquad\qquad 2\xi\omega_n = 2$$

$$\omega_n = \sqrt{10} = 3.162 \text{ rad/sec} \qquad\qquad \xi = \frac{2}{\omega_n} = \frac{1}{3.162}$$

$$\xi = 0.316$$

$$\theta = \tan^{-1}\frac{\sqrt{1-\xi^2}}{\xi} = \frac{\tan^{-1}\sqrt{1-(0.316)^2}}{0.316} = 1.249 \text{ rad}$$

$$\omega_d = \omega_n\sqrt{1-\xi^2} = 3.162\sqrt{1-(0.316)^2} = 3 \text{ rad/sec}$$

$$\text{Rise time, } t_r = \frac{\pi-\theta}{\omega_d} = \frac{\pi-1.249}{3} = 0.63 \text{ sec}$$

$$\% M_p = e^{\frac{-\xi\pi}{\sqrt{1-\xi^2}}} = e^{\frac{-0.316\pi}{\sqrt{1-(0.316)^2}}} \times 100 = 0.3512 \times 100$$

$$M_p = 35.12\%$$

$$\text{Peak overshoot} = \frac{35.12}{100} \times 12 \text{ units} = 4.2144 \text{ units}$$

$$\text{Peak time, } t_p = \frac{\pi}{\omega_d} = \frac{\pi}{3} = 1.047 \text{ sec}$$

$$\text{Time constant, } T = \frac{1}{\xi\omega_n} = \frac{1}{0.316 \times 3.162}$$

$$T = 1 \text{ sec}$$

For 5% error, settling time, $t_s = 3T = 3$ sec

For 2% error, settling time, $t_s = 4T = 4$ sec.

Example 4.13: A closed-loop is represented by the differential equation, $\frac{d^2c}{dt^2} + 8\frac{dc}{dt} = 64e$, where C is the displacement of the output, r is the displacement of the input and $e = r - C$. Determine undamped natural frequency, damping ratio and percentage maximum overshoot for unit step input.

Solution:

$$\frac{d^2c}{dt^2} + 8\frac{dc}{dt} = 64e$$

$$e = r - C$$

$$\therefore \quad \frac{d^2c}{dt^2} + 8\frac{dc}{dt} = 64(r - C)$$

Taking Laplace transform for the above equation,

$$S^2C(s) + 8S\,C(s) = 64\,[R(s) - C(s)]$$

$$S^2C(s) + 8S\,C(s) + 64\,C(s) = 64\,R(s)$$

$$C(s)[S^2 + 8S + 64] = 64\,R(s)$$

$$\frac{C(s)}{R(s)} = \frac{64}{S^2 + 8S + 64}$$

$$\therefore \quad \frac{C(s)}{R(s)} = \frac{\omega_n^2}{S^2 + 2\xi\,\omega_n S + \omega_n^2} = \frac{64}{S^2 + 8S + 64}$$

on comparing, we get

$$\omega_n^2 = 64 \qquad\qquad 2\xi\omega_n = 8$$

$$\omega_n = 8 \text{ rad/sec} \qquad\qquad 2\xi = \frac{8}{2\omega_n} = \frac{8}{2 \times 8} = 0.5$$

$$\%\,M_p = e^{\frac{-\xi\pi}{\sqrt{1-\xi^2}}} \times 100 = e^{\frac{-0.5\pi}{\sqrt{1-(0.5)^2}}} \times 100$$

$$\%\,M_p = 16.3\%.$$

Example 4.14: For a unity feedback control system the open-loop transfer function, $G(s) = \dfrac{10(S+2)}{S^2(S+1)}$.

Find (a) The position, velocity and acceleration error constants.

(b) The steady state error when the input is

$$R(s) = \frac{3}{S} - \frac{2}{S^2} + \frac{1}{3S^3}$$

Solution: To find static error constants,

$$K_p = \lim_{s\to 0} G(s)\,H(s) = \lim_{s\to 0} \frac{10(S+2)}{S^2(S+1)}$$

$$K_p = \infty$$

Velocity error constant, $K_v = \lim_{s\to 0} SG(s)\,H(s)$

$$K_v = \lim_{s\to 0} S\frac{10(S+2)}{S^2(S+1)} = \infty$$

Acceleration error constant,

$$K_a = \lim_{s\to 0} S^2G(s)\,H(s)$$

$$= \lim_{s\to 0} S^2\frac{10(S+2)}{S^2(S+1)}$$

$$K_a = 20$$

(b) To find steady state error

The error signal in S-domain,

$$E(s) = \frac{R(s)}{1 + G(s)\,H(s)}$$

$$R(s) = \frac{3}{S} - \frac{2}{S^2} + \frac{1}{3S^2}$$

$$E(s) = \frac{\frac{3}{S} - \frac{2}{S^2} + \frac{1}{3S^2}}{1 + \frac{10(S+2)}{S^2(S+1)}}$$

$$E(s) = \frac{3}{S}\left[\frac{S^2(S+1)}{S^2(S+1) + 10(S+2)}\right] - \frac{2}{S^2}\left[\frac{S^2(S+1)}{S^2(S+1) + 10(S+2)}\right] + \frac{1}{3S^3}\left[\frac{S^2(S+1)}{S^2(S+1) + 10\,(S+2)}\right]$$

The steady state error (e_{ss}) can be obtained from final value theorem.

Steady state error, $e_{ss} = \lim\limits_{s\to 0} e(t) = \lim\limits_{s\to 0} SE(S)$

$$e_{ss} = \lim_{s\to 0} S\left\{\frac{3}{S}\left(\frac{S^2(S+1)}{S^2(S+1) + 10(S+2)}\right) - \frac{2}{S^2}\left(\frac{S^2(S+1)}{S^2(S+1) + 10(S+2}\right) + \frac{1}{3S^3}\left(\frac{S^2(S+1)}{S^2(S+1) + 10(S+2)}\right)\right\}$$

$$e_{ss} = \lim_{s\to 0}\left\{\frac{3S^2(S+1)}{S^2(S+1) + 10(S+2)} - \frac{2S(S+1)}{S^2(S+1) + 10(S+2)} + \frac{(S+1)}{3S^2(S+1) + 30(S+2)}\right\}$$

$$e_{ss} = 0 - 0 + \frac{1}{60}$$

$$e_{ss} = \frac{1}{60}.$$

Example 4.15: For servomechanisms with open-loop transfer function given below explain what types of input signal give rise to a constant steady state error and calculate their values.

(i) $G(s) = \dfrac{20(S + 2)}{S(S + 1)(S + 3)}$ (ii) $G(s) = \dfrac{10}{(S + 2)(S + 3)}$

(iii) $G(s) = \dfrac{10}{S^2 (S + 1)(S + 2)}$

Solution: (i) $G(s) = \dfrac{20(S + 2)}{S(S + 1)(S + 3)}$

Let us assume $H(s) = 1$

The open-loop system has a pole at the origin. Hence, it is a type 1 system. In type 1 system, the velocity (ramp) input will give a constant steady state error. The steady state error with unit velocity input, $e_{ss} = \dfrac{1}{K_v}$.

Velocity error constant, $K_v = \lim_{s \to 0} SG(s)\,H(s)$

$$K_v = \lim_{s \to 0} S\frac{20(S + 2)}{S(S + 1)(S + 3)} = \frac{20 \times 2}{1 \times 3} = \frac{40}{3}$$

Steady state error, $e_{ss} = \dfrac{1}{K_v} = \dfrac{3}{40} = 0.075$

(ii) $G(s) = \dfrac{10}{(S + 2)(S + 3)}$

Let us assume $H(s) = 1$

There is no pole at origin ∴ It is a type O system.

In type O system, the step input will give a constant steady state error.

The steady state error with unit step input,

$$e_{ss} = \frac{1}{1 + K_p}$$

Position error constant, $K_p = \lim_{s \to 0} G(s)\,H(s)$

$$K_p = \lim_{s \to 0} \frac{10}{(S + 2)(S + 3)} = \frac{10}{6} = \frac{5}{3}$$

Steady state error, $e_{ss} = \dfrac{1}{1 + K_p} = \dfrac{1}{1 + \frac{5}{3}}$

$$e_{ss} = 0.375$$

(iii) $G(s) = \dfrac{10}{S^2(S + 1)(S + 2)}$

Let us assume $H(s) - 1$

There are two poles at the origin. Hence, it is a type 2 system. In type 2 system, the acceleration (parabolic) input will give a constant steady state error.

The steady state error with unit acceleration input,

$$e_{ss} = \frac{1}{K_a}$$

$$K_a = \lim_{s \to 0} S^2 G(s)H(s) = \lim_{s \to 0} S^2 \frac{10}{S^2(S+1)(S+2)}$$

$$K_a = \frac{10}{1 \times 2} = 5$$

Steady state error, $e_{ss} = \frac{1}{K_a} = \frac{1}{5} = 0.2$

Example 4.16: A unity feedback system has the forward transfer function, $G(S) = \dfrac{K_1(2S+1)}{S(5S+1)(1+S)^2}$.

The input, $r(t) = 1 + 6t$, is applied to the systems. Determine the minimum value of K_1 if the steady state error is to be less than 0.1.

Solution: $r(t) = 1 + 6t$

Taking Laplace transform

$$R(s) = L[r(t)] = \frac{1}{S} + \frac{6}{S^2}$$

The error signal in S-domain,

$$E(s) = \frac{R(s)}{1 + G(s)\,H(s)}$$

$$E(s) = \frac{\frac{1}{S} + \frac{6}{S^2}}{1 + \frac{K_1(2S+1)}{S(5S+1)(1+S)^2}}$$

$$E(s) = \frac{1}{S}\left[\frac{S(5S+1)(1+S)^2}{S(5S+1)(1+S)^2 + K_1(2S+1)}\right] + \frac{6}{S^2}\left[\frac{S(5S+1)(1+S)^2}{S(5S+1)(1+S)^2 + K_1(2S+1)}\right]$$

Steady state error e_{ss} can be obtained from final value theorem.

$$e_{ss} = \lim_{t \to 0} e(t) = \lim_{s \to 0} SE(s)$$

$$e_{ss} = \lim_{s\to 0} \left\{ S\left(\frac{1}{S} \left(\frac{S(5S+1)(1+S)^2}{S(5S+1)(1+S)^2 + K_1(2S+1)} \right) \right) + \frac{6}{S^2} \left[\frac{S(5S+1)(1+S)^2}{S(5S+1)(1+S)^2 + K_1(2S+1)} \right] \right\}$$

$$e_{ss} = \lim_{s\to 0} \left\{ \frac{S(5S+1)(1+S)^2}{S(5S+1)(1+S)^2 + K_1(2S+1)} + \frac{6(5S+1)\,(1+S)^2}{S(5S+1)(1+S)^2 + K_1\,(2S+1)} \right\}$$

$$e_{ss} = 0 + \frac{6}{K_1} = \frac{6}{K_1}$$

Given that $e_{ss} < 0.1$

$$\therefore \qquad 0.1 = \frac{6}{K_1}$$

or $$K_1 = \frac{6}{0.1} = 60$$

Example 4.17: Find K to limit the error of a system for input $1 + 8t + \frac{18}{2}t^2$ to 0.8 having

$$G(s)\,H(s) = \frac{K}{S^2\,(S+1)(S+4)}.$$

Solution: The input, $r(t) = 1 + 8t + \frac{18}{2}t^2$

$$R(s) = \frac{1}{S} + \frac{8}{S^2} + \frac{18}{S^3} = \frac{S^2 + 8S + 18}{S^3}$$

$$e_{ss} = \lim_{s\to 0} \frac{SR(s)}{1 + G(s)H(s)} = \lim_{s\to 0} \frac{S\left(\dfrac{S^2 + 8S + 18}{S^3}\right)}{1 + \dfrac{K}{S^2(S+1)(S+4)}}$$

$$e_{ss} = \lim_{s\to 0} \frac{(S^2 + 8S + 10)}{S^2 + \dfrac{K}{(S+1)(S+4)}} = \frac{10}{\left(\dfrac{K}{4}\right)} = \frac{40}{K}$$

$$\frac{40}{K} = 0.8$$

$$K = 50.$$

Example 4.18: For $r(t) = 0.2t$, it is required that $e_{SS} \le 0.02$. Determine the value of K for the system shown in Fig. 4.27.

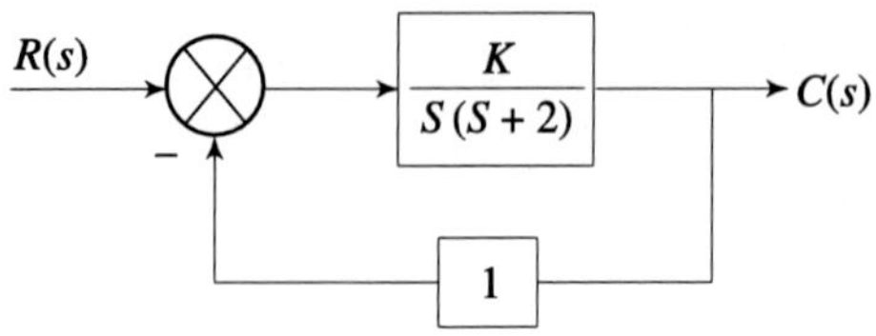

Fig. 4.27

Solution: $G(s) = \dfrac{K}{S(S+2)}, \; H(s) = 1$

$$G(s)\,H(s) = \frac{K}{S(S+2)}$$

$$r(t) = 0.2t$$

$$R(s) = 0.2t$$

$$R(s) = \frac{0.2}{S^2}$$

$$\therefore \qquad e_{ss} = \lim_{s\to 0} \frac{SR(s)}{1 + G(s)H(s)} = \lim_{s\to 0} \frac{S\left(\dfrac{0.2}{S^2}\right)}{1 + \dfrac{K}{S(S+2)}}$$

$$e_{ss} = \lim_{s\to 0} \left(\frac{0.2}{S + \dfrac{K}{S+2}}\right) = \frac{0.4}{K}$$

For $\qquad e_{ss} \le 0.2$

$$\frac{0.4}{K} \le 0.02$$

$$K \ge \frac{0.4}{0.02}$$

$$K \ge 20$$

Example 4.19: Determine the value of K, M, and J if M_p = 28%, t_p = 2.2 and output at steady state is 6 units for the feedback system shown in Fig. 4.28 having unit step input.

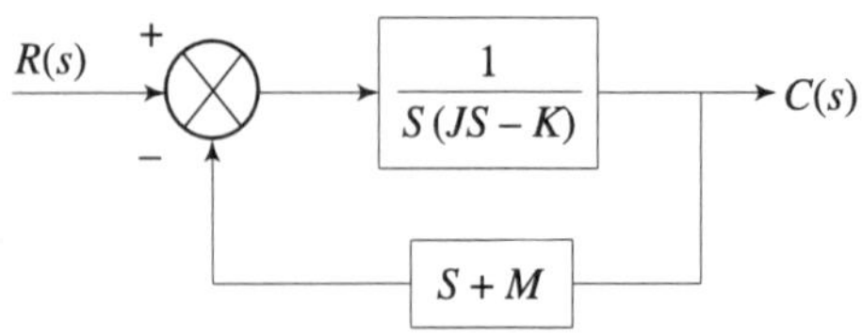

Fig. 4.28

Solution: $G(s) = \dfrac{1}{S(JS-K)}; \; H(s) = S + M$

$$\frac{C(s)}{R(s)} = \frac{G(s)}{1 + G(s)H(s)} = \frac{1}{S(JS-K) + (S+M)}$$

$$\frac{C(s)}{R(s)} = \frac{\left(\frac{1}{J}\right)}{S^2 + S\left(\frac{1-K}{J}\right) + \frac{M}{J}}$$

$$\omega_n^2 = \frac{M}{J} \qquad \text{or} \quad \omega_n = \sqrt{\frac{M}{J}}$$

$$2\xi\omega_n = \frac{1-K}{J} \qquad \text{or} \quad \xi = \frac{1}{2\omega_n}\left(\frac{1-K}{J}\right)$$

$$\xi = \frac{1-K}{2\sqrt{MJ}}$$

$$M_p = 28\%$$

$$28 = 100 \times e^{\frac{-\xi\pi}{\sqrt{1-\xi^2}}}$$

$$0.28 = e^{\frac{-\xi\pi}{\sqrt{1-\xi^2}}}$$

$$\ln(0.28) = \frac{-\xi\pi}{\sqrt{1-\xi^2}}$$

$$-1.2729 = \frac{-\xi\pi}{\sqrt{1-\xi^2}}$$

$$\therefore \qquad \xi = 0.3757$$

$$2.2 = \frac{\pi}{\omega_d}$$

$$\omega_d - \frac{\pi}{22} = 1.427 \text{ rad/sec}$$

$$\omega_d = \omega_n\sqrt{1-\xi^2}$$

$$\therefore \qquad \omega_n = \frac{\omega_d}{\sqrt{1-\xi^2}} = \frac{1.427}{\sqrt{1-(0.3757)^2}}$$

$$\omega_n = 1.5398 \text{ rad/sec}$$

The output at steady state error,

$$C_{SS} = \lim_{s\to 0} C(t) = \lim_{s\to 0} SC(s)$$

$$C_{SS} = \lim_{s\to 0} S\left[\frac{\left(\frac{1}{J}\right)}{S^2 + \left(\frac{1-K}{J}\right)S + \frac{M}{J}}\right] R(s)$$

$$R(t) = u(t)$$

$$R(s) = \frac{1}{S}$$

$$C_{SS} = \lim_{s \to 0} S\left[\frac{\left(\frac{1}{J}\right)}{S^2 + \left(\frac{1-K}{J}\right)S + \frac{M}{J}}\right] \cdot \frac{1}{S} = 6$$

$$M = \frac{1}{6} = 0.1667$$

$$\omega_n = \sqrt{\frac{M}{J}}$$

$$J = \frac{M}{\omega_n^2} = \frac{0.1667}{1.5398} = 0.10826$$

$$J = 0.10826.$$

Example 4.20: Obtain the transfer function of the second order system which has a peak overshoot of 9.5% for unit step input. The time to peak overshoot is $\frac{\pi}{12}$ seconds.

Solution:

$$t_p = \frac{\pi}{12}$$

$$M_p = 9.5\%$$

$$0.095 = e^{\frac{-\pi\xi}{\sqrt{1-\xi^2}}}$$

$$\therefore \quad \xi = 0.6$$

$$t_p = \frac{\pi}{12}$$

$$\frac{\pi}{\omega_n\sqrt{1-\xi^2}} = \frac{\pi}{12}$$

$$\therefore \quad \omega_n = 15$$

The transfer function of the system,

$$\frac{C(s)}{R(s)} = \frac{\omega_n^2}{S^2 + 2\xi\,\omega_n S + \omega_n^2} = \frac{225}{S^2 + 18S + 225}$$

Important to Remember

Second order system:

$$T(s) = \frac{C(s)}{R(s)} = \frac{\omega_n^2}{S^2 + 2\,\zeta\,\omega_n S + \omega_n^2}$$

Underdamped

$$C(t) = 1 \quad \frac{e^{-\zeta\omega_n t}}{\sqrt{1-\zeta^2}} \sin(\omega_d t + \theta),\ 0 < \zeta < 1$$

$$\theta = \tan^{-1}\frac{1-\zeta^2}{\zeta},\ \omega_d = \omega_n = \sqrt{1-\zeta^2}$$

Undamped

$$C(t) = 1 - \cos \omega_n t.$$

$$\text{Rise time } (t_r) = \frac{\pi - \theta}{\omega_d}$$

$$\text{Peak time } (t_p) = \frac{\pi}{\omega_d}$$

Maximum overshoot (M_p):

$$\% M_p = e^{\frac{-\pi\zeta}{\sqrt{1-\zeta^2}}} \times 100$$

$$\text{Settling time, } t_s = \frac{4}{\zeta\omega_n}$$

$$\text{Delay time } (t_d) = \frac{1 + 0.7\zeta}{\omega_n}$$

Example 4.21: Consider the servomechanism shown in Fig. 4.29. Determine values of K_1 and K_2 so that the maximum overshoot in unit step response is 20% and peak is 3 sec.

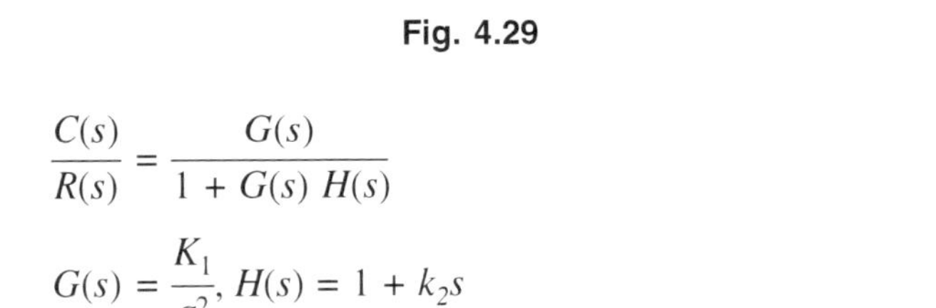

Fig. 4.29

Solution:

Here

$$\frac{C(s)}{R(s)} = \frac{G(s)}{1 + G(s)\,H(s)}$$

$$G(s) = \frac{K_1}{S^2},\ H(s) = 1 + k_2 s$$

$$\frac{C(s)}{R(s)} = \frac{\frac{K_1}{S^2}}{1 + \left(\frac{K_1}{S^2}\right)(1 + K_2 S)} = \frac{K_1}{S^2 + K_1 + K_1 K_2 S}$$

$$\frac{C(s)}{R(s)} = \frac{K_1}{S_2 + K_1K_2S + K_1}$$

The equation is in the form of

$$\frac{C(s)}{R(s)} = \frac{\omega_n^2}{S^2 + 2\zeta\omega_n S + \omega_n^2}$$

where $K_1 = \omega_n^2,\ K_1K_2 = 2\zeta\omega_n$

Here $M_p = 0.2,\ t_p = 3 \text{ sec},\ M_p = e^{\frac{-\pi\zeta}{\sqrt{1-\zeta^2}}} = 0.2$

Taking natural logarithm on both sides

$$-\frac{\pi\zeta}{\sqrt{1-\zeta^2}} = -1.609 \qquad \therefore \zeta = 0.456$$

$$\text{Peak time, } t_p = \frac{\pi}{\omega_d} = \frac{\pi}{\omega_n\sqrt{1-\zeta^2}}$$

$$\omega_n = \frac{\pi}{t_p\sqrt{1-\zeta^2}} = \frac{\pi}{3\sqrt{1-(0.456)^2}} = 1.176$$

$\therefore\ \omega_n = 1.176$ rad/sec and $\zeta = 0.456$

Substituting $K_1 = \omega_n^2 = (1.176)^2 = 1.383$

and $K_2 = \dfrac{2\zeta\omega_n}{K1} = \dfrac{2 \times 0.456 \times 1.176}{1.383}$

$\mathbf{K_2 = 0.7738.}$

Example 4.22: For the system given in Fig. 4.30, determine the values of K and K_n so that M_p is 20% and peak time, $t_p = 1$ sec. Also obtain rise time (t_r) and settling time (t_s) for these values.

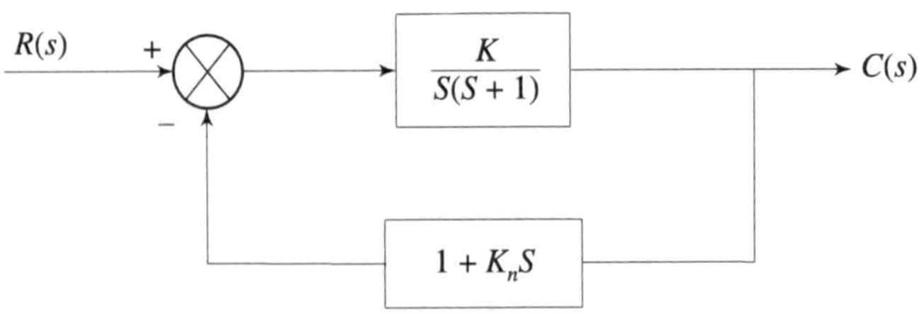

Fig. 4.30

Solution:

$$\frac{C(s)}{R(s)} = \frac{\frac{K}{S(S+1)}}{1 + \frac{K}{S(S+1)}(1 + K_nS)} = \frac{K}{S^2 + (K.K_n + 1)S + K}$$

Comparing with the transfer function of a second order system,

$$\frac{C(s)}{R(s)} = \frac{\omega_n^2}{S^2 + 2\,\zeta\omega_n S + \omega_n^2}$$

We get $K = \omega_n^2,\ 2\zeta\omega_n = K \cdot Kn + 1$

To determine ζ, $M_p = e^{-\frac{\pi\zeta}{\sqrt{1-\zeta^2}}}$

$$0.2 = e^{\frac{-\pi\zeta}{\sqrt{1-\zeta^2}}}$$

Taking natural logarithm on both sides of

$$-\frac{\pi\zeta}{\sqrt{1-\zeta^2}} = -1.6,\ \zeta = 0.456$$

$$t_p = \frac{\pi}{\omega_d} = \frac{\pi}{\omega_n\sqrt{1-\zeta^2}} \qquad \therefore\ \omega_n = \frac{\pi}{tp\sqrt{1-\zeta^2}}$$

For $t_p = 1$ sec

$$\omega_n = \frac{\pi}{\sqrt{1-\zeta^2}} = 3.53 \text{ rad/sec}$$

$$K = \omega_n^2 = 12.46 \qquad K_n = \frac{2\zeta\omega_n - 1}{K} = \frac{2(0.456)(3.53) - 1}{12.46}$$

$$K = 12.46 \qquad K_n = 0.1781$$

Example 4.23: For the system shown in Fig. 4.31 determine the value of K, so that damping ratio is 0.5. Also obtain rise time (t_p), maximum overshoot and settling time for unit step input.

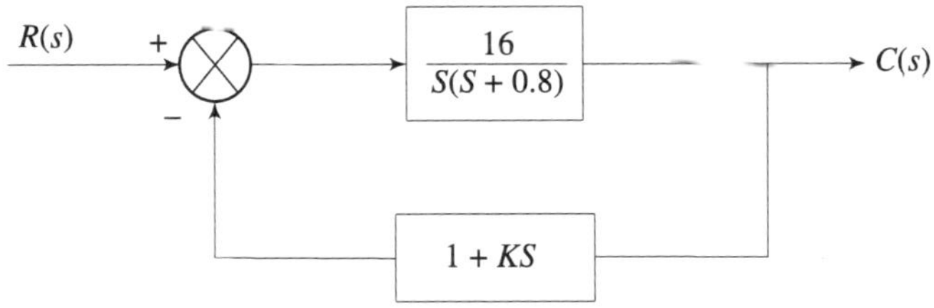

Fig. 4.31

Solution:

$$\frac{C(s)}{R(s)} = \frac{16}{S^2 + (16K + 0.8)S + 16}$$

Comparing $\frac{C(s)}{R(s)} = \frac{\omega_n^2}{S^2 + 2\zeta\,\omega_n S + \omega_n^2}$, $\omega_n^2 = 16$,

$$\omega_n = 4 \text{ rad/sec}$$

$$2\zeta\omega_n = 16K + 0.8$$

$$K = \frac{2\,\zeta\omega_n - 0.8}{16} \quad \text{and} \quad \zeta = 0.5 \text{ (given)}$$

$$K = \frac{2(0.5)4 - 0.8}{16} = 0.2$$

Rise time, $$t_r = \frac{\pi - \theta}{\omega_n\sqrt{1-\zeta^2}},\ \theta = \frac{\tan^{-1}\sqrt{1-\zeta^2}}{\zeta} = 1.047 \text{ rad}$$

$\therefore$ $$t_r = \frac{\pi - 1.047}{4\sqrt{1-(0.5)^2}}, = 0.604 \text{ sec.}$$

$$t_p = \frac{\pi}{\omega_d} = \frac{\pi}{3.464} = 0.906 \text{ sec.}$$

Max. overshoot, $$M_p = e^{-\frac{\pi\zeta}{\sqrt{1-\zeta^2}}} = e^{\frac{-\pi(0.5)}{\sqrt{1-(0.5)^2}}} = 0.163$$

For 2% criterion, $$t_s = \frac{4}{\zeta\,\omega_n} = 2 \text{ sec.}$$

For 5% criterion, $$t_s = \frac{3}{\zeta\,\omega_n} = 1.5 \text{ sec.}$$

Example 4.24: Negative feedback system has a forward path transfer function $= \dfrac{K}{S(S+1)}$ while the feedback path transfer function is $(1 + as)$. If this system is to have a peak time of 0.5 sec and also a 10% overshoot, for a unit step input, determine K and a.

Solution: $$\frac{C(s)}{R(s)} = \frac{G(s)}{1 + G(s)\,H(s)}$$

$$= \frac{\dfrac{K}{S(S+1)}}{1 + \dfrac{K}{S(S+1)}(1+as)}$$

$$= \frac{K}{S^2 + (Ka+1)S + K} \quad \text{...(4.6)}$$

Comparing Eq. (4.6) with the transfer function of a second order system,

$$\frac{C(s)}{R(s)} = \frac{\omega_n^2}{S^2 + 2\zeta\,\omega_n S + \omega_n^2}$$

$$K = \omega_n^2,\ (K_a + 1) = 2\zeta\omega_n \qquad ...(4.7)$$

To determine ζ,

$$M_p = e^{\frac{-\pi\zeta}{\sqrt{1-\zeta^2}}} \Rightarrow 0.1 = e^{\frac{-\pi\zeta}{\sqrt{1-\zeta^2}}} \qquad ...(4.8)$$

Taking natural logarithm on both sides of Eq. (4.8)

$$\frac{-\pi\zeta}{\sqrt{1-\zeta^2}} = -\ 2.303$$

$$\pi^2\zeta^2 = 5.3\ (1 - \zeta^2)$$

$$\zeta = \mathbf{0.592}$$

To determine ω_n,

$$t_p = \frac{\pi}{\omega_n\sqrt{1-\zeta^2}}$$

$$0.5 = \frac{\pi}{\omega_n\sqrt{1-(0.592)^2}}$$

$$\omega_n = \frac{\pi}{0.5\ \sqrt{(0.651)}} = 7.786 \text{ rad/sec}$$

$$K = \omega_n^2 = 60.62$$

$$K_a + 1 = 2\zeta\omega_n$$

$$K_a + 1 = 2\ (0.592)\ (7.786) = 9.22$$

$$a = \frac{9.22 - 1}{K} = \frac{8.22}{60.62} = 0.136$$

$$a = 0.136$$

Example 4.25: The closed-loop transfer function of a second order system is $\frac{C(s)}{R(s)} = \frac{25}{S^2 + 6S + 25}$. Find rise time, peak time, maximum overshoot and settling time, if the system is subjected to unit step input. Assume allowable steady state error as 2%. Find also its output response.

Solution:

$$\frac{C(s)}{R(s)} = \frac{25}{S^2 + 6S + 25}$$

$$\frac{C(s)}{R(s)} = \frac{\omega_n}{S^2 + 2\zeta\omega_n S + \omega_n^2} \qquad \Leftarrow \text{General form of second order system.}$$

Comparing both $\omega_n^2 = 25$, $\omega_n = 5$ rad/sec

$$2\zeta\omega_n = 6,\ \zeta = \frac{6}{2\omega_n} = \frac{6}{2 \times 5} = \frac{6}{10} = 0.6$$

$$\theta = \tan^{-1}\frac{\sqrt{1-\zeta^2}}{\zeta},\ \theta = 0.927 \text{ rad.}$$

$$t_r = \frac{\pi - \theta}{\omega_n\sqrt{1-\zeta^2}}$$

$$= \frac{\pi - 0.927}{5\sqrt{1-0.6^2}} = 0.5354 \text{ sec.}$$

$$t_p = \frac{\pi}{\omega d} = \frac{\pi}{\omega_n\sqrt{1-\zeta^2}} = \frac{\pi}{4} = 0.7854 \text{ sec.}$$

$$M_p = e^{\frac{-\pi\zeta}{\sqrt{1-\zeta^2}}} = 0.0948, \quad \%\ M_p = 9.48$$

For 2% criterion,

$$t_s = \frac{\pi}{\zeta\omega_n} = \frac{4}{0.6 \times 5} = 1.333 \text{ sec.}$$

$$C(t) = 1 - \frac{e^{-\zeta\omega_n t}}{\sqrt{1-\zeta^2}}\sin(\omega_d t + \theta)$$

$$C(t) = 1 - 1.25\, e^{-3t}\sin(4t + 0.927)$$

Example 4.26: Figure 4.32 shows a mechanical system and the response when 10 N of force is applied to the system. Determine the value of M, B and K.

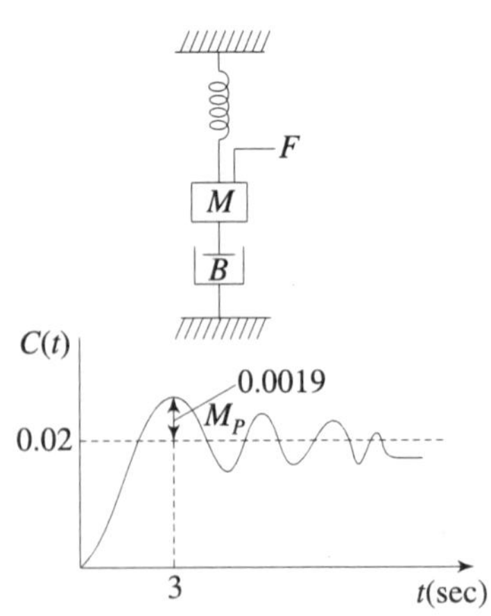

Fig. 4.32

Solution: In the given system in Fig. 4.32, when a force of 10 N is applied, the maximum overshoot is 0.00193 m and peak time is 3 sec. The steady state value is 0.02 m.

The differential equation for the system,

$$f = \frac{M \cdot d^2x}{dt^2} + B \cdot \frac{dx}{dt} + Kx \qquad \text{...(4.9)}$$

Taking Laplace transformation of Eq. (4.9)

$$F(s) = (MS^2 + BS + K)\, x(s)$$

∴ Transfer function,

$$\frac{X(s)}{F(s)} = \frac{1}{MS^2 + BS + K} \qquad \text{...(4.10)}$$

The output transform, $X(s) = \dfrac{F(s)}{MS^2 + BS + K}$

Here $F(s) = \dfrac{10}{s}$

∴ $X(s) = \dfrac{10}{S(MS^2 + BS + K)}$

To determine K,

Here steady state value, $x(\infty) = 0.02$

∴

$$X(\infty) = \lim_{s \to 0} = \frac{10}{MS^2 + BS + K} = \frac{10}{K} = 0.02$$

$$K = \frac{10}{0.02} = 500$$

$$M_p = 0.096 = e^{\frac{-\pi\zeta}{\sqrt{1 - \zeta^2}}}$$

From the above equation, $\zeta = 0.6$

From Fig. 4.32, peak time, $t_p = \dfrac{\pi}{\omega_n\sqrt{1 - \zeta^2}} = 3$ sec.

∴

$$\omega_n = \frac{\pi}{3\sqrt{1 - (0.6)^2}} = 1.31 \text{ rad/sec.}$$

By comparing Eq. (4.10) with transfer function for a second order system, only for dominator.

$$\frac{B}{M} = 2\,\zeta\omega_n \text{ and } \frac{K}{M} = \omega_n^2$$

$$M = \frac{K}{\omega_n^2} = \frac{500}{(1.3)^2} = 291.35 \text{ kg.}$$

$$B = (2\zeta\omega_n)(M) = 458 \text{ N/m/sec.}$$

Example 4.27: The step response of a second order system is shown in Fig. 4.33 for an input of $2u(t)$. Determine the open-loop and closed-loop transfer functions. Assume unity feedback system.

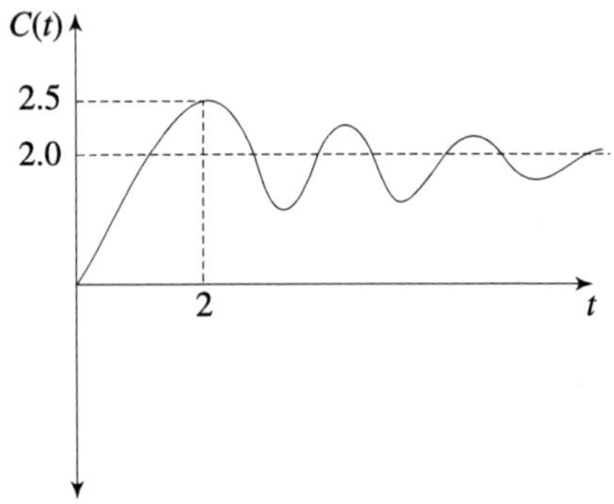

Fig. 4.33

Solution: From Fig. 4.33, $t_p = 2$ sec.

$$M_p = 0.25 \text{ at steady state}$$

$$\% M_p = \frac{C(tp) - C(\infty)}{C(\infty)}$$

$$\pi\zeta = 1.386\sqrt{1-\zeta^2}$$

Squaring on both sides

$$\pi^2\zeta^2 = 1.921(1-\zeta^2)$$

$$\zeta^2 \cdot (11.791) = 1.921$$

$$\zeta^2 = \frac{1.921}{11.79} = 0.163, \quad \zeta = \mathbf{0.4036}$$

$$t_p = \frac{\pi}{\omega_n\sqrt{1-\zeta^2}} = \frac{\pi}{\omega_d}$$

$$2 = \frac{\pi}{\omega_n\sqrt{1-(0.4036)^2}} = 2\left(\omega_n\sqrt{1-(0.4036)^2}\right) = \pi$$

$$\omega_n^2(1-0.163)\,4 = \pi^2$$

$$\omega_n^2 = \frac{\pi^2}{3.348} = 2.948$$

$$\therefore \qquad \omega_n = 1.717 \text{ rad/sec}$$

Closed-loop transfer function for standard form

$$\frac{C(s)}{R(s)} = \frac{\omega_n^2}{S^2 + 2\zeta\omega_n^2 + \omega_n^2} = \frac{(1.717)^2}{S^2 + 2 \times 403 \times 1.71 \times S + (1.71)^2}$$

$$\frac{C(s)}{R(s)} = \frac{2.94}{S^2 + 1.38S + 2.94} = \frac{G(s)}{1 + G(s) \cdot H(s)}$$

$$= \frac{\dfrac{2.94}{S^2 + 1.38S}}{1 + \dfrac{2.94}{S^2 + 1.38S}} = \frac{\dfrac{294}{S^2 + 1.38S}}{\dfrac{S^2 + 1.38S + 2.94}{S^2 + 1.38S}}$$

Divide by $(S^2 + 1.38S)$ to numerator and denominator

$$\frac{C(s)}{R(s)} = \frac{\dfrac{2.94}{S^2 + 1.38S}}{\dfrac{1 + 2.94}{S^2 + 1.38S}} = \frac{G(s)}{1 + G(s) \cdot H(s)}$$

$$\therefore \qquad G(s) = \frac{2.94}{S^2 + 1.38S} \text{ and } H(s) = 1.$$

Example 4.28: Figure 4.33 shows a mechanical vibratory system when a force of 8.9 N is applied to the system, the mass oscillates as shown in Fig. 4.34. Find the values of M, B, K.

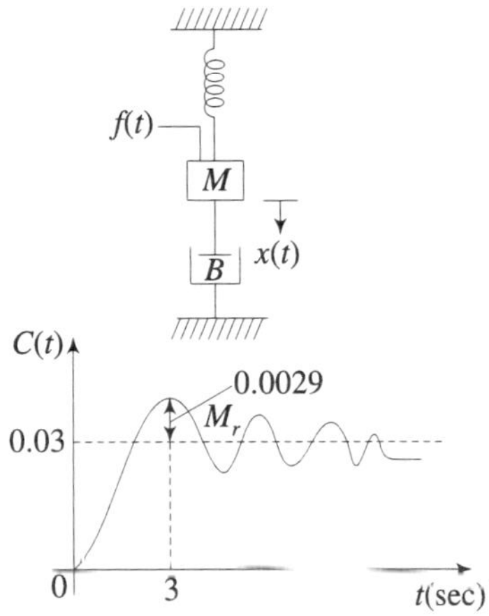

Fig. 4.34

Solution: The system equations are:

$$f(t) = \frac{M \cdot d^2\theta}{dt} + \frac{B \cdot d\theta}{dt} + K \cdot \theta(t)$$

Taking Laplace transform,

$$F(s) = M \cdot S^2 X(s) + B\, S\, X(s) + K \cdot X(s)$$

$$F(s) = X(s)\,[MS^2 + BS + K]$$

$$\therefore \qquad \frac{X(s)}{F(s)} = \frac{1}{MS^2 + BS + K}$$

Compare denominator with standard equation

$$S^2 + 2\zeta\omega_n \cdot S + \omega_n^2 = 0$$

$$\therefore \qquad \omega_n = \sqrt{\frac{K}{M}},\ 2\zeta\omega_n = \frac{B}{M}$$

Given $f(t) = 8.9$ N

$$F(s) = \frac{8.9}{S}$$

$$\frac{X(s)}{\frac{8.9}{S}} = \frac{1}{S^2 + \frac{B}{M} \cdot S + \frac{K}{M}}$$

$$X(s) = \left(\frac{8.9}{s}\right) \frac{1}{MS^2 + BS + K}$$

$$X(\infty) = \lim_{s \to 0} S \cdot X(S)$$

$$0.03 = \lim_{s \to 0} S\left(\frac{8.9}{S}\right)\left(\frac{1}{MS^2 + BS + K}\right)$$

$$0.03 = \frac{8.9}{K}$$

$$(0.03)\,(K) = 8.9$$

$$K = \frac{8.9}{0.03} = 296.67\ \omega/\text{m}$$

$$M_p = e^{\frac{-\pi\zeta}{\sqrt{1-\zeta^2}}}$$

$$(0.0029) = (0.03)\, e^{\frac{-\pi\zeta}{\sqrt{1-\zeta^2}}}$$

$$e^{\frac{-\pi\zeta}{\sqrt{1-\zeta^2}}} = \frac{0.0029}{0.03} = 0.0966$$

$$\frac{-\pi\zeta}{\sqrt{1-\zeta^2}} = -2.336$$

$$\pi^2\zeta^2 = 2.336 - \zeta^2\,(2.336)$$

$$\zeta^2 = \frac{2.336}{12.21} = 0.19138$$

$$\zeta = 0.4375$$

$$t_p = 2 \text{ sec.}$$

$$t_p = \frac{\pi}{\omega_n\sqrt{1-\zeta^2}}$$

$$\therefore \quad \omega_n = \frac{\pi}{t_p\sqrt{1-\zeta^2}} = \frac{\pi}{2\sqrt{1-\zeta^2}} = \frac{\pi}{2\sqrt{1-(0.4375)^2}} = 1.7468$$

$$\omega_n^{\;2} = \frac{K}{M}$$

$$(1.7446)^2 = \frac{296.67}{M}$$

$$M = 97.28 \text{ kg.} \qquad 2(0.4375)\,(1.7468) = \frac{B}{97.28}$$

$$2\zeta\omega_n = \frac{B}{M} \qquad B = 148.69 \text{ N/m/sec.}$$

Example 4.29: The open-loop transfer function of a unity feedback control system is given by

$G(s) = \dfrac{K}{S(ST+1)}$. Find:

(i) By what factor should K is multiplied so that damping ratio is reduced from 0.2 to 0.8.

(ii) By what factor should T be multiplied so that damping ratio is reduced from 0.6 to 0.3.

Solution:

$$G(s) = \frac{K}{S(ST+1)},\; H = 1$$

$$\frac{C}{R} = \frac{G}{1+GH} = \frac{\dfrac{K}{S(ST+1)}}{1+\dfrac{K}{S(ST+1)}} = \frac{K}{S^2T+S+K}$$

$$\frac{C}{R} = \frac{\dfrac{K}{S(ST+1)}}{\dfrac{S^2T+S+K}{S(ST+1)}} = \frac{K}{S^2T+S+K} \div \text{by } T.$$

$$\frac{C}{R} = \frac{\frac{K}{T}}{S^2 + S/T + K/T}$$

Comparing dominator $(S^2 + S/T + K/T)$ with $(S^2 + 2\zeta\omega_n S + \omega_n^2)$

$$\therefore \qquad \omega_n^2 = \frac{K}{T},\ 2\omega_n\,\zeta = \frac{1}{T}$$

Here $\qquad \zeta = \dfrac{1}{2\,\omega_n T}$. But $\omega_n = \sqrt{\dfrac{K}{T}}$

$$\zeta = \frac{1}{2\sqrt{T}\cdot\frac{\sqrt{K}}{\sqrt{T}}\cdot\sqrt{T}} = \frac{1}{2\sqrt{KT}}$$

Here $\qquad \zeta_1 = 0.2 = \dfrac{1}{2\sqrt{K_1T_1}} \quad\Rightarrow\quad 0.4 = \dfrac{1}{\sqrt{K_1T_1}}$

Now $\qquad \zeta_2 = 0.8 = \dfrac{1}{2\sqrt{K_2T_2}} \quad\Rightarrow\quad 1.6 = \dfrac{1}{\sqrt{K_2T_2}}$

But $T_2 = T_1$ squaring

$$\frac{\frac{1}{\sqrt{K_1T_1}}}{\frac{1}{\sqrt{K_2T_2}}} = \frac{0.4}{1.6} \quad\Rightarrow\quad \frac{\sqrt{K_2T_2}}{\sqrt{K_1T_1}} = 0.25 \quad\Rightarrow\quad \frac{K_2.T_2}{K_1.T_1} = 0.0625$$

$$\frac{K_2}{K_1} = 0.0625$$

Thus, $\qquad K_2 = 0.0625,\ K_1 = 1$

Thus reduce gain 0.0625 times previous value.

(ii) $\qquad \zeta = 0.6 = \dfrac{1}{2\sqrt{K_1T_1}} \quad\Rightarrow\quad 0.6 \times 2 = \dfrac{1}{\sqrt{K_1T_1}}$

$$\zeta_2 = 0.3 = \frac{1}{2\sqrt{K_2T_2}} \quad\Rightarrow\quad 0.3 \times 2 = \frac{1}{\sqrt{K_2.T_2}}\cdot\frac{T_2}{T_1} = 1$$

But now $K_1 = K_2$

$$\therefore \qquad \frac{\zeta_1}{\zeta_2} = \frac{0.6}{0.3} = \sqrt{\frac{T_2}{T_1}}$$

Example 4.30: Measurements conducted in a servomechanism show the system response to be $C(t) = 1 + 0.2\,e^{-60t} - 1.2\,e^{-10t}$ when subjected to a unit step input. Obtain the expression for closed-loop transfer function the damping ratio and undamped natural frequency of oscillation.

$$C(t) = 1 + 0.2\, e^{-60t} - 1.2 \cdot e^{-10t}$$

Solution: Taking *L.T.* of $C(t)$.

$$\frac{C(s)}{R(s)} = \frac{\frac{1}{S} + \frac{0.2}{S+60} - \frac{1.2}{S+10}}{\frac{1}{S}} = \frac{1 + 0.25}{S+60} - \frac{1.2S}{S+10}$$

$$= \frac{(S+10)(S+60) + 0.2S(S+10) - 1.25(S+60)}{(S+60)(S+60)}$$

$$= \frac{600}{S^2 + 70S + 600} \rightarrow \text{Closed-loop transfer function.}$$

$\therefore$

$$\omega_n^2 = 600$$

$$\omega_n = 24.495 \text{ rad/sec and } 2\zeta\omega_n = 70$$

$$\zeta = \frac{70}{2.\omega_n} = 1.4288$$

Example 4.31: Identify the type of the system:

(i) $$G(s) = \frac{K(1+3S)}{S^2},\ H(s) = \frac{1+5S}{S^2+6S+7}$$

$$G(s)\,H(s) = \frac{K(1+3S)(1+5S)}{S^2(S^2+6S+7)}$$

This is type 2 system.

(ii) $$G(s) = \frac{K}{S} \text{ and } H(s) = \frac{1}{S(S+2)}$$

$$G(s)\,H(s) = \frac{K(1)}{S\,(S(S+2))} = \frac{K}{S^2(S+2)}$$

This is a type 2 system.

(iii) $$G(s) = \frac{4}{S^2+6S+7} \text{ and } H(s) = (S+3)$$

$$G(s)\,H(s) = \frac{4(S+3)}{S^2+6S+7}$$

This is type zero system.

(iv) $$G(s) = \frac{5}{s} \text{ and } H(s) = \frac{1}{s}$$

$$G(s)\,H(s) = \frac{5}{S^2}$$

This is a type 2 system.

Example 4.32: Find the static error coefficients for a system having

$$G(s)\,H(s) = \frac{(S+3)}{S(1+0.60S)\,(1+0.35S)}.$$

Solution:

$$K_p = \lim_{s\to 0} G(s)\,H(s) = \lim_{s\to 0} \frac{(S+3)}{S(1+0.60S)\,(1+0.35S)}$$

$$K_p = \infty$$

$$K_v = \lim_{s\to 0} SG(s)\,H(s) = \lim_{s\to 0} \frac{S(S+3)}{S(1+0.60S)\,(1+0.35\,S)}$$

$$K_v = 3$$

$$K_a = \lim_{s\to 0} S^2G(s)\,H(s) = \lim_{s\to 0} \frac{S(S+3)}{(1+0.60S)\,(1+0.35S)}$$

$$K_a = 0.$$

Example 4.33: Find K_p, K_v, K_a for the system having $G(s) = \frac{10}{S^2}$ and $H(s) = 0.7$

Solution:

$$K_p = \lim_{s\to 0} G(s)\,H(s) = \lim_{s\to 0} \left(\frac{10}{S^2}\right)(0.7) = \infty$$

$$K_v = \lim_{s\to 0} S\,G(s)\,H(s)$$

$$= \lim_{s\to 0} \left\{S\left(\frac{10}{S^2}\right)0.7\right\}$$

$$K_v = \infty$$

$$K_a = \lim_{s\to 0} S^2\,G(s)\,H(s)$$

$$= \lim_{s\to 0} \left\{S^2\left(\frac{10}{S^2}\right)0.7\right\}$$

$$K_a = 7.$$

Example 4.34: Find K_p, K_v and K_a for the system having $G(s) = \frac{5}{S^2+3S+5}$ and $H(s) =$ 0.6.

Solution:

$$K_p = \lim_{s\to 0} G(s)\,H(s)$$

$$= \lim_{s\to 0} \left(\frac{5}{S^2+3S+5}\right)(0.6)$$

$$K_p = 0.6$$

$$K_v = \lim_{s\to 0} S\, G(s)\, H(s) = \lim_{s\to 0} \left(\frac{3S}{S^2 + 3S + 5}\right) = 0$$

$$K_a = \lim_{s\to 0} S^2\, G(s)\, H(s) = \lim_{s\to 0} \left(\frac{3S^2}{S^2 + 3S + 5}\right) = 0$$

Example 4.35: For a system having $G(s)H(s) = \dfrac{K(S+4)}{S(S^3 + 5S^2 + 6S)}$ find:

(i) type of the system, (ii) error coefficient, and (iii) error due to input $\frac{A}{2}t^2$.

Solution:

$$G(s)H(s) = \frac{K(S+4)}{S(S^3 + 5S^2 + 6S)} = \frac{K(S+4)}{S^2(S^2 + 5S + 6)}$$

(i) This is a type 2 system
Since it is type 2 system
K_p and $K_v = \infty$

(ii) $K_a = \lim_{s\to 0} S^2 G(s)\, H(s)$

$$= \lim_{s\to 0} \left\{ S^2 \frac{K(S+4)}{S^2(S^2 + 5S + 6)} \right\}$$

$$= \frac{4K}{6} = \frac{2K}{3}$$

(iii) $e_{ss} = \dfrac{A}{K_a} = \dfrac{A}{\frac{2K}{3}} = \dfrac{3A}{2K}$

Example 4.36: A unity feedback system has $G(s) = \dfrac{180}{S(S+6)}$ and $r(t) = 4t$. Determine:

(i) The steady state error
(ii) The value of K to reduce the error by 6%

Solution:

$$G(s)\, H(s) = \frac{180}{S(S+6)}$$

$$r(t) = 4t$$

$$R(s) = \frac{4}{S^2}$$

$$e_{ss} = \lim_{s\to 0} \frac{SR(s)}{1 + G(s)H(s)}$$

$$= \lim_{s\to 0} \frac{S\cdot\frac{4}{S^2}}{1 + \frac{180}{S(S+6)}} = \lim_{s\to 0} \frac{4}{S + \frac{180}{S+6}}$$

$$e_{ss} = 0.134$$

Since new error is 6% less,

$$(e_{ss})_{\text{new}} = (0.94)\, e_{ss} = 0.12596$$

$$e_{ss} = \lim_{s \to 0} \left[\frac{4}{S + \dfrac{K}{S+6}} \right] = \frac{24}{K}$$

$$\frac{24}{K} = 0.12596$$

$$\therefore \qquad K = \frac{24}{0.12596} = 190.537$$

Static error constants

Constant	Equation	Steady state error (e_{ss})
Position error constant (K_p)	$K_p = \lim_{s \to 0} S\,G(s)\,H(s)$	$e_{ss} = \dfrac{A}{1 + K_p}$
Velocity error constant (K_v)	$K_v = \lim_{s \to 0} S\,G(s)\,H(s)$	$e_{ss} = \dfrac{A}{K_v}$
Acceleration error constant (K_a)	$K_v = \lim_{s \to 0} S^2\,G(s)\,H(s)$	$e_{ss} = \dfrac{A}{K_a}$

Error for different types of input

Type	Step input		Ramp input		Parabolic input	
	K_p	e_{ss}	K_v	e_{ss}	K_a	e_{ss}
Type 0	K	$\dfrac{A}{1+K}$	0	∞	0	∞
Type 1	∞	0	K	$\dfrac{A}{K}$	0	∞
Type 2	∞	0	∞	0	K	$\dfrac{A}{K}$

S. No.	Time response specification
1.	Delay time, $t_d = \dfrac{1 + 0.7\zeta}{\omega_n}$
2.	$t_r = \dfrac{\pi - \theta}{\omega_d} \Rightarrow$ rise time where $\theta = \dfrac{\tan^{-1}\sqrt{1-\zeta^2}}{\zeta}$ in radian, $\omega_d = \omega_n\sqrt{1-\zeta^2}$

3.	Peak time, $t_p = \dfrac{\pi}{\omega_d}$
4.	Maximum Overshoot $\% M_p = e^{-\frac{\zeta\pi}{\sqrt{1-\zeta^2}}} \times 100$
5.	Settling time, $t_s = e^{-4/\zeta\omega_n}$ 2% criterion or $t_s = e^{-2\zeta\omega_n}$ for 5% criterion

Example 4.37: For a system having $\dfrac{C(s)}{R(s)} = \dfrac{20}{S^2 + 7S + 2S}$ find its time response specifications and expression for output.

Solution: $\dfrac{C(s)}{R(s)} = \dfrac{20}{25}\left[\dfrac{2S}{S^2 + 7S + 2S}\right]$ comparing with standard for

$$\frac{C(s)}{R(s)} = \frac{\omega_n^2}{S^2 + 2\zeta\omega_n S + \omega_n^2}$$

$$\omega_n^2 = 25,\ \omega_n = 5 \text{ rad/sec}$$

$$2\zeta\omega_n = 7$$

$$\zeta = \frac{7}{10} = 0.7$$

$$\omega_d = \omega_n\sqrt{1-\zeta^2} = 5\sqrt{1-(0.7)^2} = 3.57 \text{ rad/sec.}$$

$$t_d = \frac{1 + 0.7\zeta}{\omega_n} = 0.298 \text{ sec.}$$

$$t_z = \frac{\pi - \theta}{\omega_d}$$

$$\theta = \tan^{-1}\frac{\zeta}{\sqrt{1-\zeta^2}} = 0.775 \text{ rad.}$$

$$t_r = 0.662 \text{ sec.}$$

$$t_p = \frac{\pi}{\omega_d} = 0.879 \text{ sec.}$$

$$M_p = 100\, e^{-\frac{\zeta\pi}{\sqrt{1-\zeta^2}}} = 4.60\%$$

$$t_s = \frac{4}{\zeta\omega_n} = 1.1428 \text{ sec.}$$

For unit step input

$$y(t) = \frac{20}{25}\left(1 - \frac{e^{-3.5t}}{0.714}\sin(3.57t + 0.775)\right)$$

Example 4.38: Determine the time response specifications and expression for output for unit step input to the system having system equations as follows:

$$\frac{d^2y}{dt^2} + 5\frac{dy}{dt} + 16\,y = 9\,x$$

where y is output and x is input.

Solution: $\frac{d^2y}{dt} + 5\frac{dy}{dt} + 16\,y = 9\,x$

Taking LT on both sides.

$$S^2\,Y(s) + 5\,SY(s) + 16\,Y(s) = 9\,X(s)$$

$$(S^2 + 5\,S + 16)\,Y(s) = 9\,X\,(s)$$

$$\frac{Y(s)}{X(s)} = \frac{9}{S^2 + 5\,S + 16}$$

$$\omega_n^2 = 16$$

$$\omega_n = 4 \text{ rad/sec.}$$

$$2\zeta\omega_n = 5$$

$$\zeta = \frac{5}{2\omega_n} = 0.625$$

$$\omega_d = \omega_n\sqrt{1 - \zeta^2}$$

$$= 4\sqrt{1 - (0.625)^2}$$

$$= 3.1225$$

$$t_d = \frac{1 + 0.7\zeta}{\omega_n} = 0.3593 \text{ sec.}$$

$$t_r = \frac{\pi - \theta}{\omega_d}$$

$$\theta = \tan^{-1}\frac{\zeta}{\sqrt{1 - \zeta^2}} = 0.6751 \text{ rad}$$

$$t_r = \frac{3.14 - 0.6751}{3.1225} = 0.719 \text{ sec}$$

$$t_p = \frac{\pi}{\omega_d} = 1.005 \text{ sec}$$

$$\% M_p = 100 \times e^{-\frac{\zeta\pi}{\sqrt{1-\zeta^2}}} = 8.09\%$$

$$t_s = \frac{4}{\zeta\omega_n} = 1.6 \text{ sec}$$

Expression for output

$$\frac{Y(s)}{X(s)} = \frac{9}{16}\left[\frac{16}{S^2 + 5\,S + 16}\right]$$

$$C(t) = \frac{9}{16}\left[1 - \frac{e^{-\zeta\omega_n t}}{\sqrt{1-\zeta^2}} \sin(\omega_d t + \phi)\right]$$

$$C(t) = \frac{9}{16}\left[1 - 1.2810\, e^{-2.5t} \sin(3.1225t + 0.675)\right]$$

Example 4.39: Find K to limit the error of a system for input $1 + 8\,t + \frac{18}{2}\,t^2$ to 0.8 having

$$G(s)H(s) = \frac{K}{S^2(S+1)(S+4)},\ r(t) = 1 + 8\,t + \frac{18}{2}\,t^2.$$

Solution:

$$R(s) = \frac{1}{S} + \frac{8}{S^2} + \frac{18}{S^3} = \frac{S^2 + 8S + 18}{S^3}$$

$$e_{ss} = \lim_{s\to 0} \frac{SR(s)}{1 + G(s)\,H(s)}$$

$$= \lim_{s\to 0} \frac{S\left(\dfrac{S^2 + 8S + 18}{S^3}\right)}{1 + \dfrac{K}{S^2\,(S+1)\,(S+4)}}$$

$$= \lim_{s\to 0} \frac{S^2 + 8S + 18}{S^2 + \dfrac{K}{(S+1)\,(S+4)}}$$

$$= \frac{18}{\left(\dfrac{K}{4}\right)} = \frac{72}{K}$$

$$\frac{72}{K} = 0.8$$

$$K = \frac{72}{0.8} = 90.$$

Example 4.40: A system has $G(s) = \dfrac{K}{S(1 + ST)}$ with unity feedback when K and T are constant. Determine the factor by which K should be multiplied to reduce the overshoot from 85% to 35%.

Solution:

$$G(s) = \frac{K}{S(1 + ST)} \quad \text{and} \quad H(s) = 1$$

$$\frac{C(s)}{H(s)} = \frac{G(s)}{1 + G(s)\,H(s)}$$

$$= \frac{\dfrac{K}{S(1 + ST)}}{1 + \dfrac{K}{S(1 + ST)}}$$

$$= \frac{K}{S^2\,T + S + K} = \frac{\left(\dfrac{K}{T}\right)}{S^2 + \dfrac{1}{T}S + \dfrac{K}{T}}$$

$$\omega_n^2 = \frac{K}{T},\ \omega_n = \sqrt{\frac{K}{T}} \text{ rad/sec.}$$

$$2\zeta\omega_n = \frac{1}{T} \Rightarrow \frac{1}{2\omega_n T} = \xi$$

$$\zeta = \frac{1}{2\left(\sqrt{\dfrac{K}{T}}\right)T} = \frac{1}{2\sqrt{KT}}.$$

Example 4.41: The unity feedback system has $G(s) = \dfrac{K}{S(ST + 2)}$. Find the following:

The factor by which the gain K should be multiplied to increase the damping ratio from 0.15 to 0.6.

The factor by which the time constant T should be multiplied to reduce the damping ratio from 0.8 to 0.4.

Solution:

$$G(s) = \frac{K}{S(ST + 2)} \cdot H(s) = 1$$

$$\frac{C(s)}{H(s)} = \frac{G(s)}{1 + G(s)\,H(s)} = \frac{K}{S^2T + 2S + K}$$

$$= \frac{\dfrac{K}{T}}{S^2 + \dfrac{2}{T}S + \dfrac{K}{T}}$$

$$\omega_n^2 = \frac{K}{T}, \quad \omega_n = \sqrt{\frac{K}{T}}$$

$$2\zeta\omega_n = \frac{2}{T}$$

$$\zeta = \frac{2}{T}\frac{1}{2\omega_n} = \frac{1}{T\omega_n}$$

$$\zeta = \frac{1}{T\sqrt{\frac{K}{T}}} = \frac{1}{\sqrt{T}.\sqrt{T}\,\frac{\sqrt{K}}{\sqrt{T}}} = \frac{1}{\sqrt{KT}}$$

Let $\zeta = \zeta_1$ for $M_p = 85\%$ or 0.85

$$0.85 = e^{-\frac{\zeta_1\pi}{\sqrt{1-\zeta_1^2}}}$$

$$\ln(0.85) = \ln\left(e^{-\frac{\zeta_1\pi}{\sqrt{1-\zeta_1^2}}}\right)$$

$$\sqrt{1-\zeta_1^2} = (19.32)\,\zeta_1$$

$$\left(1-\zeta_1^2\right) = (373.264)\,\zeta_1^2$$

$$\zeta_1 = 0.01517 \qquad \text{...(4.11)}$$

Let $\zeta = \zeta_2$ for $M_p = 35\%$ or 0.35

$$0.35 = e^{-\frac{\zeta_2\pi}{\sqrt{1-\zeta_2^2}}}$$

$$\sqrt{1-\zeta_2^2} = 2.99\,\zeta_2$$

$$\zeta_2 = 0.317 \qquad \text{...(4.12)}$$

$$\frac{\zeta_1}{\zeta_2} = \frac{0.0517}{0.317}$$

$$\frac{\frac{1}{2\sqrt{K_1T}}}{\frac{1}{2\sqrt{K_2T}}} = \frac{0.0517}{0.317} \qquad \left\{\therefore\ \zeta = \frac{1}{2\sqrt{KT}}\right\}$$

$$K_2 = 0.02659\,K_1$$

$$\therefore \qquad \zeta = \frac{1}{\sqrt{KT}}$$

Let $$\zeta_1 = \frac{1}{\sqrt{K_1 T}}$$

$\zeta_2 = \dfrac{1}{\sqrt{K_2 T}}$ for $\zeta = 0.15$ and $\zeta = 0.6$ respectively.

$$\frac{\zeta_1}{\zeta_2} = \sqrt{\frac{K_2 T}{K_1 T}} = \frac{0.15}{0.60} = \frac{1}{4}$$

$$\frac{K_2}{K_1} = \frac{1}{16}$$

$$K_2 = \frac{K_1}{16}$$

The gain must be multiplied by factor $\frac{1}{16}$ to increase the damping ratio from 0.15 to 0.6.

Let $$\zeta_1 = \frac{1}{\sqrt{T_1 K}}$$

$\zeta_2 = \dfrac{1}{\sqrt{T_2 K}}$ for $\zeta = 0.8$ and $\zeta = 0.4$ respectively.

$$\frac{\zeta_1}{\zeta_2} = \sqrt{\frac{T_2 K}{T_1 K}} = \frac{0.8}{0.4} = 2$$

$$\frac{T_2}{T_1} = 4$$

$$T_2 = 4T_1$$

The time constant (T) must be multiplied by factor 4 to reduce the damping ratio from 0.8 to 0.4.

Example 4.42: Figure 4.35 shows step response of a second order system having unity feedback when input is $3u(t)$. Determine the open-loop and closed-loop transfer function of the system.

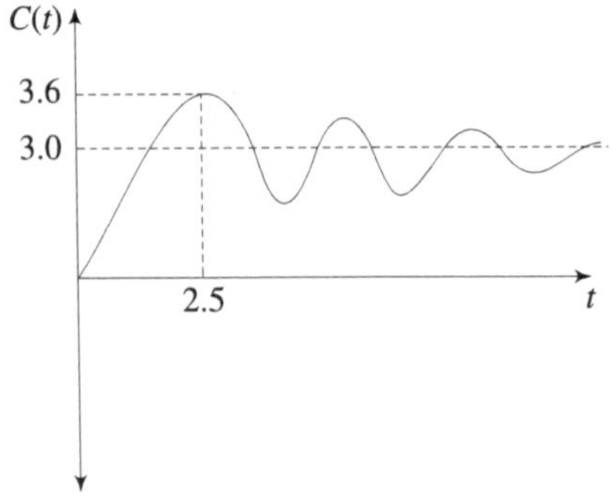

Fig. 4.35

Solution: From Fig. 4.35 $t_p = 2.5$ sec.

Overshoot = 3.6 – 3.0 = 0.6 for final output 3

$$M_p = \frac{3.6 - 3.0}{3.0} = \frac{0.6}{3} = 0.2$$

$$\% \ M_p = \frac{0.6}{3} \times 100 = 20\%$$

$$\% \ M_p = e^{-\frac{\pi\zeta}{\sqrt{1-\zeta^2}}} \times 100$$

$$0.2 = e^{-\frac{\pi\zeta}{\sqrt{1-\zeta^2}}} \times 100$$

$$-1.61 = -\frac{\pi\zeta}{\sqrt{1-\zeta^2}}$$

$$\zeta = 0.456$$

$$t_p = \frac{\pi}{\omega_n\sqrt{1-\zeta^2}}$$

$$\omega_n = 1.414 \text{ rad/sec.}$$

$$\frac{C(s)}{R(s)} = \frac{\omega_n^2}{S^2 + 2\zeta\omega_n S + \omega_n^2}$$

$$= \frac{(1.414)^2}{S^2 + 2(0.456)\,(1.414)\,S + 1.414)^2}$$

$$= \frac{(1.414)^2}{S^2 + 1.289S + (1.414)^2}$$

Let $\quad H(s) - 1$

$$G(s) = \frac{1.999}{s^2 + 1.2895}.$$

Disadvantages of static error coefficient method

(1) They do not give any information on the steady state error when the inputs are other than the standard test inputs (step, ramp and parabolic).

(2) This method does not provide precise value of error. In most of the cases the error found is either zero or infinite.

(3) It does not provide variation of error with respect to time which may be needed for design purpose.

(4) The method is applicable to stable system.

4.7 DYNAMIC ERROR COEFFICIENTS

The dynamic error coefficient method is a generalized method which includes inputs of any arbitrary functions of time. The disadvantages of the static error coefficient method are eliminated in this method.

For a unity feedback system,

$$\frac{E(s)}{R(s)} = \frac{1}{1 + G(s)} = F(s)$$

The function, $F(s)$, can be expanded as a power series in S by Taylor's series expansion.

$$F(s) = \frac{1}{1 + G(s)\,H(s)}$$

$$F(s) = \frac{E(s)}{R(s)} = \frac{1}{1 + G(s)} = C_0 + \frac{C_1 S}{1!} + \frac{C_2 S^2}{2!} + \frac{C_3 S^3}{3!} + \ldots\ldots$$

where C_0, C_1, C_2, C_3 ... are defined to be dynamic error coefficient or generalized error coefficients.

$$\therefore \qquad E(s) = C_0 R(s) + \frac{C_1 SR(s)}{1!} + \frac{C_2 S^2 R(s)}{2!} + \frac{C_3 S^3 R(s)}{3!} + \ldots\ldots$$

Taking the inverse Laplace transform, the dynamic error is given by

$$e(t) = C_0\, r(t) + C_1\, r^{\text{I}}(t) + \frac{C_2\, r^{\text{II}}(t)}{2!} + \ldots\ldots$$

The generalized error coefficients are evaluated as

$$C_0 = \lim_{s \to 0} F(s)$$

$$C_1 = \lim_{s \to 0} \frac{d}{dS} F(s)$$

$$C_2 = \lim_{s \to 0} \frac{d^2}{ds^2} F(s)$$

$$C_n = \lim_{s \to 0} \frac{d^n}{ds^n} F(s)$$

Example 4.43: For the closed-loop system $G(s) = \frac{1}{S + 5}$ and $H(s) = 5$. Calculate the generalized error coefficients and find error series.

Solution:

$$\frac{E(s)}{R(s)} = \frac{1}{1 + G(s)\,H(s)}$$

$$E(s) = F(s)\,R(s)$$

$$F(s) = \frac{1}{1 + G(s)\,H(s)}$$

$$F(s) = \frac{1}{1 + \left(\frac{1}{S+5}\right)5} = \frac{S+5}{S+10}$$

The error coefficients are

$$C_0 = \lim_{s\to 0} F(s) = \lim_{s\to 0}\left(\frac{S+5}{S+10}\right) = 0.5$$

$$C_1 = \lim_{s\to 0}\frac{d}{ds}F(s) = \lim_{s\to 0}\frac{(S+10)-(S+5)}{(S+10)^2}$$

$$= \lim_{s\to 0}\frac{5}{(5+10)^2} = 0.05$$

$$C_2 = \lim_{s\to 0}\frac{d^2F(s)}{dS} = \lim_{s\to 0}\frac{d}{dS}\left(\frac{5}{(S+10)^2}\right)$$

$$= \lim_{s\to 0}\frac{5(-2)}{(S+10)^3} = -\,0.01$$

$\therefore$ The dynamic error, $e(t) = C_0\,r(t) + C_1\frac{d}{dt}r(t) + \frac{C_2}{2!}\frac{d^2}{dt^2}r(t) + \ldots\ldots$

$$e(t) = 0.5\,r(t) + 0.05\frac{d}{dt}r(t) - \frac{0.01}{\lfloor 2}\frac{d^2}{dt^2}r(t)$$

Example 4.44: A unity feedback system has $G(s) = \dfrac{1}{s(1+S)}$

The input to the system is described by $r(t) = 4 + 6\,t + 2\,t^3$. Find the generalized error coefficients and hence the steady state error.

Solution:

$$F(s) = \frac{1}{1+G(s)} = \frac{1}{1+\frac{1}{S(1+S)}} = \frac{S^2+S}{S^2+S+1}$$

$$C_0 = \lim_{s\to 0}F(s) = \frac{0+0}{0+0+1} = 0$$

$$\frac{dF(s)}{ds} = \frac{(S^2+S+1)(2S+1)-(S^2+S)(2S+1)}{(S^2+S+1)^2}$$

$$= \frac{2S+1}{(S^2+S+1)^2}$$

$\therefore$

$$C_1 = \lim_{s\to 0}\frac{dF(s)}{ds} = 1$$

$$\frac{d^2F(s)}{ds} = \frac{2(S^2 + S + 1)^2 - 2(2S + 1)^2 (S^2 + S + 1)}{(S^2 + S + 1)^4}$$

$$= \frac{-6(S^2 + S)}{(S^2 + S + 1)^3}$$

$$C_2 = \lim_{s \to 0} \frac{d^2F(s)}{ds^2} = 0$$

$$\frac{d^3F(s)}{ds^3} = -6\left[\frac{(S^2 + S + 1)^3 (2S + 1)^2 - 3(S^2 + S +)^2 (2S + 1)(S^2 + S)}{(S^2 + S + 1)^6}\right]$$

$$C_3 = \lim_{s \to 0} \frac{d^3F(s)}{ds^3} = -6$$

$$r(t) = 4 + 6t + 2t^3$$

$$\frac{d}{dt} r(t) = 6 + 6t^2$$

$$\frac{d^2r(t)}{dt^2} = \frac{d}{dt}(6 + 6t^2) = 12t$$

$$\frac{d^3r(t)}{dt^3} = \frac{d}{dt}(12t) = 12$$

$$\frac{d^4r(t)}{dt^4} = \frac{d}{dt}(12) = 0$$

∴ The dynamic error,

$$e(t) = C_0\, r(t) + C_1 \frac{d}{dt} r(t) + \frac{C_2}{2!}\frac{d^2r(t)}{dt^2} + \frac{C_3}{3!}\frac{d^3r(t)}{dt^3} + \ldots\ldots$$

$$= 0 + 1\,(6 + 6t^2) + 0\left(\frac{-6}{6} \times 12\right) = 6 + 6t^2 - 12$$

$$= 6 + 6t^2 - 12 = 6\,(t^2 - 1)$$

Steady state error,

$$6 + 6t^2 - 12 = 6t^2 - 6$$

$$e_{ss} = \lim_{t \to \infty} e(t) = \infty$$

$$e(t) = 6(t^2 - 1)$$

$$e_{ss} = \lim_{t \to \infty} e(t) = \infty.$$

Example 4.45: Figure 4.36 (a) shows a mechanical system, when a force of 8.9 N is applied to the system, the mass oscillates as shown in Fig. 4.36 (b). Find the values of M, B and K.

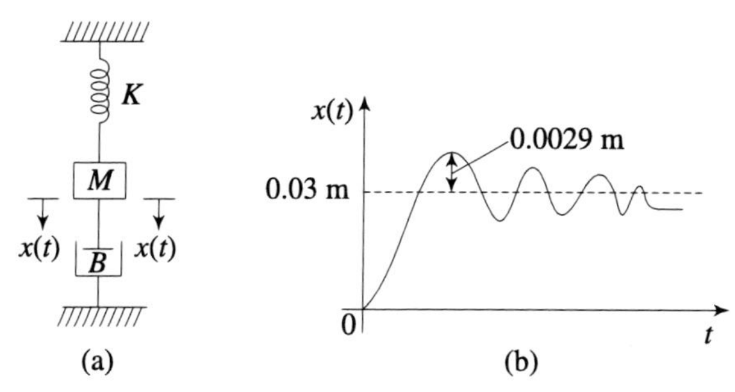

Fig. 4.36

Solution: The differential equation is

$$f(t) = M\frac{d^2x}{dt^2} + B\frac{dx}{dt} + Kx$$

Taking Laplace transform on both sides

$$F(s) = M\,S^2\,x\,(s) + B\,S\,x(s) + K\,x(s)$$

$$F(s) = x(s)\,[M\,S^2 + B\,S + K]$$

$$\frac{x(s)}{F(s)} = \frac{1}{MS^2 + BS + K} = \frac{\left(\frac{1}{m}\right)}{S^2 + \frac{B}{M}S + \frac{K}{M}}$$

$$f(t) = 8.9\text{ N}$$

$$f(s) = \frac{8.9}{S}$$

$$\therefore \quad X(s) = \frac{8.9\left(\frac{1}{m}\right)}{S\left(S^2 + \frac{B}{M}S + \frac{K}{M}\right)}$$

The final value of $x(t)$ is 0.03

$$\therefore \quad x(\infty) = 0.03$$

$$x(\infty) = \lim_{s\to 0} S\,x\,(S)$$

$$= \lim_{s\to 0} S\,\frac{89}{S}\,\frac{\left(\frac{1}{M}\right)}{S^2 + \frac{B}{M}S + \frac{K}{M}}$$

$$= \lim_{s\to 0} 89\,\frac{\left(\frac{1}{M}\right)}{0 + 0 + \frac{K}{M}}$$

$$0.03 = \frac{8.9}{K}$$

$$K = \frac{8.9}{0.03} = 296.67 \text{ N/m}$$

Comparing the transfer form with the standard form

$$\frac{X(s)}{F(s)} = \frac{\left(\frac{1}{M}\right)}{S^2 + \frac{B}{M}S + \frac{K}{M}} = \frac{\omega_n{}^2}{S^2 + 2\zeta\omega_n S + \omega_n{}^2}$$

$$\omega_n^2 = \frac{K}{M}, \ \omega_n = \sqrt{\frac{K}{M}}$$

$$2\zeta\omega_n = \frac{B}{M}$$

$$\zeta = \frac{B}{2\omega_n M} = \frac{B}{2M\sqrt{\frac{K}{M}}} = \frac{B}{2\sqrt{MK}}$$

The overshoot from Fig. 4.35(b) is 0.0029 for a final value of 0.03.

$$\therefore \quad \% \ M_p = \frac{0.0029}{0.03} \times 100$$

$$= 9.67\%$$

$$= 0.0967$$

$$\therefore \quad \% \ M_p = e^{\frac{-\pi\zeta}{\sqrt{1-\zeta^2}}} \times 100$$

$$0.0967 = e^{\frac{-\pi\zeta}{\sqrt{1-\zeta^2}}}$$

Taking natural logarithmic on both sides

$$\ln (0.0967) = \ln \left(e^{\frac{-\pi\zeta}{\sqrt{1-\zeta^2}}}\right)$$

$$-2.3361 = \frac{-\pi\zeta}{\sqrt{1-\zeta^2}}$$

$$\frac{\zeta}{\sqrt{1-\zeta^2}} = 0.7436$$

$\therefore$ Squaring both sides

$$\zeta^2 = (0.7436)^2 (1 - \zeta^2)$$

$$\zeta^2 (1 + 0.5529) = 0.5529$$

$$\zeta^2 = 0.356$$

$$\zeta^2 = 0.5967$$

Since t_p is 2 sec.

$$t_p = \frac{\pi}{\omega_n \sqrt{1-\zeta^2}} = 2$$

$$\omega_n = \frac{\pi}{2\sqrt{1-(0.5967)^2}} = 1.9575 \text{ rad/sec.}$$

$$\omega_n = \sqrt{\frac{K}{M}}$$

$$1.9575 = \frac{296.67}{M}$$

$$M = 77.42 \text{ kg.}$$

$$\zeta = \frac{B}{2\sqrt{MK}}$$

$$0.5967 = \frac{B}{2\sqrt{77.42 \times 296.67}}$$

$$B = 180.863 \text{ Nm/s}$$

Example 4.46: Determine the step, ramp and parabolic error constants of the following feedback control systems.

(i) $G(s) = \dfrac{1200}{(1 + 0.25)(1 + 115)}$, $H(s) = 1$

Ans: $K_p = 1200$, $K_v = 0$, $K_a = 0$

(ii) $G(s) = \dfrac{120}{S(S^2 + 10S + 120)}$, $H(s) = 1$

Ans: $K_p = \infty$, $K_v = 1$, $K_a = 0$

(iii) $G(s) = \dfrac{110}{S(1 + 0.2S)(1 + 12s)}$, $H(s) = 1$

Ans: $K_p = \infty$, $K_v = 110$, $K_a = 0$

(iv) $G(s) = \dfrac{120}{S(S^2 + 105 + 110)}$, $H(s)$

Ans: $K_p = \infty, K_v = \infty, K_a = 1$

(v) $G(s) = \dfrac{1100}{S(S+10)(S+110)}, H(s) = 1$

Ans: $K_p = \infty, K_v = \infty, K_a = 0$

(vi) $G(s) = \dfrac{10(1+4S)(1+6S)}{S^2(S^2+2S+1)}, H(s) = 1$

Ans: $K_p = \infty, K_v = \infty, K_a = 10$

Example 4.47: For the system shown in Fig. 4.37, find the error using the dynamic error coefficient method for input of $4 + 3t + \frac{8}{2}t^2$.

For the dynamic coefficient method,

$$e_{ss}(t) = K_0\, r(t) + K_1\, r^1(t) + K_2\, r^{11}(t) + \ldots\ldots$$

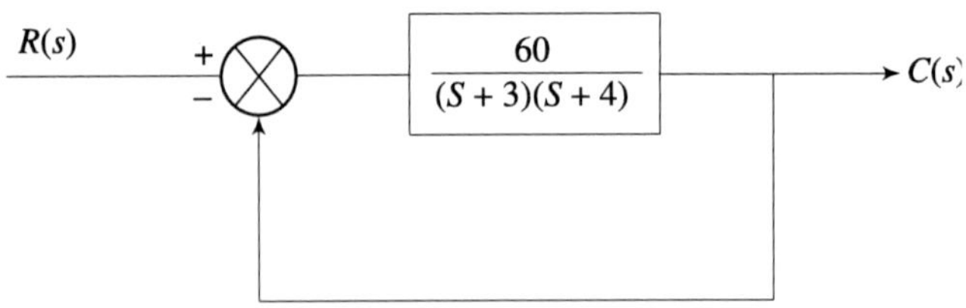

Fig. 4.37

where

$$K_0 = \lim_{s\to 0} F_1(s)$$

$$K_1 = \lim_{s\to 0} \frac{dF_1(s)}{ds}$$

$$K_2 = \lim_{s\to 0} \frac{d^2F_1(s)}{ds^2}$$

$$F_1(s) = \frac{1}{1+G(s)H(s)} = \frac{1}{1+\dfrac{60}{(s+3)(s+4)}}$$

$$= \frac{(s+3)(s+4)}{(s+3)(s+4)+60} = \frac{s^2+7s+12}{s^2+7s+72}$$

$$K_0 = \lim_{s\to 0} F_1(s) = \lim_{s\to 0} = \frac{s^2+7s+12}{s^2+7s+72} = \frac{12}{72} = 0.1667$$

$$K_1 = \lim_{s\to 0} \frac{dF_1(s)}{ds} = \lim_{s\to 0} \frac{d}{ds}\left[\frac{s^2+7s+12}{s^2+7s+72}\right]$$

$$= \lim_{s\to 0}\left[\frac{(S^2 + 7s + 72)\,(2s + 7) - (S^2 + 7s + 12)\,(2s + 7)}{(S^2 + 7s + 72)^2}\right]$$

$$= \lim_{s\to 0}\left[\frac{60(2s + 7)}{(S^2 + 7s + 72)^2}\right] = \frac{60 \times 7}{(72)^2} = 0.0810$$

$$K_2 = \lim_{s\to 0}\frac{d^2F_2(s)}{ds^2}$$

$$= \lim_{s\to 0} 60\,\frac{d}{dS}\left[\frac{2s + 7}{(S^2 + 7s + 72)^2}\right]$$

$$= 60\lim_{s\to 0}\left\{\frac{(S^2 + 7s + 72)^2(2) - 2(2s + 7)(S^2 + 7s + 72)(2s + 7)}{(S^2 + 7s + 72)^4}\right\}$$

$$K_2 = 60\lim_{s\to 0}\left\{\frac{2 \times (72)^2 - 7 \times 2 \times 72 \times 7}{(72)}\right\}$$

$$= 7.394s + 10^{-3}$$

$$e_{ss}(t) = K_0 r(t) + K_1\, r^1\,(t) + K_2\, r^{11}\,(t)$$

$$r(t) = 4 + 3t + \frac{8}{2}\,t^2$$

$$r^1(t) = 3 + 8t$$

$$r^1(t) = 8$$

$$\therefore \quad e_{ss} = 0.1667\,(4 + 3t + 4t^2) + 0.0810\,(3 + 8t) + 7.3945 \times 10^{-3} \times 8$$

$$e_{ss} = 0.968956 + 1.1481t + 0.6668\,t^2.$$

EXERCISE

Fill in the Blanks

1. Find the type for the system given by the following transfer function:

$$\frac{C(s)}{R(s)} = \frac{(10(s + 5)^2)}{(s^3(s + 2)(s + 10))}.$$

Ans: 3

2. What is the value of K for a unity feedback (negative) system with open-loop transfer function $G(s) = K/(S(S + 1)(S + 3))$ to get $e_{ss} = 0.2$ when a unit ramp input is applied?

Ans: 15

3. For a second order system, if the peak time is 2 times the rise time ($t_p = 2t_r$), then $\xi =$ __________.

Ans: 0

4. The step error coefficient of a system $G(s) = \dfrac{1}{(s+6)\,(s+1)}$ with unity feedback is ______.

Ans: 1/6

5. Consider the feedback control system with loop-transfer function

$$G(s)\,H(s) = \frac{K(1 + 0.5s)}{s(1 + s)\,(1 + 2s)}.$$ The type of closed-loop system __________.

Ans: One

6. Consider the unity feedback control system with open-loop transfer function $G(s) = \dfrac{K}{s(s+1)}$. The steady state error of the system due a unit step input is __________.

Ans: *K*

7. For a second order system with a closed-loop transfer function $T(s) = \dfrac{9}{s^2 + 4s + 9}$ the settling time for 2 per cent band, in seconds, is __________.

Ans: 4

8. Consider a system with transfer function $G(s) = \dfrac{s+6}{Ks^2 + s + 6}$. Its damping ratio will be 0.5 when the value of *K* is __________.

Ans: 3

9. Step responses of a set of three second order under damped systems all have the same percentage overshoot. Which one of the following diagrams represents the poles of three systems?

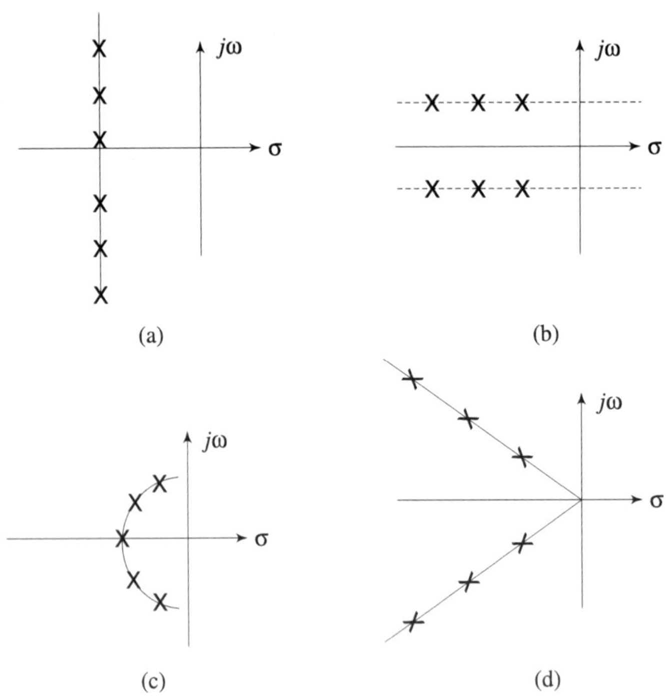

Fig. 4.38

Ans: Option (a)

10. The open-loop gain for a unity feedback system is $G(s) = \dfrac{12}{s(s + 12)}$. The steady state velocity error of the system is __________.

Ans: 1

11. Which of the following systems is generally preferred?

Ans: Underdamped

12. If the characteristic equation of a closed-loop system is $s^2 + 2s + 2 = 0$, then the system is __________.

Ans: Critically damped

13. The natural frequency of an undamped second order system is 40 rad/s. If the system is damped with a damping ratio of 0.3, the damped natural frequency in rad/s is __________.

Ans: 38.15 r/sec

14. A critically damped continuous time second order system when sampled will have __________.

Ans: two simple poles on negative real axis.

15. The steady state error of a stable type 0 unity feedback system for a unit step function is __________.

Ans: $1/1 + kp$

16. A unity feedback control system has the open-loop transfer function $G(s) = \dfrac{4(1 + 2s)}{s2(s + 2)}$. If the input to the system is a unit ramp, the steady-state error will be __________.

Ans: zero

17. The response of an LCR circuit to a step input is __________.

Ans: oscillatory if the poles are imaginary

18. The transfer function, $T(s) = \dfrac{100}{(s + 1)(s + 100)}$ for a unit step input to the system the approximate settling time for 2 per cent band in seconds, is __________.

Ans: 4 sec

19. The transfer function, $G(s) = \dfrac{1}{(s + 2)}$ is excited with $10u(t)$. The time at which the output reaches 99% of its steady state value is __________.

Ans: 2.3 sec

20. A ramp input applied to an unity feedback system results in 5% steady state error. The type number and zero frequency gain of the system are respectively __________.

Ans: 1 and 20

21. The unit step response of a system starting from rest is given by for $c(t) = 1 - e^{-2t}$ $(t \geq 0)$, the transfer function of the system is __________.

Ans: $\dfrac{2}{(s + 2)}$

22. The transfer function of a system is $T(s) = \dfrac{K}{s^3(1 + sT)}$. The type and order of the system are __________.

Ans: 3 and 4

23. A system has the following transfer function.

$$G(s) = \frac{100(S + 5)(S + 50)}{S^4(S + 10)(S^2 + 3S + 10)}$$

The type and order of the system are __________.

Ans: 4 and 7

24. For the function $X(s) = 1/S\ (S + 1)^3\ (S + 2)$, the residues associated with the simple poles @ $S = 0$ and $S = -2$ are __________ respectively.

Ans: 1/2 and 1/2

25. The open-loop system has __________ poles and __________ zero

Ans: 2 camplex poles and a zero

26. The unit step response of a system function is given by $y(t) = 1 - e^{-8t}$. The transfer function of the system is given by __________.

Ans: $8/s + 8$.

27. It is given that $G(s) = K/[s^3\ (1 + sT)]$ and the system is operated in a closed-loop with unit feedback. The order and type of closed-loop system are __________ and __________.

Ans: 4 and 0.

28. The ratio of settling time to time constant for a tolerance of 2% of a under damped system is approximate __________.

Ans: 6

29. Find the value of *M*, *D*, *K* for the mechanical vibratory system, when force applied is 93 N.

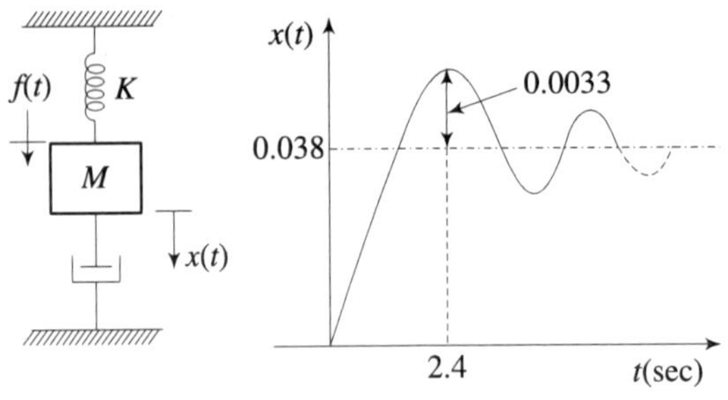

Fig. 4.39

Ans: 89.14 kg, 231.47 Nm/rad/sec, 244. 74 N/m.

30. The ratio of settling time to time constant for a tolerance of 2% of a under damped system is approximately __________.

Ans: 6

31. The overshoot due to decrease of damping factor __________.

Ans: increases

32. The ratio of damped frequency to natural frequency of a system having damping factor is __________.

Ans: $\sqrt{1-\zeta^2}$, $\sqrt{1+\zeta^2}$

33. If the gain K of the system increases the steady state error of the system __________.

Ans: decreases

34. The first derivative control can be used to __________.

Ans: eigenvalue analysis

35. For a second-order type 1 system, the resonance peak will occur when the system gain is at the __________.

Ans: critical damping value

36. The open-loop grain for a unity feedback system is $G(s) = 12/s(s + 12)$. The steady state velocity error of the system is __________.

Ans: 1

37. A linear second order system with the transfer function $G(s) = 49/s^2 + 16s + 49$ is initially at rest and is subjected to a step input signal the response of the system will exhibit a peak overshoot of __________.

Ans: Zero

38. Match the following

(a) $\delta > 1$ (overdamped)	(i) Complex conjugate
(b) $\delta < 1$ (underdamped)	(ii) Real and different
(c) $\delta = 1$ (critically damped)	(iii) Purely imaginary
(d) $\delta = 0$ (undamped)	(iv) Real and equal

Ans: (a) → (ii), (b) → (i), (c) → (iv), (d) → (iii)

39. The value of damping ratio of 0.6 in the step response of a second order system results in maximum overshoot of __________.

Ans: 9.44

40. How the system is classified depending on the value of damping.

Ans: Depending on the value of damping, the system can be classified into the following four cases:

Case 1: undamped system, $\zeta = 0$

Case 2: undamped system, $0 < \zeta < 1$

Case 3: critically damped system, $\zeta = 1$

Case 4: over damped system, $\zeta > 1$

41. Give the expression for natural frequency of oscillations, of electrical and mechanical system.

Ans: The natural frequency of oscillation of second order mechanical translational system $\left.\right\}$ $\omega_n = \sqrt{\dfrac{K}{M}}$

The natural frequency of oscillation of second order mechanical rotational system $\left.\right\}$ $\omega_n = \sqrt{\dfrac{K}{J}}$

The natural frequency of oscillation of second order electrical system $\Big\} \omega_n = \frac{1}{\sqrt{LC}}$

42. The closed-loop transfer function of second order system is $\frac{C(s)}{R(s)} = \frac{10}{S^2 + 6S + 10}$. What is the type of damping in the system?

Ans: Let us compare the given transfer function with the standard form of second order transfer function

$$\therefore \quad \frac{C(s)}{R(s)} = \frac{\omega_n^2}{S^2 + 2\zeta\omega_n s + \omega_n^2} = \frac{10}{S^2 + 6S + 10}$$

$$\therefore \quad \omega_n^2 = 10 \qquad\qquad 2\zeta\omega_n = 6$$

$$\omega_n = \sqrt{10} \qquad\qquad \zeta = \frac{6}{2 \times \omega_n} = \frac{6}{2\sqrt{10}}$$

$$= 3.1622 \text{ rad/sec} \qquad \therefore \quad \zeta = 0.95$$

Since $\zeta < 1$, the system is underdamped.

43. The closed-loop transfer function of a second order system is given by $\frac{200}{S^2 + 20S + 200}$. Determine the damping ratio and natural frequency of oscillation.

Ans: Let us compare the given transfer function to standard form of second order transfer function

$$\therefore \quad \frac{C(s)}{R(s)} = \frac{\omega_n^2}{S^2 + 2\zeta\omega_n s + \omega_n^2} = \frac{200}{S^2 + 20S + 200}$$

$$\therefore \quad \omega_n^2 = 200 \qquad\qquad 2\zeta\omega_n = 20$$

$$\omega_n = \sqrt{200} = 14.4 \text{ rad/sec} \qquad \zeta = \frac{20}{2 \times \omega_n} = \frac{20}{2 \times 14.14}$$

Damping ratio, $\zeta = 0.707$ $\qquad \therefore \zeta = 0.707$

Natural frequency of oscillation, $\omega_n = 14.14$ rad/sec

44. A second order system has a damping ratio of 0.6 and natural frequency of oscillation is 10 rad/sec. Determine the damped frequency of oscillation.

Ans: Damped frequency of oscillation, $\omega_d = \omega_n \sqrt{1 - \zeta^2} = 10\sqrt{1 - (0.6)^2} = 10 \times 0.8 = 8$ rad/sec

45. The open-loop transfer function of a unity feedback system is $G(s) = \frac{20}{s(s + 10)}$. What is the nature of response of closed-loop system for unit step input?

Ans: The closed-loop transfer function

$$\frac{C(s)}{R(s)} = \frac{G(s)}{1 + G(s)} = \frac{20(s(s + 10))}{1 + \frac{20}{s(s + 10)}} = \frac{20}{s(s + 10) + 20} = \frac{20}{s^2 + 10s + 20}$$

The standard form of second order transfer function is $\frac{C(s)}{R(s)} = \frac{\omega_n^2}{s^2 + 2\zeta\omega_n s + \omega_n^2}$

On comparing the system transfer function with standard form of second order transfer function, we get

$$\omega_n^2 = 20 \qquad\qquad 2\zeta\omega_n = 10$$

$$\therefore \quad \omega_n = \sqrt{20} = 4.47 \text{ rad/sec} \qquad \zeta = \frac{10}{2 \times \omega_n} = \frac{10}{2 \times 4.47} = 1.12$$

Since damping ratio, $\zeta > 1$, the system is overdamped and the response will be exponentially rising.

46. The damping ratio of a system is 0.75 and the natural frequency of oscillation is 12 rad/sec. Determine the peak overshoot and the peak time.

Ans: Peak overshoot, $M_p = \dfrac{-\zeta\pi}{e^{\sqrt{1-\zeta^2}}} = \dfrac{-0.75 \times \pi}{e^{\sqrt{1-(0.75)^2}}}$

$$\% M_p = 0.028 \times 100 = 2.8\%$$

Damped frequency of oscillation, $\omega_d = \omega_n \sqrt{1 - \zeta^2}$

$$= 12\sqrt{1 - (0.75)^2} = 7.94 \text{ rad/sec}$$

Peak time, $t_p = \dfrac{\pi}{\omega_d} = \dfrac{\pi}{7.94} = 0.396$ sec

47. The damping ratio of a system is 0.6 and the natural frequency of oscillation is 8 rad/sec. Determine the rise time.

Ans: Rise time, $t_r = \dfrac{\pi - \theta}{\omega_d}$

$$\theta = \tan^{-1}\sqrt{1-\zeta^2/\zeta} = \frac{\tan^{-1}\sqrt{1-(0.6)^2}}{0.6} = 53.13$$

$$= \frac{53.13}{180} \times \pi \text{ rad} = 0.927 \text{ rad}$$

$$\omega_d = \omega_n\sqrt{1-\zeta^2} = 8\sqrt{1-(0.6)^2} = 6.4 \text{ rad/sec}$$

$$\therefore \quad t_r = \frac{\pi - 0.927}{6.4} = 0.34 \text{ sec}$$

48. What is type number of a system? What is its significance?

Ans: The type number is given by the number of poles of loop-transfer function at the origin. The type number of the system decides the steady state error.

49. Distinguish between type and order of a system.

Ans: (i) Type number is specified for loop-transfer function but order can be specified for any transfer function (open-loop or closed-loop transfer function)

(ii) The type number is given by the number of poles of loop-transfer function lying at the origin of s-plane but the order is given by the number of poles of transfer function.

50. For the system with the following transfer function, determine type and order of the system.

(i) $G(s)H(s) = \dfrac{K}{s(s+1)(s^2+6s+8)}$ (ii) $G(s)H(s) = \dfrac{20(s+2)}{s^2(s+3)(s+0.5)}$

(iii) $G(s)H(s) = \dfrac{(S+4)}{(s-2)(s+0.25)}$ (iv) $G(s)H(s) = \dfrac{10}{s^3(s^2+2s+1)}$

Ans: (i) Type–1, order-4 (ii) Type–2, order-4 (iii) Type–0, order-2 (iv) Type–3, order-5

51. What is steady state error?

Ans: The steady state error is the value of error signal $e(t)$, when t tends to infinity. The steady state error is a measure of system accuracy. These errors arise from the nature of inputs, types of system and from nonlinearrity of system components.

52. What are static error constants?

Ans: The K_p, K_v and K_a are called static error constants. These constants are associated with steady state error in a particular type of system and for a standard input.

53. A unity feedback system has a open-loop transfer function of $G(s) = \frac{10}{(S+1)(S+2)}$. Determine the steady state error for unit step input.

Ans: The steady state error for unit step input, $e_{ss} = \frac{1}{1+K_p}$

where $$K_p = \underset{s\to 0}{\text{Lt}}\ G(s)H(s)$$

For unity feedback system, $H(s) = 1$

$$\therefore \quad K_p = \underset{s\to 0}{\text{Lt}}\ G(s) = \underset{s\to 0}{\text{Lt}}\ \frac{10}{(S+1)(S+2)} = 5$$

$$\therefore \quad e_{ss} = \frac{1}{1+5} = \frac{1}{6}$$

54. A unity feedback system has a open-loop transfer function of $G(s) = \frac{25(s+4)}{s(s+0.5)(s+2)}$. Determine the steady state error for unit ramp input.

Ans: The steady state error for unit ramp is $e_{ss} = \frac{1}{K_v}$,

where $K_v = \underset{s\to 0}{\text{Lt}}\ SG(s)H(s)$. For unity feedback system $H(s) = 1$

$$\therefore \quad K_v = \underset{s\to 0}{\text{Lt}}\ SG(s) = \underset{s\to 0}{\text{Lt}}\ s\left[\frac{25(S+4)}{S(S+0.5)(S+2)}\right] = \frac{25\times 4}{0.5\times 2} = 100$$

$$e_{ss} = \frac{1}{K_v} = \frac{1}{100} = 0.01$$

55. A unity feedback system has a open-loop transfer function of $G(s) = \frac{20(S+5)}{S(S+0.1)(s+3)}$. Determine the steady state error for a parabolic input.

Ans: The steady state error for unit parabolic input, $e_{ss} = \frac{1}{K_a}$

where $K_a = \underset{s\to 0}{\text{Lt}}\ s^2\, G(s)H(s)$. For unity feedback system $H(s) = 1$

$$\therefore \quad K_a = \underset{s\to 0}{\text{Lt}}\ s^2\left[\frac{20(s+5)}{s^2\,(s+0.1)(s+3)}\right] = \frac{20\times 5}{0.1\times 3} = \frac{100}{0.3} = 333.33$$

$$\therefore \quad e_{ss} = \frac{1}{K_a} = \frac{1}{333.33} = 0.003$$

56. Give the relation between generalized and static error coefficients

Ans: The following expression shows the relation between generalized and static error coefficient

$$C_0 = \frac{1}{1+K_p},\ C_1 = \frac{1}{K_v},\ C_2 = \frac{1}{K_a}$$

57. What is the unit step response of the system shown in Fig. 4.40?

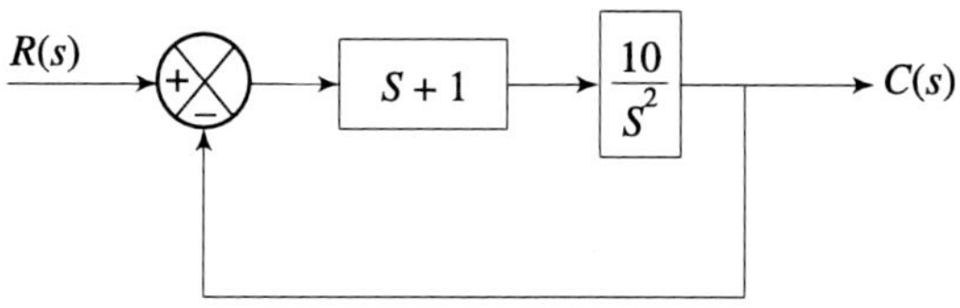

Fig. 4.40

Ans: $C(t) = -\ 1.1455\ e^{-8.87t} + 0.1455e^{-1.13t} + 1$

58. Obtain the unit step response of a unity feedback system whose open-loop transfer function is

$$G(s) = 5(S + 20)|S(S + 4.59)\ (S^2 + 3.41S + 16.35)$$

Ans: $C(t) = 1 + \dfrac{3}{8}\ e^{-t}\cos 3t - \dfrac{17}{24}\ e^{-t}\sin 3t \dfrac{-11}{8}\ e^{-3t}\cos t \dfrac{-13}{8}\ e^{-3t}\sin t$

59. The open-loop transfer function of an unity feedback control system is given by $G(s) = 100|S(S + 2)(S + 5)$. For unit step input, find the time response of the closed-loop system and determine% overshoot and the rise time.

Ans: $C(t) = [1 - 0.186\ e^{-7.45t} - 0.88e^{0.225t}\cos(3.65t - 22)]$

As t tends to infinity, $C(t)$ tends to infinity and so the system is unstable. Therefore % overshoots and rise time are not defined.

60. A servomechanism has its moment of inertia $J = 10 \times 10^{-6}$ kg-m^2 retarding friction, $B = 400 \times 10^{-6}$ N-m (rad/sec) and elasticity coefficient, $K = 0.004$ N-m/rad. Find the natural frequency and damping factor of the system.

Ans: Natural frequency, $\omega_n = 20$ rad/sec, damping ratio, $\zeta = 1$

61. For a second order system where open-loop transfer function $G(s) = 4|s(s + 2)$, determine the maximum overshoot, the time to reach the maximum overshoot when a step displacement of 18 is given to the system. Find the rise time, time constant and the settling time for an error of 7%.

Ans: Maximum overshoot = 0.16, When input is 18°, $Mp = 2.88°$

Peak time, $t_p = 1.8$ sec Rise time, $t_r = 1.2$ sec

Time constant, $T = 1$ sec, Settling time for 7% error = 2.6 sec

62. Consider the unity feedback closed-loop system where the forward transfer function is $G(s) = \dfrac{25}{S(S + 5)}$. Obtain the rise time, peak time, maximum overshoot and the settling time when the system is subjected to a unit-step input.

Ans: Rise time, $t_r = 0.5$ sec, % peak overshoot $Mp = 9.5\%$

Peak time, $t_p = 0.785$ sec, settling time, $t_s = 1.33$ sec (for 2% error) = 1 sec (for 5% error)

63. Consider the system shown in Fig. 4.41, where $\zeta = 0.6$ and $w_n = 0.5$ rad/sec. Determine the rise time, peak time, maximum overshoot and settling time, when the system is subjected to a unit step input.

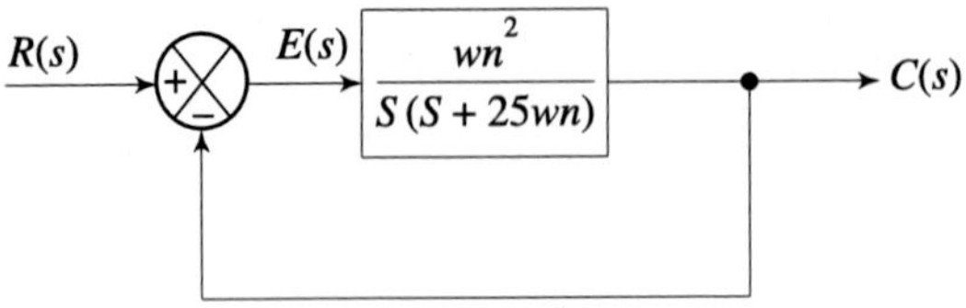

Fig. 4.41

Ans: Rise time, t_r = 0.55 sec Maximum overshoot, M_p = 0.095

Peak time, t_p = 0.785 sec Settling time, t_s = 1 sec (for 5% criterion)

64. For the system shown in Fig. 4.42 determine the values of K and K_n so that the maximum overshoot in the unit step response is 0.2 and the peak time is 1 sec. with these values of K and K_n, obtain rise time and settling time.

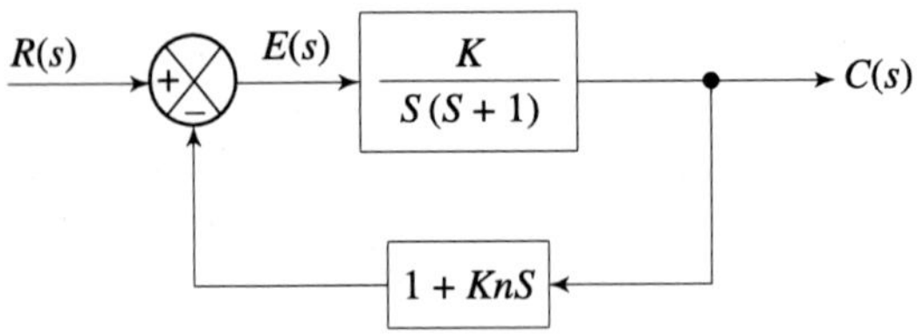

Fig. 4.42

Ans: K = 12.5 Rise time, t_r = 0.65 sec

K_n = 0.178 Settling time, t_s = 2.48 sec (for 2% error)

t_s = 1.86 sec (for 5% error)

65. The system shown in Fig. 4.43 is subjected to a unit step input. Determine the values of K and T, where the maximum overshoot of the system is 25.4% corresponding to $\zeta = 0.4$.

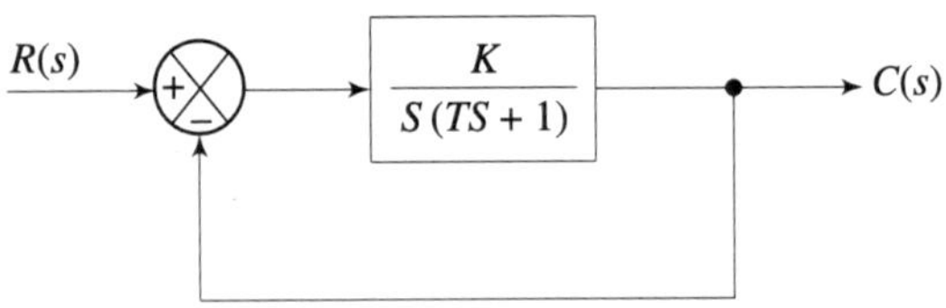

Fig. 4.43

Ans: K = 1.42 T = 1.09

66. Determine the values of K and T of the closed-loop system shown in Fig. 4.44. So that the maximum overshoot in unit step response is 25% and the peak time is 2 sec. Assume that J-1 kg-m^2.

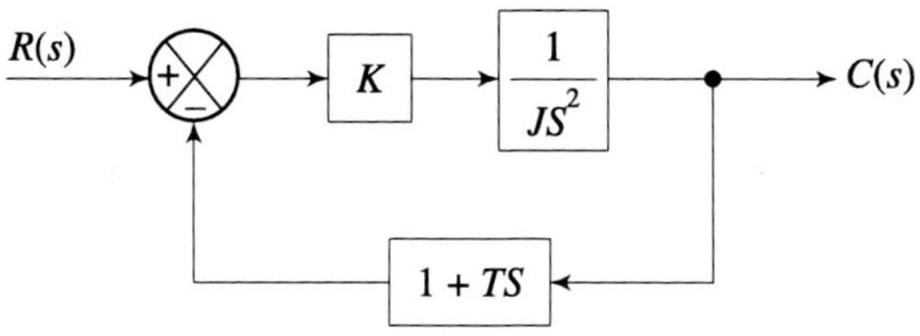

Fig. 4.44

Ans: $K = 2.95$ N-m $T = 0.471$ sec

67. A unity feedback system is characterized by the open-loop transfer function $G(s) = 1/S(0.5S + 1)(0.25 + 1)$, (a) determine the steady-state error to unit step, unit-ramp and unit parabolic inputs, (b) determine rise time, peak time, peak overshoot and settling time of the unit step response of the system.

Ans: (a) e_{ss}/unit step = 0

e_{ss}/unit ramp = 1

e_{ss}/unit parabola = ∞

(b) Rise time, $t_r = 1.91$ sec

Peak time, $t_p = 2.79$ sec

Peak overshoot $M_p = 0.1265$

Settling time, $t_s = 5.4$ sec

68. For a system whose $G(s) = 10(S(S + 1)(S + 2)$, find the steady state error when it is subjected to the inputs $r(t) = 1 + 2t + 1.5\, t^2$.

Ans: The total steady state error is ∞

69. A unity feedback system has $G(s) = 1/s\ (1 + s)$. The input to the system is described by $r(t) = 4 + 6s + 2t^3$. Find the generalized error coefficients and steady state error.

Ans: $C_0 = 0 \quad C_1 = 1 \quad e_{ss} = \infty$

$C_2 = 0 \quad C_3 = -6$

70. A unity feedback system has the forward path transfer function $G(s) = 10|(S + 1)$. Find the steady state error and the generalized error coefficients for $r(t) = -t$.

Ans: $C_0 = 1/11 \quad e_{ss} = \infty$

$C_1 = 10|12|$

71. Find out the position velocity and acceleration error coefficients for the following unity feedback systems having forward loop transfer function $G(s)$ as

(a) $100/(1 + 0.5s)(1 + 2s)$

(b) $K|S(1 + 0.15)(1 + S)$

(c) $K/S^2(S^2 + 8S + 100)$

(d) $K(1 + S)(1 + 2S)|s^2(S^2 + 4S + 20)$

Ans:

Question	K_p	K_v	K_a
(a)	100	0	0
(b)	∞	K	0
(c)	∞	∞	$K/100$
(d)	∞	∞	$K/20$

72. The open-loop transfer function of a unity feedback control system is $G(s) = 9|(s + 1)$, using the generalized error series determine the error signal and steady state error of the system when the system is excited by

(i) $r(t) = 2$ (ii) $r(t) = t$ (iii) $r(t) = 3t^2/2$ (iv) $r(t) = 1 + 2t + 3t^2/2$

Ans: (i) $e(t) = 0.2$; $e_{ss} = 0.2$ (ii) $e(t) = 0.1t + 0.09$; $e_{ss} = \infty$

(iii) $e(t) = 0.15t^2 + 0.27t - 0.054$; $e_{ss} = \infty$ (iv) $e(t) = 0.15t^2 + 0.77t + 0.226$; $e_{ss} = \infty$

73. For unity feedback system having open-loop transfer function as $G(s) = \dfrac{K(s + 2)}{S^2(S^2 + 7S + 12)}$, determine

(i) Type of system (ii) Error constant K_p, K_v, and K_a

(iii) Steady state error for parabolic input

Ans:

(i) It is type 2 system (ii) $K_p = \infty$, $K_v = \infty$, $K_a = K/6$

(iii) $e_{SS} = 6/k$

74. Derive the step input response of a second order system.

75. Draw the time domain response $C(t)$ of a typical underdamped, second order system to a step input indicate the following time domain specifications on the diagram

(i) Delay time (ii) Rise time

(iii) Peak time (iv) Maximum overshoot

(v) settling time

76. Define the steady state error and error constants with respect to unit step, unit velocity and unit acceleration inputs. How can the steady error be reduced?

77. The open-loop transfer function of a unity feedback control system is given by $G(s) = \dfrac{25}{S(S+5)}$. Obtain the maximum overshoot, peak time, rise time and settling time.

Ans: $M_p = 16.3\%$, $t_p = 0.7255S$, $t_r = 0.4836S$, $t_s = 1.6S$

78. Given the transfer function $G(s) = \dfrac{100}{S^2 + 15S + 100}$, find T_p, percent overshoot and T_S.

Ans: $T_p = 0.475S$% overshoot = 2.838 % $T_r = 0.533$ sec

79. The block diagram of a simple servosystem is shown in Fig. 4.45. Compute the values of K and T to give overshoot of 20% and peak time of 2 sec.

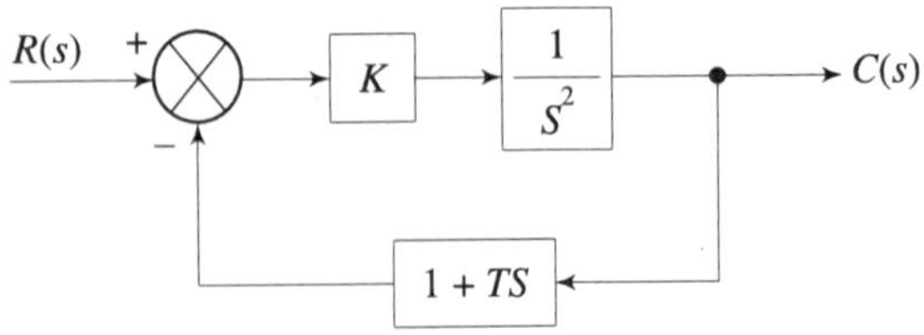

Fig. 4.45

Ans: $K = 3.1147$, $T = 0.5166$

80. The open-loop transfer function of unity feedback system is $G(s) = \dfrac{K}{S(1 + TS)}$. For this system overshoot reduces from 0.6 to 0.2 due to change in K only. Show that $\dfrac{TK_1 - 1}{TK_2 - 1} = 43.33$ where K_1 and K_2 are value of K for 0.6 and 0.2 overshoots respectively.

81. Determine the value of K and a such that the system has a damping ratio of 0.7 and an undamped natural frequency of 4 rad/sec for the system shown in Fig. 4.46.

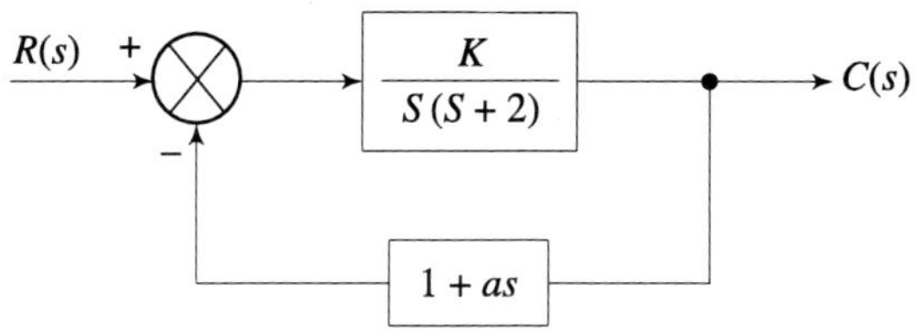

Fig. 4.46

Ans: $K = 16$, $a = 0.225$

82. The forward path transfer function of a unity feedback control system is given by $G(s) = \dfrac{2}{s(s + 3)}$. Obtain the expression for unit step response of the system.

Ans: $y(t) = 1 + e^{-2t} - 2e^{-t}$

83. Sketch the transient response of a second order system and derive the expression for rise time and peak overshoot.

84. Draw the response of a first order system subjected to unit step input.

85. The specification given on a certain second order feedback control system is that the overshoot of the step response should not exceed 25%. What are the corresponding limiting values of the damping ratio ξ and peak resonance M_r?

Ans: $\xi = 0.403$, $M_r = 1.51$

86. A unity feedback control system has an open-loop transfer function $G(s) = \dfrac{10}{s(s + 2)}$. Find t_r, M_p %, t_p, t_s and t_d for a step input of 12 units.

Ans: $t_r = 0.628$ sec, % $M_p = 36\%$, $t_p = 1.05$ sec, $t_d = 0.3849$ sec, $t_s = 4$ sec

87. Find the unit impulse response of a second order system whose transfer function is

$$G(s) = \frac{9}{S^2 + 4S + 9}$$

Ans: $C(t) = \dfrac{9}{\sqrt{5}} - e^{-2t} \sin \sqrt{5}t$

88. Determine the step, ramp and parabolic error constants of the following feedback control systems. The open-loop transfer one given below:

(i) $G(s) = \dfrac{1200}{(1 + 0.25)(1 + 11S)}$ (ii) $G(s) = \dfrac{120}{S(S^2 + 10S + 120)}$

(iii) $G(s) = \dfrac{110}{S(1 + 0.2S)(1 + 12S)}$ (iv) $G(s) = \dfrac{120}{S(S^2 + 10S + 110)}$

(v) $G(s) = \dfrac{1100}{S(S+10)(S+110)}$ (vi) $G(s) = \dfrac{10(1+4S)(1+6S)}{S^2(S^2+2S+1)}$

Ans: (i) $K_p = 1200$, $K_v = 0$, $K_a = 0$ (ii) $K_p = \infty$, $K_v = 1$, $K_a = 0$

(iii) $K_p = \infty$, $K_v = 110$, $K_a = 0$ (iv) $K_p = \infty$, $K_v = \infty$, $K_a = 1$

(v) $K_p = \infty$, $K_v = \infty$, $K_a = 0$ (vi) $K_p = \infty$, $K_v = \infty$, $K_a = 10$

89. Find type, error coefficient and steady state error of the system having unity feedback and open-loop transfer function $G(s) = \dfrac{K(s+4)}{s(s^3+8s^2+4)}$ when input is $A/2\ t^2$.

Ans: 2, ∞, ∞, $\dfrac{4K}{15}, \dfrac{15A}{5K}$.

90. Determine the range of values of K of the system shown in Fig. 4.47, so that $e_{ss} < 0.004$ when $r(t) = 0.02t$.

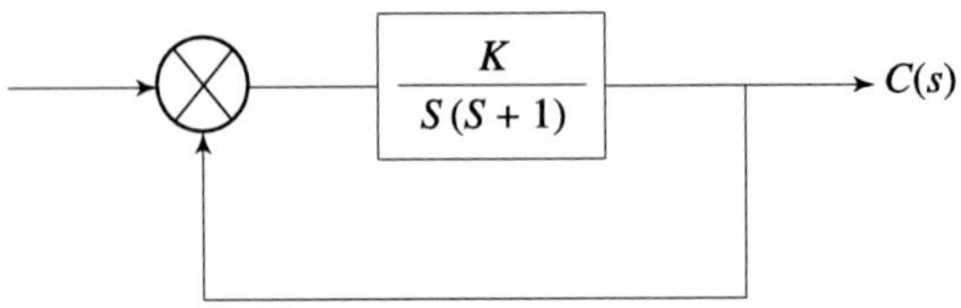

Fig. 4.47

Ans: $50 < K < \infty$.

91. A unity feedback control system has the following feedback path transfer function

$$G(s) = \frac{K(2S+1)}{S(4S+1)(S^2+2S+3)}$$

If the input $r(t) = 1 + 5t$ is applied to the system, determine the minimum value of K to get the steady state error equal to or less than 0.1.

Ans: 150

92. Determine the error co-constants, steady state error to the following inputs:

(i) A unit step input (ii) A unit velocity step velocity input

(iii) A unit velocity acceleration input

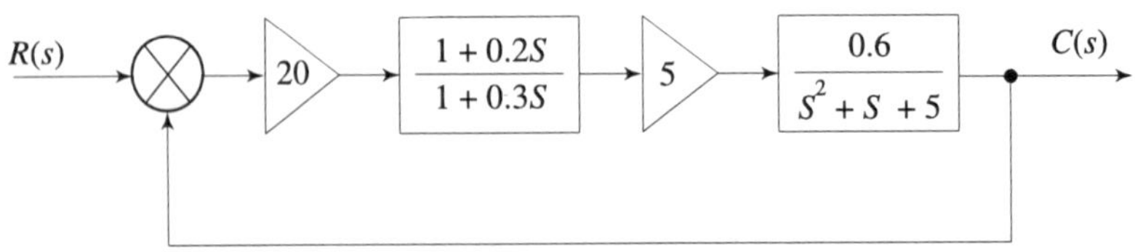

Fig. 4.48

Ans: $K_p = 12$, $K_v = \infty$, $K_a = \infty$.

(i) 0.0769 (ii) 0 (iii) 0

93. Find ω_n, ω_d, t_r, t_p, M_p and t_s and $C(t)$ for the system having transfer function $\frac{C(s)}{R(s)} = \frac{2S}{S^2 + 6S + 2S}$.

Ans: 5 rad/sec, 4 rad/sec, 0.5335 sec, 0.785, 9.84% 1.33 sec, $c(s) = 1 - 1.5635\ e^{-3t} \sin (4t + 0.9272)$

94. A second order control system is represented by a transfer function given below:
$\frac{Q_0(s)}{T(s)} = \frac{1}{Js^2 + Ds + K}$. If a step input of 12 Nm is applied to the system, calculate the value of J, D and K for the following results: (a) $M_p = 7\%$ (b) $t_p = 1.2$ sec and (c) steady state value of output is 0.6 rad.

Ans: $J = 1.7029$ N-m /rad/sec^2

$D = 7.54$ N-m /rad/sec

$K = 20$ N-m/rad

95. A unity feedback servo derive instrument has open-loop-transfer function $G(s + 2) = \frac{10}{s(s + 2)}$.

Find (i) The time domain response of a unit step input.
(ii) The natural frequency of oscillation
(iv) Maximum overshoot and peak time
(iv) Steady state error to an input $1 + 4t$

Ans: (i) $C(t) = 1 - 1/054928e^{-t} \sin (3t + 71.565°)$.

(ii) 3.16227 rad/sec

(iii) 35.09% 11.0472 sec

(iv) 0.8.

5 Stability Analysis

5.1 INTRODUCTION

5.1.1 Definition of Stability

A system is said to be stable if the output of the system is bounded (finite) for any bounded (finite) input. The term stability refers to the stable working condition of a control system. Every working system is designed to be stable. In a stable system the response or output is predictable, finite and stable for a given input (or changes in input or changes in system parameters).

5.2 CONCEPTS OF STABILITY

5.2.1 Bounded-Input Bounded-Output (BIBO) Stability

A linear relaxed system is said have BIBO stability if every bounded (finite) input results in a bounded (finite) output. A condition for BIBO stability can be obtained from convolution theorem.

The response of a system, $C(t)$, for any input is given by convolution of the input and impulse response.

The response in time domain, $C(t) = \int_{o}^{+\infty} m(\tau)r(t-\tau)d\tau$

If the input, $r(t)$, is bounded then there exists a constant A, such that $|r(t)| \leq A < \infty$. The condition for bounded output for this bounded input condition can be derived as follows

On taking the absolute value on both sides of equation, we get

$$|C(t)| = \left| \int_{o}^{+\infty} m(\tau)r(t-\tau)d\tau \right|$$

Since the absolute value of an integral is not greater than the integral of the absolute value of the integrand the equation can be written as

$$|C(t)| \leq \int_{o}^{+\infty} |m(\tau)r(t-\tau)|d\tau$$

$$\leq \int_{o}^{+\infty} |m(\tau)|\,|r(t-\tau)|\,d\tau$$

$$\leq \int_{0}^{+\infty} |m(\tau)|A \, d\tau \qquad (\because \text{ for bounded input } |r(t-\tau)| \leq A)$$

$$\leq A \int_{0}^{+\infty} |m(\tau)| \, d\tau$$

If the output, $C(t)$, is bounded then there exists a constant A_1 such that $|C(t)| \leq A_1 < \infty$.

$$\therefore \qquad A_1 \int_{0}^{+\infty} |m(\tau)| d\tau \leq A_1 < \infty$$

The above condition is satisfied if

$$\int_{0}^{+\infty} |m(\tau)| d\tau < \infty \quad \text{Hence for bounded output} \int_{0}^{\infty} |m(\tau)| d\tau < \infty$$

Therefore, we can conclude that a system with impulse response $m(\tau)$ is BIBO stable if and only if the impulse response is absolutely integrable (i.e., $\int_{0}^{\infty} |m(\tau)| d\tau$ is finite. This means that area under the absolute value curve of the impulse response $m(\tau)$ evaluated from $t = 0$ to $t = \infty$ must be finite).

5.3 TYPES OF STABILITY

The different types of the stability are the following.

1. **Absolutely stable:** If the system's response is stable for all variations of the input or parameters then it is called absolutely stable system. The locations of the roots of the characteristic equation of the system are in the left half of the S-plane (LHP). The location of poles is very close to the $j\omega$ (imaginary) axis.
2. **Unstable system:** A system is unstable if for a bounded disturbing input signal the output is of infinite amplitude or is oscillatory. The poles or the roots of the characteristic equation are in the right half of the S-plane (RHP). The system may have an unbounded output even with a bounded input, that is, no control on the response.
3. **Critically or marginally stable system:** For a bounded input signal, if the output has constant amplitude oscillations then the system may be stable or unstable under some limited constraints. Such a system is called critically stable. The response does not go increasing infinitely or does not go to zero as the time increases. The response will be same for entire time. The output oscillates in a finite range or it stays at same value. The location of poles is on $j\omega$ (imaginary) axis. If there are no repeated poles on the imaginary axis, the system is said to be stable.
4. **Conditional stable system:** If a system's output is stable for a limited range of variations of its parameters then the system is called a conditionally stable system. Stability of the system depends on the condition of the parameter of the system.
5. **Relative stable system:** The location of the poles lies far away from the $j\omega$ axis of left half of the S-plane (LHP). The response of the system takes more time to settle.

5.4 NECESSARY CONDITIONS FOR STABILITY

A system's stability can be determined by finding out the roots of the characteristic equation. The system is considered to be stable if the roots of the characteristic equation are located in left half of the *S*-plane.

The closed-loop transfer function, $M(s)$, can be expressed as a ratio of two polynomials in s. The denominator polynomial of closed-loop transfer function is called characteristic equation. The roots of characteristic equation are poles of closed-loop transfer function.

For BIBO stability the integral of the impulse response should be finite, which implies that the impulse should be finite as t tends to infinity. (*The impulse response is the inverse Laplace transform of the transfer function.*) This requirement for stability can be linked to the location of roots of characteristic equation in the *S*-plane.

The closed-loop transfer function, $M(s)$, can be expressed as a ratio of two polynomials in s as shown in the following equations.

$$M(s) = \frac{b_0 s^m + b_1 s^{m-1} + b_2 s^{m-2} + \dots + b_{m-1} s + b_m}{a_0 s^n + a_1 s^{n-1} + a_2 s^{n-2} + \dots + a_{n-1} s + a_n}$$

$$= \frac{(s + z_1)(s + z_2)(s + z_3) \dots (s + z_m)}{(s + p_1)(s + p_2)(s + p_3) \dots (s + p_n)}$$

The roots of numerator polynomial $z_1, z_2 \dots\dots z_n$ are zeros. The roots of denominator polynomial $p_1, p_2 \dots p_n$ are poles. The denominator polynomial is the characteristic equation and so the poles are roots of characteristic equation.

By partial fraction expansion we can write,

$$M(s) = \frac{A_1}{s + p_1} + \frac{A_2}{s + p_2} + \frac{A_3}{s + p_3} + \dots\dots + \frac{A_n}{s + p_n} \qquad \dots(5.1)$$

The poles $p_1, p_2 \dots p_n$ may be at the origin or lie on imaginary axis or lie on right or left half *S*-plane. The impulse response is given by inverse Laplace transform of $M(s)$. The impulse response of various terms of $M(s)$ are shown in Table 5.1.

Table 5.1 Analysis of location of roots on *S*-plane w.r.t. time domain response

Transfer function and location of roots on S-plane	*Impulse response (inverse Laplace transform and the sketch of time domain function)*
$M(s) = \frac{A}{s + a}$ 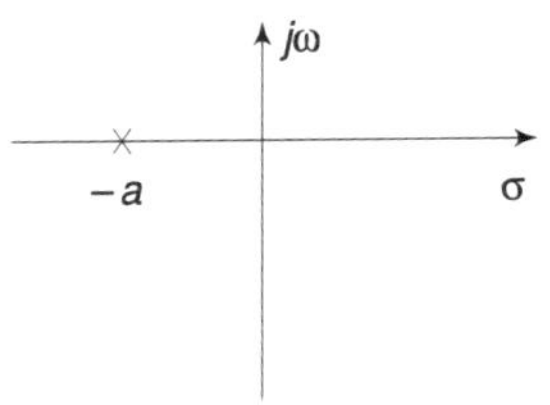 Root on negative real axis	$m(t) = L^{-1}\left[\frac{A}{s + a}\right] = A_e^{-at}$ 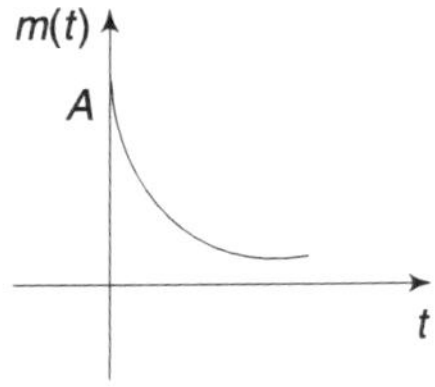 Response is exponentially decaying. Stable system.

(*Contd.*)

Table 5.1 Contd.

$M(s) = \dfrac{A}{s-a}$ 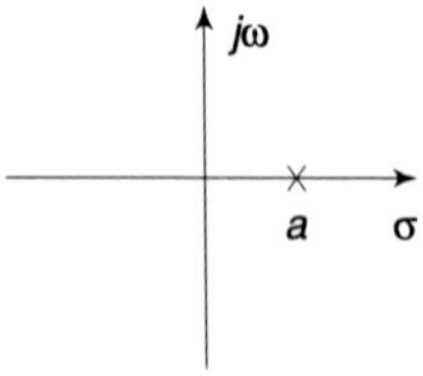 *Root on positive real axis*	$m(t) = L^{-1}\left[\dfrac{A}{s-a}\right] = Ae^{+at}$ 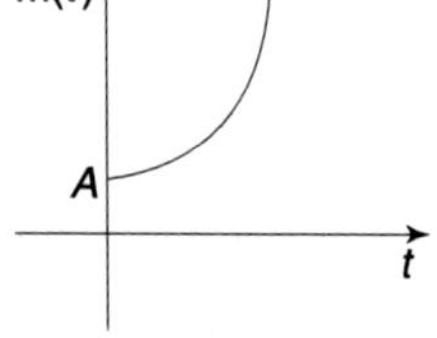 Response is exponentially increasing. Unstable system
$M(s) = \dfrac{A}{s+a+jb} + \dfrac{A^*}{s+a-jb}$ 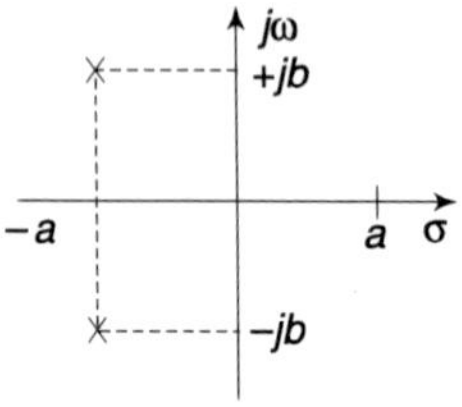 Complex conjugate root on positive roots on left half S-plane	$m(t) = L^{-1}\left[\dfrac{A}{s+a+jb} + \dfrac{A}{s+a-jb}\right]$ $= Ae^{-(a+jb)t} + Ae^{-(a-jb)t}$ $= 2Ae^{-at}\cos bt = 2Ae^{-at}\sin(bt + 90°)$ 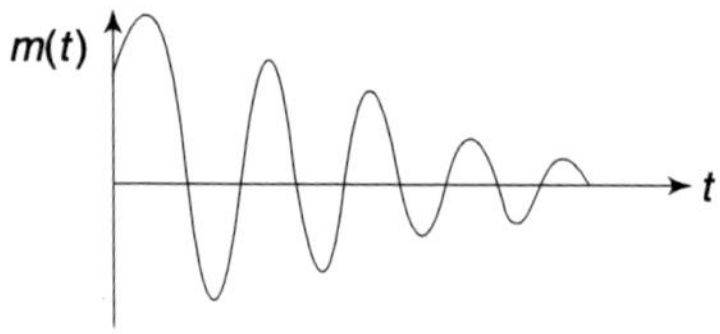 The response is damped sinusoidal
$M(s) = \dfrac{A}{s-a+jb} + \dfrac{A}{s-a-jb}$ 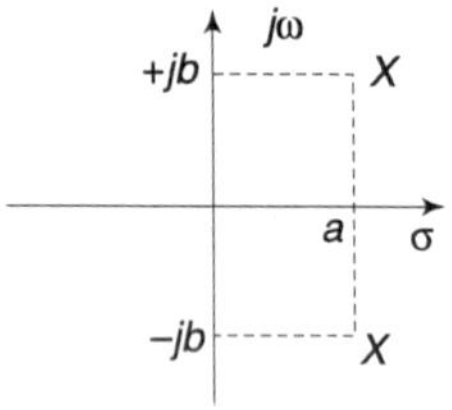 Complex conjugate root on right half S-plane	$m(t) = L^{-1}\left[\dfrac{A}{s-a+jb} + \dfrac{A}{s-a-jb}\right]$ $= Aa^{-(a+jb)t} + Ae^{-(a-jb)t}$ $= 2Ae^{at}\cos bt = 2Ae^{at}\sin(bt + 90°)$ 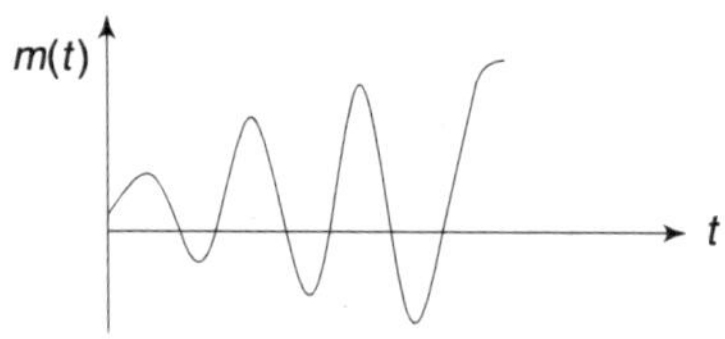 The response is exponentially increasing sinusoidally (i.e., amplitude of oscillation exponentially increases with time). Unstable system

(*Contd.*)

Table 5.1 Contd.

$M(s) = \dfrac{A}{s+jb} + \dfrac{A}{s-jb}$ 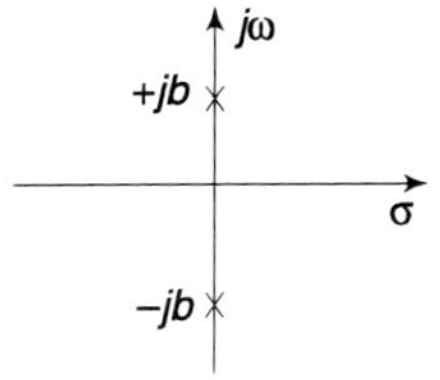 Single pair of roots on imaginary axis	$m(t) = L^{-1}\left[\dfrac{A}{s+jb} + \dfrac{A}{s-jb}\right]$ $= Ae^{-jb} + Ae^{+jb}$ $= 2A\cos bt = 2A\sin(bt + 90°)$ 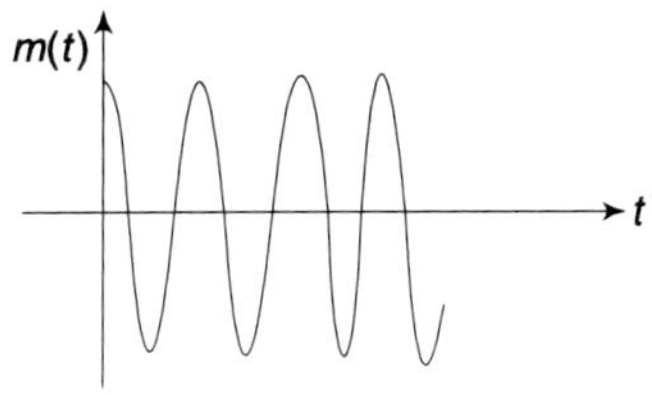 The response is oscillatory. Marginally stable system
$M(s) = \dfrac{A}{(s+jb)^2} + \dfrac{A}{(s-jb)^2}$ 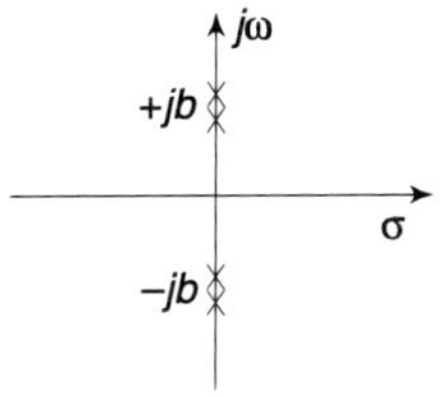 Double pair of roots on imaginary axis	$m(t) = L^{-1}\left[\dfrac{A}{(s+jb)^2} + \dfrac{A}{(s-jb)^2}\right]$ $= At\,e^{-jb} + At\,e^{+jb}$ $= 2At\,e\cos bt = 2At\sin(bt + 90°)$ 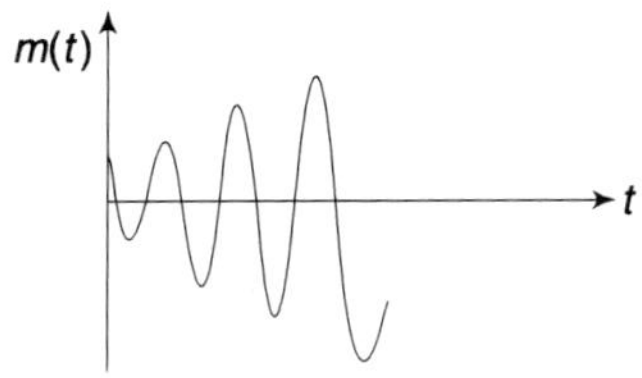 Response is linearly increasing sinusoidal (i.e., amplitude of oscillations linearly increases with time). Unstable system
$M(s) = \dfrac{A}{s}$ 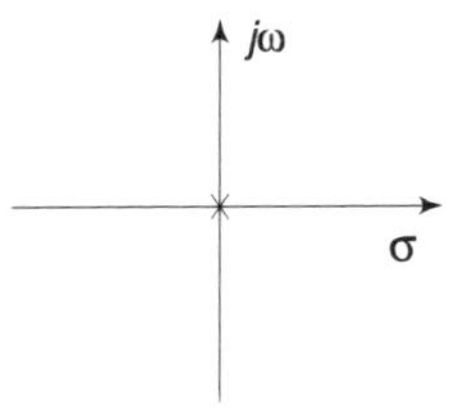 Single root at origin	$m(t) = L^{-1}\left[\dfrac{A}{s}\right] = A$ 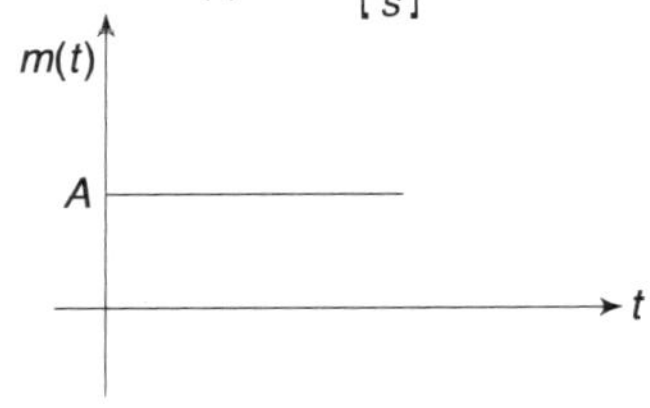 The response is constant. Marginally stable system

(*Contd.*)

Table 5.1 Contd.

$M(s) = \frac{A}{s^2}$ 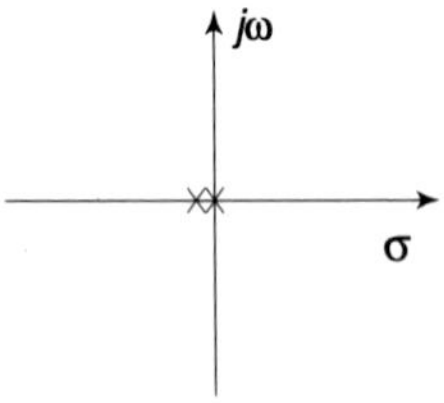 Double root at origin	$m(t) = L^{-1}\left[\frac{A}{s^2}\right] = At$ 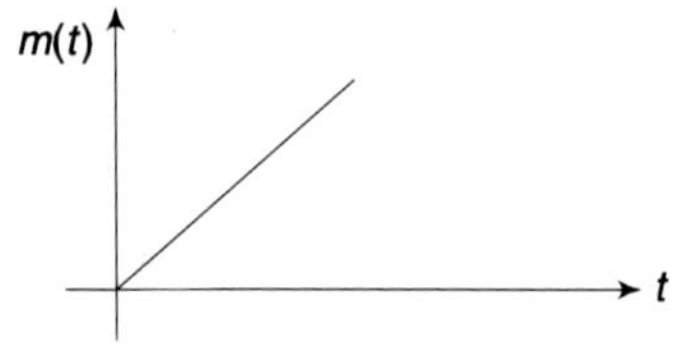 The response linearly increases with time. Unstable system

From Table 5.1, the following conclusions are drawn based on the location of roots of characteristic equation.

1. If all the roots of characteristic equation have negative real parts (i.e., lying on left half S-plane) then the impulse response is bounded (i.e., it decreases to zero as t tends to ∞). Hence, $\int_0^\infty m(\tau)d\tau$ is finite and the system is bounded-input bounded-output (BIBO) stable.
2. If any root of the characteristic equation has a positive real part (i.e., lying on right half S-plane) then the impulse response is unbounded. (i.e., it increases to ∞ as t tends to ∞). Hence, $\int_0^\infty m(\tau)d\tau$ is infinite and so the system is unstable.
3. If the characteristic equation has repeated roots on the imaginary axis then impulse response is unbounded (i.e., it increases to ∞ as t tends to ∞). Hence, $\int_0^\infty m(\tau)d\tau$ is infinite and so the system is unstable.
4. If one or more non-repeated roots of the characteristic equation are lying on the imaginary axis, then impulse is bounded (i.e., it has constant amplitude oscillations) but $\int_0^\infty m(\tau)d\tau$ is infinite and so the system is unstable.
5. If the characteristic equation has single root at the origin then the impulse response is bounded (i.e., it has constant amplitude) but $\int_0^\infty m(\tau)d\tau$ is infinite and so the system is unstable.
6. If the characteristic equation has repeated roots at the origin then the impulse response is unbounded (i.e., it linearly increases to infinity as t tends to ∞) and so the system is unstable.
7. In system with one or more repeated roots on imaginary axis or with single root at the origin, the output is bounded for bounded inputs except for the inputs having poles matching the system poles. These cases may be treated as acceptable or non-acceptable. Hence, when the system has non-repeated poles on imaginary axis or single pole at the origin, it is referred to as limitedly or marginally stable system.

In summary, the following three points may be stated regarding the stability of the system depending on the location of roots of characteristic equation.

(i) If all the roots of characteristic equation have negative real parts, then the system is stable.

(ii) If any root of the characteristic equation has a positive real part or if there is a repeated root on the imaginary axis then the system is unstable.

(iii) If the condition (i) is satisfied except for the presence of one or more non-repeated roots on the imaginary axis, then the system is limitedly or marginally stable.

In summary the following conclusions can be made about the coefficients of characteristic polynomial.

1. If all the coefficients are positive and if no coefficient is zero then all the roots are in the left half plane.
2. If any coefficient, a_i, is equal to zero then some of the roots may be on the imaginary axis or on the right half plane.
3. If any coefficient, a_i, is negative then at least one root is in the right half plane.

It can be concluded that the absence or negativeness of any of the coefficients of a characteristic polynomial indicates that the system is either unstable or at most marginally stable. Thus, the necessary condition for stability of the system is that all the coefficients of its characteristic polynomial should be positive. If any coefficient is zero/negative, we can immediately say that the system is unstable.

When all the coefficients are positive, the system is not necessarily stable, because there may be roots in the right half plane and/or on the imaginary axis. For all the roots to have negative real parts, it is necessary but not sufficient that all of the coefficients of characteristic equation are positive.

For example, consider the characteristic polynomial with all positive coefficients $s^3 + s^2 + 2s + 8 = 0$. The characteristic polynomial can be written as

$$(s^3 + s^2 + 2s + 8) = (s + 2)\left(s - \frac{1}{2} - j\frac{\sqrt{15}}{2}\right)\left(s - \frac{1}{2} + j\frac{\sqrt{15}}{2}\right) = 0$$

The coefficients of the polynomial are all positive but two roots are on the right half plane and so the system is unstable.

5.5 ROUTH-HURWITZ STABILITY CRITERION

The denominator polynomial of closed-loop transfer function is called the characteristic equation of the system.

The roots of the characteristic equation of a stable system should lie on the left of S-plane. Hence, the roots should have negative real parts. The Routh-Hurwitz stability criterion is an analytical procedure for determining whether all the roots of a polynomial have negative real parts or not.

The first step in analyzing the stability of a system is to examine its characteristic equation. The necessary condition for stability is that all the coefficients of the polynomial are positive.

If some of the coefficients are zero or negative it can be concluded that the system is not stable.

When all the coefficients are positive, the system is not necessarily stable. Even though the coefficients are positive, some of the roots may lie on the right half of *S*-plane or on the imaginary axis. In order for all the roots to have negative real parts, it is necessary but not sufficient that all of the coefficients of the characteristic equation are positive. If all the coefficients of the characteristic equation are positive, then the system may be stable and one should proceed further to examine the sufficient conditions of stability.

A. Hurwitz and E.J. Routh independently published the method of investigation the sufficient conditions of stability of a system. The Hurwitz criterion is in terms of determinants and Routh criterion is in terms of array formulation. The Routh stability criterion is presented here.

The Routh stability criterion is based on ordering the coefficients of the characteristic equation, $a_0s^n + a_1s^{n-1} + a_2s^{n-2} + \ldots\ldots + a_{n-1}s + a_n = 0$, where $a_0 > 0$ into a schedule, called the Routh array as shown below.

$$\begin{array}{lllllll}
s^n & : a_0 & a_2 & a_4 & a_6 & a_8 & \ldots\ldots \\
s^{n-1} & : a_1 & a_3 & a_5 & a_7 & a_9 & \ldots\ldots \\
s^{n-2} & : b_0 & b_1 & b_2 & b_3 & b_4 & \ldots\ldots \\
s^{n-3} & : c_0 & c_1 & c_2 & c_3 & c_4 & \ldots\ldots \\
\vdots & & & & & & \\
\vdots & & & & & & \\
s^1 & : g_0 & & & & & \\
s_0 & : h_0 & & & & &
\end{array}$$

The Routh stability criterion can be stated as follows.

"The necessary and sufficient condition for stability is that all of the elements in the first column of the Routh array are positive. If this condition is not met, the system is unstable and the number of sign changes in the elements of the first column of the Routh array corresponds to the number of roots of the characteristic equation in the right half of the S-plane."

Note: If the order of sign of first column element is +, +, –, +. Then + to – is considered one sign change and – to + as another sign change.

5.5.1 Construction of Routh Array

Let the characteristic polynomial be

$$a_0s^n + a_1s^{n-1} + a_2s^{n-2} + a_3s^{n-3} + \ldots + a_{n-1}s^1 + a_ns^0$$

The coefficients of the polynomial are arranged in two rows as shown below.

$$\begin{array}{llllll}
s^n & : a_0 & a_2 & a_4 & a_6 & \ldots\ldots \\
s^{n-1} & : a_1 & a_3 & a_5 & a_7 & \ldots\ldots
\end{array}$$

If n is even then s^n row is formed by coefficients of even order terms and s^{n-1} row is formed by coefficients of odd order terms (i.e., coefficients of odd powers of s).

If n is odd then s^n row is formed by coefficients of odd order terms and s^{n-1} row is formed by coefficients of even order terms (i.e., coefficients of even powers of s).

The other rows of Routh array upto s^0 row can be formed by the following procedure. Each row of Routh array is constructed by using the elements of previous two rows.

Consider two consecutive rows of Routh array as shown below.

$$\begin{array}{llllllll} s^{n-x} & : & x_0 & x_1 & x_2 & x_3 & x_4 & \ldots\ldots\ldots \\ s^{n-x-1} & : & y_0 & y_1 & y_2 & y_3 & y_4 & \ldots\ldots\ldots \end{array}$$

Let the next row be,

$$s^{n-x-2}: \quad z_0 \quad z_1 \quad z_2 \quad z_3 \quad \ldots\ldots\ldots$$

The elements of s^{n-x-2} row are given by,

$$z_0 = \frac{(-1)\begin{vmatrix} x_0 & x_1 \\ y_0 & y_1 \end{vmatrix}}{y_0} = \frac{y_0x_1 - y_1x_0}{y_0}$$

$$z_1 = \frac{(-1)\begin{vmatrix} x_0 & x_2 \\ y_0 & y_2 \end{vmatrix}}{y_0} = \frac{y_0x_2 - y_2x_0}{y_0}$$

$$z_2 = \frac{(-1)\begin{vmatrix} x_0 & x_3 \\ y_0 & y_3 \end{vmatrix}}{y_0} = \frac{y_0x_3 - y_3x_0}{y_0}$$

$$z_3 = \frac{(-1)\begin{vmatrix} x_0 & x_4 \\ y_0 & y_4 \end{vmatrix}}{y_0} = \frac{y_0x_4 - y_4x_0}{y_0}$$

The elements z_0, z_1, z_2, z_3 .... are computed until an element is equal to zero or for all possible computations as shown above.

In the process of constructing Routh array the missing terms are considered zeros. Also, all the elements of any row can be multiplied or divided by a positive constant to simplify the computational work.

In the construction of Routh array one may come across the following three cases.

Case I: Normal Routh array (non-zero elements in the first column of Routh array).

Case II: A row of all zeros.

Case III: First element of a row is zero but some or other elements are not zero.

Case I: Normal Routh Array

In this case there is no difficulty in forming Routh array. The Routh array can be constructed as explained above. The sign changes are noted to find the number of roots lying on the right half of S-plane and the stability of the system can be estimated.

In this case if there is no sign change in the first column of Routh array then all the roots are lying on left half of S-plane and the system is stable.

If there is sign change in the first column of Routh array, then the system is unstable and the number of roots lying on the right half of S-plane is equal to the number of sign changes. The remaining roots are lying on the left half of S-plane.

Case II: A Row of All Zeros

An all zero row indicates the existence of an even polynomial as a factor of the given characteristic equation. In an even polynomial the exponents of S are even integers or zero only. This even polynomial factor is called the auxiliary polynomial. The coefficients of the auxiliary polynomial will always be the elements of the row directly above the row of zeros in the array.

The roots of an even polynomial occur in pairs that are equal in magnitude and opposite in sign. Hence, these roots can be purely imaginary, purely real or complex. The purely imaginary and purely real roots occur in pairs. The complex roots occur in groups with respect to both the real and imaginary axes. Figure 5.1 shows the roots of an even polynomial.

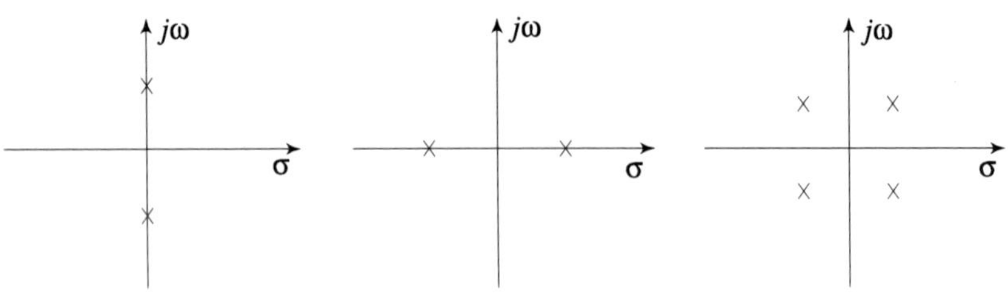

Fig. 5.1 Root of an even polynomial

Case II: Polynomial can be analyzed by any one of the following two methods.

Method 1

1. Determine the auxiliary polynomial, $A(s)$.
2. Differentiate the auxiliary polynomial with respect to s, to get $dA(s)/ds$.
3. The row of zeros is replaced with coefficients of $dA(s)/ds$.
4. Continue the construction of the array in the usual manner (as that of case I) and the array is interpreted as follows.
   (a) If there are sign changes in the first column of Routh array then the system is unstable. The number of roots lying on right half of S-plane is equal to the number of sign changes. The number of roots on imaginary axis can be estimated from the roots of auxiliary polynomial. The remaining roots are lying on the left half of S-plane.
   (b) If there are no sign changes in the first column of Routh array then the all zeros row indicates the existence of purely imaginary roots and so the system is limitedly

or marginally stable. The roots of auxiliary equation lie on imaginary axis and the remaining roots lie on left half of S-plane.

Method 2

1. Determine the auxiliary polynomial, $A(s)$.
2. Divide the characteristic equation by auxiliary polynomial.
3. Construct Routh array using the coefficients of quotient polynomial.
4. The array is interpreted as follows.
   (a) If there are sign changes in the first column of Routh array of quotient polynomial then the system is unstable. The number of roots of quotient polynomial lying on right half of S-plane is given by the number of sign changes in the first column of Routh array.

   The roots of auxiliary polynomial are directly calculated to find whether they are purely imaginary or purely real or complex.

   The total number of roots on the right half of S-plane is given by the sum of number of sign changes and the number of roots of auxiliary polynomial with positive real part. The number of roots on imaginary axis can be estimated from the roots of auxiliary polynomial. The remaining roots are lying on the left half of S-plane.

   (b) If there is no sign change in the first column of Routh array of quotient polynomial then the system is limitedly or marginally stable. Since there is no sign change all the roots of quotient polynomial are lying on the left half of S-plane.

   The roots of auxiliary polynomial are directly calculated to find whether they are purely imaginary or purely real or complex. The number of roots lying on the imaginary axis and on the right half of S-plane can be estimated from the roots of auxiliary polynomial. The remaining roots are lying on the left half of S-plane.

Case III: First Element of a Row is Zero

While constructing Routh array, if a zero is encountered as the first element of a row then all the elements of the next row will be infinite. To overcome this problem let $0 \rightarrow \varepsilon$ and complete the construction of array in the usual way (as that of case I).

Finally, let $\varepsilon \rightarrow 0$ and determine the value of the elements of the array which are functions of ε. The resultant array is interpreted as follows.

Note: If all the elements of a row are zeros then the solution is attempted by considering the polynomial as case II polynomial. Even if there is a single element zero on s^1 row, it is considered a row of all zeros.

(a) If there is no sign change in the first column of Routh array and if there is no row with all zeros, then all the roots are lying on the left half of S-plane and the system is stable.

(b) If there are sign changes in the first column of Routh array and there is no row with all zeros, then some of the roots are lying on the right half of S-plane and the system is unstable. The number of roots lying on the right half of S-plane is equal to the number of sign changes and the remaining roots are lying on the left half of S-plane.

(c) If there is a row of all zeros after letting $\varepsilon \to 0$, then there is a possibility of roots on imaginary axis. Determine the auxiliary polynomial and divide the characteristic equation by auxiliary polynomial to eliminate the imaginary roots. The Routh array is constructed using the coefficients of quotient polynomial and the characteristic equation is interpreted as explained in method-2 case-II polynomial.

Example 5.1: Determine the stability of a system whose characteristic equation is

$$S^4 + 5S^3 + 20S^2 + 40S + 50 = 0$$

Solution:

$$\begin{array}{lllll} S^4 & : & 1 & 20 & 50 \\ S^3 & : & 5 & 40 & \\ S^2 & : & 12 & 50 & \\ S^1 & : & \frac{115}{6} & & \\ S^0 & : & 50 & & \end{array}$$

↑ Column 1

$S^2 : \dfrac{5 \times 20 - 40 \times 1}{5}$

$S^2 : 12$

$S^2 : \dfrac{5 \times 50 - 1 \times 0}{5}$

$S^2 : 50$

$S^1 : \dfrac{12 \times 40 - 5 \times 50}{12}$

$S^1 : \dfrac{115}{6}$

$S^0 : \dfrac{\frac{115}{6} \times 50 - 12 \times 0}{\left(\frac{115}{6}\right)}$

$S^0 : 50$

In the first column there is no change of the sign. Therefore, the system is stable.

Example 5.2: Determine the stability of a system whose characteristic equation is $S^5 + 2S^4 + 4S^3 + 4S^2 + 3S + 8 = 0$.

Solution:

$$\begin{array}{lllll} S^5 & : & 1 & 4 & 3 \\ S^4 & : & 2 & 4 & 8 \\ S^3 & : & 2 & -1 & \\ S^2 & : & 5 & 8 & \\ S^1 & : & \frac{-21}{5} & 0 & \\ S^0 & : & 8 & & \end{array}$$

$S^3 : \dfrac{2 \times 4 - 4 \times 1}{2} = 2$

$S^3 : \dfrac{3 \times 2 - 8 \times 1}{2} = -1$

$S^2 : \dfrac{2 \times 4 - (2 \times -1)}{2} = 5$

$S^2 : \dfrac{2 \times 8 - 2 \times 0}{2} = 8$

The complete Routh array is

$$\begin{array}{lllll} S^5 & : & 1 & 4 & 3 \\ S^4 & : & 2 & 4 & 8 \\ S^3 & : & 2 & -1 & \\ S^2 & : & 5 & 8 & \\ S^1 & : & \frac{-21}{5} & & \\ S^0 & : & 8 & & \end{array}$$

↑ Column 1st

$$S^1 : \frac{5 \times -1 - 8 \times 2}{5} = -\frac{21}{5}$$

$$S^1 : 0$$

$$S^0 : \frac{\frac{-21}{5} \times 8 - 5 \times 0}{\frac{-21}{5}} = 8$$

There are two changes of sign in the first column, one from + 5 to $\frac{-21}{5}$ and the other from $\frac{-21}{5}$ to 8. Hence, two roots of characteristic equation are lying on the right half of S-plane and remaining three roots are lying on left half of S-plane. Hence, the system is unstable.

Example 5.3: Examine the stability of the system having characteristic equation using Routh's criterion

$$3S^4 + 10S^3 + 5S^2 + 5S + 3 = 0$$

Solution: The Routh array is

$$\begin{array}{llllll} S^4 & : & 3 & 5 & 3 & \\ S^3 & : & 10 & 5 & \text{divided by 5} \\ S^3 & : & 2 & 1 & & \\ S^2 & : & \frac{7}{2} & 3 & & \\ S^1 & : & \frac{-5}{7} & & & \\ S^0 & : & 3 & & & \end{array}$$

↑ Column 1

$$S^2 : \frac{2 \times 5 - 3 \times 1}{2} = \frac{7}{2}$$

$$S^2 : \frac{2 \times 3 - 3 \times 0}{2} = 3$$

$$S^1 : \frac{\frac{7}{2} \times 1 - 2 \times 3}{\frac{7}{2}} = \frac{-5}{7}$$

$$S^0 : \frac{\frac{-5}{7} \times 3 \frac{-7}{2} \times 0}{\frac{-5}{7}} = 3$$

There are two changes of sign in that first column: one from + $\frac{7}{2}$ to $-\frac{5}{7}$ and the other from $-\frac{5}{7}$ to 3. Thus, there are two roots of the characteristic polynomial in the right half of the S-plane. Hence, the system is unstable.

Example 5.4: Determine the stability of a closed-loop control system whose characteristic equation is

$$S^5 + 3S^4 + 2S^3 + 6S^2 + 6S + 9 = 0$$

Solution: The Routh array be

$$
\begin{array}{lcccc}
S^5 & : & 1 & 2 & 6 \\
S^4 & : & 3 & 6 & 9 \quad \text{divide by 3} \\
S^4 & : & 1 & 2 & 3 \\
S^3 & : & 0 & 3 & \varepsilon \\
S^3 & : & \varepsilon & 3 & \\
S^2 & : & \dfrac{2\varepsilon - 3}{\varepsilon} & 3 & \\
S^1 & : & 3 & -\dfrac{3\varepsilon^2}{2\varepsilon - 3} & 0 \\
S^0 & : & 3 & &
\end{array}
$$

$$S^3 : \frac{2-2}{1} = 0$$

$$S^3 : \frac{6-3}{1} = 3$$

$$S^2 : \frac{2\varepsilon - 3}{\varepsilon}$$

$$S^2 : \frac{3\varepsilon}{\varepsilon} = 3$$

$$S^1 : \frac{\dfrac{3(2\varepsilon - 3)}{\varepsilon} - 3\varepsilon}{\left(\dfrac{2\varepsilon - 3}{\varepsilon}\right)} = \frac{3\varepsilon^2}{2\varepsilon - 3}$$

In the S^2 row, the first term is

$$\frac{2\varepsilon - 3}{\varepsilon} = 2 - \frac{3}{\varepsilon}$$

$$\lim_{\epsilon \to 0}\left(2 - \frac{3}{\varepsilon}\right) = -\infty$$

In S^1 row, $3 - \dfrac{3\varepsilon^2}{2\varepsilon - 3}$

$$\lim_{\varepsilon \to 0}\left(3 - \frac{3\varepsilon^2}{2\varepsilon - 3}\right) = 3$$

Therefore the array is

$$
\begin{array}{lcccc}
S^5 & : & 1 & 2 & 6 \\
S^4 & : & 1 & 2 & 3 \\
S^3 & : & \varepsilon & 3 & 0 \\
S^2 & : & -\infty & 3 & \\
S^1 & : & 3 & 0 & \\
S_0 & : & 3 & &
\end{array}
$$

↑ Column 1

Therefore, there are two changes of sign one from S^3 row to S^2 and the other from S^2 row to S^1 row in the first column of the array. This means that there are two poles in the right half of the S-plane. Hence, the system is unstable.

Example 5.5: Using Routh criterion, determine the stability of the system represented by the characteristic equation $s^4 + 8s^3 + 18s^2 + 16s + 5 = 0$. Comment on the location of the roots of characteristic equation,

Solution: The characteristic equation is 4th order of the system is $s^4 + 8s^3 + 18s^2 + 16s + 5 = 0$.

The given characteristic equation and so it has 4 roots. Since the highest power of s is an even number, form the first row of Routh array using the coefficients of even powers of s and form the second row using the coefficients of odd powers of s.

s^4	:	1	18	5	... Row-1
s^3	:	8	16		... Row-2

The elements of s^3 row can be divided by 8 to simplify the computations.

s^4	:	1	18	5	... Row-1
s^3	:	1	2		... Row-2
s^4	:	16	5		... Row-3
s^3	:	1.7			... Row-4
s^3	:	5			... Row-5

↑ Column 1

$$s^2 : \frac{1 \times 18 - 2 \times 1}{1} \quad \frac{1 \times 5 - 0 \times 1}{1}$$

$$s^2 : 16 \quad 5$$

$$s^1 : \frac{16 \times 2 - 5 \times 1}{16}$$

$$s^1 : 1.6875 = 1.7$$

$$s^0 : \frac{1,7 \times 5 - 0 \times 16}{1.7}$$

$$s^0 : 5$$

On examining the elements of the first column of Routh array it is observed that all the elements are positive and there is no sign change. Hence, all the roots are lying on the left of S-plane and the system is stable.

Result

1. Stable system
2. All the four roots are lying on the left half of S-plane.

Example 5.6: Construct a Routh array and determine the stability of the system whose characteristic equation is $s^6 + 2s^5 + 8s^4 + 12s^3 + 20s^2 + 16s + 16 = 0$. Also determine the number of roots lying on the right half of S-plane, left half of S-plane and on imaginary axis.

Solution: The characteristic equation of the system is: $s^6 + 2s^5 + 8s^4 + 12s^3 + 20s^2 + 16s + 16 = 0$.

The given characteristic polynomial is 6th order equation and so it has 6 roots. Since the highest power of s is even number, form the first row of Routh array using the coefficients of even powers of s and form the second row using the coefficients of odder power of s.

s^6	:	1	8	20	16	... Row-1
s^5	:	2	12	16		... Row-2

The elements of s^5 row can be divided by 2 to simplify the calculations.

s^6	:	1	8	20	16	... Row-1
s^5	:	1	6	8		... Row-2
s^4	:	1	6	8		... Row-3
s^3	:	0	0			... Row-4
s^3	:	1	3			... Row-4
s^2	:	3	8			... Row-5
s^1	:	0.33				... Row-6
s^0	:	8				... Row-7

↑ Column 1

$$s^4 : \quad \frac{1 \times 8 - 6 \times 1}{1} \qquad \frac{1 \times 20 - 8 \times 1}{1} \qquad \frac{1 \times 16 - 0 \times 1}{1}$$

s^4 : 2 12 16

Divide by 2

s^4 : 1 6 8

The auxiliary equation is $A = s^4 + 6s^2 + 8$. On differentiating A with respect to s we get,

$$\frac{dA}{ds} = 4s^3 + 12s$$

The coefficients of $\frac{dA}{ds}$ are used to form s^3 row

s^3 : 4 12

Divide by 4

s^3 : 1 3

$$s^2 : \quad \frac{1 \times 6 - 3 \times 1}{1} \qquad \frac{1 \times 8 - 0 \times 1}{1}$$

s^2 : 3 8

$$s^1 : \quad \frac{3 \times 3 - 8 \times 1}{3}$$

s^1 : 0.33

$$s^0 : \quad \frac{0.33 \times 8 - 0 \times 3}{0.33}$$

s^0 : 8

On examining the elements of the 1st column of Routh array it is observed that there is no sign change. The row with all zeros indicates the possibility of roots on imaginary axis. Hence, the system is limitedly or marginally stable.

The auxiliary polynomial is

$$s^4 + 6s^2 + 8 = 0$$

Let, $s^2 = x$

$$\therefore \qquad x^2 + 6x + 8 = 0$$

The roots of quadratic are, $x = \frac{-6 \pm \sqrt{6^2 - 4 \times 8}}{2}$

$$= -3 \pm 1 = -2 \text{ or } -4$$

The roots of auxiliary polynomial are

$$s = \pm\sqrt{x} = \pm\sqrt{-2} \text{ and } \pm\sqrt{-4}$$
$$= +j\sqrt{2},\ -j\sqrt{2},\ +j2 \text{ and } -j2$$

The roots of auxiliary polynomial are also roots of characteristic equation. Hence 4 roots are lying on imaginary axis and the remaining two roots are lying on the left half of S-plane.

Result

1. The system is limitedly or marginally stable.
2. Four roots are lying on imaginary axis and remaining two roots are lying on left half of S-plane.

Example 5.7: Construct a Routh array and determine the stability of the system represented by the characteristic equation, $s^5 + s^4 + 2s^3 + 2s^2 + 3s + 5 = 0$. Comment on the location of the roots of characteristic equation.

Solution: The characteristic equation of the system $s^5 + s^4 + 2s^3 + 2s^2 + 3s + 5 = 0$.

The given characteristic polynomial is 5th order equation and so it has 5 roots. Since the highest power of s is odd number, form the first row of Routh array using the coefficient of odd powers of s and form the second row using the coefficients of even powers of s.

s^5 :	1	2	3	... Row-1
s^4 :	1	2	5	... Row-2
s^3 :	ε	-2		... Row-3
s^2 :	$\dfrac{2\varepsilon + 2}{\varepsilon}$	5		... Row-4
s^1 :	$\dfrac{-(5\varepsilon^2 + 4\varepsilon + 4)}{2\varepsilon + 2}$			... Row-5
s^0 :	5			... Row-6

$s^3 : \dfrac{1 \times 2 - 2 \times 1}{1} \quad \dfrac{1 \times 3 - 5 \times 1}{1}$

$s^3 : \quad 0 \quad -2$

Replace 0 by ε

$s^3 : \quad \varepsilon \quad -2$

$s^2 : \dfrac{\varepsilon \times 2 - (-2 \times 1)}{\varepsilon} \quad \dfrac{\varepsilon \times 5 - 0 \times 1}{\varepsilon}$

$s^2 : \dfrac{2\varepsilon + 2}{\varepsilon} \quad 5$

$s^1 : \dfrac{\dfrac{2\varepsilon \times 2}{\varepsilon} \times (-2) - (5 \times \varepsilon)}{\dfrac{2\varepsilon + 2}{\varepsilon}}$

$s^1 : \dfrac{-(5\varepsilon^2 + 4\varepsilon + 4)}{2\varepsilon + 2}$

$s^0 : \dfrac{\dfrac{-(5\varepsilon^2 + 4\varepsilon + 4)}{2\varepsilon + 2} \times 5 - 0 \times \dfrac{2\varepsilon + 2}{\varepsilon}}{\dfrac{-(5\varepsilon^2 + 4\varepsilon + 4)}{2\varepsilon + 2}}$

$s^1 : 5$

On letting $\varepsilon \to 0$, we get

s^5 :	1	2	3	... Row-1
s^4 :	1	2	5	... Row-2
s^3 :	0	-2		... Row-3
s^2 :	∞	5		... Row-4
s^1 :	-2			... Row-5
s^0 :	5			... Row-6

(Column 1 marked; + to − between s^2 and s^1; − to + between s^1 and s^0)

On observing the elements of the first column of Routh array, it is found that there are two sign changes. Hence, two roots are lying on the right half of S-plane.

Result

1. The system is unstable.
2. Two roots are lying on the right half of *S*-plane and three roots are lying on the left half of *S*-plane.

Example 5.8: By Routh stability criterion determine the stability of the system represented by the characteristic equation, $9s^5 - 20s^4 + 10s^3 - s^2 - 9s - 10 = 0$. Comment on the location of roots of characteristic equation.

Solution: The characteristic polynomial of the system is, $9s^5 - 20s^4 + 10s^3 - s^2 - 9s - 10 = 0$.

On examining the coefficients of the characteristic polynomial, it is found that some of the coefficients are negative and so some roots will lie on the right half of *S*-plane. Hence, the system is unstable. The Routh array can be constructed to find the number of roots lying on right half of *S*-plane.

The given characteristic polynomial is 5th order equation and so it has 5 roots. Since the highest power of *s* is odd number, form the first row of Routh array using the coefficients of odd powers of *s* and form the second row using the coefficients of even powers of *s*.

Sign change		Column 1			
+ to –	s^5 :	10	10	–9	... Row-1
– to +	s^4 :	–20	–1	–10	... Row-2
	s^3 :	9.55	–13.5		... Row-3
+ to –	s^2 :	–29.3	–10		... Row-4
	s^1 :	–16.8			... Row-5
	s^0 :	–10			... Row-6

$$s^3 : \quad \frac{-20 \times 10 - (-1) \times 9}{-20} \qquad \frac{-20 \times (-9) - (-10) \times 9}{-20}$$
$$s^3 : \quad 9.55 \qquad -13.5$$

$$s^2 : \quad \frac{9.55 \times (-1) - (-13.5) \times (-20)}{9.55} \qquad \frac{9.55 \times (-10)}{9.55}$$
$$s^2 : \quad -29.3 \qquad -10$$

$$s^1 : \quad \frac{-29.3 \times (-13.5) - (-10) \times 9.55}{29.3}$$
$$s^1 : \quad -16.8$$

$$s^0 : \quad \frac{-16.8 \times (-10)}{16.8}$$
$$s^0 : \quad -10$$

On examining the elements of the first column of Routh array it is observed that there are three sign changes and so three roots are lying on the right half of *S*-plane and the remaining two roots are lying on the left half of *S*-plane.

Result

1. The system is unstable
2. Three roots are lying on right half of *S*-plane and two roots are lying on left half of *S*-plane.

Example 5.9: The characteristic polynomial of a system is $s^7 + 9s^6 + 24s^5 + 24s^4 + 24s^3 + 24s^2 + 23s + 15 = 0$. Determine the location of roots on *S*-plane and hence the stability of the system.

Solution:

Method I: The characteristic equation is $s^7 + 9s^6 + 24s^5 + 24s^4 + 24s^3 + 24s^2 + 23s + 15 = 0$.

The given characteristic polynomial is 7th order equation and so it has 7 roots. Since the highest power of s is odd number, form the first row of Routh array using the coefficients of odd powers of s and form the second row using the coefficients of even powers of s as shown below.

s^7 : 1 24 24 23 ... Row-1

s^6 : 9 24 24 15 ... Row-2

Divide s^6 row by 3 to simplify the computations.

	Column 1				
s^7 :	1	24	24	23	... Row-1
s^6 :	3	8	8	5	... Row-2
s^5 :	1	1	1		... Row-3
s^4 :	1	1	1		... Row-4
s^3 :	0	0			... Row-5
s^3 :	2	1			... Row-5
s^2 :	0.5	1			... Row-6
s^1 :	−3				... Row-7
s^0 :	1				... Row-8

(+ to −: s^2 to s^1; − to +: s^1 to s^0)

s^5 : $\dfrac{3 \times 24 - 8 \times 1}{3}$ $\dfrac{3 \times 24 - 8 \times 1}{3}$ $\dfrac{3 \times 23 - 5 \times 1}{3}$

s^5 : 21.33 21.33 21.33

Divide by 21.33

s^5 : 1 1 1

s^4 : $\dfrac{1 \times 8 - 1 \times 3}{1}$ $\dfrac{1 \times 8 - 1 \times 3}{1}$ $\dfrac{1 \times 5 - 0 \times 3}{1}$

s^4 : 5 5 5

Divide by 5

s^4 : 1 1 1

s^3 : $\dfrac{1 \times 1 - 1 \times 1}{1}$ $\dfrac{1 \times 1 - 1 \times 1}{1}$

s^3 : 0 0

The auxiliary polynomial is $A = s^4 + s^2 + 1$

Differentiate A with respect to s.

$$\frac{dA}{ds} = 4s^3 + 2s$$

s^3 : 4 2

Divide by 2

s^3 : 2 1

s^2 : $\dfrac{2 \times 1 - 1 \times 1}{2}$ $\dfrac{2 \times 1 - 0 \times 1}{2}$

s^2 : 0.5 1

s^1 : $\dfrac{0.5 \times 1 - 1 \times 2}{0.5}$

s^1 : −3

s^0 : $\dfrac{-3 \times 1}{-3}$

s^0 : 1

On examining the first column elements of Routh array it is found that there are two sign changes. Hence, two roots are lyting on the right half of S-plane and so the system is unstable.

The row of all zeros indicates the possibility of roots on imaginary axis. This can be tested by evaluating the roots of auxiliary polynomial.

The auxiliary equation is

$$s^4 + s^2 + 1 = 0$$

Put $s^2 = x$ in the auxiliary equation,

$$\therefore \qquad s^4 + s^2 + 1 = x^2 + x + 1 = 0$$

The roots of quadratic are $x = \dfrac{-1 \pm \sqrt{1-4}}{2} = -\dfrac{1}{2} \pm j\dfrac{\sqrt{3}}{2}$

$$x = 1 \angle 120° \quad \text{or} \quad 1 \angle -120°$$

But $s^2 = x$, $\therefore$

$$\begin{aligned} s &= \pm\sqrt{x} = \pm\sqrt{1 \angle 120°} && \text{or} \quad \pm\sqrt{1 \angle -120°} \\ &= \pm\sqrt{1} \angle 120°/2 && \text{or} \quad \pm\sqrt{1} \angle -120°/2 \\ &= \pm\sqrt{1} \angle 60° && \text{or} \quad \pm\sqrt{1} \angle -60° \\ &= \pm(0.5 + j0.866) && \text{or} \quad \pm(0.5 - j0.866) \end{aligned}$$

Two roots of auxiliary polynomial are lying on the right half of S-plane and the remaining two on the left half of S-plane. The roots of auxiliary equation are also the roots of characteristic polynomial. The two roots lying on the right half of S-plane are indicated by two sign changes in the first column of Routh array. The remaining five roots are lying on the right half of S-plane. No roots are lying on imaginary axis.

Result

1. The system is unstable.
2. Two roots are lying on the right half of S-plane and five roots are lying on the left half of S-plane.

Method II: The characteristic equation is $s^7 + 9s^6 + 24s^5 + 24s^4 + 24s^3 + 24s^2 + 23s + 15 = 0$.

The given characteristic polynomial is 7th order equation and so it has 7 roots. Since the highest power of s is odd number, form the first row of Routh array using the coefficients of odd powers of s and form the second row using the coefficients of even powers of s as shown below.

$$\begin{array}{llllll} s^7 : & 1 & 24 & 24 & 23 & \ldots \text{Row-1} \\ s^6 : & 9 & 24 & 24 & 15 & \ldots \text{Row-2} \end{array}$$

Divide s^6 row by 3 to simplify the computations.

$$\begin{array}{llllll} s^7 : & 1 & 24 & 24 & 23 & \ldots \text{Row-1} \\ s^6 : & 3 & 8 & 8 & 5 & \ldots \text{Row-2} \\ s^5 : & 1 & 1 & 1 & & \ldots \text{Row-3} \\ s^4 : & 1 & 1 & 1 & & \ldots \text{Row-4} \\ s^3 : & 0 & 0 & & & \ldots \text{Row-5} \end{array}$$

Since we get a row of zeros, there exists an even polynomial, the even polynomial is nothing but the auxiliary polynomial.

$$s^5 : \quad \frac{3 \times 24 - 8 \times 1}{3} \quad \frac{3 \times 24 - 8 \times 1}{3} \quad \frac{3 \times 23 - 5 \times 1}{3}$$

$$s^5 : \quad 21.33 \quad 21.33 \quad 21.33$$

Divide by 21.33

$$s^5 : \quad 1 \quad 1 \quad 1$$

$$s^4 : \quad \frac{1 \times 8 - 1 \times 3}{1} \quad \frac{1 \times 8 - 1 \times 3}{1} \quad \frac{1 \times 5 - 0 \times 3}{1}$$

$$s^4 : \quad 5 \quad 5 \quad 5$$

Divide by 5

$$s^4 : \quad 1 \quad 1 \quad 1$$

$$s^3 : \quad \frac{1 \times 1 - 1 \times 1}{1} \quad \frac{1 \times 1 - 1 \times 1}{1}$$

$$s^3 : \quad 0 \quad 0$$

$s^3 + 9s^2 + 23s + 15$ (Quotient polynomial)

$s^4 + s^2 + 1$ (Even polynomial)

$$s^7 + 9s^6 + 24s^5 + 24s^4 + 24s^3 + 24s^2 + 23s + 15$$

$$s^7 \qquad (-) \qquad (-)s^5 \qquad (-)s^3$$

The auxiliary polynomial is

$$s^4 + s^2 + 1 = 0.$$

Divide the characteristic equation by auxiliary polynomial to get the quotient polynomial.

The characteristic polynomial can be expressed as a product of quotient polynomial and axuiliary polynomial.

$9s^6 + 24s^5 + 24s^4 + 24s^3 + 24s^3 + 23s + 15$ $9s^6_{(-)} + \quad {}_{(-)}+9s^4 \qquad {}_{(-)}+9s^2$
$23s^5 + 15s^4 + 23s^3 + 15s^2 + 23s + 15$ ${}_{(-)}23s^5 \quad {}_{(-)}+23s^3 \qquad {}_{(-)}+23s^5$
$15s^4 \quad + 15s^2 \qquad +15$ ${}_{(-)}15s^4 \quad {}_{(-)}+15s^2 \qquad {}_{(-)}+15s$ 0

$$\therefore \qquad s^7 + 9s^6 + 24s^5 + 24s^4 + 24s^3 + 24s^2 + 23s + 15 = 0$$

$$= (s^4 + s^2 + 1) \qquad (s^3 + 9s^2 + 23s + 15) = 0$$

Even polynomial Quotient polynomial

The Routh array is constructed for quotient polynomial as shown below.

$$\begin{array}{lll} s^3 : & 1 & 23 \\ s^2 : & 9 & 15 \end{array}$$

Divide s^2 row by 3,

$$\begin{array}{lll} s^3 : & 1 & 23 \\ s^2 : & 3 & 5 \\ s^1 : & 21.33 & \\ s^0 : & 5 & \end{array}$$

Column-1

$s^1 : \dfrac{3 \times 23 - 5 \times 1}{3}$ $s^1 : 21.33$
$s^0 : \dfrac{21.33 \times 5 - 0 \times 3}{21.33}$ $s^0 : 5$

The elements of column-1 of quotient polynomial are all positive and there is no sign change. Hence, all the roots of quotient polynomial are lying on the left half of S-plane. To determine the stability the roots of auxiliary polynomial should be evaluated.

The auxiliary equation is $s^4 + s^2 + 1 = 0$

Put, $s^2 = x$ in the auxiliary equation, $\therefore\ s^4 + s^2 + 1 - x^2 + x + 1 = 0$

The roots of quadratic are $x = \dfrac{-1 \pm \sqrt{1-4}}{2} = -\dfrac{1}{2} \pm j\dfrac{\sqrt{3}}{2} = 1\angle 120^\circ$ or $1\angle -120^\circ$

$$\begin{aligned} \text{But } s^2 = x, \quad \therefore s = \pm\sqrt{x} &= \pm\sqrt{1\angle 120^\circ} \quad \text{or } \pm\sqrt{1\angle -120^\circ} \\ &= \pm\sqrt{1}\angle 120^\circ/2 \quad \text{or } \pm\sqrt{1}\angle -120^\circ/2 \\ &= \pm 1\angle 60^\circ \quad \text{or } \pm 1\angle -60^\circ \\ &= \pm(0.5 + j0.866) \quad \text{or } \pm(0.5 - j0.866) \end{aligned}$$

The roots of auxiliary equation are complex and have quadrantal symmetry. Two roots of auxiliary equation are lying on the right half of S-plane and the other two on the left half of S-plane.

The roots of characteristic equation are given by the roots of auxiliary polynomial and the roots of quotient polynomial. Hence, we can conclude that two roots of characteristic equation are lying on the right half of *S*-plane and so system is unstable. The remaining five roots are lying on the left half of *S*-plane.

Result

1. The system is unstable.
2. Two roots are lying on the right half and five roots are lying on left half of *S*-plane.

Example 5.10: The characteristic polynomial of system is $s^7 + 5s^6 + 9s^5 + 9s^4 + 4s^3 + 20s^2 + 36s + 36 = 0$. Determine the location of roots on the *S*-plane and hence the stability of the system.

Solution: The characteristic equation is $s^7 + 5s^6 + 9s^5 + 9s^4 + 4s^3 + 20s^2 + 36s + 36 = 0$.

The given characteristic polynomial is 7th order equation and so it has 7 roots. Since the highest power of *s* is odd number, form the first row of array using the coefficient of odd powers of *s* and form the second row of array using the coefficients of even powers of *s* as shown below.

s^7 :	1	9	4	36	... Row-1
s^6 :	5	9	20	36	... Row-2

Divide s^6 row by 5 to simplify the computations.

s^7 :	1	9	4	36	... Row-1
s^6 :	1	1.8	4	7.2	... Row-2
s^5 :	1	0	4		... Row-3
s^4 :	1	0	4		... Row-4
s^3 :	0	0			... Row-5

$$s^5 : \frac{1 \times 9 - 1.8 \times 1}{1} \quad \frac{1 \times 4 - 4 \times 1}{1} \quad \frac{1 \times 36 - 72.1}{1}$$

s^5 : 7.2   0   28.8

Divide by 7.2

s^5 : 1   0   4

$$s^4 : \frac{1 \times 1.8 - 0 \times 1}{1} \quad \frac{1 \times 4 - 4 \times 1}{1} \quad \frac{1 \times 7.2 - 0 \times 1}{1}$$

s^4 : 1.8   0   7.2

Divide by 1.8

s^4 : 1   0   4

$$s^3 : \frac{1 \times 0 - 0 \times 1}{1} \quad \frac{1 \times 4 - 4 \times 1}{1}$$

s^3 : 0   0

The row of all zeros indicates the existence of even polynomial, which is also the auxiliary polynomial. The auxiliary polynomial is $s^4 + 4 = 0$. Divide the characteristic equation by auxiliary equation to get the quotient polynomial.

$$s^3 + 5s^2 + 9s + 9$$

$$s^4 + 4 \;\overline{\big)\; s^7 + 5s^6 + 9s^5 + 9s^4 + 4s^3 + 20s^2 + 36s + 36}$$

s^7	$4s^3$
$5s^6 + 9s^5 + 9s^4$	$20s^2 + 36s + 36$
$5s^6$	$+20s^2$
$9s^5 + 9s^4$	$+36s + 36$
$9s^5$	$+36s$
$9s^4$	$+36$
$9s^4$	$+36$
0	

The characteristic equation can be expressed as a product of quotient polynomial and auxiliary equation.

$\therefore \qquad s^7 + 5s^6 + 9s^5 + 9s^4 + 4s^3 + 20s^2 + 36s + 36 = 0$

$(s^4 + 4 = 0)$	$(s^3 + 5s^2 + 9s + 9) = 0$
Even polynomial	Quotient polynomial

The Routh array is constructed for quotient polynomial is as shown below.

		Column 1	
s^3	:	1	9
s^2	:	5	9
s^1	:	7.2	
s^0	:	9	

$$s^1 : \frac{5 \times 9 - 9 \times 1}{5}$$

$s^1 : 7.2$

$$s^0 : \frac{7.2 \times 9 - 0 \times 5}{21.33}$$

$s^0 : 9$

There is no sign change in the elements of the first column of Routh array of quotient polynomial. Hence, all the roots of quotient polynomial are lying on the left half of S-plane.

To determine the stability, the roots of auxiliary polynomial should be evaluated.

The auxiliary polynomial is $s^4 + 4 = 0$.

Put $s^4 = x$ in the auxiliary equation $\quad \therefore\ s^4 + 4 = x^2 + 4 = 0$.

$$\therefore \quad x^2 = -4 \Rightarrow x = \pm\sqrt{-4} = \pm j2 = 2\angle 90° \text{ or } 2\angle -90°$$

But $\quad s = \pm\sqrt{x} = \pm\sqrt{2\angle 90°}$ or $\pm\sqrt{2\angle -90°} = \pm\sqrt{2}\angle 90°/2$ or $\sqrt{2}\angle -90°/2$

$$= \pm\sqrt{2}\angle 45° \text{ or } \pm\sqrt{2}\angle -45° = \pm(1 + j1) \text{ or } \pm(1 - j1)$$

The roots of auxiliary equation are complex and have quadrantal symmetry. Two roots of auxiliary equation are lying on the right half of S-plane and the other two on the left half of S-plane.

The roots of characteristic equation are given by the roots of quotient polynomial and auxiliary polynomial. Hence, we can conclude that two roots of characteristic equation are lying on the right half of S-plane and so the system is unstable. The remaining five roots are lying on the left half of S-plane.

Result

1. The system is unstable.
2. Two roots are lying on the right half of S-plane and five roots are lying on the left half of S-plane.

Example 5.11: Use the Routh stability criterion to determine the location of roots on the S-plane and hence the stability for the system represented by the characteristic equation $s^5 + 4s^4 + 8s^3 + 8s^2 + 7s + 4 = 0$.

Solution: The characteristic equation of the system is $s^5 + 4s^4 + 8s^3 + 8s^2 + 7s + 4 = 0$.

The given characteristic polynomial is 5th order equation and so it has 5 roots. Since the highest power of s is odd number, form the first row of Routh array using the coefficients of odd powers of s and form the second row using the coefficients of even powers of s.

s^5 :	1	8	7	... Row-1
s^4 :	4	8	4	... Row-2

Divide s^6 row by 4 to simplify the computations.

	Column 1			
s^5 :	1	8	7	... Row-1
s^4 :	1	2	1	... Row-2
s^3 :	1	1		... Row-3
s^2 :	1	2		... Row-4
s^1 :	ε	0		... Row-5
s^0 :	1			... Row-6

$$s^3 : \quad \frac{1 \times 8 - 2 \times 1}{1} \quad \frac{1 \times 7 - 1 \times 1}{1}$$

$$s^3 : \quad 6 \quad 6$$

Divide by 6

$$s^3 : \quad 1 \quad 1$$

$$s^2 : \quad \frac{1 \times 2 - 1 \times 1}{1} \quad \frac{1 \times 1 - 0 \times 1}{1}$$

$$s^2 : \quad 1 \quad 1$$

$$s^1 : \quad \frac{1 \times 1 - 1 \times 1}{1}$$

$$s^3 : \quad 0$$

Let $0 \rightarrow \varepsilon$

$$s^1 : \quad \varepsilon$$

$$s^0 : \quad \frac{\varepsilon \times 1 - 0 \times 1}{\varepsilon}$$

$$s^0 : \quad 1$$

When $\varepsilon \rightarrow 0$, there is no sign change in the first column of Routh array. But we have a row of all zeros (s^1 row or Row-5) and so there is a possibility of roots on imaginary axis, this can be found from the roots of auxiliary polynomial. Hence, the auxiliary polynomial is given by s^2 row.

The auxiliary polynomial is $s^2 + 1 = 0 \quad \therefore s^2 = -1 \quad \text{or} \quad s = \pm\sqrt{-1} = \pm j1$.

The roots of auxiliary polynomial are $+j1$, and $-j1$, lying on imaginary axis. The roots of auxiliary polynomial are also roots of characteristic equation. Hence two roots of characteristic equation are lying on imaginary axis and so the system is limitedly or marginally stable. The remaining three roots of characteristic equation are lying on the left half of S-plane.

Result

1. The system is limitedly or marginally stable.
2. Two roots are lying on imaginary axis and three roots are lying on the left half of S-plane.

Example 5.12: Use the Routh stability criterion to determine the location of roots on the S-plane and hence the stability for the system represented by the characteristic equation $s^6 + s^5 + 3s^4 + 3s^3 + 3s^2 + 2s + 1 = 0$.

Solution: The characteristic polynomial of the system is $s^6 + s^5 + 3s^4 + 3s^3 + 3s^2 + 2s + 1 = 0$.

The given characteristic polynomial is 6th order equation and so it has 6 roots. Since the highest power of s is even number, form the first row of Routh array using the coefficient of even powers of s and form the second row using the coefficients of odd powers of s as shown below.

$$
\begin{array}{llllll}
s^6 : & 1 & 3 & 3 & 1 & \ldots \text{Row-1} \\
s^5 : & 1 & 3 & 2 & & \ldots \text{Row-2} \\
s^4 : & \varepsilon & 1 & 1 & & \ldots \text{Row-3} \\
s^3 : & \dfrac{3\varepsilon-1}{\varepsilon} & \dfrac{2\varepsilon-1}{\varepsilon} & & & \ldots \text{Row-4} \\
s^2 : & \dfrac{-2\varepsilon^2+4\varepsilon-1}{3\varepsilon-1} & 1 & & & \ldots \text{Row-5} \\
s^1 : & \dfrac{4\varepsilon^2-\varepsilon}{2\varepsilon^2-4\varepsilon+1} & & & & \ldots \text{Row-6} \\
s^0 : & 1 & & & & \ldots \text{Row-7}
\end{array}
$$

$$
s^4 : \frac{1\times 3-3\times 1}{1} \quad \frac{1\times 3-2\times 1}{1} \quad \frac{1\times 1-0\times 1}{1}
$$

$$
s^4 \quad 0 \quad 1 \quad 1
$$

$$
\text{Let} \quad 0 \to \varepsilon
$$

$$
s^4 : \quad \varepsilon \quad 1 \quad 1
$$

$$
s^3 : \frac{\varepsilon\times 3-1\times 1}{\varepsilon} \quad \frac{\varepsilon\times 2-1\times 1}{\varepsilon}
$$

$$
s^3 : \frac{3\varepsilon-1}{\varepsilon} \quad \frac{2\varepsilon-1}{\varepsilon}
$$

$$
s^2 : \frac{\dfrac{3\varepsilon-1}{\varepsilon}-\dfrac{2\varepsilon-1}{\varepsilon}\times\varepsilon}{\dfrac{3\varepsilon-1}{\varepsilon}} \quad \frac{\dfrac{3\varepsilon-1}{\varepsilon}\times 1-0\times\varepsilon}{\dfrac{3\varepsilon-1}{\varepsilon}}
$$

$$
s^2 : \frac{-2\varepsilon^2+4\varepsilon-1}{3\varepsilon-1} \quad 1
$$

On letting $\in \to 0$, we get,

$$
\begin{array}{lllll}
s^6 : & 1 & 3 & 3 & 1 \ \ldots \text{Row-1} \\
s^5 : & 1 & 3 & 2 & \ldots \text{Row-2} \\
s^4 : & 0 & 1 & 1 & \ldots \text{Row-3} \\
s^3 : & -\infty & -\infty & & \ldots \text{Row-4} \\
s^2 : & 1 & 1 & & \ldots \text{Row-5} \\
s^1 : & 0 & & & \ldots \text{Row-6} \\
s^0 : & 1 & & & \ldots \text{Row-7}
\end{array}
$$

(+ to − : from s^3 row to s^2 row; − to + : from s^2 row to s^1 row)

$$
s^1 : \frac{\dfrac{-2\varepsilon^2+4\varepsilon-1}{3\varepsilon-1}\times\dfrac{2\varepsilon-1}{\varepsilon}-\dfrac{3\varepsilon-1}{\varepsilon}\times 1}{\dfrac{-2\varepsilon^2+4\varepsilon-1}{3\varepsilon-1}}
$$

$$
s^1 : \frac{(-2\varepsilon^2+4\varepsilon-1)(2\varepsilon-1)-(3\varepsilon-1)(3c-1)}{\varepsilon(-2\varepsilon^2+4\varepsilon-1)}
$$

$$
s^1 : \frac{-4\varepsilon^3+\varepsilon^2}{\varepsilon(-2\varepsilon^2+4c \quad 1)} = \frac{4\varepsilon^2-\varepsilon}{2\varepsilon^2-4\varepsilon+1}
$$

$$
s^0 : \frac{\dfrac{4\varepsilon^2-\varepsilon}{2\varepsilon^2-4\varepsilon+1}\times 1-0\,\dfrac{-2\varepsilon^2+4\varepsilon-1}{3\varepsilon-1}}{(4\varepsilon^2-\varepsilon)/(2\varepsilon^2-4\varepsilon+1)}
$$

$$
s^1 : 1
$$

Since there is a row of all zeros (s^1 row) there is a possibility of roots on imaginary axis. The auxiliary polynomial is $s^2 + 1 = 0$.

The roots of auxiliary polynomial are $s = \pm\sqrt{-1} = \pm j1$.

The roots of auxiliary polynomial are also roots of characteristic equation. Hence two roots are lying on imaginary axis. Therefore divide the characteristic polynomial by auxiliary equation and construct the Routh array for quotient polynomial to find the roots lying on the right half of S-plane.

The characteristic polynomial can be expressed as a product of auxiliary polynomial and quotient polynomial.

$$\therefore \quad s^6 + s^5 + 3s^4 + 3s^3 + 3s^2 + 2s + 1 = 0 \Rightarrow \underbrace{(s^2 + 1)}_{\text{Even polynomial}} \quad \underbrace{(s^4 + s^3 + 2s^2 + 2s + 1)}_{\text{Quotient polynomial}} = 0$$

The Routh array for quotient polynomial is constructed as shown below.

s^4	:	1	2	1	... Row-1
s^3	:	1	2		... Row-2
s^2	:	ε	1		... Row-3
s^1	:	$\dfrac{2\varepsilon - 1}{\varepsilon}$			... Row-4
s^0	:	1			... Row-5

On letting $\varepsilon \Rightarrow 0$, we get

s^4	:	1	2	1	... Row-3
s^3	:	1	2		... Row-4
s^2	:	0	1		... Row-5
s^1	:	$-\infty$			... Row-6
s^0	:	1			... Row-7

+ to − (from s^2 to s^1); − to + (from s^1 to s^0)

Column 1

$$s^2 + 1 \overline{\big) \, s^6 + s^5 + 3s^4 + 3s^3 + 3s^2 + 2s + 1 \,} \quad s^4 + s^3 + 2s^2 + 2s + 1 \text{ Quotient polynomial}$$

$$\begin{array}{l}
s^6 \; + s^4 \quad (-)\;(-) \\
\hline
s^5 + 2s^4 + 3s^3 + 3s^2 + 2s + 1 \\
(-)s^5 \quad (-) + s^3 \\
\hline
2s^4 + 2s^3 + 3s^2 + 2s + 1 \\
(-)2s^4 \quad (-) + 2s^2 \\
\hline
2s^3 + s^2 + 2s + 1 \\
(-)2s^3 \quad (-) + 2s \\
\hline
s^2 + 1 \\
(-)s^2 \; (-) + 1 \\
\hline
0
\end{array}$$

(Even polynomial)

$$s^2 : \frac{1 \times 2 - 2 \times 1}{1} \quad \frac{1 \times 1 - 0 \times 1}{1}$$

$$s^2 : \quad 0 \quad 1$$

Let $0 \rightarrow \varepsilon$

$$s^2 : \quad \varepsilon \quad 1$$

$$s^1 : \frac{\varepsilon \times 2 - 1 \times 1}{\varepsilon}$$

$$s^1 : \frac{2\varepsilon - 1}{\varepsilon}$$

$$s^0 : \frac{\dfrac{2\varepsilon - 1}{\varepsilon} \times 1 - 0 \times \varepsilon}{(2\varepsilon - 1)/\varepsilon}$$

$$s^0 : \quad 1$$

On examining the first column of the Routh array of quotient polynomial, we found that there are two sign changes. Hence, two roots are lying on the right half of S-plane and other two roots of quotient polynomial are lying on the left half of S-plane.

The roots of characteristic equation are given by roots of auxiliary polynomal and quotient polynomial. Hence two roots are lying on imagninary axis, two roots are lying on the right half of S-plane and the remaining two roots are lying on the left half of S-plane. Hence, the system is unstable.

Result

1. The system is unstable.
2. Two roots are lying on imaginary axis, two roots as lying on the right half of S-plane and two roots are lying on the left half of S-plane.

Example 5.13: Determine the range of K for stability of unity feedback system whose open-loop transfer function is $G_s = \dfrac{K}{s(s+1)(s+2)}$.

Solution: The closed-loop transfer function,

$$\frac{C(s)}{R(s)} = \frac{G(s)}{1+G(s)} = \frac{\dfrac{K}{s(s+1)(s+2)}}{1+\dfrac{K}{s(s+1)(s+2)}} = \frac{K}{s(s+1)(s+2)+K}$$

The characteristic equation is $s(s+1)(s+2)+K=0$

$$\therefore \qquad s(s^2+3s+2)+K=0 \quad \Rightarrow \quad s^3+3s^2+2s+K=0$$

The Routh array is constructed as shown below.

The highest power of s in the characteristic polynomial is odd number. Hence, form the first row using the coefficients of odd powers of s and form the second row using the coefficients of even powers of s.

	Column 1	
s^3 :	1	2
s^2 :	3	K
s^1 :	$\dfrac{6-K}{3}$	
s^0 :	K	

$$s^1 : \frac{3\times 2 - K\times 1}{3}$$

$$s^1 : \frac{6-K}{3}$$

$$s^1 : \frac{\dfrac{6-K}{3}\times K - 0\times 3}{(6-K)/3}$$

$$s^1 : K$$

For the system to be stable there should not be any sign change in the elements of first column. Hence, choose the value of K so that the first column elements are positive.

From s^0 row, for the system to be stable, $K > 0$.

From s^1 row, for the system to be stable, $\dfrac{6-K}{3} > 0$

For $\dfrac{6-K}{3} > 0$, the value of K should be less than 6.

$\therefore$ The range of K for the system to be stable is $0 < K < 6$.

Result

The value of K is in the range of $0 < K < 6$ for the system to be stable.

Example 5.14: The open loop transfer function of a unity feedback control system is given by

$$G(s) = \frac{K}{(s+2)(s+4)(s^2+6s+25)}$$

By applying the Routh criterion, discuss the stability of the closed-loop system as a function of K. Determine the value of K which will cause sustained oscillations in the closed-loop system. What are the corresponding oscillating frequencies?

Solution:

$$\left.\begin{array}{r}\text{The closed-loop}\\ \text{transfer function}\end{array}\right\} \frac{C(s)}{R(s)} = \frac{G(s)}{1+G(s)} = \frac{\dfrac{K}{(s+2)(s+4)(s^2+6s+25)}}{1+\dfrac{K}{(s+2)(s+4)(s^2+6s+25)}}$$

$$= \frac{K}{(s+2)(s+4)(s^2+6s+25)+K}$$

The characteristic equation is given by the denominator polynomial of closed-loop transfer function.

The characteristic equation is $(s+2)(s+4)(s^2+6s+25)+K=0$

$$\therefore \quad (s+2)(s+4)(s^2+6s+25)+K=0 \Rightarrow (s^4+12s^3+69s^2+198s+200+K=0$$

The Routh array is constructed as shown below. The highest power of s in the characteristic equation is an even number. Hence, form the first row using the coefficients of even powers of s and form the second row using the coefficients of odd powers of s.

$$\begin{array}{llll}
s^4: & 1 & 69 & 200+K \quad \ldots \text{Row-1}\\
s^3: & 12 & 198 & \quad \ldots \text{Row-2}
\end{array}$$

Divide s^3 row by 12 to simplify the calculations

$$\begin{array}{llll}
s^4: & 1 & 69 & 200+K \quad \ldots \text{Row-1}\\
s^3: & 1 & 16.5 & \quad \ldots \text{Row-2}\\
s^2: & 52.5 & 200+K & \quad \ldots \text{Row-3}\\
s^1: & \dfrac{666.25-K}{52.5} & & \quad \ldots \text{Row-4}\\
s^0: & 200+K & & \quad \ldots \text{Row-5}
\end{array}$$

↑ Column 1

$$\begin{array}{lll}
s^2: & \dfrac{1\times 69-16.5\times 1}{1} & \dfrac{1\times(200+K)}{1}\\
s^2: & 52.5 & 200+K\\
\hline
s^1: & \dfrac{52.5\times 16.5-(200+K)\times 1}{52.5} & \\
s^1: & \dfrac{666.25-K}{52.5} & \\
\hline
s^0: & \dfrac{\dfrac{666.25-K}{52.5}\times(200+K)}{(666.25-K)/52.5} & \\
s^0: & 200+K &
\end{array}$$

For the system to be stable there should not be any sign change in the elements of first column. Hence, choose the value of K so that the first column elements are positive.

From s^1 row, for the system to be stable, $(666.25 - K) > 0$.

Since $(666.25 - K) > 0$, should be less than 666.25.

From s^0 row, for the system to be stable, $(200 + K) > 0$.

Since $(200 + K) > 0$, K should be greater than – 200, but practical value of K starts from 0. Hence, K should be greater than 0.

∴ The range of K for the system to be stable is $0 < K < 666.25$.

When $K = 666.25$ the s^1 row becomes zero, which indicates the possibility of roots on imaginary axis. A system will oscillate if it has roots on imaginary axis and no roots on the right half of S-plane.

When $K = 666.25$, the coefficients of auxiliary equation are given by the s^2 row.

∴ The auxiliary equation is $52.5s^2 + 200 + K = 0$

$$52.5s^2 + 200 + 666.25 = 0$$

$$s^2 = \frac{-200 - 666.25}{52.5} = -16.5$$

$$s = \pm\sqrt{-16.5} = \pm j\sqrt{16.5} = \pm j4.06$$

When $K = 666.25$, the system has roots on imaginary axis and so it oscillates. The frequency of oscillation is given by the value of root on imaginary axis.

∴ The frequency of oscillation, $\omega = 4.06$ rad/sec

Result

1. The range of K for stability is $= 0 < K < 666.25$.
2. The system oscillates when $K = 666.25$.
3. The frequency of oscillation, $\omega = 4.06$ rad/sec (when $K = 666.25$).

Example 5.15: The open-loop transfer function of a unity feedback system is given by, $G(s) = \dfrac{K(s+1)}{(s^3 + as^2 + 2s + 1)}$. Determine the value of K and a so that the system oscillates at a frequency of 2 rad/sec.

Solution:

$$\left.\begin{array}{r}\text{The closed-loop}\\ \text{transfer function}\end{array}\right\}\frac{C(s)}{R(s)} = \frac{G(s)}{1+G(s)} = \frac{\dfrac{K(s+1)}{s^3 + as^2 + 2s + 1}}{1 + \dfrac{K(s+1)}{s^3 + as^2 + 2s + 1}}$$

$$= \frac{K(s+1)}{s^3 + as^2 + 2s + 1 + K(s+1)}$$

∴ The characteristic equation is $s^3 + as^2 + 2s + 1 + K(s + 1) = 0$

$$s^3 + as^2 + 2s + 1 + K(s + 1) = 0 \Rightarrow s^3 + as^2 + (2 + K)s + 1 + K = 0$$

The Routh array of characteristic polynomial is constructed as shown below. The maximum power of s odd. Hence, the first row of Routh array is formed using coefficients of odd powers of s and the second row of Routh array is formed using coefficients of even powers of s.

If the elements of s^1 row are all zeros then there exists an even polynomial (or auxiliary polynomial). If the roots of the auxiliary polynomial are purely imaginary then the roots are lying on imaginary axis and the system oscillates. The frequency of oscillation is the root of auxiliary polynomial.

Routh Array

$$\begin{array}{lll} s^3 : & 1 & 2 + K \\ s^2 : & a & 1 + K \\ s^1 : & \dfrac{a(2 + K) - (1 + K)}{a} & \\ s^0 : & 1 + K & \end{array}$$

From s^2 row, the auxiliary polynomial is

$$as^2 + (1 + K) = 0 \Rightarrow as^2 = -(1 + K) \Rightarrow = s \pm j\sqrt{\frac{1 + K}{a}}$$

Given that, $s = + j2$,

$$\therefore \qquad \sqrt{\frac{1 + K}{a}} = 2 \Rightarrow \frac{1 + K}{a} = 4 \Rightarrow K = 4a - 1$$

From s^1 row,

$$\frac{a(2 + K) - (1 + K)}{a} = 0 \Rightarrow a(2 + K) - (1 + K) = 0 \Rightarrow 2a + Ka - 1 - K = 0$$

$$\therefore \qquad 2a - 1 + K(a - 1) = 0$$

Put $K = 4a - 1$

$$\therefore \qquad 2a - 1 + (4a - 1)(a - 1) = 0 \Rightarrow 2a - 1 + 4a^2 - 4a - a + 1 = 0$$

$$\Rightarrow 4a^2 - 3a = 0 \text{ or } a(4a - 3) = 0$$

Since $\quad a \neq 0,\ 4a - 3 = 0, \quad \therefore a = 3/4$

When $\quad a = (3/4),\ K = 4a - 1 = 4 \times (3/4) - 1 = 2$

Result

When the system oscillates at a frequency of 2 rad/sec, $K = 2$ and $a = \dfrac{3}{4}$.

Example 5.16: A feedback system has open-loop transfer function of $G(s) = \dfrac{Ke^{-s}}{s(s^2 + 5s + 9)}$. Determine the maximum value of K for stability of closed-loop system.

Solution: Generally control systems have very low bandwidth which implies that it has very low frequency range of operation. Hence, for low frequency range the term e^{-s} can be replaced by, $[1 - s]$, [(i.e., $e^{-sT} \approx 1 - sT$)].

$$\therefore \qquad G(s) = \frac{Ke^{-s}}{s(s^2 + 5s + 9)} \approx \frac{K(1-s)}{s(s^2 + 5s + 9)}$$

$$\left.\begin{array}{l}\text{The closed-loop} \\ \text{transfer function}\end{array}\right\} \frac{C(s)}{R(s)} = \frac{G(s)}{1 + G(s)} = \frac{\dfrac{K(1-s)}{s(s^2 + 5s + 9)}}{1 + \dfrac{K(1-s)}{s(s^2 + 5s + 9)}}$$

$$= \frac{K(1-s)}{s(s^2 + 5s + 9) + K(1-s)}$$

The characteristic equation is given by the denominator polynomial of closed-loop transfer function.

$\therefore$ The characteristic equation is $s(s^2 + 5s + 9) + K(1 - s) = 0$

$$\therefore \quad s(s^2 + 5s + 9) + K(1 - s) = s^3 + 5s^2 + 9s + K - Ks = 0 \Rightarrow s^3 + 5s^2 + (9 - K)s + K = 0$$

The Routh array of characteristic polynomial is constructed as shown below.

The maximum power of s in the characteristic polynomial is odd. Hence form the first row of Routh array using coefficients of odd powers of s and the second row of Routh array using coefficients of even powers of s.

s^3 :	1	9-K
s^2 :	5	K
s^1 :	$9 - 1.2K$	
s^0 :	K	

$$s^1 : \frac{5 \times (9 - K) - K \times 1}{52.55}$$

$$s^1 : \frac{45 - 5K - K}{5}$$

$$s^1 : \frac{45 - 6K}{5} \approx 9 - 1.2\,K$$

$$s^0 : \frac{(9 - 1.2\,K) \times K}{(9 - 1.2\,K)}$$

$$s^0 : K$$

From s^1 row, for stability of the system, $(9 - 1.2K) > 0$.

If $(9 - 1.2K) > 0$ then $1.2K < 9$; $K < 9/1.2 = 7.5$.

From s^0 row, for stability of the system, $K > 0$.

Finally, we can conclude that for stability of the system K should be in the range of $0 < K < 7.5$.

Result

For stability of the stem K should be in the range of, $0 < K < 7.5$.

Example 5.17: Using Routh's stability criterion determine the stability of the system which has the characteristics equation

$$S^4 + 2S^3 + 11S^2 + 18S + 18 = 0$$

Solution: The routh array is

S^4 :	1	11	18
S^3 :	2	18	← divided by 2
S^3 :	1	9	

$$
\begin{array}{lllll}
S^2 & : & \dfrac{1 \times 11 - 1 \times 9}{1} = 2 & \dfrac{1 \times 18 - 0}{1} = 18 & \leftarrow \text{dividing by 2} \\
S^2 & : & 1 & 9 & \\
S^1 & : & 0 & 0 &
\end{array}
$$

All terms in S^1 row are zero. From the auxiliary equation from the preceding row. The auxiliary equation is formed from the S^2 row.

$$S^2 + 9 = 0$$

Complete the Routh array, differentiate the above auxiliary equation

$$2S + 0 = 0$$

The coefficients of this equation in S^1 row and the array is then completed

$$
\begin{array}{l}
S^1 : 2 \\
S^0 : 9
\end{array}
$$

The complete Routh array is

$$
\begin{array}{llllll}
S^4 & : & 1 & 11 & 18 \\
S^3 & : & 1 & 9 & \\
S^2 & : & 1 & 9 & \\
S^1 & : & 2 & & \\
S^0 & : & 9 & &
\end{array}
$$

↑ Column 1

Since there are no changes of sign in the first column, no roots lie on the right half of the S-plane The presence of a zero row in the Routh array of the original characteristic equation indicates that there are roots lying on the imaginary axis. These roots can be calculated by solving auxiliary equation

$$S^2 + 9 = 0$$

$$S = \pm j3$$

Two roots lying on imaginary axis indicate that the system is marginally stable.

Example 5.18: The closed-loop transfer functions of a system is given by

$$\frac{C(s)}{R(s)} = \frac{K}{S(S^2 + S + 1)\,(S + 2) + K}$$

Determine the value of K for which the system is stable.

Solution: The characteristic polynomial is

$$S(S^2 + S + 1)(S + 2) + K = 0$$

$$S^4 + 3S^3 + 3S^2 + 2S + K = 0$$

Routh array is

$$\begin{array}{llll} S^4 & : & 1 & 3 & K \\ S^3 & : & 3 & 2 \\ S^2 & : & b_1 & b_2 \\ S^1 & : & C_1 \\ S^0 & : & d_1 \end{array}$$

$$b_1 = \frac{3 \times 3 - 1 \times 2}{3} = \frac{7}{3}$$

$$b_2 = \frac{3 \times K - 1 \times 0}{3} = K$$

$$C_1 = \frac{\frac{7}{3} \times 2 - 3 \times K}{\frac{7}{3}} = 2 - \frac{9K}{7}$$

$$d_1 = \frac{\left(2 - \frac{9K}{7}\right) K - \frac{7}{3} \times 0}{\left(2 - \frac{9K}{7}\right)} = K$$

Therefore, the routh array is

$$\begin{array}{lllll} S^4 & : & 1 & 3 & \text{K} \\ S^3 & : & 3 & 2 \\ S^2 & : & \frac{7}{3} & \text{K} \\ S^1 & : & \left(2 - \frac{9K}{7}\right) \\ S^0 & : & K \end{array}$$

For stability all the coefficients in the first column of the array must be positive

$\therefore \quad K > 0$ and $\left(2 - \frac{9K}{7}\right) > 0$

or $\quad K < \frac{14}{9}$. For $K = \frac{14}{9}$, the system gives sustained oscillations.

Example 5.19: Using Rouths stability criterion, determine the range of K for the system shown in Fig. 5.2.

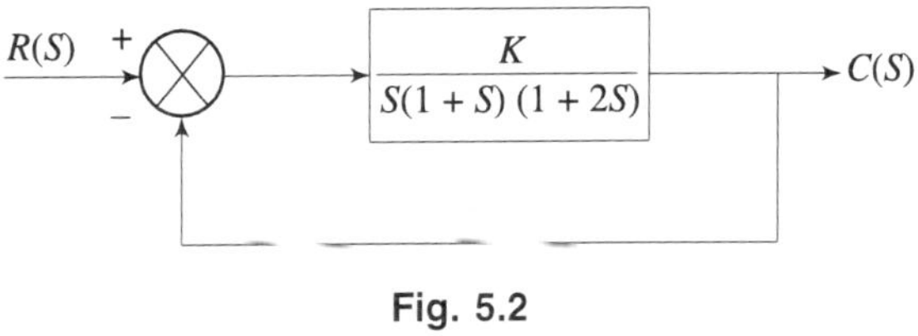

Fig. 5.2

Solution: The characteristic equation of the given system is

$$1 + \frac{K}{S(1+S)(1+2S)} = 0$$

$$S(1+S)(1+2S) + K = 0$$

$$2S^3 + 3S^2 + S + K = 0$$

The Routh array is

$$\begin{array}{lcll} S^3 & : & 2 & 1 \\ S^2 & : & 3 & K \\ S^1 & : & b_1 & \\ S^0 & : & C_1 & \end{array}$$

$$b_1 = \frac{3 \times 1 - 2 \times K}{3} = 1 - \frac{2K}{3}$$

$$C_1 = \frac{\left(1 - \frac{2K}{3}\right) K - 0}{\left(1 - \frac{2K}{3}\right)} = K$$

∴ The complete Routh array is

$$\begin{array}{lcll} S^3 & : & 2 & 1 \\ S^2 & : & 3 & K \\ S^1 & : & \left(1 - \frac{2}{3}K\right) & \\ S^0 & : & K & \end{array}$$

For stability all the coefficients of the first column must be positive

∴ $$K > 0$$

$$\left(1 - \frac{2}{3}K\right) > 0 \text{ or } K < \frac{3}{2}$$

Hence for stability

$$0 < K < \frac{3}{2}.$$

Example 5.20: For what range of K is the system given in Fig. 5.3 asymptotically stable? Assume $K \geq 0$.

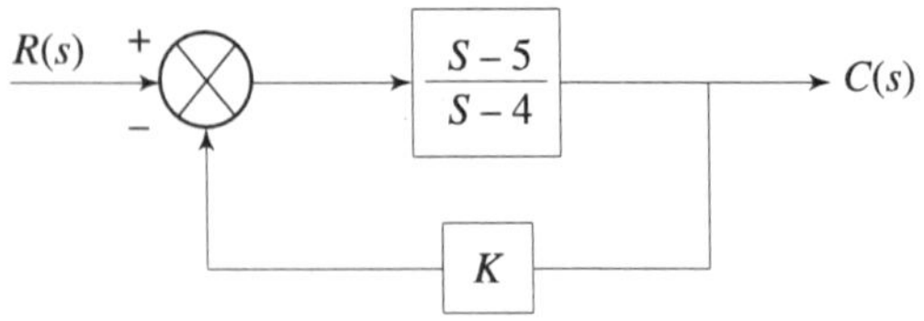

Fig. 5.3

Solution: The characteristics equation is given by

$$1 + G(s)\,H(s) = 0$$

$$1 + \frac{S-5}{S-4}K = 0$$

$$(S+4) + K(S-5) = 0$$

$$(K+1)S + 4 - 5K = 0$$

The single root is given by

$$S = \frac{5K-4}{K+1}$$

For stability

$$S = \frac{5K-4}{K+1} \leq 0$$

Since $K \geq 0$

$$5K - 4 \leq 0$$

$$K \leq \frac{4}{5}.$$

Example 5.21: The characteristic equation of a feedback control system is

$$S^4 + 20S^3 + 15S^2 + 25 + K = 0$$

(a) Determine the range of K for the system to be stable.

(b) Can the system be marginally stable? If so, find the required value of K and the frequency of sustained oscillation.

Solution: The Routh array is

$$\begin{array}{llll}
S^4 & : & 1 & 15 & K \\
S^3 & : & 20 & 2 & \rightarrow \text{divide by 2} \\
S^3 & : & 10 & 1 & \\
S^2 & : & \dfrac{10\times 10 - 1\times 1}{10} = 14.9 & \dfrac{10\times K - 0}{10} = K & \\
S^1 & : & \dfrac{14.9\times 1 - 10\times K}{14.9} = 1 - \dfrac{10}{14.9}K & & \\
S^0 & : & K & &
\end{array}$$

(a) For stability all the coefficients in the first column of the array should be positive

$$\therefore \left(1 - \frac{10}{14.9}K\right) > 0 \quad \text{or} \quad K < 1.49$$

$$K > 0$$

Combining these two requirements we conclude that the system is stable if and only if $0 < K < 1.49$

(b) For a marginally stable system

$$1 - \frac{10}{14.9} K = 0$$

$$K = \frac{14.9}{10} = 1.49$$

The Routh array becomes

$$\begin{array}{llll} S^4 & : & 1 & 15 & K \\ S^3 & : & 10 & 1 \\ S^2 & : & 14.9 & K \\ S^1 & : & 0 & 0 \end{array}$$

S^1 row becomes a zero row. The auxiliary equation is obtained from S^2 row as

$$A(s) = 14.9S^2 + K = 0$$

$$S^2 = -\frac{K}{14.9} = -\frac{1.49}{14.9} = -0.1$$

$$S = \sqrt{-0.1} = \pm j\,0.316$$

The presence of an imaginary pair of roots gives oscillatory response $S = j\omega_n$
The frequency of sustained oscillation $w_n = 0.316$ rad/sec.

Example 5.22: Consider the closed-loop feedback control system shown in Fig. 5.4 using Routh Hurwitz criterion, determine the range of K for which the system is stable. Find also the number of roots of the characteristic equation that are in the right half of S-plane for $K = 0.5$

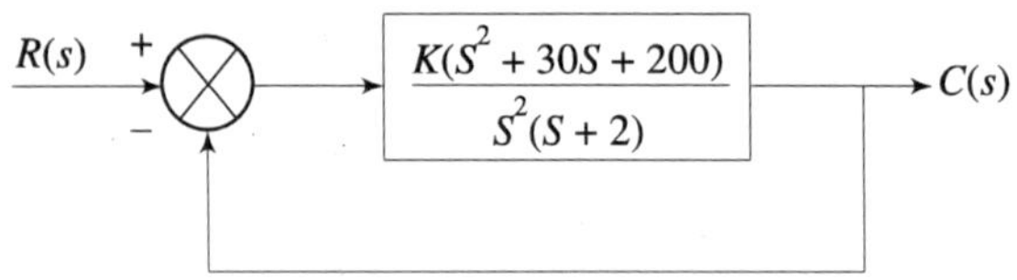

Fig. 5.4

Solution: The characteristic equation is

$$1 + G(s)\,H(s) = 0$$

$$1 + \frac{K(S^2 + 30S + 200)}{S^2\,(S + 2)} = 0$$

$$S^2\,(S + 2) + K(S^2 + 30S + 200) = 0$$

$$S^3 + (2 + K)S^2 + 30KS + 200K = 0$$

The Routh array is

$$
\begin{array}{llll}
S^3 & : & 1 & 30K \\
S^2 & : & (2+K) & 200K \\
S^1 & : & \dfrac{(2+K)\,30K - 200\,K}{2+K} = \dfrac{30K^2 - 140\,K}{2+K} & 0 \\
S^0 & : & 200K &
\end{array}
$$

For stability all the coefficients in the first column of the array should be positive. Therefore,

$$200\,K > 0$$

$$K > 0$$

$$(30\,K^2 - 140\,K) > 0$$

$$30\,K > 140$$

$$K > \frac{140}{30}$$

$$K > \frac{14}{3}$$

Therefore, the range of K for stability is

$$0 < K < \frac{14}{3}$$

If $K = 0.5$, Routh array becomes

$$
\begin{array}{llll}
S^3 & : & 1 & 15 \\
S^2 & : & (2+0.5) = 2.5 & 100 \\
S^1 & : & \dfrac{30(0.5)^2 - 140 \times 0.5}{2+0.5} = -25 & \\
S^0 & : & 200 \times 0.5 = 100 &
\end{array}
$$

Since there are two changes of sign in the first column of the array, the number of roots in the right half of the S-place is two.

Example 5.23: In a closed-loop $G(s)\,H(s) = \dfrac{K}{S(S+1)^2}$ which values of K render the closed-loop stable?

Solution: The characteristic equation is

$$1 + G(s)\,H(s) = 0$$

$$1 + \frac{K}{S(S+1)^2} = 0$$

$$S(S+1)^2 + K = 0$$

$$S^3 + 2S^2 + S + K = 0$$

The Routh array is

$$\begin{array}{llll} S^3 & : & 1 & 1 \\ S^2 & : & 2 & K \\ S^1 & : & \dfrac{2\times 1 - 1\times K}{2} = 1 - 0.5\text{ K} & 0 \\ S^0 & : & K & \end{array}$$

For stability all the coefficients in the first column of the array should be positive

$\therefore \quad (1 - 0.5\,K) > 0, \quad$ or $\quad K < 2$

$$K > 0$$

$\therefore$ The system is stable if and only if

$$0 < K < 2$$

Example 5.24: Determine the stability of a system having the following characteristic equation

$$S^5 + 5S^4 + 11S^3 + 23S^2 + 28S + 12 = 0$$

Solution: The Routh array is

$$\begin{array}{lllll} S^5 & : & 1 & 11 & 28 \\ S^4 & : & 5 & 23 & 12 \\ S^3 & : & \dfrac{5\times 11 - 1\times 23}{5} = \dfrac{32}{5} & \dfrac{5\times 28 - 1\times 12}{5} = \dfrac{128}{5} & 0 \\ S^2 & : & \dfrac{\frac{32}{5}\times 23 - \frac{128}{5}\times 5}{\frac{32}{5}} = 3 & \dfrac{\frac{32}{5}\times 12 - 5\times 0}{\frac{32}{5}} = 12 & \\ S^1 & : & \dfrac{3\times\frac{128}{5} - 12\times\frac{32}{5}}{\frac{32}{5}} = 0 & 0 & \end{array}$$

Since all the terms in S^1 row are zero, we form the auxiliary polynomial from the preceding S^2 row as

$$A(s) = 3S^2 + 12$$

$$\frac{dA(s)}{dS} = 6S$$

The S^1 row of zero is replaced by 6 and 0. The S^0 row is obtained from S^2 and S^1 row. Hence, the Routh array becomes

$$\begin{array}{lllll} S^5 & : & 1 & 11 & 28 \\ S^4 & : & 5 & 23 & 12 \\ S^3 & : & \frac{32}{5} & \frac{128}{5} & 0 \\ S^2 & : & 3 & 12 & \\ S^1 & : & 6 & 0 & \\ S^0 & : & 12 & & \end{array}$$

There are no sign changes in the first column. Hence, all the roots lie on the left of the S-plane except for a pair on the imaginary axis. The imaginary roots are calculated by making

$$3S^2 + 12 = 0$$

$$S^2 = -4$$

$$S = \pm j2$$

∴ The system is marginally stable

Example 5.25: Determine the stability of a system having the characteristic equation

$$S^6 + 5\,S^5 + 11S^4 + 25\,S^3 + 36\,S^2 + 30S + 36 = 0$$

Solution: The Routh array is

$$\begin{array}{llllll} S^6 & : & 1 & 11 & 36 & 36 \\ S^5 & : & 5 & 25 & 30 & \leftarrow \text{divide by } 5 \\ S^5 & : & 1 & 5 & 6 & \\ S^4 & : & \frac{1 \times 11 - 1 \times 5}{1} = 6 & \frac{1 \times 36 - 1 \times 6}{1} = 30 & \frac{1 \times 30 - 1 \times 0}{1} = 36 & \leftarrow \text{divide by } 6 \\ S^4 & : & 1 & 5 & 6 & \\ S^3 & : & 0 & 0 & & \end{array}$$

It is seen that all the terms in S^3 row are zero.

∴ We form the auxiliary polynomial from the preceding S^4 row. The highest power of S in S^4 row is S^4 and only even powers of S appear.

The auxiliary polynomial is

$$A(s) = S^4 + 5S^2 + 6$$

$$\frac{dA(s)}{dS} = 4S^3 + 10S$$

The zeros in the S^3 row are replaced by 4, 10 or 2, 5. Hence the Routh array becomes

$$\begin{array}{llccc}
S^6 & : & 1 & 11 & 36 & 36 \\
S^5 & : & 1 & 5 & 6 \\
S^4 & : & 1 & 5 & 6 \\
S^3 & : & 2 & 5 \\
S^2 & : & \dfrac{2 \times 5 - 1 \times 5}{2} = 2.5 & \dfrac{2 \times 6 - 1 \times 0}{2} = 6 \\
S^1 & : & 0.2 \\
S^0 & : & 6
\end{array}$$

Since there is no change in the sign in the first column there will be no root which lies on the right half the S-plane.

All the elements of a row in Routh array being zero indicates a pair of a conjugate roots on imaginary row

$$A(s) = 0$$

$$S^4 + 5S^2 + 6 = 0$$

$$S^2 = \frac{-5 + \sqrt{25 - 24}}{2} = \frac{-5 \pm 1}{2} = -3 - 2$$

$$\therefore \qquad S = \pm j\sqrt{3}, \quad S = \pm j\sqrt{2}$$

Two pairs of roots on the imaginary axis indicate a marginally stable when there is no other root which lies on the right half the S-plane

Example 5.26: A system oscillates with frequency ω if it has poles at $S = \pm j\omega$ and no poles to the right half of the S-plane. Determine the values of K and so that the system shown in Fig. 5.5 oscillates at frequency of 2 rad/sec.

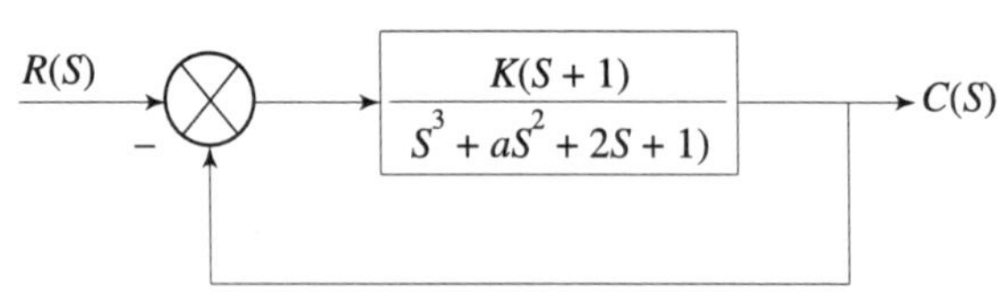

Fig. 5.5

Solution: $G(s) = \dfrac{K(S+1)}{S^2 + aS^2 + 2S + 1}, \; H(s) = 1$

The characteristic equation is

$$1 + G(s)\,H(s) = 0$$

$$1 + \frac{K(S+1)}{S^3 + aS^2 + 2S + 1} = 0$$

The characteristic equation is

$$1 + G(s)\,H(s) = 0$$

$$1 + \frac{K(S + 1)}{S^3 + aS^2 + 2S + 1} = 0$$

$$S^3 + aS^2 + (2 + K)S + (1 + K) = 0$$

$$\begin{array}{lccc} S^3 & : & 1 & (2 + K) \\ S^2 & : & a & (1 + K) \\ S^1 & : & \dfrac{a(2 + K) - (1 + K)}{a} & 0 \\ S^0 & : & (1 + K) & \end{array}$$

For stability, all elements in the first column of the Routh array should be positive.

From S^0 row, $(1 + K) > 0$

$$K > -1$$

From S^1 row, $\dfrac{a(2 + K) - (1 + K)}{a} > 0$

From S^2 row, $a > 0$

For the system to oscillate

$$a(2 + K) - (1 + K) = 0$$

$$a = \frac{1 + K}{2 + K} \qquad \text{...(5.2)}$$

Using the coefficients from the S^2 row, the auxiliary equation is given by

$$A(s) = aS^2 + (1 + K) = 0$$

The frequency of oscillation, $\omega = 2$ rad/s, $S = j\omega$

$$A(s) = a(-\omega)^2 + 1 + K = 0$$

$$-4a = -(1 + K)$$

$$a = \frac{(1 + K)}{4} \qquad \text{...(5.3)}$$

Equating the two values of a from Eqs (5.2) and (5.3), we get

$$a = \frac{1 + K}{4} = \frac{1 + K}{2 + K}$$

$$2 + K = 4 \quad \text{and} \quad K = 2$$

$$a = \frac{1 + K}{2 + K} = \frac{1 + 2}{2 + 2} = 0.75$$

$$a = 0.75.$$

Example 5.27: For the system having characteristic equation $2S^4 + 4S^2 + 1 = 0$. Find the following

(a) The number of roots in the left half of S-plane
(b) The number of roots in the right half of S-plane
(c) The number of roots on the imaginary axis use the Routh-Hurwitz stability criterion.

Solution: The Routh array is given by

$$\begin{array}{lcccc} S^4 & : & 2 & 4 & 1 \\ S^3 & : & 0 & 0 & 0 \end{array} \quad \leftarrow \text{row with all elements zero}$$

Form the auxiliary equation, using coefficients from the row just above the row of zeros

The auxiliary equation is

$$A(s) = 2S^4 + 4S^2 + 1 = 0$$

$$\frac{dA(s)}{dS} = 8S^3 + 8S + 0$$

Now replace the row of zeros with the coefficients of the resultant equation obtained by taking the derivative of the auxiliary equation. The coefficients 8, 8, 0 are used to replace the S^3 row of zeros in the Routh table.

$$\begin{array}{lcccc} S^4 & : & 2 & 4 & 1 \\ S^3 & : & 8 & 8 & 0 \quad \leftarrow \text{coefficient of } \dfrac{dA(s)}{dS} \\ S^2 & : & 2 & 1 & 0 \\ S^1 & : & 4 & 0 & \\ S^0 & : & 1 & & \end{array}$$

Since the coefficients of the first column of the table are all positive. The system is stable There is no change in sign in the first column, the number of roots in the right half of S-plane is zero. Since the powers of S on the characteristic equation are even integers and zero only, it is an even polynomial. Even polynomials have roots symmetrical about the origin. Since there are no roots in the right half of S-plane, the only possibility is that all roots must lie on the imaginary axis. Therefore, the number of roots on the imaginary axis is four.

Example 5.28: Determine the range of values for K for the system to be stable

$$S^4 + 4S^3 + 13S^2 + 36S + K = 0$$

Solution: The characteristic equation

$$S^4 + 4S^3 + 13S^2 + 36S + K = 0$$

The Routh array table

$$\begin{array}{lcccc} S^4 & : & 1 & 13 & K \\ S^3 & : & 4 & 36 & \\ S^2 & : & \dfrac{4 \times 13 - 1 \times 36}{4} = 4 & \dfrac{4 \times K - 1 \times 0}{4} = K & \end{array}$$

$$S^1 \quad : \quad \frac{4 \times 36 - 4 \times K}{4} = 36 - K \qquad\qquad 0$$

$$S^0 \quad : \quad \frac{(36 - K)K - 4 \times 0}{36 - K} = K$$

All the elements in the first column of the Routh array is positive. $\therefore$ The system is stable

$$K > 0$$

From S^1 row, $(36 - K) > 0$

$$K < 36$$

$\therefore$ The range of values of K for stability is $0 < K < 36$.

Example 5.29: A unity feedback system has the following open-loop transfer function $G(s) = \dfrac{KS(3S + 1)}{S^2 + 2S + 3}$.

Find the stability of the system in terms of the parameters K.

Solution: The characteristic equation is

$$1 + G(s)\,H(s) = 0$$

$$1 + \frac{KS(3S + 1)}{S^2 + 2S + 3} = 0$$

$$S^2 + 2S + 3 + 3KS^2 + KS = 0$$

$$S^2(3K + 1) + S(K + 2) + 3 = 0$$

This is a second order system. For the second order system to be stable, it is sufficient that all the coefficient of the characteristic equation must be real and positive.

$\therefore$
$$(3K + 1) > 0$$

$$K > \frac{1}{3}$$

$$(K + 2) > 0$$

$$K > -2$$

Since K the open loop gain, the system is stable for all positive value of K.

Example 5.30: A unity feedback control system has an open-loop transfer function consisting of two poles, two zeros and variable gain K. The zeros are located at -2 and -1 and the poles at -0.1 and 1 using the Routh stability criterion, determine the range of values of K for which the closed-loop system has 0,1 or 2 poles in the right half of the S-plane.

Solution: $G(s) = \dfrac{K(S + 1)(S + 2)}{(S + 0.1)(S - 1)}$, $H(s) = 1$

The characteristic equation is

$$1 + G(s)\,H(s) = 0$$

$$1 + \frac{K(S+1)\,(S+2)}{(S+0.1)\,(S-1)} = 0$$

$$(S+0.1)\,(S-1) + K\,(S+1)\,(S+2) = 0$$

$$(1+K)\,S^2 + (3K-0.9)\,S + (2K-0.1) = 0$$

The Routh array is formulated as

$$\begin{array}{lll} S^2 : & (1+K) & (2K-0.1) \\ S^1 : & (3K-0.9) & 0 \\ S^0 : & (2K-0.1) & \end{array}$$

For no pole to lie in the right half of the S-plane (i.e., for the system to be stable), all the terms in the first column of the Routh array must be positive.

$$(K+1) > 0$$

$$K > -1$$

$$(3K-0.9) > 0$$

$$K > 0.3$$

$$(2K-0.1) > 0$$

$$K > 0.05$$

So the condition for no pole to lie in the right half of the S-plane is $K > 0.3$.

For $K > 0.05$, two poles to lie in the right half of the S-plane, there must be two changes of sign in the elements of the first column.

Example 5.31: A unity feedback control system is characterized by the open-loop transfer function

$$G(s) = \frac{K(S+13)}{S(S+3)(S+7)}$$

Using the Routh criterion

(a) Calculate the range of value of K for the system to be stable.

(b) What is the marginal value of K for stability? Determine the frequency of oscillations.

(c) Check if for $K = 1$, all the roots of the characteristic equation of the above system have the damping factor greater than 0.5.

Solution: (a) The characteristic equation of the system

$$1 + G(s)\,H(s) = 0$$

$$1 + \frac{K(S+13)}{S(S+3)(S+7)} = 0$$

$$S(S + 3)(S + 7) + K(S + 13) = 0$$

$$S^3 + 10S^2 + (21 + K)S + 13K = 0$$

The Routh table as

$$\begin{array}{lll} S^3 : & 1 & (21 + K) \\ S^2 : & 10 & 13K \\ S^1 : & \dfrac{10 \times (21 + K) - 13K}{10} = \dfrac{210 - 3K}{10} & \\ S^0 : & 13K & \end{array}$$

For the system to be stable, all the elements in the first column of the Routh array must be positive.

For S^0 row, $13K < 0$

$$K > 0$$

From the S^1 row, $(210 - 3K) > 0$

$$3K < 210$$

$$K < 70$$

Hence, the range of values of K for stability of the system is $0 < K < 70$.

(b) The marginal value of K for stability is $K = 70$.

The frequency of sustained oscillations is given by the solution of the auxiliary equation formed by using the coefficients of the row just above the row from which the marginal value of K is determined.

$$10S^2 + 13K = 0$$

$$10S^2 + 13 \times 70 = 0$$

$$S = \pm j\sqrt{\frac{70 \times 13}{10}} = \pm j9.53$$

∴ The frequency of sustained oscillation

$$\omega = 9.53 \text{ rad/sec}$$

(c) Putting $K = 1$ in the equation $S^3 + 10S^2 + (21 + K)S + 13K = 0$

The characteristic equation is

$$S^3 + 10S^2 + 22S + 13 = 0$$

Factorizing the above equation

$$(S + 1)(S^2 + 9S + 13) = 0$$

$$(S + 1)(S + 4.5 + j5.38)(S + 4.5 - j5.38) = 0$$

∴ The roots of the characteristic equation are

$$S = -1,\ S = -4.5 - j5.38 \quad \text{and} \quad S = -4.5 + j5.38$$

Now for root at $S = -1$, $\theta = 0°$ and so the damping factor $\xi = \cos\theta = \cos 0° = 1$

For roots at $S = -4.5 \pm j5.38$

$$Q = \tan^{-1}\frac{5.38}{4.5} = 50.08$$

The damping factor $\xi = \cos\theta = \cos 50.08 = 0.64$

Thus, the damping factors for the roots are 1 and 0.64 both greater than 0.5. That means, all of roots of the characteristic equation of the above system have damping factor greater than 0.5.

The damping factor associated with $S^2 + 9S + 13$ can be determined by comparing it with the standard equation

$$S^2 + 2\xi\,\omega_n\,S + \omega_n^2 = 0$$

$$\omega_n^2 = 13,\ \omega_n = \sqrt{13} = 3.6$$

$$2\xi\,\omega_n = 9$$

$$\xi = \frac{9}{2\omega_n} = \frac{9}{2 \times 3.6} = 1.25$$

EXERCISE

Fill in the Blanks

1. The number of roots of $s^3 + 5s^2 + 7s + 3 = 0$ in the left half of the S-plane is __________ .

Ans: 3

2. The open-loop transfer function of an unity feedback open loop system is $G(s) = \dfrac{2s^2 + 6s + 5}{(s+1)^2(s+2)}$. The characteristic equation of the closed-loop system is __________ .

Ans: $2s^2 + 6s + 5 + (s + 1)^2(s + 2) = 0$

3. A system described by the transfer function $H(s) = \dfrac{1}{s^3 + as^2 + Ks + 3}$ is stable. The constraints on α and K are ____________.

Ans: $\alpha > 0,\ \alpha K > 3$

4. Consider a characteristic equation given by $s^4 + 3s^3 + 5s^2 + 6s + K + 10 = 0$. The condition for stability is __________.

Ans: $-10 < K < -4$

5. An electromechanical closed-loop control system has the following characteristic equation given by $s^3 + 6Ks^2 + (2 + K)s + 8 = 0_3$ where K is the forward gain of the system. The condition for closed-loop stability is __________.

Ans: $K = 0.528$

6. If $s^3 + 3s^2 + 4s + A = 0$, then all the roots of this equation are in the left half plane provided that __________.

Ans: $0 < A < 12$

7. The characteristic polynomial of a system $q(s) = 2s^5 + s^4 + 4s^3 + 2s^2 + 2s + 1$. The system is __________.

Ans: unstable

8. For the polynomial of a system $P(s) = s^5 + s^4 + 2s^3 + 2s^2 + 3s + 15$, the number of roots which lie in the right half of the S-plane is __________.

Ans: 2

9. The open-loop transfer function of a unity feedback system is $G(s) = \dfrac{K}{s(s^2 + s + 2)(s + 3)}$, the range of K for which the system is stable.

Ans: $21/4 > K > 0$

10. The number of open right half plane poles of $G(s) = \dfrac{10}{s^5 + 2s^4 + 3s^3 + 6s^2 + 5s + 3}$ is _____.

Ans: 2

11. The characteristic equation of an LTI system is given by $F(s) = s^5 + 2s^4 + 3s^3 + 6s^2 - 4s - 8 = 0$. The number of roots that lie strictly in the left half S-plane is __________.

Ans: 2

Unsolved Questions

12. Defind BIBO stability.

Ans: A linear released system is said to have BIBO stability if every bounded (finite) input results in a bounded (finite) output.

13. What is impulse response?

Ans: The impulse response of a system is the inverse Laplace transform of the system transfer function.

14. What is the requirement for BIBO stability?

Ans: The requirement for BIBO stability is that $\int_0^\infty m(\tau)dt < \infty$, where $m(\tau)$ is the impulse response of the system

15. What is a characteristic equation?

Ans: The denominator polynomial of $C(s)$ $R(s)$ is the characteristic equation of the system.

16. How the roots of characteristic equation are related to stability?

Ans: If the roots of characteristic equation have positive real part then the impulse response of the system is not bounded (the impulse response will be infinite as $t \to \infty$). Hence, the system will be unstable. If the roots has negative real parts then the impulse response is bounded (the impulse response becomes 0 as $t \to \infty$). Hence, the system will be stable.

17. What is the necessary condition for stability?

Ans: The necessary condition for stability is that all the coefficients of the characteristic polynomial are positive.

18. What is the relation between stability and coefficient of characteristic polynomial?

Ans: If the coefficients of characteristic polynomial are negative or zero, then some of the roots lie on the right half of S-plane. Hence, the system is unstable. If the coefficients of characteristic polynomial are positive and if no coefficient is zero then there is a possibility of the system to be stable provided all the roots are lying on the left half of S-plane.

19. What will be the nature of impulse response when the roots of characteristic equation are lying on imaginary axis?

Ans: If the roots of characteristic equation lie on the imaginary axis the nature of impulse response is oscillatory.

20. What will be the nature of impulse response if the roots of characteristic equation are lying on the right half *S*-plane?

Ans: When the roots are lying on the real axis on the right half of *S*-plane. The response is exponentially increasing. When the roots are complex conjugate and lie on the right half of *S*-plane, the response is oscillatory with exponentially increasing amplitude.

21. What is principle of argument?

Ans: The principle of argument states that let $F(s)$ be an analytic function and if an arbitrary closed contour in the clockwise direction is chosen in the *S*-plane so that $F(s)$ is analytic at every point of the contour. Then the corresponding $F(s)$-plane contour mapped in the $F(s)$ plane will encircle the origin N times in the anticlockwise direction where N is the difference between the number of poles and zeros of $F(s)$, that are encircled by the chosen closed contour in the *S*-plane (i.e., $N = P - Z$).

22. What is the necessary and sufficient condition for stability.

Ans: The necessary and sufficient condition for stability is that all the elements in the first column of the Routh array should be positive.

23. What is Routh stability criterion?

Ans: Routh criterion states that the necessary and sufficient condition for stability is that all of the elements in the first column of the Routh array are positive. If this condition is not met, the system is unstable ς the number of sign changes in the elements of the first column of Routh array corresponds to the numbers of roots of characteristic equation in the right half of the *S*-plane.

24. What is auxiliary polynomial?

Ans: In the construction of Routh array a row of all zeros indicates the existence of an even polynomial as a factor of the given characteristic equation. In an even polynomial the exponents of s are even integers or zero only. This even polynomial factor is called auxiliary polynomial. The coefficients of auxiliary polynomial are given by the elements of the row just above the row of all zeros.

25. What is quadrantal symmetry?

Ans: The symmetry of roots with respect to both real and imaginary axis is called quadrantal symmetry.

26. In Routh array what conclusion you can make when there is a row of all zeros?

Ans: All zero row in Routh array indicates the existence of an even polynomial as a factor of the given characteristic equation. The even polynomial may have roots on imaginary axis.

27. What is limitedly stable system?

Ans: For a bounded input signal, if the output has constant amplitude oscillations then the system may be stable or unstable under some limited constraints, such a system is called limitedly stable system.

28. How to find the crossing point of root locus in imaginary axis.

Ans: Method (i) By Routh-Hurwitz Criterion

Method (ii) By letting $s = j\omega$ in the characteristic equation and separate the real and imaginary parts. These two equations are equated to zero. Solve the two equations for ω and k. The value of ω gives the point where the root locus crosses imaginary axis and the value of k is the gain corresponding to the crossing point.

29. For the system represented by the following characteristic equation say whether the necessary condition for stability is satisfied or not.

(i) $s^4 + 3s^3 + 4s^2 + 5s + 10 = 0$ (ii) $s^3 - 8s^2 + 7s + 6 = 0$

(iii) $s^6 - 2s^5 + s^3 + s^2 + s + 6 = 0$ (iv) $s^5 + 4s^4 - 5s^3 - 4s^2 + 2s + 1 = 0$

Ans: In equation (i) all the coefficient are positive and so the necessary condition for stability is satisfied.

In equations (ii), (iii) and (iv) some of the coefficients are negative and some of the coefficients are missing, hence, the necessary condition for stability is not satisfied.

30. Using Routh criterion determine the locations of the roots of the following characteristic equations and comment on the stability of the systems.

(a) $2s^5 + 2s^4 + 5s^3 + 5s^2 + 3s + 5 = 0$

(b) $3s^4 + 10s^3 + 5s^2 + 5s + 3 = 0$

(c) $2s^6 + 4s^5 + s^4 - 32s^3 + 51s^2 + 3s + 15 = 0$

(d) $s^5 + s^4 + 4s^3 + 24s^2 + 3s + 63 = 0$

(e) $s^5 + 2s^4 + 24s^3 + 48s^2 - 25s - 50 = 0$

(f) $s^6 + 3s^5 + 5s^4 + 9s^3 + 8s^2 + 6s + 4 = 0$

31. The characteristic equations for certain feedback control systems are given below. In each case, determine the range of values of K, for which the system is stable.

(a) $s^4 + 3s^3 + 3s^2 + S + K = 0$ (b) $s^5 + s^4 + ks^3 + s^2 + s + 1 = 0$

(c) $s^5 + s^4 + s^2 + s + k = 0$ (d) $s^4 + s^3 + 3ks^2 + (k + 2)s + 4 = 0$

(e) $s^4 + s^3 + 3(k + 1)s^2 + (7k + 5)s + (4k + 7) = 0$

32. Open-loop transfer function of certain unity, feedback systems are given below. In each case determine the location of closed-loop poles in the S plane using Routh criterion comment on the stability of closed-loop system.

(a) $G(s) = \dfrac{200(1 + s)}{s(1 + 0.1s)(1 + 0.2s)(1 + 0.5s)}$ (b) $G(s) = \dfrac{10}{(s + 2)(s + 4)(s^2 + 6s + 25)}$

(c) $G(s) = \dfrac{(s + 1)}{s(s - 1)(s^2 + 4s + 16)}$ (d) $G(s) = \dfrac{2.5}{s(s + 5)(0.1s + 1)}$

33. Open-loop transfer function of certain unity feedback systems are given below. In each case determine the range of values of K for which the system is stable.

(a) $G(s) = \dfrac{K(s + 13)}{s(s + 3)(s + 7)}$ (b) $G(s) = \dfrac{K(s + 1)}{s(s - 1)(s + 6)}$

(c) $G(s) = \dfrac{K(s + 2)}{s(s + 5)(s^2 + 2s + 5)}$ (d) $G(s) = \dfrac{K(s - 1)}{s(s + 2)}$

34. The open-loop transfer function of a unity feedback control system is given by $G(s) = K(s + 2)|s(s - 2)(s^2 + 5s + 16)$. Determine the value of k which will cause sustained oscillation in the closed-loop system and what are the corresponding oscillation frequencies?

Short Answer Questions

35. Determine the stability of the following system using Routh criterion

$$G(s)H(s) = \frac{1}{(s + 2)(s + 4)}$$

Ans: Stable

36. Using Routh criterion determine the stability of the following system $G(s)H(s) = \dfrac{9}{s^2 (s + 2)}$

Ans: Unstable

37. Define the following terms

(i) Stable system (ii) Critically stable system

(iii) Conditionally stable system

38. Determine the number of roots that lie in the right half of the S-plane for the characteristic equation

$$2s^4 + s^3 + 3s^2 + 5s + 7 = 0.$$

Ans: 2

39. Using Routh criterion determine the stability of the system represented by the characteristic equation

$$S^4 + s^3 + 3s^2 + s + 1 = 0.$$

Ans: Stable

40. Use Routh criterion to determine the stability of the system whose characteristic equation is

$$s^5 + 5s^4 + 12s^3 + 13s^2 + 3s + 6 = 0.$$

Ans: Unstable

41. Determine the number of roots that lie in the right half of the S-plane for a system representation by the characteristic equation

$$s^3 + s^2 + 3s + 8 = 0.$$

Ans: 2

42. Use Routh criterion to determine the number of roots of the following characteristic equation which lie in the right hand of S-plane.

$$s^6 + s^5 + 2s^4 + s^3 + 2s^2 + 5s + 6 = 0.$$

Ans: 2

43. For a system with $F(s) = s^4 + 22s^3 + 10s^2 + s + k = 0$. Obtain the marginal value of K and the frequency of oscillation of the value of K.

Ans: $K = 0.452$, $\omega = 0.213$ rad/s

44. For a unity feedback system having forward path transfer function

$$G(s) = \frac{K}{s(1 + 0.6s)(1 + 0.4s)}$$

Determine (i) The range of value of K for stability

(ii) Marginal value of K

(iii) Frequency of sustained oscillations

Ans: (i) $0 < K < 4.17$ (ii) $K = 4.17$ (iii) $w = 2.04$ rad/sec

45. The characteristic equation of a closed-loop system is given by $s^4 + 2s^3 + 3s^2 + 4s + 5 = 0$. Determine the number of roots that lie in the right half of the S-plane.

Ans: 2

46. Determine the range of values of K for which the unity feedback system whose open-loop transfer function $G(s) = \dfrac{K}{s(s^2 + s + 1)(s + 4)}$ is stable and determine the frequency of sustained oscillations

Ans: $0 < K < 3.36$, $w = 0.89$ rad/sec

47. The open loop transfer function of a unity feedback control system is given by

$$G(s) = \frac{K}{(s + 2)(s + 4)\,(s^2 + 6s + 25)}$$

By applying the Routh criterion, discuss the stability of the closed-loop system as a function of K.

Ans: $-200 < K < 666.25$

48. For a closed-loop system shown in Fig. 5.6, find the characteristic equation and the range for K to have a stable system

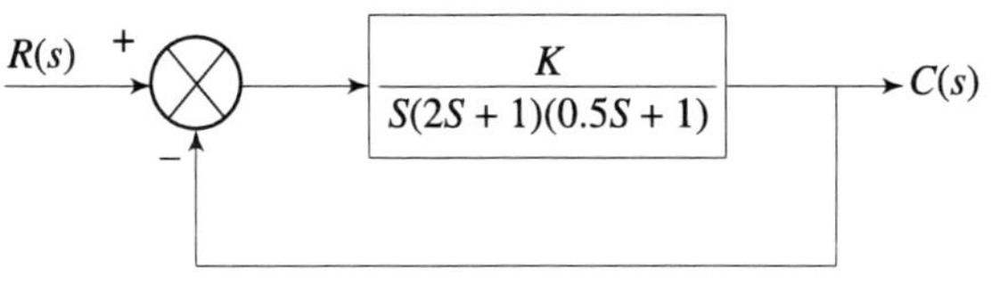

Fig. 5.6

Ans: $S^3 + 2.5S^2 + S + K = 0$. For the system to be stable $0 < K < 2.5$.

49. Using Routh criterion, determine the range of values of K for stability for the following system

$$\frac{C(s)}{R(s)} = \frac{K}{S(S^2 + S + 1)(S + 4) + K}$$

Ans: $0 < K < 3.36$

50. Find the range of K so that the system shown in Fig. 5.7 is stable where

$$G_p(S) = \frac{2}{S^3 + 4S^2 + 5S + 2}$$

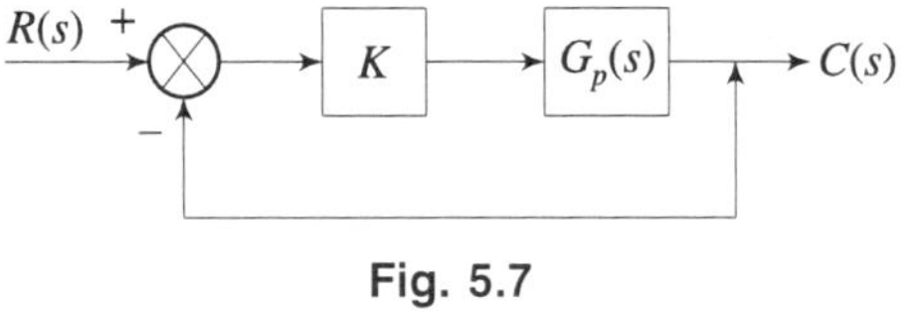

Fig. 5.7

Ans: For the system to be stable $-1 < K < 9$

51. For the block diagram shown in Fig. 5.8, the limiting value of K for stability of the inner loop is found to be $X < K < Y$. The overall system will be stable of and only if __________.

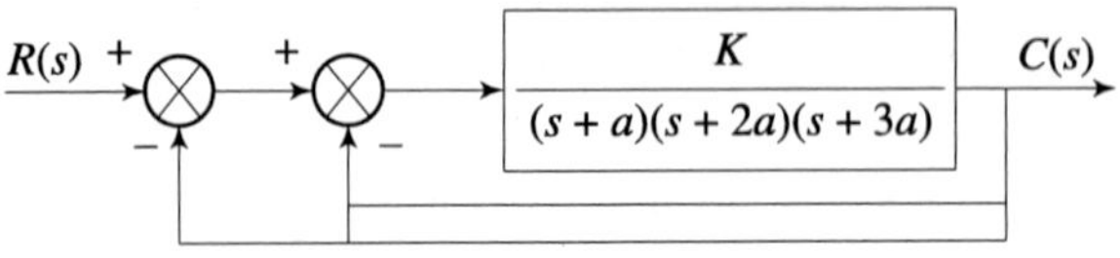

Fig. 5.8

Ans: $X/2 < K < Y/2$

52. The system shown below has an internal rate feedback loop. The closed-loop system for open and closed condition of the switch will be respectively __________.

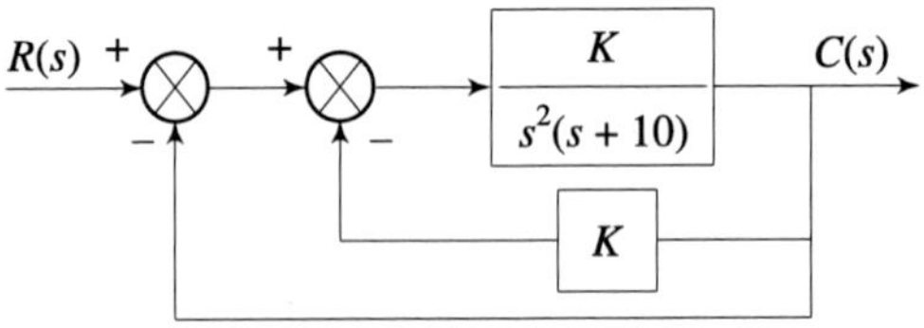

Fig. 5.9

Ans: Unstable & stable

53. The characteristic equation of a unity feedback system is given by $s^3 + s^2 + 4s + 4 = 0$. The system has __________ poles in the RH S-plane.

Ans: no poles

54. A unity feedback servo-driven instrument has and open-loop TF $G(s) = \dfrac{10}{s(s + 2)(s^2 + 3)}$. Find the absolute stability of the CL system.

Ans: Unstable

55. A unity feedback control system has an open OLTF $G(s) = \dfrac{Ke^{-S}}{s(s^2 + 5s + 9}$. Determine the maximum value of K for the CL system be stable.

Ans: 7.5

56. Investigate the stability of the system with characteristic equation $s^5 + 2s^4 + 24s^3 + 48s^2 - 25s - 50 = 0$. Find all the roots of this equation.

Ans: Unstable -1, $-j5$ and -2

57. A unity feedback control system is characterized by the loop TF $G(s) = \dfrac{K(s + 13)}{s(s + 3)(s + 7)}$. Find the range of values of K for which this system is stable.

Ans: $0 < K < 70$.

58. If the unity feedback system having $G(s) = \dfrac{5}{s(s^2 + PS + k)}$ is marginally stable and oscillates with a frequency 5 rad/sec, calculate K_{map} and p.

Ans: $K_{map} = 25/3$ p = 0.2

59. For the unity feedback system $G(s) = k/(s + 1)^3(s + 4)$ determine the following
    (i) Range of K for stability, and (ii) frequency of oscillation when this system is marginally stable

Ans: (i) $-4 < K < 20.41$ (ii) 1.3628 rad/sec.

6 Root Locus Technique

6.1 INTRODUCTION

The stability of any closed-loop system depends on the location of the closed-loop poles in the S-plane. If some parameters of the closed-loop system are varied, there is movement of its poles. Such movement of the poles with small variations of the parameters is remarkably useful in the design of the closed-loop system. The root locus technique was introduced by **W. R. Evans** in 1948 for the analysis of control systems which provides knowledge of such movement of poles by a graphical method in the S-plane. The root locus technique is a powerful tool for adjusting the location of closed-loop poles to achieve the desired system performance by varying one or more system parameters.

Consider the open-loop transfer function of system, $G(s) = \dfrac{K}{s(s + p_1)(s + p_2)}$

The closed-loop transfer function of the system with unity feedback is given by

$$\frac{C(s)}{R(s)} = \frac{G(s)}{1 + G(s)} \; \frac{\dfrac{K}{s(s + p_1)(s + p_2)}}{1 + \dfrac{K}{s(s + p_1)(s + p_2)}} = \frac{K}{s(s + p_1)(s + p_2) + K}$$

The denominator polynomial of $C(s)/R(s)$ is the characteristic equation of the system. The characteristic equation is given by $s(s + p_1)(s + p_2) + K = 0$.

The root of characteristic equation is a function of open-loop gain K. (In other words the roots of characteristic equation depend on open-loop gain K). When the gain, K, is varied from 0 to ∞, the roots of characteristic equation take different values. When $K = 0$, the roots are given by open-loop poles. When $K \to \infty$, the root will take the value of open-loop zeros.

The path taken by the roots of characteristic equation when open-loop gain K is varied from 0 to ∞ is called root loci (or the path taken by a root of characteristic equation when open-loop gain K is varied from 0 to ∞ is called root locus).

Note: In general, the root of characteristic equation can be varied by varying any other system parameter other than gain.

6.2 ROOT LOCUS CONCEPTS

For the single loop system shown in Fig. 6.1.

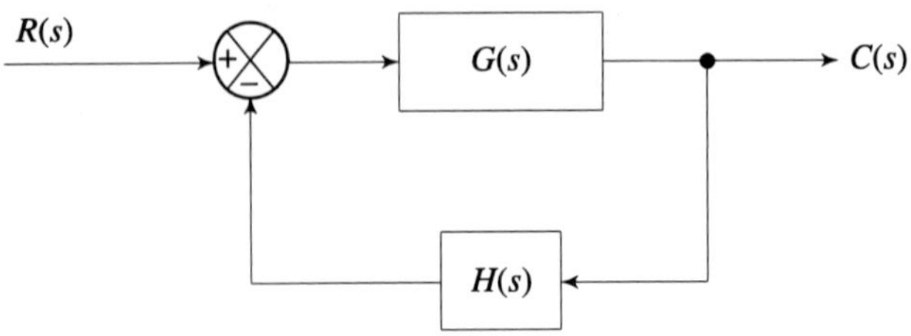

Fig. 6.1

$$\frac{C(s)}{R(s)} = \frac{G(s)}{1 + G(s)H(s)}$$

the characteristic equation is $1 + G(s)H(s) = 0$

Let $\quad D(s) = G(s)H(s)$

Therefore, $\quad 1 + G(s)H(s) = 1 + D(s) = 0$

or $\quad D(s) = -1$

From the above equation it can be concluded that the roots of characteristic equation occur only for those values of s for which $D(s) = -1$.

The equation can be converted to two Evan's conditions given below:

$$|D(s)| = 1 \text{ and } \angle D(s) = \pm\, 180^\circ\, (2q + 1) \text{ where } q = 0, 1, 2, 3.....$$

The equation $|D(s)| = 1$ is called magnitude criterion and equation $\angle D(s) = \pm\, 180^\circ\, (2q + 1)$ is called angle criterion. The magnitude criterion states that $s = s_a$ will be a point on root locus if for that values of $S|D(s)| = |G(s)H(s)| = 1$. The angle criterion states that $s = s_a$ will be a point on root locus if for that value of s, $\angle D(s) = \angle G(s)H(s)$ is equal to an odd multiple of 180°.

The function $D(s)$ can be expressed as a ratio of two polynomials in s as shown below.

$$D(s) = G(s)H(s) = K\,\frac{(s + z_1)(s + z_2)(s + z_3)\;....}{(s + p_1)(s + p_2)(s + p_3)\;....}$$

Therefore, $$|D(s)| = K\,\frac{|s + z_1|\cdot|s + z_2|\cdot|s + z_3|\;....}{|s + p_1|\cdot|s + p_2|\cdot|s + p_3|\;....} = K\,\frac{\prod_{i=1}^{m}|s + zi|}{\prod_{i=1}^{n}|s + pi|}$$

where, m = number of zeros, n = number of poles

The magnitude criterion states that $|D(s)| = 1$.

The closed-loop poles, that is, the roots of the characteristic equation are dependent upon the value of K. If the gain K is varied from $-\infty$ to $+\infty$, a separate set of locations of the roots of the characteristic equation will be obtained. The root locus is obtained by joining all such locations. Thus, we can define root locus as the path of the closed-loop poles traced out in the S-plane as the system open-loop gain K is varied from $-\infty$ to $+\infty$. The general, shape of the root locus gives an idea of the type of controller or compensator needed to meet the particular design criterion.

6.3 CONSTRUCTION OF ROOT LOCUS

The exact root locus is sketched by trial and error procedure. In this method, the poles and zeros of $G(s)$ $H(s)$ are located on the S-plane on a graph sheet. Determine the angles of vectors drawn from poles and zeros to the trial point. From the angle criterion, determine the angle to be contributed by these vectors to make the trial point as a point on root locus. Shift the trial point suitably so that the angle criterion is satisfied.

A number of points are determined using the above procedure. Join the points by a smooth curve which is the root locus. The value of K for a particular root can be obtained from the magnitude criterion.

The trial and error procedure for sketching root locus is tedious. A set of rules has been developed to reduce the task involved in sketching the root locus and to develop a quick approximate sketch. From the approximate sketch, a more accurate root locus can be obtained by a few trials.

6.3.1 Rules for Construction of Root Locus

Rule 1: The root locus is symmetrical about the real axis.

Rule 2: Each branch of the root locus originates from an open-loop pole corresponding to $K = 0$ and terminates at either on a finite open-loop zero (or open-loop zero at infinity) corresponding to $K = \infty$. The number of branches of the root locus terminating on infinity is equal to $(n–m)$, i.e., the number of open-loop poles minus the number of finite zeros.

Rule 3: Segments of the real axis having an odd number of real axis open-loop poles plus zeros to their right are parts of the root locus.

Rule 4: The $(n–m)$ root locus branches that tend to infinity, do so along straight line asymptotes making angles with the real axis given by $\phi_A = \dfrac{\pm 180°(2q + 1)}{n-m}$, where $q = 0, 1, 2, 3...(n–m)$.

Rule 5: The points of intersection of the asymptotes with the real axis are at $s = \sigma_A$ where $\sigma_A = \dfrac{\text{sum of poles} - \text{sum of zeros}}{n-m}$.

Rule 6: The breakaway and breakin points of the root locus are determined from the roots of the equation $dK/ds = 0$. If r number of branche of root locus meet a point, then they break away at an angle of $\pm 180°/r$.

Rule 7: The angle of departure from a complex open-loop pole is given by

$\Phi_P = \pm 180°\ (2q + 1) + \Phi;\ q = 0, 1, 2, \ldots$

where Φ is the net angle contribution at the pole by all other open-loop poles and zeros. Similarly, the angle of arrival at a complex open-loop is given by

$\Phi_z = \pm 180°\ (2q + 1) + \Phi;\ q = 0, 1, 2, \ldots$

where Φ is the net angle contribution at zero by all other open-loop poles and zeros.

Rule 8: The intersection of root locus branches with the imaginary axis can be determined by use of the Routh criterion, or by letting $s = j\omega$ in the characteristic equation and equating the real part and imaginary part to zero, to solve for ω and K. The value of ω is the intersection point on imaginary axis and K is the value of gain at the intersection point.

Rule 9: The open-loop gain K at many point $s = s_a$ on the root locus is given by

$$K \frac{\prod_{i=1}^{n} |s_a + pi|}{\prod_{i=1}^{m} |s_a + zi|} = \frac{\text{Product of vector lengths from open-loop poles to the point, } s_a}{\text{Product of vector lengths from open-loop zeros to the point, } s_a}$$

Note: The length of vector should be measured to scale. If there is no finite zero then the product of vector lengths from zeros is equal to 1.

6.3.2 Typical Sketches of Root Locus Plots

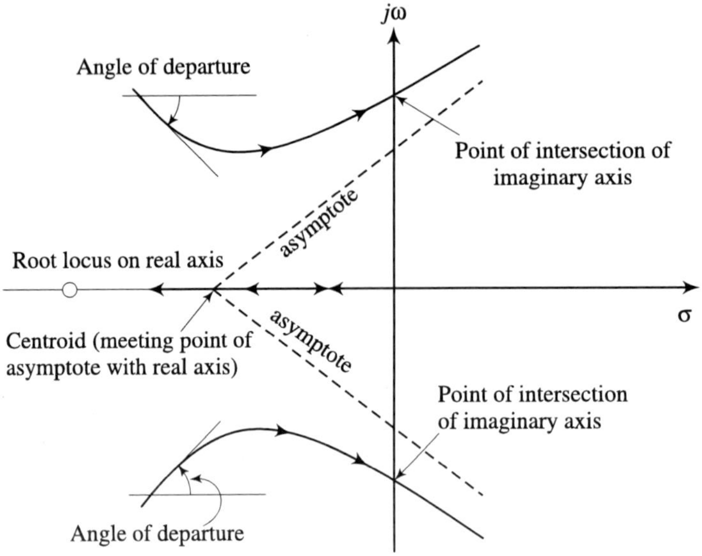

(a)

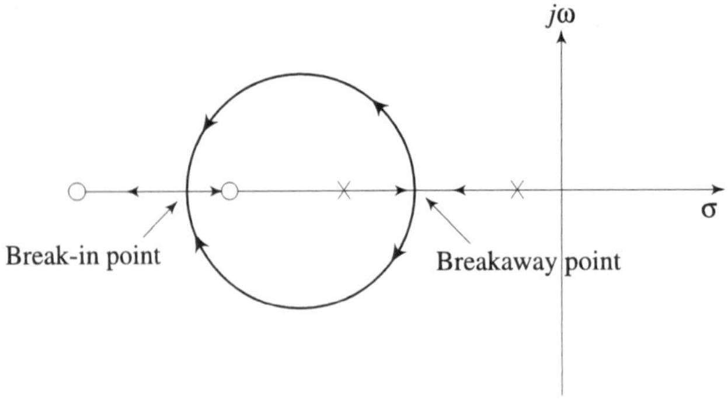

(b)

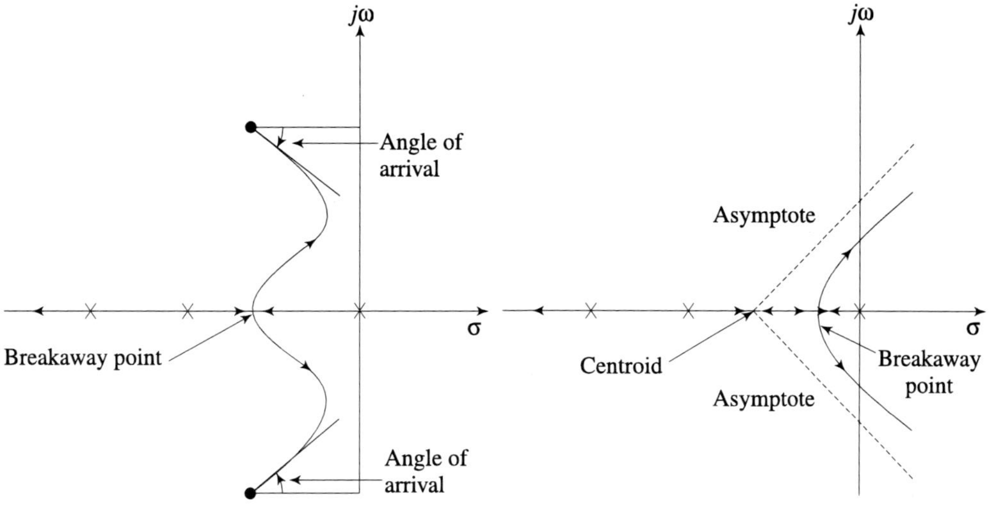

(c)

Fig. 6.2

6.3.3 Procedure for Constructing Root Locus

Step 1: Locate the poles and zeros of $G(s)H(s)$ on the S-plane. The root locus branch starts from open-loop poles and terminates at zeros.

Step 2: Determine the root locus on real axis.

Step 3: Determine the asymptotes of root locus branches and meeting point of asymptotes with real axis.

Step 4: Find the breakaway and break-in points.

Step 5: If there is a complex pole then determine the angle of departure from the complex pole. If there is a complex zero then determine the angle of arrival at the complex zero.

Step 6: Find the points where the root loci may cross the imaginary axis.

Step 7: Take a series of test points in the broad neighbourhood of the origin of the S-plane and adjust the test points to satisfy angle criterion. Sketch the root locus by joining the test points by smooth curve.

Step 8: The value of gain K at any point on the locus can be determined from magnitude condition.

The magnitude condition is given by

Gain K at a point $(s = s_a)$

$$= \frac{\text{Product of length of vectors from poles to the point } (s = s_a)}{\text{Product of length of vectors from finite zeros to the point } (s = s_a)}$$

Note: When there is no finite zero, the denominator is taken as unity. The length of vectors should be measured to scale.

6.3.4 Explanation for the Various Steps in the Procedure for Constructing Root Locus

Step 1: ***Location of poles and zeros:*** Draw the real and imaginary axis on an ordinary graph sheet and choose same scales both on real and imaginary axis.

The poles are marked by cross **X** and zeros are marked by small circle "o". The number of root locus branches is equal to the number of poles of open-loop transfer function. The origin of a root locus is at a pole and the end is at a zero.

If n = number of poles and m = number of finite zeros, then $\boldsymbol{m}$ root locus branches end at finite zeros. The remaining $(\boldsymbol{n} - \boldsymbol{m})$ root locus branches will end at zeros or at infinity.

Step 2: ***Root locus on real axis:*** To decide the part of root locus on real axis, take a test point on real axis. If the total number of poles and zeros on the real axis to the right of this test point is an odd number then the test point lies on the root locus. If it is even then the test point does not lie on the root locus.

Step 3: ***Angles of asymptotes and centroid:*** If n = number of poles and m = number of zeros, then $(n - m)$ root locus branches will terminate at zeros or at infinity. These root locus branches will go along an asymptotic path and meet the asymptotes at infinity. Hence, the number of asymptotes is equal to the number of root locus branches going to infinity. The angles of asymptotes and the centroid are given by the following formulae.

$$\text{Angle of asymptotes} = \varnothing_A = \frac{\pm 180^\circ(2q + 1)}{n - m}, \text{ where } q = 0, 1, 2, 3\ldots(n - m)$$

$$\text{Centroid (meeting point of asymptotes with real axis)} = \frac{\text{sum of poles} - \text{sum of zeros}}{n - m}$$

Step 4: ***Breakaway and break-in points:*** The breakaway or break-in points either lie on real axis or exist as complex conjugate pairs. If there is a root locus on real axis between 2 poles then there exists a breakaway point. If there is a root locus on real axis between 2 zeros then there exists a break-in point. If there is a root locus on real axis between pole and zero then there may or may not be breakaway or break-in point.

Let the characteristic equation be in the form

$$B(s) + KA(s) = 0$$

$$K = -\frac{B(s)}{A(s)}$$

The breakaway and break-in points are given by roots of the equation $dK/ds = 0$. The roots of $dK/ds = 0$ are actual breakaway or break-in points provided for this value of the root, the gain K should be positive and real.

Step 5: ***Angle of departure and angle of arrival***

$$\left\{\begin{array}{c}\text{Angle of departure}\\ \text{from a complex pole } A\end{array}\right\} = 180^\circ - \left[\begin{array}{c}\text{Sum of angles of vector to the}\\ \text{complex pole } A \text{ from other poles}\end{array}\right] + \left[\begin{array}{c}\text{Sum of angles of vectors to the}\\ \text{complex pole } A \text{ from zeros}\end{array}\right]$$

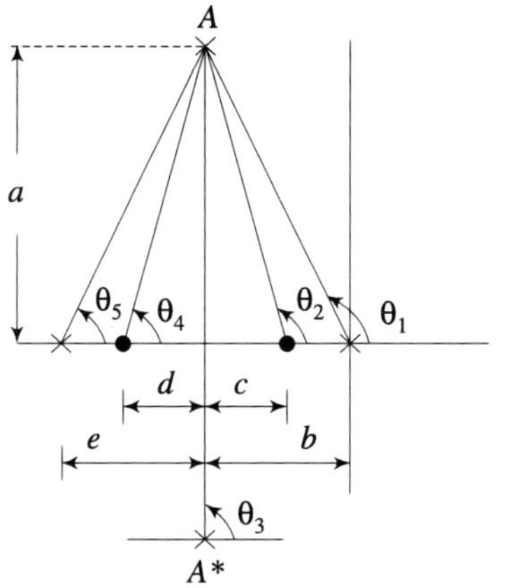

$$\theta_1 = 180^\circ - \tan^{-1}\frac{a}{b}$$

$$\theta_2 = 180^\circ - \tan^{-1}\frac{a}{c}$$

$$\theta_1 = 90^\circ$$

$$\theta_1 = \tan^{-1}\frac{a}{d}$$

$$\theta_1 = \tan^{-1}\frac{a}{e}$$

Fig. 6.3 Calculation of angle of depature

If poles are complex then they exist only as conjugate pairs. Consider the two complex conjugate poles A and A^* (shown in Fig. 6.3).

$$\left\{\begin{array}{c}\text{Angle of departure}\\ \text{at pole } A\end{array}\right\} = 180^\circ - (\theta_1 + \theta_3 + \theta_5) + (\theta_2 + \theta_4)$$

$$\left\{\begin{array}{c}\text{Angle of departure}\\ \text{at pole } A^*\end{array}\right\} = -[\text{Angle of departure at pole } A]$$

Note: The angle can be calculated as shown in Fig. 6.3 or they can be measured using protractor.

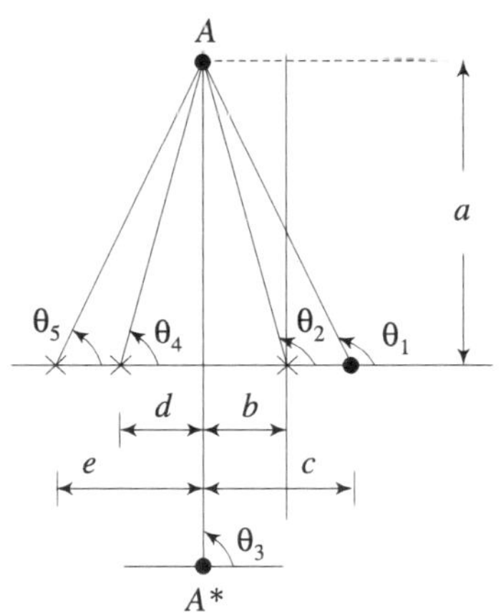

$$\theta_1 = 180^\circ - \tan^{-1}\frac{a}{c}$$

$$\theta_2 = 180^\circ - \tan^{-1}\frac{a}{b}$$

$$\theta_3 = 90^\circ$$

$$\theta_4 = \tan^{-1}\frac{a}{d}$$

$$\theta_5 = \tan^{-1}\frac{a}{e}$$

Fig. 6.4 Calculation of angle of arrival

If zeros are complex conjugate. Consider the two complex conjugate zeros A and A^* shown in Fig. 6.4

$$\left\{\begin{matrix}\text{Angle of arrival}\\ \text{at zero } A\end{matrix}\right\} = 180° - (\theta_1 + \theta_3) + (\theta_2 + \theta_4 + \theta_5)$$

$$\left\{\begin{matrix}\text{Angle of arrival}\\ \text{at zero } A^*\end{matrix}\right\} = -\text{ [Angle of arrival at zero } A]$$

Note: The angles can be calculated as shown in Fig. 6.4 or they can be measured using protractor.

Step 6: *Point of intersection of root locus with imaginary axis:* The point where the root loci intersects the imaginary axis can be found by following three methods.

1. By Routh-Hurwitz array
2. By trial and error approach
3. Letting $s = j\omega$ in the characteristic equation and separate the real part and imaginary part. Two equations are obtained one by equating real part to zero and the other by equating imaginary part to zero. Solve the two equations for ω and K. The value of ω gives the point where the root locus crosses imaginary axis. The value of K gives the value of gain, K, at this crossing point. Also this value of K is the limiting value of K for stability of the system.

Step 7: *Test points and root locus:* Choose a test point. Using a protractor roughly estimate the angle of vectors drawn to this point and adjust the point to satisfy angle criterion. Repeat the procedure for a few more test points. Sketch the root locus from the knowledge of typical sketches and the information obtained from steps 1 through 6.

Note: In practice the approximate root locus can be sketched from the informations obtained from steps 1 through 6 and from the knowledge of typical sketches of root locus.

6.4 DETERMINATION OF OPEN-LOOP GAIN FOR A SPECIFIED DAMPING OF THE DOMINANT ROOTS

The dominant pole is a pair of complex conjugate poles which decides the transient response of the system. In higher order systems the dominant poles are given by the poles which are very close to the origin, provided all other poles are lying far away from the dominant poles. The poles which are far away from the origin will have less effect on the transient response of the system.

The transfer function of higher order systems can be approximated to a second order transfer function, whose standard form of closed-loop transfer function is

$\dfrac{C(s)}{R(s)} = \dfrac{\omega_n^2}{s^2 + 2\xi\omega_n s + \omega_n^2}$ the dominant poles (s_d and s_d^*) are given by the roots of quadratic factor, $(s^2 + 2\xi\omega_n s + \omega_n^2) = 0$

$$\therefore \qquad Sd = \frac{-2\xi\omega_n \pm \sqrt{4\xi^2\omega_n^2 - 4\omega_n^2}}{2} = -\xi\omega_n \pm \omega_n\sqrt{1-\xi^2}$$

The dominant pole can be plotted on the S-plane as shown in Fig. 6.5.

In the right angle triangle OAP,

$$\cos\alpha = \frac{\xi\omega_n}{\omega_n} = \xi$$

$$\therefore \qquad \alpha = \cos^{-1}\xi$$

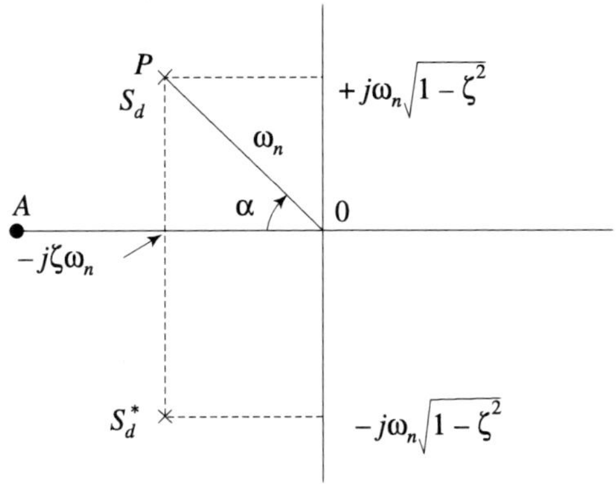

Dominant pole, S_d

Fig. 6.5

To fix a dominant pole on the root locus, draw a line at an angle of $\cos^{-1}\xi$ with respect to negative real axis. The meeting point of this line with root locus will give the location of dominant pole. The value of K corresponding to dominant pole can be obtained from magnitude condition.

The gain K corresponding to dominant pole Sd

$$K_{sd} = \frac{\text{Product of length of vectors from open-loop poles to dominant pole}}{\text{Product of length of vectors from open-loop zeros to dominant pole}}$$

6.4.1 Importance of Root Locus

The root locus technique is an important tool in designing control system with desired performance characteristics. The desired performance of the system can be achieved by adjusting the location of its closed-loop poles in the S-plane by varying one or more system parameters.

The root locus can be plotted in the S-plane by varying a system parameter (usually gain, K) over the complete range of values. The roots corresponding to a particular value of the system parameter can then be located on the locus or the value of the parameter for a desired root location can be determined from the locus.

The root locus technique is also used for stability analysis. Using root locus the range of values of K, for a stable system can be determined. It is also easier to study the relative stability of the system from the knowledge of location of closed-loop poles. The dominant roots are used to estimate the damping ratio and natural frequency of oscillation of the system. From ξ and ω_n the time domain specifications can be calculated.

Example 6.1: A unity feedback control system has an open-loop transfer function $G(s) = \dfrac{K}{s(s^2 + 4s + 13)}$ sketch the root locus.

Solution:

Step 1: ***To locate poles and zeros.*** The poles of open-loop transfer function are the roots of the equation, $s(s^2 + 4s + 13) = 0$. The roots of the quadratic are:

$$s = \frac{-4 \pm \sqrt{4^2 - 4 \times 13}}{2} = -2 \pm j3.$$

$\therefore$ The poles are 0, $-2 + j3$ and $-2 - j3$ and the poles are marked by X (cross) as shown in Fig. 6.6.

Step 2: ***To find the root locus on real axis.*** There is only one pole on real axis at the origin. Hence, if we choose any test point on the negative real axis then to the right of that Point the total numbers of real poles and zeros is one, which is an odd number. Hence, the entire negative real axis will be part of root locus. The root locus on real axis is shown as a bold line in Fig. 6.6.

Note: For the given transfer function one root locus branch will start at the pole at the origin and travel through the negative real axis to meet the zero at infinity.

Step 3: ***To find angle of asymptotes and centroid.*** Since there are 3 poles, the number of root locus branches are three. There is no finite zero. Hence, all the three root locus branches end at zeros or at infinity. The number of asymptotes required are three.

$$\text{Angles of asymptotes, } \varnothing_A = \frac{\pm 180^\circ (2q + 1)}{n - m}, \ q = 0, 1 \ldots\ldots (n - m)$$

Here $n = 3$ and $m = 0$, therefore, $q = 0, 1, 2, 3$

$$\text{If } q = 0, \text{ angles} = \pm\frac{180^\circ}{3} = \pm 60^\circ$$

If $q = 1$, angles $= \pm 180^\circ \times 3/3 = \pm 180^\circ$

If $q = 2$, angles $= \pm 180^\circ \times 5/3 = \pm 300^\circ \pm 60^\circ$

If $q = 3$, angles $= \pm 180^\circ \times 7/3 = \pm 420^\circ \pm 60^\circ$

Note: It is enough if you calculate the required number of angles. Here it is given by the first three values of angles. The remaining values will be repetitions of the previous values.

$$\text{Centroid} = \frac{\text{Sum of poles} - \text{Sum of zeros}}{n - m} = \frac{0 - 2 + j3 - 2 - j3 - 0}{3} = \frac{-4}{3} = -1.33$$

The centroid is marked on real axis and from the centroid the angles of asymptotes are marked using a protractor. The asymptotes are drawn as dotted lines as shown in Fig. 6.6.

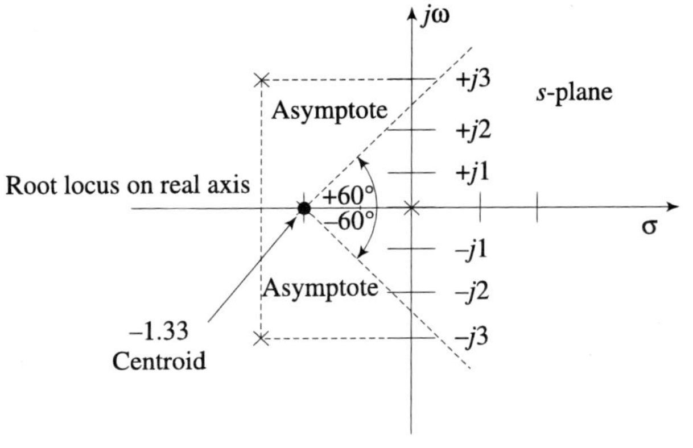

Fig. 6.6

Step 4: ***To find the breakaway and break-in points.*** The closed-loop transfer function $= \frac{C(s)}{R(s)} = \frac{G(s)}{1 + G(s)}$

$$\therefore \qquad \frac{C(s)}{R(s)} = \frac{\frac{K}{s(s^2 + 4s + 13)}}{1 + \frac{K}{s(s^2 + 4s + 13)}} = \frac{K}{s(s^2 + 4s + 13) + K}$$

The characteristic equation is $s(s^2 + 4s + 13) + K = 0$

$$s^3 + 4s^2 + 13s + K = 0$$

$$\therefore \qquad K = -s^3 - 4s^2 - 13s$$

On differentiating the equation of K with respect to s, we get

$$\frac{dK}{ds} = -(3s^2 + 8s + 13)$$

Put $$\frac{dK}{ds} = 0$$

$$\therefore \qquad -(3s^2 + 8s + 13) = 0$$

$$(3s^2 + 8s + 13) = 0$$

$$\therefore \qquad s = \frac{-8 \pm \sqrt{8^2 - 4 \times 13 \times 3}}{2 \times 3} = -1.33 \pm j1.6$$

Check for K,

When $s = (-1.33 + j1.6)$, the value of K is given by $K = -(s^3 + 4s^2 + 13s)$

$$= -[(-1.33 + j1.6)^3 + (-1.33 + j1.6)^2 + 13(-1.33 + j1.6)]$$

$$\neq \text{Positive and real}$$

Also when $s = -1.33 - j1.6$ the value of K is not equal to real and positive.

Since the value of K for $s = (-1.33 \pm j1.6)$ is not real and positive, the points are not an actual breakway or break-in points. The root locus has neither breakaway nor break-in point.

Step 5: ***To find the angle of departure.*** Let us consider the complex pole A shown in Fig. 6.7. Draw, vectors from all other poles to the pole A as shown in Fig. 6.7. Let the angles of these vectors be θ_1 and θ_2.

Here, $\qquad \theta_1 = 180° - \tan^{-1}(3/2) = 123.7°$

$$\theta_2 = 90°$$

Angle of departure from the complex pole, $A = 180° - (\theta_1 + \theta_2)$

$$= 180° - (123.7° + 90°) = -33.7°$$

The angle of departure at complex pole A^* is negative of the angle of departure at complex pole A.

$\therefore$ Angle of departure at pole, $A^* = +33.7°$

Mark the angles of departure at complex poles using protractor.

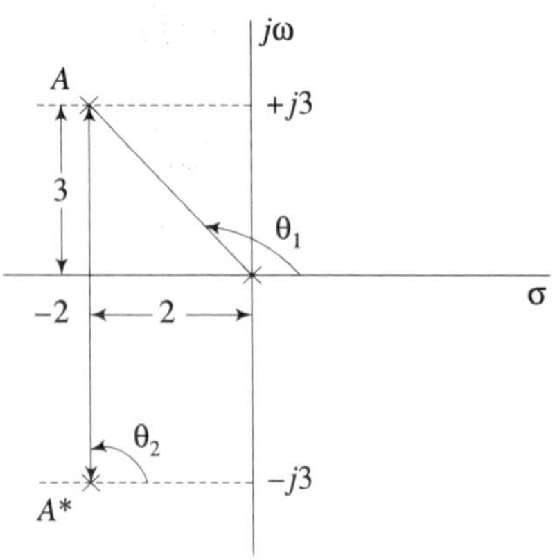

Fig. 6.7

Step 6: ***To find the crossing point on imaginary axis.*** The characteristic equation is given by

$$s^3 + 4s^2 + 13s + K = 0$$

Put $s = j\omega$

$$(j\omega)^3 + 4(j\omega)^2 + 13(j\omega) + K = 0$$

$$-j\omega^3 - 4\omega^2 + 13j\omega + K = 0$$

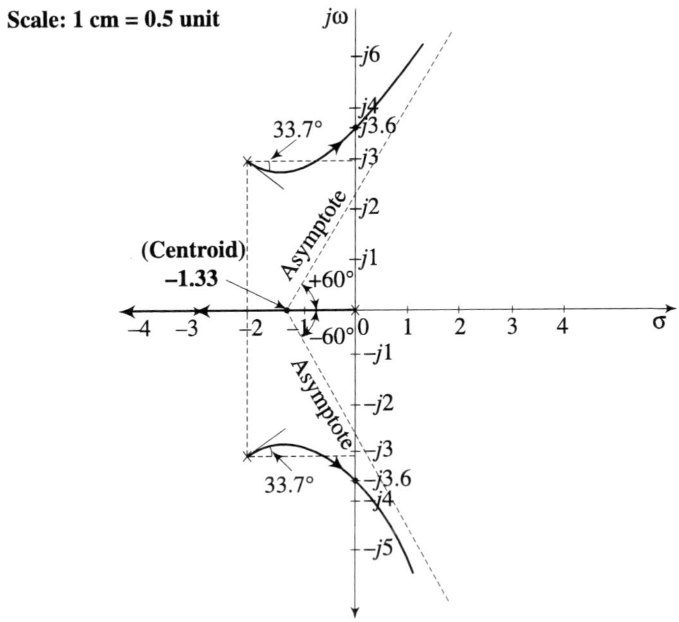

Fig. 6.8 Root locus sketch of $1 + G(s) = 1 + \dfrac{K}{s(s^2 + 4s + 13)}$

On equating imaginary part to zero

$$-\omega^3 + 13\omega = 0$$

$$-\omega^3 = -13\omega$$

$$\omega = \pm\sqrt{13} = \pm 3.6$$

On equating real part to zero

$$-4\omega^2 + K = 0$$

$$4\omega^2 = K$$

$$K = 4(13) = 52$$

The crossing point of root locus is $\pm j3.6$. The value of K at this crossing point is $K = 52$. (this is the limiting value of K for the stability of the system).

The complete root locus sketch is shown Fig. 6.8. The root locus has three branches, one branch starts at the pole at the origin and travels through negative real axis to meet the zero at infinity. The other two root locus branches start at complex poles (along the angle of departure), crosses the imaginary axis at $\pm j3.6$ and travel-parallel to asymptotes to meet the zeros at infinity.

Example 6.2: Sketch the root of the system whose open-loop transfer function is $G(s) = \dfrac{K}{s(s+2)(s+4)}$. Find the value of K so that the damping ratio of the closed-loop system is 0.5

Solution:

Step 1: *To locate poles and zeros.* The poles of open-loop transfer function are the roots of the equation, $s(s + 2)(s + 4) = 0$

$\therefore$ The poles are $s = 0, -2, -4$.

Step 2: *To find the root locus on real axis.* There are three poles on the real axis. Choose a test point on real axis between $s = 0$ and $s = -2$. To the right of this point the total number of real poles and zeros is one, which is an odd number. Hence, the real axis between $s = 0$ and $s = -2$ will be a part of root locus.

Choose a test point on real axis between $s = -2$ and $s = -4$. To the right of this point, the total number of real poles and zeros is two which is an even number, hence the real axis between $s = -2$ and $s = -4$ will not be a part of root locus.

Choose a test point on real axis to the left of $s = -4$. To the right of this point, the total number of real poles and zeros is three, which is an odd number. Hence, the entire negative real axis from $s = -4$ to $-\infty$ will be a part of root locus.

Step 3: *To find asymptotes and centroid.* Since there are three poles the number of root locus branches is three. There is no finite zero. Hence, all the three root locus branches end at zeros at infinity. The number of asymptotes required is three.

$$\text{Angles of asymptotes, } \varnothing_A \ \frac{\pm 180^\circ(2q+1)}{n-m} \quad q = 0, 1 \ldots\ldots\ldots (n-m)$$

Here $\qquad n = 3$ and $m = 0$

If $q = 0$, angle $= \pm 180^\circ/3 = \pm 60^\circ$

If $q = 1$, angle $= \pm 180^\circ \times 3/3 = \pm 180^\circ$

If $q = 2$, angle $= \pm 180^\circ \times 5/3 = \pm 300^\circ \pm 60^\circ$

Note: It is enough if you calculate the required number of angles. Here it is given by the first three values of angles. The remaining values will be repetitions of the previous values.

$$\text{Centroid} = \frac{\text{Sum of poles} - \text{Sum of zeros}}{n-m} = 0 - 2 - 4 - 0/3 = -2$$

The centroid is marked on real axis and from the centroid the angles of asymptotes are marked using a protractor. The asymptotes are drawn as dotted lines as shown in Fig. 6.9.

Step 4: *To find the breakaway and break-in points.* The closed-loop transfer function $= \dfrac{C(s)}{R(s)} = \dfrac{G(s)}{1 + G(s)}$.

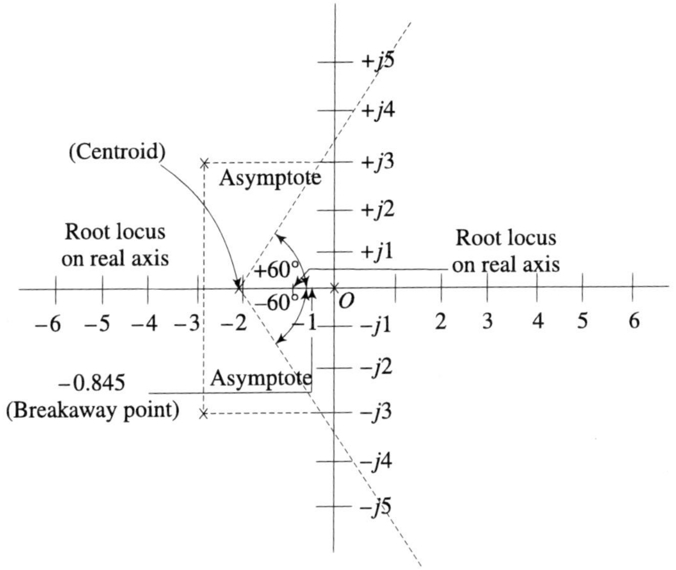

Fig. 6.9 The asymptotes, root locus on real axis and location of poles and breakaway points

$$\therefore \qquad \frac{C(s)}{R(s)} = \frac{\dfrac{K}{s(s+2)(s+4)}}{1 + \dfrac{K}{s(s+2)(s+4)}} = \frac{K}{s(s+2)(s+4) + K}$$

The characteristic equation is $s(s + 2)(s + 4) + K = 0$

$$S\,(s^2 + 6s + 8) + K = 0$$

$$s^3 + 6s^2 + 8s + K = 0$$

$$\therefore \qquad K = -s^3 - 6s^2 - 8s$$

On differentiating the equation of K with respect to s, we get

$$\frac{dK}{ds} = -(3s^2 + 12s + 8)$$

Put $$\frac{dK}{ds} = 0$$

$\therefore$ $$-(3s^2 + 12s + 8) = 0$$

$$(3s^2 + 12s + 8) = 0$$

$\therefore$ $$s = \frac{-12 \pm \sqrt{12^2 - 4 \times 3 \times 8}}{2 \times 3} = -0.845 \text{ or } -3.154$$

Check for K,

when $s = -0.845$, the value of K is given by

$$K = -[(-0.845)3 + 6(-0.845) + 8(-0.845)]$$

$$= 3.08$$

Since the value of K is positive and real for $s = -0.845$, this point is the actual breakaway point.

The breakaway point is marked on the negative real axis as shown in Fig. 6.10

Step 5: ***To find the angle of departure.*** Since there are no complex pole or zero, we only need to find the angle of departure or arrival.

Step 6: ***To find the crossing point on imaginary axis.*** The characteristic equation is given by

$$s^3 + 6s^2 + 8s + K = 0$$

Put $$s = j\omega$$

$$(j\omega)^3 + 6(j\omega)^2 + 8j\omega + K = 0$$

$$-j\omega^3 - 6\omega^2 + 8j\omega + K = 0$$

On equating imaginary part to zero

$$-j\omega^3 + 8j\omega = 0$$

$$-j\omega^3 = -j8\omega$$

$$\omega = \pm\sqrt{8} = \pm 2.8$$

On equating real part to zero

$$-6\omega^2 + K = 0$$

$$6\omega^2 = K$$

$$K = 6(8) = 48$$

The crossing point of root locus is $\pm j2.8$. The value of K at this crossing point is $K = 48$. (This is the limiting value of K for the stability of the system.)

The complete root locus sketch is shown in Fig. 6.10. The root locus has three branches. One branch starts at the pole at $s = -4$ and travels through negative real axis to meet the zero at infinity. The other two root locus branches start at $s = 0$ and $s = -2$ and travel through negative real axis, break away from real axis it $s = -0.845$, then cross imaginary axis at $\pm j2.8$ and travel parallel to asymptotes to meet the zeros at infinity.

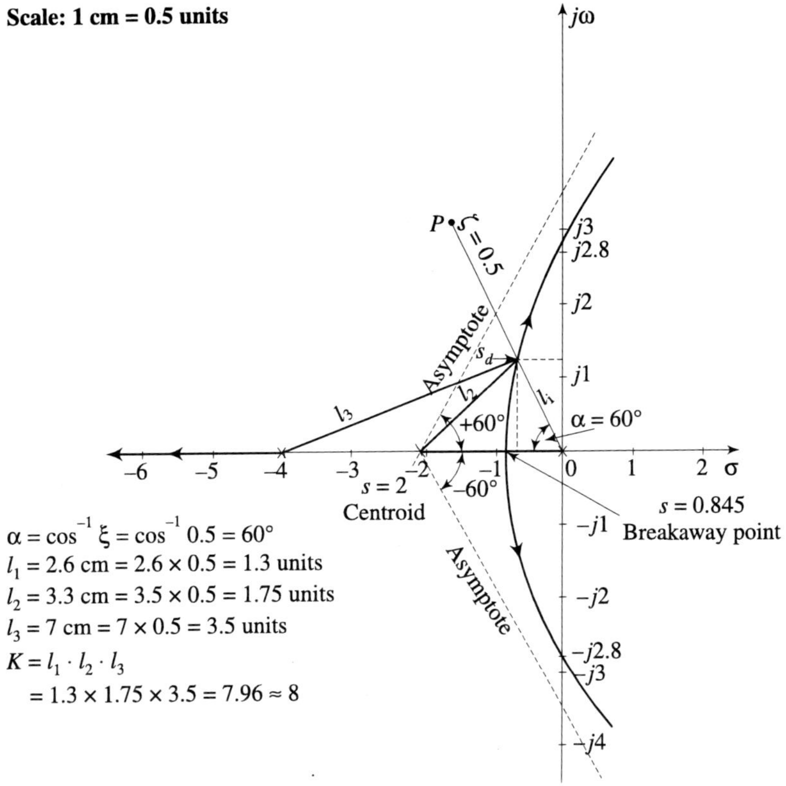

Fig. 6.10 Root locus sketch of $1 + G(s) = 1 + \dfrac{1}{s(s+2)(s+4)}$

To find the value of *K* corresponding to $\xi = 0.5$

Given that $\quad \xi = 0.5$

Let $\quad \alpha = \cos^{-1}\xi = \cos^{-1}(0.5) = 60°$

Draw a line *OP*, such that the angle between line *OP* and negative real axis 60° ($\alpha = 60°$) as shown in Fig. 6.10. The meeting point of the line *OP* and root locus give the dominant poles, s_d.

The value of *K* corresponding to the point, $s = s_d$

$$K_{sd} = \frac{\text{Product of length of vector from all poles to the point, } s = s_d}{\text{Product of length of vectors from all zeros to the point, } s = s_d}$$

$$= \frac{l_1, l_2, l_3}{1} = 1.3 \times 1.75 \times 3.5 = 7.96 \approx 8$$

Note: The length of vectors is measured to scale.

Example 6.3: The open-loop transfer function of a unity feedback system is given by $G(s) = \dfrac{K(s+9)}{s(s^2+4s+11}$. Sketch the root locus of the system.

Solution:

Step 1: ***To locate poles and zeros.*** The poles of open-loop transfer function are the roots of the equation, $s(s^2 + 4s + 11) = 0$.

The roots of the quadratic $s = \dfrac{-4 \pm \sqrt{4^2 - 4 \times 11}}{2} = -2 \pm j2.64$

$\therefore$ The poles are lying at $s = 0, -2 + j2.64, -2 - j2.64$.

The zeros are lying at $s = -9$ and infinity.

Let us denote the poles as p_1, p_2, p_3 and zero by z_1

$$p_1 = 0,\ p_2 = -2 + j2.64,\ p_3 = -2 - j2.64 \text{ and } z_1 = -9$$

Step 2: ***To find the root locus on real axis.*** One pole and one zero lie on real axis. Choose a test point to the left of $s = 0$, then to the right of this point, the total number of poles and zeros is one which is an odd number. Hence, the portion of real axis from $s = 0$ to $s = -9$ will be a part of root locus. If we choose a test point to the left of $s = -9$ then to the right of this point, the total number of pole and zeros is two, which is an even number. Hence, the real axis from $s = -9$ to $-\infty$ will not be a part of root locus. The root locus on real axis is shown in Fig. 6.11.

Step 3: ***To find asymptotes and centroid.*** Since there are three poles, the number of root locus branches is three. One root locus branch starts at the pole at origin and travels along negative real axis to meet the zero at $s = -9$. The other two root locus branches meet the zeros at infinity. The number of asymptotes required is two.

$$\text{Angles of asymptotes, } \varnothing_A = \frac{\pm 180°(2q+1)}{n-m} \quad q = 0, 1 \ldots\ldots\ldots (n-m)$$

Here $n = 3$ and $m = 0$

$$\text{If } q = 0, \text{ angles} = \pm\frac{180°}{2} = \pm 90°$$

$$\text{If } q = 1, \text{ angles} = \pm 180° \times 3/2 = \pm 270° \ \pm 90°$$

$$\text{If } q = 2, \text{ angles} = \pm 180° \times 5/2 = \pm 450° \ \pm 90°$$

Note: It is enough if you calculate the required number of angles. Here it is given by the first two values of angles. The remaining values will be repetitions of the previous values.

$$\text{Centroid} = \frac{\text{Sum of poles} - \text{Sum of zeros}}{n-m} = \frac{(0 - 2 + j2.64 - 2 - j2.64 - (-9))}{2} = 2.5$$

The centroid is marked and from the centroid the angles of asymptotes are marked using a protractor. The asymptotes are drawn as dotted lines as shown in Fig. 6.12.

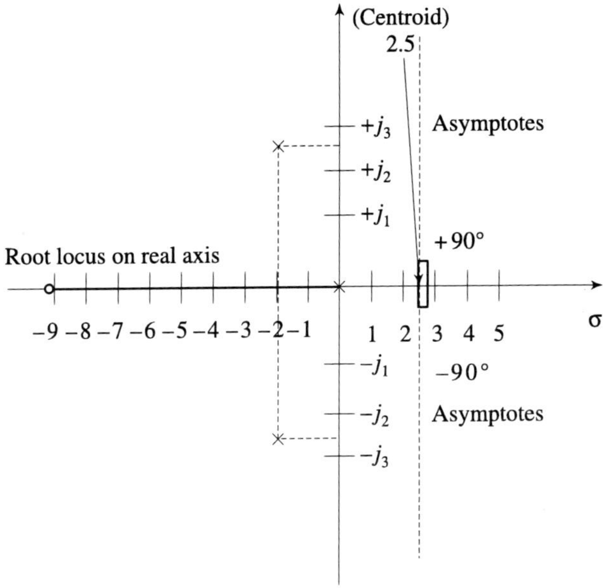

Fig. 6.11 Asymptotes root locus on real axis and location of pole and centroid

Step 4: ***To find the breakaway and break-in points.*** From the location of poles and zero and from the knowledge of typical sketches of root locus, it can be concluded that there is no responsibility of breakaway or break-in points.

Step 5: ***To find the angle of departure.*** Let us consider the complex pole A as shown in Fig. 6.12. Draw vectors from all other poles and zero to the pole A. Let the angles of these vectors be θ_1, θ_2 and θ_3.

Here $$\theta_1 = 180° - \tan^{-1}\frac{2.64}{2} = 127.1°$$

$$\theta_2 = 90°,\ \theta_3 = \tan^{-1}\frac{2.64}{7} = 20.7°$$

Angle of departure from the complex pole $A = 180 - (\theta_1 + \theta_2) + \theta_3$

$$= 180 - (127.1 + 90) + 20.7 = -16.4°$$

Angle of departure at the complex pole A^* is negative of the angle of departure at complex pole A.

$\therefore$ Angle of departure from the complex pole $A^* = -(-16.4) = +16.4°$.

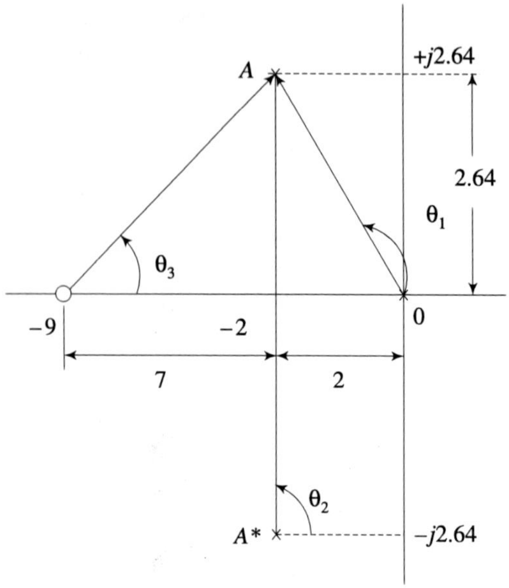

Fig. 6.12

Step 6: ***To find the crossing point on imaginary axis.*** The closed-loop transfer function

$$= \frac{C(s)}{R(s)} = \frac{G(s)}{1+G(s)} = \frac{\dfrac{K(s+9)}{s(s^2+4s+11)}}{1+\dfrac{K(s+9)}{s(s^2+4s+11)}} = \frac{K(s+9)}{s(s^2+4s+11)+K(s+9)}$$

The characteristic equation is the denominator polynomial of $C(s)/R(s)$

∴ The characteristic equation is

$$s(s^2+4s+11)+K(s+9)=0$$

$$(s^3+4s^2+11s)+Ks+9K=0$$

Put $s = j\omega$

$$(j\omega)^3+4(j\omega)^2+11(j\omega)+K(j\omega)+9K=0$$

$$-j\omega^3-4\omega^2+11\omega+jK\omega+9K=0$$

On equating imaginary part to zero

$$-j\omega^3+11j\omega+jK\omega=0$$

$$-j\omega^3=-j11\omega-jK\omega$$

$$\omega^2=11+K$$

Put $K = 8.8$

$$\therefore \quad \omega^2=\pm\sqrt{19.8}=\pm 4.4$$

On equating real part to zero

$$-4\omega^2+9K=0$$

$$4\omega^2=9K$$

Put $\omega^2=11+K$

$$\therefore \quad 9K=4(11+K)$$

$$9K-4K=44;\ 5K=44$$

$$K=44/5=8.8$$

The crossing point of root locus is $\pm j4.4$. The value of K at this crossing point is $K = 8.8$. (This is the limiting value of K for the stability of the system.)

The complete root locus sketch is shown Fig. 6.13. The root locus has three branches. One branch starts at the pole at the origin and travels through negative real axis to meet the zero at $s = -9$.

The other two root locus branches start at complex poles (along the angle of departure) cross the imaginary axis at $\pm j4.4$ and travel parallel to asymptotes to meet the zeros at infinity.

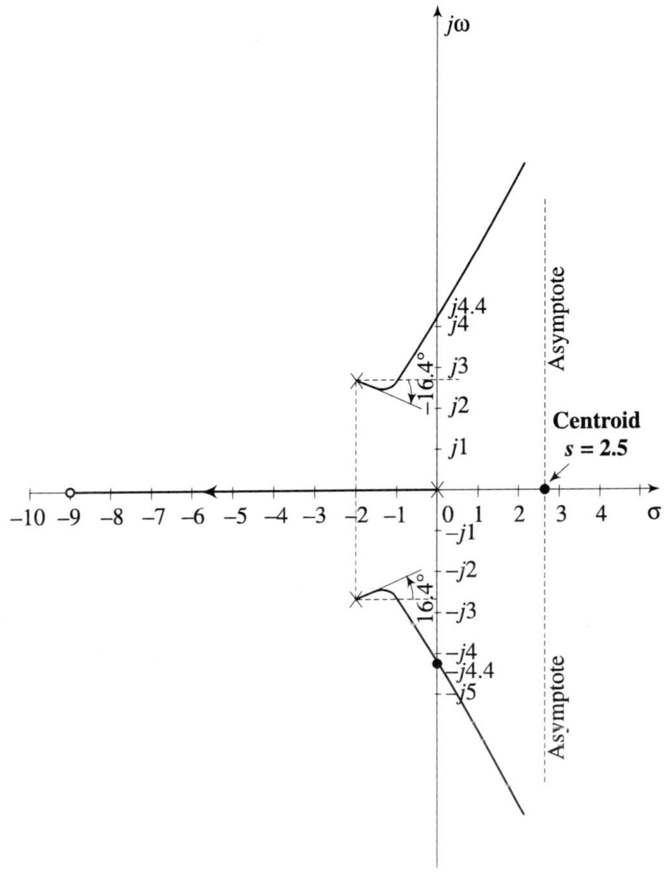

Fig. 6.13 Root locus sketch for $1 + G(s) = 1 + \dfrac{K(s+9)}{s(s^2+4s+11)}$

Example 6.4: Sketch the root locus for the unity feedback system whose loop-transfer function is $G(s)H(s) = \dfrac{K}{s(s+4)(s^2+4s+20)}$.

Solution:

Step 1: ***To locate poles and zeros.*** The poles of open-loop transfer function are the roots of the equation,

$$s(s + 4)(s^2 + 4s + 20) = 0$$

The roots of the quadratic are $s = \dfrac{-4 \pm \sqrt{4^2 - 4 \times 1 \times 20}}{2} = -2 \pm j4$

$\therefore$ The poles are lying at $s = 0, -4, -2 + j4$ and $-2 - j4$

The zeros are lying at infinity.

Step 2: ***To find the root locus on real axis.*** There are two poles on the real axis. Choose a test point on real axis between $s = 0$ and $s = -4$. To the right of this point, the total number of real poles is one which is an odd number. Hence, the real axis between $s = 0$ and $s = -4$ will be a part of root locus. Choose a test point to the left of $s = -4$, now to the right of this test point the total number of poles and zeros is two which is an even number. Hence, the real axis from $s = -4$ to $s = -\infty$ will not be a part of root locus. The root locus on real axis is as shown in Fig. 6.14.

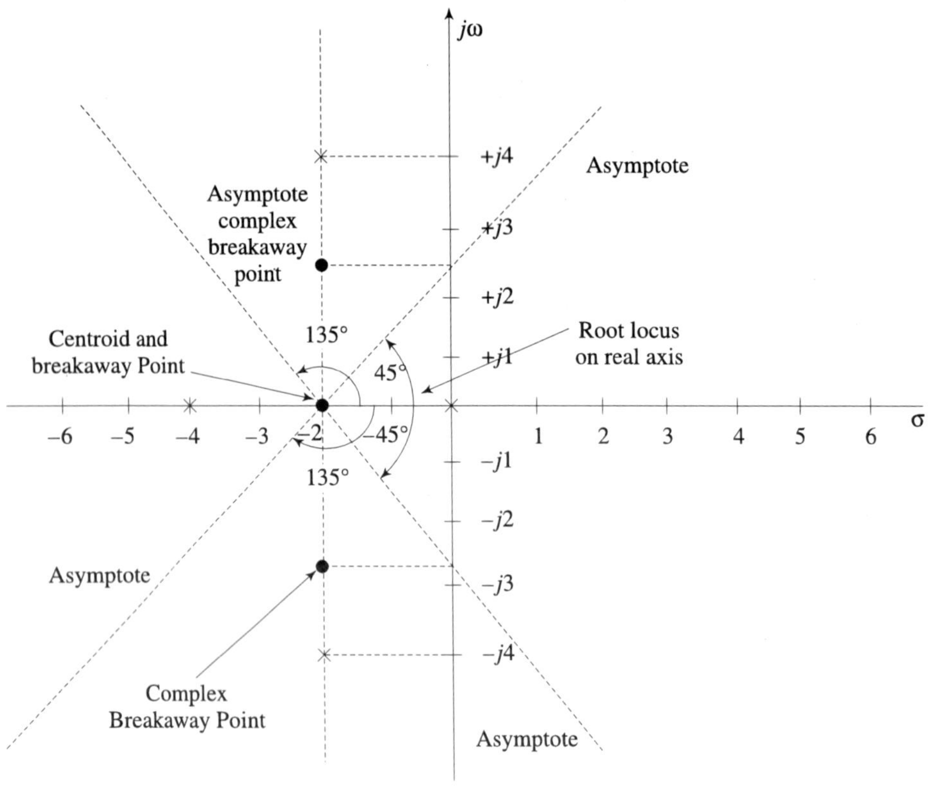

Fig. 6.14 Asymptotes, root locus on real axis and location of poles, centroid and breakaway points

Step 3: ***To find angles of asymptotes and centroid.*** Since there are four poles, the number of root locus branches is four. There is no finite zero. Hence, all the four root locus branches end at zeros at infinity. Hence, the number of asymptotes required is four.

$$\text{Angles of asymptotes, } \varnothing_A = \frac{\pm 180°(2q + 1)}{n - m} \quad q = 0, 1 \ldots\ldots\ldots (n - m)$$

Here $n = 4$ and $m = 0$

If $q = 0$, Angles $= \pm 180° / 4 = \pm 45°$

If $q = 1$, Angles $= \pm 180° \times 3/4 = \pm 135°$

If $q = 2$, Angles $= \pm 180° \times 5/4 = \pm 225° \pm 135°$

If $q = 3$, Angles $= \pm 180° \times 7/4 = \pm 315° \pm 45°$

Note: It is enough if you calculate the required number of angles. Here it is given by the first four values of angles. The remaining values will be repetitions of the previous values.

$$\text{Centroid} = \frac{\text{Sum of poles} - \text{Sum of zeros}}{n - m} = \frac{(0 - 4 - 2 + j4 - 2 - j4 - 0)}{(4 - 0)} = -\frac{8}{4}$$

$$= -2.$$

The centroid is marked on real axis and from the centroid the angles of asymptotes are marked using a protractor. The asymptotes are drawn as dotted lines as shown in Fig. 6.15.

Step 4: ***To find the breakaway and break-in points.*** The closed-loop transfer function $= \dfrac{C(s)}{R(s)} = \dfrac{G(s)}{1 + G(s)}$.

$$\therefore \quad \frac{C(s)}{R(s)} = \frac{\dfrac{K}{s(s + 4)(s^2 + 4s + 20)}}{1 + \dfrac{K}{s(s + 4)(s^2 + 4s + 20)}} = \frac{K}{s(s + 4)(s^2 + 4s + 20) + K}$$

The characteristic equation is $s(s + 4)(s^2 + 4s + 20) + K = 0$

$$\therefore \quad K = -s(s + 4)(s^2 + 4s + 20) = (s^2 + 4s)(s^2 + 4s + 20)$$

$$K = -(s^4 + 8s^3 + 36s^2 + 80s)$$

On differentiating the equation of K with respect to s we get,

$$\frac{dK}{ds} = -(4s^3 + 24s^2 + 72s + 80)$$

Put $$\frac{dK}{ds} = 0$$

$$\therefore \quad (-4s^3 + 24s^2 + 72s + 80) = 0$$

$$(4s^3 + 24s^2 + 72s + 80) = 0$$

On dividing by 4 we get $s^3 + 6s^2 + 18s + 20 = 0$

The equation $s^3 + 6s^2 + 18s + 20 = 0$ will have at least one real root. By trial and error, the real root is found to be $s = -2$.

To find the real root of $\frac{dK}{ds} = 0$.

The polynomial $s^3 + 6s^2 + 18s + 20 = 0$ can be expressed as

$$s^3 + 6s^2 + 18s + 20 = (s + 2)(s^2 + 4s + 10) = 0$$

The root of the quadratic $(s^2 + 4s + 10) = 0$ is given by

$$s = \frac{-4 \pm \sqrt{4^2 - 4 \times 10}}{2} = -2 \pm j2.45$$

Check for K,

When $s = -2$, the value of K is given by $K = -(s^4 + 8s^3 + 36s^2 + 80s)$

$$= -[(-2)^4 + 8(-2)^3 + 36(-2)^2 + 80(-2)]$$

$$= -[-64] = 64$$

When, $s = -2 \pm j2.45 = 3.16\angle\pm129°$,

$$K = -[s^4 + 8s^3 + 36s^2 + 80s]$$

$$= -3.16\angle\pm129°)^4 + 8(3.16\angle\pm129°)^3 + 36(3.16\angle\pm129°)^2 + 80(3.16\angle\pm129°)]$$

$$= -[99.7\angle\pm156° + 252.4\angle\pm27° + 359.5\angle\pm258° + 252.8\angle\pm129°]$$

For positive values of angles,

$$K = -[-91 + j40 + 225 + j115 - 75 - j351 - 159 + j196]$$

$$= -[-100] = 100$$

For negative values of angles,

$$K = -[-91 - j40 + 225 + j115 - 75 - j351 - 159 + j196]$$

$$= -[-100] = 100$$

For all the roots of the equation $dK/ds = 0$, the value of K is positive and real. Hence, all the three roots are actual breakaway points. The breakaway points are shown in Fig. 6.15.

Step 5: ***To find the angle of departure.*** Let us consider the complex pole A shown in Fig. 6.15.

Draw vectors from all other poles to the pole A as shown in Fig. 6.15. Let angles of these vectors be θ_1, θ_2 and θ_3.

$$\theta_1 = 180° - \tan^{-1}\frac{4}{2} = 117°$$

$$\theta_2 = 90°$$

$$\theta_3 = \tan^{-1}\frac{4}{2} = 63°$$

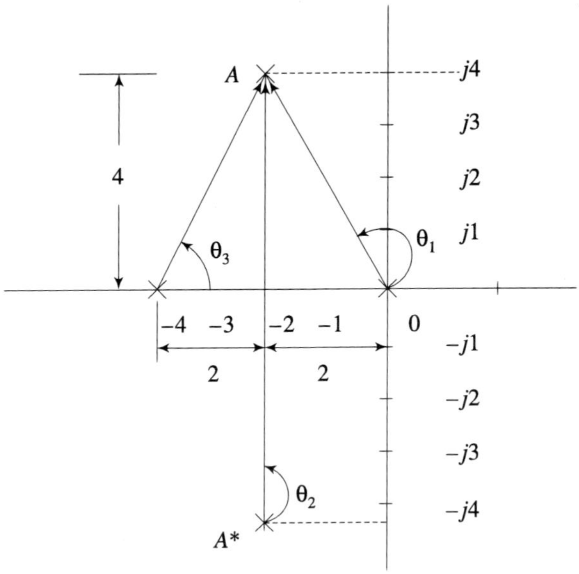

Fig. 6.15

Angle of departure from complex pole,

$$A = 180 - (\theta_1 + \theta_2 + \theta_3) = 180 - (117 + 90 + 63) = -90°$$

The angle of departure at complex pole

A^* is negative of the angle of departure at complex pole A.

∴ The angle of departure from complex pole $A^* = +90°$

Step 6: *To find the crossing point on imaginary axis.* The characteristic equation is given by

$$s^4 + 8s^3 + 36s^2 + 80s + K = 0$$

Put $s = j\omega$

$$(j\omega)^4 + 8(j\omega)^3 + 36(j\omega)^2 + 80(j\omega) + K = 0$$

$$\omega^4 - j8\omega^3 - 36\omega^2 + 80j\omega + K = 0$$

On equating imaginary part to zero

$$-j8\omega^3 + 80j\omega = 0$$

$$-j8\omega^3 = -j80\omega;\ \omega^2 = 10$$

$$\omega = \pm\sqrt{10} = \pm 3.2$$

On equating real part to zero

$$\omega^4 - 36\,\omega^2 + K = 0$$

$$K = -\,\omega^4 + 36\,\omega^2$$

Put $\omega^2 = 10$ $K = -\,(10)^2 + (36 \times 10) = 260$

The crossing point of root locus is $\pm j3.2$. The value of K at this crossing point is $K = 260$. (This is the limiting value of K for the stability of the system.)

The complete root locus sketch is shown Fig. 6.16. The root locus has four branches. All the root locus branches go to infinity along the asymptotes to meet the zeros at infinity.

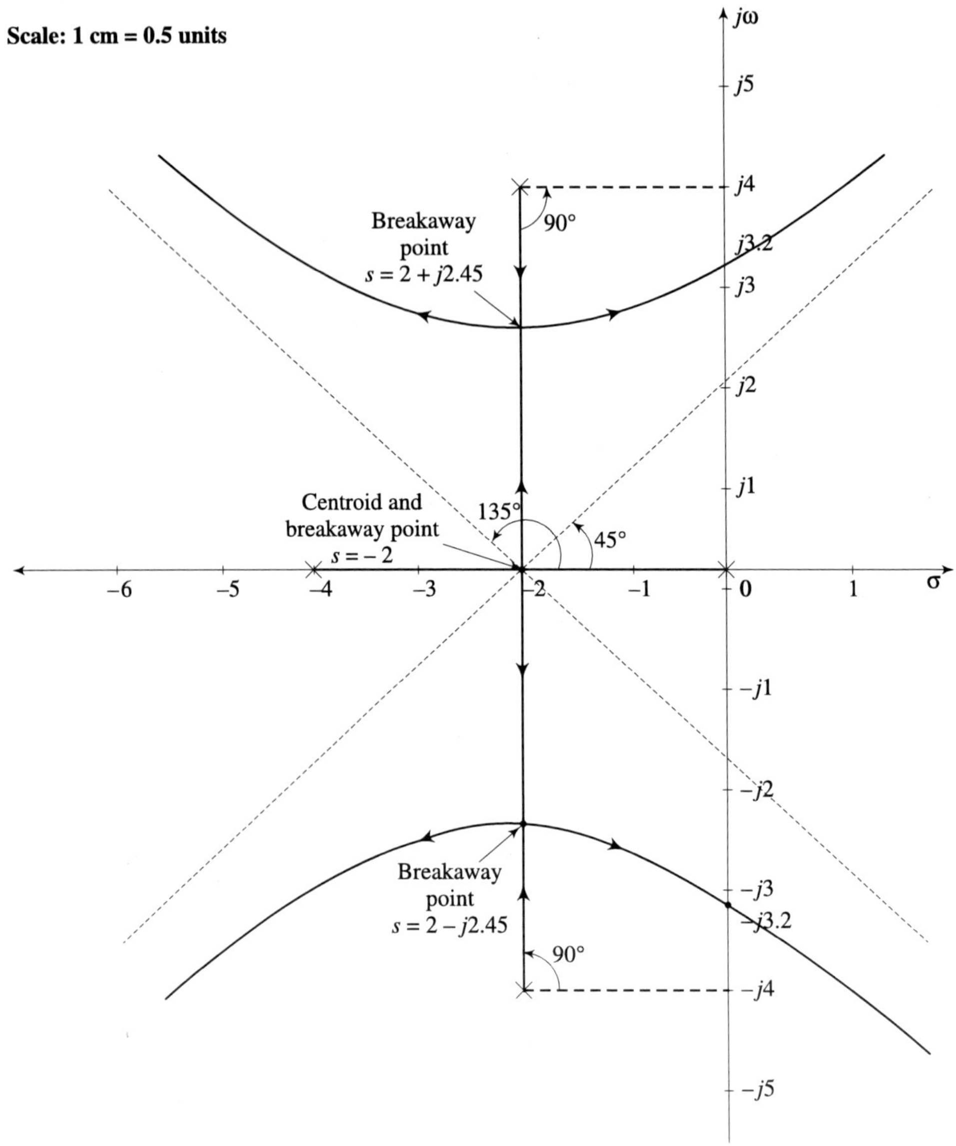

Fig. 6.16 Root locus sketch for $1 + G(s) = 1 + \dfrac{K}{s(s + 4)(s^2 + 4s + 20)}$

Example 6.5: Sketch the root locus for the infinity feedback system whose open-loop transfer function is

$$G(s)H(s) = \frac{K(s + 1.5)}{s(s + 1)\ (s + 5)}.$$

Solution:

Step 1: ***To locate poles and zeros.*** The poles of open-loop transfer function are the roots of the equation, $s(s + 1)(s + 5) = 0$ and the zeros are the roots of the equation $(s + 1.5) = 0$.

The poles are lying at $s = 0, -1, -5$.

The zeros are lying at $s = -1.5$ and infinity.

Step 2: ***To find the root locus on real axis.*** The segment of real axis between $s = 0$ and $s = -1$ and the segment of real axis between $s = -1.5$ and $s = -5$ will be part of root locus. Because if we choose a test point in the segment then to the right of this point we have odd number of real poles and zeros. The root locus on real axis are as shown in Fig. 6.17.

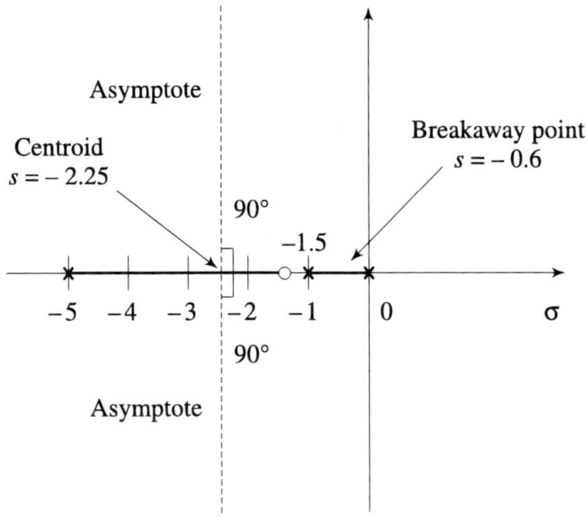

Fig. 6.17 The asymptotes, root locus on real axis and location of poles, zero, centroid and breakaway points

Step 3: ***To find angles of asymptotes and centroid.*** Since there are three poles, the numbers of root locus branches are three. There is one finite zero, so one root locus branch will end at finite zero. The other two branches will meet the zeros at infinity. Hence, the number of asymptotes required is two.

$$\text{Angles of asymptotes, } = \varnothing_A = \frac{\pm 180°(2q + 1)}{n - m} \quad q = 0, 1 \;\dots\;\dots\;\dots\; (n - m)$$

Here $\qquad n = 3$ and $m = 1$

If $\qquad q = 0, \text{ Angles} = \dfrac{\pm 180°}{2} = \pm 90°$

$$\text{Centroid} = \frac{\text{Sum of poles} - \text{Sum of zeros}}{n - m} = \frac{0 - 1 - 5 - (-15)}{2} = -2.25$$

The centroid is marked on real axis and from the centroid the angles of asymptotes are marked using a protractor.

Step 4: ***To find the breakaway and break-in points.*** The closed loop-transfer function $= \frac{C(s)}{R(s)} = \frac{G(s)}{1 + G(s)}$.

$$\therefore \qquad \frac{C(s)}{R(s)} = \frac{\dfrac{K(s + 1.5)}{s(s + 1)(s + 5)}}{1 + \dfrac{K(s + 1.5)}{s(s + 1)(s + 5)}} = \frac{K(s + 1.5)}{s(s + 1)(s + 5)) + K(s + 1.5)}$$

The characteristic equation is $s(s + 1)(s + 5) + K(s + 1.5) = 0$

$$\therefore \qquad K = -\frac{-s(s + 1)(s + 5)}{s + 1.5} = \frac{-s(s^2 + 6s + 5)}{s + 1.5} = \frac{-(s^3 + 6s^2 + 5s)}{s + 1.5}$$

On differentiating the equation of K with respect to s, we get

$$\frac{dK}{ds} = \frac{-(3s^2 + 12s + 5)(s + 1.5) - [-(s^3 + 6s^2 + 5s)](1)}{(s + 1.5)^2}$$

$$= \frac{(-3s^3 - 4.5s^2 - 12s^2 - 18s - 5s - 7.5 + s^3 + 6s^2 + 5s)}{(s + 1.5)^2}$$

$$= \frac{(-2s^3 - 10.5s^2 - 18s - 7.5)}{(s + 1.5)^2}$$

$$= \frac{(-2(s^3 + 5.25s^2 + 9s + 3.75))}{(s + 1.5)^2}$$

Put $\qquad \frac{dK}{ds} = 0$

$$\therefore \qquad s^3 + 5.25s^2 + 9s + 3.75) = 0$$

The third order polynomials have one real root. The real root of the above polynomial can be determined by Lin's method.

The polynomial can be expressed $s^3 + 5.25s^2 + 9s + 3.75 = (s + 0.6)(s^2 + 4.56s + 6.2) = 0$

The roots of the quadratic $(s^2 + 4.56s + 6.2)$ are

$$s = \frac{-4.65 \pm \sqrt{4.65^2 - 4 \times 6.2}}{2} = -2.3 \pm j0.89$$

Check for K

When $\qquad s = -0.6,\ K = \frac{-(s^3 + 6s^2 + 5s)}{s + 1.5} = \frac{-[(-0.6)^3 + 6(-0.6)^2 + 5(-0.6)]}{(-0.6 + 1.5)} = 1.17.$

For $s = -0.6$, the value of K is positive and real and so it is the actual breakaway point. It can be shown that for $s = -2.3 \pm j0.86$ the value of K is not positive and real and so they cannot breakaway points. The actual breakaway point is as shown in Fig. 6.18.

Step 5: ***To find the angle of departure.*** Since there are no complex poles or zeros we need not find angle of departure or arrival.

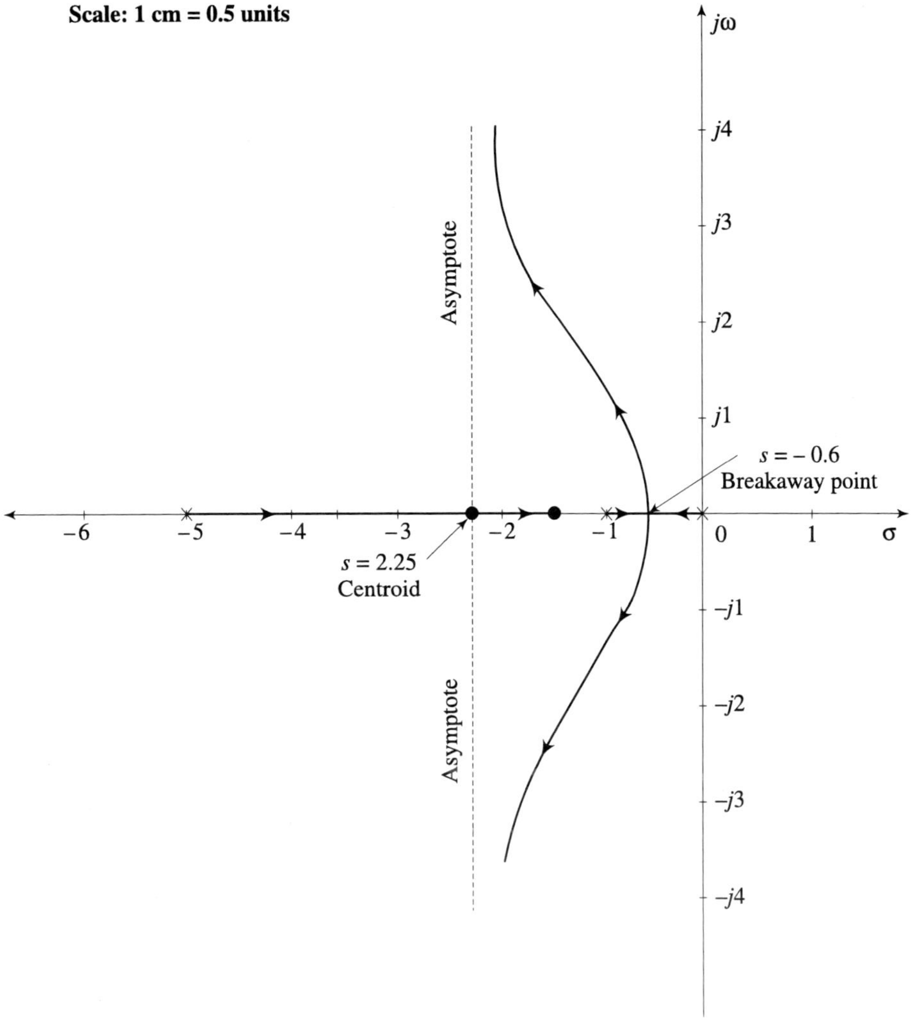

Fig. 6.18 Root locus sketch for $1 + G(s) = 1 + \dfrac{K(s + 1.5)}{s(s + 1)\,(s + 5)}$

Step 6: ***To find the crossing point on imaginary axis.*** The characteristic equation is given by

$$s(s + 1)(s + 5) + K(s + 1.5) = 0$$

$$s(s^2 + 6s + 5) + Ks + 1.5K = 0$$

$$s^3 + 6s^2 + 5s + Ks + 1.5K = 0$$

Put $s = j\omega$

$$(j\omega)^3 + 6(j\omega)^2 + 5(j\omega) + K(j\omega) + 1.5K = 0$$

$$-j\omega^3 - 6\omega^2 + 5j\omega + jK\omega + 1.5K = 0$$

On equating imaginary part to zero

$$-j\omega^3 + j5\omega + jK\omega = 0$$

$$-j\omega^3 = -j5\omega - jK\omega$$

$$\omega^2 = 5 + K$$

On equating real part to zero

$$-6\omega^2 + 15K = 0$$

Put $\omega^2 = 5 + K$

$$-6(5 + K) + 1.5K = 0$$

$$-30 - 4.5K = 0$$

$$-4.5K = 30$$

$$K = -30/4.5 = -6.67.$$

Since the value of K is negative, there is no crossing point on imaginary axis, or for any positive values of K the root locus will not cross imaginary axis. The complete root locus sketch is shown in [Fig. 6.18. The root locus has three branches. One branch starts at $s = -5$ and ends at finite zero at $s = -1.5$. The other two root loci start at $s = 0$ and $s = -1$ and breakaway from axis at $s = -0.6$, then travel parallel to asymptotes to meet the zeros at infinity.

Example 6.6: Sketch the root locus for the unity feedback system whose open-loop transfer function is

$$G(s) = \frac{K(s^2 + 6s + 25)}{s(s + 1)(s + 2)}.$$

Solution:

Step 1: ***To locate poles and zeros.*** The poles of open-loop transfer function are the roots of the equations $(s + 1)(s + 2) = 0$ and the zeros are the roots of the equation $(s^2 + 6s + 25) = 0$.

The roots of quadratic are $s = \dfrac{-6 \pm \sqrt{6^2 - 4 \times 25}}{2} = -3 \pm j4$

The poles are lying at $s = 0, -1, -2$.

The zeros are lying at $s = -3 + j4, -3 - j4$.

Step 2: ***To find the root locus on real axis.*** The segment of real axis between $s = 0$ and $s = -1$ and the entire negative real axis from $s = -2$ will be a part of root locus. Because if we choose a test point in this segment then to the right of this point we have odd number of real poles and zeros. The root locus on real axis is shown in Fig. 6.19.

Step 3: ***To find asymptotes and centroid.*** Since there are three poles the number of root locus branches is three. There are two finite zeros, so two root locus branches will end at finite zeros. The third root locus will meet the zeros at infinity by travelling through negative real axis. Here the number of asymptotes is one and the angle of asymptotes is $\pm 180°$.

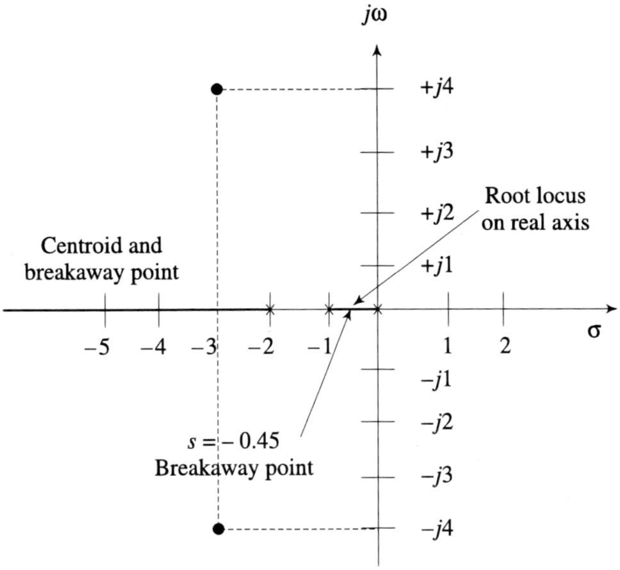

Fig. 6.19 The root locus on real axis and location of poles, zeros and breakaway points

Step 4: ***To find the breakaway and break-in points.*** The closed-loop transfer function $= \dfrac{C(s)}{R(s)} = \dfrac{G(s)}{1 + G(s)}$.

$$\therefore \qquad \frac{C(s)}{R(s)} = \frac{\dfrac{K(s^2 + 6s + 25)}{s(s + 1)(s + 2)}}{1 + K\dfrac{(s^2 + 6s + 25)}{s(s + 1)(s + 2)}}$$

The characteristic equation is $s(s + 1)(s + 2) + K(s^2 + 6s + 25) = 0$

$$\therefore \qquad K = \frac{-s(s + 1)(s + 2)}{(s^2 + 6s + 25)} = \frac{-s(s^2 + 3s + 2)}{s^2 + 6s + 25} = \frac{-s^3 - 3s^2 - 2s}{s^2 + 6s + 25}$$

On differentiating the equation of K with respect to s, we get

$$\frac{dK}{ds} = \frac{-(3s^2 - 6s - 2)(s^2 + 6s + 25) - (-s^3 - 3s^2 - 2s)(2s + 6)}{(s^2 + 6s + 25)^2}$$

$$= \frac{-(s^4 + 12s^3 + 91s^2 + 150s + 50)}{(s^2 + 6s + 25)^2}$$

Put $\dfrac{dK}{ds} = 0$, the numerator should be zero.

$$\therefore \qquad s^4 + 12s^3 + 91s^3 + 150s + 50 = 0$$

The fourth order polynomial can be split into two quadratic equations. The two quadratic factors can be obtained by Lin's method.

The roots of the quadratic $(s^2 + 2s + 0.7)$ are

$$s = \frac{-2 \pm \sqrt{2^2 - 4 \times 0.7}}{2} = -0.45, -1.55$$

The roots of the quadratic $(s^2 + 10s + 70.3) = 0$ are

$$s = \frac{-10 \pm \sqrt{10^2 - 4 \times 70.3}}{2} = -5 + j6.73$$

Here, $s = -1.55$ is not a point on root locus, hence it cannot be a breakaway point. Check the other three values for actual breakaway point.

$$\text{When} \quad s = -0.45,\ K = \frac{-s^3 - 3s^2 - 2s}{s^2 + 6s + 25} = \frac{-(-0.45)^3 - 3(-0.45)^2 - 2(-0.45)}{(-0.45)^2 + 6(-0.45) + 25} = 0.017$$

For $s = -0.45$, the value of K is positive and real and so it is the actual breakaway point. It can be shown that for $s = -5 \pm j6.73$ the value of K is not positive and real and so cannot be breakway points. The actual breakaway point is as shown in Fig. 6.19.

Step 5: ***To find the angle of arrival.*** Let us consider the complex zero A shown in Fig. 6.20. Draw vectors from all other poles to the pole A as shown in Fig. 6.20. Let the angles of these vectors be $\theta_{1,}\theta_{2,}$ and θ_3 and θ_4

Here
$$\theta_1 = 180° - \tan^{-1}\frac{4}{3} = 126.9°$$

$$\theta_2 = 180° - \tan^{-1}\frac{4}{2} = 116.6°$$

$$\theta_3 = 180° - \tan^{-1}\frac{4}{1} = 104°$$

$$\theta_4 = 90°$$

Angle of arrival at complex zero

$$\begin{aligned} A &= 180 - (\theta 4) + (\theta_1 + \theta_2 + \theta_3) \\ &= 180 - 90 + 126.9 + 116.6 + 104 \\ &= 437.5° = 77.5° \end{aligned}$$

The angle of arrival at complex zero A^* is negative of the angle of arrival at complex zero A.

$\therefore$ The angle of arrival at complex zero $A^* = -77.5°$

Mark the angles of arrival at complex zeros using protractor.

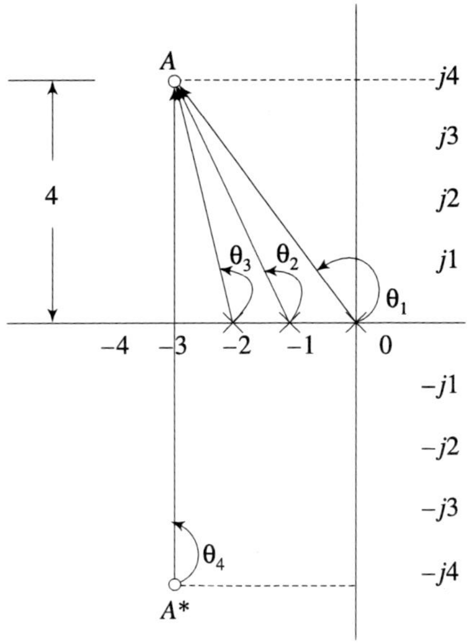

Fig. 6.20

Step 6: ***To find the crossing point on imaginary axis.*** The characteristic equation is given by

$$s(s + 1)(s + 2) + K(s^2 + 6s + 25) = 0$$

$$s(s^2 + 3s + 2) + Ks^2 + 6Ks + 25K = 0$$

$$s^3 + 3s^2 + 2s + Ks^2 + 6Ks + 25K = 0$$

$$s^3 + (3 + K)s^2 + (2 + 6K)s + 25K = 0$$

Put $s = j\omega$

$$(j\omega)^3 + (3 + K)(j\omega)^2 + (2 + 6K)j\omega + 25K = 0$$

$$-j\omega^3 - (3 + K)\omega^2 + j(2 + 6K)\omega + 25K = 0$$

On equating imaginary part to zero

$$-j\omega^3 + j(2 + 6K)\omega = 0$$

$$-j\omega^3 = -j(2 + 6K)\omega$$

$$\omega^2 = (2 + 6K)$$

On equating real part to zero

$$-(3 + K)\omega^2 + 25K = 0$$

Put $\omega^2 = 2 + 6K$

$$-(3 + K)(2 + 6K) + 25K = 0$$

$$-(6 + 18K + 2K + 6K^2) + 25K = 0$$

$$-6K^2 + 5K - 6 = 0$$

$$K = \frac{-5 \pm \sqrt{5^2 - 4\times(-6)(-6)}}{2(-6)} = 0.4 \pm j0.9$$

Since the value of K is not real and positive, there is no crossing point on imaginary axis, or for any positive values of K the root locus will not cross imaginary axis.

Step 7: *To find points on root locus.* Choose test points a, b, c, d on the S-plane and adjust the test points to satisfy angle criterion. The test points are shown in Fig. 6.21.

On the upper half of S-plane the root locus is sketched through the test points a, b, c and d. The root locus on the lower half of the plane is the mirror image of the root locus on the upper half of S-plane.

The complete root locus sketch is as shown in Fig. 6.21. The root locus has three branches. One branch starts at $s = -2$ and goes to infinity along negative real axis. The other two root locus branches start at $s = 0$ and $s = -1$ and break from real axis at $s = -0.45$, then meet the complex zeros.

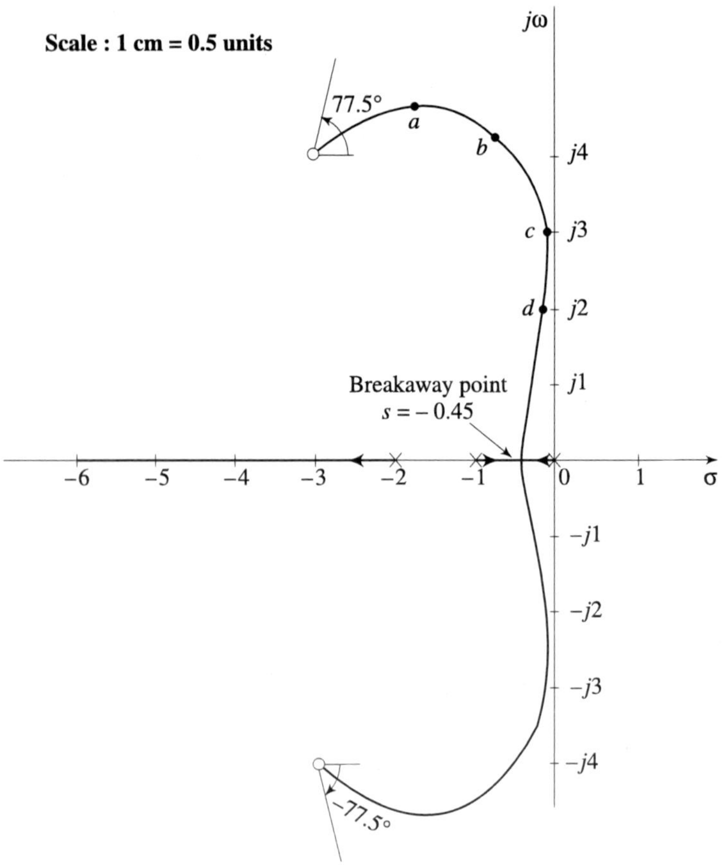

Fig. 6.21 Root locus sketch for $1 + G(s) = 1 + \dfrac{K(s^2 + 6s + 25)}{s(s+1)(s+2)}$

6.5 EFFECTS OF ADDING POLES AND ZEROS ON THE ROOT LOCI IN TERMS OF STABILITY

6.5.1 Effect of Addition of Poles

The effects of addition of poles are as follows:

- There is change in shape of the root locus and it shifts towards the imaginary axis.
- The intercept on the $j\omega$ axis occurs for a lower value of K because of asymptote's angle being lower down.
- System becomes oscillatory.
- Gain margin and relative stability decrease.
- There is reduction in the range of K.
- Settling time increases.

6.5.2 Effects of Addition of Zeros

The effects of addition of zeros are as follows.

- There is change in shape of the root locus and it shifts towards the left of the S-plane.
- Stability of the system is enhanced.
- Range K increases.
- Settling time speeds up.

Example 6.7: A unity feedback system has an open-loop transfer function

$$G(s) = \frac{K(S+1)}{S(S-1)}$$

Sketch the root locus plot with K as a variable parameter and show that the loci of complex roots are part of a circle with (– 1, 0) as centre and radius $\sqrt{2}$. Is the system stable for all values of K? If not, determine the range of K for stable system operation. Find also the marginal value of K, which causes sustained oscillations and the frequency of their oscillations. From the root locus plot, determine the value of K such that the resulting system has a settling time of 4 seconds. What are the corresponding values of the roots?

Solution: $G(s)\,H(s) = \dfrac{K(S+1)}{S(S-1)}$

The open-loop poles are at $S = 0$, $S = 1$ $\quad \therefore n = 2$

Open-loop zero is at $S = -1$ $\quad \therefore m = 1$

The number of asymptotes $= n - m = 2 - 1 = 1$

Root locus starts at the open-loop poles, $S = 0$, $S = 1$ and terminate on the open-loop zeros at $S = -1$, $S = \infty$

The angle of asymptote, $\theta_A = \dfrac{(2q+1)\times 180^\circ}{n-m}$

$$q = 0, \quad \theta_A = \frac{(2 \times 0 - 1)}{2 - 1} \times 180° = 180°$$

The centroid is given by

$$\sigma = \frac{\text{Sum of poles} - \text{Sum of zero}}{n - m}$$

$$\sigma = \frac{(0 + 1) - (-1)}{2 - 1} = 2$$

Mark the centroid on the real axis

Breakaway and break-in points are obtained by solving the equation $\frac{dK}{dS} = 0$

The characteristic equation is

$$1 + G(s)\,H(s) = 0$$

$$1 + \frac{K(S + 1)}{S(S - 1)} = 0$$

$$K = -\frac{S(S - 1)}{S + 1}$$

$$\frac{dK}{dS} = \frac{d}{dS}\left[-\frac{S(S - 1)}{S + 1}\right] = \frac{(S + 1)(-2S + 1) - S(S - 1)}{(S + 1)^2}$$

$$\frac{dK}{dS} = \frac{-(S^2 + 2S - 1)}{(S + 1)^2}$$

$$\frac{dK}{DS} = 0 = \frac{-(S^2 + 2S - 1)}{(S + 1)^2}$$

$$S^2 + 2S - 1 = 0$$

$$\therefore \qquad S = \frac{-2 \pm \sqrt{4 + 4}}{2} = 0.414 \quad \text{and} \quad -2.414$$

Hence, $S = 0.414$ is the breakaway point and $S = -2.414$ is the break-in point.

Since the open-loop transfer function has no complex poles or complex zeros, there is no need to calculate the angle of departure and the angle of arrival. The point of intersection of the root locus, with the imaginary axis and the critical value of K are determined by using Routh-Hurwitz criterion.

The characteristic equation is

$$1 + G(s)\,H(s) = 0$$

$$1 + \frac{K(S + 1)}{S - 1} = 0$$

$$S^2 + (K - 1)\,S + K = 0$$

The Routh array is

$$\begin{array}{lll} S^2 & : & 1 \quad K \\ S^1 & : & (K-1) \\ S^0 & : & K \end{array}$$

For stability, all the elements, in the first column of the Routh array must be positive

$$\therefore \quad K > 0$$

$$(K - 1) > 0$$

$$K > 1$$

The system is not stable for all values of K. The marginal value of K for stable system $K = 1$. The frequency of sustained oscillations is given by the solution of the auxiliary equation

$$S^2 + K = 0$$

$$S^2 + 1 = 0$$

$$S = \pm j1$$

$\therefore$ The root locus intersects the imaginary axis at $S = \pm j1$

The frequency of sustained oscillation is $\omega = 1$ rad/sec

The complete root locus

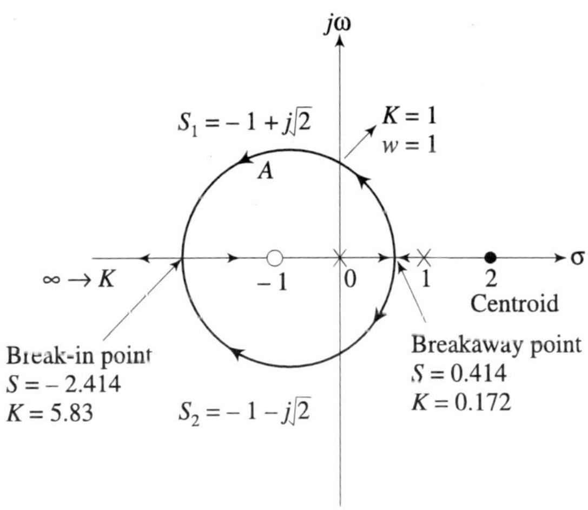

Fig. 6.22

Example 6.8: To show that the loci of complex roots are part of a circle

$$G(s)\,H(s) = \frac{K(S + 1)}{S(S - 1)}$$

Solution:

$$G(s)\,H(s)\Big|_{S=\sigma+j\omega} = \frac{K(\sigma+j\omega+1)}{(\sigma+j\omega)(\sigma+j\omega-1)}$$

$$\angle G(s)\,H(s)\Big|_{S=\sigma+j\omega} = 180^\circ$$

$$\angle \frac{K(\sigma+j\omega+1)}{(\sigma+j\omega)(\sigma+j\omega-1)} = 180^\circ$$

$$\angle K + \angle \sigma+j\omega+1 - \angle \sigma+j\omega - \angle \sigma+j\omega-1 = 180^\circ$$

$$0^\circ + \tan^{-1}\frac{\omega}{\sigma+1} - \tan^{-1}\frac{\omega}{\sigma} - \tan^{-1}\frac{\omega}{\sigma-1} = 180^\circ$$

$$\tan^{-1}\frac{\omega}{\sigma+1} - \tan^{-1}\frac{\omega}{\sigma} = 180 + \tan^{-1}\frac{\omega}{\sigma+1}$$

Taking the tangent on both sides of this equation

$$\tan\left[\tan^{-1}\frac{\omega}{\sigma+1} - \tan^{-1}\frac{\omega}{\sigma}\right] = \tan\left[180 + \tan^{-1}\frac{\omega}{\sigma-1}\right]$$

Using the relation

$$\tan(\alpha-\beta) = \frac{\tan\alpha - \tan\beta}{1+\tan\alpha\tan\beta}$$

$$\frac{\dfrac{\omega}{\sigma+1} - \dfrac{\omega}{\sigma}}{1+\dfrac{\omega}{\sigma+1}\cdot\dfrac{\omega}{\sigma}} = \frac{\omega}{\sigma-1}$$

$$\frac{\omega(\sigma-\sigma-1)}{\sigma^2+\sigma+\omega^2} = \frac{\omega}{\sigma-1}$$

$$1-\sigma = \sigma^2+\sigma+\omega^2$$

$$\sigma^2+2\sigma+\omega^2 = 1$$

$$\sigma^2+2\sigma-1+\omega^2+1-1 = 0$$

$$(\sigma+1)^2+(\omega-0)^2 = (\sqrt{2})^2$$

This equation represents a circle with centre at (– 1, 0) and radius $\sqrt{2}$

The value of K for settling time of 4 seconds

$$t_s = \frac{4}{\xi\omega_n}$$

$$4 = \frac{4}{\xi\omega_n}$$

$$\xi\omega_n = 1$$

The point has been marked on the root locus. The coordinates of A are

$$S = -1 \pm j\sqrt{2}$$

The value of K at this point is given by

$$[KG(s)H(s)]\big|_{s=-1\pm\sqrt{2}} = -1$$

$$\frac{K(-1 \pm j\sqrt{2} + 1)}{(-1 \pm j\sqrt{2})(-1 \pm j\sqrt{2} - 1)} = -1$$

$$\frac{K(\pm j\sqrt{2})}{(-1 \pm j\sqrt{2})\,(-2 \pm \sqrt{2})} = -1$$

Considering the positive value, we get

$$\frac{K(+j\sqrt{2})}{(-1 + j\sqrt{2})\,(-2 + j\sqrt{2})} = -1$$

$$K = \frac{(-1 + j\sqrt{2})(-2 + j\sqrt{2})}{j\sqrt{2}} = 3$$

$$K = 3$$

Example 6.9: For a unity feedback system, the open-loop transfer function is given by $G(s) = \dfrac{K(S+3)}{S(S+2)}$.

Show that a part of the root locus of the system is a circle having centre (–3, 0) and radius $\sqrt{3}$.

Solution: For all points on the root locus, the angle condition should be satisfied

$$G(s)\,H(s) = \frac{K(S+3)}{S(S+2)}$$

Put

$$S = \sigma + j\omega$$

$$G(s)H(s)\big|_{s=\sigma+j\omega} = \frac{K(\sigma + j\omega + 3)}{(\sigma + j\omega)(\sigma + j\omega + 2)}$$

$$\underline{|G(s)H(s)}\big|_{s=\sigma+j\omega} = 180^\circ$$

$$\underline{\left|\frac{K(\sigma + j\omega + 3)}{(\sigma + j\omega)(s + j\omega + 2)}\right.} = 180^\circ$$

$$\underline{|K} + \underline{|\sigma + j\omega + 3} - \underline{|(\sigma + j\omega)} - \underline{|\sigma + j\omega + 2} = 180^\circ$$

$$0^\circ + \tan^{-1}\frac{\omega}{\sigma + 3} - \tan^{-1}\frac{\omega}{\sigma} - \tan^{-1}\frac{\omega}{\sigma + 2} = 180^\circ$$

$$\tan^{-1}\left(\frac{\omega}{\sigma+3}\right) - \tan^{-1}\left(\frac{\omega}{\sigma}\right) = 180 + \tan^{-1}\left(\frac{\omega}{\sigma+2}\right)$$

Taking the tangent on both sides of this equation

$$\tan\left[\tan^{-1}\frac{\omega}{\sigma+3} - \tan^{-1}\frac{\omega}{\sigma}\right] = \tan\left[180 + \tan^{-1}\frac{\omega}{\sigma+2}\right] = \frac{\omega}{\sigma+2}$$

Using the relation

$$\tan(\alpha - \beta) = \frac{\tan\alpha - \tan\beta}{1 + \tan\alpha\tan\beta}$$

$$\frac{\dfrac{\omega}{\sigma+3} - \dfrac{\omega}{\sigma}}{1 + \dfrac{\omega}{\sigma+3}\dfrac{\omega}{\sigma}} = \frac{\omega}{\sigma+2}$$

$$\frac{\omega(\sigma - \sigma - 3)}{\sigma^2 + 3\sigma + \omega^2} = \frac{\omega}{\sigma+2}$$

$$\sigma^2 + 6\sigma + \omega^2 + 6 = 0$$

$$(\sigma + 3)^2 + \omega^2 = (\sqrt{3})^2$$

This is the equation of a circle with centre at (–3, 0) and radius $\sqrt{3}$.

Example 6.10: Sketch the root loci of open-loop transfer function of the feedback control system given as

$$G(s)\,H(s) = \frac{K(S+3)}{S(S+2)}.$$

Solution: $G(s)H(s) = \dfrac{K(S+3)}{S(S+2)}$

Open-loop poles are at $S = 0$ and $S = -2$ $\therefore n = 2$

Open-loop zero is at $S = -3$, $\therefore m = 1$

No. of branches of root locus, $n = 2$

Number of asymptotes, $n - m = 2 - 1 = 1$

The angle of asymptote is given by

$$\theta_A = \frac{(2q+1)\,180^\circ}{n-m}, \quad q = 0$$

$$\theta_A = \frac{180}{2-1} = 180^\circ$$

The centroid is given by

$$\sigma_A = \frac{\text{Sum of poles} - \text{Sum of zeros}}{\text{Number of poles} - \text{Number of zeros}}$$

$$\sigma_A = \frac{(-2)-(-3)}{2-1} = 1$$ →It is the point of intersection of the asymptote on the real axis.

The breakaway points are found by solving the equation

$$\frac{dK}{dS} = 0$$

$$|G(s)\ H(s)| = \left|\frac{K(S+3)}{S(S+2)}\right| = 1$$

$$\therefore \quad K = -\frac{S(S+2)}{S+3}$$

$$\frac{dK}{dS} = \frac{d}{ds}\left[\frac{S(S+2)}{S+3}\right] = 0$$

$$(S+3)\ (2S+2) - S(S+2) = 0$$

$$2S^2 + 2S + 6S + 6 - S^2 - 2S = 0$$

$$S^2 + 6S + 6 = 0$$

Breakaway points are at

$$S = \frac{-6 \pm \sqrt{36-24}}{2} = -3 \pm \sqrt{3}$$

$$S_1 = -3 + \sqrt{3} = -1.268$$

$$S_2 = -3 - \sqrt{3} = -4.732$$

There is no need to find the angles of departure and arrival as there are no complex poles or complex zeros. The root locus does not cross the imaginary axis, therefore, there is no need to apply the Routh criterion.

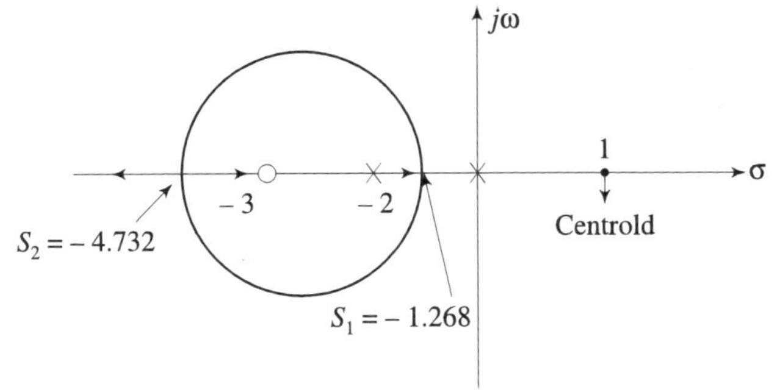

Fig. 6.23

Example 6.11: For a unity feedback system the open-loop transfer function is given by

$$G(S) = \frac{K}{S(S+2)(S^2+6S+2s)}$$

(a) Sketch the root locus for $0 \le K \le \infty$.
(b) At what value of K the system becomes unstable?
(c) At this point of instability, determine the frequency of oscillation of the system.

Solution: $G(s)\,H(s) = \dfrac{K}{S(S+2)(S^2+6S+25)}$

The open-loop poles are at

$$S_1 = 0,\ S_2 = -2$$

S_3 and S_4 are obtained by solving $S^2 + 6S + 25 = 0$

$$S = \frac{-6 \pm \sqrt{36-100}}{2} = -3 + j4$$

$$S_3 = -3 + j4$$

$$S_4 = -3 - j4$$

Number of open-loop poles, $n = 4$

There are no open zeros, $m = 0$

Number of asymptotes $= n - m = 4 - 0 = 4$

The angle of asymptotes are

$$\theta_A = \frac{(2q+1)}{n-m} \times 180°,\ q = 0, 1, 2, 3$$

For $$q = 0,\ \theta_1 = \frac{2\times 0+1}{4} \times 180 = 45°$$

$$q = 1,\ \theta_2 = \frac{2\times 1+1}{4} \times 180 = 135°$$

$$q = 2,\ \theta_3 = \frac{2\times 2+1}{4} \times 180 = 225°$$

$$q = 3,\ \theta_4 = \frac{2\times 3+1}{4} \times 180 = 315°$$

Centroid: The point of intersection of the asymptotes on the real axis is given by

$$\sigma_A = \frac{(\text{Sum of real parts of poles}) - (\text{Sum of real parts of zeros})}{\text{Number of poles} - \text{Number of zeros}}$$

$$\sigma_A = \frac{(0-2-3-3)-(0)}{4} = -2$$

Breakaway points are found from the solution of $\frac{dK}{dS} = 0$

$$G(s)H(s) = \left|\frac{K}{S(S+2)(S^2+6S+25)}\right| = 1$$

$$K = S(S+2)\,(S^2+6S+25)$$

$$K = S^4 + 8S^3 + 37S^2 + 50S$$

$$\frac{dK}{dS} = \frac{d}{ds}\,(S^4 + 8S^3 + 37S^2 + 50S) = 0$$

$$4S^3 + 24S^2 + 74S + 50 = 0$$

$$2S^2 + 12S^2 + 37S + 25 = 0$$

By trial and error method, breakaway point is at $S = -0.9$

The remaining two roots can be found by dividing

$$(2S^3 + 12S^2 + 37S + 25) \text{ by } (S + 0.9)$$

$$\frac{2S^3 + 12S^2 + 37S + 25}{(S+0.9)} = 2S^2 + 10.2S + 27.82$$

The roots are

$$S = \frac{-10.2 \pm \sqrt{(10.2)^2 - 4 \times 2 \times 27.82}}{2 \times 2}$$

$$S = -2.55 \pm j2.72$$

$$S = -2.55 + j2.72, \quad S = -2.55 - j\,2.72$$

These two points do not lie on the root locus, hence ignored. The only breakaway point is at $S = -0.9$.

The angle of departure from the complex poles at $S = -3 + j4$ is given by

$$\phi_d = 180 + \phi$$

$$\phi = -(-\theta_1 + \theta_2 + \theta_3)$$

The angles θ_1, θ_2 and θ_3 from the poles at $S = 0$, $S = -2$, $S = -3 - j4$ are formed from Fig. 6.24.

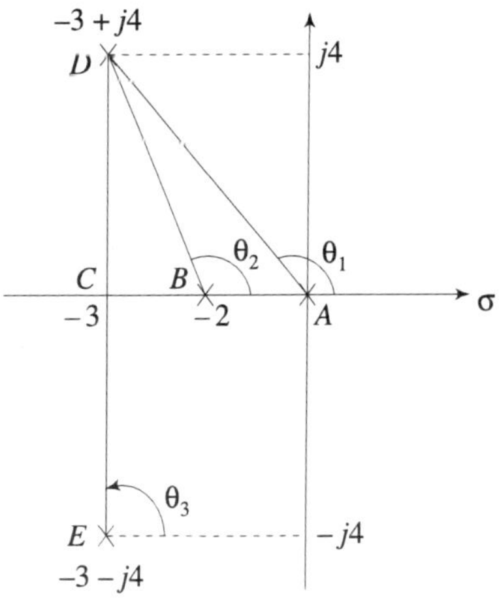

Fig. 6.24

$$\theta_1 = 180 - \tan^{-1}\frac{DC}{CA} = 180 - \tan^{-1}\frac{4}{3}$$

$$\theta_1 = 180 - 53 = 127^\circ$$

$$\theta_2 = 180 - \tan^{-1}\frac{DC}{CB} = 180 - \tan^{-1}\frac{4}{1}$$

$$\theta_2 = 180 - 75.96 = 104$$

$$\theta_3 = 90 \text{ (by inspection)}$$

$$\phi = -(\theta_1 + \theta_2 + \theta_3)$$

$$= -(127 + 104 + 90) = -321$$

$$\phi_{d_1} = 180 + \phi = 180 + (-321) = -141$$

The angle of departure from the complex pole at $S = -3 - j4$ is

$$\phi_{d_2} = -\phi_{d_1} = -(-141) = 141$$

The point of intersection of the root locus with the imaginary axis and the critical value of K can be found by using Routh criterion.

The characteristic equation is

$$1 + G(s)\,H(s) = 1 + \frac{K}{S(S+2)(S^2 + 6S + 25)} = 0$$

$$S^4 + 6S^3 + 10S^2 + 8S + K = 0$$

S^4	:	1	10	K
S^3	:	6	8	
S^2	:	$\frac{60-8}{6} = 8.66$	K	
S^1	:	$\frac{8.66 \times 8 - 6K}{8.66}$		
S^0	:	K		

For the system to be stable, all the elements in the first column of the Routh array must be positive.

$$\therefore \qquad K > 0, \frac{8.66 \times 8 - 6\,K}{8.66} > 0$$

$$6\,K < 8.66 \times 8$$

$$K < \frac{8.66 \times 8}{6}$$

$$K < 11.55$$

$\therefore$ The range of values of K for stability is $0 < K < 11.55$

The marginal value of K for stability $K = 11.55$

The frequency of oscillation at the point of instability is given by the solution of the auxiliary equation.

Auxiliary equation is given by

$$8.66\ S^2 + K = 0$$

$$8.66\ S^2 + 11.55 = 0$$

$$S^2 = \frac{-11.55}{8.66} = -1.3$$

$$S = \sqrt{-1.3} = \pm j\sqrt{1.3} = \pm j1.16$$

$$S = j\omega = j1.16$$

$$\therefore \qquad \omega = 1.16 \text{ rad/sec.}$$

Frequency of sustained oscillation ω = 1.16 rad/sec

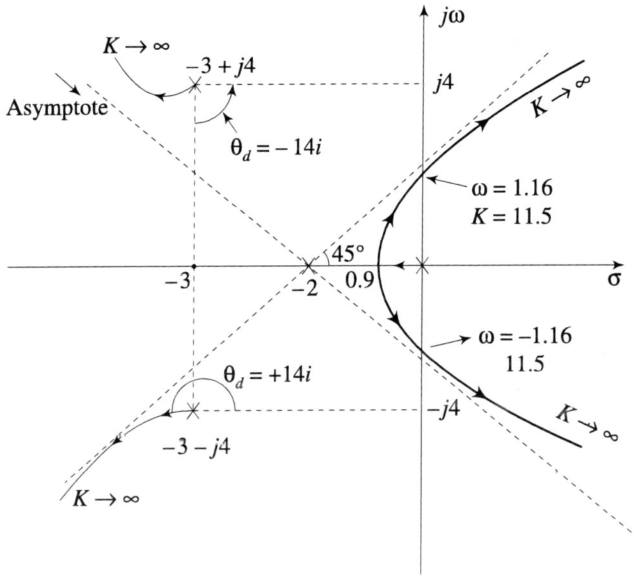

Fig. 6.25

Example 6.12: For a unity feedback system, the open-loop transfer function is given by

$$G(s) = \frac{K(S+2)(S+3)}{S(S+1)}$$

Sketch the root loci for the system when K is varied from 0 to ∞.

Solution: $G(s)H(s) = \dfrac{K(S+2)(S+3)}{S(S+1)}$

Open-loop poles are at $S = 0$, $S = -1$ $\qquad \therefore n = 2$

Open-loop zeros are at $S = -2$, $S = -3$ $\quad \therefore m = 2$

The number of asymptotes $= n - m = 2 - 2 = 0$

The function $G(s)H(s)$ contains no complex poles or complex zeros there is no need to calculate the angles of departure and arrival.

The breakaway points are found from the solution of the equation $\frac{dK}{dS} = 0$

The characteristic equation for the system is

$$1 + G(s)\,H(s) = 0$$

$$1 + \frac{K(S+2)(S+3)}{S(S+1)} = 0$$

$$K = -\frac{S(S+1)}{(S+2)(S+3)}$$

$$\frac{dK}{dS} = \frac{(S+2)(S+3)(2S+1) - S(S+1)(2S+5)}{[(S+2)(S+3)]^2} = 0$$

$$4S^2 + 12S + 6 = 0$$

$$S = \frac{-12 \pm \sqrt{144 - 4 \times 4 \times 6}}{2 \times 4} = -0.634,\ -2.366$$

Both these points lie on the root locus.

The root locus does not cross the $j\omega$ axis. There is no need to apply the Routh-Hurwitz criterion.

The system is stable for all positive values of K

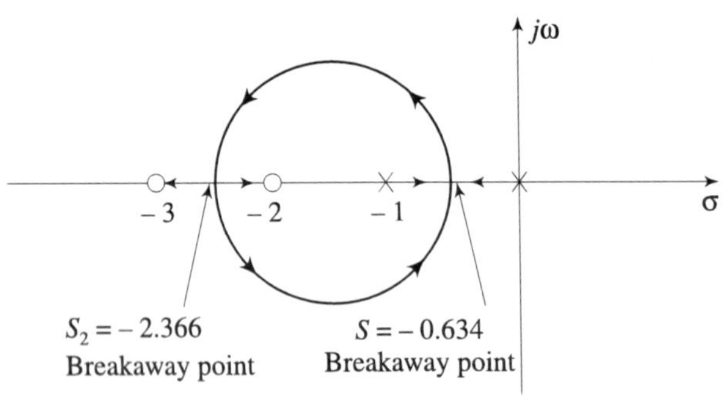

Fig. 6.26

Example 6.13: The open-loop transfer function of a system is given by $G(s)H(s) = \frac{K(S+2)}{S^2(S+20)}$.

Sketch the root loci for the system when K is varied from 0 to ∞.

Solution: The open-loop poles are at $S = 0$, $S = 0$, $S = -20$

$\therefore \quad n = 3$

The open-loop zero is at $S = -12$, $m = 1$

The number of asymptotes, $n - m = 3 - 1 = 2$

The angle of asymptotes are

$$\theta_A = \frac{(2q+1)}{n-m} \times 180, \; q = 0,1$$

For
$$q = 0, \; \theta_1 = \frac{2 \times 0 + 1}{3 - 1} \times 180 = 90^\circ$$

$$q = 1, \; \theta_2 = \frac{2 \times 1 + 1}{3 - 1} \times 180 = 270^\circ$$

The centroid is given by

$$\sigma = \frac{(\text{Sum of real parts of poles}) - (\text{Sum of real parts of zeros})}{\text{Number of poles} - \text{Number of zeros}}$$

$$\sigma = \frac{0 + 0 - 20 + 12}{3 - 1} = -4$$

Mark the centroid on the real axis.

Breakaway points are found from the solution

$$\frac{dK}{dS} = 0$$

The characteristic equation is given by

$$1 + \frac{K(S+12)}{S^2(S+20)} = 0$$

$$K = -\frac{S^2(S+20)}{(S+12)}$$

$$\frac{dK}{dS} = \frac{d}{ds}\left[-\frac{S^2(S+20)}{S+12}\right] = 0$$

$$S^3 + 28S^2 + 240S = 0$$

$$S(S^2 + 28S + 240) = 0$$

$$\therefore \quad S = 0$$

$$S = \frac{-28 \pm \sqrt{(28)^2 - 4 \times 240}}{2}$$

$S = -14 \pm j6.63$ are neither breakaway or break-in points, because the corresponding gain values of K become complex quantities.

$\therefore$ The breakaway point is $S = 0$

Since there are no complex poles or complex zeros, there is no need of calculating the angles of departure and arrival.

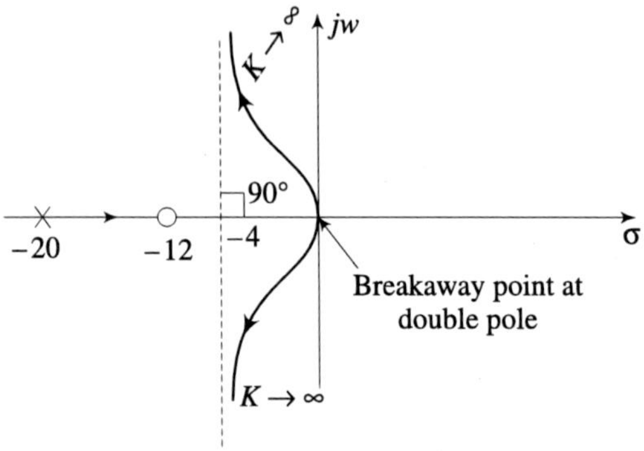

Fig. 6.27

EXERCISE

Fill in the Blanks

1. If the open-loop transfer function is a ratio of a numerator polynomial of degree m and a denominator polynomial of degree n, then the integer $(n–m)$ represents the number of __________ .

Ans: asymptotes

2. Which of the following points is NOT on the root locus of a system with the open-loop transfer function $G(s)H(s) = \dfrac{K}{s(s+1)(s+3)}$ __________ .

Ans: $s = -1.5$

3. Given $G(s)H(s) = \dfrac{K}{s(s+1)(s+3)}$, the point of intersection of the asymptotes of the root loci with the real axis is __________ .

Ans: -1.33

4. The root locus plot for a system is given below. The open-loop transfer function corresponding to this plot is given by __________ . (GATE 2011)

Ans: $G(s)H(s) = \dfrac{k(s+1)}{s(s+2)(s+3)^2}$

5. A unity negative feedback system has the open-loop transfer function $G(s)H(s) = \dfrac{K}{s(s+2)(s+3)}$.

   The value of the gain $K(> 0)$ at which the root locus crosses the imaginary axis is __________ . (GATE 2015)

Ans: 1

6. The characteristic equation of a feedback control system is given by $s^3 + 5s^2 + (K + 6)s + K = 0$ where $K > 0$ is a scalar variable parameter. In the root locus diagram of the system the asymptotes of the root loci for large values of K meet at a point in the S-plane, whose coordinators are __________ .

Ans: (–2, 0)

7. The root locus of the system $G(s)H(s) = \dfrac{K}{s(s + 2)(s + 3)}$ has breakaway point located at __________ .

Ans: (–0.784,0)

8. A unity feedback control system has an open-loop transfer function $G(s) = \dfrac{K}{s(s^2 + 7s + 12)}$.

   The gain K for which $s = -1 + j_1$ will lie on the root locus of this system is __________ .

Ans: 10

9. The open-loop transfer function of a plant in a unity feedback configuration is given as $G(s) = \dfrac{K(s + 4)}{(s + 8)(s^2 - 9)}$. The value of the gain $K > 0$ for which $-1 + j_2$ lies on the root locus is __________ . (GATE 2015)

Ans: 25.5

10. The characteristic equation of a control system is given by $s(s + 4)(s^2 + 2s + 5) + K(s + 1) = 0$ considering $K \geq 0$ what are the angles of the asymptotes for the root loci.

Ans: Angles of asymptotes are 60°, 180°, 300°

11. $G(s) = \dfrac{K(s - 2)}{s(s + 1)(s + 4)^2}$ what is the number of asymptotes and the centroid of the asymptotes of the root loci of closed-loop system.

Ans: $\dfrac{11}{3}$

12. Consider a unity feedback system with open-loop transfer function given by $G(s) = \dfrac{K(s + 2)}{s(s + 1)}$, the breakaway point is __________ .

Ans: $s = -2 + \sqrt{2}$ breakaway point and break-in point $-\sqrt{2} - \sqrt{2}$.

13. $G(s)H(s) = \dfrac{K(s + 8)}{s(s + 4)(s + 12)(2 + 20)}$, if a closed-loop pole is located at $s = -10$ then the value of K is __________ .

Ans: $K = 600$

14. The forward path transfer function of a unity feedback system is $G(s) = \dfrac{K(s + 2)}{(s + 3)(s^2 + 2s + 2)}$. The angles of departure from the complex poles are __________ .

Ans: $\theta d = \mp 108.435°$

15. The root locus of a system with loop transfer function $G(s)H(s) = \dfrac{K(s^2 + 8s + 64)}{s(s + 4)(s^2 + 4s + 5)}$ have __________ number of breakaway and break-in points.

Ans: $s = -0.837, -1.95, -3.24, -4.99 \mp j9.82$

16. A unity negative feedback system has the open-loop transfer function $G(s) = \dfrac{k}{s(s+1)(s+3)}$. The value of the gain at which the root locus crosses the imaginary axis.

Ans: $K = 12$

17. The open-loop transfer function of a feedback system is $G(s)H(s) = \dfrac{K(s+1)(s+3)}{(s^2+4s+8)}$, the angle of departure is __________ .

Ans: 90°

18. The starting point of a root locus is __________.

Ans: Open-loop poles

19. If the gain (K) of a system become zero, the root will __________.

Ans: coincide with the poles.

20. If the characteristic equation of a system is $1 + K/[s(s+1)(s+4)] = 0$, the centroid of the asymptotes in the root locus is given by __________.

Ans: Breakaway points

21. The breakaway points of the root locus occurs at __________.

Ans: multiple root of characteristic equations

22. Loop transfer function zeros are the same as

Ans: zeros of $G(s)H(s)$

23. In the root locus analysis, the angle between adjacent asymptotes is __________.

Ans: $360°/n - m$

24. The root locus of a unity feedback system is shown in Fig. 6.28. The-open-loop transfer function of the system is __________.

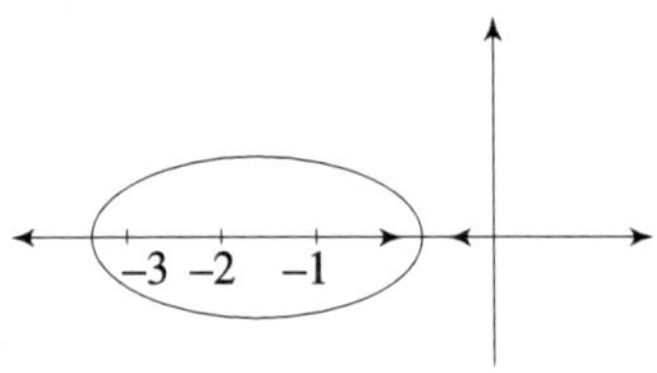

Fig. 6.28

Ans: $\dfrac{K(s+3)}{S(s+1)}$

25. Match the following *(Types of loci)*

List I	*List II*
A. Separate	1. One
B. Loci on the real axis	2. Two
C. Asymptotes	3. Three
D. Breakaway points	4. Five

Ans: A → 4, B → 3, C → 2, D → 1

26. What is root locus?

Ans: The path taken by a root of characteristic equation when open-loop gain K is varied from 0 to ∞ is called root locus.

27. How will you find root locus on real axis?

Ans: To find the root locus on real axis, choose a test point on real axis. If the total number of poles and zeros on the real axis to the right of this test point is an odd number then the test point lies on the root locus. If it is even then the test point does not lie on the root locus.

28. What are asymptotes? How will you find the angle of asymptotes?

Ans: Asymptotes are straight lines which are parallel to root locus going to infinity and meet the root locus at infinity.

$$\text{Angle of asymptotes} = \frac{\pm 180(2q+1)}{n-m};\ q = 0, 1, 2 \ldots (n-m)$$

29. What is centroid? How is a centroid calculated?

Ans: The meeting point of asymptotes with real axis is called centroid. The centroid is given by:

$$\text{Centroid} = \frac{\text{Sum of poles} - \text{Sum of zeros}}{n-m}$$

30. What are breakaway and break-in points? How to determine them?

Ans: At breakaway point the root locus breaks from the real axis to enter into the complex plane. At break-in point the root locus enters the real axis from the complex plane.

To find the breakaway or break-in points, from on equation for K from the characteristic equation and differentiate the equation of K with respect to S. Then find the roots of equation $dK/dS = 0$. The roots of $dK/dS = 0$ are breakaway or break-in points provided for this value of root the gain K should be positive and real.

31. How will you find a dominant pole on root locus and find the gain K corresponding to the dominant pole?

Ans: The dominant poles are given by roots of a quadratic factor

$$(S^2 + 2\zeta\omega_n S + \omega_n^2 = 0)$$

$$\therefore \quad S = \frac{-2\zeta\omega_n \pm \sqrt{4\zeta\omega_n^2 - 4\omega_n^2}}{2}$$

$$= -\zeta\omega_n \pm j\omega_n\sqrt{1-\zeta^2}$$

The dominant poles can be plotted on the S-plane as shown in Fig. 6.29. In the right angle triangle OAP.

$$\cos\alpha = \frac{\zeta\omega_n}{\omega_n} = \zeta \quad \therefore \alpha = \cos^{-1}\zeta$$

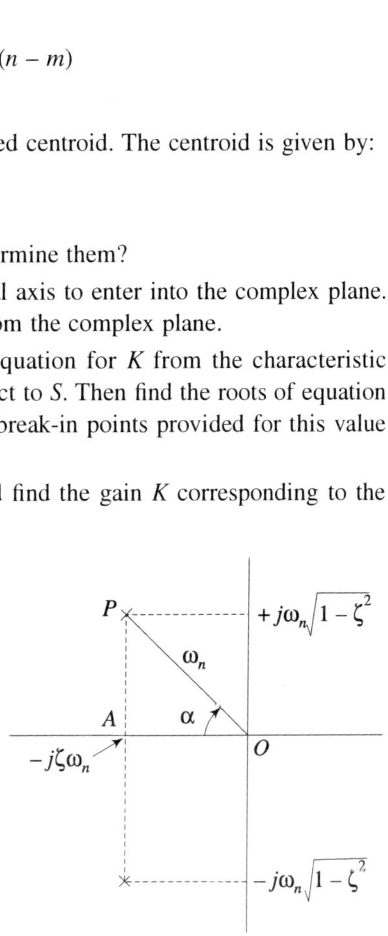

Fig. 6.29

32. The open-loop transfer functions of certain unity feedback systems are given below, sketch the root locus of each system.

(a) $G(s) = \dfrac{K(s+2)}{(s+3)^2(s^2+2s+17)}$

(b) $G(s) = \dfrac{K}{s(s+3)(s^2+2s+2)}$

(c) $G(s) = \dfrac{K(s^2+2s+10)}{s(s+2)(s+4)}$

(d) $G(s) = \dfrac{K(s+1)}{s^2(s+12)}$

(e) $G(s) = \dfrac{K(S + 4)}{s(s + 0.5)(s + 2)}$ (f) $G(s) = \dfrac{K}{s(s^2 + 8s + 20)}$

(g) $G(s) = \dfrac{K}{s(s + 2)(s^2 + 2s + s)}$ (h) $G(s) = \dfrac{K(s^2 + 1)}{S(s + 2)}$

33. A unity feedback system has an open-loop transfer function $G(s) = K/S\ (s^2 + 8s + 32)$. Sketch the root locus and determine the dominant closed-loop poles with $\zeta = 0.5$. Determine the value of K at this point.
34. Draw the root locus plot for a unity feedback system having forward path transfer function, $G(s) = K|s(s + 1)(s + 5)$. (a) Determine the value of K which gives continuous oscillation and the frequency of oscillation. (b) Determine the value of K corresponding to a dominant closed-loop pole with damping ratio, $\zeta = 0.7$.
35. What do you mean by root locus? What are the advantages, applications and limitations of root locus?
36. What are the conditions to be satisfied for the root locus to exist at any point in the S-plane?
37. Define the following terms
    (i) Asymptote (ii) Centroid (iii) Breakaway and break-in points
    (iv) Angle of departure (v) Angle of arrival
38. How do you determine the point of intersection of the root locus with $j\omega$ axis and the critical value of K?
39. How do you determine the value of K at any point on the root locus?
40. How do you determine the closed-loop transfer function for a given value of K from the root locus plot?
41. How do you determine the breakaway points on the root locus?
42. How do you determine the value of K for a given damping ratio from the root locus.
43. Sketch the root locus of a unity feedback system with forward path transfer function

    $G(s) = \dfrac{K}{s(s + 4)(s + 5)}$ where K is varied from zero to infinity.

Ans:

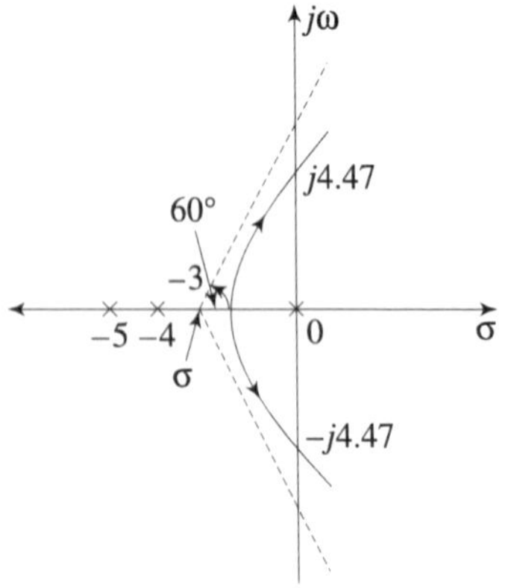

Fig. 6.30

44. The open-loop transfer function of a unity feedback system is $G(s) = \dfrac{K(s + a)}{s(s + b)}$

(i) Prove that breakaway and break-in points will exist only when $|a| > |b|$.

(ii) Prove that the complex points on the root locus form a circle with centre $(-a, 0)$ and radius $\sqrt{a^2 - ab}$.

45. The block diagram of a system shown in Fig. 6.31. Use root locus method to determine the value of α so that the damping ratio of the closed-loop poles is 0.6.

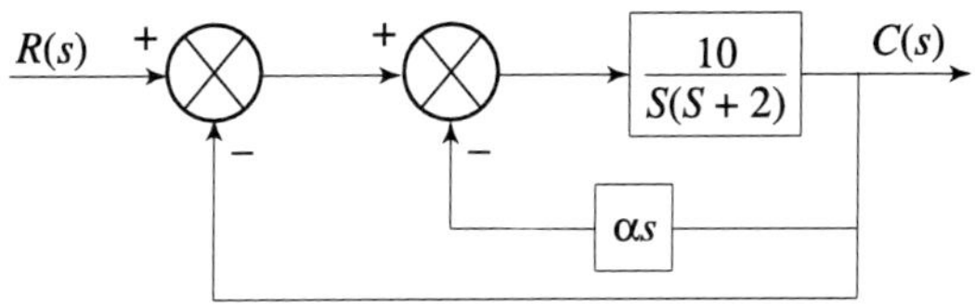

Fig. 6.31

Ans: $\alpha = 0.1162$

46. For a unity feedback system, the open-loop transfer function is given by $G(s) = \dfrac{K}{s(s^2 + 6s + 25)}$. Sketch the root loci for the system when K is varied from 0 to ∞.

Ans:

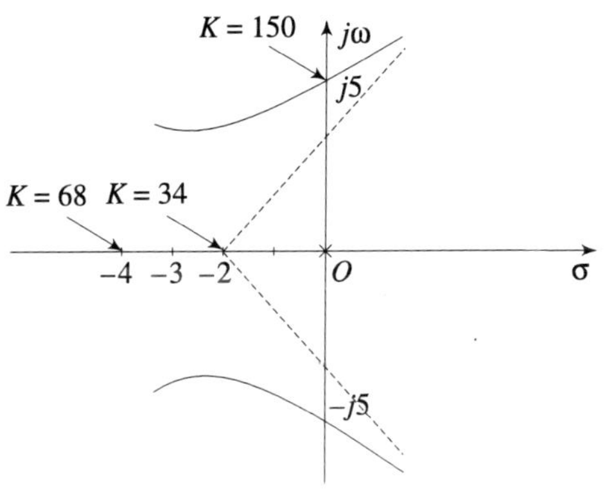

Fig. 6.32

47. For a unity feedback system the open-loop transfer function is $G(s) = \dfrac{K}{s(s + 1)(s^2 + 4s + 13)}$

Sketch the root loci for the system when K is varied from zero to infinity.

Ans:

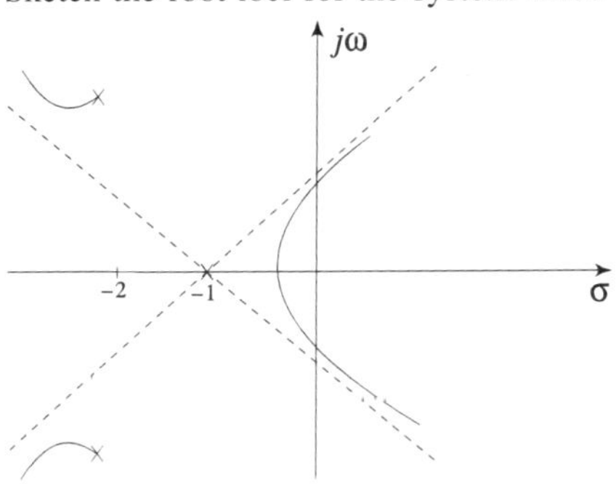

Fig. 6.33

48. What do you mean by root locus?

Ans: Defined as the locus of the plot of the closed-loop poles as a function of open-loop gain (K), when K is varied from $-\infty$ to $+\infty$. If K is varied from 0 to $+\infty$, it is known as direct root locus. If K is vaired from $-\infty$ to 0, it is known as inverse root locus.

49. State the angle and magnitude condition of root locus?

Ans: Magnitude condition: $|G(s)H(s)| = 1$

Angle condition $\angle G(s)H(s) = \pm (2q + 1)\ 180°$

50. Define the number of loci.

Ans: Let the number of OLTF poles of OLFT zeros be n and m respectively. If $n > m$ is the number of OLTF poles is greater than the number of OLTF zeros. The number of loci is n. Each locus will start from OLTF pole and ends m OLFT zero. Otherwise loci will end at infinity.

51. What do you mean by real axis loci?

Ans: Part of the loci will lie on the real axis. A point on the real axis will lie on the root locus if and only if the sum of the OLTF poles and zeros to the right of the point is odd.

52. Write the expression for angle of asymptotes.

Ans: $\theta = \dfrac{2q + 1}{n - m} \times 180°$ for $k > 0$

where $q = 0, 1, 2 \ldots n - m - 1$.

$\theta = \dfrac{2q}{n - m} \times 180°$ for $k < 0$, where $\theta = 0, 1, 2\ n - m - 1$.

53. State the effect of addition of zeros.

Ans: Range of K increases

Setting time speeds up

Stability of the system is enhanced.

54. Using the root-locus method find the gain (k) for the system having $G(s)H(s) = \dfrac{K}{s(s + 2)(s + 3)}$ when $\xi = 0.341$.

Ans: $k = 8.28$

55. Using the root locus method find the gain (K) for the system having

$$G(s)H(s) = \frac{K(s + 2)(s + 4)}{(s + 1)(s + 3)(s + 5)} \text{ where } \xi = 0.246.$$

Ans: 33.2

7 Frequency Response Analysis

7.1 INTRODUCTION

Using time domain characteristics it is possible to obtain an idea about the performance of a given system. If the order of the system increases, it is very difficult to analyse the system by the time domain approach. In design problem, there is no single method to determine the design solution which obeys the various time domain specifications such as rise time, maximum overshoot, etc. The root locus method (i.e., time domain method) was developed during 1940s. The frequency response method was developed earlier than the time domain method. The frequency response of a system is defined as the steady state response of the system to a sinusoidal input signal. The response of a linear constant coefficient system to a sinusoidal input signal is an output sinusoidal signal at the same frequency as the input. However, the magnitude and phase of the output signal differ from those of the input sinusoidal signal, and the amount of difference is a function of the input frequency. Frequency domain analysis is another analytic method that can be used for analysis and design.

7.1.1 Frequency Response

It is defined as the steady state response of a system due to sinusoidal input. Figure 7.1 shows a linear system with a sinusoidal input.

Let us study its output and derive the relation between its input and output.

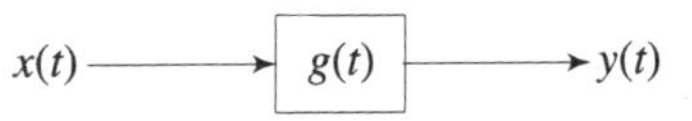

Fig. 7.1 A linear system with sinusoidal input

In Fig. 7.1, $x(t)$ is the sinusoidal input function, $y(t)$ is the output function, and $g(t)$ is the transfer function. Let $x(s)$, $y(s)$ and $G(s)$ be the Laplace transforms of $x(t)$, $y(t)$ and $g(t)$ respectively. $G(s)$ is the transfer function of the system and it is defined for the zero initial condition.

$$\therefore \qquad G(s) = \frac{N(s)}{D(s)} = \frac{N(s)}{(s+a)(s+b)(s+c)} \qquad \ldots(7.1)$$

In Eq. (7.1), $N(s)$ is a polynomial of s and the poles of $G(s)$ are at $-a$, $-b$, $-c$, etc.,

$$G(s) = \frac{y(s)}{x(s)} \qquad ...(7.2)$$

Using Eq. (7.1) and (7.2), we have,

$$y(s) = \frac{N(s)x(s)}{(s+a)(s+b)(s+c)} \qquad ...(7.3)$$

Let $x(t) = A \sin \omega t$ be the input of the system. The Laplace transform of $x(t)$ is given by

$$X(s) = \frac{A\omega}{s^2 + \omega^2} \qquad ...(7.4)$$

Using Eqs. (7.4), (7.3) can be written as follows

$$Y(s) = \left[\frac{N(s)}{(s+a)(s+b)(s+c)}\right]\left(\frac{Aw}{s^2+\omega^2}\right)$$

$$= \frac{A_1}{(s+a)} + \frac{A_2}{(s+b)} + \frac{A_3}{(s+c)} + \ldots + \frac{B_1}{(s+j\omega)} + \frac{B_2}{(s-j\omega)}$$

$$y(t) = A_1 e^{-at} + A_2 e^{-bt} + A_3 e^{-ct} + \ldots + B_1 e^{-j\omega t} + B_2 e^{j\omega t} \qquad ...(7.5)$$

The terms with coefficients A_1 to A_n will decay to zero as time tends to infinity, because these terms contribute to decaying exponents.

∴ These terms constitute transient response. The terms with coefficient B_1 and B_2 will contribute steady state response. The steady state response of Eq. (7.5) is given by

$$y_{ss}(t) = B_1 e^{-j\omega t} + B_2 e^{j\omega t} \qquad ...(7.6)$$

$$B_1 = \left[G(s)\frac{A\omega}{s^2-\omega^2}(s+j\omega)\right]_{s=-j\omega} = -\frac{A}{2j}G(-j\omega) \qquad ...(7.7)$$

$$B_2 = \left[G(s)\frac{Aw}{s^2-\omega^2}(s-j\omega)\right]_{s=j\omega} = \frac{A}{2j}G(+j\omega) \qquad ...(7.8)$$

$G(j\omega)$ is a complex quantity and we can write $G(j\omega)$ as follows

$$G(j\omega) = |G(j\omega)|e^{j\varnothing} \qquad ...(7.9)$$

where $|G(j\omega)|$ is the magnitude of $G(j\omega)$ using Eqs. (7.7) and (7.8), we have equation (7.6).

$$y_{ss}(t) = -\frac{A}{2j}|G(j\omega)|e^{-j\varnothing}e^{-j\omega t} + \frac{A^*}{2j}|G(j\omega)|e^{j\varnothing}e^{j\omega t}$$

$$y_{ss}(t) = A|G(j\omega)|\left\{\frac{e^{j(\omega t+\varnothing)} - e^{-j(\omega t+\varnothing)}}{2j}\right\}$$

$$y_{ss}(t) = A|G(j\omega)|\sin(\omega t + \varnothing)$$

$$y_{ss}(t) = |Y(j\omega)| < \varnothing \qquad ...(7.10)$$

where $$|Y(j\omega)| = A|G(j\omega)| \qquad ...(7.11)$$

From Eq. (7.11)

$$Y(j\omega) = |G(j\omega)|\ |X(j\omega)|$$

$$|G(j\omega)| = \frac{|Y(j\omega)|}{|X(j\omega)|}$$

$$\underline{|Y(j\omega)|} = \varnothing = \varnothing + 0^\circ$$

$$\underline{|Y(j\omega)|} = \underline{|G(j\omega)|} + \underline{|X(j\omega)|}$$

$$\underline{|G(j\omega)|} = \underline{|Y(j\omega)|} - \underline{|X(j\omega)|}$$

∴ The frequency of sinusoidal output is the same as the input and there is merely any change in the amplitude and phase of the output with respect to input.

7.1.2 Method in Frequency Response

The magnitude function and phase function are the two plots in the frequency response characteristic. The angular frequency (ω) is the common variable for both these plots. The following curves are commonly used in frequency response analysis.

(i) **Polar plot:** The plot of the magnitude versus phase angle in polar coordinate where is varied from zero to ∞ is called Polar plot.

(ii) **Bode plot:** The plots of both (a) magnitude and (b) phase versus frequency (ω) in logarithmic values are called Bode plots.

(iii) **Magnitude versus phase plot:** The magnitude versus phase plot on rectangular coordinate as varying ω is known as magnitude versus phase plot.

7.1.3 Definitions of Frequency Domain Specifications

The closed-loop transfer function is given by

$$\frac{C(s)}{R(s)} = \frac{G(s)}{1 + G(s)H(s)} = M(s)$$

or $$M(j\omega) = \frac{G(j\omega)}{1 + G(j\omega)H(j\omega)}.$$

7.1.4 Frequency Response Specification

The basic objectives of control system are to meet the performance specifications. These specifications are the constraints or limitations put on the mathematical functions describing the system characteristics. The frequency response specifications are described in Fig. 7.2.

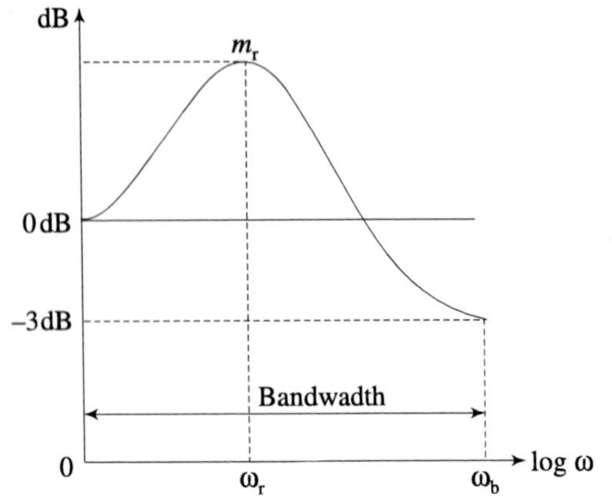

Fig. 7.2

(i) Bandwidth (BW): It is defined as the range of frequencies over which the system will respond satisfactorily or it is the range of frequencies in which the magnitude response is almost flat in nature The frequency at which the magnitude of $M(j\omega)$ drops by 30 dB below a certain specified level, generally 0 dB level. If bandwidth is large, higher frequencies will pass through the system. This indicates a faster rise in time. Therefore, bandwidth gives an idea about transient response. Small bandwidth means lower frequencies are passed and hence response is sluggish. Larger bandwidth indicates susceptibility to noise. Bandwidth also implies the sensitivity to parameter variations.

Hence, it is defined as the range of frequencies over the magnitude of closed-loop response, i.e., $|C(j\omega)/R(j\omega)|$ does not drop by more than 3 dB from its zero frequency value. Bandwidth (BW) indicates the speed of the response.

(ii) Cutoff frequency (ω_b): The frequency at which the magnitude of the closed-loop response is 3 dB down from its zero frequency value is called cut of frequency. The range 0 to ω_b is bandwidth of the system. BW α $1/t_r$

(iii) Cutoff rate: The slope of the resultant magnitude (log magnitude) curve near the cutoff frequency is called cutoff rate. The ability of the system to distinguish a signal from noise can be determined by cutoff rate.

(iv) Resonant peak (M_r): The maximum value of magnitude $M(j\omega)$ as ω is varied is termed resonance peak (M_r). Larger the value of resonant peak more is the value of peak overshoot of the system for step input. It is a measure of relative stability of the system. Generally, M_r is between 1.1 and 1.5.

(v) Resonance frequency (ω_r): The frequency at which peak resonance (M_r) occurs is known as resonance frequency.

(vi) Phase margin: If ϕ is the phase angle of a system at unity gain, the phase margin is given by $(180 + \phi)$. The frequency (ω_1) at which gain is unity is known as gain crossover frequency. The increment in the system angle that causes the system to be unstable is indicated by phase margin. It is a measure of relative stability.

(vii) Gain margin: The frequency (ω_1) at which the phase angle of the transfer function is $-180°$ is known as phase crossover frequency. The gain at this frequency ω_2 is $|G(j\omega_1) H(j\omega_1)|$ and gain margin is given by $\frac{1}{|G(j\omega_1)H(j\omega_1)|}$. The increment in gain required to cause the system to become unstable is indicated by gain margin.

7.2 CORRECTION BETWEEN TIME AND FREQUENCY DOMAIN SPECIFICATIONS FOR A SECOND ORDER SYSTEM

The relationship between frequency domain specifications and time domain specifications is important and should be thoroughly understood before designing a system.

For a Second Order System

$$G(s) = \frac{C(s)}{R(s)} = \frac{\omega_n^2}{s^2 + 2\xi\,\omega_n s + \omega_n^2}$$

$$H(s) = 1$$

$$G(j\omega) = \frac{C(j\omega)}{R(j\omega)} = \frac{\omega_n^2}{(j\omega)^2 + 2\xi\,\omega_n(j\omega) + \omega_n^2}$$

$$G(j\omega) = \frac{1}{1 + 2j\xi\left(\frac{\omega}{\omega_n}\right) - \left(\frac{\omega}{\omega_n}\right)^2}$$

Let

$$u = \frac{\omega}{\omega_n}$$

$$G(j\omega) = \frac{1}{1 + 2ju\,\xi - u^2}$$

or

$$|G(j\omega)| = M = \frac{1}{(1 - u^2)^2 + (2u\xi)^2}$$

$$\text{Ang } G(j\omega) = -\tan^{-1}\left(\frac{2u\xi}{1 - u^2}\right).$$

Putting $\frac{dM}{du} = 0$, resonant frequency (ω_r) can be obtained.

$$\left|\frac{dM}{du}\right|_{u=u_r} = -\frac{1}{2}\,\frac{[2(1 - u_r^2)(-2u_r) + 8\xi^2 u_r]}{[(1 - u_r^2)^2 + (2u_r\xi)^2]^{3/2}} = 0$$

or

$$4u_r - 8u_r\xi^2 - 4u_r^3 = 0$$

$$4u_r(1 - 2\xi^2 - u_r^2) = 0$$

$\therefore$

$$u_r = 0$$

$$1 - 2\xi^2 - 4u_r^2 = 0$$

$$(1 - 2\xi^2 - 4u_r^2) = 0$$

$$u_r^2 = 1 - 2\xi^2 \quad \text{or} \quad u_r = \sqrt{1 - 2\xi^2} \qquad ...(7.12)$$

or

$$\frac{\omega_r}{\omega_n} = \sqrt{1 - 2\xi^2}$$

$$\omega_r = \omega_n \sqrt{1 - 2\xi^2} \qquad ...(7.13)$$

$$Mr = G(j\omega)|_{u=ur} = \frac{1}{\sqrt{(1 - u_r^2)^2 (2u_r\xi)^2}} = \frac{1}{\sqrt{(2\xi^2)^2 + 4\xi^2(1 - 2\xi^2)}}$$

$$= \frac{1}{\sqrt{4\xi^4 + 4\xi^2 - 8\xi^4}} = \frac{1}{\sqrt{4\xi^2 - 4\xi^4}} = \frac{1}{\sqrt{4\xi^2(1 - \xi^2)}}$$

$$M_r = \frac{1}{2\xi\sqrt{1 - \xi^2}} \qquad ...(7.14)$$

From Eq. (7.12), u_r is valid as long as $(1 - 2\xi^2) \geq 0$, i.e., $\xi \leq 0.707$ and Eq. (7.13) shows that M_r is a function of ξ.

We have $\phi_r = G(ju)|_{u = u_r} = -\tan^{-1} \dfrac{\left(\sqrt{1 - 2\xi}\right)}{\xi}$

Equation (7.13) shows that ω_r is a function of ξ and ω_n from Eqs. (7.13) and (7.14), the following conclusion can be drawn.

(i) M_r approaches infinity and ω_r approaches ω_n when ξ approaches zero.

(ii) M_r is greater than unity and ω_r is less than ω_n for $0 < \xi < 0.707$.

(iii) There is no resonant peak and the maximum value of M_r is unity and $\omega_r = 0$ when $\xi > 0.707$.

To determine cutoff rate and bandwidth let us consider Fig. 7.3.

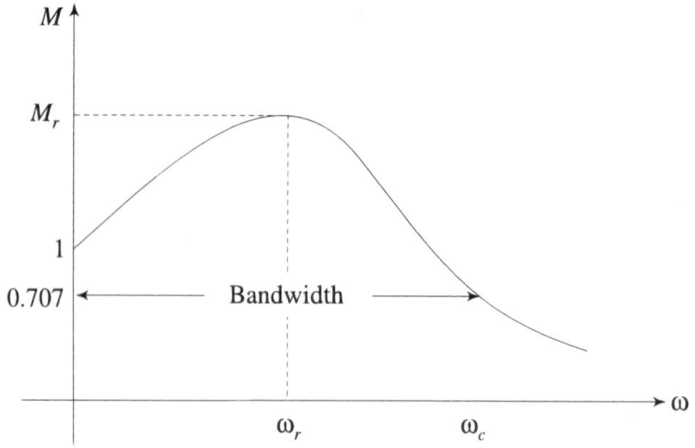

Plot of M_r vs ω

Fig. 7.3

Figure 7.3 shows the plot of M_r versus ω known as the typical magnification curve of feedback control.

Figure 7.4 shows the plot of bandwidth versus damping factor.

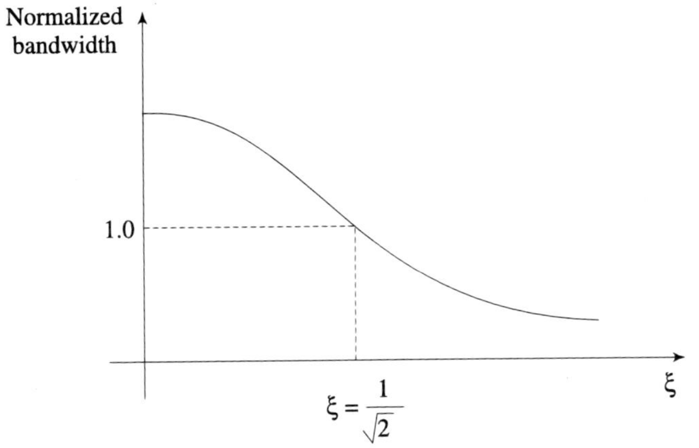

Fig. 7.4 Bandwidth Versus and ξ

Figure 7.5 shows the plot of M_r and M_p versus ξ.

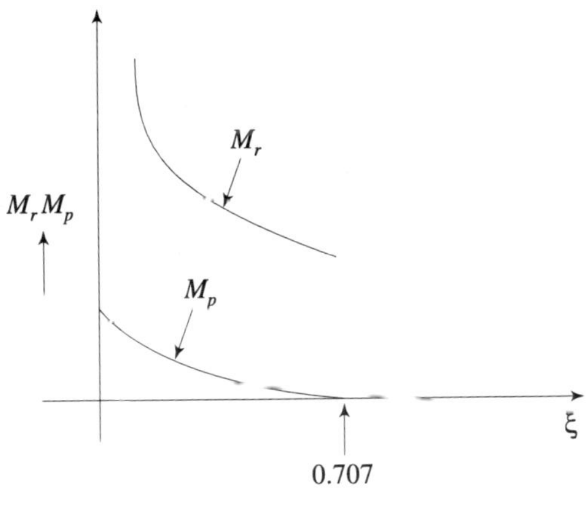

Fig. 7.5 M_r and M_p Versus ξ

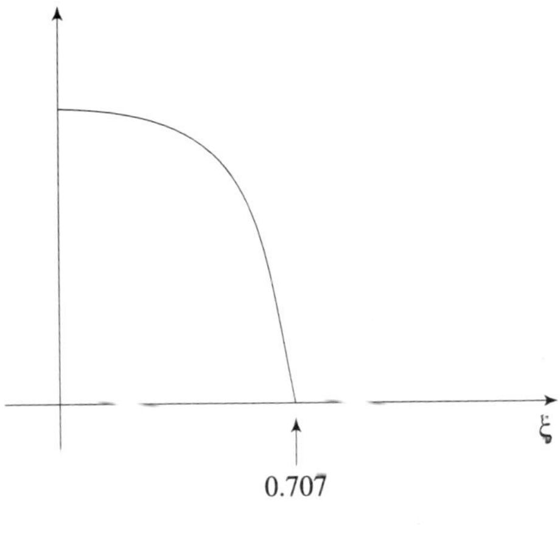

Fig. 7.6 Correlation between ω_r and ω_d

Figure 7.6 represents the correlation between ω_r and ω_d.

At cutoff frequency (ω_c) the value of M is 0.707. Generally, control systems are low-pass filters and at zero frequency, $M = 1$. Therefore, bandwidth becomes equal to ω_c.

WKT $$|G(j\omega)| = |G(u)| = M = \frac{1}{(1 - u^2)^2 + (2u\,\xi)^2}$$

Taking $M = 0.707$ and $u = u_c$ in the above equation,

We get
$$0.707 = \frac{1}{[(1 - u_c^2)^2 + (2u_c\xi)^2]^{1/2}}$$

where
$$u_c = \frac{\omega_c}{\omega_n}$$

$$(1 - u_c^2)^2 + (2u_c\xi)^2 = 2$$

Solving, we have

$$u_c^2 = (1 - 2\xi^2) \pm \sqrt{4\xi^2 - 4\xi^2 + 2}$$

$$\omega_c = \omega_n \left\{(1 - 2\xi^2) \pm \sqrt{4\xi^4 + 4\xi^2 + 2}\right\}^{1/2}$$

Bandwidth (BW) = ω_c.

The bandwidth is a function of ω_n and ξ. The following important conclusions can be drawn.

- Bandwidth α ω_n
- Bandwidth decreases as ξ increases
- For $0 < \xi < 0.707$, bandwidth and M_r are directly proportional.
- Bandwidth decreases due to increase in rise time. M_p and M_r are functions of and only for $0 < \xi < \frac{1}{\sqrt{2}}$. M_p is well damped and M_r does not exist when $\xi > \frac{1}{\sqrt{2}}$. The relation between ω_r and ω_d is given by
$$\frac{\omega_d}{\omega_r} = \frac{1 - \xi^2}{\sqrt{1 - 2\xi^2}}.$$

7.3 BODE PLOT

Basic of any frequency response is to plot magnitude M and angle Φ against input frequency ω when ω is varied from 0 to ∞ there is wide range of variations in M and Φ. Hence, it becomes difficult to accommodate all such variations with linear scale.

Hence, H. W. Bode suggested the method in which logarithmic values of magnitude are to be plotted against logarithmic values of frequencies. It is a graphical method to determine the frequency response of a circuit. Such plots are called logarithmic plots.

In general, Bode plot consists of two plots which are:

(i) Magnitude expressed in logarithmic values against logarithmic values of frequency is called magnitude plot.

(ii) Phase angle in degrees against logarithmic values of frequency is called phase angle plot.

7.3.1 Advantages of Bode Plot

(i) Multiplication and division of magnitude can be converted into addition and subtraction respectively, as logarithmic representation is used.

(ii) Both low frequency and high frequency portions can be represented on the same graph (semilog sheets).

(iii) Approximate plot using straight line asymptotes can be drawn very quickly and easily.

(iv) Both absolute and relative stability of a system can be determined using Bode plots.

Note: The logarithmic scale used for frequency in Bode plots is nonlinear, the point at $\omega = 0$ cannot be located on the log scale as $\log 0 = -\infty$.

In Bode diagrams, frequency ratios are expressed in terms of octaves and decades.

An octave is a frequency band from ω_1 to $2\,\omega_1$.

A decade is a frequency band from ω_1 to $10\,\omega_1$.

Magnitude is given as $|G(j\omega)| = 20\log_{10}|G(j\omega)|_{dB}$.

Bode plot consists of two separate plots. One shows how the amplitude of $G(j\omega)$ varies, with frequency and the other shows how the phase angle of $G(j\omega)$ varies with frequency. Bode plots are made in semilog graph paper for greater accuracy in representing the wide range of frequency values. In both the amplitude and phase plots, the frequency is plotted on the horizontal log scale and the amplitude and the phase angle are plotted on the linear vertical scale. A graph paper with horizontal axis in logarithmic scale and vertical axis in linear scale is called semilog paper.

It is the plot of the logarithm of the magnitude of a sinusoidal transfer function $G(j\omega)$ versus low ω where ω is the radian frequency. The standard representation of logarithmic magnitude in $20 \log |G(j\omega)|$. The unit is decibel (dB).

Bode phase angle plot: It is the plot of the phase angle in degree of the sinusoidal transfer function versus $\log_{10}\omega$.

Bode plots are also called Bode diagrams or logarithmic plots. Also known as corner plot or an asymptotic plot, because the Bode plot can be constructed by using straight line approximations that are asymptotic to the actual plot.

In the logarithmic representation, the curves are drawn on semilog paper. In such a paper the horizontal axis is in logarithmic scale (i.e., $\log\omega$) which is nonlinear. The vertical axis is divided in linear scale for either magnitude in decibels or phase angle in degrees. Hence, such a paper with horizontal axis in logarithmic scale and vertical axis in linear scale is called semilog paper.

Basic principle

In general, the OLTF of a control system is of the form:

$$G(s)H(s) = \frac{K(1 + TaS)(1 + TbS)\ \dots\ (1 + TmS)}{S^N\,(1 + T_1 s)(1 + T_2 s)\ \dots\ \left[1 + 2\xi\,\frac{s}{\omega_n} + \left(\frac{s}{\omega_n}\right)^2\right]} \qquad \dots(7.15)$$

This form of transfer function is known as time constant form for drawing Bode plots. The given transfer function has to be written in 'time constant form'.

The sinusoidal transfer function is obtained by setting $s = j\omega$ in Eq. (7.15)

$$G(j\omega)\,H(j\omega) = \frac{K(1 + j\omega Ta)(1 + j\omega Tb)\,\ldots\,(1 + j\omega Tm)}{(j\omega)^N (1 + j\omega T_1)(1 + j\omega T_2)\,\ldots\left[1 + 2\xi\frac{j\omega}{\omega_n} + \left(\frac{j\omega}{\omega_n}\right)^2\right]} \qquad \ldots(7.16)$$

In order to obtain the magnitude plot

$$|G(j\omega)H(j\omega)| = 20\log_{10}|G(j\omega)H(j\omega)|_{dB}$$

$$= 20\log|K| + 20\log|(1 + j\omega T_a)| + 20\log|(1 + j\omega T_b)| + \ldots$$

$$- (20N\log j\omega + 20\log|(1 + j\omega T_1)| + 20\log|(1 + j\omega T_2)|$$

$$+ 20\log\left|\left(1 + j2\xi\frac{\omega}{\omega_n} - \left(\frac{\omega}{\omega_n}\right)^2\right)\right| \qquad \ldots(7.17)$$

For plotting the magnitude plot, each term of Eq. (7.17) is considered and the magnitude plots of individual factors are plotted separately.

The resultant magnitude plot of the transfer function is obtained by adding the magnitude plots of individual factors.

Phase angle plot

The phase angle of the transfer function is:

$$\angle G(j\omega)H(j\omega) = \tan^{-1}(\omega T_a) + \tan^{-1}(\omega T_b) + \ldots -N(90^\circ) - \tan^{-1}(\omega T_1)$$

$$- \tan^{-1}(\omega T_2)\,\ldots - \tan^{-1}\frac{2\xi\frac{\omega}{\omega_n}}{1 - \left(\frac{\omega}{\omega_n}\right)^2} \qquad \ldots(7.18)$$

Equation (7.18) can be used to calculate the phase angle for different frequencies ranging from low to high value.

Plots of individual factors

The various factors in Eq. (7.16) belong to one of the following forms:

- A constant gain K.
- Plots at the origin, $\frac{1}{j\omega}$ at zeros at origin $j\omega$.
- Plots on real axis $\frac{1}{1 + j\omega T}$ at zeros on real axis $(1 + j\omega T)$
- Complex poles $\frac{1}{\left(1 + j2\xi\frac{\omega}{\omega_n} - \left(\frac{\omega}{\omega_n}\right)^2\right)}$ at complex zeros $\left[1 + j2\xi\frac{\omega}{\omega_n} - \left(\frac{\omega}{\omega_n}\right)^2\right]$

(i) Gain K at constant K:

Let
$$G(s)H(s) = K$$
$$G(j\omega)H(j\omega) = K + j0$$
$$|G(j\omega)H(j\omega)| = \sqrt{K^2} = K$$
$$|G(j\omega)H(j\omega)| \text{ in dB} = 20 \log_{10}K \text{ dB}$$

As K is constant, 20 $\log_{10}K$ is always constant through ω is varied from 0 to ∞. Therefore, its magnitude plot will be a straight line parallel to X-axis.

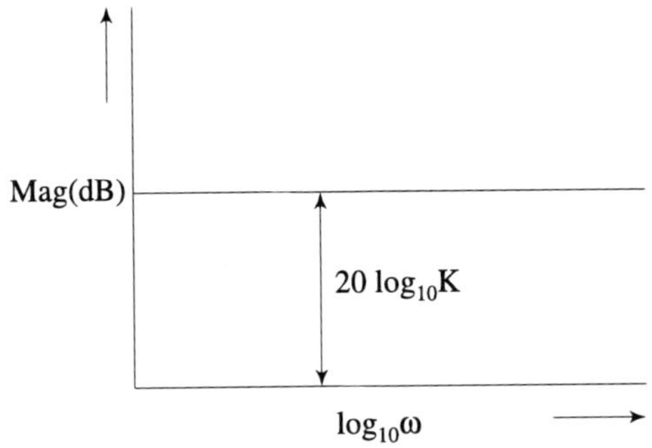

Fig. 7.7

Phase angle $\angle G(j\omega)H(j\omega) = \angle K = 0°$

Therefore, it does not affect the phase angle plot as its contribution to phase angle is 0°

$$G(j\omega)H(j\omega) = K + j0$$

$$\Phi = \tan^{-1}\frac{\text{Imaginary}}{\text{Real}} = \tan^{-1}\frac{0}{K} = 0$$

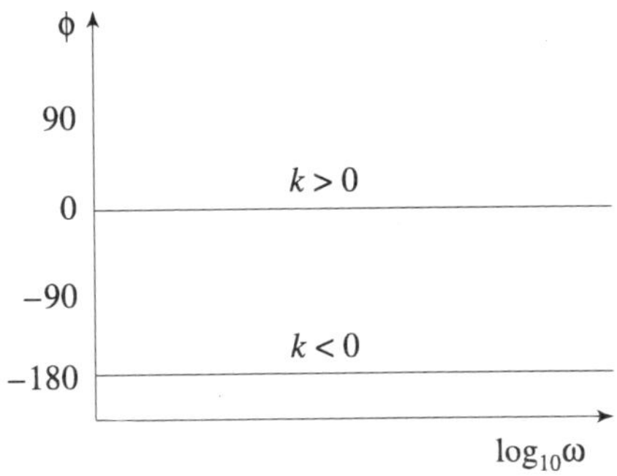

Fig. 7.8

(ii) Factors of the form, $\frac{1}{j\omega}$ that is pole at the origin

$$G(s)H(s) = \frac{1}{s} \quad \text{Put } s = j\omega$$

Let
$$G(s)H(s) = \frac{1}{j\omega}$$

$$G(j\omega)H(j\omega) = \frac{1}{j\omega} = \frac{1}{0+j\omega}$$

$$|G(j\omega)H(j\omega)| = \frac{1}{\sqrt{(0^2+\omega^2)}} = \frac{1}{\omega}$$

Therefore, $|G(j\omega)H(j\omega)| = 20\log_{10}\left(\frac{1}{\omega}\right)$ dB $= -20 \log \omega$ dB

The above equation is similar to $y = mx$. That one pole at the origin contributes to the magnitude plot, i.e., straight line of slope -20 dB

$$1 \text{ pole} = -20 \text{ dB/dec}$$

When $\omega = 0;\ |G(j\omega)H(j\omega)| = -\infty,$

When $\omega = 1;\ |G(j\omega)H(j\omega)| = 0\,\text{dB},$

When $\omega = 10;\ |G(j\omega)H(j\omega)| = -20\,\text{dB},$

When $\omega = 100;\ |G(j\omega)H(j\omega)| = -40\,\text{dB}$

Therefore, the slope of the straight line is -20 dB per unit change in log ω. A unit change means changes by a factor of 10.

Phase Angle $\angle G(j\omega)H(j\omega) = \angle \frac{1}{j\omega} = -\tan^{-1}\frac{\omega}{0} = -90^\circ$

Therefore, at all frequencies this factor contributes a phase angle of -90°

(iii) Zero at the origin $(j\omega)$

Let $G(j\omega)H(j\omega) = j\omega + 0$

$$|G(j\omega)H(j\omega)| = \sqrt{\omega^2} = \omega$$

$$|G(j\omega)H(j\omega)| = 20 \log \omega \text{ dB}$$

When $\omega = 1;\ |G(j\omega)H(j\omega)| = 0$ dB

When $\omega = 10;\ |G(j\omega)H(j\omega)| = 20$ dB

The slope of the straight line is 20 dB per unit change in ω.

Phase angle $\angle G(j\omega)H(j\omega) = \angle j\omega = -\tan^{-1}\frac{\omega}{0} = 90^\circ$

Therefore, at all frequencies this factor contributes a phase angle of 90°.

(iv) Multiple poles at the origin:

Let $$G(j\omega)H(j\omega) = \frac{1}{(j\omega)^N}$$

$$|G(j\omega)H(j\omega)| = \frac{1}{(\sqrt{\omega^N})^2} = \frac{1}{(\omega)^N}$$

$$|G(j\omega)H(j\omega)| = -20 \log(\omega^N) = -20\,\mathrm{N} \log \omega$$

When $\omega = 1;\ |G(j\omega)H(j\omega)| = 0$ dB,

When $\omega = 10;\ |G(j\omega)H(j\omega)| = -20\mathrm{N}$ dB

If $N = 2$ that is $\frac{1}{(j\omega)^2}$ then, $|G(j\omega)H(j\omega)| = -20(2) = -40$ dB

Phase angle $$\angle G(j\omega)H(j\omega) = \angle \frac{1}{(j\omega)^N} = -N - \tan^{-1}\left(\frac{\omega}{0}\right) = -90^\circ(\mathrm{N})$$

$$\frac{1}{(j\omega)^2} = \left(\frac{1}{j\omega}\right)\left(\frac{1}{j\omega}\right)$$

$$\angle \frac{1}{(j\omega)^2} = \left(\frac{1}{90^\circ}\right)\left(\frac{1}{90^\circ}\right) = -90^\circ(2) = -180^\circ$$

Similarly, multiple zeros at origin $|G(j\omega)H(j\omega)| = +20(N)$ dB, $\angle G(j\omega)H(j\omega) = 90\ (N)$

One pole at origin	*One zero at origin*
$G(j\omega)H(j\omega) = \frac{1}{j\omega}$	$G(j\omega)H(j\omega) = j\omega$
$|G(j\omega)H(j\omega)| = \frac{1}{\sqrt{\omega^2}} = \frac{1}{\omega} = -20\log\omega\,\mathrm{dB}$ $\omega = 0.1;\ |G(j\omega)H(j\omega)| = -20\log 0.1 = 20\,\mathrm{dB}$ $\omega = 1;\ |G(j\omega)H(j\omega)| = 0\,\mathrm{dB}$ $\omega = 10;\ |G(j\omega)H(j\omega)| = -20\,\mathrm{dB}$ 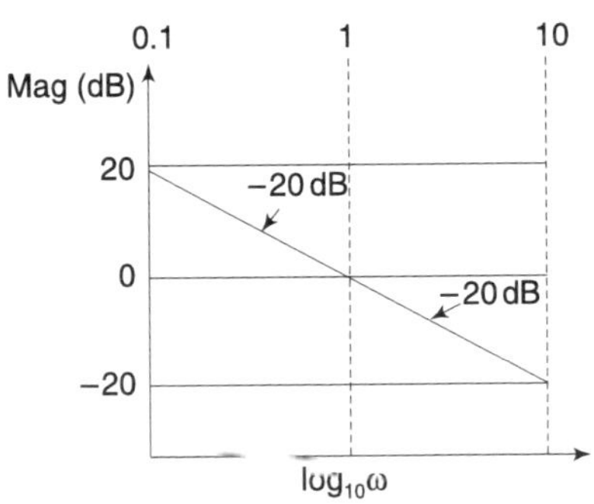 	$|G(j\omega)H(j\omega)| = \sqrt{(\omega^2)} = \omega = 20\log\omega$ $\omega = 0.1;\ |G(j\omega)H(j\omega)| = -20\,\mathrm{dB}$ $\omega = 1;\ |G(j\omega)H(j\omega)| = 0\,\mathrm{dB}$ $\omega = 10;\ |G(j\omega)H(j\omega)| = 20\,\mathrm{dB}$ 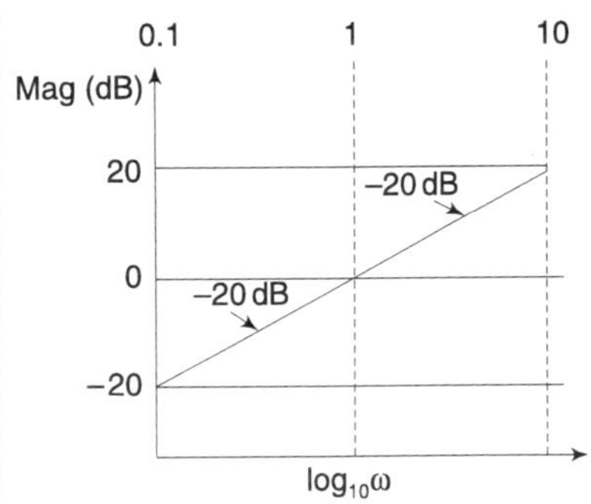

Multiple poles at origin	*Multiple zeros at origin*
$\left(\frac{1}{j\omega}\right)^4$	$(j\omega)^4 = G(j\omega)H(j\omega)$
$\vert -G(j\omega)H(j\omega) = -20(4)\log\omega = -80\log\omega$ $\omega = 0.1;\ G(j\omega)H(j\omega) = 20(4) = 80$ dB $\omega = 1;\ G(j\omega)H(j\omega) = 0(4) = 0$ $\omega = 10;\ G(j\omega)H(j\omega) = -20(4) = -80$ dB	$\vert G(j\omega)H(j\omega)\vert = 20(4)\log\omega = 80\log\omega$ $\omega = 0.1;\ G(j\omega)H(j\omega) = -20(4) = -80$ dB $\omega = 1;\ G(j\omega)H(j\omega) = 0(4) = 0$ $\omega = 10;\ G(j\omega)H(j\omega) = 20(4) = 80$ dB
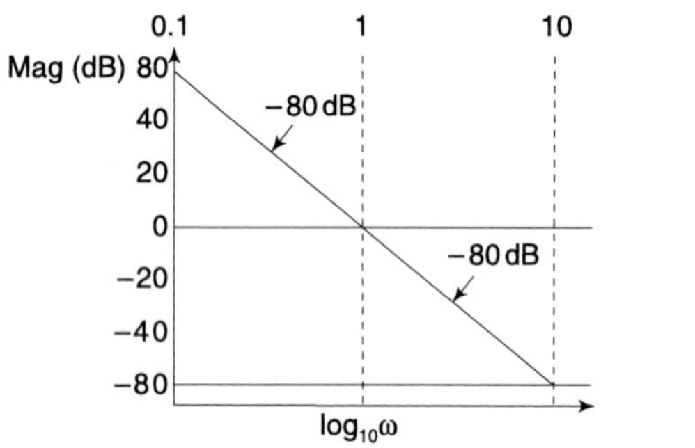	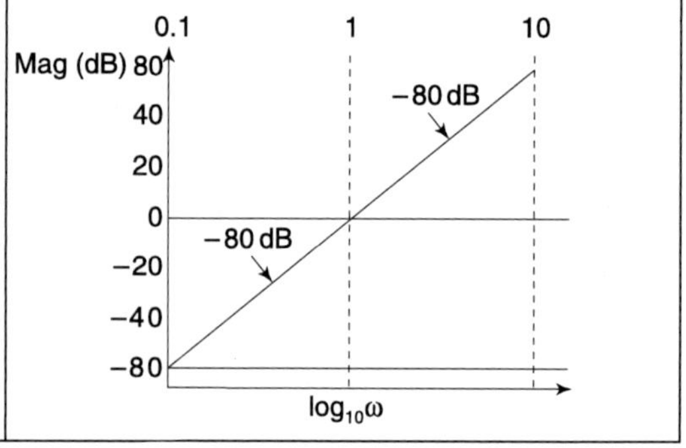

(v) First order factors of the form $\dfrac{1}{1 + j\omega T}$ (poles on real axis)

$$G(j\omega)H(j\omega) = \frac{1}{1 + j\omega T}$$

$$|G(j\omega)H(j\omega)| = \frac{1}{\sqrt{[1^2 + \omega^2 T^2]}} = [\sqrt{(1 + (\omega T)^2}]^{-1} = [1 + (\omega T)^2]^{-1/2}$$

$$= 20 \log (1 + \omega^2 T^2)^{-1/2} = -10 \log (1 + \omega^2 T^2) \text{ dB}$$

For low frequencies, $\omega << 1/T$, i.e., $\omega^2 T^2 << 1$ hence can be neglected.

Therefore, Mag(dB) = $-10 \log (1) = 0$ dB, i.e., straight line of 0 dB only.

For high frequencies, $\omega >> 1/T$, i.e., $\omega^2 T^2 >> 1$

Therefore, Mag (dB) = $-10 \log (\omega^2 T^2) = -20 \log \omega T$ dB

Hence, at high frequencies the log magnitude of this factor is a straight line of slope −20 dB/unit change in ω.

At, $\omega = \dfrac{1}{T}$, the slope is 0 dB $\left(-20\log \dfrac{1}{T} \cdot T = 0 \text{ dB}\right)$, i.e., straight line passes through 0 dB.

The log magnitude of $\dfrac{1}{1 + j\omega T}$ is represented approximately by two straight line asymptotes.

(a) For $\omega < 1/T$, a straight line of 0 dB

(b) For $\omega > 1/T$, a straight line of -20 dB/dec

These two asymptotes intersect at $\omega = \frac{1}{T}$ and this frequency is known as **corner frequency.**

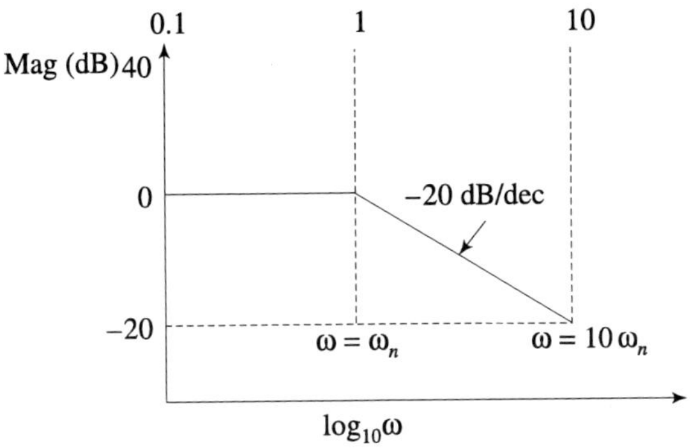

Fig. 7.9

Magnitude plot of $\frac{1}{1 + j\omega T}$

Phase angle of this factor $\frac{1}{1 + j\omega T}$: $\quad \Phi = \angle \frac{1}{1 + j\omega T} = \frac{1}{\tan^{-1} \frac{\omega T}{1}} = -\tan^{-1} \omega T$

At, $\omega = 0$; $\Phi = 0°$, $\omega = \frac{1}{T}$; $\Phi = -45°$, $\omega = \infty$; $\Phi = -90°$

To calculate the error between the exact plot and the asymptotic plot:

For low frequency: $\omega < \frac{1}{T} \Rightarrow |G(j\omega)H(j\omega)| = \frac{1}{\sqrt{[1^2 + \omega T^2]}} = (1 + \omega^2 T^2)^{-1/2}$

$$= -10 \log(1 + \omega^2 T^2)$$

At corner frequency $\omega = \frac{1}{T}$

$$\text{Mag(dB)} = -10 \log (1 + 1) = -10 \log 2 = -3.03 \text{ dB}$$

Similarly, at $\qquad \omega = \frac{2}{T}$

$$\text{Actual Mag(dB)} = -10 \log\left(1 + \frac{4}{T^2} \cdot T^2\right) = -10 \log 5 = -7 \text{ dB}$$

Approximating Mag(dB) $= -10 \log (1 + \omega^2 T^2)$ when $\omega < \frac{1}{T}$

$$= -10 \log (\omega^2 T^2)$$

At $\qquad \omega = \frac{2}{T}$

Approximating Mag(dB) $= -10 \log (4) = 6$ dB

The error at $\omega = \frac{2}{T}$ is 7dB – 6dB = –1 dB

Approximating Mag(dB) = $-10\log(\omega^2T^2)$

General values are shown as tabular form:

Frequency	$\omega = \omega c = \frac{1}{T}$	$\omega = 2\omega c$	$\omega = \frac{\omega c}{2}$
Error	3dB down	1dB down	1dB down

(vi) Factors of the form $\frac{1}{(1+j\omega T)^N}$

This factor is represented by two straight line asymptotes, one of 0 dB at $\omega < \frac{1}{T}$ and of slope of –20 N dB/dec for $\omega > \frac{1}{T}$.

Phase angle: $\Phi(\omega) = \angle\frac{1}{(1+j\omega T)^N} = -N\tan^{-1}\omega T$

(vii) Factors of the form $(1 + j\omega T)$

$$|G(j\omega)H(j\omega)| = \sqrt{1^2+\omega^2T^2} = [(1+(\omega^2T^2)^{1/2} = 10\log(1+\omega^2T^2)\text{ dB}$$

For low frequency $\omega < \frac{1}{T}$, Mag (dB) = 10 log (1) = 0dB

For high frequency $\omega > \frac{1}{T}$, Mag (dB) = $10\log(\omega^2T^2) = 20\log\omega T$

At $\omega = \frac{1}{T}$, 20 log dB/dec

Phase angle

$$\Phi(\omega) = \angle(1+j\omega T) = \tan^{-1}\omega T$$

At $\omega = 0$; $\Phi(\omega) = 0°$

At $\omega = \frac{1}{T}$; $\Phi(\omega) = 45°$

At $\omega = \infty$; $\Phi(\omega) = 90°$

(viii) Factors of the form $\frac{1}{1+j2\xi\frac{\omega}{\omega_n}-\frac{\omega^2}{\omega_n^2}}$ (complex conjugate poles)

For $\xi > 1$, the quadratic factor is factored into two factors of the form $\frac{1}{1+j\omega T}$

For $\xi < 1$, $20\log\frac{1}{\left(1+j2\xi\frac{\omega}{\omega_n}\frac{\omega^2}{\omega_n^2}\right)} = 20\log\frac{1}{\sqrt{\left[\left(1-\frac{\omega^2}{\omega_n^2}\right)^2+\left(2\xi\frac{\omega}{\omega_n}\right)^2\right]}}$

$$\text{Mag (dB)} = -10\log\left[\left(1-\frac{\omega^2}{\omega_n^2}\right)^2 + \left(2\xi\frac{\omega}{\omega_n}\right)^2\right]$$

For low frequency $\omega < \omega_n$, $\dfrac{\omega}{\omega_n} << 1$

$$\text{Mag (dB)} = -10\log[1] = 0\,\text{dB}$$

For high frequency $\omega > \omega_n$, Mag (dB) $= -10\log\left(\dfrac{\omega^2}{\omega_n^2}\right)^2 = -40\log\dfrac{\omega}{\omega_n}$

$$= -40\log\omega + 40\log\omega_n$$

The log magnitude of $G(s) = \dfrac{1}{1 + j2\xi\dfrac{\omega}{\omega_n} - \dfrac{\omega^2}{\omega_n^2}}$ is represented by two straight line asymptotes

(i) For $\omega < \omega_n$, a straight line of value, 0 dB

(ii) For $\omega > \omega_n$, a straight line of value, -40 dB/dec

Centre frequency is $\omega_c = \omega_n$, where these two asymptotes would meet.

The phase angle is $\Phi(\omega) = -\tan^{-1}\left[\dfrac{\dfrac{2\xi\,\omega}{\omega_n}}{1-\dfrac{\omega^2}{\omega_n^2}}\right] = -\tan^{-1}\left(\dfrac{2\xi\omega_n}{\omega_n^2-\omega^2}\right)$

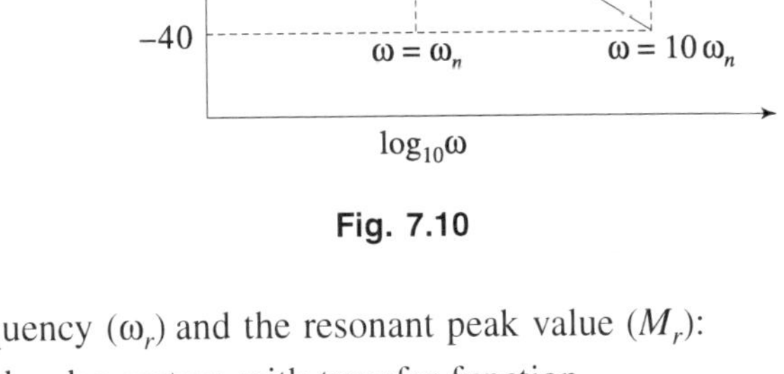

Fig. 7.10

The resonant frequency (ω_r) and the resonant peak value (M_r):

Consider a second order system with transfer function,

$$G(s) = \frac{1}{1 + 2\xi\dfrac{s}{\omega_n} + \dfrac{s^2}{\omega_n^2}}$$

The sinusoidal transfer function is: $G(j\omega) = \dfrac{1}{1 + j2\xi \dfrac{\omega}{\omega_n} - \dfrac{\omega^2}{\omega_n^2}}$

$$|G(j\omega)| \frac{1}{\sqrt{1\left[\left(1 - \dfrac{\omega^2}{\omega_n^2}\right)^2 + \left(2\xi \dfrac{\omega}{\omega_n}\right)^2\right]}} \quad ...(7.19)$$

The magnitude $G(j\omega)$ is a function of frequency and the magnitude has a peak value at a frequency known as resonant frequency (ω_r).

To determine resonant frequency (ω_r): The magnitude of $|G(j\omega)$ has a peak when the denominator of Eq. (7.19) has a minimum value.

Let $$f(\omega) = \left(1 - \frac{\omega^2}{\omega_n^2}\right)^2 + \left(2\xi \frac{\omega}{\omega_n}\right)^2$$

$$\frac{df(\omega)}{d\omega} = 2\left(1 - \frac{\omega^2}{\omega_n^2}\right)\left(\frac{-2\omega}{\omega_n^2}\right) + \frac{4\xi^2(2\omega)}{\omega_n^2}$$

At $$\omega = \omega_r, \quad \frac{df(\omega)}{d\omega} = 0$$

$$\frac{df(\omega)}{d\omega} = 2\left(1 - \frac{\omega_r^2}{\omega_n^2}\right)\left(-\frac{2\omega_r}{\omega_n^2}\right) + \frac{4\xi^2(2\omega_r)}{\omega_n^2} = 0$$

$$\Rightarrow \quad \frac{4\omega_r}{\omega_n^2}\left\{-\left(1 - \frac{\omega_r^2}{\omega_n^2}\right) + 2\xi^2\right\} = 0.$$

$$\Rightarrow \quad -\left(1 - \frac{\omega_r^2}{\omega_n^2}\right) + 2\xi^2 = 0$$

$$\Rightarrow \quad \left(1 - \frac{\omega_r^2}{\omega_n^2}\right) = 2\xi^2$$

$$\Rightarrow \quad \frac{\omega_r^2}{\omega_n^2} = 1 - 2\xi^2 \Rightarrow \omega_r^2 = \omega_n^2(1 - 2\xi^2)$$

$$\Rightarrow \quad \omega_r = \omega_n\sqrt{(1 - 2\xi^2)}$$

Therefore, the resonance peak occurs at: $\omega_r = \omega_n \boldsymbol{\sqrt{(1 - 2\xi^2)}}$ at $\xi < 0.707$

The magnitude at $\omega = \omega_r$ is: $M_r = |G(j\omega)|_{\omega = \omega_r} = \dfrac{1}{\sqrt{\left(1 - \dfrac{\omega_r^2}{\omega_n^2}\right)^2 + \left(2\xi\dfrac{\omega_r}{\omega_n}\right)^2}} = \dfrac{1}{2\xi\sqrt{(1 - \xi^2)}}$

Example 7.1: Determine the corner frequencies of the following transfer functions:

(a) $G(s) = \dfrac{5}{1 + 2S}$ (b) $G(j\omega) = \dfrac{2}{1 + j5\omega}$

(c) $G(j\omega) = 10\ (1 + j4\omega)^2$

Solution: The break frequency or corner frequency is that value of ω which makes the *j* term equal to 1. The corner frequency is denoted by ω_c. Therefore, at corner frequency

$$\omega = \omega_c$$

(a)
$$G(S) = \frac{5}{1 + 2S}$$

Put
$$S = j\omega$$

$$G(j\omega) = \frac{5}{1 + j\omega 2}$$

$$2\omega_c = 1$$

$$\omega_c = \frac{1}{2} = 0.5 \text{ rad/sec}$$

(b)
$$G(j\omega) = \frac{2}{1 + j5\omega}$$

$$5\omega_c = 1$$

$$\omega_c = \frac{1}{5} = 0.2 \text{ rad/sec}$$

(c)
$$G(j\omega) = 10\ (1 + j4\omega)^2$$

$$4\omega_c = 1$$

$$\omega_c = \frac{1}{4} = 0.25 \text{ rad/s}$$

Example 7.2: Sketch the asymptotic Bode plot for the system having $G(s) = \dfrac{20}{S(1 + 0.1S)}$.

Solution: The given function in time constant form

$$G(s) = \frac{20}{S(1 + 0.1S)}$$

Put $S = j\omega$ in the above equation

$$G(s)\big|_{S = j\omega} = G(j\omega) = \frac{20}{j\omega\ (1 + j\ 0.1\omega)}$$

Identify factors: The factors of the transfer function in order of their occurrence as frequency increases are as follows.

- Constant gain, $K = 20$
- Pole at the origin $= \dfrac{1}{j\omega}$
- Factor $\dfrac{1}{1 + j0 \cdot 1\omega}$, pole at $S = -10$

Factors in tabular form in ascending order of corner frequencies.

Factor	Corner frequency	Slope	Resultant slope
$K = 20$	None	0 dB/dec	The magnitude $20 \log K = 20 \log 20 = 26$ dB This is represented by a horizontal line of slope 0 dB/dec i.e., parallel to 0 dB axis and starting from 26 dB
$\frac{1}{j\omega}$	None	−20 dB/dec	−20 dB/dec Straight line of constant slope −20 dB/dec through $w = 1$ rad/sec
$\frac{1}{1 + j\,0.1\omega}$	$\omega_c = 10$	−20 dB/dec	−20 dB/dec −20 dB/dec = −40 dB/dec

Magnitude plote The magnitude of the first constant term $K = 20$ is $20 \log K = 20 \log 20 = 26$ dB.

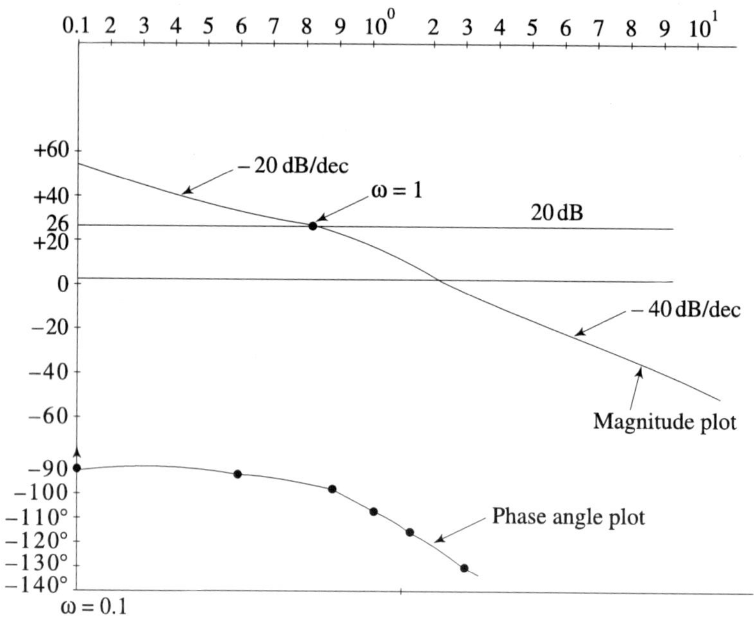

Fig 7.11

Phase angle plot $\quad G(j\omega) = \dfrac{20}{j\omega\,(1 + j\,0.1\omega)}$

The resultant phase angle is

$$\phi(\omega) = \angle G(j\omega) = \angle \frac{1}{j\omega} - \angle \frac{1}{1 + j\,0.1\omega}$$

$$\phi(\omega) = -90 - \tan^{-1} 0.1\,\omega$$

ω rad/s	φ(ω)
0.1	– 90.57°
0.5	– 92.9°
1	– 95.7°
2	– 101.3°
3	– 106.7°
5	– 116.6°
10	– 135°

Example 7.3: Construct the Bode plots for a unity feedback control system having $G(s) = \dfrac{2000}{S(S+1)(S+100)}$.

From the Bode plots determine:

(a) Gain crossover frequency
(b) Phase crossover frequency
(c) Gain margin
(d) Phase margin

Comment on the stability of the system.

Solution: $G(s) = \dfrac{2000}{S(S+1)(S+100)}$

The given transfer function is in pole zero form. Convert it into time constant form.

$$G(s) = \frac{2000}{S(S+1)\,100\left(1+\dfrac{S}{100}\right)} = \frac{20}{S(1+S)(1+0.01S)}$$

Put $S = j\omega$

$$G(j\omega) = \frac{20}{j\omega(1+j\omega)(1+j\,0.01\omega)}$$

Arrange the factors in order of increasing frequency:

- Constant gain, $K = 20$, the corner frequency is none.
- For the factor $\dfrac{1}{j\omega}$, the corner frequency is none.
- For the factor $\dfrac{1}{1+j\omega}$, the corner frequency is 1 rad/sec.
- For the factor $\dfrac{1}{1+j\,0.01\,\omega}$, the corner frequency is 100 rad/sec.

Magnitude plot

Factor	Corner frequency	slope	Resultant slope
$K = 20$	None	0 dB/dec	20 log K = 20 log 20 = 26 dB
$\frac{1}{j\omega}$	None	−20 dB/dec	−20 dB/dec
$\frac{1}{1 + j\omega}$	$\omega_c = 1$ rad/sec	−20 dB/dec	−20 − 20 = −40 dB/dec
$\frac{1}{1 + j0.01\,\omega}$	$\omega_c = 100$	−20 dB	−40 dB −20 dB = −60 dB/dec

Phase angle plot

$$G(j\omega) = \frac{20}{j\omega(1 + j\omega)(1 + j\,0.01\omega)}$$

$$\phi(\omega) = \underline{|G(j\omega)} = \underline{|0^\circ} + \underline{\left|\frac{1}{j\omega}\right.} + \underline{\left|\frac{1}{1+j\omega}\right.} + \underline{\left|\frac{1}{1+j0.01\omega}\right.}$$

$$\phi(\omega) = 0^\circ - 90 - \tan^{-1}\omega - \tan^{-1}(0.01)\,\omega$$

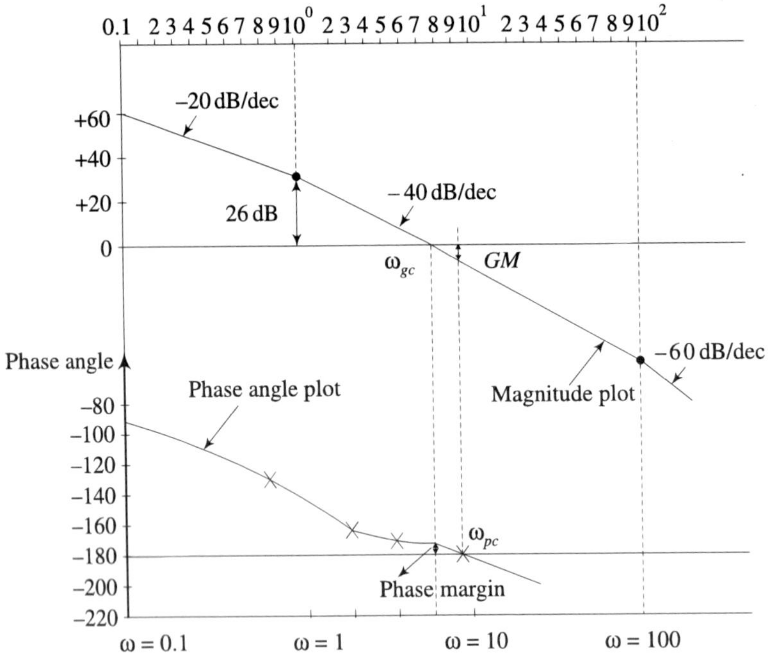

Fig. 7.12

The value of phase angle ϕ for different frequencies

ω	$\phi(\omega)$
0.1	– 95.75
1.0	– 135.57
3	– 163.28
5	– 171.6°
10	– 180°
50	– 205.41°

The complete magnitude and phase angle Bode plots.

Gain crossover frequency ω_{gc} and phase margin *PM*.

To determine the gain crossover frequency, ω_{gc}, find the point at which the magnitude plot crosses the 0 dB line. From the plot it is found that

$$\omega_{gc} = (4.5) \text{ rad/sec}$$

From this point draw a vertical line up to the phase angle plot which cuts the phase curve at – 172°

The phase margin is $PM = 180° - 172° = 8°$

Phase crossover frequency (ω_{pc}) and gain margin (*GM*). To determine the phase crossover frequency (ω_{pc}) find the point at which the phase angle plot crosses the 180° line. From the plot it is found that

$$\omega_{pc} = 10 \text{ rad/sec}$$

At this frequency draw a vertical line up to the magnitude curve.

This cuts the magnitude curve at magnitude – 14 dB

∴ The gain margin is + 14 dB

Stability: Since both phase margin and gain are positive, the system is absolutely stable.

Example 7.4: Draw the Bode plot for the transfer function

$$G(s) = \frac{36\,(1 + 0.2S)}{S^2(1 + 0.05S)\,(1 + 0.01S)}$$

From the Bode plot determine

(a) Phase crossover frequency (b) Gain crossover frequency

(c) Gain margin (d) Phase margin

Comment on the stability of the system.

Solution:

$$G(s) = \frac{36(1 + 0.2S)}{S^2\,(1 + 0.05S)\,(1 + 0.01S)}$$

Put $$S = j\omega$$

$$G(s)\big|_{S=j\omega} = G(j\omega) = \frac{36(1+j0.2\,\omega)}{(j\omega)^2\,(1+j0.05\,\omega)\,(1+j0.0\,1\,\omega)}$$

Magnitude plot

Factor	Corner frequency	Slope	Resultant slope
$\frac{36}{(j\omega)^2}$	None	−40 dB/dec	−40 dB/dec
$(1 + 0.2\,j\omega)$	5	+20 dB/dec	−40 + 20 = −20 dB/dec
$\frac{1}{1+j0.05\,\omega}$	20	−20 dB/dec	−20 − 20 = −40 dB/dec
$\frac{1}{1+j0.01\,\omega}$	100	−20 dB/dec	−40 − 20 = −60 dB/dec

The system is type 2, hence the initial slope of the Bode magnitude plot is −40 dB/dec intersects at $\omega = \sqrt{K} = \sqrt{36} = 6$ rad/sec.

$$20 \log K = 20 \log 36 = 31.12 \text{ dB}$$

Phase angle plot: The resultant phase angle is given by

$$\phi(\omega) = \underline{|G(j\omega)}$$

$$= \underline{|0} + \underline{|1 + j\,0.2\,\omega} + \underline{\left|\left(\frac{1}{j\omega}\right)^2\right.} + \underline{\left|\frac{1}{1+j\,0.05\,\omega}\right.} + \underline{\left|\frac{1}{1+j\,0.001\,\omega}\right.}$$

$$\phi(\omega) = 0 - \tan^{-1} 0.02\,\omega - 180^\circ - \tan^{-1} 0.05\,\omega - \tan^{-1} 0.001\,\omega$$

ω	$\phi(\omega)$
1	− 172.1°
5	− 151.9°
10	− 148.8°
30	− 172.5°
50	− 190.4°
100	− 216.6°

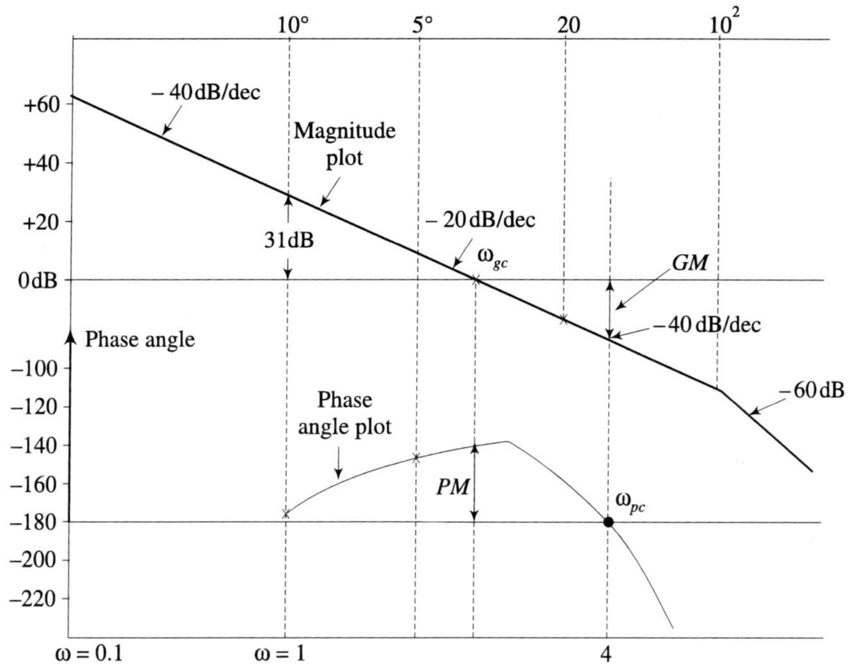

Fig. 7.13

Gain crossover frequency (ω_{gc}) and phase margin *PM*).

To determine the gain crossover frequency (ω_{gc}) find the point at which the magnitude plot crosses the 0 dB axis. From the plot it is found that ω_{gc} = 7.2 rad/sec.

From the vertical line up to the phase angle plot which cuts the phase curve at – 150°

∴ Phase margin is $PM = 180 - 150 = 30°$

Phase crossover frequency (ω_{pc}) and gain margin (*GM*): To determine the phase crossover frequency (ω_{pc}), find a point at which the phase angle plot crosses the 180° line. From the plot it is found that ω_{pc} = 37 rad/sec.

At this frequency draw a vertical line up to the marginal curve

This cuts magnitude curve at magnitude – 20 dB

∴ The gain margin is + 20 dB

Stability: Since both phase margin and gain margin are positive, the system is absolutely stable.

Example 7.5: The open-loop transfer function of a unity feedback system is

$$G(s) = \frac{K}{S(S+1)(S+10)}$$

Draw the Bode plot and determine

(a) Limting value of K for the system to be stable

(b) The value K for gain margin of 7 dB

(c) The value of K for phase margin of 40°.

Solution: $G(s) = \dfrac{K}{S(S+1)(S+10)}$

Convert it into time constant form

$$G(s) = \frac{K}{S(1+S)10(1+0.1S)} = \frac{0.1\,K}{S(1+S)(1+0.1S)}$$

Let $0.1\,K = 1$, then

$$G(s) = \frac{1}{S(1+S)(1+0.1S)}$$

Put $S = j\omega$

$$G(s)\big|_{S=j\omega} = G(j\omega) = \frac{1}{j\omega(1+j\omega)(1+j0.1\,\omega)}$$

Magnitude plot

Factor	*Corner frequency*	*Slope*	*Resultant slope*
$\dfrac{1}{j\omega}$	None	−20 dB/dec	−20 dB/dec
$\dfrac{1}{1+j\omega}$	$\omega_c = 1$	−20 dB/dec	−20 − 20 = −40 dB/dec
$\dfrac{1}{1+j0.1\,\omega}$	$\omega_c = 10$	−20 dB/dec	−40 − 20 = −60 dB/dec

Phase angle plot

$$\phi(\omega) = \angle G(j\omega)$$

$$= 0° + \angle\frac{1}{j\omega} + \angle\frac{1}{1+j\omega} + \angle\frac{1}{1+j0.1\omega}$$

$$\phi(\omega) = -90 - \tan^{-1}\omega - \tan^{-1}(0.1\,\omega)$$

ω	$\phi(\omega)$
1	− 140.7°
2	− 164.7°
3	− 178.2°
4	− 187.8°
10	− 219.3°

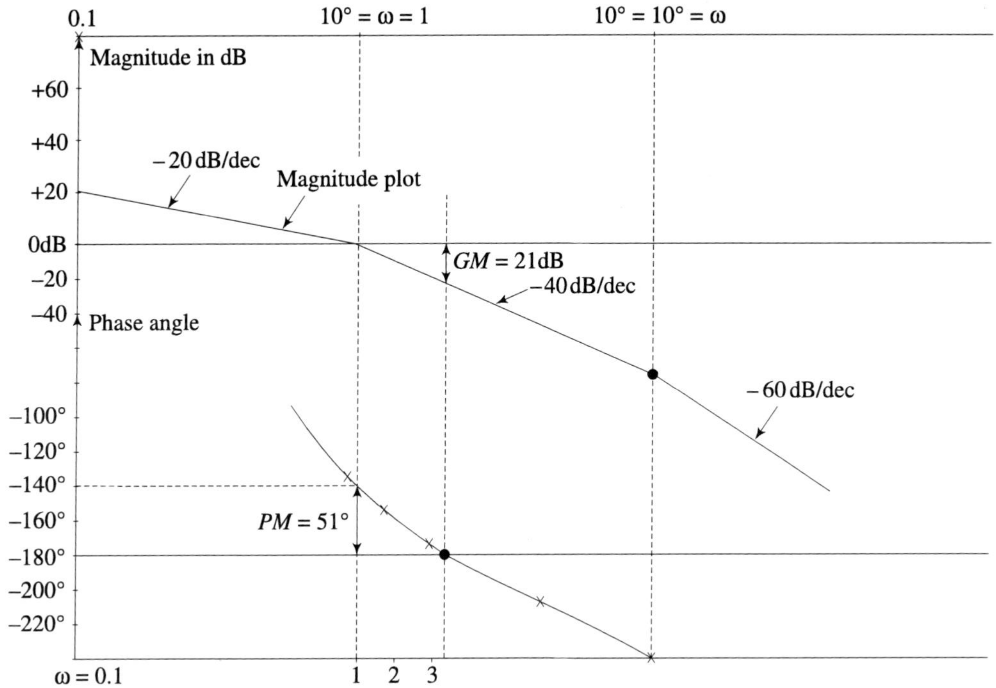

Fig. 7.14

(a) From the Bode plot, the gain margin is 21 dB

$$20 \log K = 21$$

$$K = 11.22$$

$$0.1\, K = 11.22$$

$$K = 112.2$$

(b) For the gain margin to be 7 dB, the graph is to be lifted by

$$21 - 7 = 14 \text{ dB}$$

$$20 \log K = 14$$

$$K = 5$$

$$0.1\, K = 5$$

$$K = 50$$

(c) For the phase margin to be 40°, the value of ω at (– 180 + 40) = – 140° is 3.5 rad/sec gain margin at 3.5 rad/sec is 4 dB.

∴ In order to have a phase margin of 40°, magnitude plot has to be lifted by 4 dB so that the gain crossover frequency is 3.5 rad/s

$$20 \log K = 4$$

$$K = 1.58$$
$$0.1\ K = 1.58$$
$$K = 15.8$$

Example 7.6: Draw the Bode magnitude (asymptotic) plot and phase angle plot for the open-loop transfer function.

$$G(s)H(s) = \frac{48(s + 10)}{s(s + 20)(s^2 + 2.4s + 16)}.$$

Solution: Rewrite the transfer function in time constant form.

$$G(s)H(s) = \frac{48\left(\frac{s}{10} + 1\right) \times 10}{s\left(\frac{s}{20} + 1\right) 20 \left(\frac{s^2}{16} + \frac{2.4}{16} s + 1\right)16}$$

$$G(s)H(s) = \frac{1.5\left(1 + \frac{s}{10}\right)}{s\left(1 + \frac{s}{20}\right)\left(1 + \frac{2.4}{16} s + \frac{s^2}{16}\right)}$$

Obtain the sinusoidal transfer function by replacing s by $j\omega$

$$G(j\omega)H(j\omega) = \frac{1.5\left(1 + \frac{j\omega}{10}\right)}{j\omega\left(1 + \frac{j\omega}{20}\right)\left(1 + \frac{j2.4\omega}{16} - \frac{\omega^2}{16}\right)}$$

List the factors and their corner frequencies:

Factor	*Corner frequencies*
(a) Constant gain, $K = 1.5$	None
(b) $\frac{1}{j\omega}$	None
(c) $1 + \frac{j\omega}{10}$	$\omega_c = 10$
(d) $\frac{1}{1 + \frac{j\omega}{20}}$	$\omega_c = 20$
(e) $\frac{1}{1 + \frac{j2.4\omega}{16} - \frac{\omega^2}{16}}$ $[\omega_n^2 = 16,\ \omega_n = 4]$	$\omega_c = 4$

Factors in ascending order of corner frequencies.

Factor	*Corner frequency (ωc)*	*Asymptotic log magnitude characteristic*	*Resultant slope*
$\frac{1.5}{j\omega}$	None	Straight line of slope = –20 dB/dec passing through 20 log 1.5 = 3.52 dB at ω = 1. It passes through 0 dB at ω = 1.5	–20 dB/dec
$\frac{1}{1+\frac{j2.4\omega}{16}-\frac{\omega^2}{16}}$	$\omega_1 = 4$	Straight line of 0 dB for $\omega < \omega_1$. Straight line of slope –40 dB/dec for $\omega > \omega_1$.	–60 dB/dec
$1+\frac{j\omega}{10}$	$\omega_2 = 10$	Straight line of 0 dB for $\omega < \omega_2$. Straight line of slope +20 dB/dec for $\omega > \omega_2$.	–40 dB/dec
$1+\frac{j\omega}{20}$	$\omega_3 = 20$	Straight line of 0 dB for $\omega < \omega_3$. Straight line of slope – 20 dB/dec for $\omega > \omega_3$.	–60 dB/dec

7.3.2 Procedure to Draw Asymptotic Log Magnitude Plot

(a) At frequencies less than ω_1, only the factor $\frac{1.5}{j\omega}$ is effective. All other factors have zero value. Therefore, the slope of the resultant plot for $\omega < \omega_1$ is –20 dB/dec. At $\omega = 1$ the resultant plot has a value 3.52 dB and $\omega = 1.5$ the resultant plot has a value 0 dB.

(b) Above ω_1, the factor $\frac{1}{1+\frac{j2.4\omega}{16}-\frac{\omega^2}{16}}$ has a slope of –40 dB/dec and this slope is added to the previous slope to get resultant slope of the plot between ω_1 and ω_2. The resultant slope is –20 + (–40) = –60 dB/dec.

(c) Above ω_2, the factor $1+\frac{j\omega}{10}$ has a slope of +20 dB/dec and this slope is added to the previous slope, (i.e., – 60 dB/dec) to get resultant slope of the plot between ω_2 and ω_3. The resultant slope is –60 + 20) = –40 dB/dec.

(d) Above ω_3, the factor $\frac{1}{1+\frac{j\omega}{20}}$ has a slope of –20 dB/dec and this slope is added to the previous slope, (i.e., –40 dB/dec) to get resultant slope of the plot above ω_3. The resultant slope is –40 + (–20) = –60 dB/dec.

The asymptotic log magnitude curve is drawn on semilog sheet. The starting frequency on frequency scale should be less than the lowest corner frequency (i.e., ω_1). If ω_0 is the starting frequency then the magnitude at ω_0 is calculated by considering $\left|\frac{K}{j\omega}\right|$ factor.

Here ω_0 is selected as 0.1 rad/sec

Therefore, magnitude at ω_0 is $20\log|G(j\omega)| = 20\log K - 20\log\omega_0$

$$= 20\log 1.5 - 20\log 0.1 = 23.5\,\text{dB}.$$

Phase angle curve

The phase angle of the sinusoidal transfer function is given by

$$\angle G(j\omega)H(j\omega) = \tan^{-1}\frac{\omega}{10} - 90 - \tan^{-1}\frac{\omega}{20} - \tan^{-1}\left(\frac{\dfrac{2.4\omega}{16}}{1-\dfrac{\omega^2}{16}}\right)$$

$$\angle G(j\omega)H(j\omega) = \tan^{-1} 0.1\omega - 90^\circ - \tan^{-1} 0.05 - \tan^{-1}\frac{2.4\omega}{16-\omega^2}.$$

Phase angle is calculated at various frequencies

Ω	$\angle G(j\omega)H(j\omega)$
1	−96.24°
2	−106.2°
4	−169.5°
5	−204.3°
8	−233°
10	−235.6°
20	−244.4°

Note: For $\omega > 4$, (i.e., $\omega > \omega_n$).

$$\tan^{-1}\left(\frac{2.4\omega}{16-\omega^2}\right) = \tan^{-1}\left(\frac{-2.4\omega}{\omega^2-16}\right) + 180^\circ$$

By using the above data, phase angle curve is drawn.

Bode magnitude and phase angle plots are shown in Fig. 7.15.

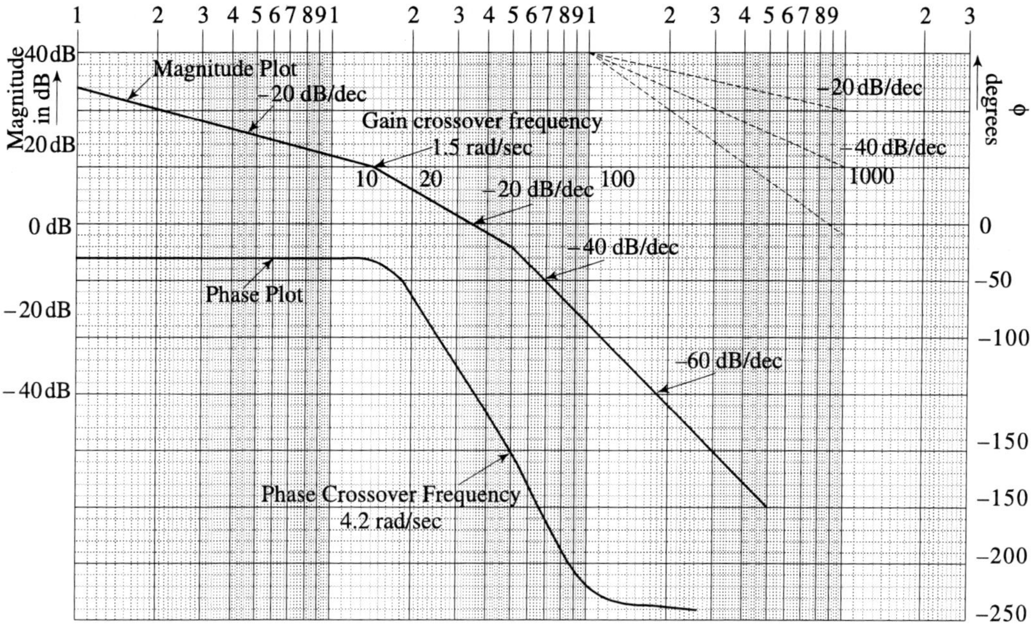

Fig. 7.15

To Determine Gain Margin and Phase Margin

The crossing of zero dB line by magnitude plot is known as gain crossover frequency, ω_g. From the plot, $\omega_g = 1.5$ rad/sec.

At ω_g draw a perpendicular line which intersects the phase plot and obtain the phase angle at ω_g, from the plot it is $-100°$.

Phase margin (PM) = 180° + Phase angle at gain crossover frequency

$$= 180° - 100° = 80°$$

The frequency at which phase angle is $-180°$ is known as phase crossover frequency, ω_p.

From the plot, $\omega_p = 4.2$ rad/sec.

At this point draw a perpendicular line which intersects the magnitude plot. The point of intersection gives the magnitude at phase crossover frequency. From the plot it is -10 dB.

Gain margin (GM) = − [Magnitude at crossover frequency]

$$= -(-10) = 10 \text{ dB}$$

Example 7.7: The open-loop transfer function of a unity feedback system is given by

$$G(s) = \frac{K(1 + 0.5s)}{s(1 + 2s)(1 + 0.05s + 0.125^2s^2)}$$

Taking $K = 4$, draw the Bode plot (both magnitude and phase plots) using the asymptote approximation. Determine the phase margin and gain margin.

Solution: The transfer function is in time constant form

$$G(s) = \frac{K(1 + 0.5s)}{s(1 + 2s)(1 + 0.05s + 0.125^2s^2)}$$

$$= \frac{4\left(1 + \frac{s}{2}\right)}{s\left(1 + \frac{s}{0.5}\right)\left(1 + \frac{0.05s(8)}{8} + \frac{s^2}{8^2}\right)}$$

$$= \frac{4\left(1 + \frac{s}{2}\right)}{s\left(1 + \frac{s}{0.5}\right)\left(1 + 2\,\frac{0.2s}{8} + \frac{s^2}{8^2}\right)}$$

Obtain the transfer function by replacing s by $j\omega$

$$G(j\omega)H(j\omega) = \frac{4\left(1 + \frac{j\omega}{2}\right)}{j\omega\left(1 + \frac{j\omega}{0.5}\right)\left(1 + \frac{j2(0.2)\omega}{8} - \frac{\omega^2}{8^2}\right)}.$$

List the factors and their corner frequency

Factor	Corner frequency (ωc)
Constant gain, $K = 4$	None
$\frac{1}{j\omega}$	None
$1 + \frac{j\omega}{2}$	$\omega_c = 2$
$\frac{1}{1 + \frac{j\omega}{0.5}}$	$\omega_c = 0.5$
$\frac{1}{1 + j2(0.2)\frac{\omega}{8} - \frac{\omega^2}{8^2}}$	$\omega_c = 8$

Factors in ascending order of corner frequencies

Factor	Corner frequency (ωc)	Slope of the factor for $\omega > \omega c$*	Resultant slope
$\frac{4}{j\omega}$	None	−20 dB/dec	−20 dB/dec
$\frac{1}{1+\frac{j\omega}{0.5}}$	$\omega_1 = 0.5$	−20 dB/dec	−40 dB/dec
$1+\frac{j\omega}{2}$	$\omega_2 = 2$	20 dB/dec	−20 dB/dec
$\frac{1}{1+j2(0.2)\frac{\omega}{8}-\frac{\omega^2}{8^2}}$	$\omega_3 = 8$	−40 dB/dec	−60 dB/dec

*All factors have a value of 0 dB for $\omega < \omega_c$ except $\frac{4}{j\omega}$.

Starting frequency ω_o is selected as 0.1 rad/sec

Magnitude at $\omega = 0.1$ is $20\log|G(j\omega)| = 20\log K - 20\log \omega_o$

$$= 20\log 4 - 20\log 0.1 = 32 \text{ dB}$$

The asymptotic log magnitude curve is drawn.

The phase angle of the transfer function is:

$$G(j\omega) = -90^\circ + \tan^{-1}(0.5\omega) - \tan^{-1}(2\omega) - \tan^{-1}\left(\frac{2(0.2)\left(\frac{\omega}{8}\right)}{1-\frac{\omega^2}{8^2}}\right)$$

$$= -90^\circ + \tan^{-1}(0.5\omega) - \tan^{-1}(2\omega) - \tan^{1}\left(\frac{3.2\omega}{64-\frac{\omega^2}{1}}\right)$$

ω	$\Phi(\omega)$
0.1	−104.4°
0.5	−122.4°
1.0	−129.7°
2.0	−127.1°
3.0	−124°
5	−128.4°
7	−158.1°
8	−190.4°
9	−219.9°
10	−236.8°

The magnitude and phase plots are shown in Fig. 7.16.

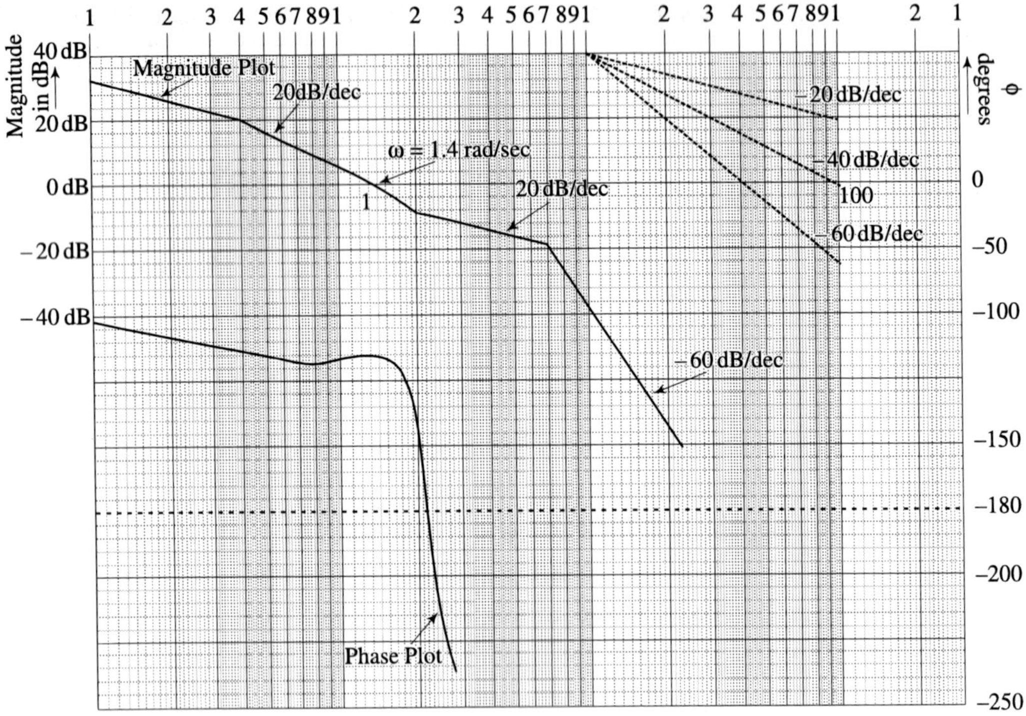

Fig. 7.16

To determine gain margin and phase margin:

Here ω_p = 7.8 rad/sec. the magnitude at phase crossover frequency is –18 dB

Therefore, gain margin = – [magnitude at phase crossover frequency] = 18 dB

Here ω_g = 1.4 rad/sec. the phase angle at gain crossover frequency is – 129°

Therefore, phase margin = 180° + phase angle at gain crossover frequency

$$= 180° - 129° = 51°.$$

Example 7.8: The open-loop transfer function of a unity feedback control system is $G(s) = \dfrac{90(1 + 0.5s)}{(1 + 0.1s)(1 + 2s)(1 + 0.02s)}$. Draw the Bode plot (both magnitude and phase plots). Determine phase and gain margins.

Solution:

$$G(s) = \frac{90(1 + 0.5s)}{(1 + 0.1s)(1 + 2s)(1 + 0.02s)}$$

$$G(j\omega)H(j\omega) = \frac{90(1 + j0.5\omega)}{(1 + j0.1\omega)(1 + j2\omega)(1 + j0.02\omega)}$$

$$G(j\omega)H(j\omega) = \frac{90\left(1+\frac{j\omega}{2}\right)}{\left(1+\frac{j\omega}{10}\right)\left(1+\frac{j\omega}{0.5}\right)\left(1+\frac{j\omega}{50}\right)}$$

The corner frequencies (CF) are at $\omega = 0.5$, 10, 50 (poles) and $\omega = 2$ (zero). Assume the origin as $\omega_o = 0.1$ rad/sec.

Magnitude at origin 20 log 90 = 39.08 dB

CF	*Factor*	*Slope at CF in dB/dec*	*Resultant slope in dB/dec*
None	90	0 dB/dec	0 dB/dec
0.5	$\frac{1}{1+j2\omega}$	−20 dB/dec	−20 dB/dec
2	$(1+j0.5\omega)$	20 dB/dec	0 dB/dec
10	$\frac{1}{1+j0.1\omega}$	−20 dB/dec	−20 dB/dec
50	$\frac{1}{1+j0.02\omega}$	−20 dB/dec	−40 dB/dec

$$\Phi = \tan^{-1} 0.5\omega - \tan^{-1} 0.1\omega - \tan^{-1} 2\omega - \tan^{-1} 0.02\omega$$

ω	$\Phi(\omega)$
0.1	−9°
1	−43.8°
5	−48.4°
10	−64.8°
20	−89.6°
50	−125.6°
100	−148.6°
200	−163.5°
500	−173°
1000	−176°

Magnitude and phase plots are shown in Fig. 7.10.

At $\omega_g = 105$ rad/sec, phase angle = −150°. Therefore, PM = 180° − 150° = 30°

Since $\Phi = 180°$ occurs at $\omega = \infty$, the gain margin is always positive

Hence, the system is stable.

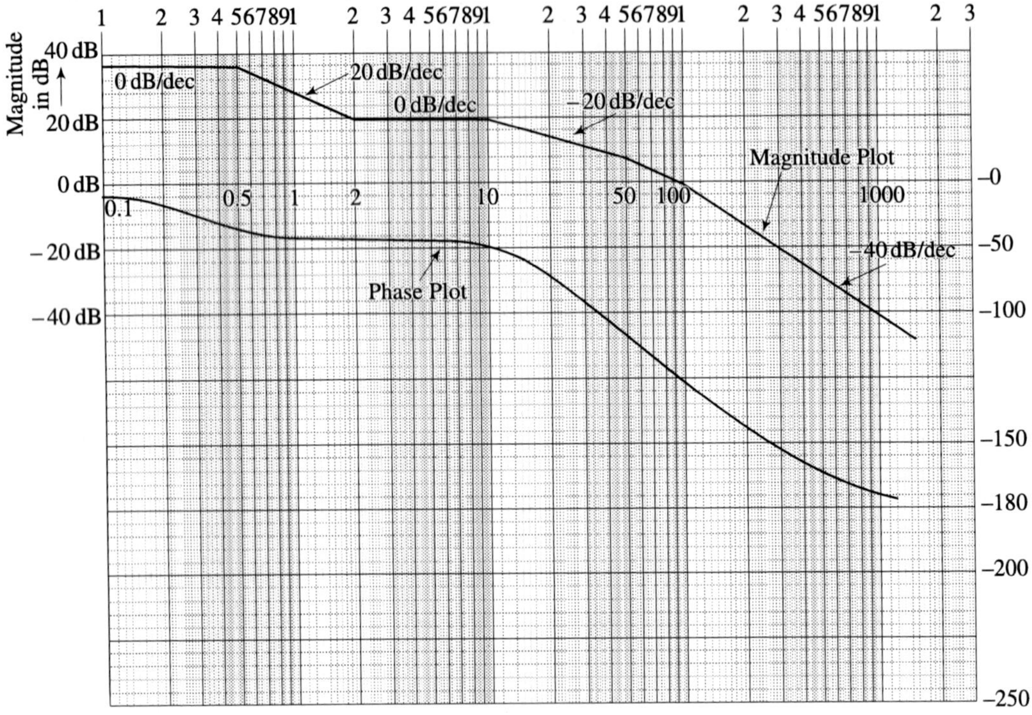

Fig. 7.17

Example 7.9: Determine the open-loop transfer function of a system whose approximate plot is shown in Fig. 7.18.

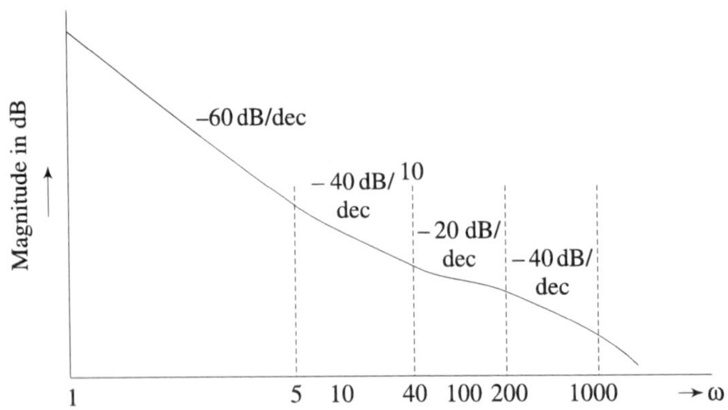

Fig. 7.18

Solution:

1. Initial slope is -60 dB/dec. This indicates factor of the form, $\frac{K}{(j\omega)^3}$, i.e., 3 poles at the origin. A single pole at the origin, $\left(\text{i.e., } \frac{1}{j\omega}\right)$ contributes a slope of -20 dB/dec. Since the slope is -60 dB/dec, there are 3 poles at the origin.

2. At the first corner frequency, $\omega_1 = 5$, the resultant slope changes by $-40 - (-60) = 20$ dB/dec. This is due to a factor of the form $\left(1 + \frac{j\omega}{5}\right)$, i.e., a zero on the real axis.

3. At the second corner frequency, $\omega_2 = 40$, the resultant slope changes by $-20-(-40) = 20$ dB/dec. This is due to a factor of the form $\left(1 + \frac{j\omega}{40}\right)$.

4. At the third corner frequency, $\omega_3 = 200$, the resultant slope changes by $-40-(-20) = -20$ dB/dec. This is due to a factor of the form $\frac{1}{\left(1 + \frac{j\omega}{200}\right)}$, i.e., a pole on real axis.

5. At the corner frequency, $\omega_4 = 1000$, the resultant slope changes by $-60-(-40) = -20$ dB/dec. This is due to a factor of the form $\frac{1}{\left(1 + \frac{j\omega}{1000}\right)}$.

The transfer function is of the form $GH(j\omega) = \dfrac{K\left(1 + \frac{j\omega}{5}\right)\left(1 + \frac{j\omega}{40}\right)}{(j\omega)^3\left(1 + \frac{j\omega}{200}\right)\left(1 + \frac{j\omega}{1000}\right)}$.

To determine K:

Asymptote for $\omega < 5$ is given by the equation, $20 \log |G(j\omega)| = 20 \log K - 20 \log \omega^3$

At $\qquad \omega = 1,\ 20 \log |G(j\omega)| = -60$ dB

Therefore, $\quad 20 \log K - 0 = -60$ dB

$$K = 0.001$$

Therefore,
$$G(s) = \frac{0.001\left(1 + \frac{s}{5}\right)\left(1 + \frac{s}{40}\right)}{s^3 + \left(1 + \frac{s}{200}\right)\left(1 + \frac{s}{1000}\right)}$$

Example 7.10: Determine the open-loop transfer function of a system whose approximate plot is shown in Fig. 7.19.

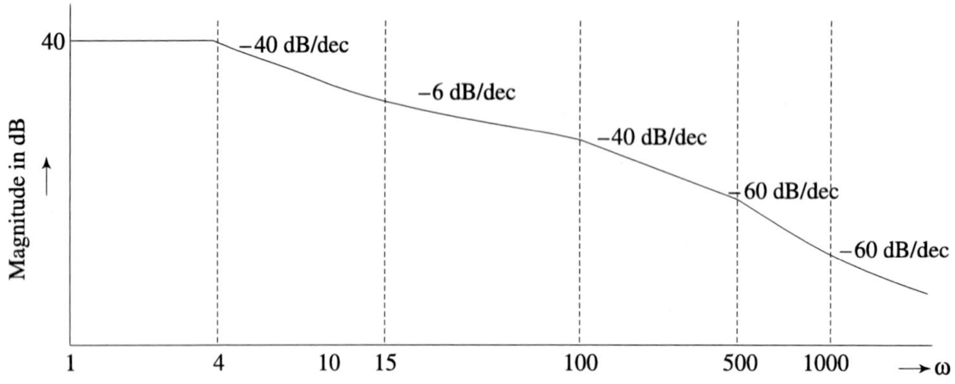

Fig. 7.19

Solution:

1. Initial slope is 0 dB/dec. This indicates absence of poles or zeros at the origin. Only constant term is effective for $\omega < 4$.
2. At the first corner frequency, $\omega_1 = 4$, the resultant slope changes by $-40 - 0 = -40$ dB/dec. This is due to multiple poles on the real axis. The factors, $\dfrac{1}{\left(1+\dfrac{j\omega}{200}\right)^2}$.
3. Between $\omega_2 = 15$ and $\omega_3 = 100$, the resultant slope is -6 dB/octave. This is equivalent to -20 dB/dec. Hence, at $\omega_1 = 15$, the resultant slope changes by $-20-(-40) =$ 20 dB/dec. This is due to a factor of the form $\left(1+\dfrac{j\omega}{15}\right)$.
4. At the third corner frequency, $\omega_3 = 100$, the resultant slope changes by $-40-(-20)$ $= -20$ dB/dec. This is due to a factor of the form $\dfrac{1}{\left(1+\dfrac{j\omega}{100}\right)}$.
5. At the fourth corner frequency, $\omega_4 = 500$, the resultant slope changes by $-60-(-40)$ $= -20$ dB/dec. This is due to a factor of the form $\dfrac{1}{\left(1+\dfrac{j\omega}{500}\right)}$.
6. At the fifth corner frequency, $\omega_5 = 1000$, the resultant slope changes by $-40-(-60)$ $= +20$ dB/dec. This is due to a factor of the form $\dfrac{1}{\left(1+\dfrac{j\omega}{1000}\right)}$.

The transfer function is of the form $GH(j\omega) = \dfrac{K\left(1+\dfrac{j\omega}{15}\right)\left(1+\dfrac{j\omega}{1000}\right)}{\left(1+\dfrac{j\omega}{100}\right)^2\left(1+\dfrac{j\omega}{100}\right)\left(1+\dfrac{j\omega}{500}\right)}$

To determine K:

For $\omega < \omega_1$, is given by the equation, $20 \log |G(j\omega)| = 20 \log K = 40$. Therefore, $K = 100$.

Therefore, $$G(s) = \frac{100\left(1+\frac{s}{15}\right)\left(1+\frac{s}{1000}\right)}{\left(1+\frac{s}{4}\right)^2\left(1+\frac{s}{100}\right)\left(1+\frac{s}{500}\right)}.$$

Example 7.11: Determine the open-loop transfer function of a system whose approximate plot is shown in Fig. 7.20.

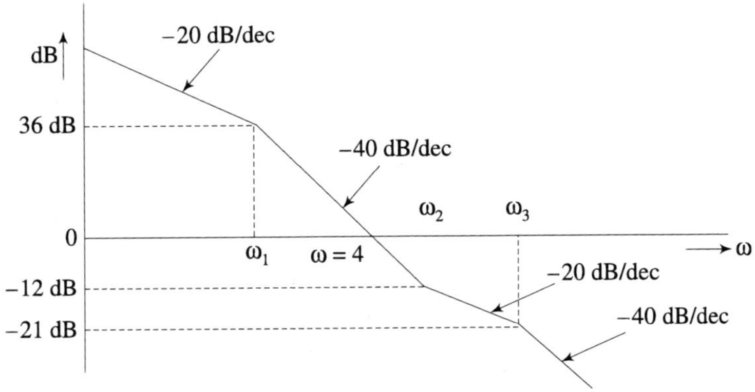

Fig. 7.20

Solution: The corner frequencies are ω_1, $\omega_2 = 8$, ω_3

The transfer function is of the form: $$GH(j\omega) = \frac{K\left(1+\frac{j\omega}{\omega_2}\right)}{j\omega\left(1+\frac{j\omega}{\omega_1}\right)\left(1+\frac{j\omega}{\omega_3}\right)}$$

1. For $\omega < \omega_1$, the factor contributing to the magnitude plot is $\frac{K}{j\omega}$

   The equation of asymptote is: $20 \log|G(j\omega)| = 20 \log K - 20 \log \omega$ ...(7.20)

   At ω_1, $20 \log \left|\frac{K}{j\omega}\right| = 36$ dB

   Therefore, $20 \log K - 20 \log \omega_1 = 36$ ...(7.21)

2. Between ω_1 and ω_2, the factors contributing to the magnitude plot are $\frac{K}{j\omega}$ and $\frac{1}{\left(1+\frac{j\omega}{\omega_1}\right)}$.

The equation of asymptote in this region is:

$$20 \log |G(j\omega)| = 20 \log K - 20 \log \omega - [20 \log \omega - 20 \log \omega_1]$$

$$= 20 \log K - 40 \log \omega + 20 \log \omega_1 \qquad ...(7.22)$$

At $\qquad \omega = 4,\ 20 \log |G(j\omega)| = 0 \text{ dB}$

Therefore, $\qquad 20 \log K - 40 \log 4 + 20 \log \omega_1 = 0 \qquad ...(7.23)$

From Eqs. (7.21) and (7.22), K and ω_1 are determined. Equation (7.21) + Eq. (7.22) gives:

$$40 \log K - 40 \log 4 = 36$$

$$40 \log K = 36 + 40 \log 4$$

Therefore, $K = 31.77$

Substituting K in Eq. (7.21), ω_1 is determined. $\omega_1 = 0.5$

To determine ω_3:

For $\omega > \omega_2$, the factor $\left(1 + \frac{j\omega}{\omega_2}\right)$ contributes to the magnitude plot.

Therefore, for $\omega_2 < \omega < \omega_3$,

$$20 \log|G(j\omega)| = 20 \log K - 20 \log \omega - [20 \log \omega - 20 \log \omega_1] + [20 \log \omega - 20 \log \omega_2]$$

$$= 20 \log K - 20 \log \omega + 20 \log \omega_1 - 20 \log \omega_2$$

At $\qquad \omega = \omega_3,\ 20 \log |G(j\omega)| = -21 \text{ db.}$

$$-21 = 20 \log K - 20 \log \omega_3 + 20 \log \omega_1 - 20 \log \omega_2$$

$$-21 = 20 \log 31.77 - 20 \log \omega_3 + 20 \log 0.5 - 20 \log 8$$

$$20 \log \omega_3 = 26.95$$

Therefore, $\omega_3 = 22.27$

$$\text{Therefore, } G(j\omega) = \frac{31.77\left(1 + \frac{j\omega}{8}\right)}{j\omega\left(1 + \frac{j\omega}{0.5}\right)\left(1 + \frac{j\omega}{22.27}\right)} = \frac{31.77\left(1 + \frac{j\omega}{8}\right)}{j\omega(1 + j2\omega)(1 + j0.04\omega)}$$

$$G(s) = \frac{31.77(1 + 0.125s)}{s(1 + 2s)(1 + 0.044s)}.$$

Example 7.12: Determine the open-loop transfer function of a system whose approximate plot is as shown in Fig. 7.21.

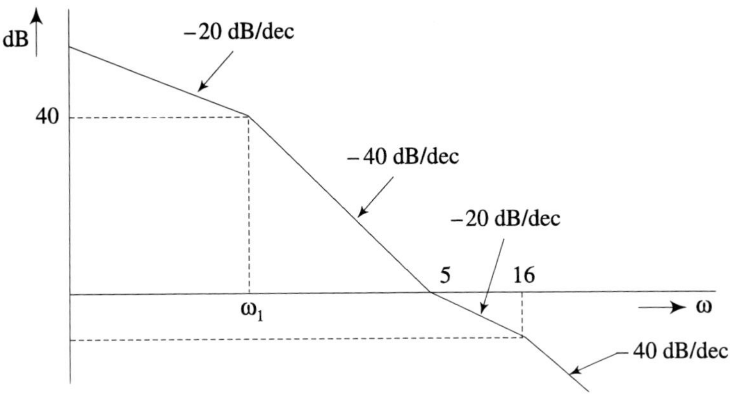

Fig. 7.21

Solution: The corner frequencies are ω_1, $\omega_2 = 5$ and $\omega_3 = 16$

The transfer function is of the form $G(j\omega) = \dfrac{K\left(1 + \dfrac{j\omega}{\omega_2}\right)}{j\omega\left(1 + \dfrac{j\omega}{\omega_1}\right)\left(1 + \dfrac{j\omega}{\omega_3}\right)}$

1. For $\omega < \omega_1$, the factor contributing to the magnitude plot is $\dfrac{K}{j\omega}$
   The equation of asymptote in this region is

$$20\log|G(j\omega)| = 20\log K - 20\log\omega$$

At $\omega = \omega_1$

$$20\log|G(j\omega)| = 40\text{ dB}$$

$$\therefore \quad 40 = 20\log K - 20\log\omega_1$$

2. For $\omega > \omega_1$, $\dfrac{1}{1 + \dfrac{j\omega}{\omega_1}}$ contributes to the magnitude plot.

$$\therefore \quad 20\log|G(j\omega)| = 20\log K - 20\log\omega - [20\log\omega - 20\log\omega_1]$$

$$= 20\log K - 40\log\omega + 20\log\omega_1$$

At $\omega = 5$,

$$20\log|G(j\omega)| = 0\text{ dB}$$

$$0 = 20\log K - 40\log 5 + 20\log\omega_1$$

Adding

$$40 = 20\log K - 20\log\omega_1 \text{ and } 0 = 20\log K - 40\log 5 + 20\log\omega_1$$

$$40 = 40\log K - 40\log 5 \quad \therefore K = 50$$

Substituting $K = 50$ in the above equation

$$40 = 20\log 50 - 20\log \omega_1$$

$$\omega_1 = 0.5$$

$$\therefore \qquad G(j\omega) = \frac{50\left(1 + \frac{j\omega}{5}\right)}{j\omega\left(1 + \frac{j\omega}{0.5}\right)\left(1 + \frac{j\omega}{16}\right)}$$

$$G(s) = \frac{50\left(1 + \frac{s}{5}\right)}{s\left(1 + \frac{s}{0.5}\right)\left(1 + \frac{s}{16}\right)}$$

$$G(s) = \frac{50(1 + 0.2s)}{s(1 + 2s)(1 + 0.0625s)}$$

7.4 EFFECTS OF ADDING A ZERO TO THE FORWARD PATH

If a zero is added in the forward path of the second order system, the following effects occur.

- Rise time decreases due to increase of bandwidth.
- Settling time will increase.
- System will be more stable.

7.4.1 Effects of Adding a Pole to the Forward Path

If a pole is added in the forward path of a second order system, the following effects occur.

- Rise time will increase and bandwidth will decrease.
- M_r will increase.
- System will be less stable.

EXERCISE

Fill in the Blanks

1. In the Bode plot of a unity feedback control system, the value of phase of $G(j\omega)$ at the gain crossover frequency is –125°. The phase margin of the system is __________ .

Ans: 55°

2. The gain margin of a system having a loop transfer function is $G(s)H(s) = \dfrac{\sqrt{2}}{s(s + 1)}$ is __________ .

Ans: ∞

3. The phase margin (in degree) of a system having a loop-transfer function

$$G(s)H(s) = \frac{\sqrt{2}}{s(s+1)} \text{ is } \_\_\_\_\_\_\_\_\_\_ .$$

Ans: 300

4. The phase margin of a system with the open-loop transfer function $G(s)H(s) = \dfrac{(1-s)}{(1+s)(2+s)}$ is __________ .

Ans: ∞

5. The gain margin for the system with the open-loop transfer function $G(s)H(s) = \dfrac{2(1+s)}{s^2}$ is __________ .

Ans: 0

6. The open-loop transfer function of a unity gain feedback control system is given by $G(s) = \dfrac{K}{(s+1)(s+2)}$, the gain margin of the system in dB is given by __________ .

Ans: ∞

7. A system has fourteen poles and two zeros. Its high frequency asymptote in its magnitude plot having a slop of __________ .

Ans: –240 dB/decade

8. Advantage of frequency domain analysis is __________.

Ans: to visualize the effect of noise disturbance and parameters variations.

9. A system has a transfer function $(1 - s)/(1 + s)$. It is called __________.

Ans: all pass system.

10. The OLTF of a unity feedback system is $G(s) = K/s(0.02s + 1)\,(1 + 0.04s)$.
If the gain K produces a phase margin of 45°. Find the corresponding gain margin.

Ans: $K = 12.125$, 12 dB

11. For a type –3 system, the asymptote at a lower frequency will have a slope of __________.

Ans: –18 dB/dec

12. The pole factor $1/1 + j\omega T$ has a slope of __________.

Ans: –20 dB/dec

13. Gain crossover frequency is the one at which $|G(j\omega)H(j\omega)|$ is __________.

Ans: equal to 1

14. The open-loop transfer function of a unity feedback control system is $10(s + 5)^3$. The gain margin is __________.

Ans: 40 dB

15. A decade frequency range is specified by

Ans: $\omega_2/\omega_1 = 10$

16. The slope of a Bode plot within the margin $5 < \omega < 20$ for the open-loop system function $G(s) = [s(s + 5)(s + 20)]$ is __________.

Ans: –12 dB/octane

17. For a type-3 system, the asymptote at a lower frequency will have a slope of ________.

Ans: –24 dB/octane

18. Define bandwidh?

Ans: Bandwidh is the range of frequencies to which the system gain is more than –3 dB.

19. Define gain margin?

Ans: The gain margin is defined as the reciprocal of the magnitude of open-loop transfer function, at phase crossover frequency, ω_{pc}

$$\text{Gain margin} = \frac{1}{|G(j\omega)|_{\omega=\omega_{pc}}}$$

When expressed in decibels, it is given by the negative of dB magnitude of $G(j\omega)$ at phase crossover frequency

$$\text{Gain margin is dB} = 20\log\frac{1}{|G(j\omega)|_{\omega=\omega_{pc}}} = -20\log|G(j\omega)|_{\omega=\omega_{pc}}$$

20. Define phase margin?

Ans: The phase margin, γ is the amount of additional phase lag at the gain crossover frequency, ω_{gc} required to bring the system to the verge of instability. It is given by $180° + \phi_{gc}$, where ϕ_{gc} is the phase of $G(j\omega)$ at the gain cross over frequency.

$$\text{Phase margin, } \gamma = 180° + \phi_{gc}$$

$$\text{where, } \phi_{gc} = Arg\ [G(j\omega)]|_{\omega=\omega_{gc}}$$

21. What is phase and gain crossover frequency?

Ans: The gain crossover frequency is the frequency at which the magnitude of the open-loop transfer function is unity.

The phase crossover frequency is the frequency at which the phase of the open-loop transfer function is 180°.

22. Write the expression to resonant peak and resonant frequency.

Ans:

$$\text{Resonant peak, } M_r = \frac{1}{2\zeta\sqrt{1-\xi^2}}$$

$$\text{Resonant frequency, } \omega_r = \omega_n\sqrt{1-2\xi^2}$$

23. Write a short note on the correlation between the time and frequency response?

Ans: There exist a correlation between time and frequency response of first or second order systems. The frequency domain specification can be expressed in terms of the time domain parameters ζ and ω_n. For a peak overshoot in time domain there is a corresponding resonant peak in frequency domain.

For higher order systems there is no explicit correlation between time and frequency response. But if there is a pair of dominant complex conjugate poles, then the system can be approximated to second order system and the correlation between time and frequency response can be estimated.

24. The damping ratio and natural frequency of oscillation of a second order system is 0.5 and 8 rad/sec respectively, calculate the resonant peak and resonant frequency?

Ans: Resonant peak, $M_r = \dfrac{1}{2\zeta\sqrt{1-\zeta^2}} = \dfrac{1}{2\times(0.5)\sqrt{1-(0.5)^2}} = 1.154$

Resonant frequency, $\omega_r = \omega_n \sqrt{1 - 2\zeta^2}$

$$= 8 \times \sqrt{1 - 2(0.5)^2} = 5.657 \text{ rad/sec}$$

25. Sketch the Bode plot of the following open-loop transfer function and from the plot determine the phase margin and gain margin.
    (a) $G(s) = 100(1 + 0.1S)|S(1 + 0.2S)(1 + 0.5S)$
    (b) $G(s) = 50(1 + 0.1S)|(1 + 0.01S)(1 + S)$
    (c) $G(s) = 30(1 + 0.1S)|S(1 + 0.01S)(1 + S)$
    (d) $G(s) = S^2(S + 10)|(s + S)^2 (S + 0.1)$
    (e) $G(s) = 40(Hs)|(1 + 5S)(S^2 + 4S + 4)$
    (f) $G(s) = 10(1 + S)e^{-0.1S}[S(1 + 0.2S)]$
26. The open-loop transfer function of a system is given by $G(s) = K/s(1 + 0.5S)(1 + 0.2S)$. Using Bode plot find the value of K so that (i) the gain margin of the system is 6 dB, and (ii) the phase margin of the system is 25.
27. Draw the Bode plot of, $G(s) = \dfrac{K}{S^n}$

Ans: Let $S = j\omega$

$\therefore \quad G(j\omega) = \dfrac{K}{(j\omega)^n}$

The magnitude of $G(j\omega)$ is unity when $\omega = k^{1/n}$

The magnitude plot is a straight line with slope of -20 dB/dec and passing through $\omega = k^{1/n}$.

The phase plot is straight line parallel to x-axis at $-90°n$

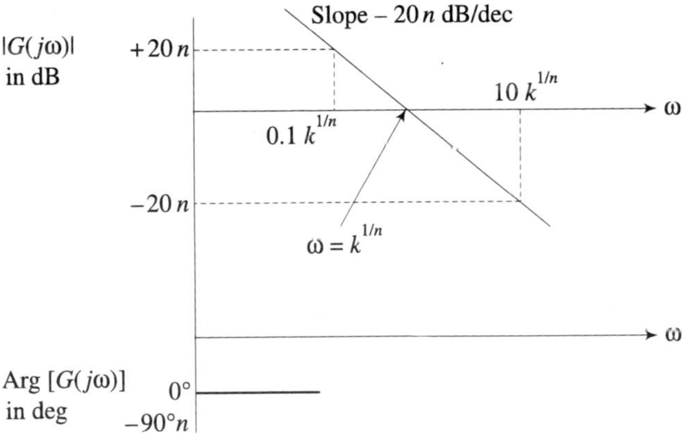

Fig. 7.22

28. Sketch the bode plot of $G(s) = 1|(1 + ST)$

Ans: Let $s = j\omega$

$\therefore\ G(j\omega) = \dfrac{1}{1 + j\omega T}$; The corner frequency, $\omega_c = \dfrac{1}{T}$

The magnitude plot is approximated by two straight lines: one straight line at 0 dB in the frequency range 0 to ω_c and the other straight line with the slope of -20 dB/dec in the frequency range ω_c to ∞.

The phase of $G(j\omega)$ varies from 0 to $-90°$ as ω is varied from 0 to ∞. Hence, the phase plot is a curve passing through $-45°$ at the corner frequency.

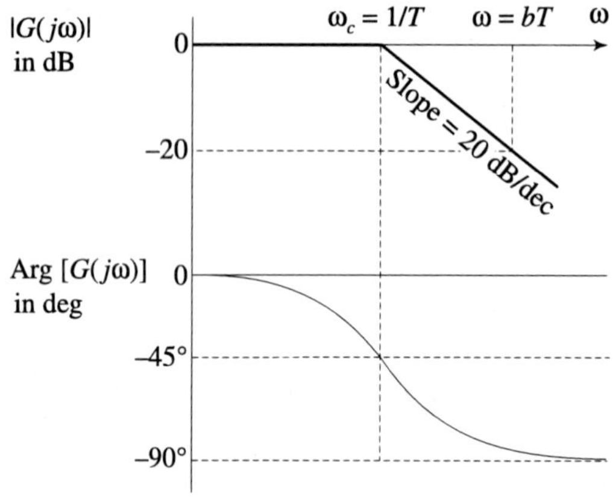

Fig. 7.23

29. What are Bode plots? State the advantages of Bode plots
30. Define the following terms
    (i) Corner frequency (ii) Gain crossover frequency
    (iii) Phase corner frequency (iv) Gain margin
    (v) Phase margin
31. Explain the concept of gain margin and phase margin. Explain how these values help in studying relative stability.
32. (i) How do we determine the gain crossover frequency and phase crossover frequency from the Bode plot.
    (ii) How do you determine the gain margin and phase margin of a system from the Bode plot.
33. Explain the nature of Bode plots for
    (i) Poles at origin (ii) Simple pole (iii) Simple zero
34. Construct the Bode plot for the system having $G(s) = \dfrac{80}{s(s+2)(s+20)}$

    From the Bode plots determine (i) gain and phase crossover frequencies, (ii) gain and phasemargins, and (iii) comment on the stability of the system.

Ans: (i) $\omega_{gc} = 2.1$ rad/sec, $\omega_{pc} = 6.35$ rad/sec
(ii) Gain margin = 21 dB, phase margin = 38°
(iii) The system is absolutely stable.

35. The open-loop transfer function of a unity feedback system is

$$\frac{K(1+s)}{s(1+0.1s)(1+0.4s)}$$

Using the straight line approximations, draw the Bode diagram and hence find
(i) the value of k for a gain margin of 22 dB, and
(ii) the value of k for a phase margin of 45°.

Ans: (i) 31.6 (ii) 10

36. The open-loop transfer function of a unity feedback system is $G(s) = \dfrac{K}{s(s+2)(s+20)}$
    Construct Bode plot and determine
    (i) limiting value of K for system to be stable,
    (ii) value of K for gain margin to be 10 dB,
    (iii) value of K for phase margin to be 50°, and

Ans: (i) $K = 798$ (ii) $K = 252$ (iii) $K = 75.32$.

37. What are the advantages and limitations of frequency response analysis.
38. Write a short note on the correlation between time and frequency response.
39. What are frequency domain specifications?
40. Define the following terms
    (i) Resonant peak (ii) Resonant frequency (iii) Bandwith
    (iv) Cut-off rate (v) Gain margin (vi) Phase margin
41. Find frequency domain specifications for a second order system with unity feedback and
    $G(s) = \dfrac{169}{s(s+7)}$.

Ans: $M_r = 1.920$ $\omega_r = 12.02$ rad/sec $\omega_c = 28.22$ rad/sec B.W. = 28.22 rad/sec

42. Find the frequency domain specification for a second order system with unity feedback and $G(s) = 121/[S(S + 11)]$,

Ans: $M_r = 1.155$ $\omega_r = 7.778$ rad/sec, $\omega_c = 17.798$ rad/sec B.W. = 17.798 rad/sec

43. Find K and a for the feedback system. So that $M_r = 1.04$ and $\omega_r = 11.55$ rad/sec will be satisfied Determine bandwidth also.

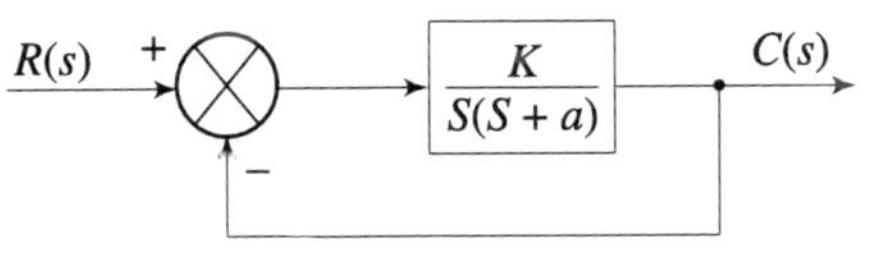

Fig. 7.24

Ans: $K = 491$, $a = 26.77$ B.W. = 38.62 rad/sec

44. What will happen if a pole is added in the forward path of a second order system?

Ans: The system will be less stable
M_r will increase
Rise time will increase and B.W. will decrease.

45. State disadvantages of frequency response methods

Ans: Time conserving process applicable only for linear systems.

8 Stability in Frequency Domain

8.1 INTRODUCTION

In earlier chapters, it has been shown that the stability of a system depends upon the location of the roots of its characteristic equation in the s plane. For a system to be stable, the roots of the characteristic equation should not lie in the right half of the S-plane or on the imaginary axis, for such roots lead to instability or sustained oscillations of the system. The root locus technique is an approach to determine the location of the roots in the S-plane, given the open-loop pole zero configurations. In root locus technique, the computation of closed-loop poles is not necessary in this case and the stability study can be carried out graphically from the open-loop frequency response. An important and useful technique, the Nyquist stability criterion, which relates the location of the roots of the characteristic equation to the open-loop frequency response of a system. Unlike the root locus technique, computation of closed-loop poles is not necessary in this case and the stability study can be carried out graphically from the open-loop frequency response. Also polar plot is useful in determining the stability of a closed-loop system from its open-loop frequency response.

8.2 POLAR PLOTS

The polar plot of function of a complex variable s, $G(s)$ is a plot of the magnitude of $G(j\omega)$ versus the phase angle of $G(j\omega)$ on polar coordinates as ω is varied from zero to infinity.

Thus, the polar plot is the locus of phasors $|G(j\omega)|\angle G(j\omega)$ as ω is varied from zero to infinity.

From the mathematical viewpoint, the process can be regarded as mapping of the positive half of the imaginary axis of the S-plane on the $G(j\omega)$ plane. The polar plot of $G(j\omega)$ is called the direct polar plot and the polar plot of $\dfrac{1}{G(j\omega)}$ is called the inverse polar plot. The polar plot is strictly for the frequency range $0 < \omega < \infty$, while the Nyquist plot is in the frequency range $-\infty < \omega < \infty$.

In polar plots, a positive phase angle is measured counterclockwise from the positive real axis while negative phase angle is measured clockwise from the positive real axis.

Advantages

- It depicts the frequency response characteristics over the entire range of frequencies in a single plot.
- Polar plots can be used for stability studies of control systems.

Limitations

- The addition of poles or zeros to an existing system requires the recalculation of the frequency response.
- The calculation of the frequency response by this method is tedious and does not indicate the effect of the individual poles or zeros of the transfer function.

8.2.1 Procedure to Sketch the Polar Plot

A rough sketch of the polar plot of the transfer function $G(j\omega)$ is quite adequate for stability analysis in the frequency domain.

Step 1: Determine the transfer function, $G(s)$, of the system.

Step 2: Put $s = j\omega$ in the transfer function to obtain $G(j\omega)$.

Step 3: Calculate the magnitudes, $G(j\omega)$ at limiting frequencies of $\omega \to 0$ and $\omega \to \infty$ as follows.

$$|G(j\omega)|_{\omega = 0} = \lim_{\omega \to 0} |G(j\omega)|$$

$$|G(j\omega)|_{\omega = \infty} = \lim_{\omega \to \infty} |G(j\omega)|$$

Step 4: Calculate the phase angles of $G(j\omega)$ at limiting frequencies of $\omega \to 0$ and $\omega \to \infty$ as follows

$$\angle G(j\omega) = \lim_{\omega \to 0} \angle G(j\omega)$$

$$\angle G(j\omega) = \lim_{\omega \to \infty} \angle G(j\omega)$$

Step 5: To find the frequencies at which the polar plot crosses the real and imaginary axes, we rationalize the complex frequency function $G(j\omega)$, multiply numerator and denominator of $G(j\omega)$ with complex conjugates of the denominator factors separate the real and imaginary parts.

Step 6: Equate the imaginary part of $G(j\omega)$ to zero to determine the values of the frequency, ω, at which the polar plot intersects the real axis, calculate the value of $G(j\omega)$ at the intersecting points by substituting the determined values of frequency, ω, in the rationalized expressing $G(j\omega)$.

Step 7: Equate the real part of $G(j\omega)$ to zero and determine the value of the frequency ω, at which the plot intersects the imaginary axis. Calculate the values of $G(j\omega)$ at the points of intersection by substituting the determined values of frequency, ω, in the rationalized expression of $G(j\omega)$.

Step 8: Sketch the polar plot from the above information.

The above procedure is used to sketch a rough polar plot. However, a detailed plot can be drawn by calculating the magnitude (M) and phase (ϕ) of the given transfer function at different frequencies in the range $0 < \omega < \infty$.

8.2.2 Typical Sketches of Poles Plot

Type 0, order 1

$$G(s) = \frac{1}{1 + ST}$$

$$G(s) = \frac{1}{1 + j\omega T}$$

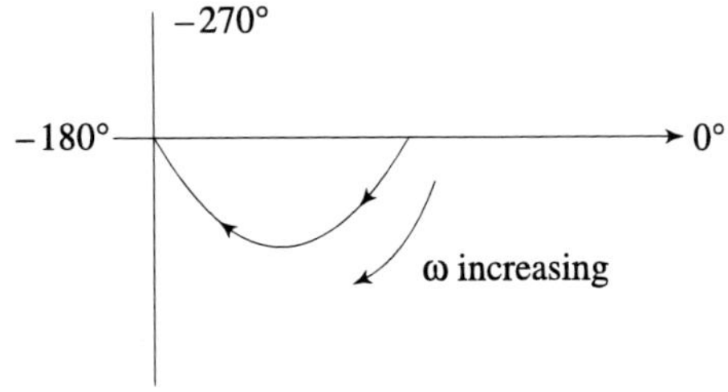

Type 1, order 2

$$G(s) = \frac{1}{S(1 + ST)}$$

$$G(j\omega) = \frac{1}{j\omega(1 + j\omega T)}$$

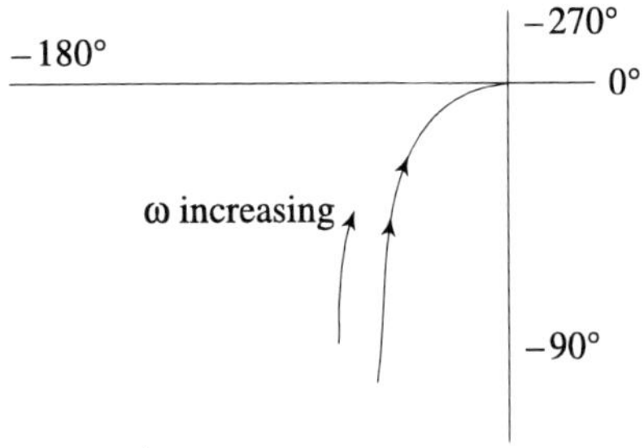

Type 0, order 2

$$G(s) = \frac{1}{(1 + ST_1)\,(1 + ST_2)}$$

$$G(j\omega) = \frac{1}{(1 + j\omega T_1)\,(1 + j\omega T_2)}$$

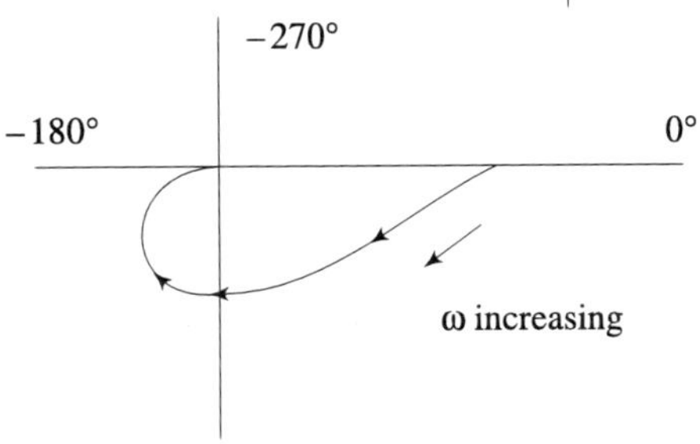

Type 0, order 3

$$G(s) = \frac{1}{(1 + ST_1)\,(1 + ST_2)\,(1 + ST_3)}$$

$$G(j\omega) = \frac{1}{(1 + j\omega T_1)\,(1 + j\omega T_2)\,(1 + j\omega T_3)}$$

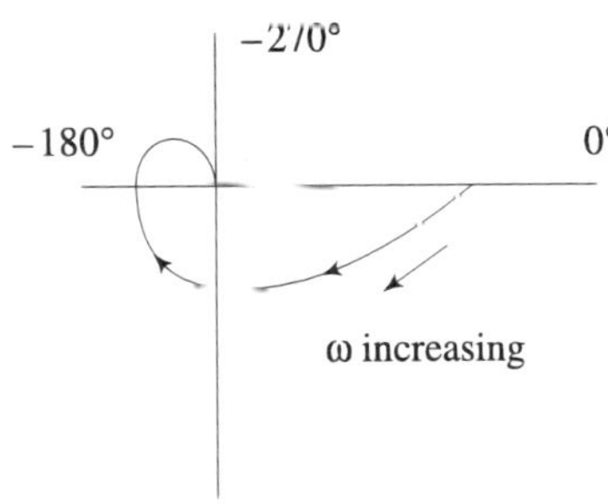

Type 1, order 3

$$G(s) = \frac{1}{S(1 + ST_1)\,(1 + ST_2)}$$

$$G(j\omega) = \frac{1}{j\omega\,(1 + j\omega T_1)(1 + j\omega T_2)}$$

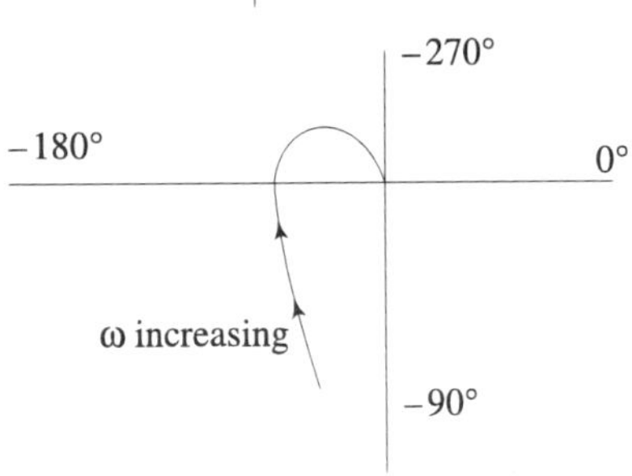

Type 2, order 4

$$G(S) = \frac{1}{s^2 (1 + ST_1)(1 + ST_2)}$$

$$G(j\omega) = \frac{1}{(j\omega)^2 (1 + j\omega T_1)(1 + j\omega T_2)}$$

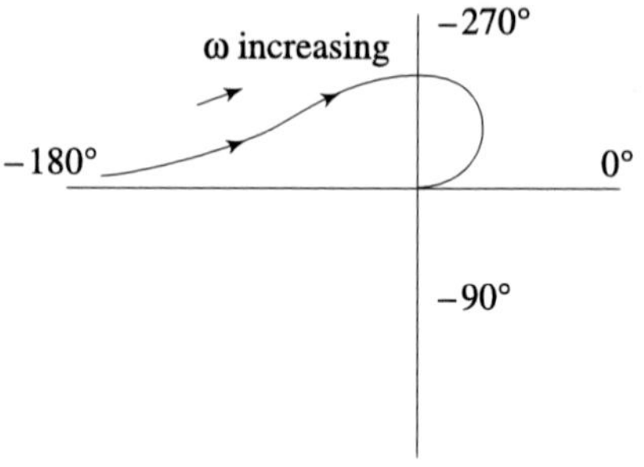

Type 2, order 5

$$G(s) = \frac{1}{s^2 (1 + ST_1)(1 + ST_2)(1 + ST_3)}$$

$$G(j\omega) = \frac{1}{(j\omega)^2 (1 + j\omega T_1)(1 + j\omega T_2)(1 + j\omega T_3)}$$

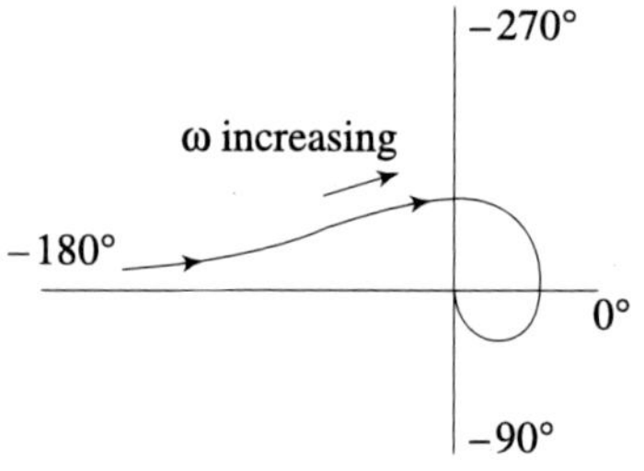

Example 8.1: Sketch the polar plot of $G(s) = \frac{1}{s}$.

Solution:

$$G(j\omega) = \frac{1}{j\omega} + \frac{1 + j0}{(0 + j\omega)}$$

$$M = |G(j\omega)| = \frac{1}{\omega}$$

$$\phi = |G(j\omega) = \frac{\tan^{-1}\frac{0}{1}}{\tan^{-1}\frac{\omega}{0}} = \frac{0}{\underline{|90^\circ}} = -90^\circ$$

$$\lim_{\omega \to 0} |G(j\omega)| = \lim_{\omega \to \infty} \frac{1}{\omega} = \infty$$

$$\lim_{\omega \to \infty} |G(j\omega)| = \lim_{\omega \to \infty} \frac{1}{\omega} = 0$$

$$G(j\omega)|_{\omega = 0} = \infty \underline{|-90^\circ} \qquad ...(8.1)$$

$$G(j\omega)|_{\omega = \infty} = 0 \underline{|-90^\circ} \qquad ...(8.2)$$

It is seen from Eq. (8.1) that the polar plot starts at $M = \infty$ at an angle -90°, where $\omega = 0$. The end point of the plot is the origin 0 where $\omega = \infty$.

The polar plot is shown in Fig. 8.1.

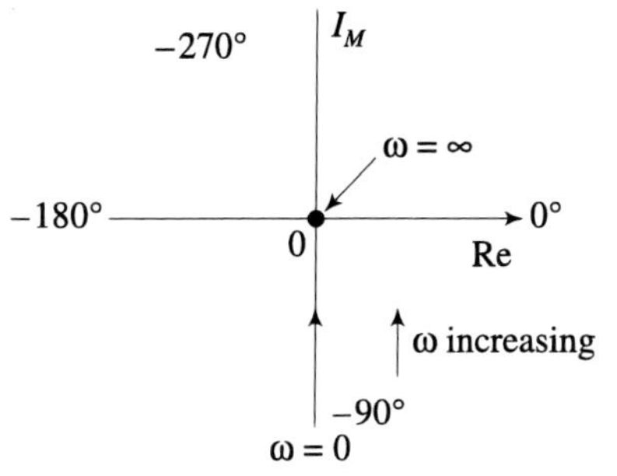

Fig. 8.1

Example 8.2: Sketch the polar plot of $G(s) = 1 + sT$

Solution: Put $s = j\omega$

$$G(j\omega) = 1 + j\omega T = \sqrt{1 + \omega^2T^2}\ \underline{|\tan^{-1}\omega t}$$

At $\omega = 0,\quad M = 1,\ \phi = 0°$

At $\omega = \infty,\quad M = \infty,\ \phi = 90°$

The polar plot of the function $G(j\omega) = (1 + j\omega T)$ is simply the upper half of the straight line passing through the point $(1, j0)$ on the complex $G(j\omega)$ plane and parallel to the imaginary axis.

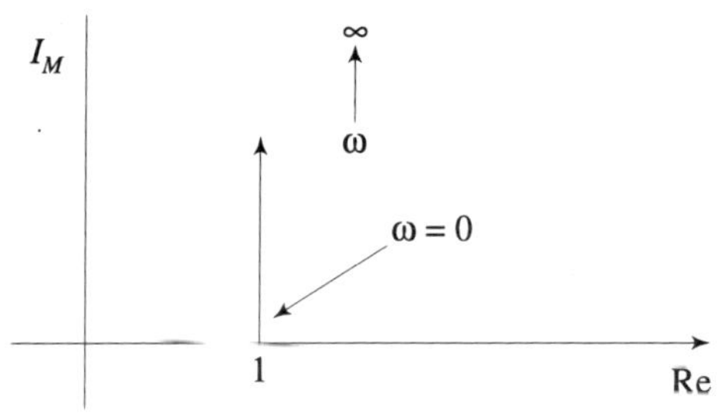

Fig. 8.2

Example 8.3: Sketch the polar plot of the transfer function $G(s) = \dfrac{1}{1 + sT}$.

Solution: For sinusoidal steady state putting $s = j\omega$

$$G(j\omega) = \frac{1}{1 + j\omega T} = \frac{1 + j0}{1 + j\omega T}$$

$$G(j\omega) = \frac{1}{\sqrt{1 + \omega^2T^2}}\ \underline{|\tan^{-1}\omega T}$$

$$G(j\omega) = |G(j\omega)|\ \underline{|G(j\omega)}$$

$$G(j\omega) = M\underline{|\phi}$$

$$M = |G(j\omega)| = \frac{1}{\sqrt{1+\omega^2T^2}}$$

$$\phi = \underline{|G(j\omega)} = \frac{\tan^{-1}\left(\frac{0}{1}\right)}{\tan^{-1}\left(\frac{\omega T}{1}\right)} = \frac{0^\circ}{\tan^{-1}\omega T} = -\tan^{-1}\omega T.$$

The value of M and ϕ can be obtained by varying ω from 0 to ∞.

Table 8.1 Values of ω and the corresponding values of M and ϕ.

ω	$M = \lvert G(j\omega)\rvert$	$\varnothing = \underline{\lvert G(j\omega)}$
0	1	0°
$\frac{1}{T}$	$\frac{1}{\sqrt{2}}$	–45°
∞	0	–90°

Thus, at

$$\omega = 0,\ G(j\omega) = 1\underline{|0^\circ} \text{ and at } \omega = \frac{1}{T},\ G(j\omega) = \frac{1}{\sqrt{2}}\underline{|-45^\circ}$$

If ω approaches infinity, the magnitude of $G(j\omega)$ approaches zero and the phase angle approaches -90°.

The polar plot of $G(j\omega)$ is a semicircle as the frequency ω is varied from zero to infinity as shown below,

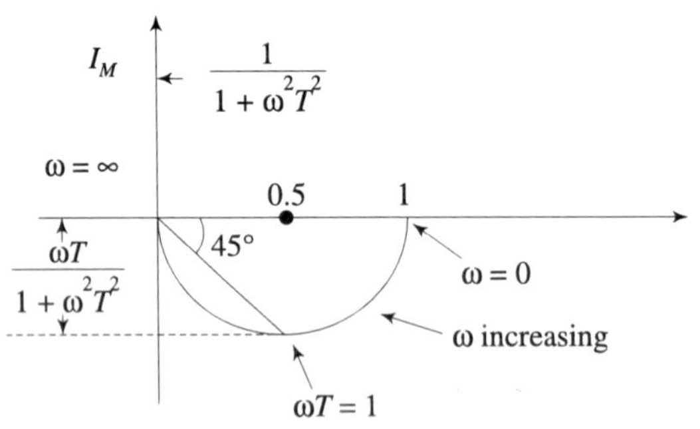

Fig. 8.3

Example 8.4: Sketch the polar plot for the following type 0 system.

$$G(s) = \frac{6}{(s+1)(s+2)}.$$

Solution:
$$G(s) = \frac{6}{(s+1)(s+2)}$$

The sinusoidal transfer function is obtained by putting $s = j\omega$.

$$G(j\omega) = \frac{6}{(1+j\omega)(2+j\omega)}$$

$$G(j\omega) = \frac{6+j0}{(1+j\omega)(2+j\omega)}$$

$$G(j\omega) = |G(j\omega)| \underline{|G(j\omega)} = M\underline{|\phi}$$

$$M = |G(j\omega)| = \frac{6}{\sqrt{(1+\omega^2)}\sqrt{(4+\omega^2)}}$$

$$\phi = \underline{|G(j\omega)} = \frac{\tan^{-1}\frac{0}{6}}{\tan^{-1}\left(\frac{\omega}{1}\right)\tan^{-1}\left(\frac{\omega}{2}\right)}$$

$$\phi = 0 - \tan^{-1}\omega - \tan^{-1}\left(\frac{\omega}{2}\right)$$

The magnitude of the transfer function $G(j\omega)$ at limiting frequencies of $\omega \to 0$ and $\omega \to \infty$.

$$M|_{\omega\to 0} = \lim_{\omega\to 0}|G(j\omega)| = \lim_{\omega\to 0}\frac{6}{\sqrt{(1+\omega^2)}\sqrt{(4+\omega^2)}}$$

$$M|_{\omega\to 0} = \frac{6}{\sqrt{(1+0)}\sqrt{(4+0)}} = 3$$

$$M|_{\omega\to 0} = \lim_{\omega\to\infty}\frac{6}{\sqrt{(1+\omega^2)}+\sqrt{(4+\omega^2)}} = 0$$

The phase angles of $G(j\omega)$ at limiting frequencies can be formed as follows. We get

$$\phi|_{\omega\to 0} = \lim_{\omega\to\infty}\left(-\tan^{-1}\omega - \tan^{-1}\left(\frac{\omega}{2}\right)\right) = 0^\circ$$

$$\phi|_{\omega\to\infty} = \lim_{\omega\to\infty}\left(-\tan^{-1}\omega - \tan^{-1}\left(\frac{\omega}{2}\right)\right)$$

$$\phi|_{\omega\to\infty} = -90^\circ - 90^\circ - 180^\circ$$

Multiplying the numerator and denominator of $G(j\omega)$ with complex conjugates of the denominator factors, we get

$$G(j\omega) = \frac{6(1-j\omega)(2-j\omega)}{(1+j\omega)(2+j\omega)(1-j\omega)(2-j\omega)}$$

$$G(j\omega) = \frac{6(2-\omega^2) - j18\omega}{4 + 5\omega^2 + \omega^4}$$

$$G(j\omega) = \frac{6(2-\omega^2)}{4 + 5\omega^2 + \omega^4} - j\frac{18\omega}{4 + 5\omega^2 + \omega^4}$$

Equating the imaginary part to zero to determine frequency at which the polar plot intersects the real axis, we get

$$\frac{-18\omega}{4 + 5\omega^2 + \omega^4} = 0$$

$$\therefore \qquad \omega = 0 \text{ and } \omega = \pm\,\infty$$

For positive values of frequency ω, the polar plot intersects the real axis at $\omega = 0$ and $\omega = \infty$, for which

$$G(j\omega) = 3\,\angle 0^\circ \text{ and } 0\,\angle 0^\circ$$

Equating the real part to zero to determine the frequency at which the polar plot intersects the imaginary axis, we get

$$\frac{6(2-\omega^2)}{4 + 5\omega^2 + \omega^4} = 0$$

$$\therefore \qquad \omega^2 = 2,\ \omega = \pm\sqrt{2},\ \omega = \pm\,\infty$$

For positive frequencies, the polar plot intersects the imaginary axis at $\omega = \pm\sqrt{2}$, $\omega = \infty$ for which $G(j\omega) = \sqrt{2}\,\angle{-90^\circ}$ and $0\,\angle{-180^\circ}$.

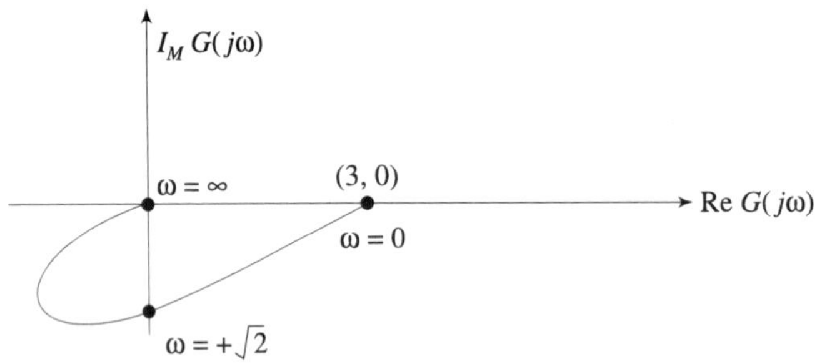

Fig. 8.4

Example 8.5: Sketch the polar plot of $G(s) = \dfrac{e^{-sL}}{1 + sT}$.

Solution: Put $\qquad s = j\omega$

$$G(j\omega) = \frac{e^{-j\omega L}}{1 + j\omega T}$$

$$e^{-j\omega L} = \cos \omega L - j \sin \omega L$$

$$G(j\omega) = (\cos \omega L - j \sin \omega L) \left(\frac{1}{1 + j\omega T} \right)$$

$$|G(j\omega)| = |\cos \omega L - j \sin \omega L) \left| \frac{1}{1 + j\omega T} \right|$$

$$|G(j\omega)| = (\cos^2 \omega L + \sin^2 \omega L) \frac{1}{\sqrt{1 + \omega^2 T^2}}$$

or $$|G(j\omega)| = \frac{1}{\sqrt{1 + \omega^2 T^2}}$$

$$\angle G(j\omega) = \angle e^{-j\omega L} + \angle \frac{1}{1 + j\omega T}$$

$$\angle G(j\omega) = -\omega L - \tan^{-1} \omega T$$

At $\omega = 0,\ G(j\omega) = 1 \angle 0^\circ$

At $\omega = \infty,\ |G(j\omega) = 0$

Since the magnitude decreases monotonically from unity and phase angle also decreases monotonically with increasing ω the polar plot is a spiral shown in Fig. 8.5.

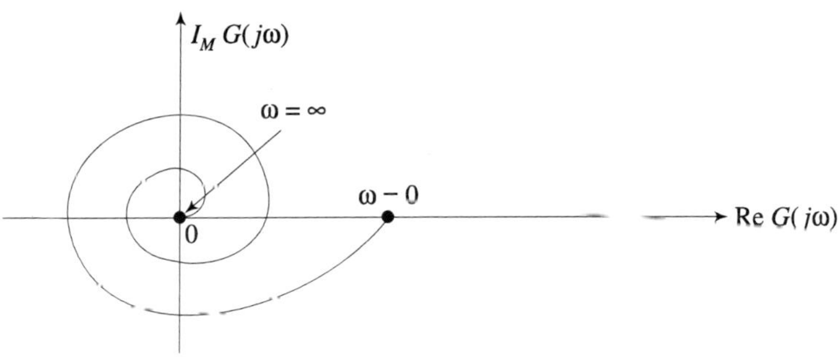

Fig. 8.5

8.3 MATHEMATICAL PRELIMINARIES

Let $F(s)$ be a function of S, which is expressed as a ratio of two polynomials in S, as shown in Eq. (8.3)

$$F(s) = \frac{(s - Z_1)\ (s - Z_2)\ \dots\ (s - Z_m)}{(s - P_1)\ (s - P_2)\ \dots\ (s - P_n)} \qquad \dots(8.3)$$

The roots of numerators polynomial are zeros and the roots of denominators polynomial are poles. The function has m number of zeros and n number of poles.

S is a complex variable expressed as $S = \sigma + j\omega$. For a particular value of σ and ω, S will represent a point in the S-plane. $F(s)$ can also be expressed as $F(s) = u + jv$, where u is the real part of $F(s)$ and v is the imaginary part of $F(s)$.

Since any number of points of analytic in the S-plane can be mapped into the $F(s)$ plane it can be concluded that for a contour in the S-plane which does not go through any singular point, there exists a corresponding contour in the $F(s)$ plane shown in Fig. 8.6.

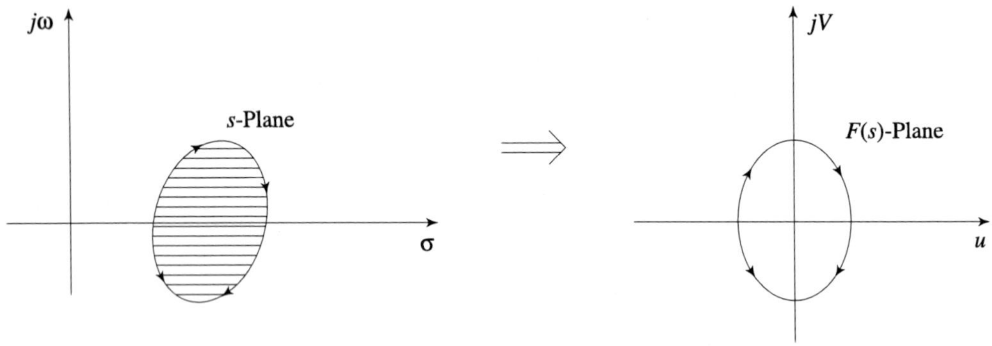

Fig. 8.6

Normally the direction of arbitrary contour in S-plane is chosen as clockwise. Here zeros are marked by small circles (o) and poles by (x). The S-plane plane contours and the corresponding $F(s)$ plane contours are shown in Fig. 8.7. It can be proved that there exists a relationship between the enclosure of poles and zeros by the S-plane closed contour and number of encirclements of the origin of $F(s)$ plane by the corresponding $F(s)$-plane contour.

S-plane contour **$F(s)$-plane contour**

$F(s) = s^2 - 2S + 6 = (S - 1 - j2)(S - 1 + j2)$

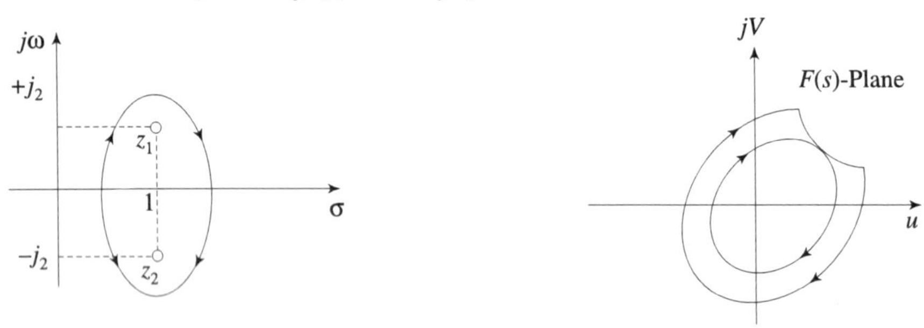

Fig. 8.7

$$\text{Zeros} = z_1 = 1 + j2,\ z_2 = 1 - j2$$

$$F(s) = \frac{1}{S^2 - 4S + 8} = \frac{1}{(S - 2 - j2)(S - 2 + j2)}$$

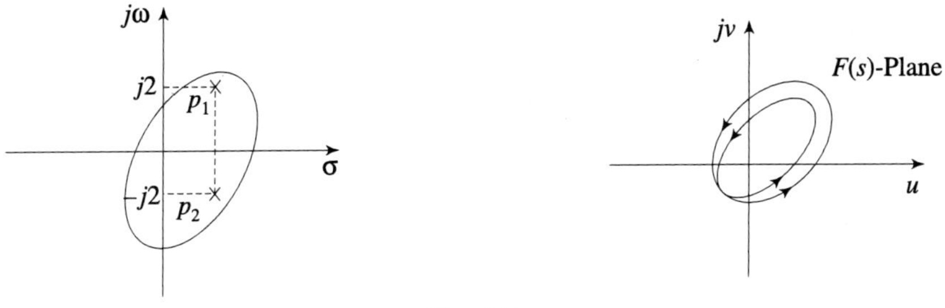

Fig. 8.8

Poles: $P_1 = 2 + j2,\ P_2 = 2 - j2$

$$F(s) = \frac{s - 1}{(s - 2) + (s - 4)}$$

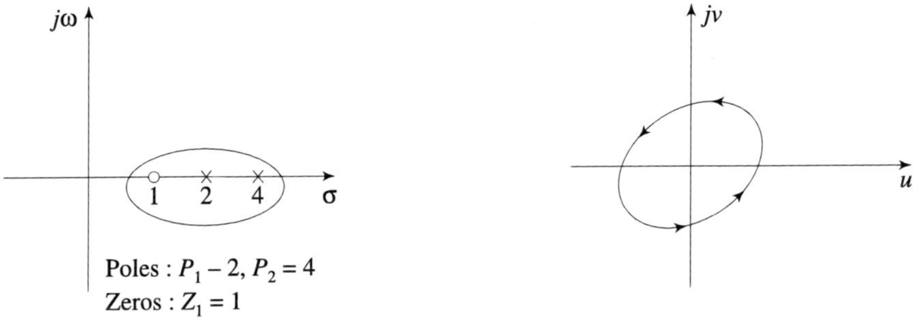

Fig. 8.9

$$F(s) = \frac{S^2 - 2S + 6}{S - 2} = \frac{(s - 1 - j2)(S - 1 + j2)}{S - 2}$$

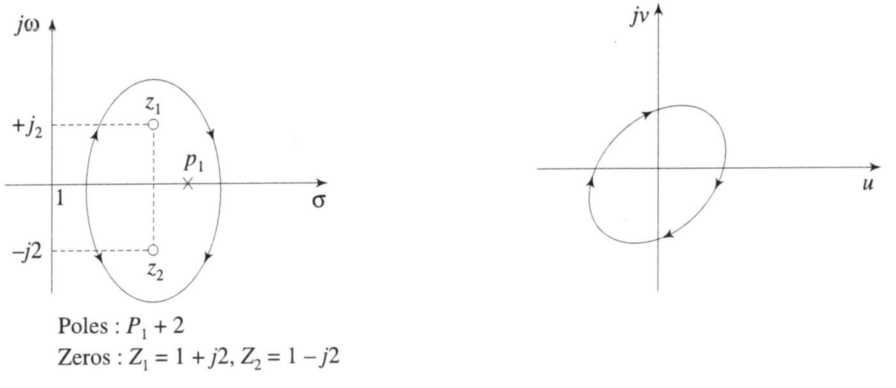

Fig. 8.10

$$F(s) = \frac{S^2 - 2S + 6}{S^2 - 4S + 8} = \frac{(S - 1 - j2)(S - 1 + j2)}{(S - 2 - j2)(S - 2 + j2)}$$

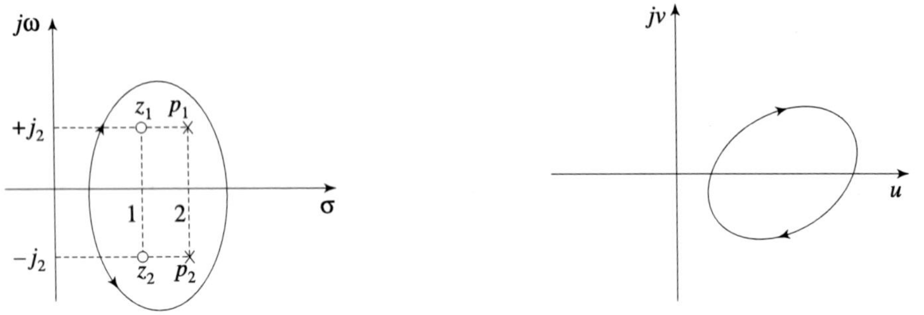

Fig. 8.11

$$F(s) = \frac{S - 2}{(S - 1)(S - 3)}$$

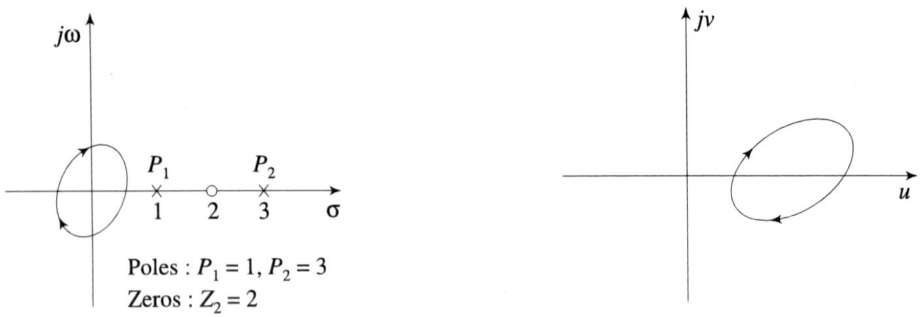

Fig. 8.12

8.3.1 Concept of Encircled and Enclosed

It is important to distinguish between the concept of encircled and enclosed which are frequently used to apply Nyquist criterion.

Encircled: A point is said to be encircled by a closed path if it is found inside the path.

In Fig. 8.13 the point *A* is encircled in the clockwise direction and the point *B* is not encircled.

Any point or region is said to be enclosed by a closed path, if it is found to lie to the right of the path. When the path is traversed in the prescribed direction

In Fig. 8.12 the point *A* is enclosed by closed path and the point *B* is not enclosed.

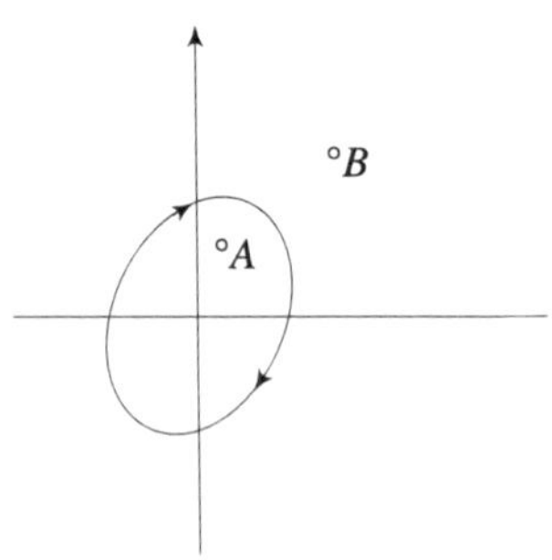

Fig. 8.13

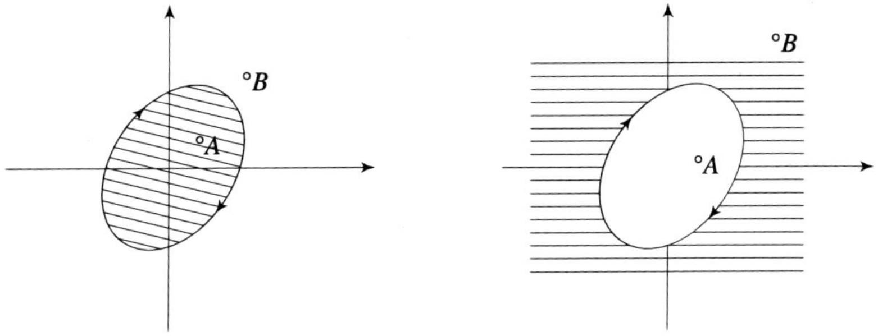

Fig. 8.14

In Fig. 8.14 the point A is not enclosed by closed path but point B is enclosed.

The principle of argument is stated as follows: Let $F(s)$ be a single valued rational function and is analytic in a given region in the S-plane except at some points. Now an arbitrary closed contour is chosen in the S-plane, so that $F(s)$ is analytic at every point on the closed contour in S-plane then the corresponding $F(s)$ plane contour mapped in the $F(s)$ plane will encircle the origin N times in anticlockwise direction where N is the difference between the number of poles and number of zeros of $F(s)$ that are encircled by the chosen closed contour in S-plane.

Mathematically, it can be expressed as $N = P - Z$,

where N = number of encirclement of origin of $F(s)$ plane, made by $F(s)$ contour.

Z = number of zeros of $F(s)$ lying on right half of S-plane and enclosed by S-plane closed contour.

P = number of poles of $F(s)$ lying on right half of S-plane and enclosed by the S-plane closed contour.

The value of N can be positive, zero or negative. Based on the sign of N, following conclusion can be made.

- If N is positive, then direction of encirclement of origin of $F(s)$ plane will be anticlockwise.
- If N is zero, then there will be no encirclement of origin of $F(s)$–plane.
- If N is negative, then direction of encirclement of origin of $F(s)$ plane will be clockwise.

8.4 NYQUIST STABILITY CRITERION

The Nyquist criterion is a semigraphical method that determines the stability of a closed-loop system by investing the properties of the frequency domain plot of the transfer function $G(s)H(s)$.

The Nyquist method has the following features that make it suitable for the analysis and design of control systems.

- It provides the same information on the absolute stability of a control system as does the R–H criterion.
- In addition to absolute system stability, the Nyquist criterion provides information for the relative stability of a system. It also gives indication on how the system stability may be improved.
- The frequency domain plots of loop gain $G(j\omega)H(j\omega)$ gives the information on the frequency domain characteristics such as M_r, ω_r and bandwidth.
- This stability criterion was developed by H. Nyquist in 1932. The method determines the location of the characteristic equation roots in the left half and right half of the S-plane. Unlike the root locus method the Nyquist criterion does not give the exact location of the roots of the characteristic equation. This is the drawback of the Nyquist stability criterion.

8.4.1 Nyquist Path or Nyquist Contour

In order to apply Nyquist criterion, we make a speical choice of the closed contour in the S-plane. This contour is known as the Nyquist contour or Nyquist path. This path is shown in Fig. 8.15.

The Nyquist contour is so chosen that it encloses the entire right half side of the imaginary $j\omega$ axis in the S-plane. It encloses all the right half S-plane poles and zeros of $G(s)H(s)$ except the singular points. The Nyquist path shown is composed of two segments C_1 and C_2. The first segment is the whole imaginary axis from $-j\infty$ to $+j\infty$. The segment C_2 is a semicircle of radius infinity that encloses the entire right half of S-plane.

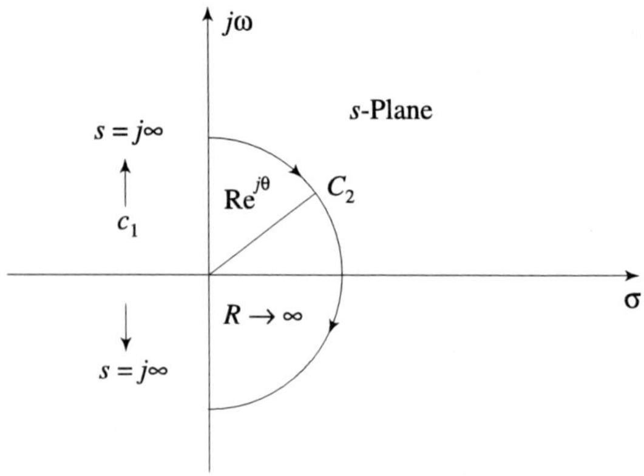

Fig. 8.15

8.4.2 Nyquist Stability Criterion

In order to apply the Nyquist criterion a special contour called the Nyquist contour is chosen so that it encloses the entire right half side of the $j\omega$ axis of the S-plane. When the S-plane contour C is the Nyquist contour, the Nyquist stability criterion is given by $Z = N + P$.

where P = number of poles of $G(s)H(s)$ or $F(s)$ in the right half of the S-plane.

N = number of clockwise encirclements of $(-1, j0)$ point in the GH plane or the origin in $F(s)$ plane by the Nyquist path.

Z = number of zeros of $F(s) = 1 + G(s)H(s)$ in the right half of the S-plane.

We know that to make a closed-loop system stable, there must not be any zero in the right half S-plane. Therefore, $Z = 0$, $N = -P$.

If $P \neq 0$, then for closed-loop system to be stable, $N = -P$, which means that we must have contour clockwise encirclements of the $(-1, j0)$ point, that is, the locus, C_{GH}, will encircle the point $(-1, j0)$ N times in the direction opposite to the direction of the Nyquist contour C in the S-plane.

If $P = 0$, that is the open-loop transfer function has no poles on the right hand side of the S-plane, then $N = 0$. Thus, for a closed-loop system to be stable, there must not be the encirclements of the $(-1, j0)$ point by the $G(s)H(s)$ locus.

The Nyquist stability criterion can be stated as following:

If the contour C_{GH} of the open-loop transfer function $G(s)H(s)$ in the $G(s)H(s)$ plane corresponding to a Nyquist contour in the S-plane encircles, the point $(-1, j0)$ in the counterclockwise direction as many times as the right half S-plane poles of $G(s)H(s)$, then the closed-loop system is stable.

8.4.3 Procedure for Investigating Stability Using Nyquist Criterion

The following procedure can be followed to investigate the stability of a closed-loop system from the knowledge of the open-loop system using Nyquist stability criterion.

1. Substitute $s = j\omega$ in the open-loop transfer function $G(s)H(s)$. This converts the transfer function into a frequency domain function $G(j\omega)H(j\omega)$.
2. Choose a Nyquist contour in S-plane shown in Fig. 8.16. This contour can be divided into four sections C_1, C_2, C_3 and C_4. The mapping of the four sections into C_{GH} contour in $G(s)H(s)$ plane can be done sectionwise and then combined together to get entire $G(s)H(s)$ contour.

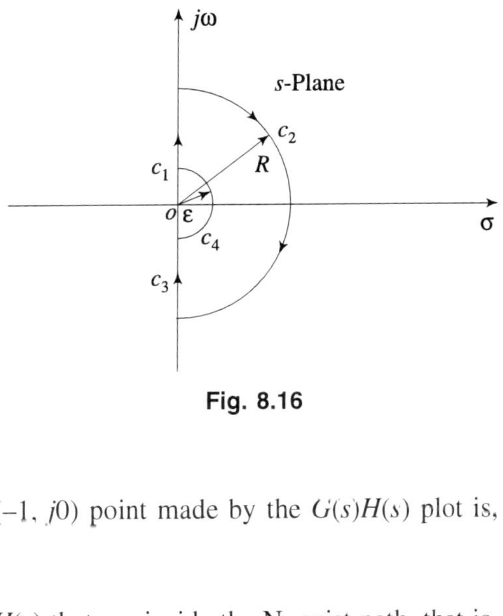

Fig. 8.16

3. The number of encirclements N of the $(-1, j0)$ point made by the $G(s)H(s)$ plot is, observed.
4. The Nyquist criterion gives $N = Z - P$.

   where P = number of poles of $1 + G(s)H(s)$ that are inside the Nyquist path, that is, the right half S-plane.

   N = number of encirclements of $(-1 + j0)$ point made by the $G(s)H(s)$ plot

   Z = number of zeros of $1 + G(s)H(s)$ that are inside the Nyquist path, that is the right half of the S-plane.

It is to be noted that the poles of $1 + G(s)H(s)$ are the same as that of $G(s)H(s)$. The stability requirements for the two types of stability in terms of Z and P are as follows:

For closed-loop stability, Z must be equal to zero.

For open-loop stability, P must be equal to zero.

Example 8.6: Sketch the Nyquist plot for the open-loop transfer function $G(s)H(s) = \dfrac{10}{(s+2)(s+4)}$. Determine the stability of the closed-loop system by Nyquist criterion.

Solution: The given transfer function by putting $s = j\omega$ in $G(s)H(s)$

$$G(s)H(s)|_{s=j\omega} = \frac{10}{(2+j\omega)(4+j\omega)} = \frac{10\angle 0^\circ}{\left(\tan^{-1}\frac{\omega}{2}\right)\left(\sqrt{(4+\omega^2)}\right)\left(\tan^{-1}\frac{\omega}{4}\right)\left(\sqrt{(16+\omega^2)}\right)}$$

$$|G(j\omega)H(j\omega)| = M = \frac{10}{\left(\sqrt{(4+\omega^2)}\right)\left(\sqrt{(16+\omega^2)}\right)}$$

$$\angle G(j\omega)H(j\omega) = \Phi = 0^\circ - \tan^{-1}\frac{\omega}{2} - \tan^{-1}\frac{\omega}{2}$$

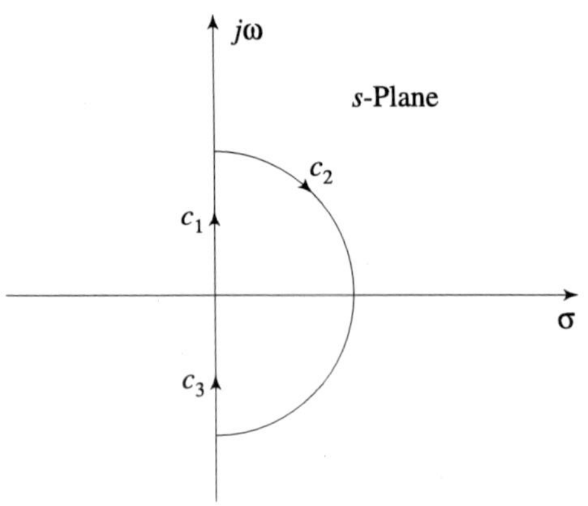

Fig. 8.17

Nyquist path is divided into three sections C_1, C_2 and C_3 mapping each section from S-plane to G-H plane

(i) *Mapping of section C_1:* In this section of the Nyquist contour on the $j\omega$ axis, S varies from 0 to $+\infty$.

At $\omega = 0^+$

$$\lim_{\omega\to 0} |G(j\omega)H(j\omega)| = \frac{10}{2\times 4} = 1.25$$

$$\angle G(j\omega)H(j\omega) = \lim_{\omega\to 0} \angle -\left(\tan^{-1}\frac{\omega}{2}\right) - \left(\tan^{-1}\frac{\omega}{4}\right)$$

$$= 0 - 0 = 0^\circ$$

At $\omega = +\infty$

$$\lim_{\omega \to 0} |G(j\omega)H(j\omega)| = \frac{10}{(\sqrt{4+\infty^2})(\sqrt{16+\infty^2})} = 0$$

$$\underline{|G(j\omega)\,H(j\omega)} = \lim_{\omega \to 0} \underline{\left| -\left(\tan^{-1}\frac{\infty}{2}\right) - \left(\tan^{-1}\frac{\infty}{4}\right)\right.}$$

$$= -90 - 90 = -180°$$

As the exact shape of $G(s)H(s)$ is not required, we can draw this part of G-H contour by noting that the angle of G-H changes from 0 to $-180°$ as ω changes from 0 to ∞.

The mapping of section C_1 from S-plane contour to G-H plane contour is shown in figure

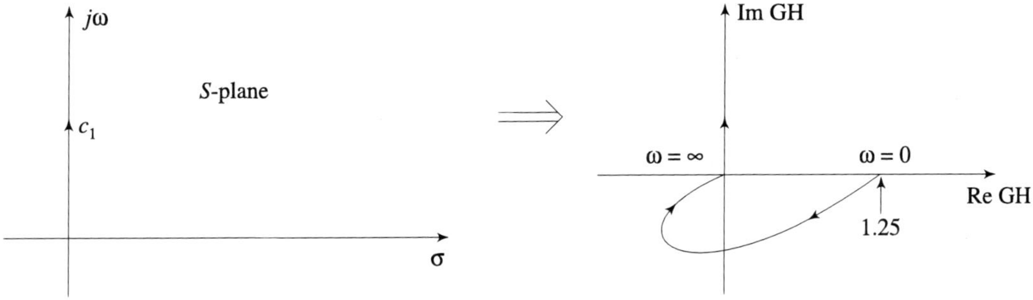

Fig. 8.18

(ii) *Mapping of section C_2:* Section C_2 in S-plane is a semicircle of infinite radius. It can be mapped in $G(s)H(s)$ plane by substituting

$$s = \lim_{R \to \infty} Re^{j\theta}$$

In $G(s)H(s)$ and varying θ from $+90°$ to $-90°$ through $0°$

$$G(s)H(s) = \frac{10}{(2 + Re^{j\theta})(4 + Re^{j\theta})}$$

As $\qquad R \to \infty$, $Re^{j\theta} >> 2$ and $Re^{j\theta} >> 4$

$\therefore$ $(Re^{j\theta} + 2)$ is approximated to $Re^{j\theta}$ and $(Re^{j\theta} + 4)$ is approximated to $Re^{j\theta}$

$$\lim_{R \to \infty} G(s)H(s) = \lim_{R \to \infty} \frac{10}{((2 + Re^{j\theta})(4 + Re^{j\theta}))}$$

$$= \lim_{R \to \infty} \frac{10}{(R^2 e^{j2\theta})} = 0\,e^{-j2\theta}$$

Thus, the entire semicircle C_2 in S-plane is mapped into a circular arc of radius zero at the origin in the GH plane.

(iii) *Mapping of section C_3:* In this section ω varies from $-\infty$ to 0. The mapping will be mirror image about the real axis in GH plane corresponding to section C_1.

This is shown in Fig. 8.19

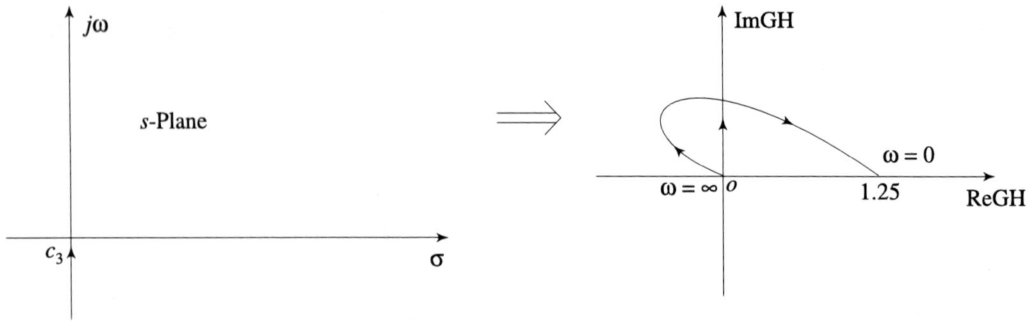

Fig. 8.19

In order to get the complete Nyquist plot we combine the mapping for sections C_1, C_2 and C_3. The complete Nyquist plot is shown in Fig. 8.20.

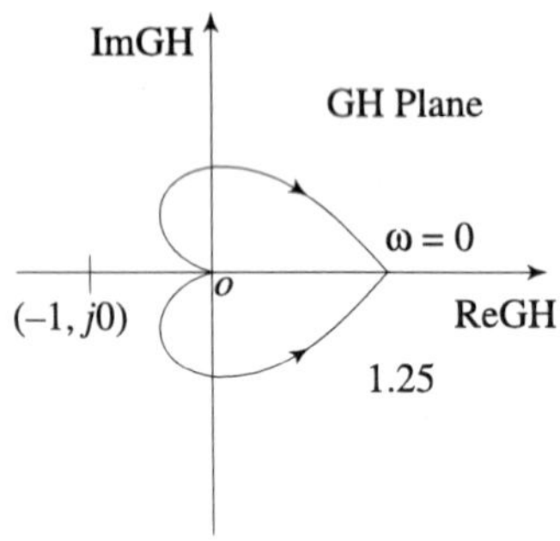

Fig. 8.20

8.4.4 Determination of Stability

To determine stability, we need to know whether the Nyquist plot encircled the critical point $(-1, j0)$ or not. It is seen that the Nyquist plot does not encircle the point $(-1 + j0)$. Therefore, $N = 0$ since there are no poles of $G(s)H(s)$ in the right half of S-plane $P = 0$, $Z = N + P = 0 + 0 = 0$.

Thus, there are no zeros of $1 + G(s)H(s)$. There is no pole of the closed-loop system in the right hand side of the S-plane. Hence, according to Nyquist criterion the closed-loop system is stable.

Example 8.7: The open-loop transfer function of a control system is given by $G(s)H(s) = \dfrac{K(s+1)}{s^2(s+2)(s+4)}$ using Nyquist criterion, determine the range of K for which the system is stable.

Solution: Putting $s = j\omega$ in $G(s)H(s)$

$$G(s)H(s)|_{s=j\omega} = G(j\omega)H(j\omega) = \frac{K(j\omega + 1)}{j\omega^2\,(j\omega + 2)(j\omega + 4)}$$

$$G(j\omega)H(j\omega) = \frac{K\left(\sqrt{(1+\omega^2)}\right) \angle \tan^{-1}\omega}{(\omega^2 \angle 180^\circ)\left(\sqrt{(4+\omega^2)}\ \angle\tan^{-1}\frac{\omega}{2}\right)\left(\sqrt{(16+\omega^2)}\ \angle\tan^{-1}\frac{\omega}{4}\right)}$$

$$[G(j\omega)H(j\omega)] = M = \frac{K\left(\sqrt{(1+\omega^2)}\right)}{(\omega^2)\left(\sqrt{(4+\omega^2)}\right)\left(\sqrt{(16+\omega^2}\right)}$$

$$\angle G(j\omega)H(j\omega) = \Phi = \tan^{-1}\omega - 180^\circ - \tan^{-1}(\omega/2) - \tan^{-1}(\omega/4)$$

There is a double pole at the origin. Nyquist path is shown in Fig. 8.21. It has four sections C_1, C_2, C_3 and C_4. We map each section of the Nyquist path.

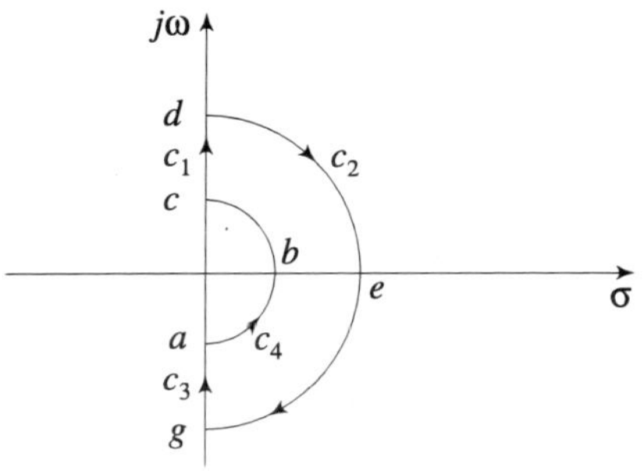

Fig. 8.21

(i) Mapping of section C_1

(ii) Along the section C_1 of the Nyquist path on the $j\omega$ axis, ω varies from $\omega = 0^+$ to $\omega = +\infty$

$$|G(j\omega)H(j\omega)| = \frac{K(0+1)}{(j0)^2\,(0+2)(0+4)} = \frac{K}{(j0)^2)} = \infty$$

$$\angle G(j\omega)H(j\omega) = \Phi = \tan^{-1}(0) - 180^\circ - \tan^{-1}\left(\frac{0}{2}\right) - \tan^{-1}\left(\frac{0}{4}\right)$$

$$= 0^\circ - 180^\circ \quad 0^{\circ -} - 0^\circ = -180^\circ$$

$$\therefore \quad G(j\omega)H(j\omega) = \infty \angle -180^\circ$$

At $\quad \omega = \infty$

$$|G(j\omega)H(j\omega)| = \frac{K(j\infty + 1)}{(j\infty)^2(j\infty + 2)(j\infty + 4)} = \frac{K(j\infty)}{(j\infty)^2(j\infty)(j\infty)} = 0$$

$$\angle G(j\omega)H(j\omega) = \Phi - \tan^{-1}(\infty) - 180^\circ - \tan^{-1}\left(\frac{\infty}{2}\right) - \tan^{-1}\left(\frac{\infty}{4}\right)$$

$$= 90^\circ - 180^\circ - 90^\circ - 90^\circ = -270^\circ$$

$$G(j\omega)H(j\omega) = 0 \angle -270^\circ$$

Therefore, the mapping of section C_1 in S-plane is represented by CD in GH plane, where the point C is at $\infty \angle -180^\circ$ and D is at $0 \angle -270^\circ$ as shown in Fig. 8.22. It is seen that the Nyquist plot crosses the negative real axis at E. At this point the imaginary part of $G(j\omega)H(j\omega)$ should be zero or angle of $G(j\omega)H(j\omega)$ should be $\pm 180^\circ$.

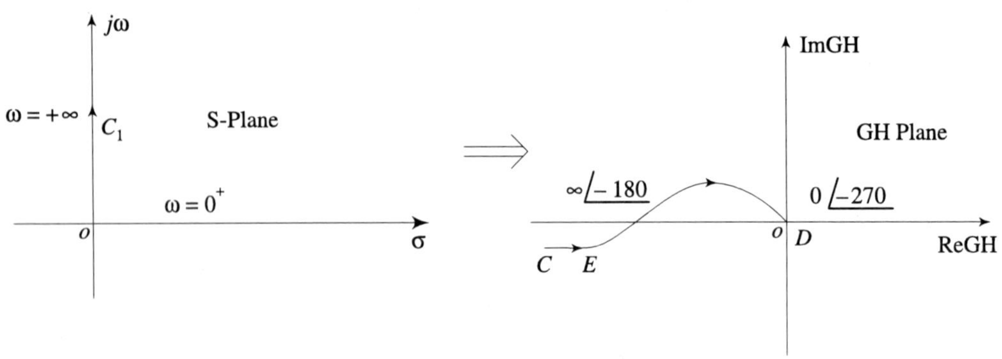

Fig. 8.22

$$\angle G(j\omega)H(j\omega) = -180^\circ$$

$$\tan^{-1}(\omega) - 180^\circ - \tan^{-1}\left(\frac{\omega}{2}\right) - \tan^{-1}\left(\frac{\omega}{4}\right) = -180^\circ$$

$$\tan^{-1}(\omega) = \tan^{-1}\left(\frac{\omega}{2}\right) + \tan^{-1}\left(\frac{\omega}{4}\right)$$

Taking tangent of angles on both sides, we get

$$\tan(\tan^{-1}(\omega) = \tan\left[\tan^{-1}\left(\frac{\omega}{2}\right) + \tan^{-1}\left(\frac{\omega}{4}\right)\right]$$

$$\omega = \frac{\left(\frac{\omega}{2}\right) + \left(\frac{\omega}{4}\right)}{1 - \left(\frac{\omega}{2}\right) + \left(\frac{\omega}{4}\right)}$$

$$1 - \frac{\omega^2}{8} = \frac{1}{2} + \frac{1}{4}$$

$$\frac{\omega^2}{8} = 1 - \frac{1}{2} - \frac{1}{4}$$

$$\omega^2 = 2, \ \omega = \sqrt{2} \text{ rad/s}$$

$$|G(j\omega)H(j\omega)| = \frac{K(\sqrt{1 + \omega^2})}{(\omega)^2 \sqrt{(\omega^2 + 4)} \sqrt{(\omega^2 + 16)}}$$

At this value of ω,

$$|G(j\omega)H(j\omega)| = \frac{K(\sqrt{1 + 2})}{2\sqrt{(2 + 4)} \sqrt{(2 + 16)}} = \frac{K}{12} = 0.0833K$$

Therefore, the GH plot crosses the negative real axis at -0.0833 K for $\omega = \sqrt{2}$

(i) *Mapping of section C_2:* Section C_2 in S-plane is a semicircle degree of infinite radius. It is mapped in $G(s)H(s)$ plane by substituting

$$s = \lim_{R \to \infty} \text{Re}^{j\theta} \text{ in } G(s)H(s) \text{ and varying } \theta \text{ from } +90^\circ \text{ to } -90^\circ \text{ through } 0^\circ.$$

$$\lim_{R \to \infty} G(s)H(s)\Big|_{s = \text{Re}^{j\theta}} = \lim_{R \to \infty} \frac{K(\text{Re}^{j\theta} + 1)}{\left(\text{Re}^{j\theta})^2\right) (\text{Re}^{j\theta} + 2)(\text{Re}^{j\theta} + 4)} = 0$$

Thus, the infinite semicircle degree in S-plane maps into the origin of the GH plane.

(ii) *Mapping of section C_3:* In section C_3 ω varies from $-\infty$ to 0^-. The mapping will be mirror image about real axis in GH plane of the plot for $\omega = 0^-$ to $\omega = \infty$ as shown in Fig. 8.23. This follows from the fact that $G(-j\omega)H(-j\omega)$ is the complex conjugate of $G(j\omega)H(j\omega)$.

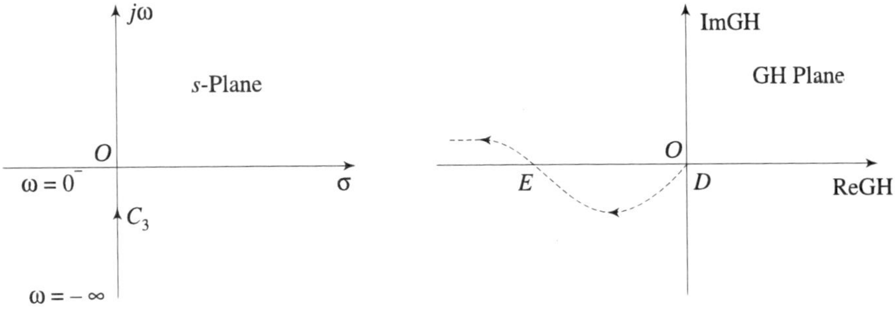

Fig. 8.23

(iii) *Mapping of section C_4:* The circle of section C_4 in S-plane can be mapped in $G(s)$ $H(s)$ plane by substituting

$$s = \lim_{\varepsilon \to \infty} \varepsilon e^{j\theta}$$

$$G(s)H(s)\big|_{s = \varepsilon e^{j\theta}} \frac{K(\in e^{j\theta} + 1)}{\left((\in e^{j\theta})^2\right)(\in e^{j\theta} + 2)(\in e^{j\theta} + 4)}$$

$$\lim_{\varepsilon \to \infty} G(\varepsilon e^{j\theta})H(\varepsilon e^{j\theta}) = \lim_{\varepsilon \to \infty} \frac{K}{(8\,\varepsilon^2 e^{j2\theta})} = \infty e^{-j2\theta}$$

Therefore, section C_4 is mapped into a semicircle of infinity radius. In section C_4 of S-plane, angle θ varies from $-90°$ to $+90°$. The corresponding Nyquist plot in GH plane is moving from $+180°$ to $-180°$ in the clockwise direction. This is shown in Fig. 8.24.

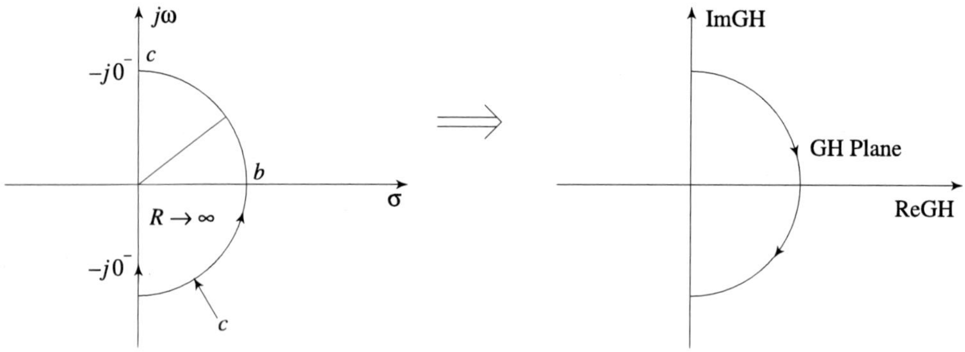

Fig. 8.24

The complete Nyquist plot is obtained by adding different sections of Nyquist plot as shown in Fig. 8.25.

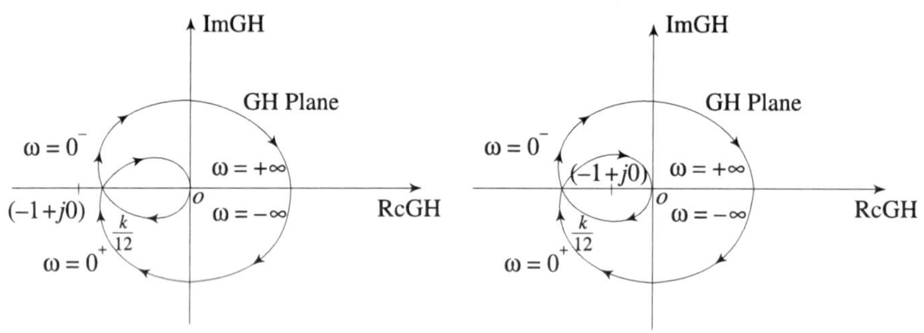

Fig. 8.25

Determination of stability

(a) When $K < 12$, the Nyquist plot is shown. In this case the plot $(-1, j0)$ is not encircled by the GH plot and therefore, $N = 0$. Since there is no pole in the right half of S-plane, $P = 0$.

By Nyquist criterion

$$Z = N + P = 0 + 0 = 0$$

∴ The system is stable for $K < 12$.

(b) When $K < 12$, the Nyquist plot that the point $(-1, j0)$ is encircled twice in clockwise direction. Thus, $N = 2$,

$$Z = N + P = 2 + 0 = 2$$

∴ The system is unstable for $K > 12$

The range of value of K for which system is stable is given by $0 < K < 12$.

Example 8.8: The open-loop transfer function of a system is $G(s)\,H(s) = \dfrac{(1 + 4S)}{S^2(1 + S)(1 + 2S)}$.

Determine the stability of closed-loop system. If the closed-loop system is not stable then find the number of closed-loop poles lying on the right half of S-plane.

Solution: Given that $G(s)\,H(s) = \dfrac{(1 + 4S)}{S^2\,(1 + S)(1 + 2S)}$

The open-loop transfer function has two poles at the origin. Hence, choose the Nyquist contour on S-plane enclosing the entire right half plane except the origin as shown in Fig. 8.26.

The open-loop transfer function has two poles at origin. Hence, choose the Nyquist contour on S-plane enclosing the entire right half plane except the origin. The Nyquist contour has 4 sections C_1, C_2, C_3 and C_4. The mapping of each section is performed separately and the overall Nyquist plot is obtained by combining the individual sections.

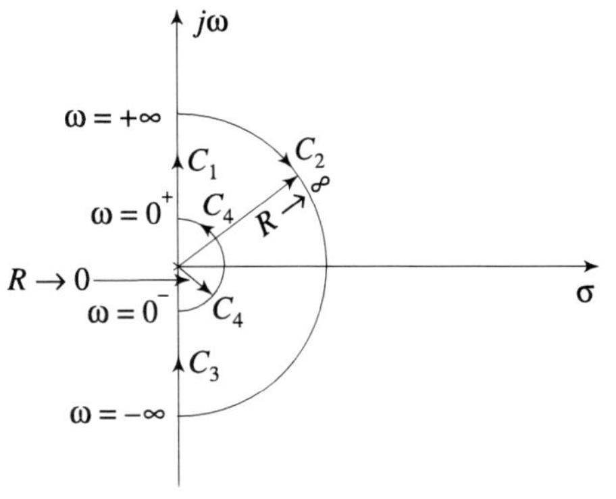

Fig. 8.26

8.4.4.1 Mapping of Section C_1

In section C_1, ω varies from 0 to + ∞. The mapping of section C_1 is given by the locus of $G(j\omega)\,H(j\omega)$ as ω is varied from 0 to ∞. This locus is the polar plot of $G(j\omega)\,H(j\omega)$.

$$G(s)\,H(s) = \frac{(1 + 4S)}{S^2(1 + S)(1 + 2S)}$$

Let $\quad S = j\omega$

$$G(j\omega)\,H(j\omega) = \frac{(1 + j4\omega)}{(j\omega)^2\,(1 + j\omega)(1 + j2\omega)}$$

$$= \frac{\sqrt{1+16\omega^2}\;\angle\tan^{-1}4\omega}{\omega^2\angle 180\;\sqrt{1+\omega^2}\;\angle\tan^{-1}\omega\;\sqrt{1+4\omega^2}\;\angle\tan^{-1}2\omega}$$

$$= \frac{\sqrt{1 + 16\omega^2}}{\omega^2\sqrt{1 + \omega^2}\ \sqrt{1 + 4\omega^2}} \underline{|(\tan^{-1} 4\omega - 180° - \tan^{-1} \omega - \tan^{-1} 2\omega)}$$

$$G(j\omega)\ H(j\omega) = \frac{\sqrt{1 + 16\omega^2}}{\omega^2\ \sqrt{1 + \omega^2}\ \sqrt{1 + 4\omega^2}}$$

$\underline{|G(j\omega)\,H(j\omega)} = \tan^{-1} 4\omega - 180 - \tan^{-1} \omega - \tan^{-1} 2\omega$ when the $G(j\omega)\ H(j\omega)$ locus crosses real axis, the phase will be -180 and the corresponding frequency is the phase crossover frequency ω_{pc}

∴ At $\omega = \omega_{pc}$, $\underline{|G(j\omega)\,H(j\omega)} = -180$

$$\tan^{-1} 4\,\omega_{pc} - 180 - \tan^{-1}\ \omega_{pc} - \tan^{-1} 2\,\omega_{pc} = -180$$

$$\tan^{-1} 4\,\omega_{pc} - 180 - \tan^{-1}\ \omega_{pc} - \tan^{-1} 2\,\omega_{pc} = -180$$

$$\tan^{-1} 4\,\omega_{pc} = \tan^{-1}\ \omega_{pc} + \tan^{-1} 2\,\omega_{pc}$$

On taking tan on both sides, we get

$$\tan\,[\tan^{-1} 4\,\omega_{pc}] = \tan\,[\tan^{-1}\ \omega_{pc} + \tan^{-1}\ \omega_{pc}]$$

$$4\,\omega_{pc} = \frac{\tan\,[\tan^{-1}\ \omega_{pc}] + \tan\,[\tan^{-1}\ 2\,\omega_{pc}]}{1 - \tan\,[\tan^{-1}\ \omega_{pc}]\ \tan\,[\tan^{-1}\ 2\,\omega_{pc}]}$$

$$\tan\,(A + B) = \frac{\tan A + \tan B}{1 - \tan A \tan B}$$

$$4\,\omega_{pc} = \frac{\omega_{pc} + 2\omega_{pc}}{1 - 2\,\omega^2_{pc}}$$

$$1 - 2\,\omega^2_{pc} = \frac{3\,\omega_{pc}}{4\,\omega_{pc}}$$

$$-\,2\,\omega^2 pc = \frac{3}{4} - 1$$

$$\omega_{pc} = \sqrt{\frac{-0.25}{-2}} = 0.354 \text{ rad/sec}$$

At $\omega = \omega_{pc} = 0.354$ rad/sec

$$|G(j\omega)\ H(j\omega)| = \frac{\sqrt{1 + 16\,\omega^2}}{\omega^2\ \sqrt{1 + \omega^2}\ \sqrt{1 + 4\,\omega^2}}$$

$$= \frac{\sqrt{1 + 16 \times (0.354)^2}}{(0.354)^2\ \sqrt{1 + (0.354)^2}\ \sqrt{1 + 4 \times (0.354)^2}} = 10.64$$

$G(s)\ H(j\omega)$ locus crosses the real axis at -10.64

At $\qquad \omega \to 0,\ G(j\omega)\ H(j\omega) \to \infty \angle -180^\circ$

At $\qquad \omega \to \infty,\ G(j\omega)\ H(j\omega) \to 0 \angle -270^\circ$

From equations we can say that the polar plot starts at –180 axis at infinity travels in third quadrant and crosses real axis at -10.64 to enter the second quadrant and then ends at the origin in the second quadrant. The section C_1 and its mapping are shown in Fig. 8.27.

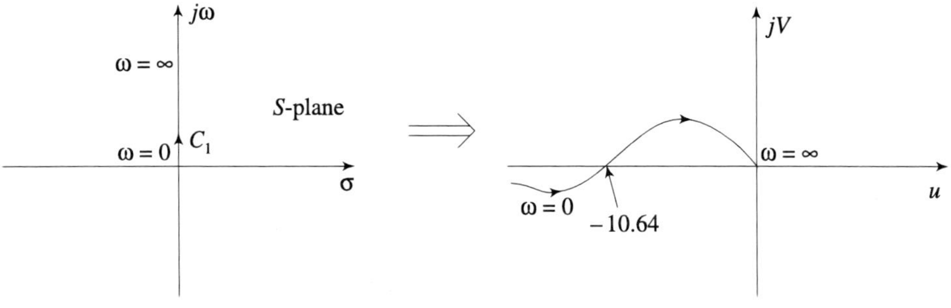

Fig. 8.27

8.4.4.2 Mapping of Section C_2

The mapping of section C_2 from S-plane to $G(s)\ H(s)$ plane is obtained by letting $S = \lim_{R\to\infty} Re^{j\theta}$ in $G(s)\ H(s)$ and varying θ from $+\frac{\pi}{2}$ to $-\frac{\pi}{2}$. Since $S \to Re^{j\theta}$ and $R \to \infty$, the $G(s)\ H(s)$ can be approximated as shown, i.e., $(1 + ST) \simeq ST$

$$G(s)\ H(s) = \frac{(1+4S)}{S^2(1+S)(1+2S)} \simeq \frac{4S}{S^2.S(2S)} = \frac{2}{S^3}$$

Let $\qquad S = \lim_{R\to\infty} Re^{j\theta}$

$$\therefore \qquad G(s)\ H(s)\Big|_{S = \lim_{R\to\infty} Re^{j\theta}} = \frac{2}{S^3}\Bigg|_{S = \lim_{R\to\infty} Re^{j\theta}} = \frac{2}{\lim_{R\to\infty}(Re^{j\theta})^3} = 0e^{-j3\theta}$$

where $\qquad \theta = \frac{\pi}{2},\ G(s)\ H(s) = 0\ e^{-i3\pi/2}$

where $\qquad \theta = -\frac{\pi}{2},\ G(s)\ H(s) = 0\ e^{i3\pi/2}$

From the above equations we can say that section C_2 in S-plane is mapped as circular arc of zero radius around origin in $G(s)\ H(s)$ plane with argument (phase) varying from $\frac{-3\pi}{2}$ to $\frac{3\pi}{2}$ as shown in Fig. 8.28.

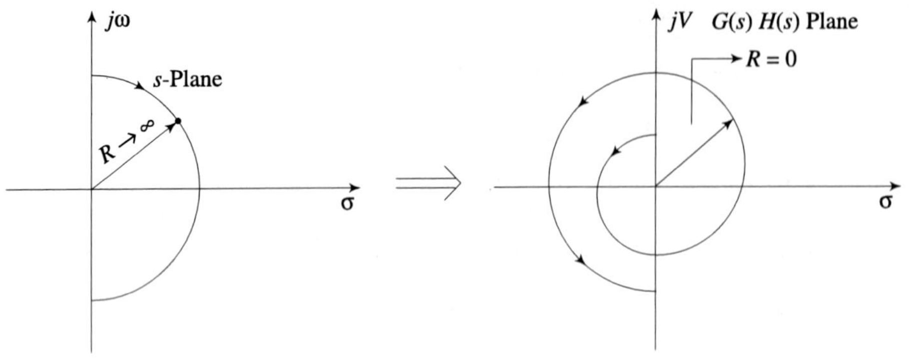

Fig. 8.28

8.4.4.3 Mapping of Section C_3

In section C_3, ω varies from $-\infty$ to 0. The mapping of section C_3 is given by the locus of $G(j\omega)\ H(j\omega)$ as ω is varied from $-\infty$ to 0. This locus is the inverse polar plot of $G(j\omega)H(j\omega)$. The inverse polar plot is given by the mirror image of polar plot with respect to real axis. The section C_3 in S-plane and its corresponding contour in $G(s)\ H(s)$ plane are shown in Fig. 8.29.

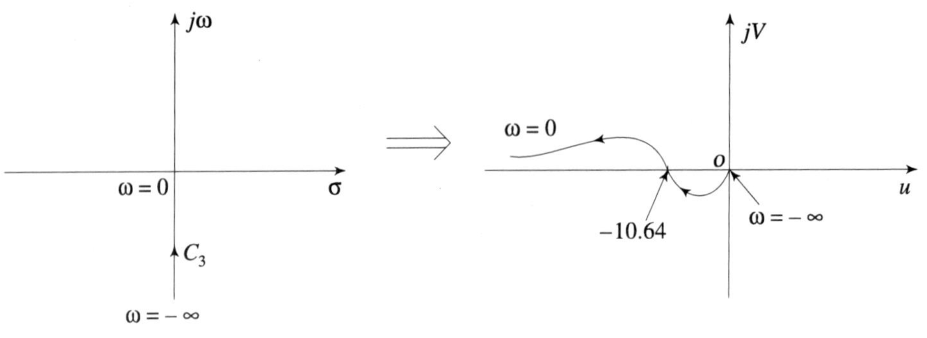

Fig. 8.29

8.4.4.4 Mapping of Section C_4

The mapping of section C_4 from S-plane to $G(s)\ H(s)$ plane is obtained by letting $S = \lim_{R\to\infty} Re^{j\theta}$ in $G(s)\ H(s)$ and varying θ from $-\frac{\pi}{2}$ to $\frac{\pi}{2}$. Since $S \to Re^{j\theta}$ and $R \to 0$, the $G(s)\ H(s)$ can be approximated as $(1 + ST) \simeq 1$

$$G(s)\ H(s) = \frac{(1 + 4S)}{S^2(1 + S)(1 + 2S)} = \frac{1}{S^2 \times 1 \times 1} = \frac{1}{S^2}$$

Let $$S = \lim_{R\to\infty} Re^{j\theta}$$

$$G(s)\,H(s)\Big|_{S = \lim\limits_{R\to\infty} \mathrm{Re}^{j\theta}} = \frac{1}{S^2}\Big|_{S = \lim\limits_{R\to\infty} \mathrm{Re}^{j\theta}}$$

$$= \frac{1}{\lim\limits_{R\to\infty} (\mathrm{Re}^{j\theta})^2} = \infty\, e^{-j2\theta}$$

where $\theta = -\dfrac{\pi}{2},\; G(s)\,H(s) = \infty\, e^{j\pi}$

where $\theta = \dfrac{\pi}{2},\; G(s)\,H(s) = \infty\, e^{-j\pi}$

From the equations, we can say that section C_4 in S-plane is mapped as a circle of infinite radius with argument (phase) varying from $-\pi$ to π as shown in Fig. 8.30.

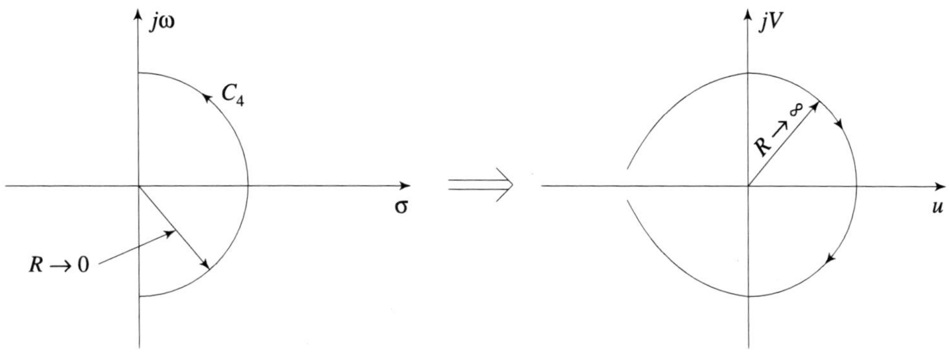

Fig. 8.30

Complete Nyquist plot

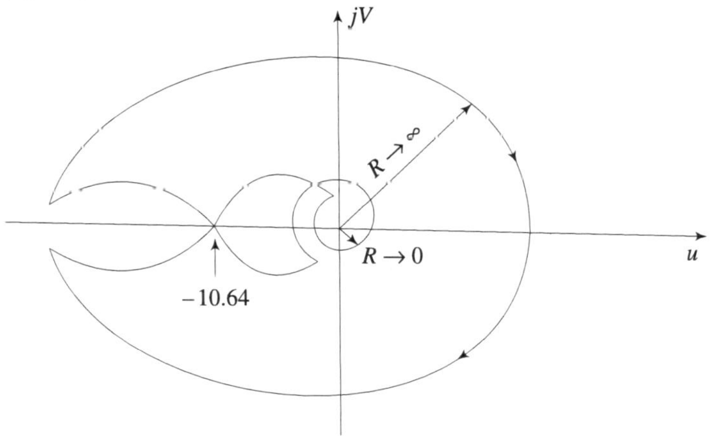

Fig. 8.31

Stability: On travelling through Nyquist contour in $G(s)\,H(s)$ plane it is observed that $(-1+j0)$ point is encircled in clockwise direction two times. Therefore, the closed-loop system is unstable. Since the $(-1+j0)$ is encircled twice in clockwise direction two poles of closed-loop system are lying on the right half S-plane.

Example 8.9: Sketch the Nyquist plot for a system with the open-loop transfer function

$$G(s)\,H(s) = \frac{K(1+0.5S)(1+S)}{(1+10S)\,(S-1)}$$

Determine the range of values of K for which the system is stable.

Solution: $G(s)\,H(s) = \dfrac{K(1+0.5S)(1+S)}{(1+10S)(S-1)}$

The open-loop transfer function does not have a pole at origin. Hence, choose the Nyquist contour on S-plane enclosing the entire right half plane as shown in Fig. 8.32.

The Nyquist contour has three sections C_1, C_2, and C_3. The mapping of each section is performed separately and the overalls Nyquist plot is obtained by combing the individuals sections.

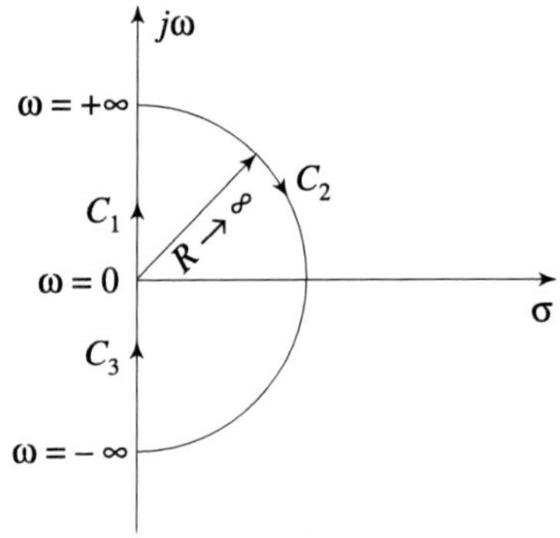

Fig. 8.32

8.4.4.5 Mapping of Section C_1

In section C_1, ω varies from 0 to + ∞. The mapping of section C_1 is given by the locus of $G(j\omega)\,H(j\omega)$ as ω is varied from 0 to ∞. This locus is the polar plot of $G(j\omega)\,H(j\omega)$.

$$G(s)\,H(s) = \frac{K(1+0.5S)\,(1+S)}{(1+10S)\,(S-1)}$$

Let $\qquad S = j\omega$

$$G(j\omega)H(j\omega) = \frac{K(1+j0.5\,\omega)(1+j\omega)}{(1+j10\,\omega)(-1+j\omega)}$$

$$= \frac{K(1+j1.5\,\omega-0.5\,\omega^2)}{-1-j9\,\omega-10\,\omega^2} = \frac{K(1-0.5\,\omega^2)+j1.5\,\omega K}{-(1+10\,\omega^2)-j9\,\omega}$$

On multiplying the numerators and denominator by the complex conjugate of denominator, we get

$$G(j\omega)\,H(j\omega) = \frac{K(1-0.5\,\omega^2)+j\,1.5\,\omega k}{-(1+10\,\omega^2)-j\,9\,\omega} \times \frac{-(1+10\,\omega^2)+j\,9\,\omega}{-(1+10\,\omega)^2+j\,9\,\omega}$$

$$G(j\omega)\,H(j\omega) = \frac{\begin{array}{l}-\;K(1-0.5\,\omega^2)\,(1+10\,\omega^2)-13.5\,\omega^2K\\ +\,j[9\omega K\,(1-0.5\omega^2)-1.5\omega K\,(1+10\omega^2)]\end{array}}{(1+10\omega^2)^2+(9\omega)^2}$$

When the $G(j\omega)\, H(j\omega)$ locus crosses the real axis the imaginary term is zero and the corresponding frequency is the phase crossover frequency.

At $\omega = \omega_{pc}$, $9\,\omega_{pc} K\,(1 - 0.5\,\omega_{pc}^2) - 1.5\,\omega_{pc} K(1 + 10\,\omega_{pc}^2) = 0$

$$9\,\omega_{pc} K(1 - 0.5\,\omega_{pc}^2) = 1.5\,\omega_{pc} K\,(1 + 10\,\omega_{pc}^2)$$

$$1 - 0.5\,\omega_{pc}^2 = \frac{1.5}{9}\,(1 + 10\,\omega_{pc}^2)$$

$$1 - 0.5\,\omega_{pc}^2 = 0.167 + 1.67\,\omega_{pc}^2$$

$$2.17\,\omega_{pc}^2 = 0.833$$

$$\omega_{pc} = \sqrt{\frac{0.833}{2.17}} = 0.62 \text{ rad/sec}$$

At $\omega = \omega_{pc} = 0.62$ rad/sec

$$G(j\omega)\, H(j\omega) = \frac{-K(1 - 0.5\,\omega^2)(1 + 10\,\omega^2) - 13.5\,\omega^2 K}{(1 + 10\,\omega^2)^2 + (9\,\omega)^2}$$

$$= \frac{-K[1 - 0.5 \times (0.62)^2 (1 + 10\,(0.62)^2 + 13.5\,(0.62)^2]}{(1 + 10\,(0.62)^2)^2 + (9\,(0.62))^2]}$$

$$= -K\left[\frac{3.913 + 5.189}{23.464 + 31.136}\right] = -\,0.1667\,K$$

The $G(j\omega)\, H(j\omega)$ locus crosses real axis at a point $-\,0.1667\,K$

The exact shape of $G(j\omega)\, H(j\omega)$ locus is determined by calculating the magnitude and phase of $G(j\omega)\, H(j\omega)$ for various values of ω.

$$G(j\omega)\, H(j\omega) = \frac{K(1 + j0.5\,\omega)(1 + \omega)}{(1 + j10\,\omega)\,(-1 + j\omega)}$$

$$= \frac{K\sqrt{1+(0.5\omega)^2}\,\underline{|\tan^{-1} 0.5\omega}\,\sqrt{1+\omega^2}\,\underline{|\tan^{-1}\omega}}{\sqrt{1+(10\omega)^2}\,\underline{|\tan^{-1} 10\omega}\,\sqrt{1+\omega^2}\,\underline{|180° - \tan^{-1}\omega}}$$

$$= K\frac{\sqrt{1 + 0.25\,\omega^2}}{\sqrt{1 + 100\,\omega^2}}\,\underline{|\tan^{-1}}\, 0.5\omega + 2\tan^{-1}\omega - \tan^{-1} 10\,\omega - 180°)$$

$$|G(j\omega)\, H(j\omega)| = K\frac{\sqrt{1 + 0.25\,\omega^2}}{\sqrt{1 + 100\,\omega^2}}$$

$$\underline{|G(j\omega)\, H(j\omega)} = \tan^{-1} 0.5\,\omega + 2\tan^{-1}\omega - \tan^{-1} 10\,\omega - 180$$

ω rad/sec	0	0.1	0.5	1.5	2.0	5.0	∞
$\lvert G(j\omega)\ H(j\omega)\rvert$	K	0.707 K	0.202 K	0.083 K	0.07K	0.054K	0
$\angle G(j\omega)\ 1 + (j\omega)$ deg	– 180	– 210	– 191	– 116	– 95	– 43	0

From the above analysis, the following conclusion are made:

- The locus of $G(j\omega)H(j\omega)$ starts at $K\angle{-180^\circ}$ when $\omega = 0$ and travels in the second quadrant.
- The locus crosses real axis at – 0.1667 K and enters the third quadrant.
- Then the locus crosses negative imaginary axis and enters the fourth quadrant.
- Finally, the locus ends at origin when $\omega = \infty$.

The section C_1 in S-plane and its corresponding mapping in $G(s)\ H(s)$ plane are shown in Fig. 8.33.

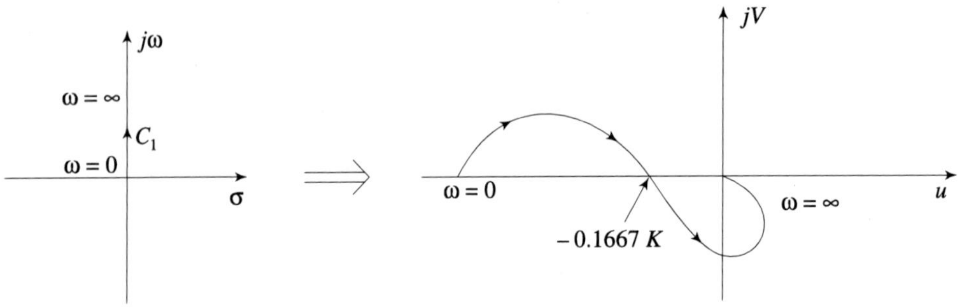

Fig. 8.33

8.4.4.6 Mapping of Section C_2

The mapping of section C_2 from S-plane to $G(s)\ H(s)$ is obtained by letting $S = \lim_{R\to\infty} Re^{j\theta}$ in $G(s)H(s)$ and varying θ from $-\frac{\pi}{2}$ to $\frac{\pi}{2}$. Since $S \to Re^{j\theta}$ and $R \to G(s)\ H(s)$ can be approximated as

$$(1 + ST) \simeq ST \text{ and } (S - 1) = S$$

$$G(s)\ H(s) = \frac{K(1 + 0.5S)\ (1 + S)}{(1 + 10S)(S - 1)} \simeq \frac{K(0.5S)(S)}{10S(S)}$$

$$= 0.05\ K$$

The approximate $G(s)\ H(s)$ is independent of S and so the contour of section C_2 in S-plane is mapped as a point at 0.05 K in $G(s)\ H(s)$ plane.

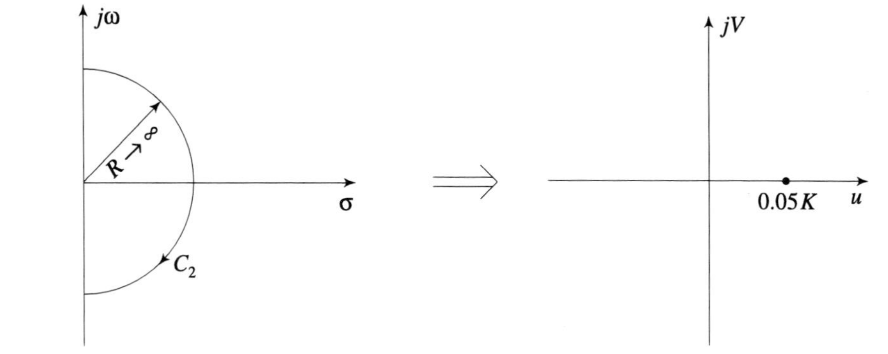

Fig. 8.34

8.4.4.7 Mapping of Section C_3

In section C_3, ω varies from $-\infty$ to 0. The mapping of section of C_3 is given by the locus of $G(j\omega)\ H(j\omega)$ as ω is varied from $-\infty$ to 0. This locus is the inverse polar plot of $G(j\omega)H(j\omega)$.

The inverse polar plot is given by the mirror image of polar plot with respect to real axis. The section C_3 in S-plane and its corresponding contour in $G(s)\ H(s)$ plane are shown in Fig. 8.35.

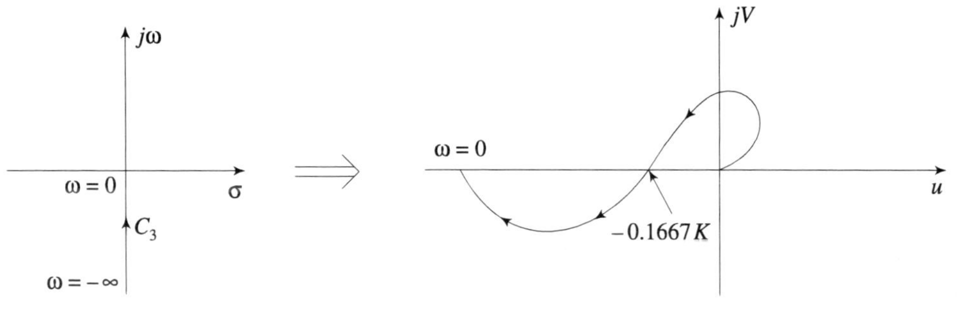

The complete Nyquist plot

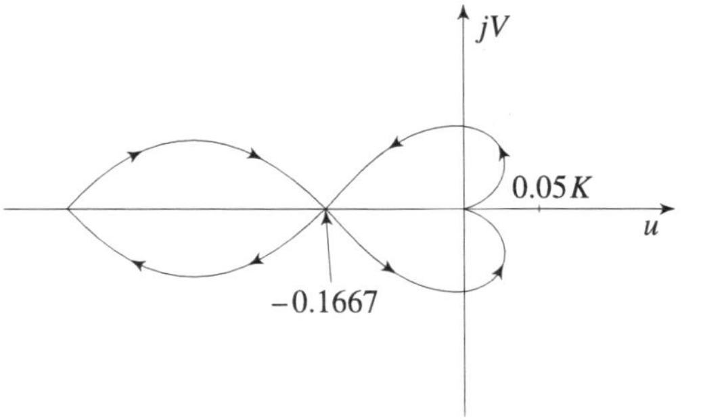

Fig. 8.35

8.5 SENSITIVITY ANALYSIS IN FREQUENCY DOMAIN

Sensitivity considerations often play an important role in the design of control systems. All physical elements have properties that change with environment and age. A good control system should be very insensitive to these parameter variations while being able to follow the command responsively. The sensitivity of a system closed-loop transfer function T with respect to the characteristic K of a given element is

$S_K^T = \frac{\frac{\partial T}{T}}{\frac{\partial K}{K}}$. Note that the sensitivity is a function of frequency.

Sensitivity of T to variations in G: The sensitivity of closed-loop transfer function $T(s)$ to variations in open-loop transfer function $G(s)$ is given by the function

$$S_K^T\,(j\omega) = \frac{1}{1 + G(j\omega)H(j\omega)}.$$

The function $S_K^T\,(j\omega)$ may conveniently be evaluated from the Nyquist plot of $G(j\omega)$ $H(j\omega)$.

8.5.1 Stability Analysis

When $-\,0.1667\,K = -\,1$, the contour passes through -1 to $j0$ point and this value of K is the limiting value for stability.

The limiting value of $K = \frac{1}{0.1667} = 6.$

When $K < 6$, the locus crosses real axis between 0 and -1. On travelling through the locus it is observed that the $-1 + j0$ point is encircled clockwise and so the closed-loop system is unstable.

When $K > 6$, the locus crosses real axis between -1 and $-\infty$ on travelling through the locus it is observed that $-1 + j0$ point is encircled anticlockwise on time. Also the open-loop system has one pole at the right half S-plane. Hence, the system is stable.

Example 8.10: Construct Nyquist plot for a feedback control system whose open-loop transfer function is given by $G(s)\,H(s) = \frac{5}{S(1+S)}$. Comment on the stability of open-loop and closed-loop systems.

Solution: $G(s)\,H(s) = \frac{5}{S(1+S)}$.

The open-loop transfer functions has a pole at origin. Hence, choose the Nyquist contour on S-plane enclosing the entire right half S-plane except the origin as shown in Fig. 8.36.

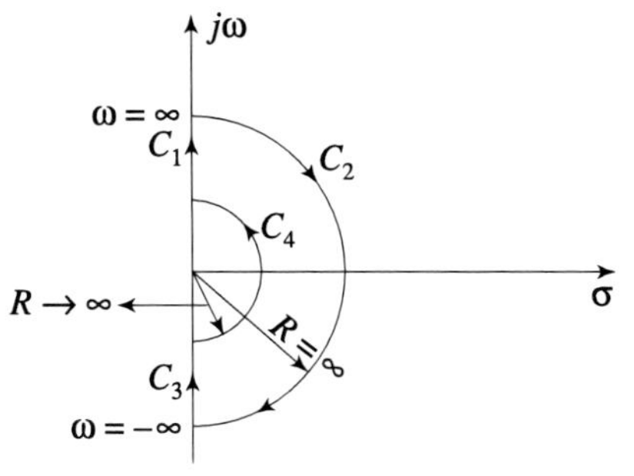

Fig. 8.36

The Nyquist contour has four sections C_1, C_2, C_3 and C_4. The mapping of each section is performed separately and the overall Nyquist plot is obtained by combining the individual sections.

Mapping of Section C_1

In section C_1, ω varies from 0 to + ∞. The mapping of section C_1 is given by the locus of $G(j\omega)\ H(j\omega)$ as ω is varies from 0 to ∞. This locus is the polar plot of $G(j\omega)\ H(j\omega)$.

Note: $(1 - j\omega)$ represents a point in the fourth quadrant

$$G(s)\ H(s) = \frac{5}{S(1 - S)}$$

Let $$S = j\omega$$

$$\therefore \quad G(j\omega)\ H(j\omega) = \frac{5}{j\omega(1 - j\omega)} = \frac{5}{\omega\angle 90^\circ \sqrt{1+\omega^2}\ \angle -\tan^{-1}\omega}$$

$$= \frac{5}{\omega\sqrt{1 + \omega^2}}\ \angle -90 + \tan^{-1}\omega$$

$$\therefore \quad |G(j\omega)\ H(j\omega)| = \frac{5}{\omega\sqrt{1 + \omega^2}}$$

$$\angle G(jw)\ H(jw) = 90 + \tan^{-1}\omega.$$

The exact shape of $G(j\omega)\ H(j\omega)$ locus is determined by calculating the magnitude and phase of $G(j\omega)H(j\omega)$ for various value of ω.

ω rad/sec	0	0.6	1.0	2.0	10.0	∞
$\lvert G(j\omega)H(j\omega)\rvert$	∞	7.15	3.53	1.12	0.05	0
$\angle G(j\omega)\ H(j\omega)$ deg	−90	−59	−45	−26	−5	0

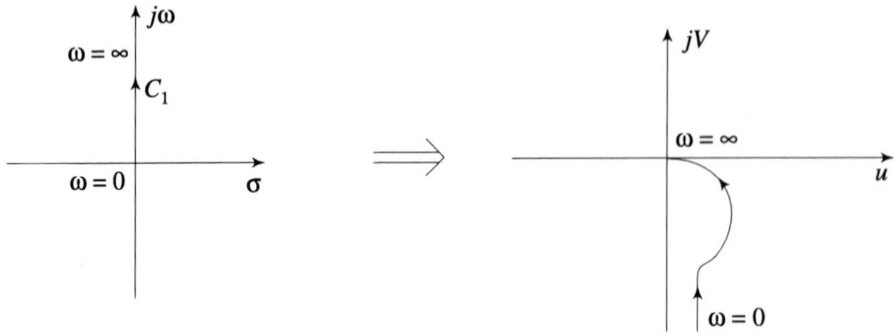

Fig. 8.37

From the above analysis we can conclude that $G(j\omega)\ H(j\omega)$ locus starts at -90 axis at infinity for $\omega = 0$ and meets the origin along 0 axis when $\omega = \infty$.

Mapping of Section C_2

The mapping of section C_2 from S-plane to $G(s)\ H(s)$ plane is obtained by letting $S = \lim_{R \to \infty} \text{Re}^{j\theta}$ in $G(s)\ H(s)$ and varying θ from $+\frac{\pi}{2}$ to $\frac{\pi}{2}$. Since $S \to \text{Re}^{j\theta}$ and $R \to \infty$, the $G(s)\ H(s)$ can be approximated as $(1 - s) \simeq -S$

$$G(s)\ H(s) = \frac{5}{S(1-S)} \simeq \frac{5}{S(-S)} = \frac{5}{S^2\ e^{j\pi}} \quad \because e^{j\pi} = -1$$

Let $$S = \lim_{R \to \infty} \text{Re}^{j\theta}$$

$$G(s)\ H(s)\Big|_{S = \lim\limits_{R\to\infty} \text{Re}^{j\theta}} = \frac{5}{\lim\limits_{R \to \infty} (\text{Re}^{j\theta})\ 2\, e^{j\pi}} = 0\, e^{-j(2\theta + \pi)}$$

When $$\theta = \frac{\pi}{2},\ G(s)\ H(s) = 0\, e^{-j2\pi}$$

When $$\theta = \frac{-\pi}{2},\ G(s)\ H(s) = 0\, e^{j0}$$

Section C_2 in S-plane is mapped as circular arc of zero radius around origin in $G(s)\ H(s)$ plane with argument varying from -2π to $+0$ as shown in Fig. 8.38.

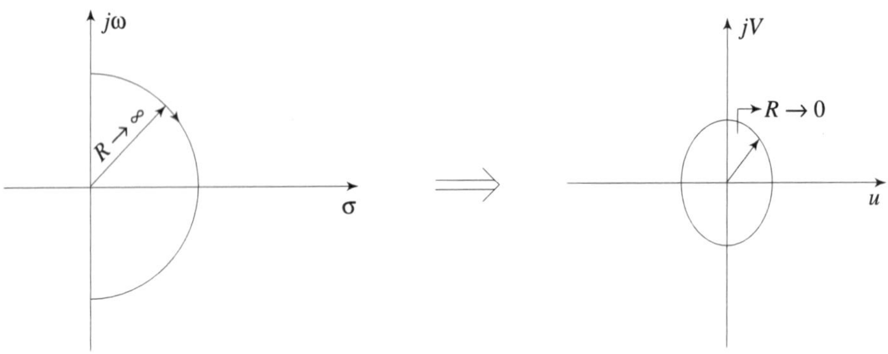

Fig. 8.38

Mapping of Section C_3

It section C_3, ω varies from $-\infty$ to 0. The mapping of section C_3 is given by the locus of $G(j\omega)$ $H(j\omega)$ as ω is varied from $-\infty$ to 0. This locus is the inverse polar plot of $G(j\omega)$ $H(j\omega)$.

The inverse polar plot is given by the mirror image of polar plot with respect to real axis. The section C_3 in S-plane and its corresponding contour in $G(s)$ $H(s)$ plane are shown in Fig. 8.39.

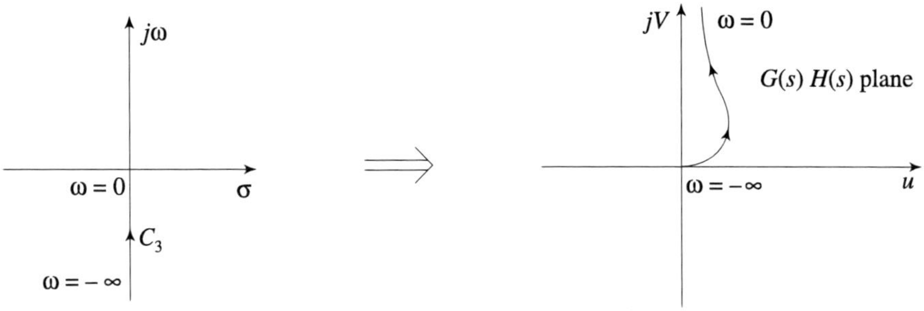

Fig. 8.39

Mapping of Section C_4

The mapping of section C_4 from S-plane to $G(s)$ $H(s)$ plane is obtained by letting $S = \lim\limits_{R\to\infty} \text{Re}^{j\theta}$ in $G(s)$ $H(s)$ and varying θ from $\frac{-\pi}{2}$ to $\frac{\pi}{2}$. Since $S \to \text{Re}^{j\theta}$, $R \to 0$ $G(s)$ $H(s)$ can be approximate $(1 - s) \simeq 1$.

$$G(s)\,H(s) = \frac{5}{S(1-S)} \simeq \frac{5}{S \times 1} = \frac{5}{S}$$

Let $$S = \lim_{R\to\infty} \text{Re}^{j\theta}$$

$$\therefore \quad G(s)\,H(s)\Big|_{\lim\limits_{R\to\infty} \text{Re}^{j\theta}} = \frac{5}{\lim\limits_{R\to\infty} \text{Re}^{j\theta}} = \infty\, e^{-j\theta}$$

When $$\theta = \frac{-\pi}{2},\; G(s)\,H(s) = \infty\, e^{j\pi/2}$$

When $$\theta = \frac{\pi}{2},\; G(s)\,H(s) = \infty\, e^{-j\pi/2}$$

The section C_4 in S-plane is mapped as a circular arc of infinite radius with argument varying from $\pi/2$ to $-\pi/2$ as shown in Fig. 8.40.

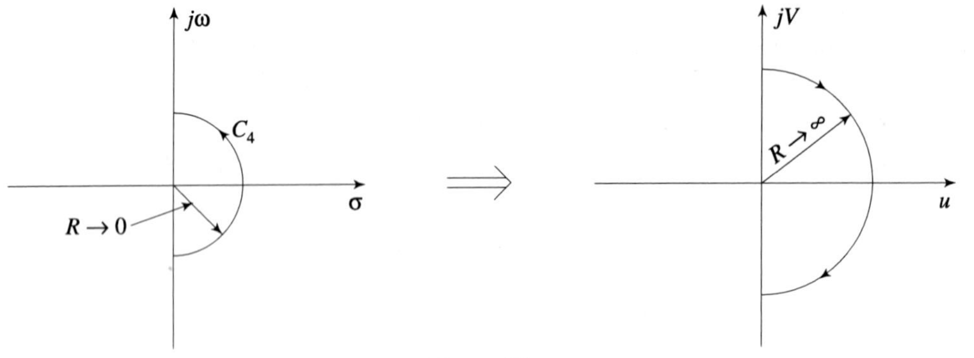

Fig. 8.40

Complete Nyquist plot

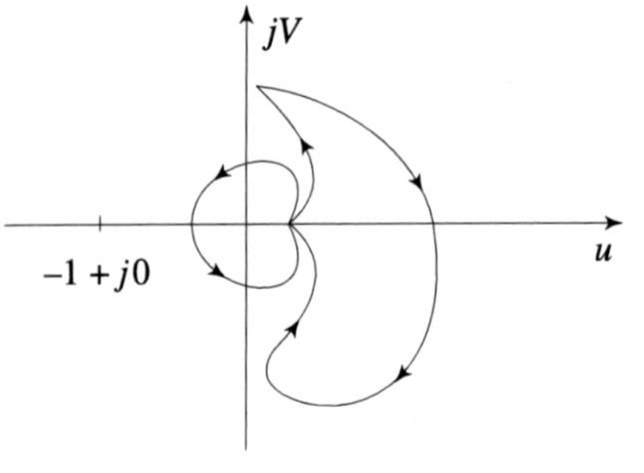

Fig. 8.41

Stability Analysis

The Nyquist contour in $G(s)\,H(s)$ plane does not encircle the point $(-1 + j0)$ but the open-loop transfer function has one pole on the right half S-plane. Therefore, the system is unstable.

Example 8.11: By Nyquist stability criterion determine the stability of closed-loop system, whose open-loop transfer function is given by $G(s)\,H(s) = \dfrac{S+2}{(S+1)(S-1)}$. Comment on the stability of open-loop and closed-loop system

Solution: $G(s)\,H(s) = \dfrac{(S+2)}{(S+1)(S-1)}$

The open-loop transfer function does not have a pole at the origin. Hence, choose the Nyquist contour on S-plane enclosing the entire right half plane as shown in Fig. 8.42. The Nyquist contour has three sections C_1, C_2 and C_3. The mapping of each section is performed.

Separately and the overall Nyquist plot is obtained by combining the individual sections.

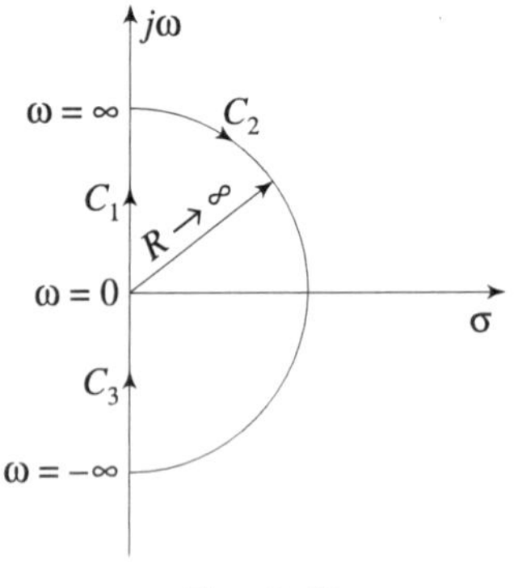

Fig. 8.42

Mapping of Section C_1

In section C_1, ω varies from 0 to +∞. The mapping of section C_1 is given by the locus of $G(j\omega)\ H(j\omega)$ as ω is varied from 0 to ∞. This locus is the polar plot of $G(j\omega)\ H(j\omega)$.

$$G(s)\ H(s) = \frac{S+2}{(S+1)(S-2)} = \frac{2(1+0.5S)}{(1+S)(-1+S)}$$

Let $\qquad S = j\omega$

$$G(j\omega)\ H(j\omega) = \frac{2(1+j0.5\,\omega)}{(1+j\omega)(-1+j\omega)} = \frac{2\sqrt{1+0.25\omega^2}\ \angle\tan^{-1}0.5\omega}{\sqrt{1+\omega^2}\ \angle\tan^{-1}\omega\ \sqrt{1+\omega^2}\ \angle 180-\tan^{-1}\omega}$$

$$= \frac{2\sqrt{1+0.25\,\omega^2}}{1+\omega^2}\ \angle -180+\tan^{-1}0.5\omega$$

$$|G(j\omega)\ H(j\omega)| = \frac{2\sqrt{1+0.25\,\omega^2}}{1+\omega^2}$$

$$\angle G(j\omega)\ H(j\omega) = -\ 180 + \tan^{-1}\ 0.5\ \omega$$

The exact shape of $G(j\omega)\ H(j\omega)$ locus is determined by calculating the magnitude and phase of $G(j\omega)\ H(j\omega)$ for varied values of ω.

ω rad/sec	0	0.4	1.0	2.0	10.0	∞
$\|G(j\omega)\ H(j\omega)\|$	2	1.76	1.12	0.57	0.1	0
$\angle G(j\omega)\ H(j\omega)$ deg	−180	−168	−153	−135	−101	−90

We can conclude that $G(j\omega)\ H(j\omega)$ locus starts at − 180 axis at a magnitude of − 2 for ω = 0 and meets the origin along −90 axis when ω = ∞.

The section C_1 in S-plane and its corresponding mapping in $G(s)\ H(s)$ plane are shown in Fig. 8.43

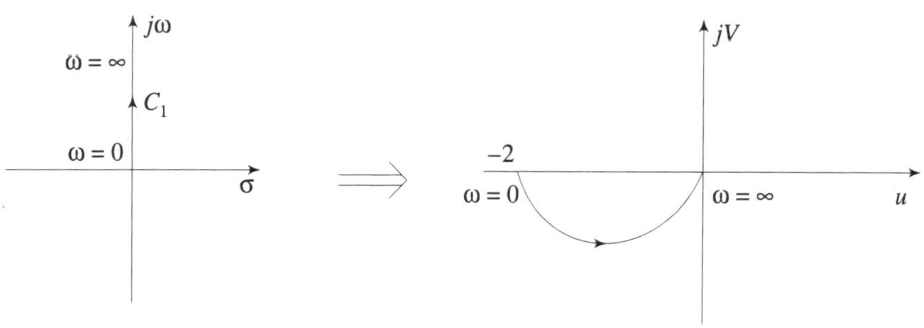

Fig. 8.43

Mapping of Section C_2

The mapping of section C_2 from S-plane to $G(s)\ H(s)$ plane is obtained by letting $S = \lim_{R\to\infty} Re^{j\theta}$ in $G(s)\ H(s)$ and varying θ from $+\frac{\pi}{2}$ to $-\frac{\pi}{2}$. Since $s = \lim_{R\to\infty} Re^{j\theta}$ and $R \to \infty$, $G(s)\ H(s)$ can be approximated $(1 + ST) \simeq ST$.

$$G(s)\ H(s) = \frac{2(1 + 0.5S)}{(1 + S)(-1 + S)} \simeq \frac{2\times 0.5S}{S\times S} = \frac{1}{S}$$

Let $$S = \lim_{R\to\infty} Re^{j\theta}$$

$$\therefore \quad G(s)\ H(s)\Big|_{\lim_{R\to\infty} Re^{j\theta}} = \frac{1}{\lim_{R\to\infty} Re^{j\theta}} = 0\,e^{-j\theta}$$

When $$\theta = \frac{\pi}{2},\ G(s)\ H(s) = 0\,e^{-j\pi/2}$$

When $$\theta = -\frac{\pi}{2}\ G(s)\ H(s) = 0\,e^{j\pi/2}$$

From the equation, section C_2 in S-plane is mapped as circular arc of zero radius around origin in $G(s)\ H(s)$ plane with argument varying from $-\pi/2$ to $\pi/2$.

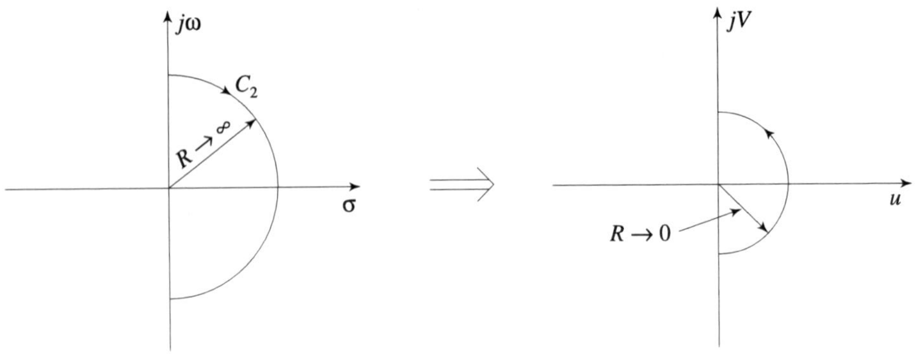

Fig. 8.44

Mapping of Section C_3

In section C_3, ω varies from $-\infty$ to 0. The mapping of section C_3 is given by the locus of $G(j\omega)\ H(j\omega)$ as ω is varied from $-\infty$ to 0. This locus is the inverse polar plot of $G(j\omega)\ H(j\omega)$.

The inverse polar plot is given by the mirror image of polar plot with respect to real axis. The section C_3 in S-plane and its corresponding contour in $G(s)\ H(s)$ plane are shown in Fig. 8.45.

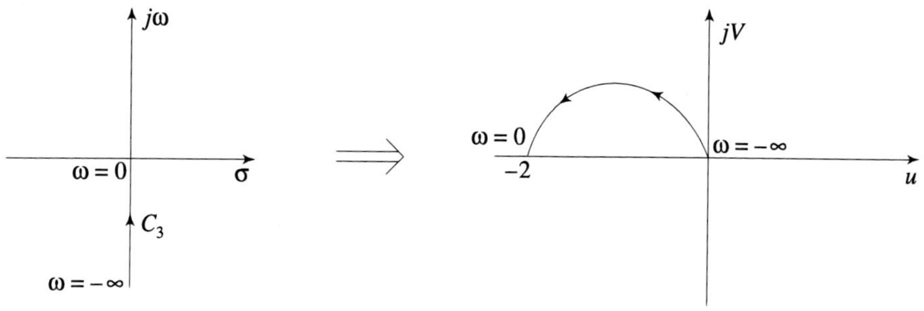

Fig. 8.45

Complete Nyquist plot

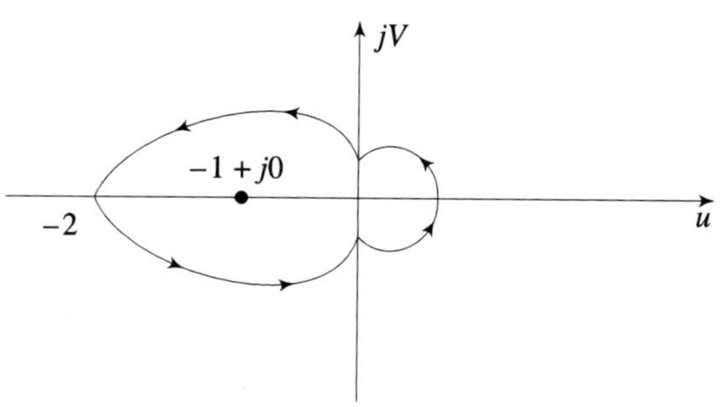

Fig. 8.46

Stability analysis: Nyquist contour is observed that $-1+j0$ point is encircled in anticlockwise direction one time. Also the open-loop transfer function has one pole at right half S-plane. Since the number of anticlockwise encirclement is equal to number of open-loop poles on right half S-plane, the closed-loop system is stable.

EXERCISE

Fill in the Blanks

1. The Nyquist plot of a loop transfer function $G(j\omega)H(j\omega)$ of a system encloses the $(-1, j0)$ point. The gain margin of the system is __________.

Ans: less than zero dB

2. The polar plot of the transfer function $G(s) = \dfrac{10(s+1)}{(s+10)}$ for $0 \le \omega \le \infty$ will be in the __________. (GATE 2015)

Ans: first quadrant

3. The polar plot of $G(s) = \dfrac{10}{s(s+1)^2}$ intercepts real axis at $\omega = \omega_o$ then, the real part and ω_o are respectively given by __________.

Ans: -5, 1

4. The open-loop transfer function of a feedback control system is $G(s)H(s) = \dfrac{s}{(s+1)^3}$. The gain margin of the system is ________.

Ans: 8

5. The gain margin and the phase margin of a feedback system with $G(s)H(s) = \dfrac{s}{(s+100)^3}$ are ________.

Ans: ∞, ∞

6. The Nyquist plot of $G(j\omega)H(j\omega)$ for a closed-loop system, passes through (–1, 0) point in the GH plane. The gain margin of the system in dB is equal to ________.

Ans: zero

7. The value of "a" so that the system has a phase margin equal to $\dfrac{\pi}{4}$ is approximately equal to ________.

Ans: 0.84

8. The magnitude of frequency response of an underdamped second order system is at 5 at 0 rad/sec and peaks to $\dfrac{10}{\sqrt{3}}$ at $5\sqrt{2}$ rad/sec. The transfer function of the system is ________.

Ans: $\dfrac{500}{s^2 + 10s + 100}$

9. Sketch the polar plot of

$$G(s) = \frac{1}{(s+2)}$$

Ans.

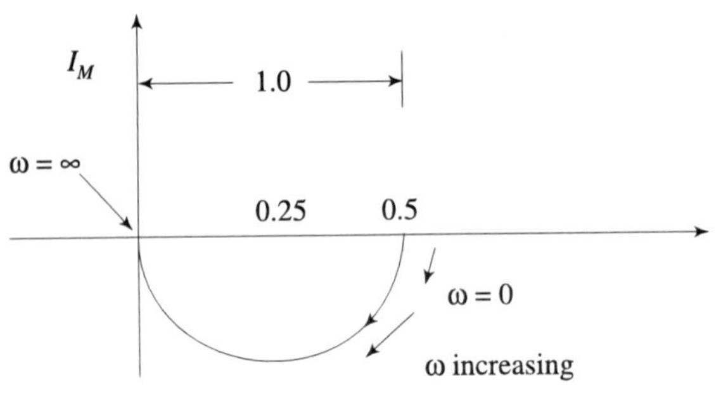

Fig. 8.47

10. Sketch the polar plot of

$$G(s) = \frac{1}{(1+s)(1+2s)}$$

Ans.

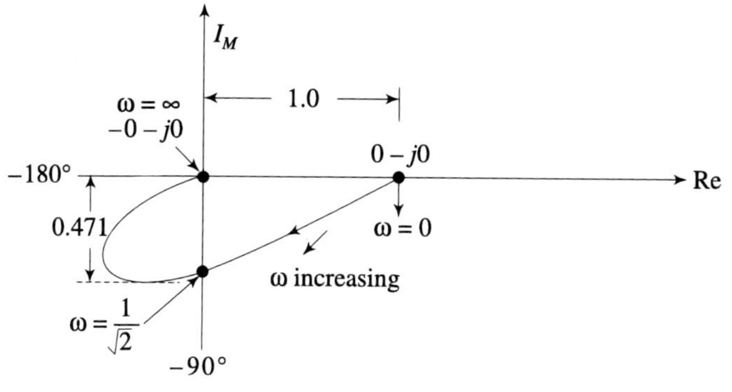

Fig. 8.48

11. Sketch the polar plot of

$$G(s) = \frac{10}{s(s+1)(s+2)}$$

Ans.

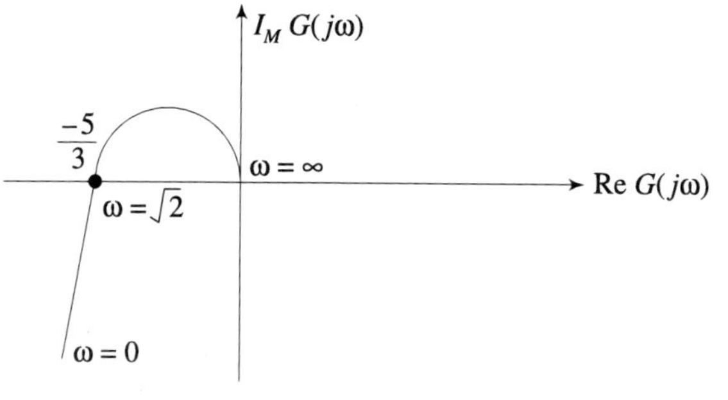

Fig. 8.49

12. What is polar plot?

Ans. The polar plot of a sinusoidal transfer function $G(j\omega)$ is a plot of the magnitude of $G(j\omega)$ versus the phase angle/argument of $G(j\omega)$ on polar on rectangular coordinates as ω is varied from zero to infinity.

13. What is minimum phase system?

Ans. The minimum phase system are systems with minimum phase transfer functions. In minimum phase transfer function all poles and zeros will lie on the left half of *S*-plane.

14. What is all-pass systems?

Ans. The all-pass systems are systems with all pass transfer functions. In all-pass transfer functions, the magnitude is unity at all frequencies and the transfer function will have anti-symmetric pole zero pattern (i.e., for very pole in the left half *S*-plane, there is a zero in the mirror image position with respect to imaginary axis.

15. What is non-minimum phase transfer function?

Ans. A transfer function which has one or more zeros in the right half *S*-plane is known as non-minimum phase transfer function.

16. In minimum phase system, how the stand and end of polar plot are identified?

Ans. For minimum phase transfer functions with only poles, the type number of the system determines the quadrant in which the polar plot starts and the order of a system determine the quadrant in which the polar plot ends

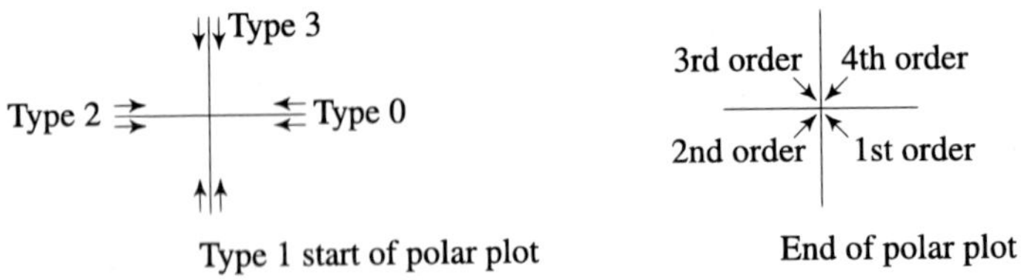

Fig. 8.50

17. Draw the polar plot of $G(s) = 1|(1 + ST)$

Ans. Let $s = j\omega$

$$\therefore\ G(j\omega) = \frac{1}{1 + j\omega T}$$

The given statement is of type–0 and order = 1. Hence, $G(j\omega)$ locus starts in fourth quadrant at 0° and ends in fourth quadrant along 90° axis.

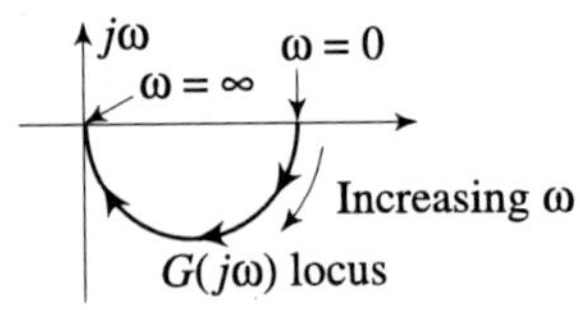

Fig. 8.51

18. Sketch the polar plot of $G(s) = \dfrac{1}{s^2(1 + ST_1)(1 + ST_2)(1 + ST_3)}$

Ans. The given system in all pole minimum phase system the type number of the system is 2 and the order is 5. Hence, the polar plot starts in second quadrant and ends in the fourth quadrant.

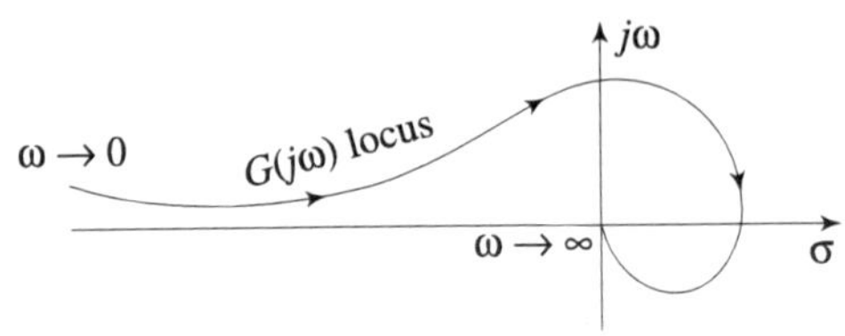

Fig. 8.52

19. Sketch the polar plot of the following transfer functions and from the plot, determine the phase margin and gain margin
    (a) $G(s) = 10(s + 1)|(s + 10)^2$ (b) $G(s) = 200\ (s + 2)|s(s^2 + 10s + 100)$
    (c) $e^{-0.1s}|s(s + 1)(s + 5)$ (d) $1|s(s + 4)(s + 8)$
20. The open-loop transfer function of a system is given by $G(s) = K|s(s^2 + s + 4)$ using polar plot, determine the value of *K*, so that phase margin is 50°. What is the corresponding value of gain margin?
21. What is Nyquist stability criterion?

Ans. If $G(s)H(s)$ contour in the $G(s)H(s)$ plane corresponding to Nyquist contour in S-plane encircle the point $-1 + j10$ in the anticlockwise direction as many times as the number of right half-plane poles of $G(s)H(s)$. Then the closed-loop s/m is stable.

22. The open-loop transfer function of certain unity feedback system are given below. In each case, sketch the Nyquist plot and determine the stability of the system.

(a) $G(s) = \dfrac{K(s+3)}{s(s-1)}$ (b) $G(s) = \dfrac{-1}{2s(1-20s)}$

(c) $G(s) = \dfrac{K(1+2s)}{s(1+s)(1+s+s^2)}$ (d) $G(s) = \dfrac{K(s+5)(s+40)}{s^3(s+200)(s+1000)}$

(e) $G(s) = \dfrac{K}{(s+1)(s+1.5)(s+2)}$ (f) $G(s) = \dfrac{K}{s(s+1)(s^2+2s+2)}$

23. Determine the phase margin and gain margin of the system with the following transfer functions.

(a) $G(s) = \dfrac{2}{(s+1)^2}$ (b) $\dfrac{20}{s(s+1)(s^2+2s+2)}$

24. The open-loop transfer function of a unity feedback system in given by $G(s) = K/s(s + 0.5s)(1 + s)$ (a) determine the value of K so that the gain margin of the system is 6 dB and (b) determine the value of K so that the phase margin of the system is 30°.

25. What do you mean by a polar plot? What is the advantage of polar plots.

26. Sketch the polar plot of $G(s) = \dfrac{1}{S+2}$.

Ans:

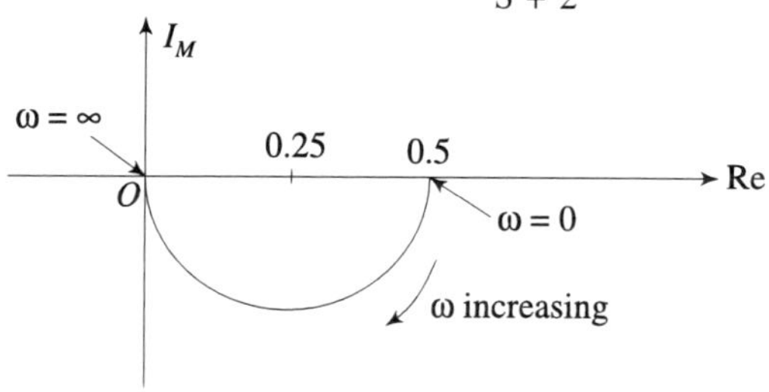

Fig. 8.53

27. Sketch the polar plot of $G(s) = \dfrac{1}{(1+S)(1+2S)}$

Ans:

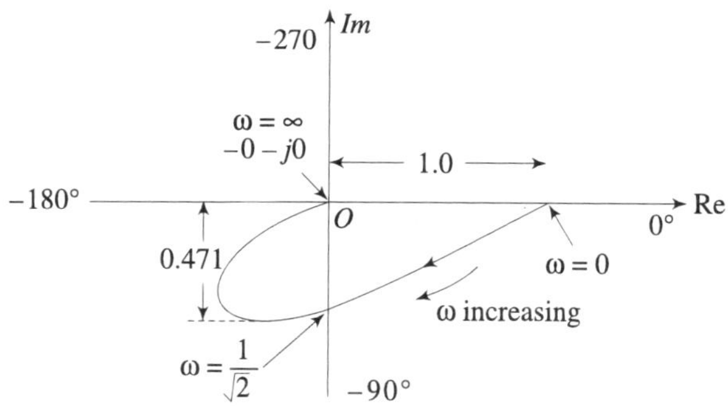

Fig. 8.54

28. Sketch the polar plot of $G(s) = \dfrac{10}{S(S+1)(S+2)}$

Ans:

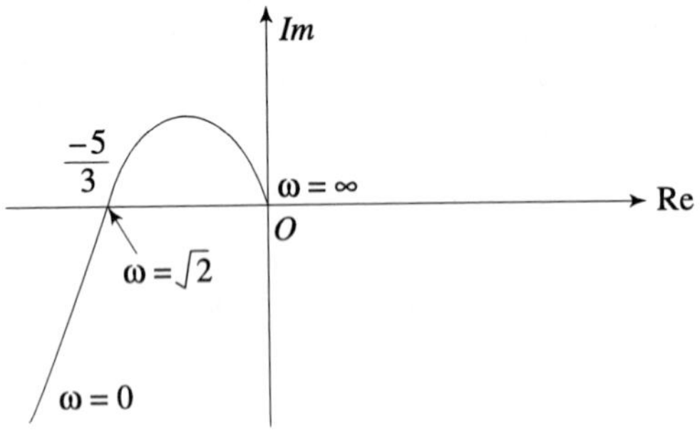

Fig. 8.55

29. Draw the polar plot of the function $G(s) = \dfrac{1}{1+2S}$

Ans:

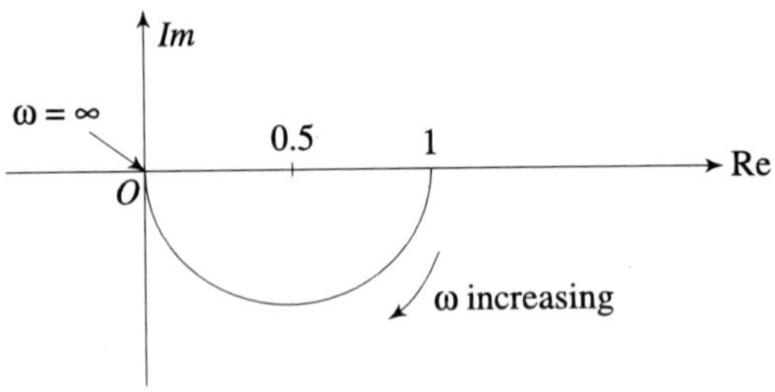

Fig. 8.56

30. Draw the polar plot of the function $G(s) = \dfrac{8}{(S+1)(S+2)}$.

Ans:

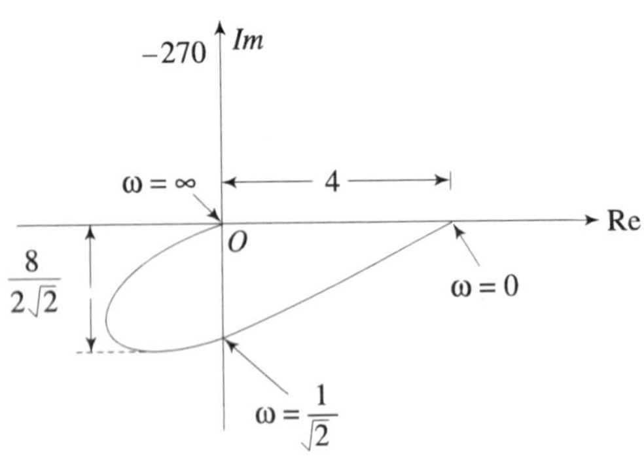

Fig. 8.57

31. Sketch the polar plot of the function $G(s) = \dfrac{14}{S(S+1)(S+2)}$

Ans:

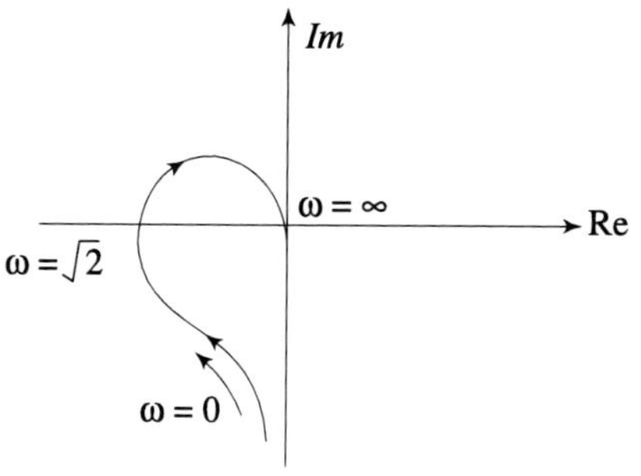

Fig. 8.58

32. Sketch the polar plot of the transfer function $G(s) = \dfrac{1}{(1 + ST_1)(1 + ST_2)(1 + ST_3)}$

Ans:

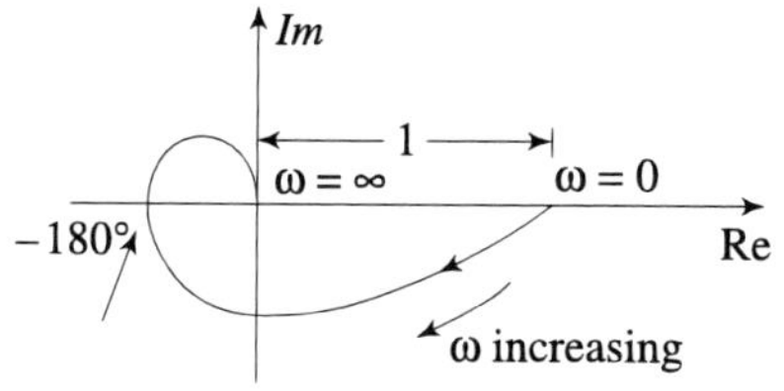

Fig. 8.59

33. State the advantages and limitations of the Nyquist stability criterion.
34. Explain the following terms
    (i) Analytic function (ii) Singular points
    (iii) Enclosure (iv) Encirclement
    (v) Principle of argument
35. State and explain the Nyquist stability criterion what is the effect of addition of a pole at $s = 0$ to $G(s)\ H(s)$ on the Nyquist plot?
36. Sketch the Nyquist plot for the open-loop system transfer function $G(s)H(s) = \dfrac{6}{(S + 1)(S + 2)}$. Determine the stability of the closed-loop system by Nyquist criterion.

Ans: 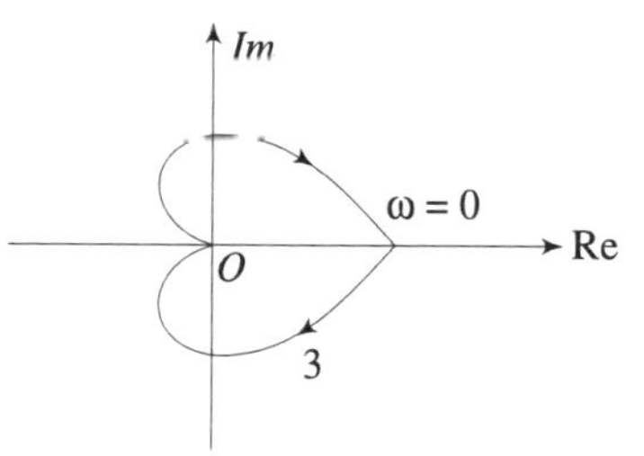

stable.

Fig. 8.60

37. The open-loop transfer function of a control system is given by

$$G(s)H(s) = \frac{60}{(S + 1)(S + 2)(S + 5)}$$

Using Nyquist stability criterion determine whether the closed-loop system is stable or not.

Ans:

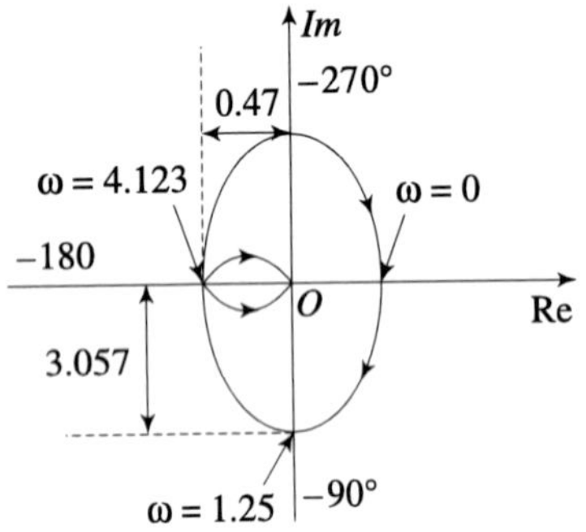

Fig. 8.61

38. Sketch the Nyquist plot and comment on the stability of the system represented by

$$G(s)H(s) = \frac{K}{(1 + ST_1)(1 + ST_2)}$$

Ans: The system is stable for all positive values of K, T, and T_2

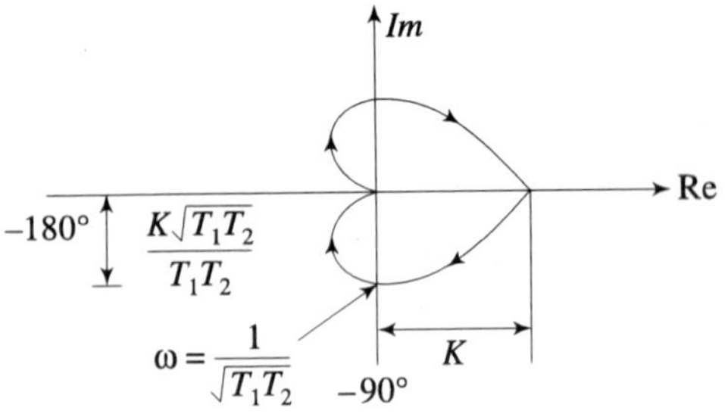

Fig. 8.62

39. Sketch the polar plot for the following function

(i) $G(s) = \dfrac{sa}{1 + sa}$ (ii) $G(s) = \dfrac{1 + sa}{1 + sb}$

Ans:

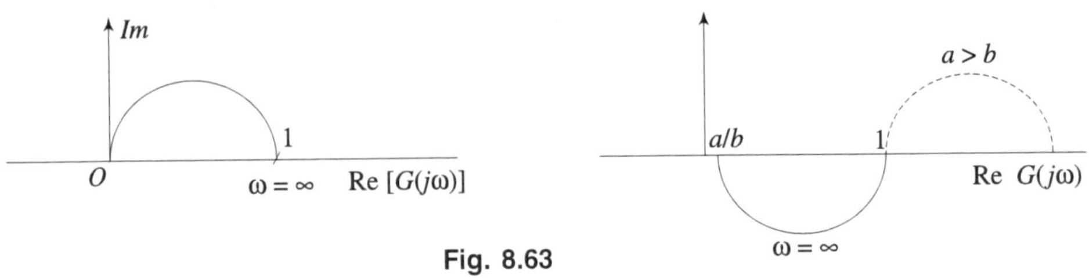

Fig. 8.63

40. Sketch the polar plot of $G(s) = 12/s(s + 1)$

Ans:

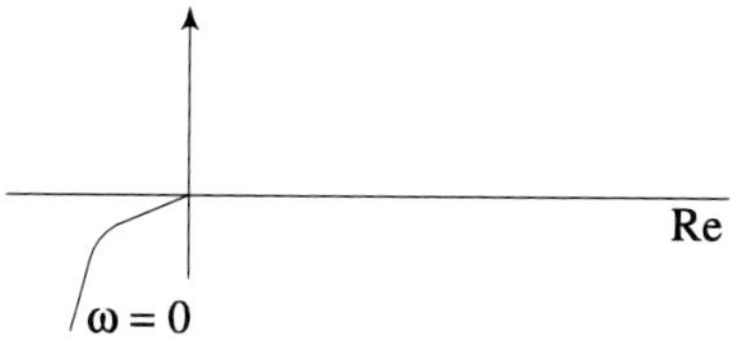

Fig. 8.64

41. Sketch the plot for $G(s) = \dfrac{1}{s(1 + sb)(1 + sb)(1 + sc)}$

Ans:

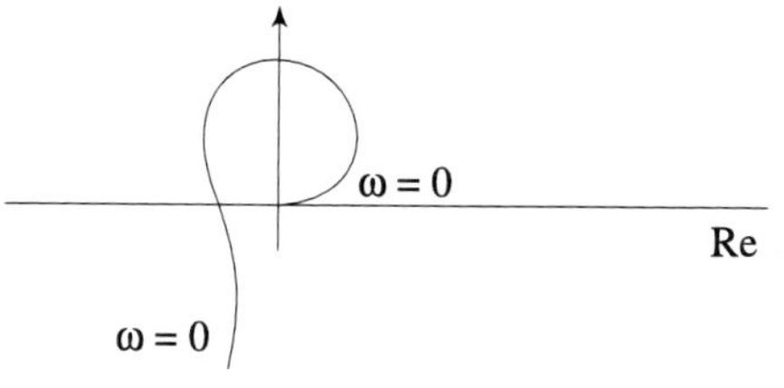

Fig. 8.65

42. Find the plot for $G(s) = \dfrac{1}{s(sb + 1)(sb + 1)(sc + 1)}$

Ans:

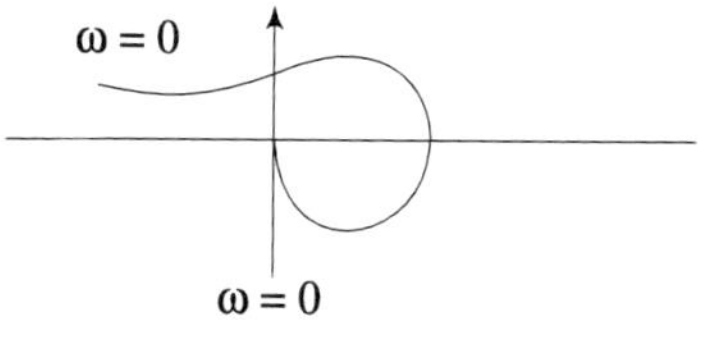

Fig. 8.66

43. Polar plot is plot of magnitude versus phase angle varying frequency from 0 to ∞ __________. True/False

Ans: True

44. What are disadvantages of polar plots?

Ans:
- Plot is cramped at high frequencies
- Calculations and time consuming
- Difficult to calculate gain & phase margin

45. The polar plot of $G(s) = (1 + s)/1 + 4s$ for $0 \leq w \leq \infty$ in G-plane is __________.

Ans:

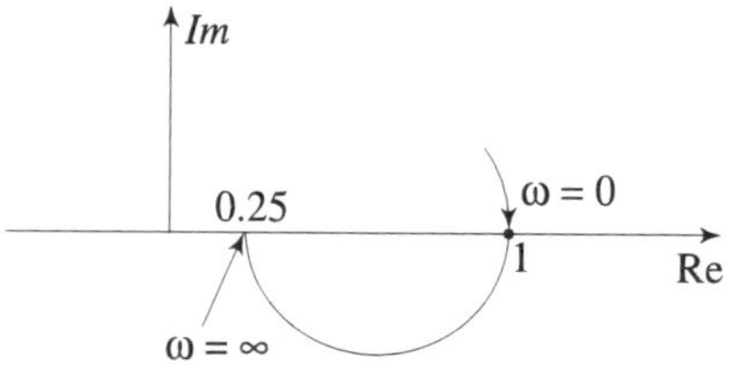

Fig. 8.67

46. Match the following:

(a) $\frac{s^2+1}{s^3}$

(i)

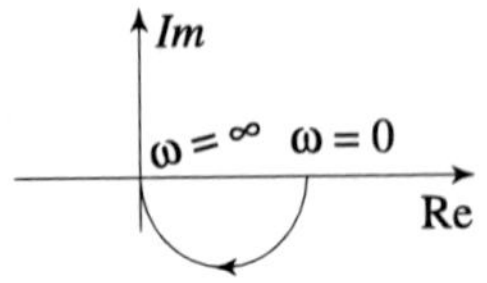

(b) $\frac{1}{s+6}$

(ii)

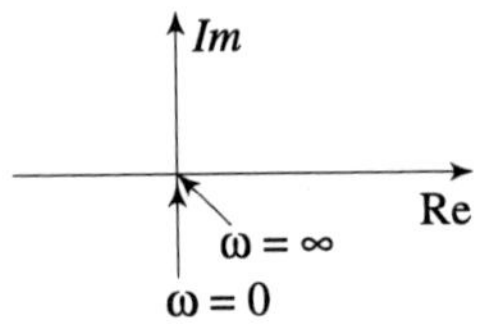

(c) $\frac{s}{(s+1)(s+2)}$

(iii)

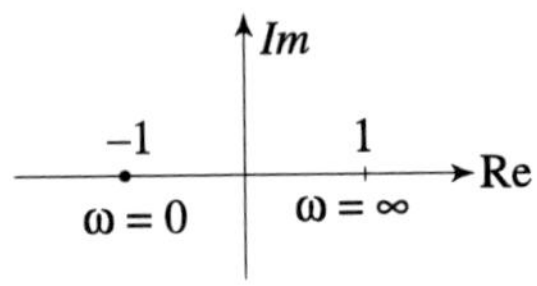

(d) $\frac{s^2-1}{s^2+1}$

(iv)

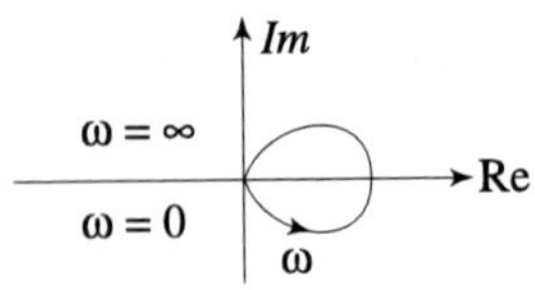

Fig. 8.68

Ans: (a) → (ii) (b) → (i) (c) → (iv) (d) → (iii)

9 Introduction to Design

9.1 INTRODUCTION

Every control systems designed for a specific application has to meet certain performance specifications.

Two methods of specifying the performance of a control system:

1. Set of specifications in time domain/frequency domain such as peak overshoot, settling time, gain margin, phase margin, and steady state error.
2. Optimality of a certain function Example, integral function

Introducing some kind of corrective sub-systems to force the chosen plant to meet the given specifications. These sub-systems are known as compensators and their job is to compensate for the deficiency in the performance of the plant. "Given plant and a set of specifications, design suitable compensators so that the overall system will meet the given specifications".

The performance of a feedback control system is primary important. A well designed control system should have the following properties:

- It should be stable.
- The controlled output should follow the changes in the reference input without oscillations or overshoots. That is, it should have suitable damping.
- It should be less sensitive to system parameter changes.
- It should be able to reduce the effect of undesirable disturbances.
- It should have as little error as possible.

A feedback control system that provides an optimum performance without any necessary adjustments is rare. Usually, it is necessary to compromise among the many conflicting and demanding specifications. Some adjustments in system parameters are to be made to provide suitable and acceptable performance when it is not possible to obtain all the desired optimum specifications. The alteration or adjustment of a control system in order to provide a suitable performance is called compensation. That is, compensation is the adjustment of a system in order to make up for inadequacies. An additional component or circuit called compensator is inserted into the system to compensate for efficient performance.

The compensating device may be electric, mechanical, hydraulic, pneumatic, or some other type of device or network. Electrical networks are often used as compensators in many control systems.

Performance of a control system is specified by a set of specifications in time domain and/or in frequency domain such as damping ratio, peak overshoot, rise time, settling time, gain margin, phase margin, bandwidth, steady state error, etc.

9.1.1 Compensating Networks

Although many types of compensators can be used, the most common compensating electrical networks are as follows:

- Lead network or lead compensator
- Lag network or lag compensator
- Lag lead network or lag lead compensator

If a sinusoidal input is applied to a network, a sinusoidal output is obtained.

If the steady state output leads the input, the network is called a lead network.

If the steady state output lags behind the input, the network is called a lag network.

Both phase lag and lead occur in the lag lead network in different frequency regions. The phase lag occurs in the low frequency region and phase lead occurs in the high frequency region.

9.1.2 Types of Compensation

A compensator is inserted into the system at a suitable location depending upon the required compensation is usually classified as follows:

- Series compensation or cascade compensation
- Parallel compensation or feedback compensation
- Series parallel compensation

9.1.3 Series or Cascade Compensation

If the compensator is placed in the forward path, the scheme is known as the series compensation or cascade compensation. The transfer function of the compensator is denoted by $G_c(s)$ and the transfer function of the original process is denoted by $G(s)$. This is shown in Fig. 9.1.

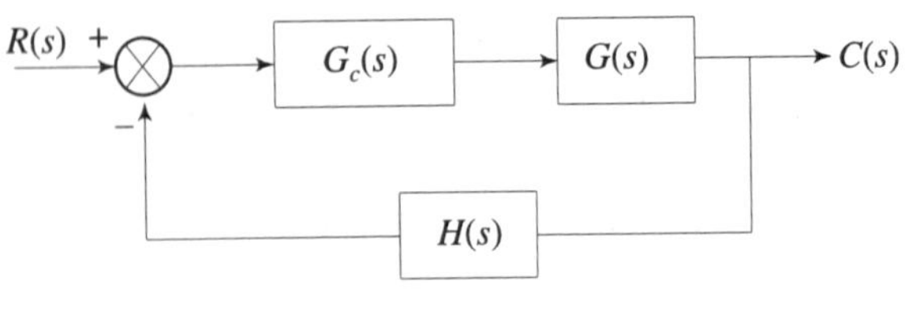

Fig. 9.1

9.1.4 Parallel Compensation or Feedback Compensation

If the compensator is placed in the feedback path, the scheme is known as parallel compensation or feedback compensation. Figure 9.2 shows the parallel or feedback compensated system.

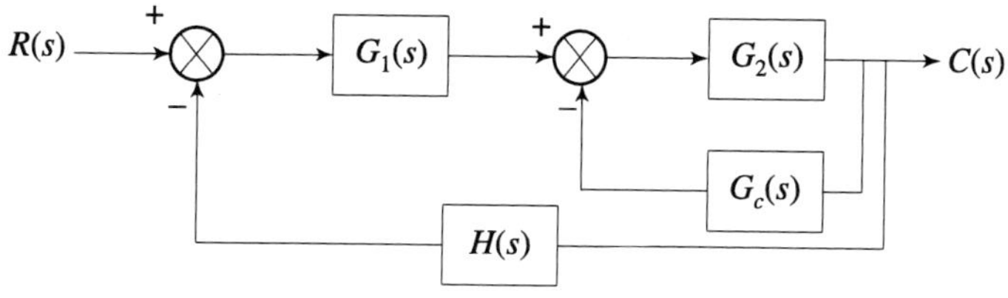

Fig. 9.2

9.1.5 Combined Cascade and Feedback Compensation or Series Parallel Compensation

Figure 9.3 shows a combined cascade and feedback compensation or series parallel compensation.

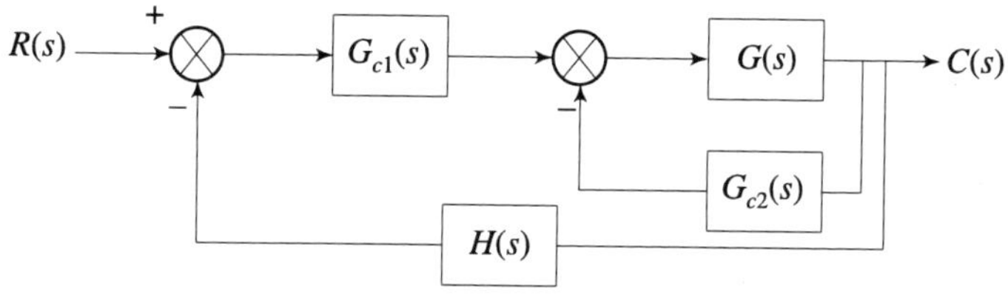

Fig. 9.3

9.1.6 Design Procedure of a Phase Lead

9.1.6.1 Compensator Using the Bode Diagram

Usually in a lead compensator design there are three primary design parameters.

- The crossover frequency, ω_c, which determines bandwidth, rise time and settling time.
- The phase margin (PM), which determines the damping coefficient (ξ) and the overshoot, M_p
- The low frequency gain, which determines the steady state characteristics.

Consider the system shown in Fig. 9.4. Let $G_c(s)$ be the transfer function of the lead compensator for the given uncompensated system with transfer function, $G(s)$.

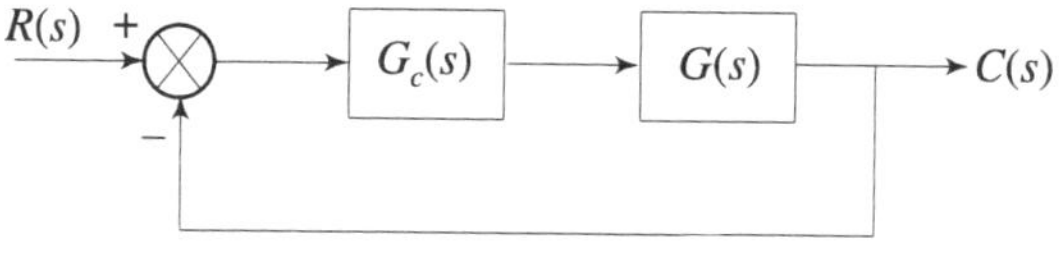

Fig. 9.4

Assume that the performance specifications are given in terms of phase margin, gain margin static velocity error constant, etc.

The lead compensator is basically a high-pass filter.

The procedure for designing a lead compensator by frequency response approach is as follows. This procedure is quite general and applies to any type and order of a system.

Consider the general transfer function of the lead compensator.

$$G_c(s) = K_c \propto \frac{1 + T_S}{1 + \alpha\, T_S}$$

$$= K_c \frac{S + \dfrac{1}{T}}{S + \dfrac{1}{\alpha T}}, \quad 0 < \alpha << 1$$

Let $\quad K_c \propto = K$

The $\quad G_c(s) = K \dfrac{1 + T_S}{1 + \alpha T_S}$

The open-loop transfer function of the compensated system is

$$G_c(s)\, G(s) = K \frac{(1 + T_S)}{(1 + \alpha T_S)}$$

$$= K \frac{(1 + T_S)}{1 + \alpha T_S} + G(s) = \frac{1 + T_S}{1 + \alpha T_S} G(s)\, K$$

$$= \frac{1 + T_S}{1 + \alpha T_S} G_1(s)$$

where $\quad G_1(s) = KG(s)$

Usually, one of the error constants is given as specification. Determine the gain K to satisfy the requirements of the given error constant or bandwidth.

The Bode plot of $G_1(j\omega)$ is drawn using this value of gain K determined in the above step. The phase margin (ϕ_1) of the uncompensated system is found from the Bode plot.

Usually, phase margin (PM) is specified for a design problem. Let the given phase margin be ϕ_S. The necessary phase lead, ϕ_L, required to be added to the system is found from $\phi_L = \phi_S - \phi_L + \varepsilon$,
where ε is the margin of safety as the crossover frequency may increase due to compensation. The margin of safety (ξ) is from 5° to 15°.

Let $\quad \phi_m = \phi_L$

The value of attenuation factors (α) is found from the equation $\alpha = \dfrac{\sin \phi_m + 1}{1 - \sin \phi_m}$.

Determine the frequency where the magnitude of the uncompensated system $G_1(j\omega)$ is equal to $\left(-20 \log \frac{1}{\sqrt{\alpha}}\right)$ dB or 10 log ∝ dB.

Since the compensation network provides a gain of 10 log ∝ dB at ω_n, this frequency is the new zero dB crossover frequency and ω_n simultaneously. Select this frequency as the new gain crossover frequency. This frequency corresponds to

$$\omega_m = \frac{1}{T\sqrt{\alpha}}$$

The maximum phase shift (ϕ_m) occurs at this frequency. This is the frequency of the compensated system since α is known $\frac{1}{T}$ can be determined compute the two corner frequencies of the compensators as

$\omega_{c_1} = \frac{1}{T} = \omega_m \sqrt{\alpha}$ = zero of lead compensator

$\omega_{c_2} = \frac{1}{\alpha T} = \frac{\omega_m}{\sqrt{\alpha}}$ = pole of lead compensator with these values of ω_{c_1} and ω_{c_2}, the design is complete and the lead compensator transfer function can be written as

$$G_c(s) = \frac{1 + T_S}{1 + \alpha\, T_S}$$

Draw the magnitude and phase plots of the compensated system and check the resulting phase margin. If the phase margin is still low increase the value of ξ and repeat the design. When satisfactory result is obtained check any additional specification of system performance, for example, bandwidth. If the specifications are not satisfied.

Compensation using root locus technique. We know that the root locus is the locus of poles for variation of closed-loop parameters from zero to infinity. Design of control system using root locus is reshaping of root locus to meet the desired specification of the system so that the new root locus is passed through the desired closed-loop poles. If the root locus of the system before compensation is passing through the desired poles, then the problem of design is reduced for selecting gain, *K*. If the uncompensated system root locus does not pass through the desired poles, then a subsystem is added to the original system so that the new root locus passes through the desired poles. The control system design using root locus is addition of a subsystem to the original system so that the new compensated system has poles and zeros at the desired location. It is therefore important to know the effects of addition of poles and zeros.

Design procedure of a phase lead compensator using the root locus method. The root locus approach to design is very effective when the specifications are given in terms of time domain quantities, such as the damping ratio and undamped natural frequency of the desired dominant closed-loop poles, maximum overshoot, rise time and settling time.

Consider the system, $G_c(s)$ is the transfer function of the lead compensator for the given uncompensated system with transfer function $G(s)$. The procedure for designing a phase lead compensator by the root locus method is as follows.

Steps to design a phase lead compensator by the root locus method:

- From the performance specifications, determine the desired location for the dominant closed-loop poles.
- Check whether the root locus plot of the uncompensated system passes through the dominant closed-loop pole.
- If so, find the gain K at the dominant closed-loop pole.
- If the root locus does not pass through the dominant closed-loop pole, then design a suitable lead compensator to be inserted in a system.
- To design a lead compensator, calculate the angle deficiency, ϕ. This angle must be contributed by the lead compensator if the new root locus is to pass through the desired locations for the dominant closed-loop poles.
- Assume the lead compensator to be

$$G_c(s) = K_c \propto \frac{T_S + 1}{\alpha T_S + 1} = \frac{K_c\left(S + \frac{1}{T}\right)}{S + \frac{1}{\alpha T}}$$

where α and T are determined from the angle deficiency. The value of K_c is determined from the requirement of the open-loop gain.

- If static error constants are not specified determine the location of the pole and zero of the lead compensator so that the lead compensator will contribute the necessary angle ϕ.
- Determine the open-loop gain of the compensated system from the magnitude condition.
- If all performance specifications are met, the design is complete.

9.1.7 Effects of Phase Lead Compensation

- The damping of the closed-loop system is increased.
- The overshoot is reduced and the transient response is improved.
- The bandwidth of the system is increased. This corresponds to faster time response.
- The slope of the magnitude curve of the Bode plot is reduced at the gain crossover frequency, therefore, the relative stability is improved.
- The steady state error of the system is not affected.
- The phase margin of the closed-loop system is increased.

The disadvantage is mainly due to the increased bandwidth with raises noise level at the system output.

9.1.7.1 Design Procedure of a Phase Lag Compensator Using Bode Diagram

The main objective of designing a phase lag compensator is to provide attenuation in the high frequency range to give a system sufficient phase margin. The phase log characteristic of a lag compensator is of no use for compensation purposes.

The procedure to design a phase lag compensator by frequency response approach is as follows.

- Assume the general transfer function of a lag compensator as

$$G_c(s) = \frac{E_o(s)}{E_i(s)} = \frac{K_C}{\beta}\left[\frac{S + \frac{1}{T}}{S + \frac{1}{\beta T}}\right]$$

$$= K_C \frac{(1 + T_S)}{(1 + \beta\, T_S)},\ (\beta > 1)$$

The open-loop transfer function of the compensated system is

$$G_c(s)\, G(s) = K \frac{(1 + T_S)}{(1 + \beta T_S)}\, G(s)$$

$$= \frac{1 + T_S}{1 + \beta T_S} KG(s) = \frac{1 + T_S}{1 + \beta T_S} G_1(s)$$

where $\quad G_1(s) = KG(s)$

Determine the value of K from the given static error constant.

- Draw the Bode plot of $G_1(j\omega)$ using the value of K determined in the above, equation check the gain margin and phase margin from this plot. If the specification of gain and phase margins are not satisfied then choose the new gain crossover frequency point where the phase angle of the loop transfer function is equal to – 180° and specified phase margin, + ε. Here ε is the margin of safety. The value of ε is 5° to 12°. It compensates the phase lag of the compensator. Let the new gain crossover frequency be ω_2
- For ω_2 as the new gain crossover frequency, the attenuation necessary to shift the magnitude curve up or down to zero dB is determined.

  This attenuation is

$$20 \log \frac{1}{\beta} \quad \text{or} \quad -20 \log \beta$$

  Downshift is taken negative while upshift is taken positive. From this the value of β is calculated.

- Usually, the upper corner frequency, $\frac{1}{T}$, of the lag compensator is approximately one octane or one decade below the new gain crossover frequency, ω_c.

  That is, $\quad \omega_{c_2} = \frac{1}{T} = \frac{\omega_2}{2} \quad \text{or} \quad \frac{\omega_2}{10}$

  This determines the value of T.

  For log compensator the other corner frequency is $\omega_c = \frac{1}{\beta T}$.

- The Bode plot of the lag compensated system is drawn. If the specifications are not satisfied, repeat the design by modifying the pole zero locations till a satisfactory result is obtained.

Design procedure for phase lag compensation by the root locus method: Phase lag compensator is used when the uncompensated system gives satisfactory transient response but unsatisfactory steady state response. In order to improve the steady state response, we have to increase the open-loop gain without appreciably changing the transient response.

This means that the root locus in the neighbourhood of the dominant closed-loop poles should not be changed appreciably, but the open-loop gain should be increased as much as needed. This can be achieved if a lag compensator is put in cascade with the given forward path transfer function. To avoid an appreciable change in the root loci, the angle contribution of the lag compensators should be limited to a small amount about 5°. To achieve this, place the pole and zero of the lag network relatively closer together and near the origin of the S-plane. Thus, the closed-loop poles of the compensated system will be shifted only slightly from their origin locations. Hence, the transient response characteristic will be changed only slightly.

It is to be noted that if the pole and zero of the lag network are placed very close to each other, then $\left(S_1 + \frac{1}{T}\right)$ and $\left(S + \frac{1}{\beta T}\right)$ are almost equal, where S_1 is the closed-loop pole. Thus,

$$\left|\frac{1}{\beta}\frac{S + \frac{1}{T}}{S + \frac{1}{\beta T}}\right| = \left|\frac{1}{\beta}\frac{S_1 + \frac{1}{T}}{S_1 + \frac{1}{\beta T}}\right| = \frac{1}{\beta}$$

This implies that the open-loop gain can be increased approximately by a factor β without altering the transient response characteristics. If the pole and zero are placed very close to the origin, the value of β_1 can be made large, usually $1 < \beta < 15$. An increase in gain means an increase in the static error coefficients. If the open-loop transfer function of the uncompensated system is $G(s)$, then the static velocity error coefficient, $K_V = \lim\limits_{S \to 0} SG(s)\, H(s)$.

If the transfer function of the lag compensator

$$G_c(s) = K_c \frac{(T_S + 1)}{\beta T_S + 1} = \frac{K_c\left(S + \frac{1}{T}\right)}{\beta\left(1 + \frac{1}{\beta T}\right)}$$

Then for the compensated system, the open-loop transfer function is $G_c(s)\, G(s)\, H(s)$ and the static velocity error coefficient (K_c') becomes zero

$$K_v' = \lim_{S \to 0} SG_c(s)\, G(s)\, H(s)$$

$$= \lim_{S \to 0} G_c(s)\, K_v \text{ as } H(s) = 1$$

$$K'_v = K_c K_v$$

Thus, the static velocity error coefficient of the compensated system is increased by a factor K_c.

The procedure for designing a phase lag compensator by the root locus method is as follows:

- Draw the root locus for the uncompensated system, whose open-loop transfer function is $G(s)$. Based on the transient response specifications locate the dominant closed-loop poles on the uncompensated root locus.
- Assume that the transfer function of lag compensator is

$$G_c(s) = K_c\beta \frac{T_S + 1}{\beta T_S + 1} = K_c \frac{S + \frac{1}{T}}{1 + \frac{1}{\beta T}}$$

where $\frac{1}{T}$ and $\frac{1}{\beta T}$ are zero and pole of the compensator respectively and $\beta > 1$. Then the open-loop transfer function of the compensated system becomes $G_c(s)\ G(s)$.

- Determine the static error coefficient specified in the problem.
- Determine the factor β by which the static error constant necessary to satisfy the specifications is to be increased.

That is,

$$\beta = \frac{\text{New static error constant}}{\text{Original static error constant}}$$

Determine the pole and zero of the lag compensator such that

$$\beta = \frac{\text{Length of the line joining the zero from the origin}}{\text{Length of the line joining the pole from the origin}}$$

- Draw the new locus plot for the compensated system, and locate the desired dominant closed-loop poles on the root locus.
- Check the angle contributed by the lag compensator to the dominating pole. In an ideal situation the net angle contributed by lag compensator is zero. The angle contribution less than 5° is permissible.
- Find the parameters K_c, β and T of the lag compensator.
- Adjust gain K_c of the compensator from the magnitude condition so that the dominant closed-loop poles lie at the desired location. Check the steady state error constant. If the given specifications are not satisfied, then repeat the procedure from the step β.

9.1.8 Effects and Limitations of Lag Compensation

The general effects and limitations of lag compensation are as follows:

- Lag compensator has gain at low frequencies and low gain at higher frequencies, therefore, a lag compensator acts as a low-pass filter.
- It attenuates high frequencies, so it acts as a high frequency noise filter.
- Lag compensator increases the order of the system by one. The transient response becomes sluggish.
- The phase lag characteristic of a lag compensator is of no use for compensation purposes.
- Since the attenuation due to lag compensation decreases, the gain crossover frequency, the bandwidth of the system are reduced. The rise time and settling time are increased and the response becomes slower.
- High gain at low frequencies improves steady state performance and low gain at high frequencies improves phase margin.
- For a given relative stability. the value of error constant is increased.

9.1.9 Design Procedure of Lag Lead Compensator

- First adjust the error constant to the specified value. For the specified error constant, determine the open-loop gain, K.
- Draw the magnitude and phase Bode plots of the uncompensated system with this value of K and determine the phase margin and gain crossover frequency.
- First design the lag section to provide only partial compensation of phase margin. Choose the given crossover frequency such that it is higher than gain crossover frequency if the system was fully lag compensated.
- Determine the value of β required such that the high frequency attenuation provided by the lag network is equal to the magnitude of the uncompensated system at this frequency.
- Choose the value of T_2 such that the upper corner frequency, $1/T_2$, of the lag network is approximately one decade below the new gain crossover frequency, ω_2.

  That is, $$\omega_{c_2} = \frac{1}{T_2} = \frac{\omega_2}{10}$$

- This determines the value of T_2. For lag compensator the other corner frequency is $\omega_{c_1} = \frac{1}{\beta T_2}$. Thus, the lag network design is complete. The magnitude and phase Bode plots are drawn for the lag compensated system. The gain crossover frequency and the phase margin are determined from these plots.
- For the lead section design, we choose α and β such that $\alpha = \frac{1}{\beta}$, so that the design procedure becomes simple. Calculate the maximum phase lead provided by the lead section using the formula

$$\phi_m = \sin^{-1}\left[\frac{1-\alpha}{1+\alpha}\right] = \sin^{-1}\left[\frac{1-\frac{1}{\beta}}{1+\frac{1}{\beta}}\right]$$

- Choose the compensated crossover frequency to coincide with ϕ_m. This is the frequency where the lag section compensated system has a gain of $(-10 \log \beta)$ dB. Read this from the Bode plot. Therefore, $\omega_n = \frac{1}{T_1 \sqrt{\alpha}}$.

  Therefore, T_1 and αT_1 can be calculated. The lead compensator transfer function is found.
- Combine the transfer function of the lag and lead section to get the lag lead compensator transfer function. Draw the Bode plotes and determine the phase margin.
- If the specifications are not satisfied, repeat the design by modifying the values of β and T.

9.2 APPROACHES TO THE DESIGN PROBLEM

There are two approaches to the control system design problem.

1. We select the configuration of the overall system by introducing compensators and then choose the parameters of the compensators to meet the given specification on performance.
2. For a given plant, we find an overall system that meets the given specifications and then compute the necessary compensators.

In a fixed system configuration and performance specified in terms of peak overshoot, settling time, gain margin, phase margin, etc. Root locus method and frequency domain methods of design are referred to as the classical methods of design.

9.3 PRELIMINARY CONSIDERATION OF CLASSICAL DESIGN

Proper selection of performance specifications is the most important step in control system design. The desired behaviour is specified in terms of transient response measures and steady state error.

Steady state error is usually specified in terms of error constant for specific inputs.

Transient response measures relative stability and speed of response may be specified in time or frequency domain or in both.

In time domain, the measure of relative stability is damping factor (ζ) or peak overshoot (M_p) speed of response is measured in terms of rise time, settling time or natural frequency.

In frequency domain, the measure of relative stability is resonant peak (M_r) or phase margin. Measure of speed of response is resonant frequency (ω_r) or bandwidth (ω_b).

The nature of compensation depends upon the given plant. The compensators may be an electrical, mechanical, hydraulic, pneumatic or other type of device or electrical network.

The compensator's transfer function may be placed in cascade or series with the plant transfer function or in the feedback path (feedback or parallel compensation).

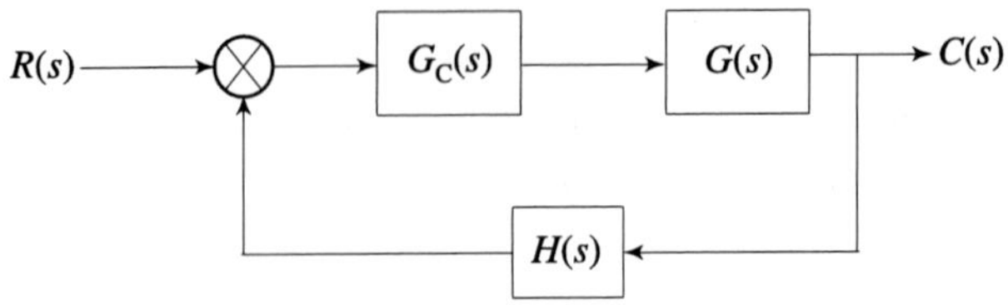

(a) Cascade compensators

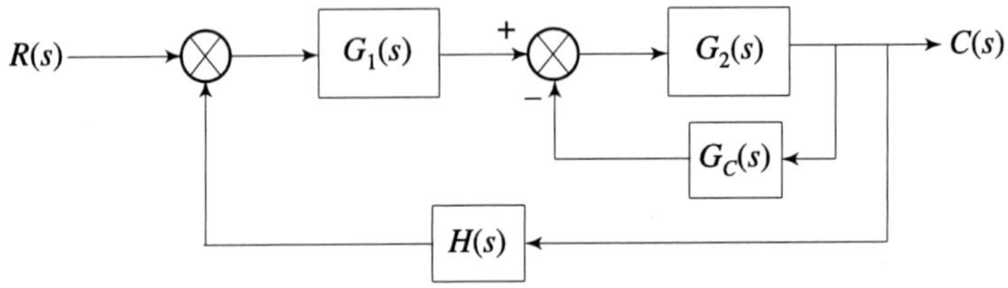

(b) Feedback compensation

Fig. 9.5

Once the configuration of the overall system has been selected, next step is to select the structure of the compensator and then to obtain the parameters of the compensators that satisfy the selected specifications in the best possible manner.

Design example: Consider the system shown in Fig. 9.6, which has an open-loop transfer function.

$$G(s) = \frac{5K_A}{5\left(\frac{S}{2}+1\right)\left(\frac{S}{6}+1\right)}$$

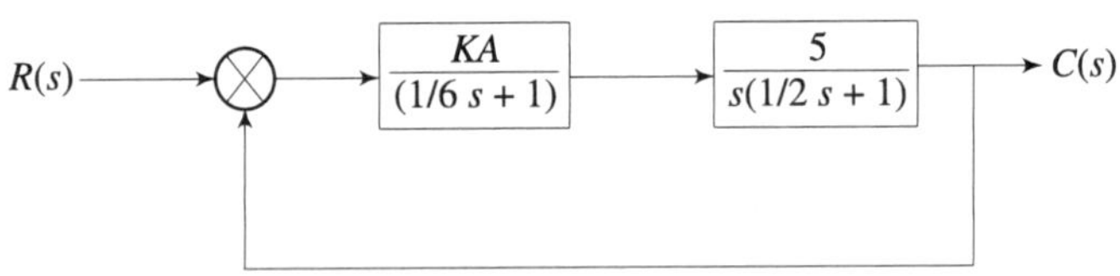

Fig. 9.6

$$K_v = \text{Velocity error constant of the system}$$

$$K_v = \lim_{s \to 0} S\, G\,(s) = 5K_A$$

For $K_A = 1$, therefore, $K_v = 5$, the steady state error to unit velocity input ($e_{ss} = 1/K_v$) is 0.2.

From the equation we have

$$G(s) = \frac{60 K_A}{s(s+2)(s+6)} = \frac{K}{s(s+2)(s+6)}$$

The root locus plot of the uncompensated system corresponding to a value $K = 60$ (i.e., $K_v = 5$) in the root locus plot, the roots is found to be $S_{1,2} = -0.3 \pm j2.8$.

For this root pair, the damping ratio (ζ) is 0.105 and ω_n is 2.85.

The settling time = $\frac{4}{\xi \omega_n}$ for the response is about 13.36 sec.

On investigation it is found that the system has poor relative stability and large settling time.

Assume that the specifications on the transient performance are $\xi = 0.6$, settling time < 4 sec.

A line corresponding to $\xi = 0.6$. The intersection of this line with root locus branch determines the value of K which yields the specified ξ.

Therefore, we can find

$$K = 10.5 \left[K_v = \frac{10.5}{2 \times 6} = 0.875, \text{ i.e., } K_A = 0.175 \right]$$

$$\omega_n = 1.26, \text{ settling time} = 5.3 \text{ second}, e_{ss} = 1.14$$

But the steady state error increases beyond the tolerable limit of 0.2, settling time specification is not satisfied. Conclude that our purpose is not served by gain adjustments. In order to meet the specification we need to increase ω_n keeping ξ constant (0.6).

Let us consider the proposition of adding a compensating zero somewhere between the non-zero poles of the open-loop system.

From the locus (Fig. 9.7) we observe the fall effect of the compensating zero.

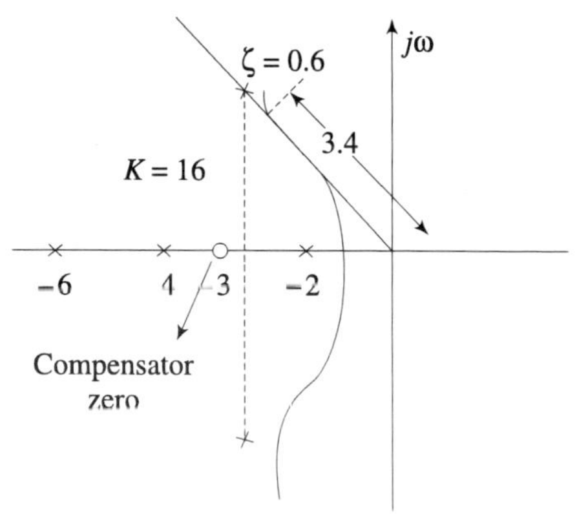

Fig. 9.7

1. All branches of the root locus now lie completely in the left half S-plane, so that K can be adjusted to any positive value without causing instability.
2. Complex root branches of the root locus are now bent away from the $j\omega$-axis It is found that for $\xi = 0.6$

$$K = 16 \left[K_v = \frac{(16 \times 3)}{(2 \times 6)} = 4 \right]$$

$$\omega_n = 3.4$$

$$\text{Settling time} = \frac{4}{(0.6 \times 3.4)} = 1.96 < 4$$

$$e_{ss} = 0.25$$

Thus, the addition of compensating zero helps satisfy the settling time specification for $\xi = 0.6$. By suitable readjustment of the compensating zero along the negative real axis, it is possible to bring the steady state error within the specified limit.

The addition of a zero in the open-loop transfer function can be achieved by adding a compensator with the transfer function.

$$G_c(s) = (S + z_c) = \left(S + \frac{1}{\tau_c}\right)$$

In cascade with the forward transfer function. Since an isolated zero is not physically realizable we must add a pole along with the compensating zero so as to achieve physical realizability. This pole must be added far away from the $j\omega$-axis such that it has relatively negligible effect on the root locus.

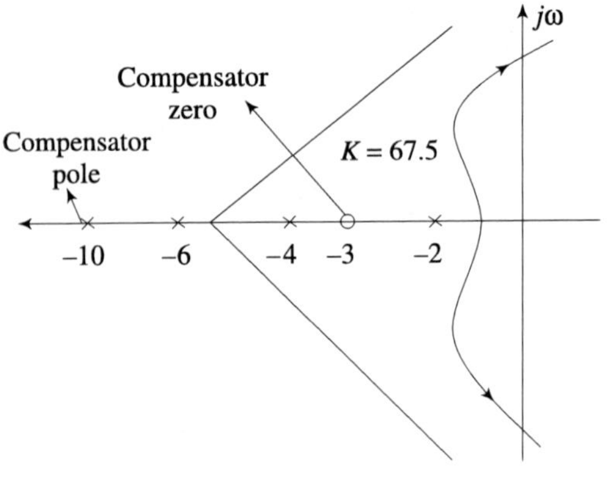

Fig. 9.8

Root locus plot in the presence of both a compensating zero and pole $\xi = 0.6$ is shown in Fig. 9.8.

We obtain from root locus,

$$K = 67.5 \left(K_v = \frac{(67.3 \times 3)}{(2 \times 6 \times 10)} = 1.7\right), \quad \omega_n = 2$$

$$\text{Settling time} = \frac{2}{2 \times 0.6} = 3.33 < 4 \text{ seconds.}$$

By adjusting the compensating zero and pole, it may be possible to raise the K_v to the specified value but satisfying the transient response specifications.

∴ Transfer function of such compensator then becomes

$$G_c(s) = \frac{(S + Z_c)}{(S + P_c)} = \frac{(S + 1/\tau)}{(1 + 1/\propto\tau)} \quad \text{...(9.1)}$$

$$\alpha = \frac{z_c}{p_c}, < 1, \tau > 0$$

Note: The $\alpha < 1$ ensures that the pole is located to the left of the zero.

The compensator having a transfer function of the form given in Eq. (9.1) is known as lead compensator.

The block diagram of lead compensator is given in Fig. 9.9.

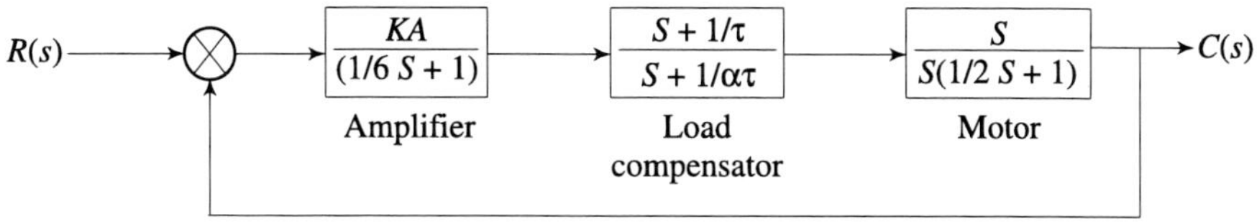

Fig. 9.9

Lead compensator speeds up the transient response and increases the margin of stability of a system. It also helps to increase the system error constant though to a limited extent.

Let us consider the case, satisfactory transient response but its steady state error is quite large. Consider the proposition of adding a pole at origin.

The velocity error constant of the compensated system is then given by:

$$K_v = \lim_{s \to 0} \frac{SG(s)}{S} \text{ tends to infinity.}$$

Adding pole at the origin, error constant has been increased to ∞. Therefore, the system has excellent steady state response.

It is to be noted that addition of a compensating pole at the origin to a type 0 system would not change the transient response so drastically as in the case of type 1 system.

Compensated system with compensation pole at the origin and compensating zero at -0.2.

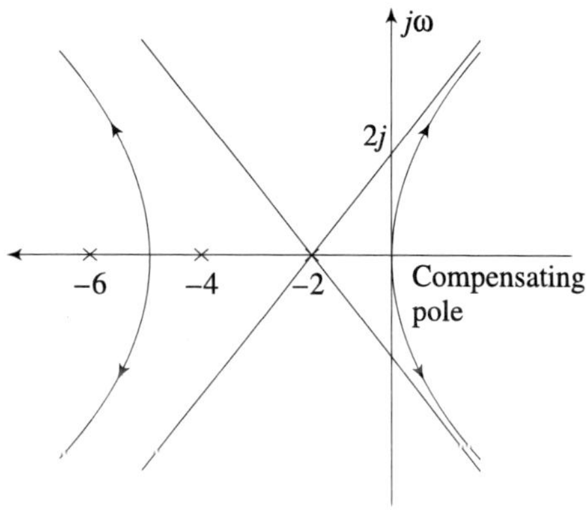

Fig. 9.10

The addition of the compensating pole and zero

$$G_c(s) = \frac{(S + Z_c)}{s}$$

Physical realizability requires that the pole at the origin shifted on the real axis slightly to the left of origin.

$$G_c(s) = \frac{(S + Z_c)}{(S + P_c)}, \frac{Z_c}{P_c} = \beta > 1 \qquad ...(9.2)$$

$\beta > 1$ ensures that pole is to the right of zero, i.e., nearer the origin than zero. The compensator having a transfer function of the form given in Eq. (9.2) is called a lag compensator.

Lag compensator improves the steady state behaviour of a system, while nearly preserving its transient response.

When both the transient and steady state responses require improvement, a lag-lead compensator is required, this is basically a lag and a lead compensator connected in series.

9.3.1 Selection of a Compensator

There are two situations in which compensation is required,

1. System is absolutely unstable and compensation is required to stabilize it as well as to achieve a specified performance.
2. System is stable but compensation is required to obtain the desired performance, systems which are type 2 or higher, are usually absolutely unstable.

For these type of systems lead compensator is required because only the lead compensator increases the margin of stability.

In type 1 and type 0 systems, stable operation is always possible if the gain is sufficiently reduced. Three compensators that is lag, lead, lag-lead may be used to obtain the desired performance.

9.4 REALIZATION OF BASIC COMPENSATORS

Compensators may be realized by electrical, mechanical, pneumatic, hydraulic, or other components. The choice of the type of components to be used depends upon the system's structure. Realization by electrical components is quite common.

Lead compensator

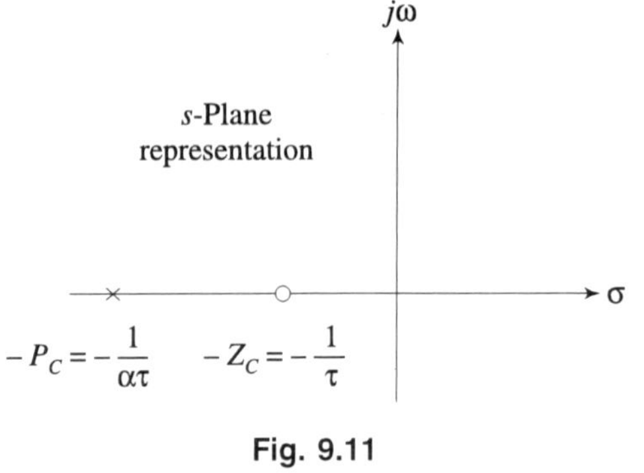

Fig. 9.11

It has zero at $S = \frac{1}{\tau}$ and pole at $S = -\frac{1}{\alpha\tau}$ with the zero closer to the origin than the pole, the general form of the lead compensator is

$$G_c(s) = \frac{(S + Z_c)}{(S + P_c)} = \frac{\left(s + \frac{1}{\tau}\right)}{\left(s + \frac{1}{\alpha\tau}\right)}, \quad \alpha = \frac{Z_c}{P_c} < 1 \quad \tau > 0 \qquad ...(9.3)$$

$$= \alpha\left(\frac{\tau S + 1}{\alpha\tau\, S + 1}\right) \qquad ...(9.4)$$

The lead compensator with the transfer function in Eq. (9.3) can be realized by an electrical lead network shown in Fig. 9.12.

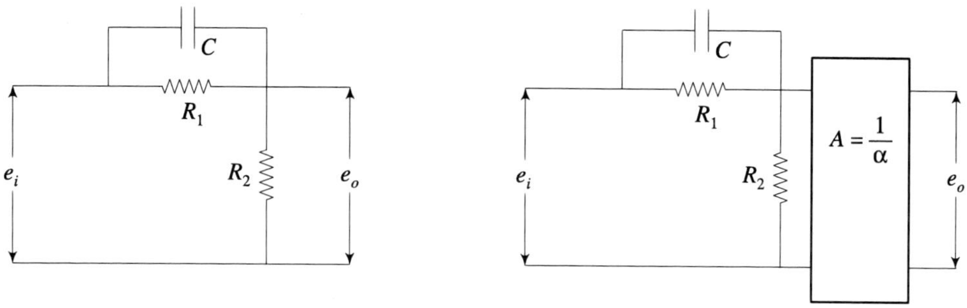

(a) Electrical load network

(b) Phase load network with amplifier

Fig. 9.12

Let us derive the transfer function of this lead network, assuming the impedance of the source to be zero and output load impedance to be infinite.

From Fig. 9.12(a)

$$\frac{E_o(s)}{E_i(s)} = \frac{R_2}{R_2 + \dfrac{\dfrac{R_1}{sC}}{R_1 + \dfrac{1}{sC}}} = \frac{\left(s + \dfrac{1}{R_1 C}\right)}{s + \dfrac{1}{\left(\dfrac{R_2}{(R_1 + R_2)}\right) R_1 C}} \qquad ...(9.5)$$

Defining $\qquad \tau - R_1 C \quad \text{and} \quad \alpha = \dfrac{R_2}{(R_1 + R_2)} < 1 \qquad ...(9.6)$

Network transfer function in Eq. (9.5) has some form as that of lead compensator in Eq. (9.3). The values of the 3 network components R_1, R_2 and C are to be determined from the two lead compensator parameters, τ and α, using Eq. (9.6).

Consider the sinusoidal transfer function of the lead network from Eq. (9.4) as

$$G_c(j\omega) = \alpha\left(\frac{1 + j\omega\tau}{1 + j\omega\tau\alpha}\right), \ \alpha < 1$$

At zero frequency the network has gain of $\alpha < 1$ or an attenuation of $\frac{1}{\alpha}$.

In the frequency domain compensation techniques, it is convenient to cancel the DC attenuation of the network with an amplification $A = \frac{1}{\alpha}$. Since $\alpha < 1$, the network output leads the sinusoidal input under steady state and hence the name lead compensator.

The sinusoidal transfer function of the lead compensator is then given by

$$G_c(j\omega) = \left(\frac{1 + j\omega\tau}{1 + j\omega\tau\alpha}\right), \ \alpha < 1$$

The Bode diagram of the lead compensator $\left(\text{with amplifier of gain } A = \frac{1}{\alpha}\right)$.

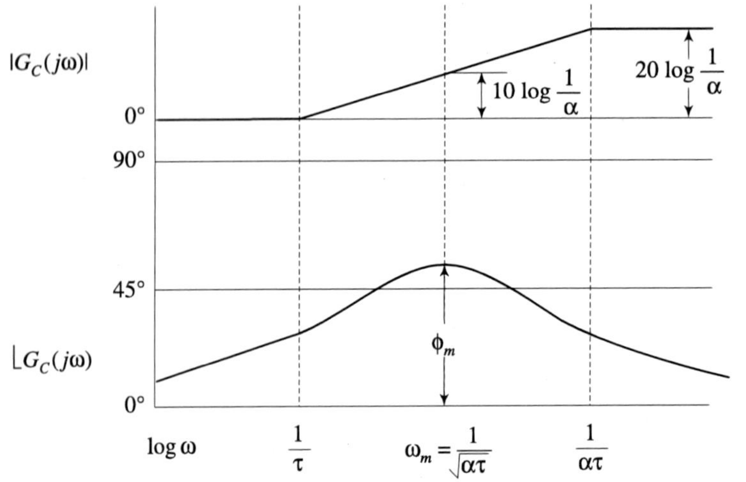

Fig. 9.13

The phase lead of compensator at any frequency ω is given by

$$\phi = \tan^{-1} \omega\tau - \tan^{-1} \alpha\omega\tau$$

or $$\tan \phi = \frac{+\,\omega\tau(1 - \alpha)}{1 + \alpha\omega^2\tau^2}$$

Using the condition $\frac{d\phi}{d\omega} = 0$. Maximum phase lead occurs at the frequency,

$$\omega_m = \frac{1}{\tau\sqrt{\alpha}} = \sqrt{\left(\frac{1}{\tau}\right)\left(\frac{1}{\alpha\tau}\right)}$$

ω_m = Geometric mean of two corner frequencies of the compensator.

At $\omega = \omega_m$, the maximum phase lead (ϕ_m) is

$$\tan \phi_m = \frac{(1 - \alpha)}{2\sqrt{\alpha}}$$

$$\sin \phi_m = \frac{(1 - \alpha)}{(1 + \alpha)}$$

This gives α in terms of ϕ_m as

$$\alpha = \frac{1 - \sin \phi m}{1 + \sin \phi m}$$

The sinusoidal transfer function of the lead compensator

$$|G_c(j\omega_m)| = \frac{1 + j\omega\tau}{1 + j\alpha\omega\tau}, \ \alpha < 1 \qquad ...(9.7)$$

Since $\alpha < 1$, the network output leads the sinusoidal input under steady state. From Eq. (9.7), the phase lead of the compensator at any frequency ω is given by

$$\Phi = \tan^{-1} \omega\tau - \tan^{-1} \alpha\, \omega\tau,$$

or

$$\tan \phi = \frac{\omega\tau(1 - \alpha)}{1 + \alpha\omega^2\tau^2}$$

Condition $\frac{d\phi}{d\omega} = 0$, find that the maximum phase lead occurs at the frequency,

$$\omega_m = \frac{1}{\tau\sqrt{\alpha}} = \sqrt{\frac{1}{\tau} \cdot \frac{1}{\alpha\tau}} \qquad ...(9.8)$$

From Eq. (9.8), ω_m is the geometric mean of the two corner frequencies of the compensator.

At $\omega = \omega_m$, the maximum phase lead Φ_m is: $\tan \Phi_m = \frac{1 - \alpha}{2\sqrt{\alpha}}$ or $\sin \Phi = \frac{1 - \alpha}{1 + \alpha}$

This gives α in terms of Φ_m as: $\alpha = \frac{1 - \sin \Phi_m}{1 + \sin \Phi_m}$ ...(9.9)

The magnitude of $G_c(j\omega)$ at $\omega = \omega_m$, the frequency of maximum phase lead is

$$|G_c(j\omega_m)| = \left| \frac{1 + j\omega_m\tau}{1 + j\alpha\omega_m\tau} \right| = \frac{1}{\sqrt{\alpha}} \qquad ...(9.10)$$

Equation (9.9) is useful in computing the α parameter of the network from the required maximum phase lead.

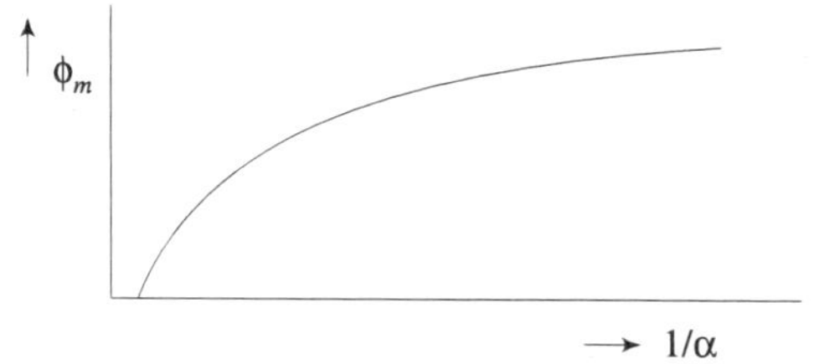

Fig. 9.14

Lag compensator: The lag compensator is shown in Fig. 9.11.

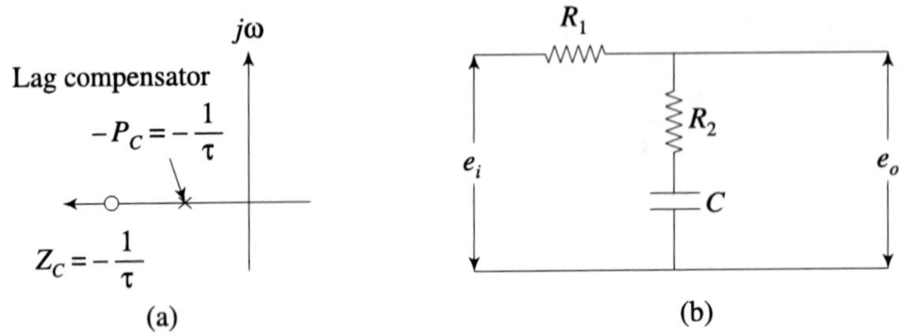

Fig. 9.15

It has a pole at $-\frac{1}{\beta\tau}$ and zero $-\frac{1}{\tau}$ with the zero located to the left of the pole on the negative real axis.

The general form of the transfer function of the lag compensator is

$$G_c(s) = \frac{s + z_c}{s + p_c} = \frac{s + \frac{1}{\tau}}{s + \frac{1}{\tau\beta}}, \ \beta = \frac{z_c}{p_c}, > 1, \tau > 0 \qquad \text{...(9.11)}$$

The realization of the transfer function of Eq. (9.11) is achieved with an electric lag network as shown in Fig. 9.15 (b)

$$\frac{E_0(s)}{E_1(s)} = \frac{R_2 + \frac{1}{sc}}{R_1 + R_2 + \frac{1}{sc}} = \frac{1}{\frac{R_1 + R_2}{R_2}} \left[\frac{s + \frac{1}{R_2 c}}{s + \frac{1}{\left(\frac{R_1 + R_2}{R_2}\right) R_2 c}} \right] \qquad \text{...(9.12)}$$

Comparing Eqs. (9.11) and (9.12), we get

$$\tau = R_2 c, \ \beta = \frac{R_1 + R_2}{R_2} > 1$$

∴ The transfer function of the network becomes

$$G_c(s) = \frac{1}{\beta} \left(\frac{s + \frac{1}{\tau}}{s + \frac{1}{\beta\tau}} \right) = \frac{1}{\beta} \left(\frac{s + z_c}{s + p_c} \right), \ \beta = \frac{z_c}{p_c} > 1$$

$$= \frac{\tau s + 1}{\beta\tau s + 1}$$

The sinusoidal transfer function of the lag network is given by

$$G_c(j\omega) = \frac{1 + j\omega\tau}{1 + j\beta\omega\tau}.$$

Since $\beta > 1$, the steady state output has a lagging phase angle with respect to the sinusoidal input. Therefore, the name lag network.

The maximum phase lag (Φ_m) and the corresponding frequency (ω_m) are:

$$\beta = \frac{1 - \sin\Phi_m}{1 + \sin\Phi_m},\ \tan\Phi_m = \frac{1-\beta}{2\sqrt{\beta}}$$

$$\sin\Phi_m = \frac{1-\beta}{1+\beta}$$

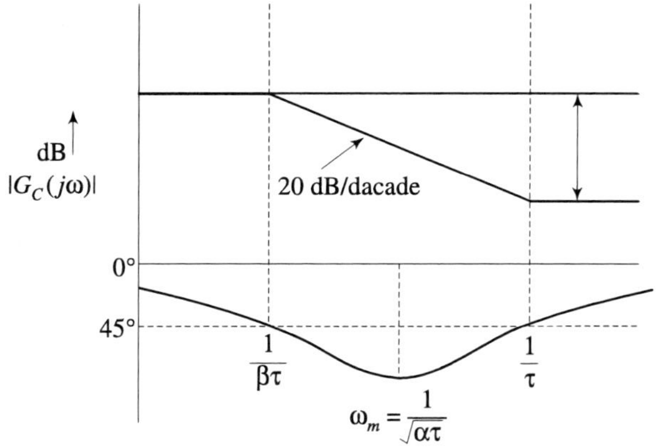

Fig. 9.16

It offers a high frequency gain of $\frac{1}{\beta}$. Since $\beta > 1$, it means that high frequency noise in attenuators is passing through the network. A typical choice of β is 10.

Lag-lead compensator: It is a combination of a lag compensator and a lead compensator. The lag section has one real pole and one real zero with the pole to the right of zero. The lead section also has one real pole and one real zero but the zero is to the right of the pole.

The general form

$$G_c(s) = \left(\frac{s + \frac{1}{\tau_1}}{s + \frac{1}{\beta\tau_1}}\right)\left(\frac{s + \frac{1}{\tau_2}}{s + \frac{1}{\alpha\tau_2}}\right),\ \beta > 1,\ \alpha < 1 \qquad ...(9.13)$$

Equation (9.13) can be realized by a single electric lag lead network shown in Fig. 9.17.

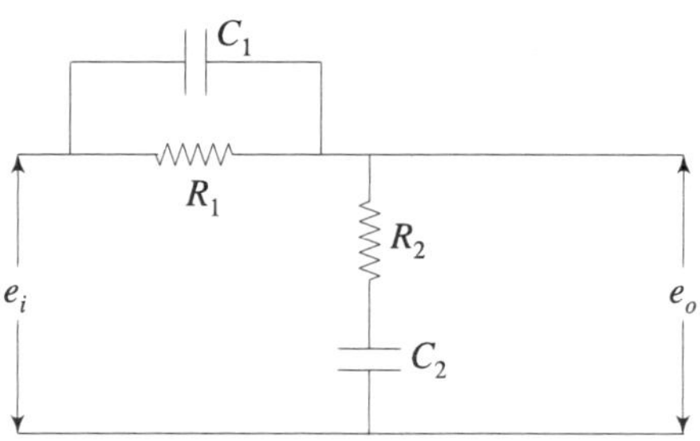

Fig. 9.17

The transfer function of the network is

$$\frac{E_0(s)}{E_1(s)} = \frac{R_2 + \frac{1}{sC_2}}{R_2 + \frac{1}{sC_2} + \frac{\frac{R_1}{sC_1}}{R_1 + \frac{1}{sC_1}}}$$

$$= \left[\frac{\left(s + \frac{1}{R_1C_1}\right)\left(s + \frac{1}{R_2C_2}\right)}{s^2 + \left(\frac{1}{R_1C_1} + \frac{1}{R_2C_1} + \frac{1}{R_2C_2}\right)s + \frac{1}{R_1R_2C_1C_2}} \right] \quad ...(9.14)$$

Comparing Eqs. (9.13) and (9.14), we have

$$R_1C_1 = \tau_1 \quad ...(9.15)$$

$$R_2C_2 = \tau_2 \quad ...(9.16)$$

$$R_1R_2C_1C_2 = \alpha\beta\, \tau_1\, \tau_2 \quad ...(9.17)$$

$$\frac{1}{R_1C_1} + \frac{1}{R_2C_1} + \frac{1}{R_2C_2} = \frac{1}{\beta\tau_1} + \frac{1}{\alpha\tau_2} \quad ...(9.18)$$

From Eqs. (9.15), (9.16) and (9.17), it is found that $\alpha\beta = 1$

It means that a single lag lead network does not permit an independent choice of α and β.

The transfer function of lag lead compensator is

$$G_c(s) = \left(\frac{s + \frac{1}{\tau_1}}{s + \frac{1}{\beta\tau_1}}\right)\left(\frac{s + \frac{1}{\tau_2}}{s + \frac{\beta}{\tau_2}}\right) \beta > 1$$

$$= \frac{(s + z_{C_1})(s + z_{C_2})}{(s + p_{C_1})(s + p_{C_2})},$$

$$\beta = \frac{z_{C_1}}{p_{C_1}} = \frac{p_{C_2}}{z_{C_2}} > 1$$

where $\tau_1 = R_1C_1$, $\tau_2 = R_2C_2$ and $\beta > 1$

$$\frac{1}{R_1C_1} + \frac{1}{R_2C_1} + \frac{1}{R_2C_2} = \frac{1}{\beta\tau_1} + \frac{\beta}{\tau_2}$$

The sinusoidal transfer function of lag lead compensator is

$$G_c(j\omega) = \frac{(1 + j\omega\tau_1)(1 + j\omega\tau_2)}{(1 + j\omega\beta\tau_1)\left(1 + \frac{j\omega\tau_2}{\beta}\right)}$$

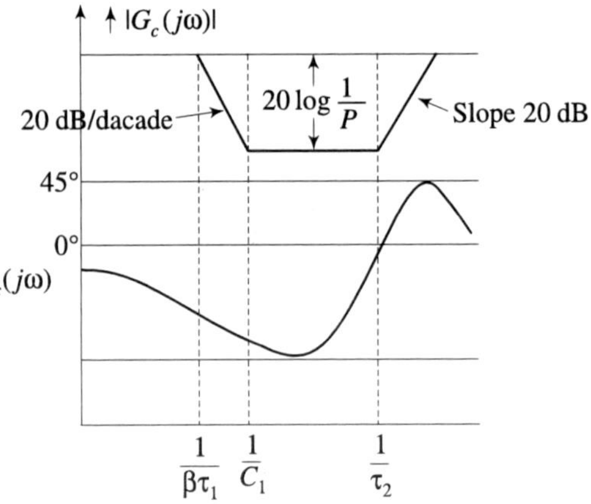

Fig. 9.18

Example 9.1: Consider the system shown in Fig. 9.19 It is desired to modify the closed-loop poles so that the undamped natural

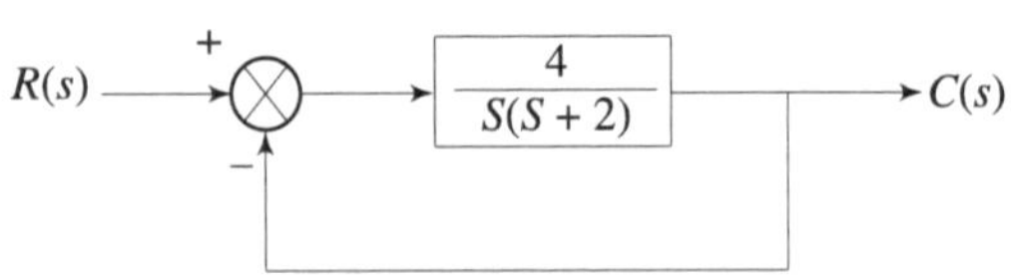

Fig. 9.19

frequency, ω_n = 4 rad/s, is obtained without changing the value of the damping ratio. Design a suitable compensator for the system.

Solution: $G(s) = \dfrac{4}{S(S+2)}$, $H(s) = 1$

The closed-loop transfer function

$$\frac{C(s)}{R(s)} = \frac{G(s)}{1+G(s)H(s)} = \frac{\dfrac{4}{S(S+2)}}{1+\dfrac{4}{S(S+2)}} = \frac{4}{S^2+2S+4}$$

$$\frac{C(s)}{R(s)} = \frac{4}{S^2+2S+4} = \frac{\omega_n^2}{S^2+2\xi\,\omega_n S+\omega_n^2}$$

$$\omega_n^2 = 4,\ \omega_n = 2 \text{ rad/sec}$$

$$2\xi\omega_n S = 2$$

$$\xi = \frac{2}{2\omega_n} = \frac{2}{2\times 2} = 0.5$$

The damping ratio of the closed-loop poles is $\xi = 0.5$. It is desired to modify the closed-loop poles so that undamped natural frequency, ω_n = 4 rad/sec, is obtained without changing the value of the damping ratio, $\xi = 0.5$.

We know that the radial distance of the dominant pole from the origin is equal to ω_n at an angle of $\cos^{-1}\xi$ with the negative real axis.

$$\xi = \cos\theta = 0.5$$

This gives $\theta = \pm 60$

The damping ratio of 0.5 requires that the complex conjugate poles lie on the lines drawn through the origin making angles of $\pm 60°$ with the negative real axis.

The desired locations of the complex conjugated poles are at P_1 and P_2 as shown in Fig. 9.20. The location of the dominant pole P_1 is given by

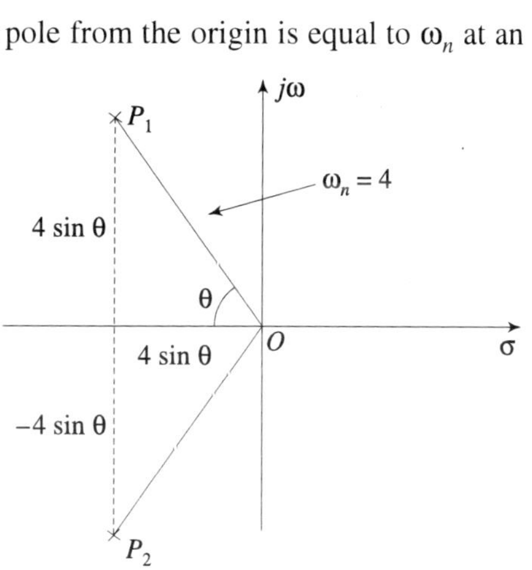

Fig. 9.20

$$S_1 = -4\cos\theta + j4\sin\theta$$

$$= -4\cos 60° + j4\sin 60°$$

$$= -4\frac{1}{2} + j4\frac{\sqrt{3}}{2}$$

$$S_1 = -2 + j2\sqrt{3}$$

The location of the dominant pole P_2 is

$$S_2 = -4 \cos \theta - j4 \sin \theta = -4 \cos 60° - j4 \sin 60°$$

$$S_2 = -2 - j2\sqrt{3}$$

For the given system, the angle of $G(s)$ at the desired closed-loop dominant pole P_1 is

$$\left.\angle\frac{4}{S(s+2)}\right|_{\sigma=-2+j2\sqrt{3}} = \angle\frac{4}{(-2+j2\sqrt{3})(-2+j2\sqrt{3}+2)}$$

$$= \angle\frac{1}{-3-j\sqrt{3}} = -210$$

Hence, a lead compensator is needed in the forward path to provide a leading angle of at least (210 – 180) = 30°.

Thus, if the root locus is to pass through the desired closed-loop pole, the lead compensator must contribute $\phi = 30°$ at this point.

A general procedure for determining the lead compensator is as follows.

First, find the sum of the angles at the desired location of one of the dominant closed-loop poles with the open-loop poles and zeros of the original system. Then determine the necessary angle ϕ to be added so that the total sum of the angles is equal to $\pm 180°$ $(2K + 1)$. The lead compensator must contribute this angle.

If the original system has the open-loop transfer function $G(s)$ then the compensated system will have the transfer function.

$$G_c(s)\,G(s) = \left[K_c \frac{S + \frac{1}{T}}{S + \frac{1}{\alpha T}}\right] G(s)$$

where $\left(\alpha \frac{TS+1}{\alpha TS + 1}\right)$ corresponds to the lead compensation, K is the gain of the amplifier and $G(s)$ is the original open-loop transfer function

$$G(S) = K_c\,\alpha \frac{TS+1}{\alpha TS+1} = K_c \frac{S + \frac{1}{T}}{S + \frac{1}{\alpha T}} \quad (0 < \alpha < 1)$$

It is to be noted that there are many possible values for T and α that will yield the necessary angle contribution at the desired closed-loop poles. The next step is to determine the locations of the zero and pole of the lead compensators. There are many possibilities for the choice of such locations. For a better performance of the system we choose the largest possible value for α. For this purpose, we introduce the following procedure.

First draw a horizontal line passing through point P, the desired location for one of the dominant closed-loop poles at $(-2 + j2\sqrt{3})$. This is shown as line PA in Fig. 9.21.

Then draw a line *PO* connecting point *P* and the origin. Bisect the angle between the lines *PA* and *PO*.

Draw two lines *PC* and *PD* which make angle $\pm \frac{\phi}{2}$ with the bisector *PB*. The intersection of PC and PD with the negative real axis gives the necessary locations for the pole and zero of lead network. The compensator thus designed will make point *P*, a point on the root locus of the compensated system.

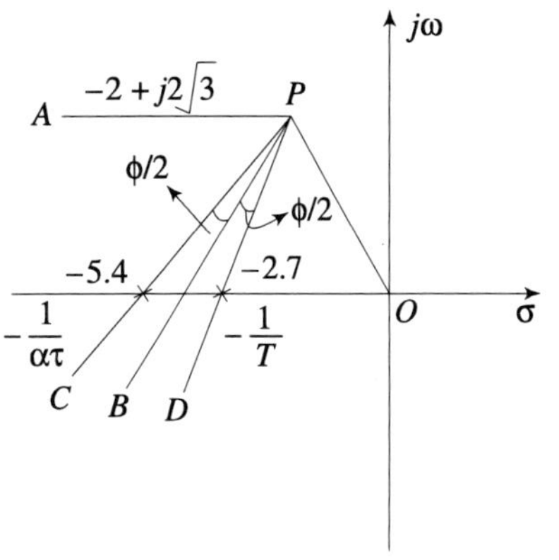

Fig. 9.21

The pole and zero of the lead compensator from the Fig. 9.21.

$$\text{Zero at } S = -2.9$$

$$\text{Pole at } S = -5.4$$

Hence, the transfer function of the lead compensator is

$$G_c(s) = \frac{S + 2.9}{S + 5.4} K_C$$

where the gain K_c of the amplifier is for gain adjustment for satisfying the magnitude condition of root locus approach.

Thus, the open-loop transfer function of the compensated system becomes

$$G_c(s)\, G(s) = \frac{S + 2.9}{S + 5.4} K_C \frac{4}{S(S + 2)}$$

$$= \frac{K(S + 2.9)}{S(S + 2)(S + 5.4)}$$

The gain *K* is evaluated from the magnitude condition as follows:

$$\left| \frac{K(S + 2.9)}{S(S + 2)(S + 5.4)} \right|_{S = -2 + j2\sqrt{3}} = 1$$

$$\left| \frac{K(-2 + j2\sqrt{3} + 2.9)}{(-2 + j2\sqrt{3})\,(-2 + j2\sqrt{3})\,(-2 + j2\sqrt{3} + 5.4)} \right| = 1$$

$$\left| \frac{K(0.9 + j2\sqrt{3})}{(2)^2 + (2\sqrt{3})^2\,(2\sqrt{3})\,\sqrt{(3.4)^2 + (2\sqrt{3})^2}} \right| = 1$$

$$\frac{K(3.579)}{4 \times 2\sqrt{3} \times 4.854} = 1$$

$$K(0.0532) = 1$$

$$K = \frac{1}{0.0532} = 18.79$$

$$G_c(s)\, G(s) = \frac{18.79\,(S + 2.9)}{S(S + 2)(S + 5.4)}$$

The velocity error constant is given by

$$K_v = \lim_{S\to 0} SG_C(s)\, G(s)$$

$$= \lim_{S\to 0} S\cdot \frac{18.79(S+2.9)}{S(S+2)(S+5.4)}$$

$$K_v = \frac{18.79 \times 2.9}{2 \times 5.4} = 5.04$$

The constant K_C of the lead compensator is

$$K_v = \frac{K}{4} = \frac{18.79}{4} = 4.68$$

The transfer function of the lead compensator is

$$G_c(s) = 4.68\,\frac{(S+2.9)}{(S+5.4)}$$

Example 9.2: The open-loop transfer function of a unity feedback system is $G(s)\,H(s) = \dfrac{K}{S(S+4)(S+5)}$.

It is desired that the velocity constant should be at least 5 and the damping ratio should be 0.707. Design a suitable lag compensator to meet these requirements using root locus technique.

Solution: There are three open-loop poles at

$$S = 0,\ S = -4\ S = -5$$

Angle of asymptotes,

$$\theta_A = \frac{(2q+1)180°}{3-0},\ q = 0, 1, 2$$

$$\theta_1 = 60°,\ \theta_2 = 180°,\ \theta_3 = 300$$

Centroid:

$$\sigma = \frac{\text{Sum of poles} - \text{Sum of zeros}}{\text{Number of poles} - \text{Number of zeros}}$$

$$\sigma = \frac{-0-4-5}{3-0} = -3$$

The root locus exists on the real axis between $S = 0$, $S = -4$ and from $S = -5$ to ∞

Calculation of the breakaway points

The characteristics equation is

$$1 + G(s)H(s) = 0$$

$$1 + \frac{K}{S(S+4)(S+5)} = 0$$

$$S^3 + 9S^2 + 20S + K = 0$$

$$K = -S^3 - 9S^2 - 20S$$

$$\frac{dK}{dS} = -3S^2 - 18S - 20 = 0$$

$$3S^2 + 18S + 20 = 0$$

$$S = \frac{-18 \pm \sqrt{(18)^2 - 4 \times 3 \times 20}}{2 \times 3}$$

$$S_1 = -1.47,\ S_2 = -4.5$$

Since $S = -4$ to $S = -5$ is not the part of the root locus, therefore, we take $S = -1.47$ as the breakaway point.

Determine the point of intersection of the root locus with the imaginary axis.

The characteristic equation is

$$S^3 + 9S^2 + 20S + K = 0$$

Routh array is

S^3	:	1	20
S^2	:	9	K
S^1	:	$\frac{180 - K}{9}$	
S^0	:	K	

For stability all the elements in the first column must be positive

$$K > 0 \text{ and } \left(\frac{180 - K}{9}\right) > 0 \text{ or } K < 180.$$

The marginal value of K is

$$K = 180$$

The auxiliary equation is

$$A(s) = 9S^2 + K = 0$$

$$9S^2 + 180 = 0$$

$$S^2 = \frac{-180}{9} = -20$$

$$S = \pm i\sqrt{20} \pm j4.47$$

The complete root locus is shown in Fig. 9.22.

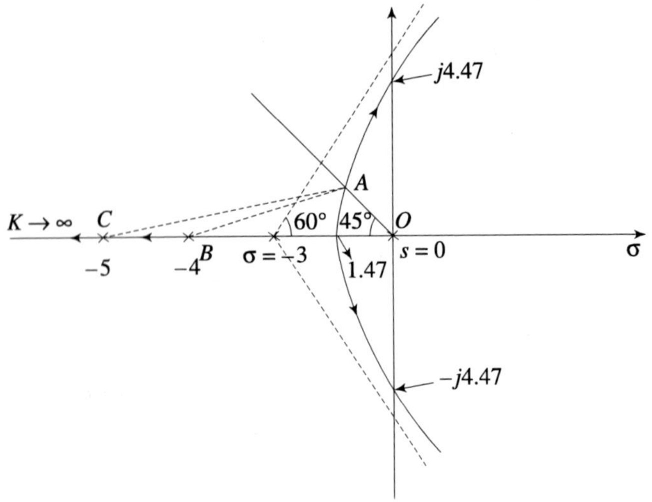

Fig. 9.22

Determination of gain K for $\xi = 0.707$. Draw a line making an angle $\cos^{-1}(0.707) = 45°$ with the real axis as shown in root locus.

This line cuts the root locus at point A. Measure the distance of all poles from point A.

$$OA = 1.8,\ AB = 3.2,\ AC = 4.1$$

Hence, gain K for required damping is

$$K = OA \times AB \times AC$$

$$= 1.8 \times 3.2 \times 4.1$$

$$K = 23.6$$

$$G(s)\,H(s) = \frac{23.6}{S(S+4)\,(S+5)}$$

Assume that the transfer function of the lag compensator is

$$G_c(s) = K_c\beta\,\frac{(1 + T_S)}{(1 + \beta T_S)} = K_c\,\frac{\left(S + \dfrac{1}{T}\right)}{\left(S + \dfrac{1}{\beta T}\right)}$$

where $\frac{1}{T}$ and $\frac{1}{\beta T}$ are zero and pole of the compensator respectively.

As the system has adjustable gain K, so we assume the gain K of the compensator $K_c = 1$.

The transfer function of the compensator is

$$G_c(s) = \frac{\beta(1 + T_S)}{(1 + \beta T_S)}$$

The transfer function of the compensated system is

$$G_c(s)G(s)\ H(s) = \frac{\beta(1 + T_S)}{1 + \beta T_S} = \frac{23.6}{S(S + 4)(S + 5)}$$

The velocity error constant is given by

$$K_v = \lim_{S\to 0} S\frac{\beta(1 + T_S)}{1 + \beta T_S} = \frac{23.6}{S(S + 4)(S + 5)}$$

$$5 = \frac{\beta \times 23.6}{4 \times 5}$$

$$\beta = 4.24$$

Location of pole and zero of compensator. The lag compensator should be selected such that the root locus is not disturbed by putting the compensating pole and zero near the origin.

Usually, 10% of the distance between the origin and first pole on the negative axis is taken for the location of compensator zero.

Zero at $-\frac{1}{T}$ is at 10% of (– 4)

$$-\frac{1}{T} = \frac{10}{100}(-4)$$

$$\frac{1}{T} = 0.4,\ T = 2.5$$

Pole of the compensator is at $\frac{1}{\beta T}$

$$\therefore \qquad \frac{1}{\beta T} = \frac{1}{4.24} \times 0.4 = 0.094$$

$$\beta T - 10.6$$

The transfer function of the compensator is

$$G_c(s) = \frac{\beta(1 + T_S)}{(1 + \beta T_S)} = \frac{4.24(1 + 2.5S)}{(1 + 10.6S)} = \frac{S + 0.4}{S + 0.094}$$

The overall transfer function of the compensated system

$$G_c(s)\ G(s)\ H(s) = \frac{23.6(S + 0.4)}{S(S + 4)(S + 5)(S + 0.094)}$$

Example 9.3: Consider a lag lead compensator $G_c(s)$ defined by

$$G_c(S) = K_c \frac{\left(S + \dfrac{1}{T_1}\right)\left(S + \dfrac{1}{T_2}\right)}{\left(S + \dfrac{\beta}{T_1}\right)\left(S + \dfrac{1}{\beta T_2}\right)}$$

Show that at frequency ω_1 where $\omega_1 = \dfrac{1}{\sqrt{T_1 T_2}}$, the phase angle of $G_c(j\omega)$ becomes zero.

Solution: Putting $S = j\omega$ in the expression for $G_c(s)$, we get

$$G_c(j\omega) = K_c \frac{\left(j\omega + \dfrac{1}{T_1}\right)\left(j\omega + \dfrac{1}{T_2}\right)}{\left(j\omega + \dfrac{\beta}{T_1}\right)\left(j\omega + \dfrac{1}{\beta T_2}\right)}$$

The angle of $G_c(j\omega)$ is given by

$$\angle G_c(j\omega) = \angle\left(j\omega + \frac{1}{T_2}\right) + \angle\left(j\omega + \frac{1}{T_2}\right) - \angle\left(j\omega + \frac{\beta}{T_1}\right) - \angle\left(j\omega + \frac{1}{\beta T_2}\right)$$

$$\angle G_c(j\omega) = \tan^{-1} \omega T_1 + \tan^{-1} \omega T_2 - \tan^{-1} \frac{\omega T_1}{\beta} - \tan^{-1} \omega T_2 \beta$$

At $\omega = \omega_1 = \dfrac{1}{\sqrt{T_1 T_2}}$, we have

$$G_c(j\omega) = \tan^{-1} \frac{T_1}{\sqrt{T_1 T_2}} + \tan^{-1} \frac{T_2}{\sqrt{T_1 T_2}} - \tan^{-1} \frac{T_1}{\beta} \frac{1}{\sqrt{T_1 T_2}} - \tan^{-1} \frac{T_2}{\sqrt{T_1 T_2}} \beta$$

$$= \tan^{-1} \sqrt{\frac{T_1}{T_2}} + \tan^{-1} \sqrt{\frac{T_2}{T_1}} - \tan^{-1} \frac{1}{\beta} \sqrt{\frac{T_1}{T_2}} - \tan^{-1} \beta \sqrt{\frac{T_2}{T_1}}$$

$$\frac{\tan\left(\tan^{-1} \sqrt{\dfrac{T_1}{T_2}} + \tan^{-1} \sqrt{\dfrac{T_2}{T_1}}\right) = \sqrt{\dfrac{T_1}{T_2}} + \sqrt{\dfrac{T_2}{T_1}}}{1 - \sqrt{\dfrac{T_1}{T_2}} + \sqrt{\dfrac{T_2}{T_1}}}$$

$$= \infty = \tan 90$$

$$\therefore \qquad \tan^{-1} \sqrt{\frac{T_1}{T_2}} + \tan^{-1} \sqrt{\frac{T_2}{T_1}} = 90^\circ$$

$$\tan\left(\tan^{-1} \frac{1}{\beta} \sqrt{\frac{T_1}{T_2}} + \tan^{-1} \sqrt{\frac{T_2}{T_1}}\right) \frac{\dfrac{1}{\beta} \sqrt{\dfrac{T_1}{T_2}} + \beta \sqrt{\dfrac{T_2}{T_1}}}{1 - \dfrac{1}{\beta} \sqrt{\dfrac{T_1}{T_2}}\, \beta \sqrt{\dfrac{T_2}{T_1}}}$$

$$= \infty = \tan 90^\circ$$

$$\tan^{-1}\frac{1}{\beta}\sqrt{\frac{T_1}{T_2}} + \tan^{-1}\beta\sqrt{\frac{T_2}{T_1}} = 90°$$

$$\underline{|G_c(j\omega_1)} = 0°$$

Thus the angle of $G_C(j\omega_1)$ becomes 0° at $\omega = \omega, = \frac{1}{\sqrt{T_1 T_2}}$.

Example 9.4: The control system shown in Fig. 9.23 employs proportional plus error rate control. Determine the value of error rate constant, K_e, so that the damping ratio is 0.6.

(a) Determine the value of settling time and maximum overshoot. Find the steady state error if the input is a unit ramp.

(b) What will be those values without error rate control.

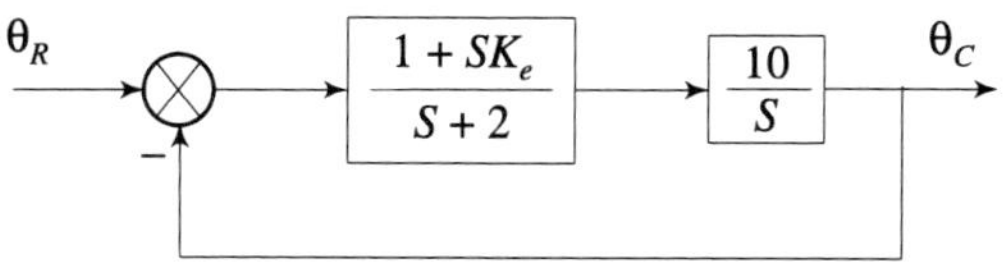

Fig. 9.23

Solution:

(a) With error rate control

$$G(s) = \frac{10(1 + SK_e)}{S(S + 2)}$$

The transfer function is

$$\frac{\theta_c(s)}{\theta_R(s)} = \frac{\dfrac{10(1 + SK_e)}{S(S + 2)}}{1 + \dfrac{10(1 + SK_e)}{S(S + 2)}} = \frac{10 + 10\,SK_e}{S^2 + S(2 + 10K_e) + 10}$$

Comparing the characteristics equation

$S^2 + S(2 + 10K_e) + 10 = 0$ with the standard form of characteristic equation of a second order system $S^2 + 2\xi\omega_n S + \omega_n^2 = 0$

$$\omega_n^2 = 10,\ \omega_n = \sqrt{10} = 3.16 \text{ rad/sec}$$

$$2\xi\omega_n = 2 + 10K_e$$

$$K_e = \frac{2\xi\omega_n - 2}{10} = \frac{2 \times 0.6 \times 3.16 - 2}{10} = 0.18$$

The settling time

$$t_s = \frac{4}{\xi\omega_n} = \frac{4}{0.6 \times 3.16} = 2.11S$$

The peak overshoot

$$M_p = e^{\frac{-\pi\xi}{\sqrt{1-\xi^2}}} = e^{\frac{-\pi \times 0.6}{\sqrt{1-0.6^2}}} = 0.0949$$

$$M_p\,\% = 9.49\%$$

Steady state error

$$e_{ss} = \frac{1}{K_v} = \frac{1}{K_v} = \frac{1}{\lim\limits_{S\to 0} SG(s)}$$

$$e_{ss} = \frac{1}{\lim\limits_{S\to 0} S\,\dfrac{10(1+SK_e)}{S(S+2)}} = \frac{1}{5} = 0.2$$

(b) Without error rate control

When $K_e = 0$

$$G(s) = \frac{10}{S(S+2)}$$

The closed-loop transfer function is

$$\frac{\theta_c(s)}{\theta_R(s)} = \frac{\dfrac{10}{S(S+2)}}{1 + \dfrac{10}{S(S+2)}} = \frac{10}{S^2 + 2S + 10}$$

Comparing $\dfrac{\omega_n^2}{s^2 + 2\xi\,\omega_n S + \omega_n^2}$

$$\omega_n^2 = 10,\ \omega_n = \sqrt{10} = 3.16 \text{ rad/sec}$$

$$2\xi\omega_n = 2$$

$$\xi = \frac{2}{2\omega_n} = \frac{2}{2 \times 3.16} = 0.32$$

Settling time

$$t_s = \frac{4}{\xi\,\omega_n} = \frac{4}{(0.32)(3.16)} = 45$$

$$M_p = e^{\frac{-\pi\xi}{\sqrt{1-\xi^2}}} = e^{\frac{-\pi\xi}{\sqrt{1-\xi^2}}} = e^{\frac{-\pi \times 0.32}{\sqrt{1-(0.32)^2}}} = 0.351$$

Steady state errors

$$e_{ss}(t) = \frac{1}{K_V} = \frac{1}{\lim\limits_{S\to 0} SG(s)} = \frac{1}{\lim\limits_{S\to 0} S\,\dfrac{10}{S(S+2)}}$$

$$e_{ss}(t) = \frac{1}{5} = 0.2.$$

Example 9.5: A closed-loop control system with unity feedback is shown in Fig. 9.24. By using derivative control, the damping ratio is to be made 0.75. Determine the value of T_d. Also determine the rise time, peak time and peak overshoot without derivative control and with derivative control. The input to the system is a unit step.

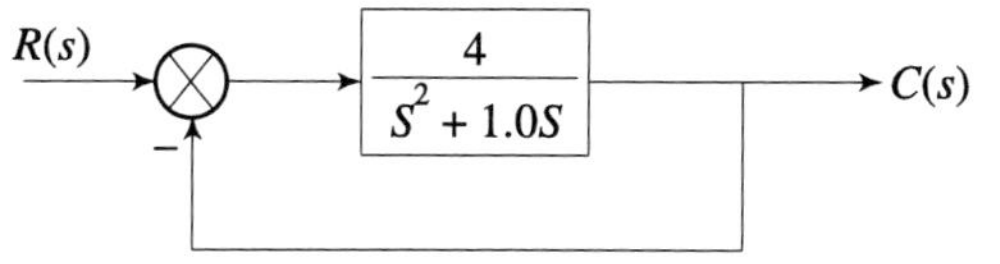

Fig. 9.24

Solution: The closed-loop transfer function of the system is

$$\frac{C(s)}{R(s)} = \frac{\dfrac{4}{S^2 + 1.0S}}{1 + \dfrac{4}{S^2 + 1.0S}} = \frac{4}{S^2 + 1.0S + 4}$$

The characteristic equation is

$$S^2 + 1.0S + 4 = 0$$

Comparing with $S^2 + 2\xi\,\omega_n S + \omega_n^2 = 0$

$$\omega_n^2 = 4$$

$$\omega_n = 2 \text{ rad/sec}$$

$$2\xi\omega_n = 1$$

$$\xi = \frac{1}{2\omega_n} = \frac{1}{2 \times 2} = 0.25$$

The damping ratio with derivative control is

$$\xi' = \xi + \frac{\omega_n t_d}{2}$$

Since the damping ratio with derivative control is to be 0.75 and $\omega_n = 2$ we have

$$0.75 = 0.25 + \frac{2 \times T_d}{2}$$

$$t_d = 0.5$$

The block diagram with derivative control is as shown below.

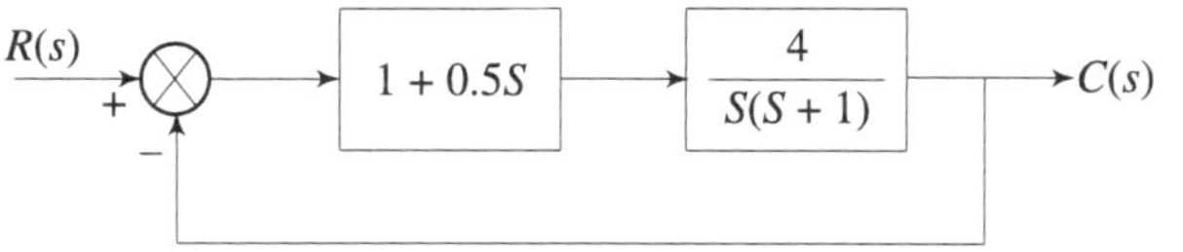

Fig. 9.25

Without derivative control the rise time

$$t_r = \frac{\pi - \theta}{\omega_d} = \frac{\pi - \tan^{-1}\dfrac{\sqrt{1-\xi^2}}{\xi}}{\omega_n\sqrt{1-\xi^2}}$$

$$= \frac{\pi - \tan^{-1}\dfrac{\sqrt{1-(0.25)^2}}{0.25}}{2\sqrt{1-(0.25)^2}} = \frac{1.823}{1.936} = 0.942 \text{ sec}$$

$$t_r = 0.942 \text{ sec}$$

The peak time

$$t_p = \frac{\pi}{\omega_d} = \frac{\pi}{\omega_d\sqrt{1-\xi^2}}$$

$$t_p = \frac{\pi}{2\sqrt{1-(0.25)^2}} = 1.62S$$

The peak overshoot is given by

$$M_p = e^{\frac{-\pi\xi}{\sqrt{1-\xi^2}}} = e^{\frac{-\pi \times 0.25}{\sqrt{1-(0.25)^2}}}$$

$$M_p = 0.4443$$

The overall transfer function with derivative control is

$$\frac{C(s)}{R(s)} = \frac{\dfrac{4(1+0.5S)}{S(S+1)}}{1+\dfrac{4(1+0.5S)}{S(S+1)}} = \frac{4(1+0.5S)}{S^2+3S+4}$$

$$= \frac{4+2S}{S^2+3S+4}$$

For a unit step input, $r(t) = 1$

$$R(s) = \frac{1}{S}$$

$$C(s) = \frac{4+2S}{S(S^2+3S+4)} = \frac{A}{S} + \frac{BS+C}{S^2+3S+4}$$

$$= \frac{1}{S} - \frac{(S+1)}{S^2+3S+4}$$

$$C(s) = \frac{1}{S} - \frac{S+1}{(S+1.5)^2 + 1.75}$$

$$= \frac{1}{S} - \frac{(S+1)}{(S+1.5)^2 + (1.32)^2} = \frac{1}{S} - \frac{(S+1.5) - 0.5}{(S+1.5)^2 + (1.32)^2}$$

$$= \frac{1}{S} - \frac{(S+1.5)}{(S+1.5)^2 + (1.32)^2} + \frac{0.5}{1.32}\frac{1.32}{(S+1.5)^2 + (1.32)^2}$$

$$C(t) = 1 - e^{-1.5t}\cos 1.32\, t + 0.378\, e^{-1.5t}\sin 1.32\, t$$

$$= 1 - e^{-1.5t}(\cos 1.32\, t - 0.378 \sin 1.32\, t)$$

$$t = t_r$$

$$C(t) = 1$$

$$C(t_r) = 1$$

$$1 = 1 - e^{-1.5t_r}(\cos 1.32\, t_r - 0.378 \sin 1.32\, t_r)$$

$$\cos(1.32\, t_r) = 0.378 \sin(1.32\, t_r)$$

$$\tan(1.32\, t_r) = \frac{1}{0.378} = 2.645$$

$$1.32\, t_r = \tan^{-1} 2.645 = 69.28 = 1.209 \text{ rad}$$

$$t_r = \frac{1.209}{1.32} = 0.916 \text{ sec}$$

For peak time, differentiating the output equation with respect to t and equating to zero.

$$\frac{d}{dt}C(t)\bigg|_{t=t_p} = \frac{d}{dt}[1 - e^{-1.5t}](\cos 1.32t - 0.378 \sin 1.32\, t)\bigg|_{t=t_p} = 0$$

$$-e^{-1.5t}[-\sin 1.32t\,(1.32) - 0.378 \times 1.32 \cos(1.32t)$$

$$-e^{-1.5t}(-1.5)[\cos(1.32t) - 0.378 \sin 1.32\, t]\bigg|_{t=t_p} = 0$$

$$-1.32 \sin 1.32\, t_p - 0.498 \cos 1.32\, t_p + 0.567 \sin 1.32\, t_p$$

$$-1.5 \cos 1.32\, t_p = 0$$

$$-0.753 \sin 1.32\, t_p - 1.998 \cos 1.32\, t_p = 0$$

$$\tan 1.32\, t_p = -\frac{1.998}{0.753} = -2.653$$

$$t_p = \frac{\tan^{-1}(-2.63)}{1.32} = \frac{\pi - \tan^{-1} 2.63}{1.32}$$

$$t_p = \frac{\pi - 1.207}{1.32} = 1.46 \text{ sec}$$

Peak output occurs at $t = t_p$

$$C(t_p) = 1 - e^{-1.5\,t_p}(\cos 1.32\,t_p - 0.378 \sin 1.32\,t_p)$$

$$= 1 - e^{-1.5(1.46)}(\cos(1.32 \times 1.46) - 0.378 \sin(1.32 \times 1.46))$$

$$C(t_p) = 1 - e^{-2.19}(\cos(1.927) - 0.378 \sin(1.927))$$

$$= 1 - 0.119(-0.348 - 0.354)$$

$$C(t_p) = 1.08$$

∴ The peak overshoot

$$M_p = C(t_p) - C(\infty)$$

$$= 1.08 - 1 = 0.08$$

$$M_p\,\% = 8\%.$$

Example 9.6: The system shown in Fig. 9.26 is a unity feedback control system with minor feedback loop (output derivative feedback).

(a) In the absence of derivative feedback ($K_t = 0$) determine the damping factor and natural frequency for $K_A = 15$. Also determine the steady state error resulting from a unit ramp input.

(b) Determine the derivative feedback constant K_t which will increase the damping factor of the system to 0.7. What is the steady state error to a unity ramp input with this setting of the derivative feedback constant.

(c) Find suitable values of the parameters K_A and K_t so that the damping ratio of the system is increased to 0.8 without affecting the steady state error as obtained in part (a).

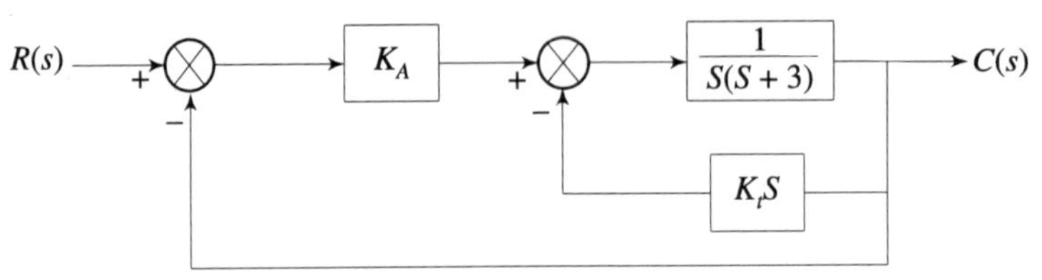

Fig. 9.26

Solution: (a) In the absence of derivative feedback, $K_t = 0$

$$G(s) = \frac{K_A}{S(S+3)} = \frac{15}{S(S+3)}, \quad H(s) = 1$$

The closed-loop transfer function is

$$\frac{C(s)}{R(s)} = \frac{G(s)}{1 + G(s)} = \frac{\dfrac{15}{S(S+3)}}{1 + \dfrac{15}{S(S+3)}} = \frac{15}{S^2 + 3S + 15}$$

Comparing with $S^2 + 2\xi\, \omega_n S + \omega_n^2 = 0$

$$\omega_n^2 = 15,\ \omega_n = \sqrt{15} = 3.873 \text{ rad/sec}$$

$$2\xi\, \omega_n = 3$$

$$\xi = \frac{3}{2\omega_n} = \frac{3}{2 \times 3.873} = 0.387$$

The system velocity error constant

$$K_v = \lim_{S \to 0} SG(s)$$

$$K_v = \lim_{S \to 0} S \frac{15}{S(S + 3)} = 5$$

∴ The steady state error for a ramp input is

$$e_{ss} = \frac{1}{K_v} = \frac{1}{5} = 0.2$$

(b) With derivative feedback

$$G(s) = \frac{15 \dfrac{1}{S(S+3)}}{1 + \dfrac{1}{S(S+3)} K_t S} = \frac{15}{s^2 + (3 + K_t)S}$$

Overall transfer function is

$$\frac{C(s)}{R(s)} = \frac{\dfrac{15}{S^2 + S(3 + K_t)}}{1 + \dfrac{15}{S^2 + S(3 + K_t)}} = \frac{15}{S^2 + S(3 + K_t) + 15}$$

Comparing with $S^2 + 2\xi\, \omega_n S + \omega_n^2 = 0$

$$\omega_n^2 = 15,\ \omega_n = \sqrt{15} = 3.873 \text{ rad/sec}$$

$$2\xi\, \omega_n = 3 + K_t$$

$$2 \times 0.7 \times 3.873 = 3 + K_t$$

$$K_t = 2 \times 0.7 \times 3.873 - 3 = 2.422$$

$$G(s) = \frac{15}{S^2 + (3 + 2.442)S} = \frac{15}{S^2 + 5.422S}$$

For a unit ramp input

$$K_v = \lim_{S \to 0} SG(s) = \lim_{S \to 0} S \frac{15}{S(S + 5.422)}$$

$$K_v = \frac{15}{5.422} = 2.766$$

$$e_{ss} = \frac{1}{K_v} = \frac{1}{2.766} = 0.3615$$

(c) The overall transfer function of the system is

$$\frac{C(s)}{R(s)} = \frac{K_A}{S^2 + (3 + K_t)\,S + K_A}$$

Comparing with $S^2 + 2\xi\,\omega_n S + \omega_n^2 = 0$

$$\omega_n^2 = K_A$$

$$\omega_n = \sqrt{K_A}$$

$$2\xi\,\omega_n = 3 + K_t$$

$$2 \times 0.8\,\sqrt{K_A} = 3 + K_t$$

The open-loop transfer function of the system is

$$G(s) = \frac{K_A}{S(S + 3 + K_t)}$$

For a ramp input

$$K_v = \lim_{S \to 0} SG(s)$$

$$= \lim_{S \to 0} S\frac{K_A}{S(S + 3 + K_t)} = \frac{K_A}{3 + K_t}$$

Steady state error

$$e_{ss} = \frac{1}{K_V} = \frac{3 + K_t}{K_A}$$

Given that, the steady state error

$$e_{ss} = 0.2$$

$$\therefore \qquad 0.2 = \frac{3 + K_t}{K_A}$$

$$0.2\,K_A = 3 + K_t = 2\xi\,\omega_n = 2 \times 0.8\,\sqrt{K_A}$$

$$\sqrt{K_A} = \frac{2 \times 0.8}{0.2} = 8$$

$$K_A = 64$$

$$3 + K_t = 0.2\,K_A$$

$$3 + K_t = 0.2 \times 64 = 12.8$$

$$K_t = 12.8 - 3 = 9.8$$

The values of K_A and K_t for $\xi = 0.8$ for a steady state error of $e_{SS} = 0.2$ for a ramp input $K_A = 64$ and $K_t = 9.8$.

9.5 BASIC CONTROL ACTION OF CONTROLLER

Comparison of the actual value of plant output with the desired value is carried out by an automatic controller.

Industrial controllers are classified on the basic of control action as

- Proportional controllers (P)
- Integral controllers (I)
- Proportional plus integral controllers (PI)
- Proportional plus derivative controller (PD)
- Proportional plus integral plus derivative controllers (PID)

9.5.1 Proportional Controller (P)

The output and the error are related by

$$m(t) = KpE(t)$$

Taking LT

$$M(s) = K_p E(s)$$

$$\frac{M(s)}{E(s)} = K_p$$

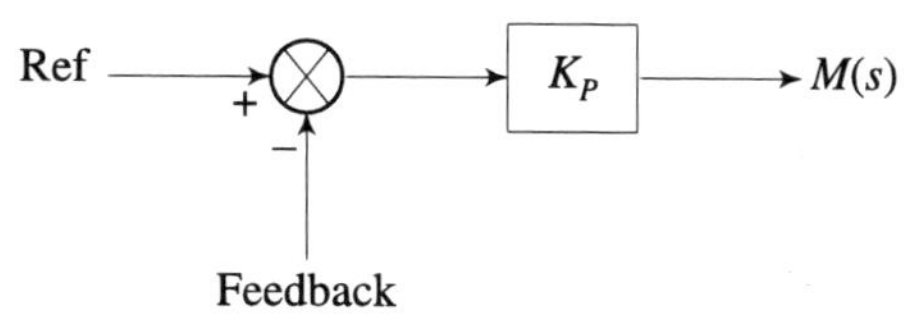

Fig. 9.27

For a 2nd order system,

$$G(s)H(s) = \frac{\omega_n^{\;2}}{s(s + 2\xi\omega_n)}$$

$$K_p = \lim_{s \to \infty} G(s)H(s) = \infty$$

$$K_v = \lim_{s \to \infty} s\ G(s)H(s) = \frac{\omega_n}{2\xi}$$

$$e_{ss} = \frac{A}{K_v}$$

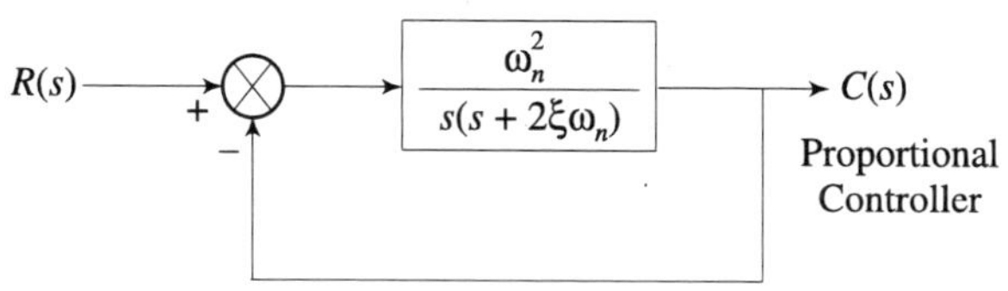

Fig. 9.28

9.5.2 Integral Controller (I)

$$\frac{d}{dt}(m(t)) = K_I e(t)$$

Taking LT

$$sM(s) = K_I E(s)$$

$$\frac{M(s)}{E(s)} = \frac{K_I}{s}$$

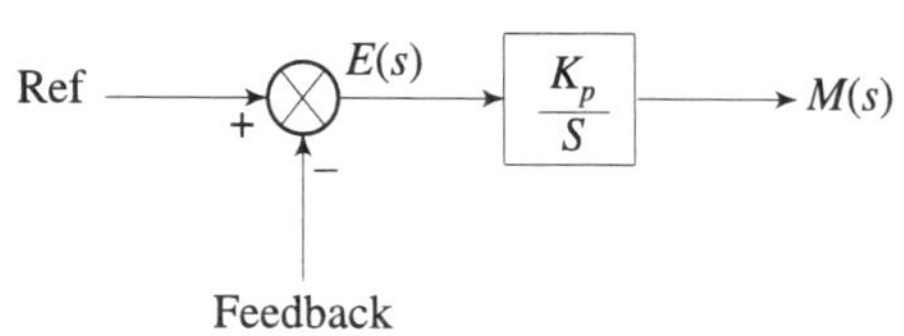

Fig. 9.29

$$m(t) = K_I \int_0^t e(t)dt$$

9.5.3 Realization of Controllers with Op-Amp

Proportional controller

$$\frac{E_0}{E_i} = \frac{-R_2}{R_1}$$

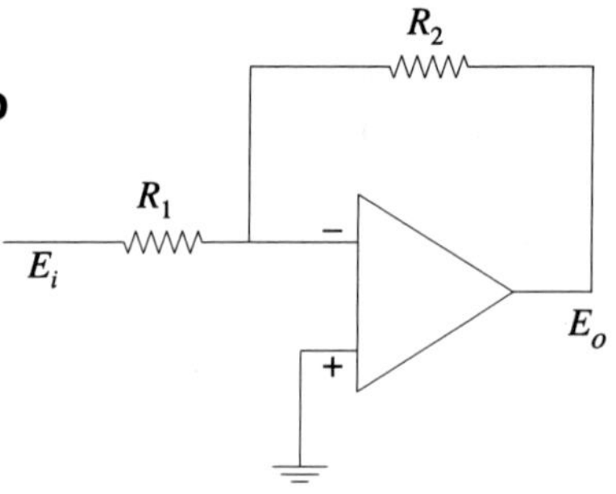

Fig. 9.30

Integral controller

$$\frac{E_0(s)}{E_i(s)} = \frac{(1 + sC_2R_2)}{sR_1C_2}$$

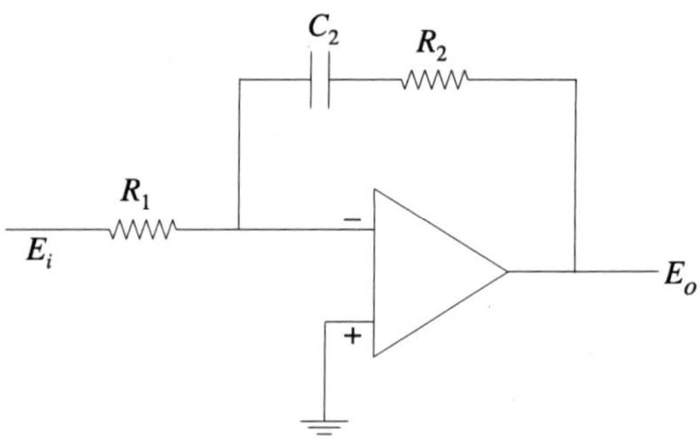

Fig. 9.31

9.5.4 Design with PD Controller

The controller has been typically a simple amplifier with a constant gain K. This type of control action is known as proportional control.

Figure 9.32 shows the block diagram of a feedback control system has a second order prototype process with transfer function.

$$G_p(s) = \frac{\omega_n^2}{s(s + 2\xi\omega_n)} \quad ...(9.19)$$

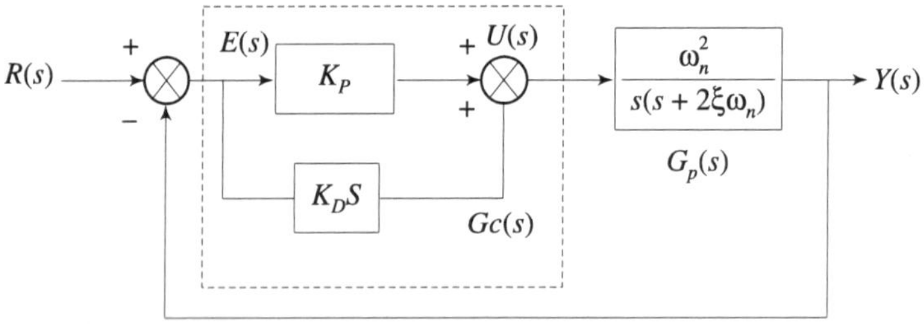

Fig. 9.32

The series controller is a proportional derivative (PD) type with the transfer function.

$$G_c(s) = K_p + K_DS \quad ...(9.20)$$

Thus, the control signal applied to the process is

$$u(t) = K_p e(t) + K_D \frac{de(t)}{dt} \qquad \text{...(9.21)}$$

where K_p and K_D are the proportional and derivative constants respectively.

Electric circuit realization of the PD controller is the transfer function of the circuit in Fig. 9.33 is

$$G_p(s) = \frac{E_0(s)}{E_{in}(s)} = \frac{R_2}{R_1} + R_2C_1s \qquad \text{...(9.22)}$$

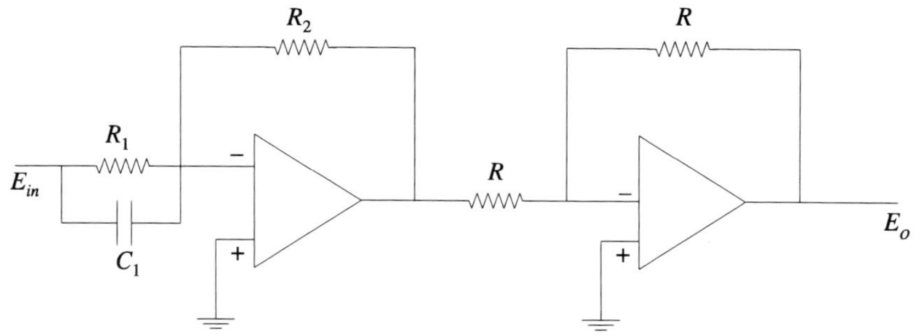

Fig. 9.33 Two op-amp circuit

Comparing Eq. (9.20) with Eq. (9.22), we have

$$K_p = \frac{R_2}{R_1},\ K_D = R_2C_1$$

Realization of PD controller using 3 op-amp circuit. The transfer function of the circuit is

$$G_p(s) = \frac{E_0(s)}{E_{in}(s)} = \frac{R_2}{R_1} + R_dC_ds \qquad \text{...(9.23)}$$

Comparing Eq. (9.20) and (9.23), we have

$$K_p = \frac{R_2}{R_1} \quad \text{and} \quad K_D = R_dC_d$$

The advantage with the circuit in Fig. 9.34 is that only 2 op-amps are used. The circuit does not allow independent selection of K_p and K_D as they are commonly dependent on R_2.

If the value of K_D is large, a large capacitor C_1 would be required. The 3 op-amp circuit, allows K_p and K_D to be controlled independently. The forward path transfer function in the compensated system is

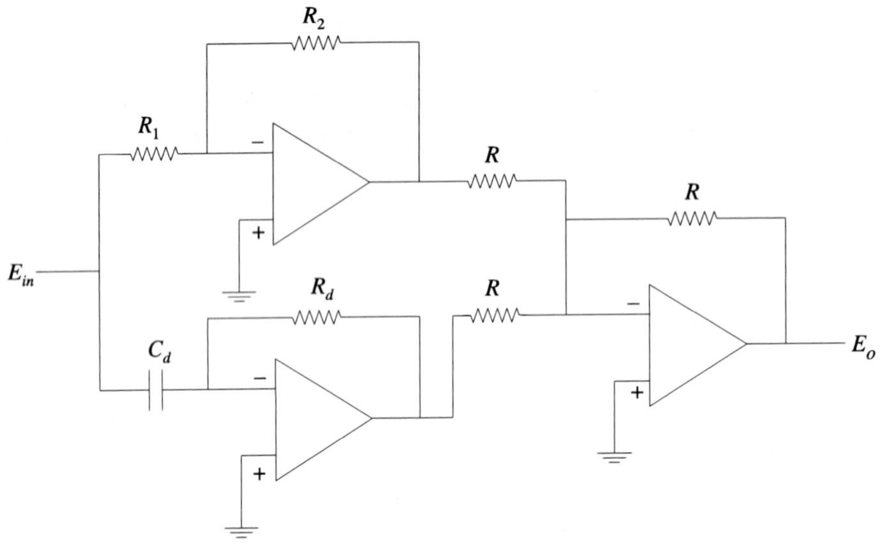

Fig. 9.34 Three op-amp circuit

$$G(s) = \frac{Y(s)}{E(s)} = G_c(s)G_p(s)$$

$$G(s) = \frac{\omega_n^2(K_p + K_D S)}{s(s + 2\xi\omega_n)}$$

which shows that the PD control is equivalent to adding simple zero at $s = \frac{-K_p}{K_D}$ to the forward path transfer function.

Properly designed PD controller will affect the performance of the control system in the following ways.

1. Improves damping and reduces maximum overshoot.
2. Reduces rise time and settling time.
3. Increases bandwidth.
4. Improves G_M, P_M and M_r.
5. May attenuate noise at higher frequencies.
6. Not effective for lightly damped or initially unstable systems.
7. May require a relatively large capacitor in circuit implementation.

PD controller can improve the damping and rise time of a control system at the expense of higher bandwidth and resonant frequency and the steady state error is not affected unless it varies with time.

Thus, PD controller may not fulfill the compensation objective in many simulations.

Design with the PI controller

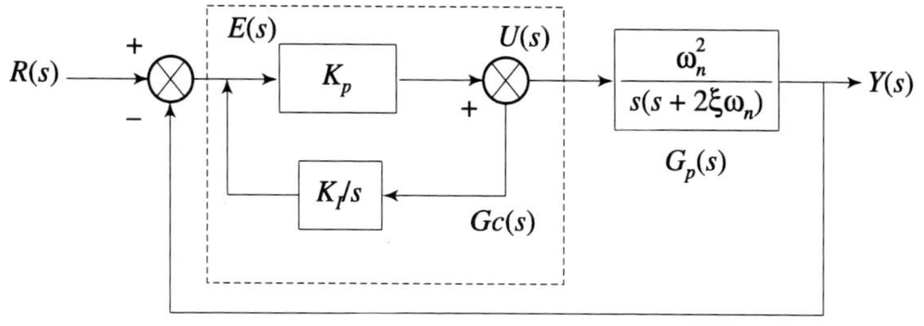

Fig. 9.35

The transfer function of the PI controller is

$$G_p(s) = K_p + \frac{K_I}{s} \qquad \text{...(9.24)}$$

Two op-amp circuit realization,

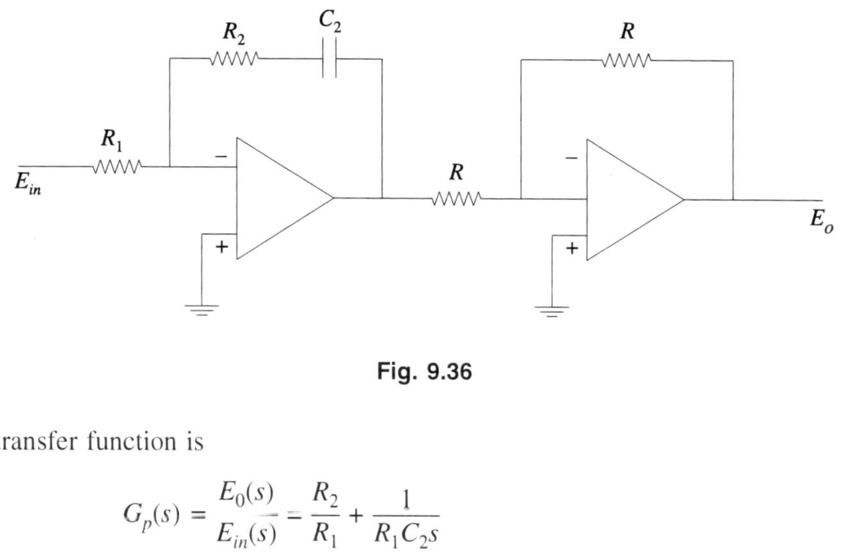

Fig. 9.36

The transfer function is

$$G_p(s) = \frac{E_0(s)}{E_{in}(s)} = \frac{R_2}{R_1} + \frac{1}{R_1 C_2 s} \qquad \text{...(9.25)}$$

Comparing Eqs. (9.24) and (9.25)

$$K_p = \frac{R_2}{R_1},\ K_I = \frac{1}{R_1 C_2}$$

Proportional Plus Derivative Controller (PD)

Figure 9.23 shows block diagram of a PD controller.

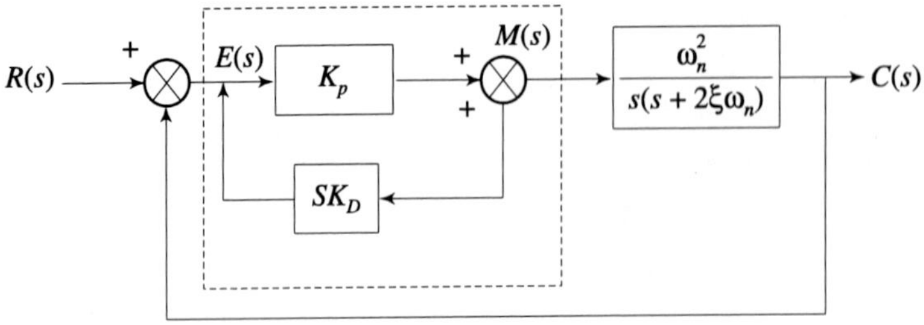

Fig. 9.37

In PD controllers output is proportional to (i) error and (ii) derivative of error.

$$M(s) = (K_p + sK_D)\, E(s)$$

Taking inverse LT

$$m(t) = K_p e(t) + K_D \frac{de(t)}{dt}$$

$$G(s) = \frac{(K_p + K_D S)\omega_n^2}{s^2 + 2\xi\omega_n s} = \frac{(K_p + K_D S)\omega_n^2}{s(s + 2\xi\omega_n)}$$

$$H(s) = 1$$

$$\frac{C(s)}{R(s)} = \frac{G(s)}{1 + G(s)H(s)} = \frac{(K_p + K_D S)\omega_n^2}{s^2 + s(K_D\omega_n^2 + 2\xi\omega_n) + K_p\omega_n^2}$$

$$2\xi\omega_n + K_D\omega_n^2 = 2\omega_n\left(\xi + \frac{K_D\omega_n}{2}\right)$$

$$= 2\xi'\omega_n$$

$$\xi' = \left(\xi + \frac{K_D\omega_n}{2}\right)$$

∴ Damping factor

$$K_v = \lim_{s \to 0} sG(s)H(s) = \frac{K_p\omega_n}{2\xi}$$

In PD controller:

(i) Damping ratio improves and maximum overshoot reduces.
(ii) Rise time and settling time are reduced.
(iii) Bandwidth increases.
(iv) Gain margin, phase margin and resonant peak improve.
(v) Noise at higher frequency is attenuated.

Realization of PD Controllers with op-amps

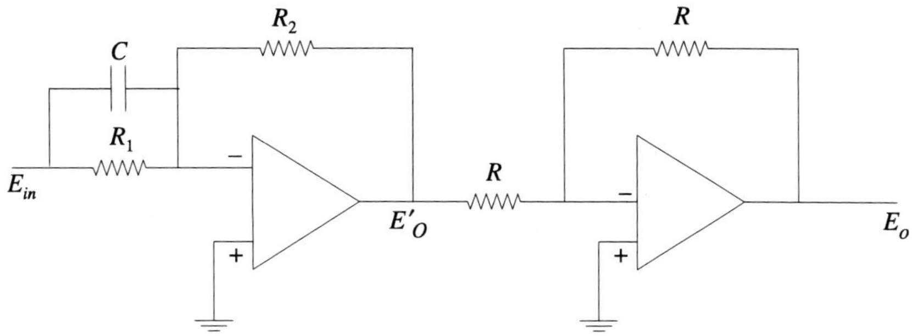

Fig. 9.38

$$\frac{-E_0'(s)}{R_2} = \frac{E_i(s)}{\dfrac{R_1\left(\dfrac{1}{sc}\right)}{R_1 + \left(\dfrac{1}{sc}\right)}}$$

$$\therefore \qquad E'_0(s) = -R_2 \frac{E_i(s)}{\dfrac{R_1\left(\dfrac{1}{sc}\right)}{R_1 + \left(\dfrac{1}{sc}\right)}}$$

$$= -R_2 \frac{E_i(s)\,(R_1 s\,C + 1)}{R_1}$$

$$E_0 = -E'_0$$

$$\therefore \qquad E_0 = \frac{R_2}{R_1}(R_1 s\,C + 1)E_i(s)$$

$$E_0(s) = \frac{R_2}{R_1}E_i(s) + sC\,R_2E_i(s)$$

$$\frac{E_0(s)}{E_i(s)} = \frac{R_2}{R_1} + s\,C\,R_2$$

The above equation shows that the output voltage consists of the following two parts:

(i) 1st part is proportional to input voltage.

(ii) 2nd part is the differential of input voltage.

Three op-amp circuit

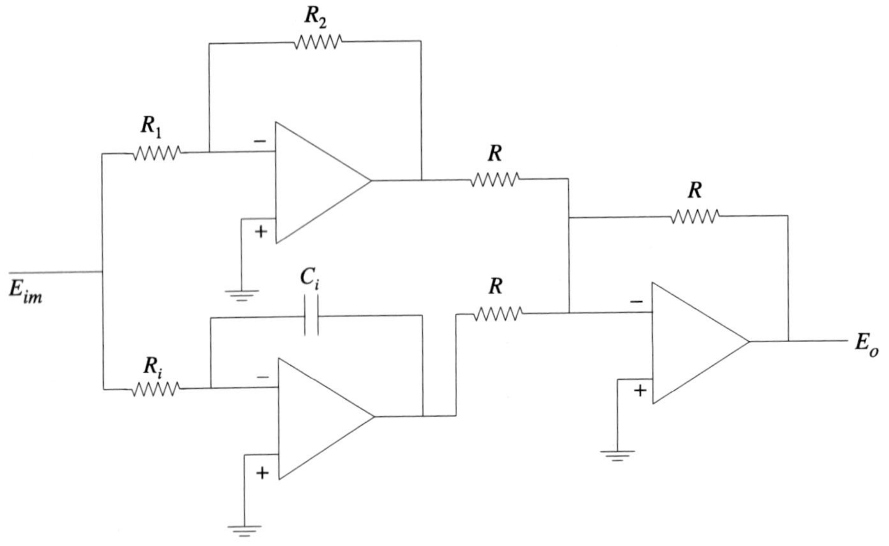

Fig. 9.39

Transfer function,

$$G_p(s) = \frac{E_0(s)}{E_{in}(s)} = \frac{R_2}{R_1} + \frac{1}{R_i C_i s}$$

The parameters of the PI controller are related to the circuit parameters as

$$K_p = \frac{R_2}{R_1},\ K_I = \frac{1}{R_i C_i}$$

The forward path transfer function of the compensated system is

$$G(s) = G_c(s)\ G_p(s) = \frac{\omega_n^2 (K_p S + K_I)}{s^2 (s + 2\xi\omega_n)}$$

1. Adds a zero at $s = \frac{-K_I}{K_p}$ to the forward path transfer function.
2. Adds a pole at $s = 0$ to the forward path transfer function. This means that the system type is increased by type 1 to a type 2 system.

Thus, the steady state error of the original system is improved by one order, if steady state error to a given input is constant, the PI control reduces it to zero.

Advantages and disadvantages of properly designed PI controller are:

1. Improves damping and reduces maximum overshoot.
2. Increases rise time.
3. Decreases bandwidth.
4. Improves gain margin, phase margin and M_r.
5. Filters out high frequency noise.
6. The problem of selecting a proper combination of K_I and K_p so that the capacitor in the circuit implementation of the controller is not excessively large is more acute than in the case of the PD controller.

Proportional integral controller (PI)

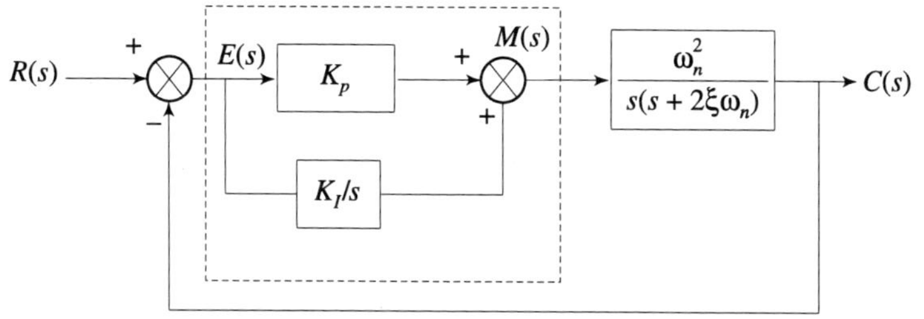

Fig. 9.40

The output in proportional to (i) error and (ii) integral of error.

$$M(s) = \left(K_p + \frac{K_I}{s}\right) E(s)$$

Taking inverse LT

$$m(t) = K_p e(t) + K_I \int_0^t e(t)dt$$

$$G(s) = \frac{\left(K_p + \frac{K_I}{s}\right)\omega_n^2}{s(s + 2\xi\omega_n)} = \frac{(K_p S + K_I)\omega_n^2}{s^2(s + 2\xi\omega_n)}$$

$$H(s) = 1$$

$$G(s)H(s) = \frac{(K_p S + K_I)\omega_n^2}{s^2(s + 2\xi\omega_n)}$$

$$\frac{C(s)}{R(s)} = \frac{G(s)}{1 + G(s)H(s)} = \frac{\dfrac{(K_p S + K_I)\omega_n^2}{s^2(s + 2\xi\omega_n)}}{1 + \dfrac{(K_p S + K_I)\omega_n^2}{s^2(s + 2\xi\omega_n)}}$$

$$= \frac{(K_p s + K_I)\omega_n^2}{s^2 + 2\xi\omega_n s^2 + K_p\omega_n s^2 + K_I\omega_n^2}$$

$$K_v = \lim_{s \to 0} s\, G(s)H(s) = \infty$$

$$\therefore \qquad e_{ss} = \frac{A}{K_v} = 0.$$

The main positive point is that system's behaviour is accurate due to reduction of steady state error (PI controller).

- Improvement of damping and reduction of overshoot.
- Bandwidth also increases.
- Noise filtered out.
- K_I must be designed properly.

Realization of PI controller using op-amp

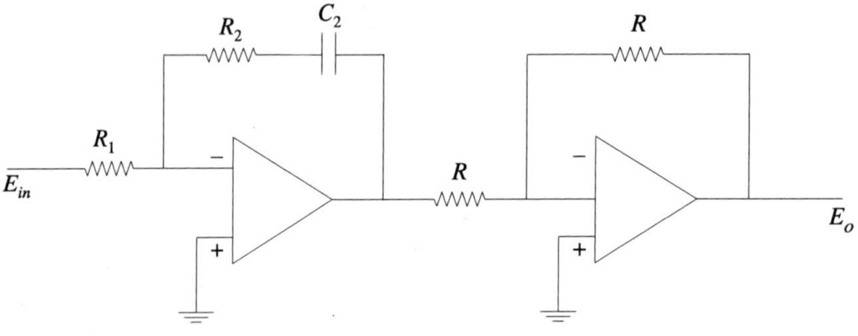

Fig. 9.41

The transfer function is

$$\frac{E_0(s)}{E_{in}(s)} = \frac{R_2}{R_1} + \frac{1}{R_1 C_2 s}$$

$$\therefore \qquad K_p = \frac{R_2}{R_1} \text{ and } K_I = \frac{1}{R_1 C_2 s}$$

Proportional plus integral plus derivative controller (PID)

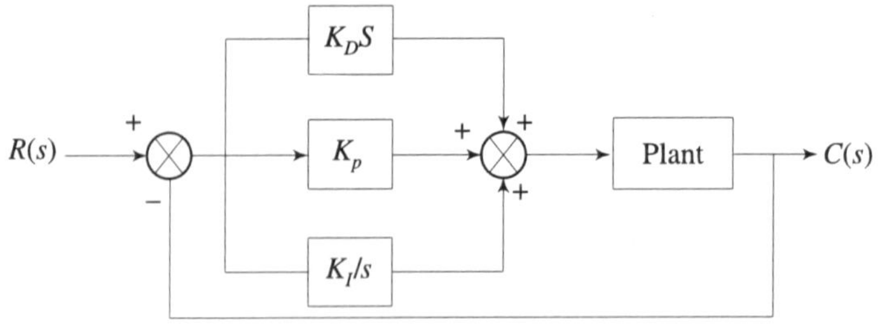

Fig. 9.42

PD controller improves the transient part and PI controller improves the steady state part combination of PD and PI improves the overall system.

Using op-amp

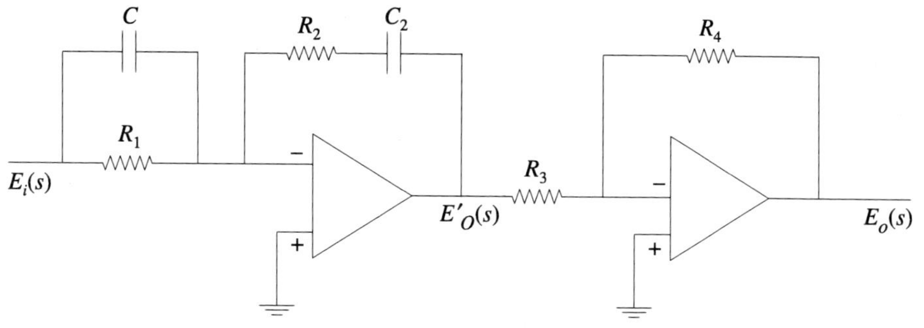

Fig. 9.43

$$\frac{\dfrac{E_i(s)}{\dfrac{R_1 \frac{1}{sc_1}}{R_1 + \frac{1}{sc_1}}}}{} = -\frac{E'_0(s)}{R_2 + \dfrac{1}{sc_2}}$$

$$\frac{E_i(s)}{E'_0(s)} = \frac{-sR_1C_2}{(1 + sR_1C_1)(sR_2C_2)} \quad \text{...(9.26)}$$

$$\frac{E'_0(s)}{R_3} = -\frac{E_0(s)}{R_4} \quad \text{...(9.27)}$$

$$E'_0(s) = -\frac{R_3}{R_4} E_0(s)$$

Substitute Eqs. (9.27) in Eqs. (9.26)

$$\frac{E_i(s)}{-\left(\dfrac{R_3}{R_4}\right) E_0(s)} = \frac{-sR_1C_2}{(1 + sR_1C_1)(sR_2C_2)}$$

$$\frac{E_i(s)}{E_0(s)} = \left(\frac{R_3}{R_4}\right)\left(\frac{sR_1C_2}{(1 + sR_1C_1)(1 + sR_2C_2)}\right)$$

$$\frac{E_0(s)}{E_i(s)} = \frac{R_4(1 + sR_1C_1)(1 + sR_2C_2)}{sR_1R_3C_2}$$

$$\frac{E_0(s)}{E_i(s)} = \frac{R_4R_2(1 + sR_1C_1)(1 + sR_2C_2)}{sR_1R_2R_3C_2}$$

$$\frac{E_0(s)}{E_i(s)} = \frac{R_2R_4}{R_1R_3}\left[\frac{1 + sR_1C_1 + sR_2C_2 + s^2R_1R_2C_1C_2}{sR_2C_2}\right]$$

$$\frac{E_0(s)}{E_i(s)} = \frac{R_2R_4}{R_1R_3}\left[\frac{sR_1C_1 + sR_2C_2}{sR_2C_2} + \frac{1}{sR_2C_2} + sR_1C_1\right]$$

$$\frac{E_0(s)}{E_i(s)} = \left(\frac{R_4(R_1C_1 + R_2C_2)}{R_1R_3C_2}\right)\left[1 + \frac{1}{(R_1C_1 + R_2C_2)s} + \frac{C_1R_1R_2C_2}{C_1R_1 + C_2R_2}\right]$$

Design with the PID controller

PD controller could add damping to a system, but the steady state response is not affected. PI controller can improve the relative stability and the steady state error at the same time, but the rise time is increased.

1. PID controller consists of a PI portion connected in cascade with a PD portion. The transfer function of the PID controller is

$$G_c(s) = K_p + K_DS + \frac{K_I}{s}$$

$$G_c(s) = (1 + K_{D1}S)\left(K_{P2} + \frac{K_{I2}}{s}\right) \quad ...(9.28)$$

The proportional constant of the PD portion is set to unity, since we need only 3 parameters in the PID controller. Equating both sides of Eq. (9.28) we have,

$$K_p = K_{P2} + K_{D1}\,K_{I2}$$

$$K_D = K_{D1} + K_{P2}$$

$$K_I = K_{I2}$$

2. Consider that the PD portion is in effect only. Select the value of K_{D1} so that a portion of the desired relative stability is achieved. In the time domain, this relative stability may be the maximum overshoot, and in the frequency domain it is phase margin.
3. Select the parameters K_{I2} and K_{P2} so that the total requirement on relative stability is satisfied.

Example 9.7: Figure 9.44 shows a PD controller. Find T_D so that the system is critically damped. Also calculate its settling time.

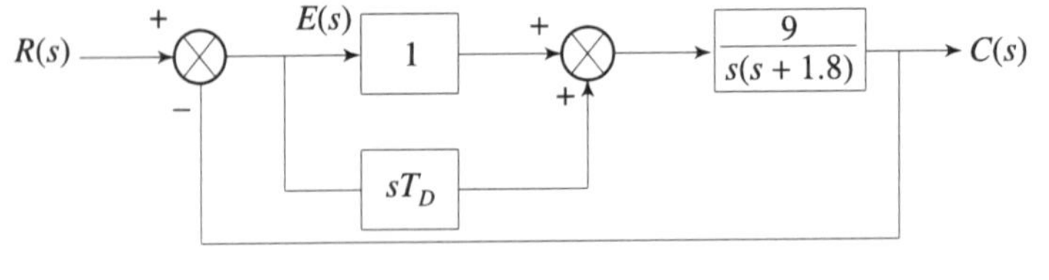

Fig. 9.44

Solution:

$$G(s) = \frac{9(1 + sT_D)}{s(s + 1.8)}$$

$$H(s) = 1$$

$$\frac{C(s)}{R(s)} = \frac{G(s)}{1 + G(s)H(s)} = \frac{\frac{9(1 + sT_D)}{s(s + 1.8)}}{1 + \frac{9(1 + sT_D)}{s(s + 1.8)}}$$

$$= \frac{9(1 + sT_D)}{s^2 + (9T_D + 1.8)s + 9}$$

$$\omega_n^2 = 9,\ \omega_n = 3 \text{ rad/sec}$$

For critically damped system, $\xi = 1$

$$2\xi\omega_n = 9T_D + 1.8$$

$$\xi = \frac{9T_D + 1.8}{2\omega_n}$$

$$\frac{9T_D + 1.8}{2\omega_n} = 1$$

$$9T_D + 1.8 = 2\omega_n$$

$$T_D = 0.467,\ \text{settling time} = \frac{4}{\xi\omega_n} = \frac{4}{1 \times 3} = 1.35 \text{ sec.}$$

Example 9.8: Refer to Fig. 9.45, calculate the following

(i) ξ and ω_n without K_D

(ii) K_D for $\xi = 0.60$ with controller

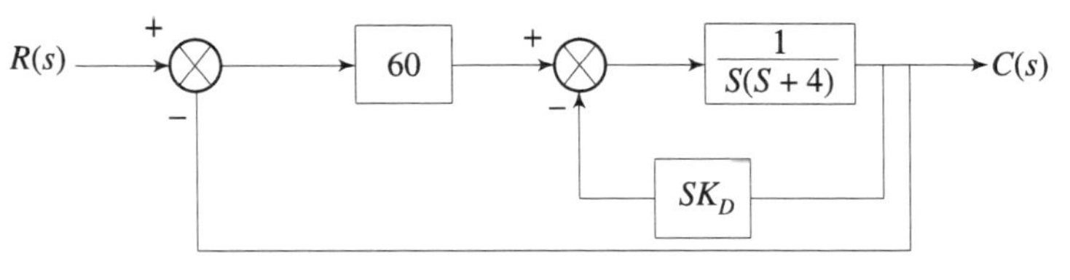

Fig. 9.45

Solution: (i) Without controller

$$G(s) = \frac{60}{S(S + 4)} \text{ and } H(s) = 1$$

$$\frac{C(s)}{R(s)} = \frac{G(s)}{1 + G(s)H(s)} = \frac{\frac{6.0}{S(S + 4)}}{1 + \frac{60}{S(S + 4)}}$$

$$\frac{C(s)}{R(s)} = \frac{60}{S^2 + 4S + 60} = \frac{\omega_n^2}{S^2 + 2\xi\omega_n + \omega_n^2}$$

$$\omega_n^2 = 60$$

$$\omega_n = 7.746 \text{ rad/sec}$$

$$2\xi\omega_n = -4$$

$$\xi = \frac{4}{2\omega_n} = \frac{4}{2 \times 7.746} = 0.258$$

(ii) With controller

$$G(s) = \frac{G(s)}{1 + G(s)\,H(s)} = \frac{\dfrac{60}{S(S+4)}}{1 + \dfrac{K_D S}{S(S+4)}}$$

$$\frac{C(s)}{R(s)} = \frac{G(s)}{1 + G(s)H(s)} = \frac{60}{S^2 + (4 + K_D)S + 60}$$

$$\omega_n^2 = 60$$

$$\omega_n = 7.746 \text{ rad/sec}$$

$$2\xi\omega_n = 4 + K_D$$

$$K_D = 2\xi\,\omega_n - 4 = 2 \times 0.6 \times 7.746 - 4$$

$$K_D = 5.2952$$

Example 9.9: For Fig. 9.46, determine ξ for a unit step input.

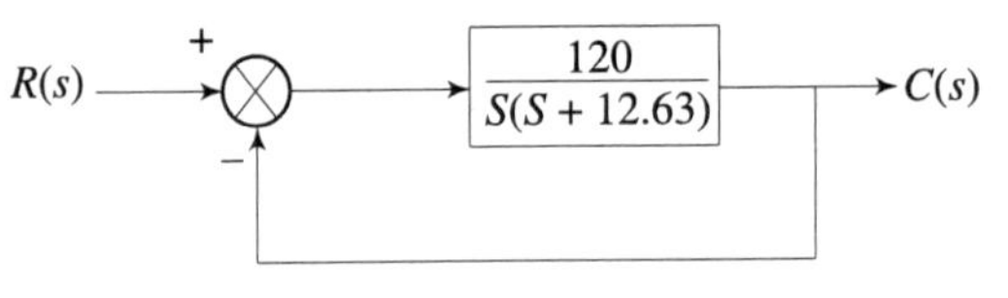

Fig. 9.46

(i) Also determine ξ, % M_p and t_s if a PD controller $K_D = \frac{1}{40}$ is introduced as shown in Fig. 9.47.

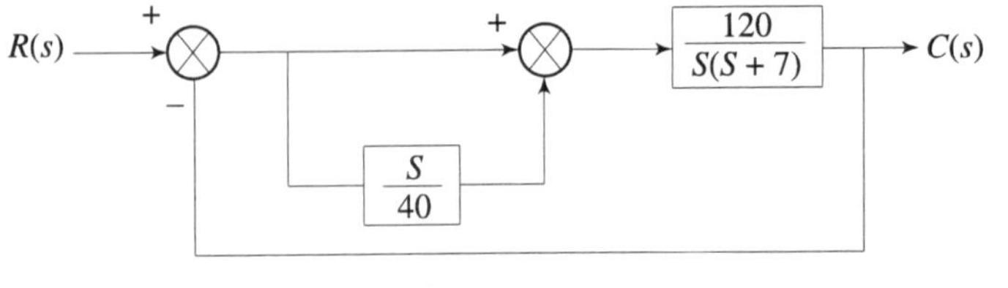

Fig. 9.47

Solution:

(i) Without controller

$$\frac{C(s)}{R(s)} = \frac{120}{S^2 + 12.63S + 120}$$

$$\omega_n^2 = 120$$

$$\omega_n = \sqrt{120} = 10.954 \text{ rad/sec}$$

$$2\xi\omega_n = 12.63$$

$$\xi = \frac{12.63}{2\omega_n} = \frac{12.63}{2 \times 10.95} = 0.5765$$

(ii) With controller

$$G(s) = \frac{120}{S(S + 12.63)}\left[1 + \frac{S}{30}\right] = \frac{4(S + 30)}{S^2 + 12.63S}$$

$$\frac{C(s)}{R(s)} = \frac{\dfrac{4(S + 30)}{S(S + 12.63)}}{1 + \dfrac{4(S + 30)}{S(S + 12.63)}} = \frac{4(S + 30)}{S^2 + 16.63S + 120}$$

$$\omega_n^2 = 120, \ \omega_n = \sqrt{120} = 10.954 \text{ rad/sec}$$

$$2\xi\omega_n = 16.63$$

$$\xi = \frac{16.63}{2\omega_n} = \frac{16.63}{2 \times 10.954} = 0.759$$

$$\% \ M_p = e^{\frac{-\xi\pi}{\sqrt{1-\xi^2}}} \times 100 = 2.57\%$$

$$t_s = \frac{4}{\xi\omega_n} = \frac{4}{0.759 \times 10.954} = 0.481 \text{ sec.}$$

EXERCISE

Objective Type Questions

1. The transfer function of a phase lead controller is $\frac{1 + 3T_S}{1 + T_S}$. The maximum value of phase provided by this controller is __________. (GATE 1998)

Ans: 30°

2. A PD controller is used to compensate a system. Compared to the uncompensated system, the compensated system has __________.

Ans: higher noise amplification

3. The transfer function of a phase lead compensator is given by $\frac{1 + 3T_S}{1 + T_S}$ where $T > 0$. The maximum phase shift provided by such a compensator is __________.

Ans: $\frac{\pi}{6}$

4. The transfer function of a first order controller is given as $G_c(s) = \frac{K(s + c)}{s + b}$ where K, a and b are positive real numbers. The condition for this controller to act as a phase lead compensator is __________.

Ans: $a < b$

5. The transfer function of a simple RC network functioning as a controller is $G_c(s) = \frac{s + z_1}{s + p_1}$. The condition for the RC network to act as a phase lead controller is __________.

Ans: $p_1 > z_1$

6. The open-loop transfer function of a plant is given as $G(s) = \frac{1}{s^2 - 1}$. If the plant is operated in a unity feedback configuration, then the lead compensator that can stabilize this control system is __________.

Ans: $\frac{2(s + 2)}{s + 10}$

7. A unity feedback closed-loop system has a plant with the transfer function $G(s) = \frac{1}{s^2 + 2s + 2}$ and a controller $G_c(s)$ in the feed forward path. For a unit step input, the transfer function of the controller that gives minimum steady state error is __________.

Ans: $G_c(s) = 1 + \frac{2}{s} + 3s$

8. A lead compensator network includes a parallel combination of R and C in the feed forward path. If the transfer function of the compensator is $G_c(s) = \frac{s + 2}{s + 4}$, the value of RC is __________.
(GATE 2015)

Ans: 0.5

9. A unity negative feedback system has an open-loop transfer function $G(s) = \frac{K}{s(s + 10)}$. The gain K for the system to have a damping ratio of 0.25 is __________.
(GATE 2015)

Ans: 400

10. To obtain high accuracy requirement we use __________.

Ans: integral error compensation

11. The term reset control refers to __________.

Ans: integral control

12. The input-output relationship is shown in Fig. 9.48. The network is an

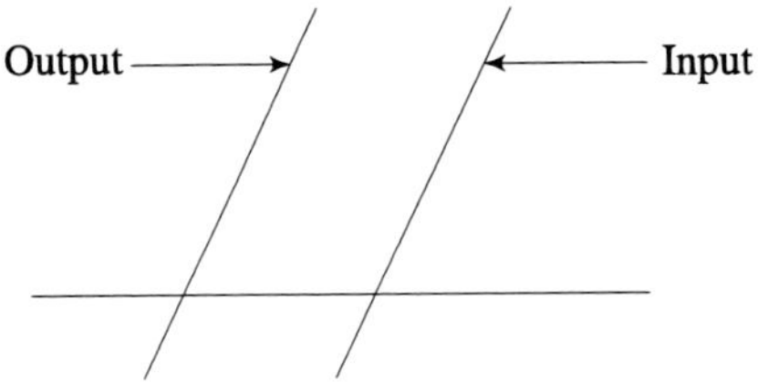

Fig. 9.48

Ans: PD controller

13. Differentiators are not used is a system due to __________.

Ans: large noise and saturation in the amplifier.

14. To eliminate the effect of disturbance $D(s)$ on $C(s)$, the transfer function $Ga(s)$ should be __________.

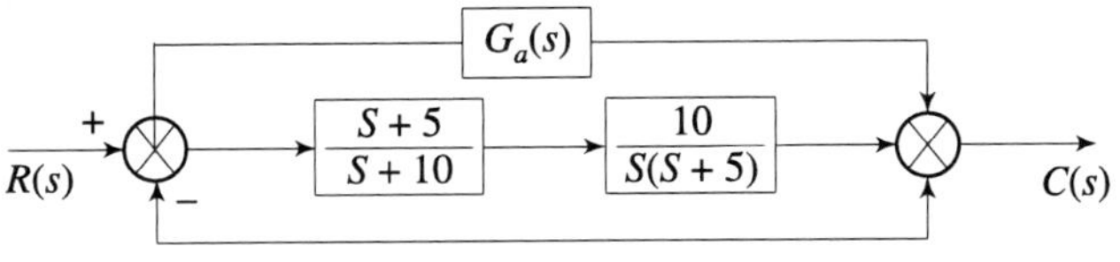

Fig. 9.49

Ans: $\frac{S(S-10)}{10}$

15. What do you mean by compensator?

Ans: The process of redesigning a system by adding an external device is called compensation of a control system. The external device is known as a compensator.

16. The circuit shown in Fig. 9.50 is __________.

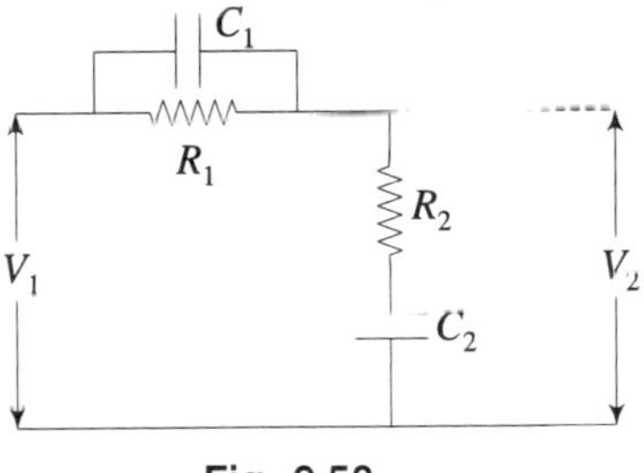

Fig. 9.50

Ans: lag-lead network

17. The circuits shown in Fig. 9.51 are—

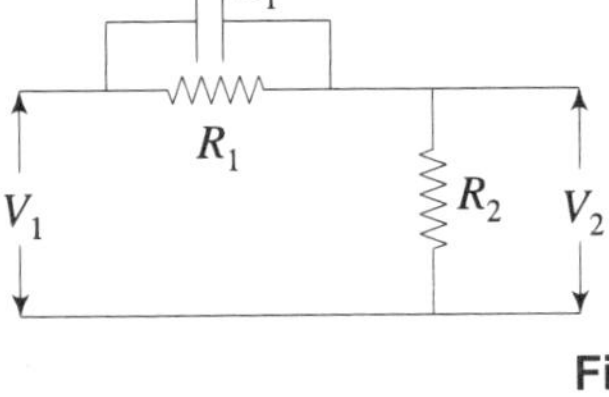

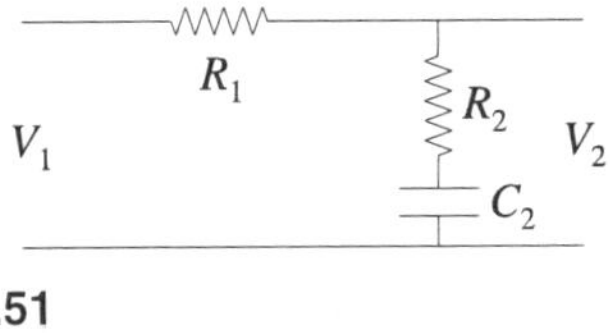

Fig. 9.51

Ans: lead network and lag network

18. Phase lag network __________ system's stability.

Ans: Increases

19. In a lag network, the inductor is used due to its __________.

Ans: time delay and hysteresis loss.

20. To increase the bandwith of a control system, we may use __________.

Ans: phase lead compensator

21. Lead-lag compensation is required for __________.

Ans: steady-state response improvement

22. The transfer function of a lead compensator $G_c(s) = \dfrac{1 + 0.12S}{1 + 0.04S}$ is __________.

Ans: 30°

23. Match the following

(i) Constant M loci	(a) Constant gain and phase shift loci of the closed-loop system
(ii) Constant N loci	(b) Plot of loop gain with variation of ω
(iii) Nichol's chart	(c) Circles of constant gain to closed-loop transfer function
(iv) Nyquist plot	(d) Circle of constant phase shift of a closed-transfer function

Ans: (i) → (c), (ii) → (d), (iii) → (a), (iv) → (b)

24. If zeros are added in a transfer function, it will cause __________.

Ans: lag compensation

25. What is the effect on a system's performance when a proportional controller is introduced to the system?

Ans: The proportional controller improves the steady state tracking accuracy, disturbance signal rejection and relative stability of the system. It also increases the loop gain of the system which results in reducing the sensitivity of the system to parameters variations.

26. What is the disadvantage in proportional controllers?

Ans: The disadvantage in proportional controlles is that it produces a constant steady state error.

27. What is the effect of PI controlles on the system's performance?

Ans: The PI controlles increase the order of the system by one, which results in reducing the steady state error but the system becomes less stable than the original system.

28. What is the effect of PD controlles on the system's performance?

Ans: The effect of PD controller is to increase the damping ratio of the system and so the peak overshoot is reduced.

29. Why a derivative controller is not used in control systems?

Ans: The derivative controlles produce a control action based on rate of change of error signal and it does not produce corrective measures to any constant error. Hence, derivative controller is not used in control systems.

30. What is the effect of addition of a pole to the forward path transfer function?
31. What is the effect of addition of a pole to the closed-loop transfer function?
32. What is the effect of addition of a zero to the closed-loop transfer function?
33. What is the effect of addition of a zero to the forward path transfer function?

34. What is meant by proportional control? What is the effect of a proportional controllers on the performance of a system? State the disadvantages of a proportional controller.
35. What is meant by PD control? State the effect of a PD controlles on the system's performance
36. What do you mean by PI control? State the effect of a PI controller on the system's performance
37. What is meant by PID control? What are the advantages of PID controller.
38. What is compensation? What are the different types of compensation.
39. What is a phase lead compensator? Derive expression for the transfer function of a phase lead compensator.
40. What are the effects of phase lead compensation. State the limitations of single stage phase lead control.
41. Write the procedure for the design of a phase lead compensator.
42. What is lag compensator? Obtain the transfer function of a lag compensator and draw the pole zero plot.
43. Explain the different steps followed for the design of a lag compensator using the Bode plot.
44. Draw the Bode plot for a phase lead network and derive an expression for the fraction α in terms of ϕ_m.
45. What are the effects and limitations of phase lag control.
46. Derive expression for the transfer function of a lag lead compensator.
47. Write the procedure for the design of a lag lead compensator.
48. A PD controller used for a system. Determine the value of T_d so that $\xi = 0.85$. Find its settling time.

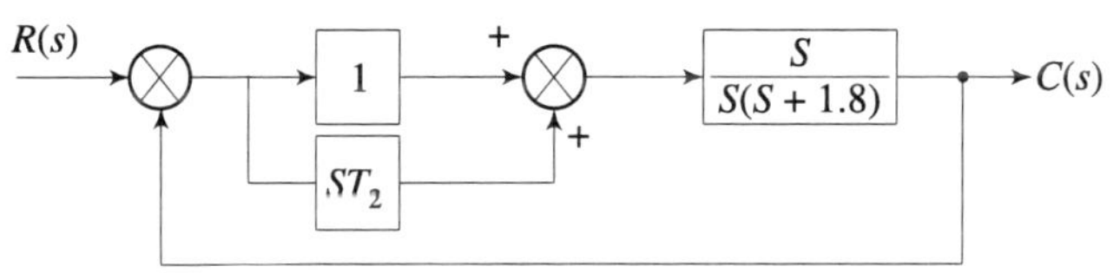

Fig. 9.52

Ans: 0.4, 1.789 sec

49 Consider a unity feedback controller

(i) In the absence of a derivative controller ($K_a = 0$), find ξ and w_n.

(ii) Also fluid K_o if $\xi = 0.56$ when controller is used.

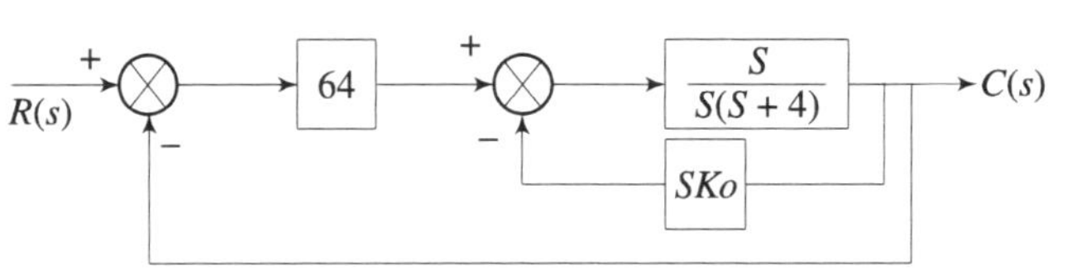

Fig. 9.53

Ans: (i) 0.25, B rad/sec (ii) 4.96

10 State Variable Analysis and Design

The conventional control theory is applicable to linear time invariant systems having single input and single output. The conventional control theory cannot be applied for time varying and/or nonlinear systems.

In the conventional control theory the main theme is to formulate the transfer function putting all initial conditions to zero.

The main purpose of state variable analysis in modern control theory automatically takes care of initial conditions and it is possible to analyze time varying or time invariant, linear or nonlinear, single or multiple input-output systems.

10.1 INTRODUCTION TO STATE VARIABLE ANALYSIS

State space analysis is very useful technique of analysing control systems. It is based on the concept of state and is applicable to linear time varying, nonlinear and MIMO systems. It employs the use of vector matrix notation. Representation of higher order system thus becomes simple. Before we proceed further, we must define state, state variable, state vector and state space.

Basic Concepts

- *State:* The state of control system at time $t = t_0$ is the smallest set of variables called state variables, such that the knowledge of these variables at time $t = t_o$ together with input for $t \geq t_o$ completely determines the behaviour of the system for any time t. The state of a dynamic system is uniquely found out by the state of the dynamic system at time $t = t_o$ and the input for $t \geq t_o$.
- *State variables:* The state variables of a dynamic system are the minimum or the smallest set of variables which describe the dynamics of a linear system. Let $x_1(t)$, $x_2(t)$ .... $x_n(t)$ be the state variable chosen to describe the dynamics of a linear system. $x_1(t_o)$, $x_2(t_o)$ .... $x_n(t_o)$ define the initial state of the system at time $t = t_o$. Once the inputs for time $t \geq t_o$ are specified and the initial state is known set of variables x_1, x_2 .... x_n have characteristics that the future behaviour of the system can be determined.
- *State vector:* The n state variables that completely desire the behaviour of a given system are said to be n components of a vector.

- *State space:* The n-dimensional space whose coordinate axis consists of the x_1 axis, x_2 axis ..... x_n axis is known as a state space. Any state can be uniquely represented by a point in the state space.

10.2 STATE MODEL

Figure 10.1 shows an nth order system having MIMO.

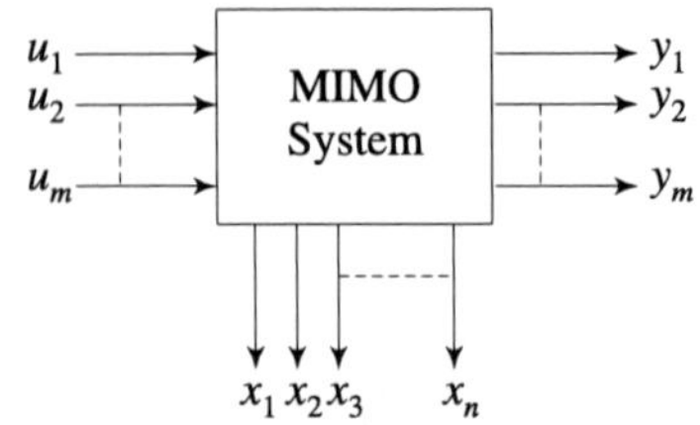

Fig. 10.1 State model

Here,

$$u(t) = \begin{bmatrix} u_1(t) \\ u_2(t) \\ M \\ u_m(t) \end{bmatrix}_{m\times 1}, \quad x(t)\begin{bmatrix} x_1(t) \\ x_2(t) \\ M \\ x_n(t) \end{bmatrix}_{n\times 1} \quad \text{and } y(t) = \begin{bmatrix} y_1(t) \\ y_2(t) \\ M \\ y_p(t) \end{bmatrix}_{P\times 1}$$

The state variable representation of such a system in the form of n first order differential equation is given below,

$$\dot{x}_1(t) = f_1\,(x_1, x_2, \dots \dots \dots x_n, u_1, u_2 \dots \dots \dots u_m)$$

$$\dot{x}_2(t) = f_2\,(x_1, x_2, \dots \dots \dots x_n, u_1, u_2 \dots \dots \dots u_m)$$

$$\vdots \qquad \vdots$$

$$\vdots \qquad \vdots$$

$$\dot{x}_n(t) = f_n\,(x_1, x_2, \dots \dots \dots x_n, u_1, u_2 \dots \dots \dots u_m)$$

where,

$$f(t) = \begin{bmatrix} f_1 \\ f_2 \\ M \\ f_n \end{bmatrix}_{n\times 1}$$

On integration, we can write

$$x_i(t) = x_i(t_o) + \int_{t_o}^{t} f_i\,(x_1, x_2, \dots \dots \dots x_n, u_1, u_2 \dots \dots \dots u_m)\,dt$$

where $i = 1, 2, \dots n$

$\therefore$ n state variables and hence state vectors at any time t can be determined uniquely. The functional form of any n-dimensional time invariant system is given by

$$\dot{x}(t) = f(x, u)$$

The output of such a system is dependent on the state of the system and the instantaneous input.

The functional output equation can be written as

$$y(t) = g(x, u)$$

where g is the functional operator. For a time invariant system we can write.

$$\dot{x}(t) = f(x, u, t) \qquad ...(10.1)$$

$$y(t) = g(x, u, t) \qquad ...(10.2)$$

Equations (10.1) and (10.2) give the state and output equations respectively. For a time invariant system the functional equations can be written in the form as follows.

$$\dot{x}(t) = A\,x(t) + B\,u(t)$$

$$y(t) = C(t)\,x(t) + D(t)\,u(t)$$

where A, B, C and D are time dependent matrices.

Figure 10.2 shows the diagrammatic representation.

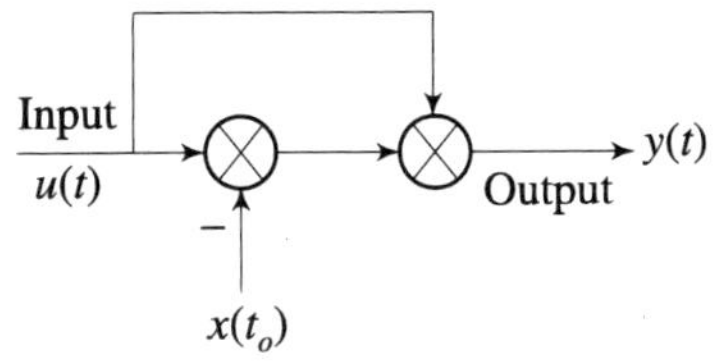

Fig. 10.2

The order of the above matrices is given below.

$A = n \times n$ known as evaluation matrix

$B = n \times m$ known as control matrix

$C = P \times n$ known as observation matrix

$D = P \times m$ known as direct transmission matrix

This is known as the state model of the given system.

Example 10.1: Obtain the state model of the parallel RLC network as shown in Fig. 10.3.

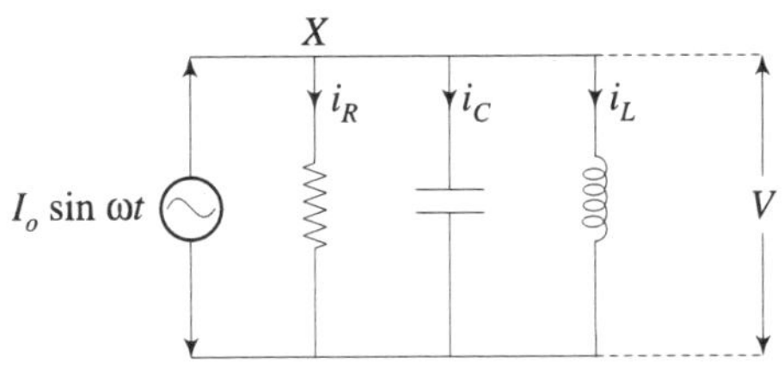

Fig. 10.3

Solution: Applying KCL at node X, we get

Let $$i(t) = i_R + i_L + i_C$$

$$i(t) = \frac{V(t)}{R} + \frac{1}{L}\int V(t)dt + C\frac{dV(t)}{dt}$$

$$\therefore \quad C\frac{d^2V(t)}{dt^2} + \frac{1}{R}\frac{dV(t)}{dt} + \frac{1}{L}\frac{di(t)}{dt} = I_m \cos\omega t$$

Let $$V(t) = x_1(t) \qquad \text{...(10.3)}$$

and $$x_1(t) = x_2(t) \qquad \text{...(10.4)}$$

$$x_2(t) = -\frac{1}{L}x_1(t) - \frac{1}{RC}x_2(t) + \frac{\omega I_m \cos\omega t}{C} \qquad \text{...(10.5)}$$

From Eqs. (10.4) and (10.5), the state equation can be written as

$$\begin{bmatrix} \dot{x}_1(t) \\ \dot{x}_2(t) \end{bmatrix} = \begin{bmatrix} 0 & 1 \\ -\frac{1}{LC} & -\frac{1}{RC} \end{bmatrix}\begin{bmatrix} x_1(t) \\ x_2(t) \end{bmatrix} + \begin{bmatrix} 0 \\ \frac{\omega I_m}{C}\cos\omega t \end{bmatrix}[u]$$

From Eq. (10.5), the output equation can be written as

$$v(t) = [1, \ 0]\begin{bmatrix} x_1(t) \\ x_2(t) \end{bmatrix}$$

10.2.1 State-Space Representation in Block Diagram Form

The state-space representation in block diagram form is shown in Fig. 10.4 below where broad arrows are used to represent vector quantities.

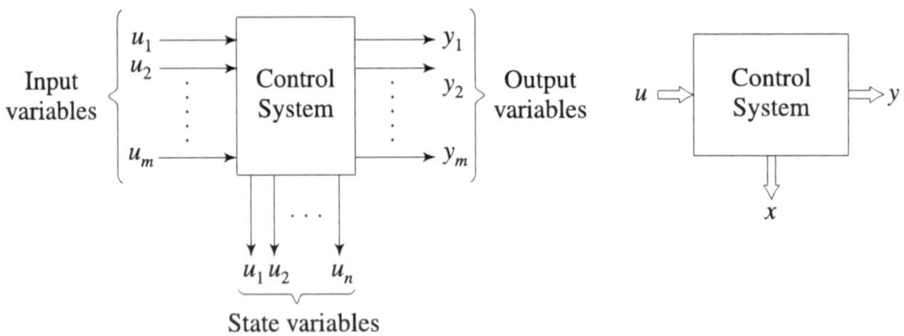

Fig. 10.4

10.3 STATE EQUATIONS

The system equations written in the form of first-order differential equation in time domain are known as state equations.

The number of state equations and state variables is equal to the order of differential equations required to model it.

The nth order time invariant differential equation with constant coefficients is represented by a set of n first-order differential equations in terms of state variables x_1, x_2 ..., x_n as follows.

$$\dot{x}_1 = a_{11}x_1 + a_{12}x_2 + ... + a_{1n}x_n + b_{11}u_1 + b_{12}u_2 + ... + b_{1m}u_m$$

$$\dot{x}_2 = a_{21}x_1 + a_{22}x_2 + ... + a_{2n}x_n + b_{21}u_1 + b_{22}u_2 + ... + b_{2m}u_m$$

$$\vdots$$

$$\dot{x}_n = a_{n1}x_1 + a_{n2}x_2 + ... + a_{nn}x_n + b_{n1}u_1 + b_{n2}u_2 + ... + b_{nm}u_m \quad ...(10.6)$$

where a_{ij}'s and b_{ij}'s are constants.

Equation (10.1) can be written in compact form as

$$\dot{x}(t) = Ax(t) + Bu(t)$$

where $$x(t) = \begin{bmatrix} x_1 \\ x_2 \\ M \\ x_n \end{bmatrix}$$

is $n \times 1$ column matrix known as state vector

$$u(t) = \begin{bmatrix} u_1 \\ u_2 \\ M \\ u_n \end{bmatrix}$$

is $m \times 1$ column matrix known as input vector

$$A = \begin{bmatrix} a_{11} & a_{12} & L & a_{1n} \\ a_2 & a_{22} & L & a_{2n} \\ M & & & M \\ a_m & a_{n2} & L & a_{nn} \end{bmatrix}$$

is $n \times n$ coefficient matrix with constant elements.

$$B = \begin{bmatrix} b_{11} & b_{12} & L & b_{1m} \\ b_2 & b_{22} & L & b_{2m} \\ M & & & M \\ b_{n1} & b_{n2} & L & b_{nn} \end{bmatrix}$$

is $n \times m$ coefficient matrix with constant elements

For a time varying system the coefficients of matrices A and B are not constants but are functions of time.

10.3.1 Output Equation

The output variables are denoted by $y_1(t), y_2(t), \ldots \ldots \ldots y_p(t)$. The output variables at time are linear combination's of input and state variables at time t.

$$y_1(t) = c_{11}x_1(t) + c_{12}x_2(t) + \ldots + c_{1n}x_n(t) + d_{11}u_1(t) + d_{12}u_2(t) + \ldots + d_{1m}u_m(t)$$
$$y_2(t) = c_{21}x_1(t) + c_{22}x_2(t) + \ldots + c_{2n}x_n(t) + d_{21}u_1(t) + d_{22}u_2(t) + \ldots + d_{2m}u_m(t)$$
$$\cdot$$
$$\cdot$$
$$\cdot$$
$$y_p(t) = c_{p1}x_1(t) + c_{p2}x_2(t) + \ldots + C_{pn}x_n(t) + d_{p1}u_1(t) + d_{p2}u_2(t) + \ldots + d_{pm}u_m(t) \quad \ldots(10.6a)$$

where the coefficients a_{ij}'s and b_{ij}'s are constants.

The set of Eq. (10.6a) written in matrix form is

$$y(t) = Cx(t) + Du(t)$$

where $y(t)$ is a $p \times 1$ output vector

C is a $p \times n$ output matrix

and D is a $p \times m$ transmission matrix.

They are defined by

$$y(t) = \begin{bmatrix} y_1(t) \\ y_2(t) \\ M \\ y_p(t) \end{bmatrix}$$

$$C = \begin{bmatrix} c_{11} & c_{12} & L & c_{1n} \\ c_{21} & c_{22} & L & c_{2n} \\ M & & & M \\ c_{pl} & c_{p2} & L & c_{pn} \end{bmatrix}$$

$$D = \begin{bmatrix} d_{11} & d_{12} & L & d_{1m} \\ d_{2} & d_{22} & L & d_{2m} \\ M & & & M \\ d_{pl} & d_{p2} & L & d_{2m} \end{bmatrix}$$

The state model of linear time-invariant systems is thus given by the following equations:

$$\dot{x}(t) = Ax(t) + Bu(t) \qquad \text{State equations}$$

$$y(t) = Cx(t) + Du(t) \qquad \text{Output equations}$$

where, A is called the system matrix

B is the input coupling matrix

C is the output coupling matrix

D is the input-output coupling matrix

The term $Du(t)$ indicates a direct coupling of the input to the output. The direct coupling of the input to the output is rare in control systems.

Therefore, $D = 0$

The state model of linear time-invariant systems is therefore, given by the following equations:

$$\dot{x}(t) = Ax(t) + Bu(t) \quad \text{State equations}$$

$$y(t) = Cx(t) \quad \text{Output equations}$$

The state equations and output equations together are called dynamic equations of the system. They are also called the state model of the system.

10.4 STATE DIAGRAM REPRESENTATION

- The dynamics of a control system may be described by a set of first-order differential equations.
- The differential equations can be depicted diagrammatically using integrators, summers and constant multipliers in the form of block diagrams or signal flow graphs.
- The integrator is shown by a rectangular box or a triangle. The sign of integration ($\int$) is put inside the box or triangle or $\frac{1}{s}$ is written inside box or triable as shown in Fig. 10.5.

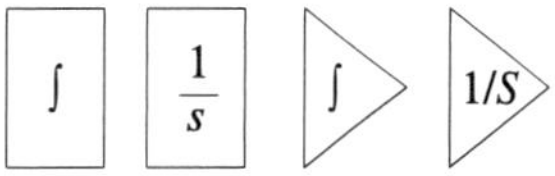

Fig. 10.5 Representation of integrator

10.5 STATE MODELS FOR LINEAR TIME-INVARIANT SYSTEM

The state model of a linear multi-input-multi-output (MIMO) system is given by the following equation.

$$\dot{x}(t) = Ax(t) + Bu(t) \quad \text{State equations}$$

$$y(t) = Cx(t) + Du(t) \quad \text{Output equations}$$

The block diagram representation of the state model is shown in Fig. 10.6.

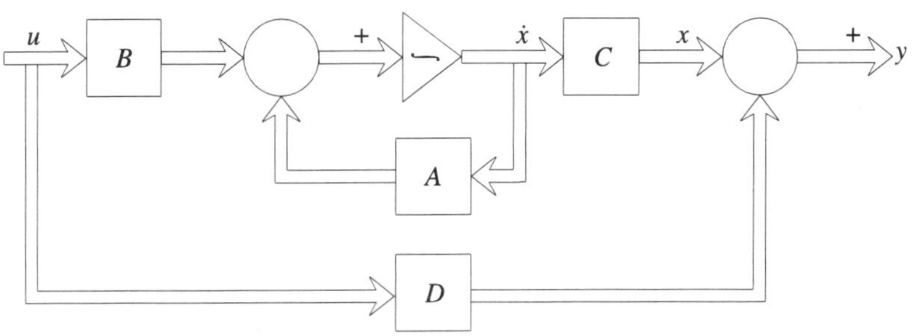

Fig. 10.6 Block diagram representation of linear multi-input-multi-output system

The state model for single-input-single-output (SISO) linear system is given by

$$\dot{x}(t) = Ax(t) + Bu(t) \quad \text{State equations}$$

$$y(t) = Cx(t) + Du(t) \quad \text{Output equations}$$

where, B is an $n \times 1$ matrix

C is a $1 \times n$ matrix

and D is a constant,

u is a scalar control variable.

The block diagram representation of this state model is shown in Fig. 10.7.

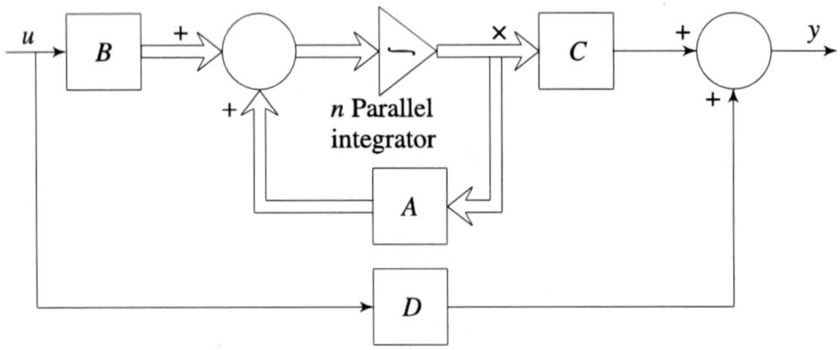

Fig. 10.7 Block diagram representation of linear single-input-single-output system

10.6 STATE-SPACE REPRESENTATION FROM TRANSFER FUNCTION

The stat-space representation provides a more complete description of the control system. The transfer function approach is used because some of the practical results of state-space approach can be most compactly and usefully presented in terms of transfer function. We shall determine the state-variable model when the transfer function is given. Since state-variable representation is not unique, there are theoretically an infinite number of ways of writing the state equations.

We shall present here some methods for deriving a set of state-variable representations from the transfer function.

10.6.1 Derivation of State Model from the Different Representations of a System

State model of linear time-invariant control system can be represented in

- Physical variable form
- Phase variable form
- Canonical variable form

State-space representation in canonical form is classified as follows:

(a) Controllable canonical form

(b) Observable canonical form

(c) Diagonal canonical form
(d) Jordan canonical form

10.6.2 State-Space Representation Using Physical Variables

The general procedure to determine the state-space models of a control system is as follows:

1. Describe the control system by formulating the equations that describe the dynamics of the control system.
2. Choose the state variables. One method is to choose physical variables as state variables that are associated with energy. The number of state variables may equal the number of energy storing elements in a control system.
   (a) In a mechanical system, potential energy and kinetic energy of a mass are functions of position and velocity, are chosen as state variables.
   (b) In an electric network, like RLC network, capacitors and inductors are energy storing elements. Therefore, the rate of change of current in an inductor and the rate of change of voltage across a capacitor can be chosen as state variables.
   (c) In chemical systems, rate of change of temperature, rate of change of pressure and rate of change of flow are usually chosen as state variables.
3. Convert the system equations into a set of first-order differential equations.
4. The state model of a linear time-invariant system is represented by the following equations.

$$\dot{x}(t) = Ax(t) + Bu(t) \quad \text{State equations}$$
$$y(t) = Cx(t) + Du(t) \quad \text{Output equations}$$

10.6.3 Procedure of Writing State Equations of RLC Networks

The following steps are involved in writing the state equations of simple RLC networks. The method is based on KCL, KVL and v-i relationships of the network elements.

Step 1: Choose all the capacitor voltages and inductor currents as the state variables.

Step 2: Write down the linearly independent KVL and KCL equations.

Step 3: Use the branch relationships to express all the network variables that are not the state variables in terms of the state variables chosen in step 1.

Step 4: Use the equations obtained in step 3 to eliminate all the variables that are not the chosen state variables in the equations obtained in step 2.

The resulting equations contain only the state variables and their derivatives and sources. Then transferring all the derivatives to the left-hand side, the resulting equations will be the desired equations in the standard form.

$$\dot{x}(t) = Ax(t) + Bu(t)$$

Example 10.2: Obtain the state model for the electric network shown in Fig. 10.8. Select i_L and v_C as state variables.

Solution: By KVL in the network of Fig. 10.8.

$$v_s = v_R + v_L + v_C$$

$$v_L = -v_R - v_C + v_S$$

$$L\frac{di_L}{dt} = -Ri_R - v_C + v_S$$

$$\frac{di_L}{dt} = -\frac{R}{L}i_R - \frac{1}{L}v_C + \frac{1}{L}v_s$$

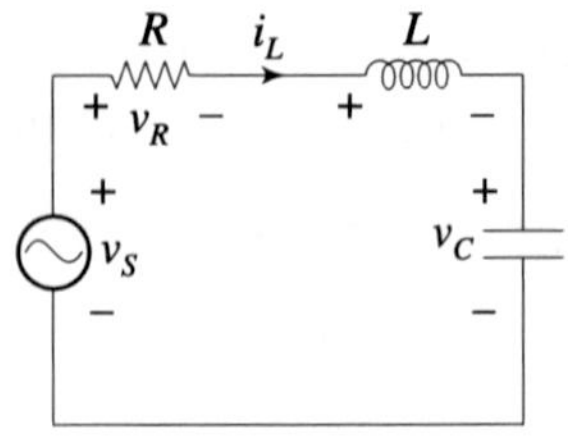

Fig. 10.8

Also $$C\frac{dv_c}{dt} = i_L \quad \text{...(10.7)}$$

$$\frac{dv_c}{dt} = \frac{1}{C}i_L \quad \text{...(10.8)}$$

Combining Eqs. (10.7) and (10.8), we obtain

$$\begin{bmatrix} \frac{di_L}{dt} \\ \frac{dv_C}{dt} \end{bmatrix} = \begin{bmatrix} -\frac{R}{L} & -\frac{1}{L} \\ \frac{1}{C} & 0 \end{bmatrix}\begin{bmatrix} i_L \\ v_C \end{bmatrix} + \begin{bmatrix} \frac{1}{L} \\ 0 \end{bmatrix}v_S \quad \text{...(10.9)}$$

Equation (10.9) is of the form

$$\frac{d}{dt}x(t) = Ax(t) + Bu(t).$$

Example 10.3: Obtain the state equations for the zero-input network as shown in Fig. 10.9.

Solution: The state variables are i_L and v_C.

Therefore, we shall express $\frac{di_L}{dt}$ and $\frac{dv_C}{dt}$ in terms of i_L and v_C

From circuit of Fig. 10.9

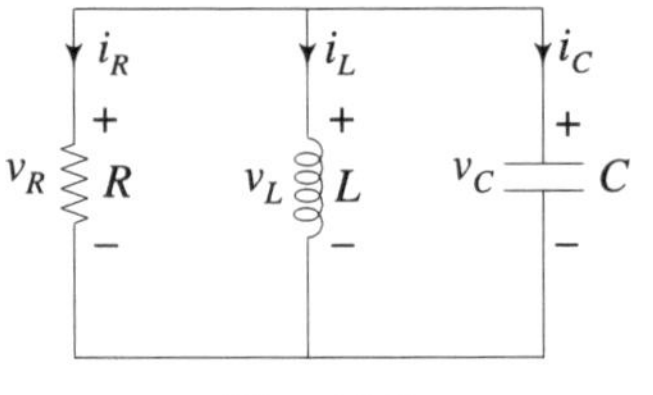

Fig. 10.9

$$v_L = v_C$$

$$L\frac{di_L}{dt} = v_C$$

$$\frac{di_L}{dt} = \frac{1}{L}v_C \quad \text{...(10.10)}$$

Now $$C\frac{dv_C}{dt} = i_C \quad \text{...(10.11)}$$

Since i_C is not a state variable, we shall eliminate it. By KCL

$$i_R + i_L + i_C = 0$$

Equation (10.11) now becomes

$$C\frac{dv_C}{dt} = -i_R - i_L \qquad ...(10.12)$$

In Eq. (10.12), v_C and i_L are state variables, but i_R is not a state variable. Therefore, i_R is to be expressed in terms of state variables.

By Ohm's law $\quad i_R = \frac{v_R}{R} = \frac{v_C}{R}$

Substituting the value of i_R in Eq. (10.12), we obtain

$$C\frac{dv_C}{dt} = -i_L - \frac{v_C}{R}$$

$$\frac{dv_c}{dt} = -\frac{1}{C}i_L - \frac{1}{RC}v_c \qquad ...(10.13)$$

Equation (10.13) is the second state equation. The two state Eqs. (10.10) and (10.11) can be combined into a single matrix state equation as

$$\begin{bmatrix} \frac{di_L}{dt} \\ \frac{dv_C}{dt} \end{bmatrix} = \begin{bmatrix} 0 & \frac{1}{L} \\ -\frac{1}{C} & -\frac{1}{RC} \end{bmatrix} \begin{bmatrix} i_L \\ v_C \end{bmatrix} \qquad ...(10.14)$$

Equation (10.14) is of the form

$$\frac{dx}{dt} = Ax(t).$$

Example 10.4: Write the state equation for the network shown in Fig. 10.10.

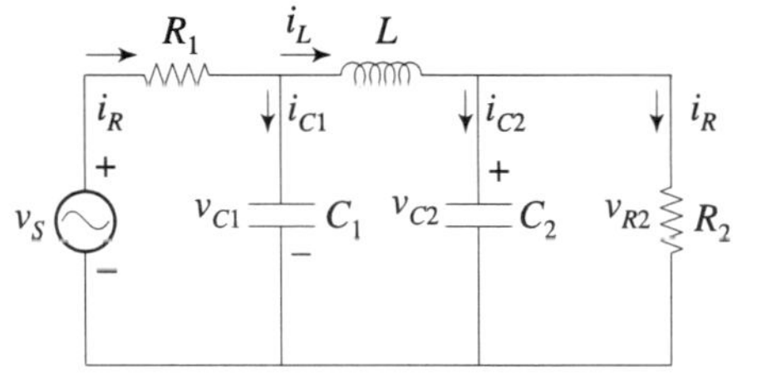

Fig. 10.10

Solution: The state variables will be i_L, v_{C1} and v_{C2}

By KVL in left-hand mesh

$$v_s - v_{R1} - v_{C1} = 0$$

But $\quad v_{R1} = R_1 i$ and $i = i_L + i_{C1}$

Therefore, $v_S - (i_L + i_{C1}) R_1 - v_{C1} = 0$

$$i_{C1} = \frac{1}{R_1}(v_S - v_{C1}) - i_L$$

$$C\frac{dv_{C1}}{dt} = \frac{1}{R_1}(v_S - v_{C1}) - i_L$$

$$\frac{dv_{C1}}{dt} = -\frac{1}{R_1C_1}v_{C1} - \frac{1}{C_1}i_L + \frac{1}{R_1C_1}v_S \qquad ...(10.15)$$

By KVL in the right-hand mesh

$$v_{C2} - v_{R2} = 0$$

But $\quad v_{R2} = R_2 i_{R2}$ and $i_L = i_{C2} + i_{R2}$

Therefore, $\quad v_{C2} = (i_L - i_{C2})R_2$

$$i_{C2} = -\frac{1}{R_2}v_{C2} + i_L$$

$$C_2\frac{dv_{C2}}{dt} = -\frac{1}{R_2}v_{C2} + i_L$$

$$\frac{dv_{C2}}{dt} = -\frac{1}{R_2C_2}v_{C2} + \frac{1}{C_2}i_L \qquad ...(10.16)$$

By KVL in the middle mesh

$$-v_L - v_{C2} + v_{C1} = 0$$

$$v_L = v_{C1} - v_{C2}$$

$$L\frac{di_L}{dt} = v_{C1} - v_{C2}$$

$$\frac{di_L}{dt} = \frac{1}{L}v_{C1} - \frac{1}{L}v_{C2} \qquad ...(10.17)$$

Equations (10.15), (10.16) and (10.17) can be written as

$$\frac{dv_{C1}}{dt} = -\frac{1}{R_1C_1}v_{C1} - 0\cdot v_{C2} - \frac{1}{C_1}i_L + \frac{1}{R_1C_1}v_S$$

$$\frac{dv_{C2}}{dt} = 0 \times v_{C1} - \frac{1}{R_2C_2}v_{C2} + \frac{1}{C_2}i_L + 0\cdot v_S$$

$$\frac{di_L}{dt} = \frac{1}{L}v_{C1} - \frac{1}{L}v_{C2} + 0\cdot i_L + 0\cdot v_S$$

These equations can be combined into a single matrix equation as

$$\begin{bmatrix} \frac{dv_{C1}}{dt} \\ \frac{dv_{C2}}{dt} \\ \frac{di_L}{dt} \end{bmatrix} = \begin{bmatrix} -\frac{1}{R_1C_1} & 0 & -\frac{1}{C_1} \\ 0 & -\frac{1}{R_2C_2} & \frac{1}{C_2} \\ \frac{1}{L} & -\frac{1}{L} & 0 \end{bmatrix} \begin{bmatrix} v_{C1} \\ v_{C2} \\ i_L \end{bmatrix} + \begin{bmatrix} \frac{1}{R_1C_1} \\ 0 \\ 0 \end{bmatrix} [v_S]$$

This equation is of the form

$$x = Ax + Bu$$

Example 10.5: Obtain a state model for the mechanical system shown in Fig. 10.11.

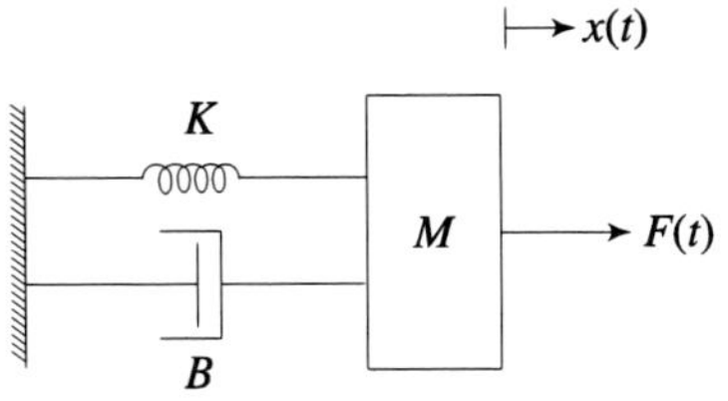

Fig. 10.11

Solution: The differential equation governing the behaviour of the given mechanical system is given by

$$\frac{M\,d^2y}{dt^2} + \frac{B\,dy}{dt} + Ky = F(t)$$

$$\frac{d^2y}{dt^2} = \frac{-B}{M}\frac{dy}{dt}\frac{-k}{M}y + \frac{1}{M}F(t)$$

The position and velocity can be chosen as the state variables of the system. Define,

$$x_1(t) = x(t)$$

$$x_2(t) = x(t)$$

Therefore, $$\frac{dx_1}{dt} = x_2$$

$$\frac{dx_2}{dt} = -\frac{K}{M}x_1 - \frac{B}{M}x_2 + \frac{1}{M}F(t)$$

The output equation is

$$y(t) = x_1(t)$$

The state model in matrix form is

$$\begin{bmatrix}\dot{x}_1\\ \dot{x}_2\end{bmatrix} = \begin{bmatrix}0 & 1\\ -\dfrac{K}{M} & -\dfrac{B}{M}\end{bmatrix}\begin{bmatrix}x_1\\ x_2\end{bmatrix} + \begin{bmatrix}0\\ \dfrac{1}{M}\end{bmatrix}F(t)$$

$$y = [1\ \ 0]\begin{bmatrix}x_1\\ x_2\end{bmatrix}.$$

Example 10.6: Obtain the state model for the mechanical system given in Fig. 10.12.

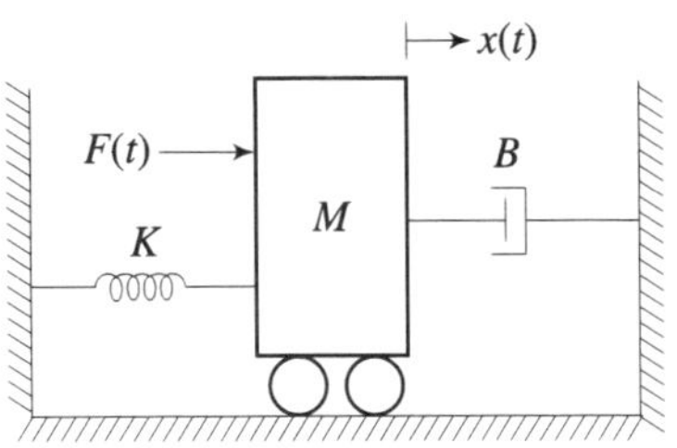

Fig. 10.12

Solution: The differential equation governing the behaviour of the given mechanical system is given by

$$M\frac{d^2x(t)}{dt^2} + B\frac{dx(t)}{dt} + Kx(t) = F(t) \quad \text{...(10.18)}$$

The position and velocity are chosen as state variables. Define

$$x_1(t) = x(t) \quad \text{...(10.19)}$$

$$x_2(t) = x(t) \quad \text{...(10.20)}$$

Thus, we have

$$\dot{x}_1(t) = \dot{x}(t) = x_2(t) \quad \text{...(10.21)}$$

$$\dot{x}_2(t) = \dot{x}(t) \quad \text{...(10.22)}$$

From Eq. (10.18), we get

$$\frac{d^2x(t)}{dt^2} = \frac{F(t)}{M} - \frac{B}{M}\frac{dx(t)}{dt} - \frac{K}{M}x(t)$$

or

$$\ddot{x} = \frac{1}{M}F(t) - \frac{B}{M}\dot{x}(t) - \frac{K}{M}x(t) \quad \text{...(10.23)}$$

Combination of Eqs. (10.19), (10.20), (10.21), (10.22) and (10.23) gives

$$\dot{x}_2(t) = \frac{1}{M}F(t) - \frac{B}{M}x_2(t) - \frac{K}{M}x_1(t)$$

Therefore, the state equations are

$$\dot{x}_1(t) = x_2(t)$$

or

$$\dot{x}_1(t) = 0 \times x_1(t) + x_2(t) + F(t) \times 0$$

$$\dot{x}_2(t) = -\frac{K}{M}x_1(t) - \frac{B}{M}x_2(t) - \frac{1}{M}F(t) \quad \text{...(10.24)}$$

These equations can be put in matrix form as

$$\begin{bmatrix}\dot{x}_1(t)\\ \dot{x}_2(t)\end{bmatrix} = \begin{bmatrix}0 & 1\\ -\frac{K}{M} & -\frac{B}{M}\end{bmatrix}\begin{bmatrix}x_1(t)\\ x_2(t)\end{bmatrix} + \begin{bmatrix}0\\ -\frac{1}{M}\end{bmatrix}F(t) \quad \text{...(10.25)}$$

We have chosen output $y = x_1$ as state variable, that is,

$$y = 1.x_1 + 0.x_2$$

The output in matrix form can be written as

$$y = [1 \ 0]\begin{bmatrix} x_1 \\ x_2 \end{bmatrix} \qquad \text{...(10.26)}$$

Equations (10.25) and (10.26) form the state model of the system.

10.7 DECOMPOSITION OF TRANSFER FUNCTION

We know that a linear system can be modelled by differential equations, transfer function, or dynamic equations. All these methods are closely related. The state diagram forms a close relationship among the equations, computer simulation, and transfer function.

The process of obtaining the state diagram of a system from its transfer is called decomposition.

In general there are three basic methods of decomposition of transfer functions. These are

- Direct decomposition
- Cascade decomposition
- Parallel decomposition

Each of these three schemes of decomposition has its own merits and is best suited for a particular purpose.

10.7.1 Direct Decomposition

Direct decomposition is applied to an input-output transfer function that is not in factored form. The advantage is that a general method and can be applied to any system. The resultant system matrix *A* is in the Bush form.

10.7.2 Cascade Decomposition

Cascade decomposition is applied to transfer functions that are written as products of simple first-order or second-order components. The state diagram is drawn for each section. The complete state diagram is obtained by combining in parallel the state diagram for each section.

The advantage of cascade decomposition is that the poles and zeros of the transfer function appear as isolated branch gains on the system diagram. Thus, the study of the effect on the system becomes easier when poles and zeros are varied.

10.7.3 Parallel Decomposition

Parallel decomposition method is based on partial-fraction approach. The following steps are used.

1. Expand the given transfer function by partial-fraction method. The expansion consists of first order terms.
2. Draw the state diagram of each factor separately.

3. The complete state diagram of the system is drawn by the parallel combination of the state diagram of each of the first-order terms obtained in the partial-fraction expansion.
4. Select the outputs of the integrators as the state variables and write the dynamic equations of the system.

Consider an nth-order system having the transfer function

$$G(s) = \frac{Y(s)}{U(s)} = \frac{b_0 s^m + b_1 s^{m-1} + \ldots + b_m}{(s + p_1)(s + p_2) \ldots + (s + p_n)} \qquad \ldots(10.26a)$$

where $p_1, p_2, \ldots, p_n$ are real and distinct poles.

Expanding Eq. (10.26a) into partial fractions, we get

$$\frac{Y(s)}{U(s)} = K_0 + \left[\frac{K_1}{s + p_1} + \frac{K_2}{s + p_2} + \ldots + \frac{K_n}{s + p_n}\right] \qquad \ldots(10.27)$$

where $K_1, K_2, \ldots, K_n$ are the residues of the transfer function at the corresponding poles $p_1, p_2, \ldots, p_n$ respectively.

The transfer function in the form of Eq. (10.27) shows that there are n first-order transfer functions in parallel and forward path with gain K_0. The state variable diagram for each term like $\frac{K_1}{s + p_1}$ is shown in Fig. 10.13.

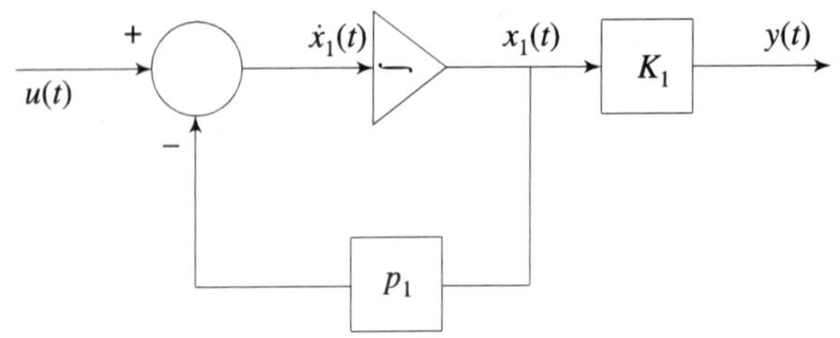

Fig. 10.13 State variable diagram for $\frac{K_1}{s + P_1}$

The state variable diagram of the system is drawn by connecting the individual state variable diagrams of the first-order terms in parallel as shown in Fig. 10.14.

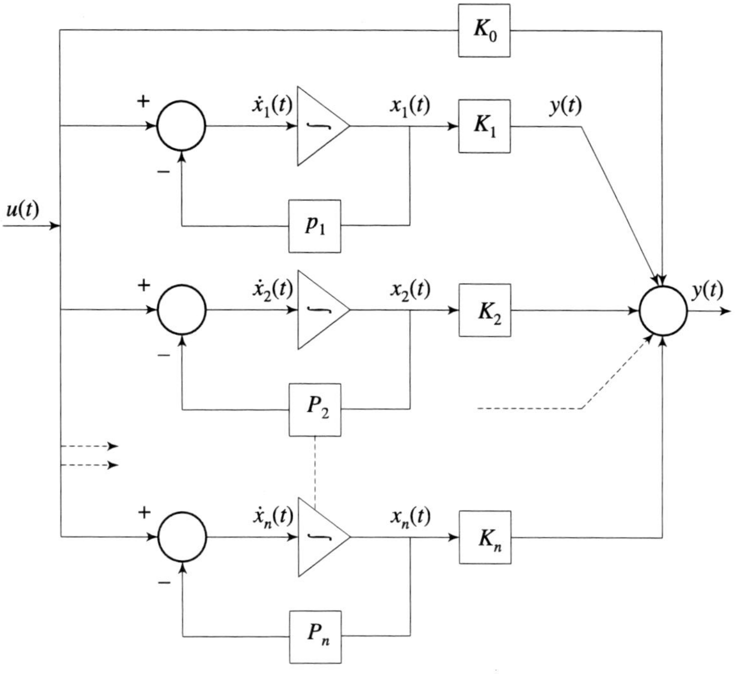

Fig. 10.14 State variable diagram by parallel decomposition

The signal flow diagram is shown in Fig. 10.15.

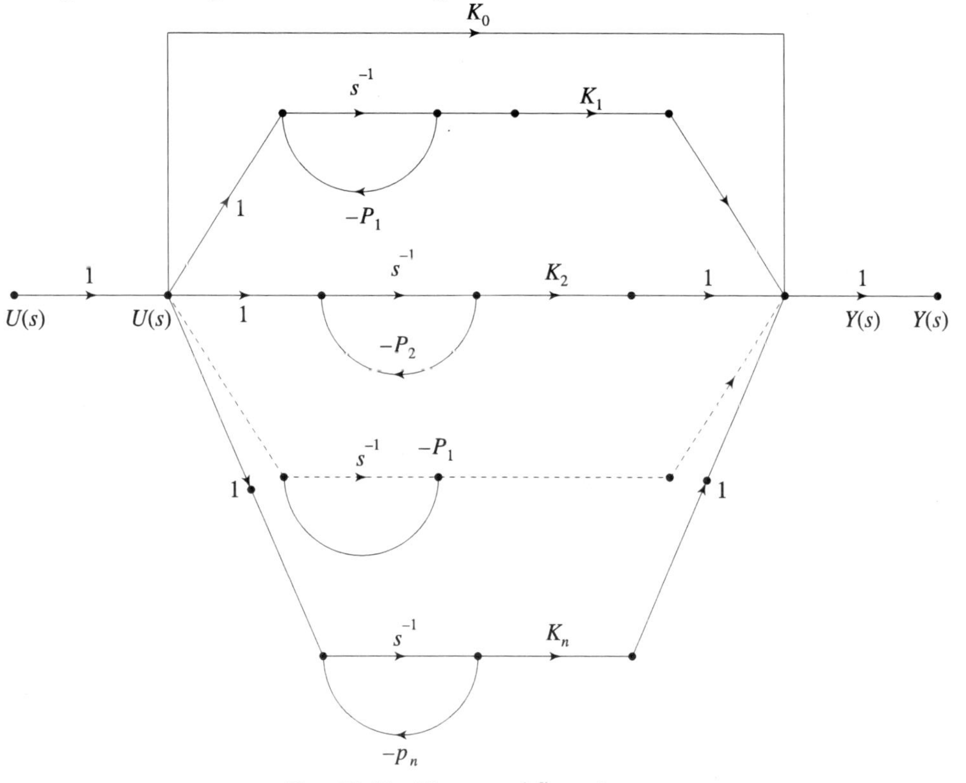

Fig. 10.15 The signal flow diagram

Selecting the outputs of the integrators as the state variable, the dynamic equations of the system in matrix form are as follows:

$$\begin{bmatrix} \dot{x}_1(t) \\ \dot{x}_2(t) \\ \vdots \\ \dot{x}_n(t) \end{bmatrix} = \begin{bmatrix} -p_1 & 0 & \cdots & 0 \\ 0 & -p_2 & \cdots & 0 \\ \vdots & \vdots & \cdots & \vdots \\ 0 & 0 & \cdots & -p_n \end{bmatrix} \begin{bmatrix} x_1(t) \\ x_2(t) \\ \vdots \\ x_n(t) \end{bmatrix} \begin{bmatrix} 1 \\ 1 \\ \vdots \\ 1 \end{bmatrix} u(t)$$

$$y(t) = [K_1 \;\; K_2 \;\; \ldots \;\; K_n] \begin{bmatrix} x_1(t) \\ x_2(t) \\ \vdots \\ x_n(t) \end{bmatrix} \qquad \ldots(10.28)$$

In compact form, the dynamic equations are

$$\dot{x}_1(t) = Ax(t) + Bu(t) \qquad \ldots(10.29)$$

$$y(t) = Cx(t) \qquad \ldots(10.30)$$

The matrix A is given by

$$A = \begin{bmatrix} -p_1 & 0 & \cdots & 0 \\ 0 & -p_2 & \cdots & 0 \\ \vdots & \vdots & \cdots & \vdots \\ 0 & 0 & \cdots & -p_n \end{bmatrix} \qquad \ldots(10.31)$$

The matrix A has the following features:

1. It is a diagonal matrix.
2. The main diagonal elements, $p_1, p_2, \ldots, p_n$, are the distinct roots (eigenvalues) of the matrix A.

This form of representation is called the **diagonal canonical form (DCF)**. It is also called the **normal form.**

As the diagonal form is most suitable for matrix algebra, this form is most important from control system analysis point of view. It has unique decoupled nature due to the fact that the n first-order differential equations are completely independent of each other. The decoupling feature is very useful in the analysis of the system.

It is to be noted that the pure diagonal form is possible only when the roots of the denominator are non-repeated roots.

10.7.4 Jordan Canonical Form (JCF)

If the transfer function has repeated poles, the matrix A will be in **Jordan canonical form (JCF).**

Suppose the transfer function has three poles at $s = -p_1$ and two poles at $s = -p_2$ and the remaining poles are distinct (non-repeated).

Then the partial-fraction expansion can be written as

$$\frac{Y(s)}{U(s)} = \left[\frac{K_1}{(s+p_1)^3} + \frac{K_2}{(s+p_1)^2} + \frac{K_3}{(s+p_1)}\right] + \left[\frac{K_4}{(s+p_2)^2} + \frac{K_5}{(s+p_2)}\right]$$

$$+ \frac{K_6}{(s+p_3)} + \ldots. + \frac{K_n}{(s+p_n)} \qquad \ldots(10.32)$$

The state equations in matrix form can be written by assigning output of each integrator as state variable.

$$\begin{bmatrix} \dot{x}_1 \\ \dot{x}_2 \\ \dot{x}_3 \\ \dot{x}_4 \\ \dot{x}_5 \\ \vdots \\ \dot{x}_n \end{bmatrix} = \begin{bmatrix} -p_1 & 1 & 0 & 0 & 0 & \cdots & 0 \\ 0 & -p_1 & 0 & 0 & 0 & \cdots & 0 \\ 0 & 0 & -p_1 & 0 & 0 & \cdots & 0 \\ 0 & 0 & 0 & -p_2 & 1 & \cdots & 0 \\ 0 & 0 & 0 & 0 & -p_2 & \cdots & 1 \\ \vdots & \vdots & \vdots & \vdots & \vdots & \cdots & \vdots \\ 0 & 0 & 0 & 0 & 0 & \cdots & -p_n \end{bmatrix} \begin{bmatrix} x_1 \\ x_2 \\ x_3 \\ x_4 \\ x_5 \\ \vdots \\ x_n \end{bmatrix} + \begin{bmatrix} K_1 \\ K_2 \\ K_3 \\ K_4 \\ K_5 \\ \vdots \\ K_n \end{bmatrix} u \qquad \ldots(10.33)$$

The state-space representation in the form given by Eq. (10.33) is said to be in the **Jordan canonical form (JCF).**

Figure 10.15 shows the signal flow diagram. The matrix A in Eq. (10.33) involves triple poles at $s = -p_1$, double poles at $s = -p_2$ and remaining poles are distinct (non-repeated).

In general, the matrix A in the JCF has the following properties:

1. The elements on the main diagonal are the eigenvalues.
2. All the elements below the main diagonal are zero.
3. The 1*s* together with the eigenvalues for typical blocks called the **Jordan blocks**. As shown in Eq. (10.33), the Jordan blocks are enclosed by dashed lines.
4. The number of Jordan blocks is equal to the number of independent eigenvectors.
5. In the Jordan block, all the diagonal elements are the same (value of the repeated pole), and the elements just above the main diagonal are all 1*s* and all other elements are 0*s*.

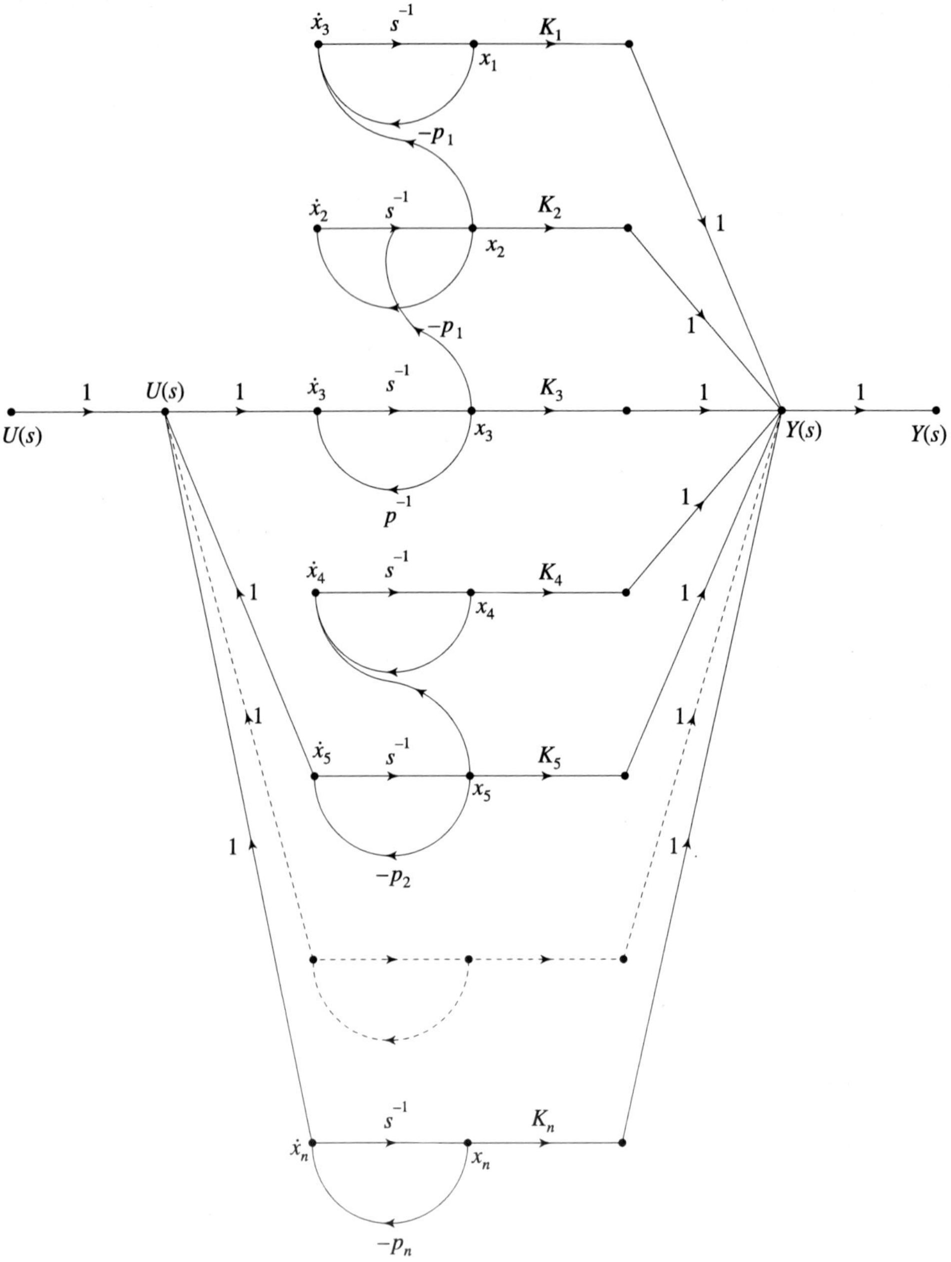

Fig. 10.16 State diagram by parallel decomposition (repeated poles present)

Example 10.7: Draw the state block diagram for the transfer function

$$\frac{Y(s)}{U(s)} = \frac{1}{s+2} \cdot \frac{1}{s+3}$$

and obtain the state model.

Solution: The given transfer function is in cascade form, that is, the system u is given to block $\frac{1}{s+2}$ and output of which is given as input to block $\frac{1}{s+3}$.

The output of block $\frac{1}{s+3}$ is the system output.

The state block diagram of the given transfer function is shown in Fig. 10.17.

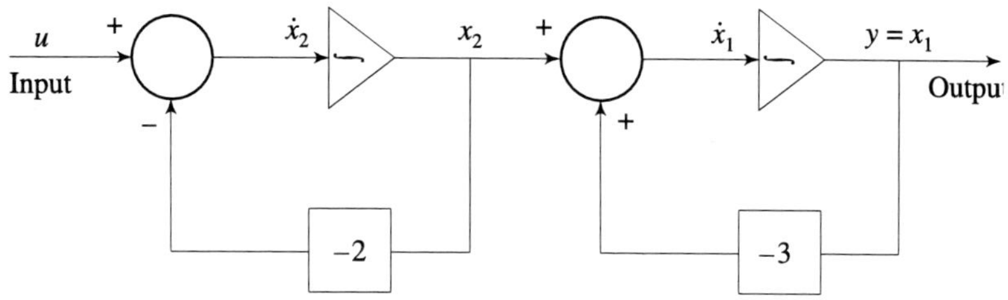

Fig. 10.17 State block diagram

The output of each integrator is assigned a state variable as shown in Fig. 10.17. The state equations are

$$\dot{x}_1 = -3x_1 + x_2$$
$$\dot{x}_2 = -2x_2 + u$$

The output equation is

$$y = x_1$$

Therefore, the state model can be written as

$$\begin{bmatrix} \dot{x}_1 \\ \dot{x}_2 \end{bmatrix} = \begin{bmatrix} -3 & 1 \\ 0 & -2 \end{bmatrix} \begin{bmatrix} x_1 \\ x_2 \end{bmatrix} + \begin{bmatrix} 0 \\ 1 \end{bmatrix} u$$

$$y = [1 \;\; 0] \begin{bmatrix} x_1 \\ x_2 \end{bmatrix}$$

Example 10.8: Draw the state block diagram for the transfer function

$$\frac{Y(s)}{U(s)} = \frac{s+5}{s^2+10s+9}$$

and obtain state equations.

Solution: The given transfer function is expanded by partial fraction expansions

$$\frac{Y(s)}{U(s)} = \frac{s+5}{s^2+10s+9}$$

$$\frac{Y(s)}{U(s)} = \frac{s+5}{(s+9)(s+1)}$$

$$\frac{Y(s)}{U(s)} = \frac{A}{s+9} + \frac{B}{s+1} = \frac{1/2}{s+9} + \frac{1/2}{s+1}$$

The transfer function expansion is in parallel form.

The input u is given simultaneously to block $\frac{1/2}{s+9}$ as well as to $\frac{1/2}{s+1}$. The outputs of these blocks are added to form system output. The block diagram of transfer function is shown in Fig. 10.18.

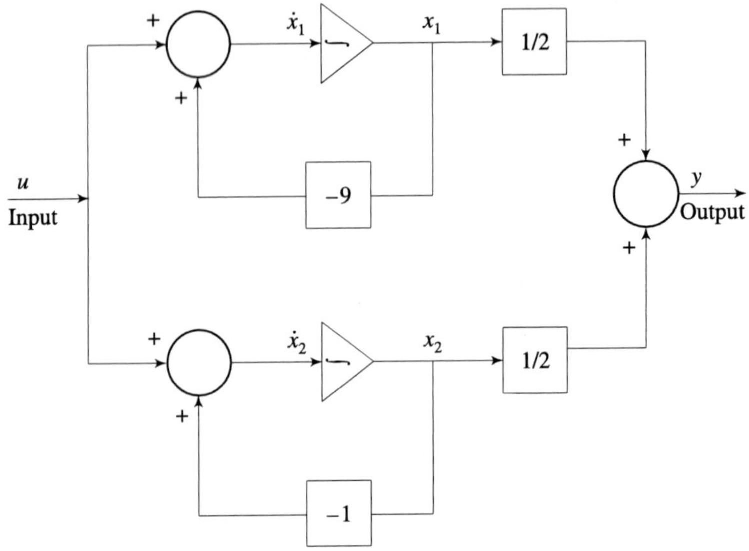

Fig. 10.18. State block diagram

The output of each integrator is assigned a state variable.

The state equations are given by

$$\dot{x}_1 = -9x + u \quad \text{and} \quad \dot{x}_2 = -x_2 + u$$

The output equation is given by

$$y = \frac{1}{2}x_1 + \frac{1}{2}x_2$$

Thus, the state model is given by

$$\begin{bmatrix} \dot{x}_1 \\ \dot{x}_2 \end{bmatrix} = \begin{bmatrix} -9 & 0 \\ 0 & -1 \end{bmatrix} \begin{bmatrix} x_1 \\ x_2 \end{bmatrix} + \begin{bmatrix} 1 \\ 1 \end{bmatrix} u$$

$$y = \begin{bmatrix} \frac{1}{2} & \frac{1}{2} \end{bmatrix} \begin{bmatrix} x_1 \\ x_2 \end{bmatrix}$$

10.8 DIAGONALIZATION

We have seen that the parallel form of realizing a system results in a diagonal system matrix A. A diagonal system matrix A has the advantage that each state equation is a function of only one state variable. Hence, each state equation can be solved independently. Therefore, the equations are said to be **decoupled**. This decoupling feature is very useful in the analysis of the system. It is therefore, desirable to make matrix **A** diagonal so as to have decoupled equations. If **A** is not diagonal, we can transform the state variables such that the resulting matrix **A** is diagonal. This technique of transforming a non-diagonal matrix to diagonal matrix is known as **diagonalization**.

10.8.1 Similarity Transformation

Consider a system defined by state model

$$\dot{x} = \mathbf{A}\mathbf{x} + \mathbf{B}\mathbf{u} \qquad ...(10.34)$$

$$\mathbf{y} = \mathbf{C}\mathbf{x} + \mathbf{D}\mathbf{u} \qquad ...(10.35)$$

We are interested in transforming these dynamic equations into another set of equations of the same dimensions.

Let us define new state vector **Z** related to the state vector x by an $n \times n$ non-singular matrix **P** such that

$$\mathbf{x} = \mathbf{P}\mathbf{z} \qquad ...(10.36)$$

Therefore, $$\dot{\mathbf{x}} = \mathbf{P}\mathbf{z} \qquad ...(10.37)$$

Substituting Eq. (10.37) in Eq. (10.34), we get

$$\mathbf{P}\dot{\mathbf{z}} = \mathbf{A}\mathbf{P}\mathbf{z} + \mathbf{B}\mathbf{u}$$

Therefore, $$\dot{\mathbf{z}} = \mathbf{P}^{-1}\mathbf{A}\mathbf{P}\mathbf{z} + \mathbf{P}^{-1}\mathbf{B}\mathbf{u} \qquad ...(10.38)$$

and $$\mathbf{y} = \mathbf{C}\mathbf{P}\mathbf{z} + \mathbf{D}\mathbf{u} \qquad ...(10.39)$$

we desire the transformation to be such that $\mathbf{P}^{-1}\mathbf{A}\mathbf{P}$ is a diagonal matrix. With this condition the matrix **P** is called **diagonalizing matrix**.

Equations (10.38) and (10.39) may be written as

$$\dot{\mathrm{z}} = \bar{A}z + \bar{B}u \qquad ...(10.40)$$

$$Y = \bar{C}z + \bar{D}u \qquad ...(10.41)$$

where $$\bar{A} = \mathbf{P}^{-1}\mathbf{A}\mathbf{P} \qquad ...(10.42)$$

$$\bar{B} = P^{-1}B \qquad ...(10.43)$$

$$\bar{C} = CP \qquad ...(10.44)$$

$$\bar{D} = D \qquad ...(10.45)$$

The matrix $\bar{\mathbf{A}}$ is a diagonal matrix,

$$\bar{A} = \begin{bmatrix} \lambda_1 & 0 & \mathrm{L} & 0 \\ 0 & \lambda_2 & \mathrm{L} & 0 \\ \mathrm{M} & & \mathrm{O} & \mathrm{M} \\ 0 & 0 & \mathrm{L} & \lambda_n \end{bmatrix} (n \times n) \qquad ...(10.46)$$

where $\lambda_1, \lambda_2, ..., \lambda_n$ are the n distinct eigenvalues of A.

The transformation given above is called **similarity transformation**, since in the transformed system, such properties as the characteristic equation, eigenvectors, eigenvalues and transfer function remain invariant by the similarity transformation.

10.8.2 Invariance of Transfer Function

It can be proved that transfer function matrix is invariant under a linear transformation.

We know that for the original state model, the transfer function matrix is

$$\mathbf{G(s)} = \mathbf{C[sI - A]^{-1}B + D}$$

For the transformed model, the transfer matrix is

$$\begin{aligned}\overline{G}(s) &= \overline{C}[\mathbf{sI - A}]^{-1}\,\overline{\mathbf{B}} + \overline{D}\\ &= \mathbf{CP}[s\mathbf{I} - \mathbf{P}^{-1}\mathbf{AP}]^{-1}\mathbf{P}^{-1}\mathbf{B} + \mathbf{D}\\ &= \mathbf{CPP}^{-1}[\mathbf{sI - A}]^{-1}\,\mathbf{PP}^{-1}\mathbf{PB} + \mathbf{D}\\ &= \mathbf{C[sI - A]^{-1} + D}\\ &= G(s)\end{aligned}$$

10.8.3 Invariance of Eigenvalues

It can also be proved that the characteristic equation, eigenvalues and eigenvectors are invariant under a linear transformation.

For the state equation of a dynamic system

$$\dot{x} = \mathbf{Ax + Bu}$$

The characteristic equation is

$$|\mathbf{sI - A}| = 0$$

Under a linear transformation $\mathbf{x = Pz}$, the state equations transformed to

$$\dot{\mathbf{z}} = \mathbf{P}^{-1}\mathbf{APz} + \mathbf{P}^{-1}\mathbf{Bu}$$

Therefore, the new characteristic equation of the system is

$$|\mathbf{sI} - \overline{A}| = 0$$

$$\begin{aligned}|sI - \overline{A}| &= |\mathbf{sI} - \mathbf{P}^{-1}\mathbf{AP}|\\ &= |\mathbf{sP}^{-1}\mathbf{P} - \mathbf{P}^{-1}\mathbf{AP}|\end{aligned}$$

We know that the determinant of a product matrix is equal to the product of the determinants of the matrices.

$$\begin{aligned}|\mathbf{sI} - \overline{A}| &= |\mathbf{P}^{-1}|\,|\mathbf{sI - A}|\,|\mathbf{P}|\\ &= |\mathbf{sI - A}|\mathbf{P}^{-1}|\,|\mathbf{P}|\\ &= |\mathbf{sI - A}|\end{aligned}$$

This shows that the characteristic equation is invariant under the linear transformation. Consequently, the eigenvalues and eigenvectors remain invariant by linear transformation.

10.9 DERIVATION OF TRANSFER FUNCTION FROM STATE MODEL

Transfer function matrix $G(s)$ relates the output vector $Y(s)$ to the input vector $U(s)$.

Let us consider the control system whose state equation and output equation are given as

$$\dot{x} = \mathbf{Ax(t)} + \mathbf{Bu(t)} \quad ...(10.47)$$

$$\mathbf{y(t)} = \mathbf{Cx(t)} + \mathbf{Du(t)} \quad ...(10.48)$$

The Laplace transform of state equation and output equation gives

$$s\mathbf{X}(s) - \mathbf{x}(0) = \mathbf{AX}(s) + \mathbf{BU}(s) \quad ...(10.49)$$

$$\mathbf{Y}(s) = \mathbf{CX}(s) + \mathbf{DU}(s) \quad ...(10.50)$$

or

$$|s\mathbf{I} - A|\, X(s) = x(0) + Bu(s)$$

$$\mathbf{X}(s) = |s\mathbf{I} - \mathbf{A}|^{-1} x(0) + (\mathbf{sI} - \mathbf{A})^{-1}\mathbf{BU}(s) \quad ...(10.51)$$

Substituting Eqs. (10.50) in (10.51), we have

$$\mathbf{Y}(s) = \mathbf{C}(s\mathbf{I} - \mathbf{A})^{-1} x(0) + \mathbf{C}(s\mathbf{I} - \mathbf{A})^{-1}\, \mathbf{BU}(s) + \mathbf{DU}(s) \quad ...(10.52)$$

Since the definition of a transfer function requires that the initial conditions be set to 0.

$$x(0) = 0$$

Thus, Eq. (10.52) becomes

$$\mathbf{Y}(s) = \{\mathbf{C}[s\mathbf{I} - \mathbf{A}]^{-1}\mathbf{B} + \mathbf{D}\}\mathbf{U}(s) \quad ...(10.53)$$

Therefore, the **transfer function matrix** $\boldsymbol{G(s)}$ for a multi-variable system is given by

$$\mathbf{G}(s) = \frac{\mathbf{Y}(s)}{\mathbf{U}(s)} = \mathbf{C}[s\mathbf{I} - \mathbf{A}]^{-1}\mathbf{B} + \mathbf{D} \quad ...(10.54)$$

Since

$$C[s\mathbf{I} - \mathbf{A}]^{-1} = \mathbf{C}\,\frac{\text{adj}[s\mathbf{I} - \mathbf{A}]}{\det[s\mathbf{I} - \mathbf{A}]} \quad ...(10.55)$$

$$\mathbf{G}(s) = \frac{\mathbf{Y}(s)}{\mathbf{U}(s)} = \mathbf{C}\,\frac{\text{adj}[s\mathbf{I} - \mathbf{A}]}{\det[s\mathbf{I} - \mathbf{A}]}\,\mathbf{B} \quad ...(10.56)$$

In the case of a single-input-single-output [SISO] system, $G(s)$ is scalar and is termed transfer function.

In general, there is no direct coupling between input and output. Therefore, $\mathbf{D} = 0$ and

$$\mathbf{G}(s) = \frac{\mathbf{Y}(s)}{\mathbf{U}(s)} = \mathbf{C}\,\frac{\text{adj}[s\mathbf{I} - \mathbf{A}]}{\det[s\mathbf{I} - \mathbf{A}]}\,\mathbf{B} \quad ...(10.57)$$

The matrix $[s\mathbf{I} - \mathbf{A}]^{-1}$ is commonly known as **resolvent matrix** and is denoted by $\mathbf{\Phi(s)}$.

Thus, $$\Phi(s) = [s\mathbf{I} - \mathbf{A}]^{-1} \quad ...(10.58)$$

10.10 CHARACTERISTIC EQUATIONS

Characteristic equations play an important role in the study of linear systems. They can be defined with respect to differential equations, transfer functions or state equations.

10.10.1 Characteristic Equation from a Differential Equation

Consider that a linear time-invariant system is described by the differential equation.

$$\frac{d^n y(t)}{dt^n} + a_{n-1}\frac{d^{n-1}y(t)}{dt^{n-1}} + \ldots + a_1\frac{y(t)}{dt} + a_0 y(t)$$

$$= b_m\frac{d^m u(t)}{dt^m} + b_{m-1}\frac{d^{m-1}u(t)}{dt^{m-1}} + \ldots + b_1\frac{du(t)}{dt} + b_0\, u(t) \quad ...(10.59)$$

where $n > m$

Let us define the operator s as

$$s^k = \frac{d^k}{dt^k}\; k = 1, 2,\ldots, n \quad ...(10.60)$$

Equation (10.59) can be written as $(s^n + a_{n-1}\, s^{n-1} + \ldots + a_1 s + a_0)\; y(t)$

$$= (b_m\, s^n + b_{m-1}\, s^{m-1} + \ldots + b_1(s) + b_0)\; u(t) \quad ...(10.61)$$

The **characteristic equation** of the system is defined as

$$s^n + a_{n-1}\, s^{n-1} + \ldots + a_1 s + a_0 = 0 \quad ...(10.62)$$

which is obtained by setting the homogeneous part of Eq. (10.61) to zero.

Example 10.9: Determine the characteristic equation of the differential equation.

$$\frac{d^3y(t)}{dt^3} + 5\frac{d^2y(t)}{dt^2} + \frac{dy(t)}{dt} + 2y(t) = u(t) \quad ...(10.63)$$

Solution: Defining the operator s as

$$s^k = \frac{d^k}{dt^k},\; k = 1, 2,\ldots, n$$

Equation (10.63) can be written as

$$(s^3 + 5s^2 + s + 2)y(t) = u(t) \quad ...(10.64)$$

The characteristic equation is obtained by setting the homogeneous part of Eq. (10.64) to zero

$$s^3 + 5s^2 + s + 2 = 0$$

Characteristic equation from a transfer function

Consider the transfer function

$$G(s) = \frac{b_m s^m + b_{m-1}s^{m-1} + \ldots + b_1 s + b_0}{s^n + a_{n-1}s^{n-1} + \ldots + a_1 s + a_0} \quad \ldots(10.65)$$

The characteristic equation is obtained by equating the denominator polynomial of the transfer function to zero.

Thus, the characteristic equation of the system described by Eq. (10.65) is

$$s^n + a_{n-1}\, s^{n-1} + \ldots. + a_1 s + a_0 = 0.$$

Example 10.10: The transfer function of a system is described by

$$\frac{Y(s)}{U(s)} = \frac{1}{s^3 + 5s^2 + s + 2}$$

What is the characteristic equation?

Solution: The characteristic equation is obtained by setting the denominator polynomial of the transfer function to zero.

Therefore, the characteristic equation is given by

$$s^3 + 5s^2 + s + 2 = 0$$

Characteristic Equation from State Function

The characteristic equation from the state equation is obtained from the relation

$$\det[sI - A] = 0$$

or $$|sI - A| = 0$$

Example 10.11: A system is represented by the following state and output equations

$$\dot{x} = \begin{bmatrix} -3 & -2 \\ -1 & -2 \end{bmatrix} x + \begin{bmatrix} 1 \\ 1 \end{bmatrix} u(t)$$

$$y = [1\ 2]\, x$$

Find the poles of the system,

Solution: The characteristic equation is given by

$$|\lambda I - A| = 0$$

$$\left| \lambda \begin{bmatrix} 1 & 0 \\ 0 & 1 \end{bmatrix} - \begin{bmatrix} -3 & -2 \\ -1 & -2 \end{bmatrix} \right| = 0$$

$$\left| \begin{bmatrix} \lambda + 3 & 2 \\ 1 & \lambda + 2 \end{bmatrix} \right| = 0$$

$$(\lambda + 3)(\lambda + 2) - 2 \times 1 = 0$$

$$\lambda^2 + 5\lambda + 6 - 2 = 0$$

$$\lambda^2 + 5\lambda + 4 = 0$$

$$(\lambda + 4)(\lambda + 1) = 0$$

$\lambda_1 = -1$, $\lambda_2 = -4$. Therefore, the poles are –1 and –4

Example 10.12: Obtain characteristic equation of the matrix

$$A = \begin{bmatrix} 0 & 1 & 0 \\ 3 & 0 & 2 \\ -12 & -7 & -6 \end{bmatrix}$$

Solution: The characteristic equation of A is

$$\det[\mathbf{sI} - \mathbf{A}] = 0$$

$$[\mathbf{sI} - \mathbf{A}] = \begin{bmatrix} s & 0 & 0 \\ 0 & s & 0 \\ 0 & 0 & s \end{bmatrix} - \begin{bmatrix} 0 & 1 & 0 \\ 3 & 0 & 2 \\ -12 & -7 & -6 \end{bmatrix}$$

$$= \begin{bmatrix} s & -1 & 0 \\ -3 & 0 & -2 \\ 12 & 7 & s+6 \end{bmatrix}$$

$$\det[\mathbf{sI} - \mathbf{A}] = |\mathbf{sI} - \mathbf{A}|$$

$$= \begin{bmatrix} s & -1 & 0 \\ -3 & 0 & -2 \\ 12 & 7 & s+6 \end{bmatrix}$$

$$= s\{s(s+6) - (-2 \times 7)\} - (-1)\{-3(s+6) - (-2)(12)\} + 0$$

$$= (s+1)(s+2)(s+3)$$

$$= s^3 + 6s^2 + 11s + 6$$

Therefore, the characteristic equation is

$$\det[\mathbf{sI} - \mathbf{A}] = 0$$

or $\quad s^3 + 6s^2 + 11s + 6 = 0$

Example 10.13: A system is represented by the following dynamic equations:

$$x = \begin{bmatrix} 0 & 1 \\ -2 & -3 \end{bmatrix} + \begin{bmatrix} x_1 \\ x_2 \end{bmatrix} + \begin{bmatrix} 0 \\ 1 \end{bmatrix} u$$

$$y = [1\ 0] \begin{bmatrix} x_1 \\ x_2 \end{bmatrix}$$

Determine the transfer function.

Solution: The transfer function is given by

$$\frac{Y(s)}{U(s)} = \mathbf{C}[s\mathbf{I} - \mathbf{A}]^{-1}\mathbf{B} + \mathbf{D}$$

$$= \mathbf{C}[s\mathbf{I} - \mathbf{A}]^{-1}\mathbf{B} + \mathbf{0}$$

$$[s\mathbf{I} - \mathbf{A}] = S\begin{bmatrix} 1 & 0 \\ 0 & 1 \end{bmatrix} - \begin{bmatrix} 0 & 1 \\ -2 & -3 \end{bmatrix}$$

$$= \begin{bmatrix} s & -1 \\ 2 & s+3 \end{bmatrix}$$

$$[s\mathbf{I} - \mathbf{A}]^{-1} = \frac{\text{adj}[s\mathbf{I} - \mathbf{A}]}{\det[s\mathbf{I} - \mathbf{A}]}$$

$$\text{adj}[s\mathbf{I} - \mathbf{A}] = [\text{cofactor matrix of } [s\mathbf{I} - \mathbf{A}]]^T$$

$$= \begin{bmatrix} C_{11} & C_{12} \\ C_{21} & C_{22} \end{bmatrix}^T = \begin{bmatrix} s+3 & -2 \\ 1 & s \end{bmatrix}^T$$

$$= \begin{bmatrix} s+3 & 1 \\ -2 & s \end{bmatrix}$$

$$\det[s\mathbf{I} - \mathbf{A}] = \begin{vmatrix} s+3 & 1 \\ -2 & s \end{vmatrix} = (s + 3)s - (1)(-2)$$

$$= s^2 + 3s + 2$$

Therefore, $[s\mathbf{I} - \mathbf{A}]^{-1} = \dfrac{\begin{bmatrix} s+3 & 1 \\ -2 & s \end{bmatrix}}{s^2 + 3s + 2}$

$$B = \begin{bmatrix} 0 \\ 1 \end{bmatrix},\ C = [1\ 0]$$

Transfer function = $\mathbf{C}[s\mathbf{I} - \mathbf{A}]^{-1}\mathbf{B}$

$$= \frac{[1\ \ 0]\begin{bmatrix} s+3 & 1 \\ -2 & s \end{bmatrix}\begin{bmatrix} 0 \\ 1 \end{bmatrix}}{s^2 + 3s + 3} = \frac{1}{s^2 + 3s + 2}.$$

Example 10.14: Obtain the transfer function for a system having state model

$$\begin{bmatrix} \dot{x}_1 \\ \dot{x}_2 \end{bmatrix} = \begin{bmatrix} -2 & -3 \\ 4 & 2 \end{bmatrix}\begin{bmatrix} x_1 \\ x_2 \end{bmatrix} + \begin{bmatrix} 3 \\ 5 \end{bmatrix}u$$

$$y = [1 \quad 1]\begin{bmatrix} x_1 \\ x_2 \end{bmatrix}, D = 0$$

Solution: The transfer function is given by

$$G(s) = \frac{Y(s)}{U(s)} = \mathbf{C}[\mathbf{sI} - \mathbf{A}]^{-1}\mathbf{B} + \mathbf{D}$$

$$= \mathbf{C}[\mathbf{sI} - \mathbf{A}]^{-1}\mathbf{B} + 0$$

$$[\mathbf{sI} - \mathbf{A}]^{-1} = \frac{\text{adj}[\mathbf{sI} - \mathbf{A}]}{|\mathbf{sI} - \mathbf{A}|}$$

$$[\mathbf{sI} - \mathbf{A}] = s\begin{bmatrix} 1 & 0 \\ 0 & 1 \end{bmatrix} - \begin{bmatrix} -2 & -3 \\ 4 & 2 \end{bmatrix}$$

$$= \begin{bmatrix} s & 0 \\ 0 & s \end{bmatrix} - \begin{bmatrix} -2 & -3 \\ 4 & 2 \end{bmatrix} = \begin{bmatrix} s+2 & 3 \\ -4 & -2 \end{bmatrix}$$

$$\text{adj}[\mathbf{sI} - \mathbf{A}] = [\text{cofactor matrix of } [\mathbf{sI} - \mathbf{A}]]^T$$

$$= \begin{bmatrix} C_{11} & C_{12} \\ C_{21} & C_{22} \end{bmatrix}^T = \begin{bmatrix} s-2 & 4 \\ -3 & s+2 \end{bmatrix}^T$$

$$= \begin{bmatrix} s-2 & -3 \\ 4 & s+2 \end{bmatrix}$$

$$|sI - A| = (s-2)(s+2) - (4)(-3)$$

$$= s^2 - 4 + 12 = s^2 + 8$$

$$\text{Therefore, } [\mathbf{sI} - \mathbf{A}]^{-1} = \frac{\begin{bmatrix} s-2 & -3 \\ 4 & s+2 \end{bmatrix}}{s^2+8}$$

$$B = \begin{bmatrix} 3 \\ 5 \end{bmatrix}, C = [1, \ 1]$$

$$\text{Transfer function} = C[\mathbf{sI} - \mathbf{A}]^{-1}\mathbf{B}$$

$$= \frac{[1 \quad 1]\begin{bmatrix} s-2 & -3 \\ 4 & s+2 \end{bmatrix}\begin{bmatrix} 3 \\ 5 \end{bmatrix}}{s^2+8}$$

$$= \frac{[1 \quad 1]\begin{bmatrix} 3s & -21 \\ 5s & 22 \end{bmatrix}}{s^2+8} = \frac{[8s+1]}{s^2+8} = \frac{8s+1}{s^2+8}.$$

10.11 STATE-SPACE REPRESENTATION FROM TRANSFER FUNCTION

We shall determine the state variable model when the transfer function is given. Since the state variable representation is not unique, there are, theoretically an infinite number of ways of writing the state equations. We shall present here some methods for deriving a set of state variable representation from the transfer function.

10.12 STATE-SPACE REPRESENTATION FROM PHASE VARIABLES

The **phase variables** are defined as those state variables, which are obtained from one of the system variables and its derivatives.

Usually, the system variable used is the system output and the remaining variables are then the derivatives of the output.

Phase variables provide a powerful method of state variable formulation.

A disadvantage of phase variable formulation is that phase variables, in general, are not physical variables of the system and therefore, not available for measurement and control purposes. However, a link is provided between transfer function approach and time domain approach while designing a control system. Thus, the design implementation of the control system becomes easy and straightforward.

10.12.1 Phase Variable Form

The **phase variable** form of state mode can be easily obtained if the system model is available in the differential equation or transfer function form.

Let us consider a general *n*th-order differential equation for a linear time-invariant system.

$$\frac{d^n}{dt^n} y(t) + a_1 \frac{d^{n-1}}{dt^{n-1}} y(t) + \ldots + a_n y(t) = b_0 \frac{d^n}{dt^n} u(t) + b_1 \frac{d^{n-1}}{dt^{n-1}} u(t) + \ldots + b_n u(t) \quad \ldots(10.66)$$

where n is an integer and $a_1, a_2, \ldots, a_n$ and $b_0, b_1, \ldots b_n$ are constants

The transfer function relating output and input in *s-domain* is

$$G(s) = \frac{Y(s)}{U(s)} = \frac{b_0 s^n + b_1 s^{n-1} + \ldots + b_n}{s^n + a_1 s^{n-1} + \ldots + a_n} \quad \ldots(10.67)$$

This is the transfer function representation of the differential Eq. (10.66).

State-space representation of an *n*th-order system in which the forcing function does not involve derivative terms

If the transfer function of the system has no zeros (that is, the differential equation of the system has no derivatives of the input), the model of the system can be obtained easily as follows:

Consider the transfer function

$$G(s) = \frac{b_0}{s^n + a_1 s^{n-1} + \ldots + a_{n-1}\, s + a_n} \qquad \ldots(10.68)$$

Corresponding to the differential equation

$$\frac{d^n}{dt^n}\, y(t) + a_1 \frac{d^{n-1}}{dt^{n-1}}\, y(t) + \ldots + a_{n-1} \frac{d}{dt}\, y(t) + a_n\, y(t) = b_0 u \qquad \ldots(10.69)$$

This is an nth-order system, s_0 n state variables are required. Let the first state variable be x_1 equal to the output variable y. Let the second state variable x_2 be equal to the derivative of the output variable and so on. That is,

$$x_1 = y$$

$$x_2 = \frac{dy}{dt} = \dot{y}$$

$$x_3 = \frac{d^2 y}{dt^2} = \ddot{y}$$

$$\vdots$$

$$x_n = \frac{d^{n-1} y}{dt^{n-1}} = y^{(n-1)}$$

The equations can be reduced to a set of first-order differential equations as follows.

$$\begin{aligned} \dot{x}_1 &= x_2 \\ \dot{x}_2 &= x_3 \\ &\vdots \\ \dot{x}_{n-1} &= \dot{x}_n \\ \dot{x}_n &= -a_n x_1 - a_{n-1} x_2 - \ldots - a_1 x_n + b_0 u \end{aligned} \qquad \ldots(10.70)$$

The last equation is obtained by equating the highest order derivative term of the output in the differential equation to all other terms.

The above set of state equations Eq. (10.70) can be written in vector-matrix form as

$$\begin{bmatrix} \dot{x}_1 \\ \dot{x}_2 \\ \vdots \\ \dot{x}_n \end{bmatrix} = \begin{bmatrix} 0 & 1 & 0 & \cdots & 0 \\ 0 & 0 & 1 & \cdots & 0 \\ \vdots & & & & \\ 0 & 0 & 0 & \cdots & 1 \\ -a_n & -a_{n-1} & -a_{n-2} & \cdots & -a_1 \end{bmatrix} \begin{bmatrix} x_1 \\ x_2 \\ \vdots \\ x_{n-1} \\ x_n \end{bmatrix} + \begin{bmatrix} 0 \\ 0 \\ \vdots \\ 0 \\ b_0 \end{bmatrix} u \qquad \ldots(10.71)$$

or $$\dot{x} = \mathbf{A}x + \mathbf{B}u \qquad \ldots(10.72)$$

where
$$A = \begin{bmatrix} 0 & 1 & 0 & \cdots & 0 \\ 0 & 0 & 1 & \cdots & 0 \\ \vdots & & & & \\ 0 & 0 & 0 & \cdots & 1 \\ -a_n & -a_1 & -a_2 & \cdots & -a_{n-1} \end{bmatrix} (n \times n) \quad \text{...(10.73)}$$

$$B = \begin{bmatrix} 0 \\ 0 \\ \vdots \\ 0 \\ 1 \end{bmatrix} (n \times n) \quad \text{...(10.74)}$$

The matrix **A** has a very special form. It has all 1*s* in the upper off diagonal. The last row of **A** contains the negative values of the coefficients of the homogeneous part of the different equation in ascending order, except for the coefficient of the high-order term, which is unity. This form of matrix **A** is known as the **bush form** or the **companion form**.

The matrix **B** is a column matrix with the last row equal to one, and the rest of the elements are all zeroes.

The state equations in Eq. (10.72) with **A** and **B** given in Eqs. (10.73) and (10.74) are known as the **phase-variable canonical form (PVCF)** or the **controllable canonical form (CCF).**

Hence, in the view of these observations, **A** and **B** and, therefore, the state equations can be written directly by inspection of the differential equation.

The output y is equal to x_1, and therefore the output equation is given by

$$y = [1 \ \ 0 \ \ 0] \begin{bmatrix} x_1 \\ x_2 \\ \vdots \\ x_n \end{bmatrix} \quad \text{...(10.75)}$$

That is, $$\mathbf{y}(t) = \mathbf{C}x(t) \quad \text{...(10.76)}$$

If the transfer function is known instead of the differential equation, we can easily obtain a differential equation form as shown.

If $$T(s) = \frac{Y(s)}{U(s)} = \frac{b_0}{s^n + a_1 s^{n-1} + \ldots + a_n} \quad \text{...(10.77)}$$

$$(s^n + a_1 s^{n-1} + \ldots + a_n)\, Y(s) = b_0 U(s)$$

or $$y^{(n)} + a_1 y^{(n-1)} + \ldots + a_n y = b_0 u \quad \text{...(10.78)}$$

where $$y^{(n)} \underline{\Delta} \frac{d^n y}{dt^n} \qquad \text{...(10.79)}$$

Equation (10.78) is same as Eq. (10.79) and hence the state-space model is again given by Eqs. (10.81) and (10.81).

Block diagram representation (state diagram for phase variable form)

A block diagram representation of the same model of Eq. (10.71) is shown in Fig. 10.19.

Each block on the forward path represents integration and the output of each integrator is taken a state variable. Figure represents the state diagram for phase variable form.

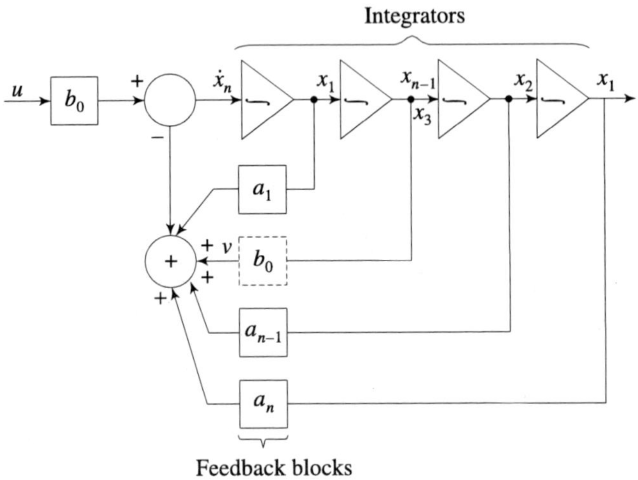

Fig. 10.19 Block diagram representation of the state model for $x = Ax + Bu$ and $y = Cx$

Signal Flow Graph Representation

A signal flow graph representation of the state model of Eq. (10.71) is shown in Fig. 10.20. In the signal flow graph the term 1/s represents pure integration.

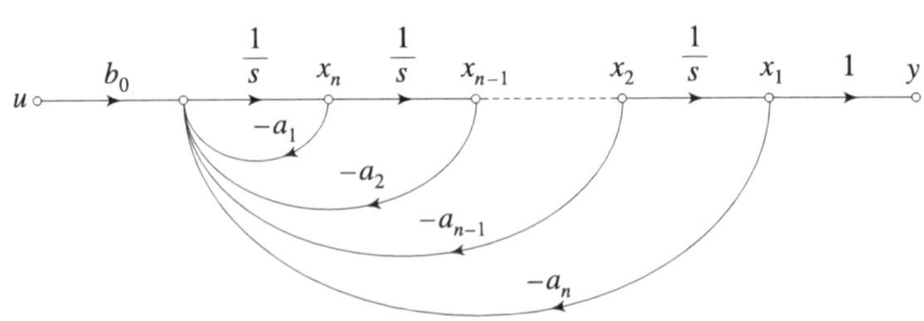

Fig. 10.20 Signal flow graph representation of the state model for $\dot{x} = Ax + Bu$ and $y = Cx$

Example 10.15: Construct the state model for a system characterized by the differential equation.

$$\ddot{y} + 5\dot{y} + 6y = u$$

Solution: The given system is of the second order and, therefore two state variables are required.

Rearranging the given differential equation so that highest order derivative term $\frac{d^2y}{dt^2}$ is equated to the rest of the terms, we have

$$\frac{d^2y}{dt^2} = -5\frac{dy}{dt} - 6y + u \quad ...(10.80)$$

The state variables are defined as

$$x_1(t) = y(t) \quad ...(10.81)$$

$$x_2(t) = \frac{dy}{dt} \quad ...(10.82)$$

The state equations are given by

$$\dot{x}_1 = x_2 \quad ...(10.83)$$

$$\dot{x}_2 = -5x_2 - 6x_1 + u \quad ...(10.84)$$

In matrix form, the sate equations can be written as

$$\begin{bmatrix} \dot{x}_1 \\ \dot{x}_2 \end{bmatrix} = \begin{bmatrix} 0 & 1 \\ -6 & -5 \end{bmatrix} \begin{bmatrix} x_1 \\ x_2 \end{bmatrix} + \begin{bmatrix} 0 \\ 1 \end{bmatrix} u \quad ...(10.85)$$

The output equation is

$$y = x_1$$

or

$$y = [1 \;\; 0] \begin{bmatrix} x_1 \\ x_2 \end{bmatrix} \quad ...(10.86)$$

The state model is given by Eqs. (10.85) and (10.86)

The block diagram representation of the state model is shown in Fig. 10.21.

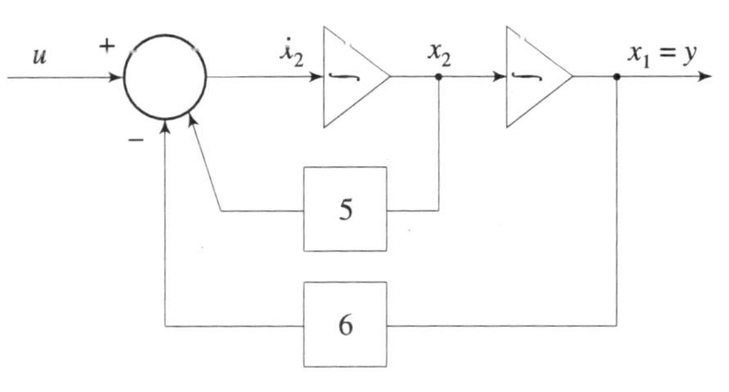

Fig. 10.21 Block diagram representation of the state model for example 10.14

10.13 CONTROLLABLE CANONICAL FORM (CCF) OF THE STATE-SPACE REPRESENTATION

Consider the following transfer function system

$$\frac{Y(s)}{U(s)} = \frac{b_0 s^n + b_1 s^{n-1} + b_{n-1}\, s + b_n}{s^n + a_1\, s^{n-1} + a_{n-1}\, s ... + a_0}$$

The given transfer function can be divided into 2 parts.

$$\frac{Y(s)}{U(s)} = \frac{Y(s)}{X_1(s)} \cdot \frac{X_1(s)}{U(s)}$$

$$= \frac{1}{s^n + a_1\, s^{n-1} + ...\, a_{n-1}\, s + a_0} \cdot (b_0 s^n + b_1 s^{n-1} + b_{n-1}\, s + b_n)$$

where
$$\frac{X_1(s)}{U(s)} = \frac{1}{s^n + a_1\, s^{n-1} + ...\, a_{n-1}\, s + a_0} \quad ...(10.87)$$

and
$$\frac{Y(s)}{X_1(s)} = b_0 s^n + b_1 s^{n-1} + ...\, b_{n-1}\, s + b_n \quad ...(10.88)$$

Consider the function

$$\frac{X_1(s)}{U(s)} = \frac{1}{s^n + a_1\, s^{n-1} + ...\, a_{n-1}\, s + a_0} \quad ...(10.89)$$

From the equations with $b_0 = 1$

Hence, its state-space representation is given by Eq. (10.91)

$$\begin{bmatrix} \dot{x}_1 \\ \dot{x}_2 \\ \vdots \\ \dot{x}_{n-1} \\ \dot{x}_n \end{bmatrix} = \begin{bmatrix} 0 & 1 & 0 & \cdots & 0 \\ 0 & 0 & 1 & \cdots & 0 \\ \vdots & & & & \\ 0 & 0 & 0 & \cdots & 1 \\ -a_n & -a_{n-1} & -a_{n-2} & \cdots & -a_1 \end{bmatrix} \begin{bmatrix} x_1 \\ x_1 \\ \vdots \\ x_{n-1} \\ x_n \end{bmatrix} + \begin{bmatrix} 0 \\ 0 \\ \vdots \\ 0 \\ 1 \end{bmatrix} u \quad ...(10.90)$$

Now consider the function

$$\frac{Y(s)}{X_1(s)} = b_0 s^n + b_1 s^{n-1} + ... + b_{n-1} s + b_n$$

By cross-multiplying we get

$$Y(s) = (b_0 s^n + b_1 s^{n-1} + ...\, b_{n-1}\, s + b_n)\, X_1(s)$$

The differential equation corresponding to this is

$$y = b_0 \frac{d^n x_1}{dt^n} + b_1 \frac{d^{n-1} x_1}{dt^{n-1}} + \ldots\ldots + b_{n-1} \frac{dx_1}{dt} + b_1 x_1 \qquad \ldots(10.91)$$

The first state variable is x_1: The second state variable x_2 is the derivative of x_1 and so on. That is,

$$x_1 = x_1$$

$$x_2 = \frac{dx_1}{dt^2} = \dot{x}_1$$

$$x_3 = \frac{d^2 x_1}{dt^2} = \dot{x}_2$$

$$\vdots$$

$$x_n = \frac{d^{n-1} x_1}{dt^{n-1}} = x_1^{(n-1)}$$

Therefore, Eq. (10.91) becomes

$$y = b_0 \dot{x}_n + b_1 x_n + \ldots.\ b_{n-1} x_2 + b_x x_1$$

Substituting $\dot{x}_n$, we get

$$y = b_0(-a_n x_1 - a_{n-2} x_2 - \ldots. -a_2 x_{n-1} - a_1 x_1 + u) + b_1 x_n + \ldots + b_{n-1} x_2 + b_n x_1$$

$$y = x_1[(b_n - b_0 a_n) + x_2(b_{n-1} - b_0 a_{n-1})] + \ldots\ldots + x_n(b_1 - b_0 a_1) + b_0 u$$

$$y = [(b_n - b_0 a_n) \vdots (b_{n-1} - b_0 a_{n-1}) \vdots \cdots \vdots (b_1 - b_0 a_1)] \begin{bmatrix} x_1 \\ x_1 \\ M \\ x_n \end{bmatrix} + b_0 u \qquad \ldots(10.92)$$

Equations (10.90) and (10.92) are said to be in the **controllable canonical form**.

10.14 OBSERVABLE CANONICAL FORM (OCF) OF THE STATE-SPACE REPRESENTATION

Consider the following transfer function system

$$\frac{Y(s)}{U(s)} = \frac{b_0 s^n + b_1 s^{n-1} + \ldots + b_{n-1} s + b_n}{s^n + a_1 s^{n-1} + \ldots + a_{n-1} s + a_n} \qquad \ldots(10.93)$$

Equation (10.93) can be modified in the following form:

$$S^n[Y(s) - b_0 U(s)] + s^{n-1}[a_1 Y(s) - b_1 U(s)] + \ldots + s[a_{n-1} Y(s) - b_{n-1} U(s)] + a_n Y(s) - b_n U(s) = 0 \qquad \ldots(10.94)$$

Dividing Eq. (10.94) by s^n and rearranging we get

$$Y(s) = b_0U(s) + \frac{1}{s}[b_1U(s) - a_1Y(s) + \ldots + \frac{1}{s^{n-1}}[b_{n-1}u(s) - a_{n-1}Y(s)]$$

$$+ \frac{1}{s^n}[b_nU(s) - a_nY(s)] \qquad \ldots(10.95)$$

Now define state variables as follows:

$$X_n(s) = \frac{1}{s}[b_1U(s) - a_1Y(s) + X_{n-1}(s)]$$

$$X_{n-1}(s) = \frac{1}{s}[b_1U(s) - a_2Y(s) + X_{n-2}(s)]$$

$$\vdots$$

$$X_2(s) = \frac{1}{s}[b_{n-1}U(s) - a_{n-2}Y(s) + X_1(s)] \qquad \ldots(10.96)$$

$$X_1(s) = \frac{1}{s}[b_nU(s) - a_nY(s)]$$

Then Eq. (10.96) can be written as

$$Y(s) = b_0U(s) + X_n(s) \qquad \ldots(10.97)$$

By substituting Eq. (10.97) into Eq. (10.96) and multiplying both sides of the equation by s, we get

$$sX_n(s) = X_{n-1}(s) - a_1X_n(s) + (b_1 - a_1b_0)U(s)$$

$$sX_{n-1}(s) = X_{n-2}(s) - a_2X_n(s) + (b_2 - a_2b_0)U(s)$$

$$\vdots \qquad \vdots \qquad \vdots$$

$$sX_2(s) = X_1(s) - a_{n-1}X_n(s) + (b_{n-1} - a_{n-1}b_0)U(s) \qquad \ldots(10.98)$$

$$sX_1(s) = -a_nX_1(s) + (b_n - a_nb_0)U(s)$$

Taking the inverse Laplace transforms of n Eq. (10.98) and writing them in the reverse order, we get

$$\dot{x}_1 = -a_nx_n + (b_n - a_nb_0)u$$

$$\dot{x}_2 = x_1 - a_{n-1}x_n + (b_{n-1} - a_{n-1}b_0)u$$

$$\vdots \qquad \vdots \qquad \vdots$$

$$\dot{x}_{n-1} = x_{n-2} - a_2x_n + (b_2 - a_2b_0)u \qquad \ldots(10.99)$$

$$\dot{x}_n = x_{n-1} - a_1x_n + (b_1 - a_1b_0)u$$

Taking the inverse Laplace transform of Eq. (10.97) we get

$$y = x_n + b_0u \qquad \ldots(10.100)$$

Writing the state and output Eq. (10.99) and (10.100) in standard vector-matrix form, we get

$$\begin{bmatrix} \dot{x}_1 \\ \dot{x}_2 \\ \vdots \\ \dot{x}_n \end{bmatrix} = \begin{bmatrix} 0 & 0 & \cdots & 0 & -a_n \\ 1 & 0 & \cdots & 0 & -a_{n-1} \\ \vdots & & & & \\ 0 & 0 & \cdots & 1 & -a_1 \end{bmatrix} \begin{bmatrix} x_1 \\ x_2 \\ \vdots \\ x_n \end{bmatrix} + \begin{bmatrix} b_n - a_n b_0 \\ b_{n-1} - a_{n-1} b_0 \\ \vdots \\ b_1 - a_1 b_0 \end{bmatrix} u \qquad ...(10.101)$$

$$y = [0 \ 0 \ \ldots \ 0 \ 1] \begin{bmatrix} x_1 \\ x_2 \\ \vdots \\ x_n \end{bmatrix} \qquad ...(10.102)$$

Equations (10.101) and (10.102) are said to be **observable canonical form (OCF)**.

Example 10.16: Obtain the phase-variable state model for a system described by the different equation

$$\frac{d^3y(t)}{dt^3} + 5\frac{d^2y(t)}{dt^2} + \frac{dy(t)}{dt} + 2y(t) = u(t).$$

Solution: The given system is of third order and therefore, $n = 3$. Three state variables are required. Rearranging the given differential equation so that the highest-order derivative term is equated to the rest of the terms, we have

$$\frac{d^3y(t)}{dt^3} = -5\frac{d^2y(t)}{dt^2} - \frac{dy(t)}{dt} + 2y(t) + u(t)$$

The state variables are defined as

$$x_1(t) = y(t)$$

$$x_2(t) = \frac{dy(t)}{dt}$$

$$x_3(t) = \frac{d^2y(t)}{dt^2}$$

Then the state equations are represented by the vector matrix equations

$$\frac{dx(t)}{dt} = Ax(t) + Bu(t)$$

For the given differential equation, $n = 3$, $a_1 = 1$, $a_2 = 1$, $a_3 = 2$, $b_0 = 1$

The state model is given by

$$\begin{bmatrix} \dot{x}_1 \\ \dot{x}_2 \\ \dot{x}_3 \end{bmatrix} = \begin{bmatrix} 0 & 1 & 0 \\ 0 & 0 & 1 \\ -a_3 & -a_2 & -a_1 \end{bmatrix} \begin{bmatrix} x_1 \\ x_2 \\ x_3 \end{bmatrix} + \begin{bmatrix} 0 \\ 0 \\ 1 \end{bmatrix} u$$

or $$\begin{bmatrix} \dot{x}_1 \\ \dot{x}_2 \\ \dot{x}_3 \end{bmatrix} = \begin{bmatrix} 0 & 1 & 0 \\ 0 & 0 & 1 \\ -2 & -1 & -5 \end{bmatrix} \begin{bmatrix} x_1 \\ x_2 \\ x_3 \end{bmatrix} + \begin{bmatrix} 0 \\ 0 \\ 1 \end{bmatrix} u$$

The output equation is

$$y(t) = x_1(t)$$

$$= [1 \;\; 0 \;\; 0] \begin{bmatrix} x_1 \\ x_2 \\ x_3 \end{bmatrix}$$

The state diagram is shown in Fig. 10.22.

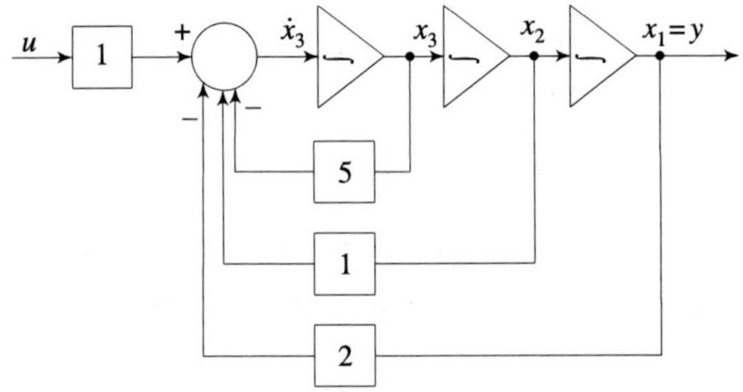

Fig. 10.22

Example 10.17: A system is described by the following differential equation:

$$\frac{d^3x}{dt^3} + 3\frac{d^2x}{dt^2} + 4\frac{dx}{dt} + 4x = u_1(t) + 3u_2(t) + 4u_3(t)$$

The outputs are: $y_1 = 4\frac{dx}{dt} + 3u_1$; $\quad y_2 = \frac{d^2x}{dt^2} + 4u_2 + u_3$

Obtain the state space representation of the system.

Solution: The given system is of third order and therefore $n = 3$. Three state variables are required. Rearranging the given differential equation so that the highest-order derivative term is equated to the rest of their terms, we have

$$\frac{d^3x}{dt^3} = -3\frac{d^2x}{dt^2} - 4\frac{dx}{dt} - 4x + u_1(t) + 3u_2(t) + 4u_3(t) \qquad ...(10.103)$$

The state variables are defined as

$$x_1 = x \qquad ...(10.104)$$

$$x_2 = \dot{x}_1 \qquad ...(10.105)$$

$$x_3 = x_2 \qquad ...(10.106)$$

From Eq. (10.103), we have

$$x_3 = -4x - 4\frac{dx}{dt} - 3\frac{d^2x}{dt^2} + u_1(t) + 3\,u_2(t) + 4u_3(t) \quad ...(10.107)$$

Therefore, Eqs. (10.105), (10.106) and (10.107) forming the set of state equations are written as

$$\dot{x}_1 = x_2 \quad ...(10.108)$$

$$\dot{x}_2 = x_3 \quad ...(10.109)$$

$$\dot{x}_3 = -4\,x_1 - 4\,x_2 - 3\,x_3 + u_1(t) + 3\,u_2(t) + 4u_3(t) \quad ...(10.110)$$

In matrix form, the set of equations (10.108), (10.109) and (10.110) can be written as

$$\begin{bmatrix} \dot{x}_1 \\ \dot{x}_2 \\ \dot{x}_3 \end{bmatrix} = \begin{bmatrix} 0 & 1 & 0 \\ 0 & 0 & 1 \\ -4 & -4 & -3 \end{bmatrix}\begin{bmatrix} x_1 \\ x_2 \\ x_3 \end{bmatrix} + \begin{bmatrix} 0 & 0 & 0 \\ 0 & 0 & 0 \\ 1 & 3 & 4 \end{bmatrix}\begin{bmatrix} u_1 \\ u_2 \\ u_3 \end{bmatrix} \quad ...(10.111)$$

The outputs are given by

$$y_1 = 4x_2 + 3u_1$$

$$y_2 = x_3 + 4u_2 + 3u_3$$

In matrix form

$$\begin{bmatrix} y_1 \\ y_2 \end{bmatrix} = \begin{bmatrix} 0 & 4 & 0 \\ 0 & 0 & 1 \end{bmatrix}\begin{bmatrix} x_1 \\ x_2 \\ x_3 \end{bmatrix} + \begin{bmatrix} 3 & 0 & 0 \\ 0 & 4 & 1 \end{bmatrix}\begin{bmatrix} u_1 \\ u_2 \\ u_3 \end{bmatrix} \quad ...(10.112)$$

The state model is given by Eqs. (10.111) and (10.112).

Example 10.18: For the given transfer function

$$T(s) = \frac{b_0}{s_3 + a_2\,s^2 + a_1\,s + a_0}$$

obtain the state model.

Solution: We can obtain the differential equation form for the given transfer function as follows:

$$T(s) = \frac{Y(s)}{U(s)} = \frac{b_0}{s^3 + a_2\,s^2 + a_1\,s + a_0} \quad ...(10.113)$$

$$(s^3 + a_2\,s^2 + a_1\,s + a_0)Y(s) = b_0U(s) \quad ...(10.114)$$

The differential equation form of Eq. (10.114) is

$$\frac{d^3y}{dt^3} + a_2\frac{d^2y}{dt^2} + a_1\frac{dy}{dt} + a_0y = b_0u \quad ...(10.115)$$

This is a third-order system, so three state variables are required.

Let the first state variable be x_1 equal to the output variable y. Let the second state variable x_2 be equal to the derivative of the output variable, and the third state variable x_3 be equal to the second derivative of the output variable. That is,

$$x_1 = y \qquad ...(10.116)$$

$$x_2 = \frac{dy}{dt} = \dot{y} \qquad ...(10.117)$$

$$x_3 = \frac{d^2y}{dt^2} = \ddot{y} \qquad ...(10.118)$$

These equations can be reduced to a set of first-order differential equations as follows:

$$\dot{x}_1 = x_2 \qquad ...(10.119)$$

$$\dot{x}_2 = x_3 \qquad ...(10.120)$$

Third order derivative term is equated to all other terms.

$$\frac{d^3y}{dt^3} = -a_0 y - a_1\frac{dy}{dt} - a_2\frac{d^2y}{dt^2} + b_0 u$$

$$= -a_0x_1 - a_1x_2 - a_2x_3 + b_0u$$

Therefore, the third state equation is given by

$$\dot{x}_3 = -\ a_0x_1 - a_1x_2 - a_2x_3 + b_0u \qquad ...(10.121)$$

The set of state equations given by Eqs. (10.119), (10.120), and (10.121) can be written in matrix form as

$$\begin{bmatrix} \dot{x}_1 \\ \dot{x}_2 \\ \dot{x}_3 \end{bmatrix} = \begin{bmatrix} 0 & 1 & 0 \\ 0 & 0 & 1 \\ -a_0 & -a_1 & -a_2 \end{bmatrix} \begin{bmatrix} x_1 \\ x_2 \\ x_3 \end{bmatrix} + \begin{bmatrix} 0 \\ 0 \\ b_0 \end{bmatrix} \begin{bmatrix} \dot{x}_1 \\ \dot{x}_2 \\ \dot{x}_3 \end{bmatrix} u \qquad ...(10.122)$$

The output y is equal to x_1 and therefore, the output equation is given by $y = x_1$

or $$y = [1\ \ 0\ \ 0] \begin{bmatrix} x_1 \\ x_2 \\ x_3 \end{bmatrix}$$

Figure 10.23 shows the block diagram representation of the state model.

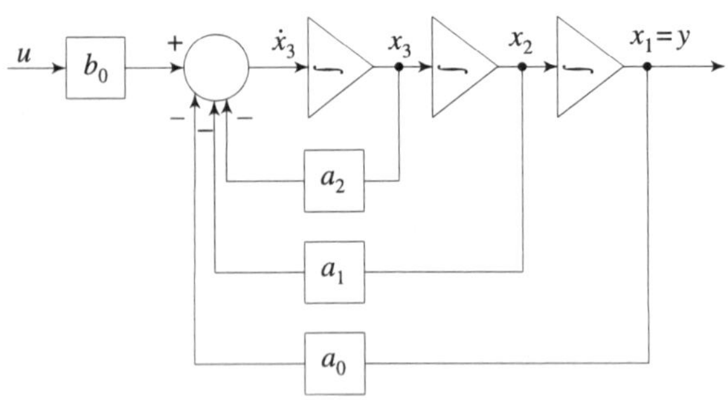

Fig. 20.23

Example 10.19: Obtain a state model for the system described by

$$T(s) = \frac{Y(s)}{U(s)} = \frac{1}{s^3 + 6s^2 + 10s + 5}$$

Solution:

$$\frac{Y(s)}{U(s)} = \frac{1}{s^3 + 6s^2 + 10s + 5} \qquad ...(10.123)$$

$$(s^3 + 6s^2 + 10s + 5)\ Y(s) = U(s) \qquad ...(10.124)$$

Taking the inverse Laplace transform on both sides, we get

$$\frac{d^3y}{dt^3} + 6\frac{d^2y}{dt^2} + 10\frac{dy}{dt} + 5y = u(t) \qquad ...(10.125)$$

Rearranging the differential Eq. (10.125) so that the highest order derivative term $\frac{d^3y}{dt^3}$ is equated to the rest of the terms, we get

$$\frac{d^3y}{dt^3} = -6\frac{d^2y}{dt^2} - 10\frac{dy}{dt} - 5y + u(t)$$

or

$$\frac{d^3y}{dt^3} = -5y - 10\frac{dy}{dt} - 6\frac{d^2y}{dt^2} + u \qquad ...(10.126)$$

The given system is of third order, therefore, three state variables are required.

The state variables are defined as

$$x_1 = y$$
$$x_2 = \dot{y} = \dot{x}_1$$
$$x_3 = \ddot{y} = \dot{x}_2$$

Therefore, the state equations can be written as

$$\begin{aligned} \dot{x}_1 &= x_2 = 0 \cdot x_1 + 1 \cdot x_2 + 0 \cdot x_3 + 0 \cdot u \\ \dot{x}_2 &= x_3 = 0 \cdot x_1 + 0 \cdot x_2 + 1 . x_3 + 0 \cdot u \\ \dot{x}_3 &= -5x_1 - 10x_2 - 6x_3 + 1 \cdot u \end{aligned} \qquad ...(10.127)$$

In matrix form, these equations can be written as

$$\begin{bmatrix} \dot{x}_1 \\ \dot{x}_2 \\ \dot{x}_3 \end{bmatrix} = \begin{bmatrix} 0 & 1 & 0 \\ 0 & 0 & 1 \\ -5 & -10 & -6 \end{bmatrix} \begin{bmatrix} x_1 \\ x_2 \\ x_3 \end{bmatrix} + \begin{bmatrix} 0 \\ 0 \\ 1 \end{bmatrix} u \qquad ...(10.128)$$

The output equation is

$$y = x_1 = 1 \cdot x_1 + 0 \cdot x_2 + 0 \cdot x_3 \qquad ...(10.129)$$

or

$$y = [1 \ \ 0 \ \ 0] \begin{bmatrix} x_1 \\ x_2 \\ x_3 \end{bmatrix} \qquad ...(10.130)$$

The state model is given by Eqs. (10.128) and (10.130).

Example 10.20: The transfer function of a control system is given by

$$\frac{Y(s)}{U(s)} = \frac{s^2 + 3s + 4}{s^3 + 2s^2 + 3s + 2}$$

Obtain a state model.

Solution: The given transfer function can be divided into two parts

$$\frac{Y(s)}{U(s)} = \frac{Y(s)}{X_1(s)} \cdot \frac{X_1(s)}{U(s)}$$

$$= \frac{1}{s^3 + 2s^2 + 3s + 2} \cdot (s^2 + 3s + 4) \qquad ...(10.131)$$

where $$\frac{X_1(s)}{U(s)} = \frac{1}{s^2 + 2s^2 + 3s + 2} \qquad ...(10.132)$$

and $$\frac{Y(s)}{X_1(s)} = s^2 + 3s + 4 \qquad ...(10.133)$$

Consider the function

$$\frac{X_1(s)}{U(s)} = \frac{1}{s^3 + 2s^2 + 3s + 2}$$

By cross multiplication,

$$(s^3 + 2s^2 + 3s + 2)\, X_1(s) = 1 \cdot U(s)$$

The differential equation corresponding to this is

$$\frac{d^3x_1}{dt^3} + 2\frac{d^2x_1}{dt^2} + 3\frac{dx_1}{dt} + 2x_1 = u \qquad ...(10.134)$$

This is a third-order system, so three state variables are required. Let the first state variable be equal to x_1. The other state variables, x_1 and x_2, are defined as follows:

$$x_2 = \dot{x}_1$$

$$x_3 = \dot{x}_2$$

Therefore, the state equations can be written as

$$\dot{x}_1 = x_2 = 0 \cdot x_1 + 1 \cdot x_2 + 0 \cdot x_3$$

$$\dot{x}_2 = x_3 = 0 \cdot x_1 + 0 \cdot x_2 + 1 \cdot x_3 \qquad ...(10.135)$$

$$\dot{x}_3 = -2x_1 - 3x_2 - 2x_3 + u$$

The last equation is obtained by equating the highest order derivative term in equation (4) to all other terms. That is,

$$\frac{d^3x_1}{dt^3} = -2\,x_1 - 3\frac{dx_1}{dt} - 2\frac{d^2x_1}{dt^2} + u = -2\,x_1 - 3\,x_2 - 2\,x_3 + u$$

The above set of state Eq. (10.135) be written in vector-matrix form as

$$\begin{bmatrix} \dot{x}_1 \\ \dot{x}_2 \\ \dot{x}_3 \end{bmatrix} = \begin{bmatrix} 0 & 1 & 0 \\ 0 & 0 & 1 \\ -2 & -3 & -2 \end{bmatrix} \begin{bmatrix} x_1 \\ x_2 \\ x_3 \end{bmatrix} + \begin{bmatrix} 0 \\ 0 \\ 1 \end{bmatrix} u \qquad ...(10.136)$$

The second expression is

$$\frac{Y(s)}{X_1(s)} = s^2 + 3s + 4$$

By cross-multiplying, we get

$$Y(s) = (s^2 + 3s + 4)\, X_1(s)$$

The differential equation corresponding to this

$$y = \frac{d^2x_1}{dt^2} + 3\frac{dx_1}{dt} + 4x_1 = x_3 + 3x_2 + 4x_1$$

or $$y = 4 \cdot x_1 + 3 \cdot x_2 + 1 \cdot x_3$$

In vector matrix form, this output equation is written as

$$y = [4 \;\; 3 \;\; 1] \begin{bmatrix} x_1 \\ x_2 \\ x_3 \end{bmatrix} \qquad ...(10.137)$$

The state model of the system is given by Eqs. (10.136) and (10.137).

Example 10.21: Obtain a state model for the system whose transfer functions is given by

$$G(s) = \frac{2s^3 + 7s^2 + 12s + 8}{s^3 + 6s^2 + 11s + 9}.$$

Solution: The given transfer function is

$$G(s) = \frac{Y(s)}{U(s)} = \frac{2s^3 + 7s^2 + 12s + 8}{s^3 + 6s^2 + 11s + 9} \qquad ...(10.138)$$

This can be written as

$$\frac{Y(s)}{U(s)} = \frac{X_1(s)}{U(s)} \cdot \frac{Y(s)}{X_1(s)} = \frac{1}{s^3 + 6s^2 + 11s + 9} \cdot (2s^3 + 7s^2 + 12s + 8)$$

where $$\frac{X_1(s)}{U(s)} = \frac{1}{s^3 + 6s^2 + 11s + 9} \qquad ...(10.139)$$

and
$$\frac{Y(s)}{X_1(s)} = 2s^3 + 7s^2 + 12s + 8 \qquad ...(10.140)$$

From Eq (10.139), by cross-multiplication

$$(s^3 + 6s^2 + 11s + 9)\, X_1(s) = 1.U(s)$$

The differential equation corresponding to this is

$$\frac{d^3x_1}{dt^3} + 6\frac{d^2x_1}{dt^2} + 11\,\frac{dx_1}{dt} + 9x_1 = u \qquad ...(10.141)$$

This is a third-order system, so three state variables are required.

Let the first state variable be equal to x_1, the other state variables are x_2 and x_3 defined as follows:

$$x_2 = \dot{x}_1$$
$$x_3 = \dot{x}_2$$

Therefore, the state equations can be written as

$$\begin{aligned} \dot{x}_1 &= x_2 = 0\cdot x_1 + 1\cdot x_2 + 0\cdot x_3 \\ \dot{x}_2 &= x_3 = 0\cdot x_1 + 0\cdot x_2 + 1\cdot x_3 \\ \dot{x}_3 &= -9x_1 - 11x_2 - 6x_3 + u \end{aligned} \qquad ...(10.142)$$

The last equation is obtained by equating the highest order derivative term in Eq. (10.142) to all other terms. That is,

$$\begin{aligned} \frac{d^3x_1}{dt^3} &= -9x_1 - 11\frac{dx_1}{dt} - 6\frac{d^2x_1}{dt^2} + u \\ &= -9x_1 - 11x_2 - 6x_3 + u \end{aligned}$$

The above set of stat Eq. (10.143) can be written in vector-matrix form as

$$\begin{bmatrix} \dot{x}_1 \\ \dot{x}_2 \\ \dot{x}_3 \end{bmatrix} = \begin{bmatrix} 0 & 1 & 0 \\ 0 & 0 & 1 \\ -9 & -11 & -6 \end{bmatrix} \begin{bmatrix} x_1 \\ x_2 \\ x_3 \end{bmatrix} + \begin{bmatrix} 0 \\ 0 \\ 1 \end{bmatrix} u \qquad ...(10.143)$$

The second expression is

$$\frac{Y(s)}{X_1(s)} = 2s^3 + 7s^2 + 12s + 8$$

By cross-multiplying, we get

$$Y(s) = (2s^3 + 7s^2 + 12s + 8)\, X_1\,(s)$$

The differential equation corresponding to this

$$y = 2\frac{d^3x_1}{dt^3} + 7\frac{d^2x_1}{dt^2} + 12\frac{dx_1}{dt} + 8x_1$$

$$= 2\dot{x}_3 + 7\dot{x}_2 + 12\dot{x}_1 + 8x_1$$

$$= 2(-9x_1 - 11x_2 - 6x_3 + u) + 7x_3 + 12x_2 + 8x_1$$

$$= -10x_1 - 10x_2 - 5x_3 + 2u$$

In vector matrix form, this output equation can be written as

$$y = [-10 \ -10 \ -5]\begin{bmatrix} x_1 \\ x_2 \\ x_3 \end{bmatrix} + 2u \qquad \text{...(10.144)}$$

The state model of the system is represented by Eqs. (10.143) and (10.144)

Example 10.22: A control system is described by the following transfer function:

$$G(s) = \frac{s+3}{(s+1)(s+2)^2}$$

Obtain the state and output equations of the system.

Solution: The given transfer function is

$$G(s) = \frac{Y(s)}{U(s)} = \frac{s+3}{(s+1)(s+2)^2} = \frac{s+3}{(s+1)(s^2+4s+4)}$$

$$= \frac{s+3}{s^3+5s^2+8s+4} \qquad \text{...(10.145)}$$

This can be divided in two parts as

$$\frac{Y(s)}{U(s)} = \frac{Y(s)}{X_1(s)} \cdot \frac{X_1(s)}{U(s)} \qquad \text{...(10.146)}$$

where $$\frac{X_1(s)}{U(s)} = \frac{1}{s^3+5s^2+8s+4} \qquad \text{...(10.147)}$$

and $$\frac{Y(s)}{X_1(s)} = s+3 \qquad \text{...(10.148)}$$

Consider the function given by Eq. (10.147)

By cross-multiplication

$$(s^3 + 5s^2 + 8s + 4)X_1(s) = U(s) \qquad \text{...(10.149)}$$

The differential equation corresponding to this is

$$\frac{d^3x_1}{dt^3} + 5\frac{d^2x_1}{dt^2} + 8\frac{dx_1}{dt} + 4x_1 = u \qquad \text{...(10.150)}$$

This is a third-order system, so three state variables are required.

Let the first state variable be equal to x_1. The other state variables, x_2 and x_3, are defined as follows:

$$x_2 = \dot{x}_1$$

$$x_3 = \dot{x}_2$$

Therefore, the state equations can be written as

$$\left.\begin{aligned} \dot{x}_1 &= x_2 = 0 \cdot x_1 + 1 \cdot x_2 + 0 \cdot x_3 \\ \dot{x}_2 &= x_3 = 0 \cdot x_1 + 0 \cdot x_2 + 1 \cdot x_3 \\ \text{and} \quad \dot{x}_3 &= -4x_1 - 8x_2 - 5x_3 + u \end{aligned}\right\} \quad \text{...(10.151)}$$

The last equation is obtained by equating the highest order derivative term in Eq. (10.149) to all other terms. That is,

$$\frac{d^3x_1}{dt^3} = -4x_1 - 8\frac{dx_1}{dt} - 5\frac{d^2x_1}{dt^2} + u$$

or

$$x_3 = -4x_1 - 8x_2 - 5x_3 + u$$

The set of state Eq. (10.151) can be written in vector-matrix form as

$$\begin{bmatrix} \dot{x}_1 \\ \dot{x}_2 \\ \dot{x}_3 \end{bmatrix} = \begin{bmatrix} 0 & 1 & 0 \\ 0 & 0 & 1 \\ -4 & -8 & -5 \end{bmatrix} \begin{bmatrix} x_1 \\ x_2 \\ x_3 \end{bmatrix} + \begin{bmatrix} 0 \\ 0 \\ 1 \end{bmatrix} u \quad \text{...(10.152)}$$

The second expression is

$$\frac{Y(s)}{X_1(s)} = s + 3$$

By cross-multiplication

$$Y(s) = (s + 3)\, X_1(s) \quad \text{...(10.153)}$$

In time domain it can be written as

$$y(t) = \frac{dx_1(t)}{dt} + 3x_1(t)$$

$$y = \dot{x}_1 + 3x_1$$

$$y = x_2 + 3x_1$$

$$y = 3x_1 + 1 \cdot x_2 + 0 \cdot x_3 \quad \text{...(10.154)}$$

In vector matrix form Eq. (10.154) can be is written as

$$y = [3 \;\; 1 \;\; 0] \begin{bmatrix} x_1 \\ x_2 \\ x_3 \end{bmatrix} \quad \text{...(10.155)}$$

Equations (10.152) and (10.155) represent the state model of the system.

Example 10.23: Obtain a state model for the system whose transfer function is given by

$$G(s) = \frac{k(c_1 s + c_2)}{s^3 + a_2 s^2 + a_1 s + a_0}.$$

Solution: The given transfer function is

$$G(s) = \frac{Y(s)}{U(s)} = \frac{K(c_1 s + c_2)}{s^3 + a_2 s^2 + a_1 s + a_0} \qquad \text{...(10.156)}$$

This can be divided into two parts as

$$\frac{Y(s)}{U(s)} = \frac{Y(s)}{X_1(s)} \cdot \frac{X_1(s)}{U(s)} \qquad \text{...(10.157)}$$

where
$$\frac{X_1(s)}{U(s)} = \frac{k}{s^3 + a_2 s^2 + a_1 s + a_0} \qquad \text{...(10.158)}$$

and
$$\frac{Y(s)}{X_1(s)} = (c_1 s + c_2) \qquad \text{...(10.159)}$$

Consider the function given by Eq. (10.159). By cross-multiplication,

$$(s^3 + a_2 s^2 + a_1 s + a_0)\, X_1(s) = kU(s)$$

The differential equation corresponding to this is

$$\frac{d^3 x_1}{dt^3} + a_2 \frac{d^2 x_1}{dt^2} + a_1 \frac{dx_1}{dt} + a_0 x_1 = ku \qquad \text{...(10.160)}$$

This is a third-order system, so three state variables are required.

Let the first state variable be equal to x_1. The other state variables, x_2 and x_3, are defined as follows:

$$x_2 = \dot{x}_1$$

$$x_3 = \dot{x}_2$$

Therefore, the state equations can be written as

$$\left.\begin{aligned} \dot{x}_1 &= x_2 = 0 \cdot x_1 + 1 \cdot x_2 + 0 \cdot x_3 \\ \dot{x}_2 &= x_3 = 0 \cdot x_1 + 0 \cdot x_2 + 1 \cdot x_3 \\ \text{and} \quad \dot{x}_3 &= -a_0 x_1 - a_1 x_2 - a_2 x_3 + ku \end{aligned}\right\} \qquad \text{...(10.161)}$$

The last equation is obtained by equating the highest order derivative term in Eq. (10.161) to all other terms. That is,

$$\frac{d^3 x_1}{dt^3} = -a_0 x_1 - a_1 \frac{dx_1}{dt} - a_2 \frac{d^2 x_1}{dt^2} + Ku$$

$$= -a_0 x_1 - a_1 x_2 - a_2 x_3 + Ku$$

The above set of state Eq, (10.161) can be written in vector-matrix form as

$$\begin{bmatrix} \dot{x}_1 \\ \dot{x}_2 \\ \dot{x}_3 \end{bmatrix} = \begin{bmatrix} 0 & 1 & 0 \\ 0 & 0 & 1 \\ -a_0 & -a_1 & -a_2 \end{bmatrix} \begin{bmatrix} x_1 \\ x_2 \\ x_3 \end{bmatrix} + \begin{bmatrix} 0 \\ 0 \\ k \end{bmatrix} u \qquad ...(10.162)$$

The second expression $\dfrac{Y(s)}{X_1(s)}$ can be written in time domain as

$$y(t) = c_1 x_2(t) + c_0 x_1(t)$$

$$= c_0 x_1(t) + c_1 x_2(t) + 0 \cdot x_3(t)$$

It can be written in vector matrix form as

$$y = [c_0 \ \ c_1 \ \ 0] \begin{bmatrix} x_1 \\ x_2 \\ x_3 \end{bmatrix} \qquad ...(10.163)$$

This is the output equation.

Equation (10.162) and (10.163) represent the state model of the system.

Example 10.24: A system is characterized by the equation.

$$\frac{Y(s)}{U(s)} = \frac{20(4s + 2)}{s^3 + 5s^2 + 8s + 2}$$

Find the state and output equations of the system and express them in vector-matrix form.

Solution: The given transfer function is

$$\frac{Y(s)}{U(s)} = \frac{20(4s + 2)}{s^3 + 5s^2 + 8s + 2} \qquad ...(10.164)$$

It can be divided into two parts as

$$\frac{Y(s)}{U(s)} = \frac{Y(s)}{X_1(s)} \cdot \frac{X_1(s)}{U(s)} \qquad ...(10.165)$$

where
$$\frac{X_1(s)}{U(s)} = \frac{20}{s^3 + 5s^2 + 8s + 2} \qquad ...(10.166)$$

and
$$\frac{Y(s)}{X_1(s)} = 4s + 2 \qquad ...(10.167)$$

Consider the function given by Eq. (10.166)

By cross-multiplication

$$(s^3 + 5s^2 + 8s + 2)\, X_1(s) = 20U(s)$$

The differential equation corresponding to this is

$$\frac{d^3x_1}{dt^3} + 5\frac{d^2x_1}{dt^2} + 8\frac{dx_1}{dt} + 2x_1 = 20u \qquad ...(10.168)$$

This is third-order system, so three state variables are required.

Let the first state variable be equal to x_1. The other state variable, x_2 and x_3, are defined as follows:

$$x_2 = \dot{x}_1$$
$$x_3 = \dot{x}_2$$

Therefore, the state equations can be written as

$$\left.\begin{aligned} \dot{x}_1 &= x_2 = 0\cdot x_1 + 1\cdot x_2 + 0\cdot x_3 \\ \dot{x}_2 &= x_3 = 0\cdot x_1 + 0\cdot x_2 + 1\cdot x_3 \\ \text{and} \quad \dot{x}_3 &= -2x_1 - 8x_2 - 5x_3 + 20u \end{aligned}\right\} \qquad ...(10.169)$$

The last equation is obtained by equating the highest order derivative term in Eq. (10.168) to all other terms.

This is

$$\frac{d^3x_1}{dt^3} = -2x_1 - 8\frac{dx_1}{dt} - 5\frac{d^2x_1}{dt^2} + 20u$$
$$= -2x_1 - 8x_2 - 5x_3 + 20u$$

The above set of state Eq. (10.169) can be written in vector-matrix form as

$$\begin{bmatrix} \dot{x}_1 \\ \dot{x}_2 \\ \dot{x}_3 \end{bmatrix} = \begin{bmatrix} 0 & 1 & 0 \\ 0 & 0 & 1 \\ -2 & -8 & 5 \end{bmatrix} \begin{bmatrix} x_1 \\ x_2 \\ x_3 \end{bmatrix} + \begin{bmatrix} 0 \\ 0 \\ 20 \end{bmatrix} u \qquad ...(10.170)$$

The second expression

$$\frac{Y(s)}{X_1(s)} = 4s + 2$$

$$Y(s) = (4s + 2)X_1(s) \qquad ...(10.171)$$

In time-domain form Eq. (10.171) can be written as

$$y(t) = 4x_2(t) + 2x_1(t)$$

or

$$y = 2\cdot x_1 + 4\cdot x_2 + 0\cdot x_3$$

This is the output equation.

It can be written in matrix form as

$$y = [2\ 4\ 0]\begin{bmatrix} x_1 \\ x_2 \\ x_3 \end{bmatrix} \quad ...(10.172)$$

The state model of the system is given by Eqs. (10.170) and (10.172).

10.15 SOLUTION OF STATE EQUATION

When a linear control system is subjected to both inputs and initial conditions, the total response, that is, the output of the system is given by:

Total response = Zero-state response + Zero-input response

1. **Zero-state response** means the response that is the output of the system due to inputs only. All the initial conditions of the system are zero.
2. **Zero-input response** means the response due to initial conditions only. All the inputs to the system are zero.

10.15.1 Homogeneous and Non-homogeneous Part of State Equation

The general form of the state equation is

$$\dot{x}(t) = Ax(t) + Bu(t) \quad ...(10.173)$$

or

$$\dot{x} = Ax + Bu$$

The term $Ax(t)$ is called the homogeneous part of the state equation.

The term $Bu(t)$ is called the non-homogeneous part of the state equation.

It is also known as the forcing function.

If the input $u(t) = 0$, the state equation becomes

$$\dot{x}(t) = Ax(t) \quad ...(10.174)$$

or

$$\dot{x} = Ax$$

The solution of the state Eq. (10.174) gives the unforced response of the control system and is governed by the initial conditions only.

If the input $u(t)$ is specified and it not equal to zero, the response obtained is known as forced response.

10.15.2 Laplace Transform Method of Solving State Equations

The state equation of a dynamic system is given by

$$\dot{x}(t) = Ax(t) + Bu(t) \quad ...(10.175)$$

Let $x(0)$ denote the initial state at $t = 0$.

Taking the Laplace transform of both sides of Eq. (10.175), we get

$$sX(s) - x(0) = AX(s) + BU(s)$$

or

$$[sI - A]X(s) = x(0) + BU(s)$$

where I is identity matrix.

$$X(s) = [sI - A]^{-1}\, x(0) + [sI - A]^{-1} + BU(s) \qquad ...(10.176)$$

The matrix $[sI - A]^{-1}$ is commonly known as resolvent matrix and is designated by $\phi(s)$.

Thus, $\phi(s) = [sI - A]^{-1}$

$$X(s) = \phi(s)\, x(0) + \phi(s) + BU(s) \qquad ...(10.177)$$

Taking inverse Laplace transform of Eq. (10.177) on both sides, we get

$$£^{-1}\, X(s) = £^{-1}[\phi(s)x(0)] + £^{-1}[\phi(s) + BU(s)]$$

or

$$x(t) = \phi(t)\, x(0) + £^{-1}[\phi(s) + BU(s)] \qquad ...(10.178)$$

In Eq. (10.178)

$$\phi(t) = £^{-1}\phi(s) = £^{-1}[sI - A]^{-1} \qquad ...(10.179)$$

The term $\phi(t)$ is known as state transition matrix (STM).

It is seen that the state-transition matrix is the inverse Laplace transform of the resolvent matrix.

Equation (10.178) gives the time response of state Eq. (10.180).

The term $\phi(t)x(0)$ is the zero-input response.

The term $£^{-1}[\phi(s)BU(s)]$ is the zero-state response.

10.15.3 Infinite Series Method for Solving Homogeneous State Equation

If the input $u(s) = 0$ the state equation becomes

$$\dot{x}(t) = Ax(t),\ x(0) = x_0 \qquad ...(10.180)$$

where $x(0)$ is the initial state at $t = 0$.

The solution of this equation gives the zero-input response.

Consider a scalar case

$$\frac{dx}{dt} = ax(t),\ x(0) = x_0 \qquad ...(10.181)$$

This equation has the solution

$$x(t) = e^{at}x_0$$

or

$$x(t) = \left(1 + at + \frac{a^2t^2}{2!} + \frac{a^3t^3}{3!} + ...\right) x_0 \qquad ...(10.182)$$

By analogy with the scalar case, the vector state equation has the solution

$$x(t) = \left(1 + At + \frac{A^2t^2}{2!} + \frac{A^3t^3}{3!} + ...\right) x_0 \qquad ...(10.183)$$

Each term inside the brackets is $n \times n$ matrix.

Equation (10.183) can be written as

$$x(t) = e^{At}x_0 \qquad ...(10.184)$$

is the solution of homogeneous state equation.

This solution of homogeneous state equation shows that the initial state $x(0)$ at $t = 0$ is driven to a state $x(t)$ at time t.

Since this transition in state is carried out by the matrix exponential e^{At}, e^{At} is known as the state transition matrix (STM).

It is denoted by $\phi(t)$, that is, $\phi(t) = e^{At}$ ...(10.185)

The term e^{At} is expressed in infinite series form as

$$e^{At} = I + At + \frac{A^2t^2}{2!} + \frac{A^3t^3}{3!} + \ldots \qquad ...(10.186)$$

10.16 STATE TRANSITION MATRIX (STM)

10.16.1 Computation of STM by Laplace Transformation

The homogeneous part of the state equation is represented by

$$\dot{x}(t) = Ax(t) \qquad ...(10.187)$$

On way of determining $\phi(t)$ is by taking the Laplace transform on both sides of Eq. (10.187); we have

$$sX(s) - x(0) = AX(s) \qquad ...(10.188)$$

Solving for $X(s)$ from Eq. (10.188), we get

$$[sI - A]\ X(s) = x(0)$$

$$X(s) = [sI - A]^{-1}\ x(0) \qquad ...(10.189)$$

$$X(s) = \phi(s)x(0) \qquad ...(10.190)$$

where it is assumed that $[sI - A]$ is a non-singular matrix.

$$\phi(s) = [sI - A]^{-1} \qquad ...(10.191)$$

It is called resolvent matrix.

Taking the inverse Laplace transform on both sides of Eq. (10.189) gives

$$x(t) = £^{-1}[sI - A]^{-1}\ x(0),\ t \geq 0 \qquad ...(10.192)$$

From Eq. (E4.5), the state transition matrix is given by

$$\phi(t) = £^{-1}\phi(s) = £^{-1}[sI - A]^{-1} = e^{At} \qquad ...(10.193)$$

10.16.2 Significance of State-Transition Matrix

Since the state-transition matrix satisfies the homogeneous state equation, it represents the free response of the system.

Thus, it governs the response that is excited by the initial conditions only.

$$\phi(t) = £^{-1}[sI - A]^{-1} \text{ and } \phi(t) = e^{At}$$

Since the state-transition matrix $\phi(t)$ is dependent only upon the matrix A, therefore, it is sometimes called the state-transition matrix of A. At the name implies, the state-transition matrix $\phi(t)$ completely defines the transition of the states from the initial time $t = 0$ any time t when the inputs are zero. Thus, the transition from any initial state $X(0)$ to any other state $X(t)$ for any onward time $t > 0$ is possible provided that $\phi(t)$ is known.

10.16.3 Properties of State-Transition Matrix

The state-transition matrix $\phi(t)$ possesses the following properties:

(1) $\phi(0)$ = identity matrix = I

Proof. We know that

$$\phi(t) = e^{At} = I + At + \frac{1}{2!}A^2t^2 + \frac{1}{3!}A^3t^3 + \ldots$$

Putting $t = 0$ gives

$$\phi(0) = I \quad \text{...(10.194)}$$

(2) $\dfrac{d\phi(t)}{dt} = A\phi(t)$

Proof. We know that

$$\dot{x}(t) = Ax(t)$$

$$x(t) = \phi(t)x(0)$$

Therefore,

$$\frac{d\phi(t)}{dt}x(0) = A\phi(t)x(0)$$

or

$$\frac{d\phi(t)}{dt} = A\phi(t) \quad \text{...(10.195)}$$

(3) $\phi^{-1}(t) = \phi(-t)$

Proof. We know that

$$\phi(t) = e^{At}$$

Multiplying both sides by e^{-At} gives

$$\phi(t)\,e^{-At} = e^{At}\,e^{-At}$$

or

$$\phi(t)\,e^{-At} = I$$

Pre-multiplying both sides by $\phi^{-1}(t)$, we get

$$\phi^{-1}(t)\phi(t)e^{-At} = \phi^{-1}(t)I$$

or

$$e^{-At} = \phi^{-1}(t)$$

or

$$\phi^{-1}(t) = e^{-At} \quad \text{...(10.196)}$$

or

$$\phi^{-1}(t) = \phi(-t) \quad \text{...(10.197)}$$

(4) $x(0) = \phi(-t)x(t)$

Proof, We know that

$$x(t) = e^{At}\, x(0)$$

or $$e^{At}\, x(0) = x(t)$$

Pre-multiplying both sides by e^{-At}

$$e^{-At}\, e^{At}\, x(0) = e^{-At}\, x(t)$$

Using Eq. (10.196), we get

$$x(0) = \phi(-t)x(t) \qquad \text{...(10.198)}$$

(5) $\phi(t_2 - t_1)\, \phi(t_1 - t_0) = \phi(t_2 - t_0)$ for any t_0, t_1, t_2

Proof, We know that

$$\phi(t) = e^{At}$$

Therefore, $$\phi(t_2 - t_1) = e^{A(t_2 - t_1)}$$

and $$\phi(t_1 - t_0) = e^{A(t_1 - t_0)}$$

Hence, $\phi(t_2 - t_1)\phi(t_1 - t_0)$

$$= e^{A(t_2 - t_1)}\, e^{A(t_1 - t_0)}$$

$$= e^{A(t_2 - t_1 + t_1 - t_0)}$$

$$= e^{A(t_2 - t_0)}$$

$$= \phi(t_2 - t_0) \qquad \text{...(10.199)}$$

This property of state-transition matrix is called the transition property. It implies that a state-transition process can be divided into a number of sequential transitions. Figure 10.24 shows that the transition for $t = t_0$ to $t = t_2$ is equal to the transition from t_0 to t_1 and then from t_1 to t_2. In general, the state-transition process can be divided into any number of parts.

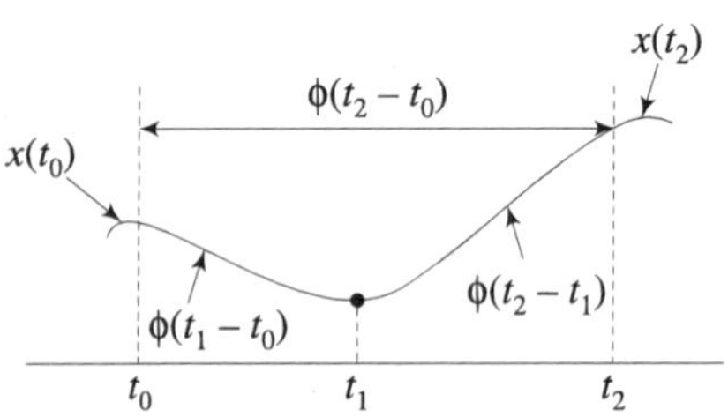

Fig. 10.24

(6) $[\phi(t)]^k = \phi(kt)$ for k = positive integer ...(10.200)

Proof. $$[\phi(t)]^k = e^{At}\, e^{At} \ldots k \text{ times}$$

$$= e^{kAt} = \phi(kt)$$

Computation of e^{At}

1. Laplace method

$$\phi(t) = e^{At} = £^{-1}[sI - A]^{-1}$$

$$= £^{-1}\left\{\frac{Adj[sI - A]}{det[sI - A]}\right\}$$

2. Power series method

$$e^{At} = I + At + \frac{A^2}{2!}t^2 + \frac{A^3}{3!}t^3 + \ldots$$

In the power series method, we have to calculate A, A^2, A^3, ... This is time consuming because it is not practicable to calculate higher powers of A.

Hence, this method is rarely used to compute e^{At}.

Laplace method is very convenient and hence generally used to compute e^{At}.

Eigenvalues

The roots of the determinant $\det|sI - A|$ are known as eigenvalues.

Eigenvalues of an $n \times n$ matrix A, also referred to a characteristic roots, are roots of the characteristic equation.

$$\det|\lambda I - A| = 0$$

or

$$|\lambda I - A| = 0$$

Eigenvalues and closed-loop poles of a system are same.

The location of eigenvalues in the S-plane indicates the system instability. For a system to be stable, eigenvalues cannot be located in right hand side of the S-plane.

Eigenvalues can be directly obtained from state equations without converting the state equations into transfer functions.

Eigenvectors

If A is an $n \times n$ matrix, there are n eigenvalues.

Any non-zero vector x_i which satisfies the matrix equation

$$(\lambda_i I - A)x_i = 0 \qquad ...(10.201)$$

where $\lambda_i = 1, 2, ..., n$ denotes the ith eigenvalue of A, is called the eigenvalues of A associated with the eigenvalue λ_i.

If A has distinct eigenvalues, the eigenvectors can be obtained directly from Eq. (10.201) by the solution of the set of n homogeneous equations written for different values of λ_i.

No eigenvector is unique. The fact at the x_i are not unique follows from the fact $|\lambda_i I - A| = 0$.

Example 10.25: Find the eigenvalues and eigenvectors of the matrix.

$$A = \begin{bmatrix} 0 & 1 \\ -2 & 3 \end{bmatrix}$$

Solution

(a) Eigenvalues

$$A = \begin{bmatrix} 0 & 1 \\ -2 & 3 \end{bmatrix}$$

$$|\lambda I - A| = \left| \lambda \begin{bmatrix} 1 & 0 \\ 0 & 1 \end{bmatrix} - \begin{bmatrix} 1 & 1 \\ -2 & 3 \end{bmatrix} \right|$$

$$= \left| \begin{bmatrix} \lambda & 0 \\ 0 & \lambda \end{bmatrix} - \begin{bmatrix} 0 & 1 \\ -2 & 3 \end{bmatrix} \right| = \begin{vmatrix} \lambda & -1 \\ +2 & \lambda - 3 \end{vmatrix}$$

The characteristic equation of A is $|\lambda I - A| = 0$

$$\lambda(\lambda - 3) - (-1)(+2) = 0$$

$$\lambda^2 - 3\lambda + 2 = 0$$

$$(\lambda - 1)(\lambda - 2) = 0$$

Therefore, the two eigenvalues are $\lambda_1 = 1$ and $\lambda_2 = 2$.

(One could also write $\lambda_1 = 2$ and $\lambda_2 = 1$.) We shall find eigenvectors for $\lambda_1 = 1$ and $\lambda_2 = 2$.

(b) Eigenvector

Let the eigenvectors be written as

$$x_1 = \begin{bmatrix} x_{11} \\ x_{21} \end{bmatrix} \; \& \; x_2 = \begin{bmatrix} x_{12} \\ x_{22} \end{bmatrix}$$

Substituting $\lambda_1 = 1$ and x_1, we get

$$\left[1 \begin{bmatrix} 1 & 0 \\ 0 & 1 \end{bmatrix} - \begin{bmatrix} 0 & 1 \\ -2 & 3 \end{bmatrix} \right] \begin{bmatrix} x_{11} \\ x_{21} \end{bmatrix} = 0$$

$$\begin{bmatrix} 1 & -1 \\ 2 & -2 \end{bmatrix} \begin{bmatrix} x_{11} \\ x_{21} \end{bmatrix} = 0$$

This gives $x_{11} - x_{21} = 0$

and $2x_{11} - 2x_{21} = 0$

That is, $x_{11} = x_{21}$

If $x_{11} = 1$, then $x_{21} = x_{11}$

Therefore, one choice of eigenvector is

$$x_1 = \begin{bmatrix} 1 \\ 1 \end{bmatrix}$$

Similarly, for $\lambda_2 = 2$

$$\left[2\begin{bmatrix} 1 & 0 \\ 0 & 1 \end{bmatrix} - \begin{bmatrix} 0 & 1 \\ -2 & 3 \end{bmatrix}\right]\begin{bmatrix} x_{12} \\ x_{22} \end{bmatrix} = 0$$

$$\begin{bmatrix} 2 & -1 \\ 2 & -1 \end{bmatrix}\begin{bmatrix} x_{12} \\ x_{22} \end{bmatrix} = 0$$

or $\quad 2x_{12} - x_{22} = 0$, that is, $x_{22} = 2x_{12}$

and $\quad 2x_{12} - x_{22} = 0$, that is, $x_{22} = 2x_{12}$

Therefore, eigenvector for $\lambda_2 = 2$ is

$$x_1 = \begin{bmatrix} x_{12} \\ x_{22} \end{bmatrix} = \begin{bmatrix} 1 \\ 2 \end{bmatrix}.$$

Example 10.26: Obtain the STM for the state model whose matrix A is given by

$$A = \begin{bmatrix} 0 & 1 \\ -2 & -3 \end{bmatrix}.$$

Solution:

$$[sI - A] = s\begin{bmatrix} 1 & 0 \\ 0 & 1 \end{bmatrix} - \begin{bmatrix} 0 & 1 \\ -2 & -3 \end{bmatrix}$$

$$= \begin{bmatrix} s & 0 \\ 0 & s \end{bmatrix} - \begin{bmatrix} 0 & 1 \\ -2 & -3 \end{bmatrix} = \begin{bmatrix} s & -1 \\ 2 & s+3 \end{bmatrix}$$

$$\phi(s) = [sI - A]^{-1} = \frac{\text{adj}[sI - A]}{|sI - A|}$$

$$\text{adj}[sI - A] = [\text{cofactor matrix of } [sI - A]]^T$$

$$\begin{bmatrix} C_{11} & C_{12} \\ C_{21} & C_{22} \end{bmatrix}^T = \begin{bmatrix} s+3 & -2 \\ 1 & s \end{bmatrix}^T$$

$$= \begin{bmatrix} s+3 & 1 \\ -2 & s \end{bmatrix}$$

$$|sI - A| = \begin{vmatrix} s+3 & 1 \\ -2 & s \end{vmatrix} = (s + 3)s - (-2)(1)$$

$$= s^2 + 3s + 2 = (s + 1)(s + 2)$$

Therefore,

$$\phi(s) = \frac{\begin{bmatrix} s+3 & 1 \\ -2 & s \end{bmatrix}}{(s+1)(s+2)}$$

The state transition matrix (STM) is given by

$$\phi(s) = £^{-1}\phi(s)£^{-1}\begin{bmatrix} \dfrac{s+3}{(s+1)(s+2)} & \dfrac{1}{(s+1)(s+2)} \\ \dfrac{-2}{(s+1)(s+2)} & \dfrac{s}{(s+1)(s+2)} \end{bmatrix}$$

$$= \begin{bmatrix} £^{-1}\dfrac{s+3}{(s+1)(s+2)} & £^{-1}\dfrac{1}{(s+1)(s+2)} \\ £^{-1}\dfrac{-2}{(s+1)(s+2)} & £^{-1}\dfrac{s}{(s+1)(s+2)} \end{bmatrix}$$

$$\frac{s+3}{(s+1)(s+2)} = \frac{2}{s+1} - \frac{1}{s+2}$$

$$£^{-1}\frac{s+3}{(s+1)(s+2)} = £^{-1}\frac{2}{s+1} - £^{-1}\frac{1}{s+2} = 2e^{-t} - e^{-2t}$$

$$\frac{1}{(s+1)(s+2)} = \frac{-1}{s+1} + \frac{1}{s+2}$$

$$£^{-1}\frac{1}{(s+1)(s+2)} = £^{-1}\frac{-1}{s+1} + £^{-1}\frac{1}{s+2} = e^{-t} - e^{-2t}$$

$$\frac{-2}{(s+2)(s+2)} = \frac{-2}{s+1} + \frac{2}{s+2}$$

$$£^{-1}\frac{-2}{(s+1)(s+2)} = £^{-1}\frac{-2}{s+1} + £^{-1}\frac{2}{s+2} = -2e^{-t} + 2e^{2t}$$

$$\frac{s}{(s+1)(s+2)} = \frac{-1}{s+1} + \frac{2}{s+2}$$

$$£^{-1}\frac{s}{(s+1)(s+2)} = £^{-1}\frac{-1}{s+1} + £^{-1}\frac{2}{s+2} = -e^{-t} + 2e^{-2t}$$

Therefore,

$$\phi(s) = \begin{bmatrix} 2e^{-t} - 2e^{-2t} & e^{-t} - e^{-2t} \\ -2e^{-t} + 2e^{-2t} & -e^{-t} + 2e^{-2t} \end{bmatrix}$$

Example 10.27: A linear time-invariant system is characterized by the state equation.

$$\begin{bmatrix} \dot{x}_1 \\ \dot{x}_2 \end{bmatrix} = \begin{bmatrix} 1 & 0 \\ 1 & 1 \end{bmatrix}\begin{bmatrix} x_1 \\ x_2 \end{bmatrix} + \begin{bmatrix} 0 \\ 1 \end{bmatrix}u$$

where u is a unit step function. The initial condition is

$$X(0) = \begin{bmatrix} 1 \\ 0 \end{bmatrix}$$

Using inverse Laplace transform method, obtain the solution of the state equation.

Solution:

$$A = \begin{bmatrix} 1 & 0 \\ 1 & 1 \end{bmatrix}, \quad B = \begin{bmatrix} 0 \\ 1 \end{bmatrix}$$

For unit step unction $U(s) = \frac{1}{s}$

The state equation of a dynamic system is

$$\dot{x} = Ax + Bu(t)$$

The solution of this equation is

$$x(t) = £^{-1}[\phi(s)x(0)] + £^{-1}[\phi(s)BU(s)]$$

$$\phi(s) = [sI - A]^{-1}$$

$$[sI - A] = s\begin{bmatrix} 1 & 0 \\ 0 & 1 \end{bmatrix} - \begin{bmatrix} 1 & 0 \\ 1 & 1 \end{bmatrix}$$

$$= \begin{bmatrix} s-1 & 0 \\ -1 & s-1 \end{bmatrix}$$

$$[sI - A]^{-1} = \frac{\text{adj}[sI - A]}{[sI - A]}$$

$$\text{adj}[sI - A] = [\text{cofactor matrix of } [sI - A]]^T$$

$$\begin{bmatrix} C_{11} & C_{12} \\ C_{21} & C_{22} \end{bmatrix}^T = \begin{bmatrix} s-1 & 1 \\ 0 & s-1 \end{bmatrix}^T$$

$$= \begin{bmatrix} s-1 & 0 \\ 1 & s-1 \end{bmatrix}$$

$$|sI - A| = \begin{bmatrix} s-1 & 0 \\ 1 & s-1 \end{bmatrix} = (s-1)(s-1) = (s-1)^2$$

Therefore,
$$[sI - A]^{-1} = \frac{\begin{bmatrix} s-1 & 0 \\ 1 & s-1 \end{bmatrix}}{(s-1)^2}$$

$$= \begin{bmatrix} \frac{1}{s-1} & 0 \\ \frac{1}{(s-1)^2} & \frac{1}{s-1} \end{bmatrix} \begin{bmatrix} 1 \\ 0 \end{bmatrix}$$

$$\phi(s)X(0) = \begin{bmatrix} \frac{1}{s-1} & 0 \\ \frac{1}{(s-1)^2} & \frac{1}{s-1} \end{bmatrix} \begin{bmatrix} 1 \\ 0 \end{bmatrix}$$

$$= \begin{bmatrix} \frac{1}{s-1} \\ \frac{1}{(s-1)^2} \end{bmatrix}$$

The zero-input response

$$£^{-1}[\phi(s)X(0)] = £^{-1} \begin{bmatrix} \frac{1}{s-1} \\ \frac{1}{(s-1)^2} \end{bmatrix}$$

$$= \begin{bmatrix} £^{-1} \frac{1}{s-1} \\ £^{-1} \frac{1}{(s-1)^2} \end{bmatrix} = \begin{bmatrix} e^t \\ te^t \end{bmatrix}$$

$$\phi(s)B = \begin{bmatrix} \frac{1}{s-1} & 0 \\ \frac{1}{(s-1)^2} & \frac{1}{s-1} \end{bmatrix} \begin{bmatrix} 1 \\ 0 \end{bmatrix} = \begin{bmatrix} 0 \\ \frac{1}{s-1} \end{bmatrix}$$

$$\phi(s)BU(s) = \begin{bmatrix} 0 \\ \frac{1}{s-1} \end{bmatrix} \begin{bmatrix} 1 \\ s \end{bmatrix} = \begin{bmatrix} 0 \\ \frac{1}{s(s-1)} \end{bmatrix}$$

Therefore, the zero-state response

$$= £^{-1}[\phi(s)BU(s)]$$

$$= £^{-1} \begin{bmatrix} 0 \\ \frac{1}{s(s-1)} \end{bmatrix} = £^{-1} \begin{bmatrix} 0 \\ -\frac{1}{s} + \frac{1}{s-1} \end{bmatrix}$$

$$= \begin{bmatrix} 0 \\ -1 + e^t \end{bmatrix}$$

The total response = zero-input response + zero-state response

$$x(t) = \begin{bmatrix} e^t \\ te^t \end{bmatrix} + \begin{bmatrix} 0 \\ -1+e^t \end{bmatrix} = \begin{bmatrix} e^t \\ -1+(t+1)e^t \end{bmatrix}.$$

10.17 CONTROLLABILITY AND OBSERVABILITY

The concepts of controllability and observability were first introduced by Kalman in 1960. These concepts are necessary to establish the validity of the classical transfer function methods.

10.17.1 Definition of Controllability

A system is said to be controllable if any initial state $x(t_0)$ can be transferred to any final state $x(t_f)$ in a finite time $t_f \geq 0$ by some control input $u(t)$.

The definition does not restrict the choice of input $u(t)$. The idea is that it is possible to move the system state to any desired destination. Thus, controllability is the ability of the control input to affect each state variable.

If every state $x(t_0)$ of the system is controllable in a finite time interval, the system is said to be completely controllable or simply controllable.

The controllability defined above is associated with the states and should not be confused with the output controllability, which is the property of the input-output relationship.

An uncontrollable system is a system which has a subsystem that is physically disconnected from the input.

10.18 KALMAN'S TEST FOR CONTROLLABILITY

A linear time-invariant system described by the state equation.

$$\frac{dx(t)}{dt} = Ax(t) + Bu(t)$$

is completely state controllable, it is necessary and sufficient that the following $n \times np$ controllability matrix has a rank of n, the order of the system.

$$Q_c = [B \vdots AB \vdots A^2B \vdots \cdots \vdots A^{n-1}\, B]$$

The controllability matrix consists of the columns of B followed by the columns of AB, and so on.

Since the matrices A and B are involved, sometimes we say that the pair $[A, B]$ is controllable which implies that Q_c is of rank n.

For a single-input system Q_c is an $n \times n$ square matrix.

The matrix Q_c is non-singular, that is, its determinant is non-zero.

Therefore, in the SISO case, the system is controllable if and only if the determinant of Q_c is non-zero.

For complete state controllability, all the columns of Q_c must be linearly independent.

10.19 RANK OF A MATRIX

A square matrix of order $n \times n$ is said to have a rank r, if and only if, it has at least one non-singular sub-matrix of order r but has no non-singular sub-matrix of order more than r.

Therefore, we find out the largest square matrix in the given matrix whose determinant is not zero.

Example 10.28: Test the following system for controllability using Kalman's test

$$\begin{bmatrix} \dot{x}_1 \\ \dot{x}_2 \end{bmatrix} = \begin{bmatrix} 2 & 3 \\ 6 & -1 \end{bmatrix} \begin{bmatrix} x_1 \\ x_2 \end{bmatrix} + \begin{bmatrix} 1 \\ -2 \end{bmatrix} u$$

Solution:

$$A = \begin{bmatrix} 2 & 3 \\ 6 & -1 \end{bmatrix}, \; B = \begin{bmatrix} 1 \\ -2 \end{bmatrix}$$

A is 2×2, so the controllability matrix is

$$Q_c = [B \,\vdots\, AB]$$

$$AB = \begin{bmatrix} 2 & 3 \\ 6 & -1 \end{bmatrix} \begin{bmatrix} 1 \\ -2 \end{bmatrix} = \begin{bmatrix} -4 \\ 8 \end{bmatrix}$$

Therefore,

$$Q_c = \begin{bmatrix} 1 & -4 \\ -2 & 8 \end{bmatrix}$$

Which is not of rank 2 (order of A), since

$$\begin{vmatrix} 1 & -4 \\ -2 & 8 \end{vmatrix} = 1 \times 8 - (-4)(-2) = 0$$

Therefore, the system is not completely controllable.

Example 10.29: For the following system, test that controllability.

$$\begin{bmatrix} \dot{x}_1 \\ \dot{x}_2 \\ \dot{x}_3 \end{bmatrix} = \begin{bmatrix} -2 & 1 & 2 \\ 4 & 0 & 3 \\ 1 & -1 & 0 \end{bmatrix} \begin{bmatrix} x_1 \\ x_2 \\ x_3 \end{bmatrix} + \begin{bmatrix} 0 & 4 \\ -5 & 0 \\ 0 & 0 \end{bmatrix} \begin{bmatrix} u_1 \\ u_2 \end{bmatrix}.$$

Solution:

$$A = \begin{bmatrix} -2 & 1 & 2 \\ 4 & 0 & 3 \\ 1 & -1 & 0 \end{bmatrix}, \; B = \begin{bmatrix} 0 & 4 \\ -5 & 0 \\ 0 & 0 \end{bmatrix}$$

A is 3×3 so the controllability matrix is given by

$$Q_c = [B \,\vdots\, AB \,\vdots\, A^2B]$$

$$AB = \begin{bmatrix} -2 & 1 & 2 \\ 4 & 0 & 3 \\ 1 & -1 & 0 \end{bmatrix}\begin{bmatrix} 0 & 4 \\ -5 & 0 \\ 0 & 0 \end{bmatrix} = \begin{bmatrix} -5 & -8 \\ 0 & 16 \\ 5 & 4 \end{bmatrix}$$

$$A^2B = [A][AB]$$

$$= \begin{bmatrix} -2 & 1 & 2 \\ 4 & 0 & 3 \\ 1 & -1 & 0 \end{bmatrix}\begin{bmatrix} -5 & -8 \\ 0 & 16 \\ 5 & 4 \end{bmatrix} = \begin{bmatrix} 20 & 40 \\ -5 & -20 \\ -5 & -24 \end{bmatrix}$$

Therefore,

$$Q_c = \begin{bmatrix} 0 & 4 & -5 & -8 & 20 & 40 \\ -5 & 0 & 0 & 16 & -5 & -20 \\ 0 & 0 & 5 & 4 & -5 & -24 \end{bmatrix}$$

To be of full rank, the controllability matrix must have three linearly independent equations. We have

$$\begin{vmatrix} 0 & 4 & -5 \\ -5 & 0 & 0 \\ 0 & 0 & 5 \end{vmatrix} = 0 - 4(5)(-5) - 0 = 100 \neq 0$$

Since the determinant is non-zero, the system is completely controllable.

Example 10.30: A system is described by the dynamic equations

$$\begin{bmatrix} \dot{x}_1 \\ \dot{x}_2 \end{bmatrix} = \begin{bmatrix} -1 & 1 \\ 0 & -1 \end{bmatrix}\begin{bmatrix} x_1 \\ x_2 \end{bmatrix} + \begin{bmatrix} 1 & 0 \\ 0 & 1 \end{bmatrix}u \text{ and } \begin{bmatrix} 1 & 0 \\ 0 & 1 \end{bmatrix}x$$

Comment on the controllability of the system.

Solution:

$$A = \begin{bmatrix} -1 & 1 \\ 0 & -1 \end{bmatrix}, \; B = \begin{bmatrix} 1 & 0 \\ 0 & 1 \end{bmatrix}$$

A is 2×2, so the controllability matrix is

$$Q_c = [B \; \vdots \; AB]$$

$$AB = \begin{bmatrix} -1 & 1 \\ 0 & 1 \end{bmatrix}\begin{bmatrix} 1 & 0 \\ 0 & 1 \end{bmatrix} = \begin{bmatrix} -1 & 1 \\ 0 & -1 \end{bmatrix}$$

$$Q_c = \begin{bmatrix} 1 & 0 & \vdots & -1 & 1 \\ 0 & 1 & \vdots & 0 & -1 \end{bmatrix}$$

Determinant of sub-matrix $\begin{bmatrix} -1 & 1 \\ 0 & -1 \end{bmatrix}$ is $\begin{bmatrix} -1 & 1 \\ 0 & -1 \end{bmatrix}$

Now, $\begin{bmatrix} -1 & 1 \\ 0 & -1 \end{bmatrix} = 1$, that is, the determinant is not equal to zero, and the rank of Q_c is 2 (order of the system).

Hence, the system is state controllable.

10.20 OBSERVABILITY

10.20.1 Definition

A system is completely observable if every state $x(t_0)$ can be exactly determined from measurements of the output $y(t)$ over a finite interval of time $t_0 \leq t \leq t_f$.

The concept of observability relates to the condition of observing or estimating the state variables from the output variables, which are generally measurable.

Observability is concerned with the ability of each state variable to influence the output of system.

There are no restrictions placed on the output. The definition indicates that any earlier value of the state vector $x(t)$ is determinable by watching the output $y(t)$.

10.20.2 Kalman's Test for Observability

A MIMO linear time-invariant system described by the dynamic equations

$$\frac{dx(t)}{dt} = Ax(t) + B\dot{u}(t)$$

$$y(t) = Cx(t)$$

to be completely observable, it is necessary and sufficient that the following $n \times nq$ observability matrix Q_0 has a rank of n, the order of the system.

$$Q_0 = \begin{bmatrix} C \\ \text{L} \\ CA \\ \text{L} \\ CA^2 \\ \text{L} \\ \text{M} \\ CA^{n-1} \end{bmatrix} = [C^T \; A^T C^T \; A^{2T} C^T \; \cdots \; A^{n-1} C^T]$$

This condition is also referred to as the pair $[A, C]$ being observable.

In particular, if the system has only one output C is a $1 \times n$ row matrix Q_c is an $(n \times n)$ square matrix. The system is completely observable if Q_c is non-singular, that is determinant is not zero.

10.20.3 Duality Property (Duality Theorem)

The principle of duality was introduced by Kalman. We have seen the conditions of controllability and observability.

The pair $[A, B]$ is controllable if the rank of

$$Q_c = [B \ AB \ \cdots \ A^{n-1}B] \text{ is } n$$

The pair $[A, C]$ is observable if the rank of

$$Q_0 = \left[C^T \ A^T C^T \ A^{2T} C^T \ \cdots \ (A^{n-1})^T C^T\right] \text{ is } n$$

Thus, there is a link between the two concepts. The duality theorem states that if the dynamics of a system is described in state space with the help of matrices $(A, B, C, D,)$ then the system is completely state controllable if and only if the dual system $(A^T, B^T, C^T, D^T,)$ is completely observable.

The duality principle specifies that

- Pair $[A, B]$ is controllable also means that the pair $[A^T, B^T,]$ is observable.
- Pair $[A, C]$ is observable also means that the pair $[A^T, C^T,]$ is controllable.

With the use of this theorem state controllability of a given system can be checked by performing observability test of its dual. Similarly, the observability of a given system can be checked by performing state controllability test of its dual.

10.20.4 Controllability, Observability, and Pole-zero Cancellation

The concepts of controllability and observability are closely related to the properties of the transfer function.

Let the input-output relationship of an nth-order system having distinct eigenvalues (roots) be represented as

$$\frac{Y(s)}{U(s)} = \frac{k(s - z_1)(s - z_2) \dots (s - z_m)}{(s - p_1)(s - p_2) \dots (s - p_n)}, \quad m < n$$

$$= \frac{c_1}{s - p_1} + \frac{c_2}{s - p_2} + \dots + \frac{c_n}{s - p_n}$$

where $c_1, c_2, \dots, c_n$ are the residues of the transfer function at the corresponding poles $p_1, p_2, \dots, p_n$.

Let us assume that the state-space representations of the control system are of the following forms:

First form

$$\begin{bmatrix} \dot{x}_1 \\ \dot{x}_2 \\ \vdots \\ \dot{x}_n \end{bmatrix} = \begin{bmatrix} p_1 & 0 & \cdots & 0 \\ 0 & p_2 & \cdots & 0 \\ \vdots & \vdots & \cdots & \vdots \\ 0 & 0 & \cdots & p_n \end{bmatrix} \begin{bmatrix} x_1 \\ x_2 \\ \vdots \\ x_n \end{bmatrix} \begin{bmatrix} c_1 \\ c_2 \\ \vdots \\ c_n \end{bmatrix} u$$

$$y = [1 \;\; 1 \;\; \ldots \;\; 1] \begin{bmatrix} x_1 \\ x_2 \\ \vdots \\ x_n \end{bmatrix} + Du$$

Second form

$$\begin{bmatrix} \dot{x}_1 \\ \dot{x}_2 \\ \vdots \\ \dot{x}_n \end{bmatrix} = \begin{bmatrix} p_1 & 0 & \cdots & 0 \\ 0 & p_2 & \cdots & 0 \\ \vdots & \vdots & \cdots & \vdots \\ 0 & 0 & \cdots & p_n \end{bmatrix} \begin{bmatrix} x_1 \\ x_2 \\ \vdots \\ x_n \end{bmatrix} \begin{bmatrix} 1 \\ 1 \\ \vdots \\ 1 \end{bmatrix} u$$

$$y = [c_1 \;\; c_2 \;\; \ldots \;\; c_n] \begin{bmatrix} x_1 \\ x_2 \\ \vdots \\ x_n \end{bmatrix} + Du$$

Suppose that the transfer function has identical pole and zero. That is, $p_i = z_i$ which makes $c_i = 0$.

In such a case, if the state-space is represented in the 'first form' then the matrix B will have the element $c_i = 0$, making the state variable x_i uncontrollable.

If the state-spaces, representation is in the 'Second Form', then the matrix C will have the element $c_i = 0$, making the state variable x_i unobservable.

Therefore, depending upon how the state variables are defined and state-space represented, the pole-zero cancellation due to identical pole and zero, will make the system either not state controllable or unobservable or both; otherwise the state-space model representation can be completely controllable and observable. The controllable or unobservable modes correspond to cancelled poles.

10.20.5 Theorem

If the input-output transfer function of a linear system has a pole-zero cancellation, the system will be either uncontrollable or unobservable, or both, depending on how the state variables are defined. On the other hand, if the input-output function does not have pole-zero cancellation, the system can always be represented by dynamic equations as a completely controllable and observable system.

The importance of this theorem is that if a linear system is modelled by a transfer function with no pole-zero cancellation, then it is a controllable and observable system, no matter how the state-variable model is derived.

Example 10.31: A linear time-invariant system is described by the following state variable model:

$$\begin{bmatrix} \dot{x}_1 \\ \dot{x}_2 \end{bmatrix} = \begin{bmatrix} -1 & 0 \\ 0 & -2 \end{bmatrix} \begin{bmatrix} x_1 \\ x_2 \end{bmatrix} + \begin{bmatrix} 0 \\ 1 \end{bmatrix} u$$

$$y(t) = [1 \quad 2] \begin{bmatrix} x_1 \\ x_2 \end{bmatrix}$$

Comment on the controllability and observability of the system.

Solution: For the given state model

$$A = \begin{bmatrix} -1 & 0 \\ 0 & -2 \end{bmatrix}, \; B = \begin{bmatrix} 0 \\ 1 \end{bmatrix}, \; C = [1 \quad 2], \; D = 0$$

Since A is 2×2 matrix, the controllability matrix can be written as

$$Q_c = [B \; \vdots \; AB]$$

$$AB = \begin{bmatrix} -1 & 0 \\ 0 & -2 \end{bmatrix} \begin{bmatrix} 0 \\ 1 \end{bmatrix} = \begin{bmatrix} 0 \\ -2 \end{bmatrix}$$

Therefore, $$Q_c = \begin{bmatrix} 0 & 0 \\ 1 & -2 \end{bmatrix}$$

The determinant of $Q_c = \begin{vmatrix} 0 & 0 \\ 1 & -2 \end{vmatrix} = 0$

Therefore, Q_c is not of rank 2 (order of A). Hence, the given system is not completely controllable.

10.20.6 Observability Test

Since A is 2×2 matrix, the observability matrix can be written as

$$Q_0 = \begin{bmatrix} C \\ L \\ CA \end{bmatrix}$$

$$CA = [1 \quad 2] \begin{bmatrix} -1 & 0 \\ 0 & -2 \end{bmatrix} = [-1 \quad -4]$$

Therefore, $$Q_0 = \begin{bmatrix} 1 & 2 \\ -1 & -4 \end{bmatrix}$$

The determinant of Q_c is

$$\begin{vmatrix} 1 & 2 \\ -1 & -4 \end{vmatrix} = 1(-4) - (-1)(2) = -2$$

Since the determinant of Q_c is non-zero, the rank of Q_c is 2 (order of A).

Hence the system is completely observable.

Example 10.32: A linear time-invariant system is described by the following differential equations.

$$\frac{dx_1(t)}{dt} = -2x_1(t) + 4x_2(t)$$

$$\frac{dx_2(t)}{dt} = -2x_1(t) - x_2(t) + u(t)$$

Comment on the controllability and stability of the system.

Solution: The state-space representation of the system can be written as

$$\begin{bmatrix} \dot{x}_1 \\ \dot{x}_2 \end{bmatrix} = \begin{bmatrix} -2 & 4 \\ 2 & -1 \end{bmatrix} \begin{bmatrix} x_1 \\ x_2 \end{bmatrix} + \begin{bmatrix} 0 \\ 1 \end{bmatrix} u(t)$$

Here, $$A = \begin{bmatrix} -2 & 4 \\ 2 & -1 \end{bmatrix}, \quad \text{and } B \begin{bmatrix} 0 \\ 1 \end{bmatrix}$$

Since A is 2×2 matrix, the controllability matrix can be written as

$$Q_c = [B \ \vdots \ AB]$$

$$AB = \begin{bmatrix} -2 & 4 \\ 2 & -1 \end{bmatrix} \begin{bmatrix} 0 \\ 1 \end{bmatrix} = \begin{bmatrix} 4 \\ -1 \end{bmatrix}$$

Therefore, $$Q_c = \begin{bmatrix} 0 & 4 \\ 1 & -1 \end{bmatrix}$$

The determinant of Q_c is

$$\begin{vmatrix} 0 & 4 \\ 1 & -1 \end{vmatrix} = 0(-1) - (1)(4) = -4$$

Since the determinant of Q_c is non-zero, the rank of Q_c is 2 (order of A).

Hence, the system is state controllable.

The characteristic equation of the system is given by

$$|sI - A| = 0$$

$$sI = s\begin{bmatrix} 1 & 0 \\ 0 & 1 \end{bmatrix} = \begin{bmatrix} s & 0 \\ 0 & s \end{bmatrix}$$

$$|sI - A| = \begin{bmatrix} s & 0 \\ 0 & s \end{bmatrix} - \begin{bmatrix} -2 & 4 \\ 2 & -1 \end{bmatrix}$$

$$|sI - A| = \begin{bmatrix} s+2 & -4 \\ -2 & s+1 \end{bmatrix}$$

Therefore, the characteristic equation of the system becomes

$$\begin{bmatrix} s+2 & -4 \\ -2 & s+1 \end{bmatrix} = 0$$

or $$(s + 2)(s + 1) - (-2)(-4) = 0$$

$$s^2 + 3s + 2 - 8 = 0$$

$$s^2 + 3s - 6 = 0$$

The roots of this equation are given by

$$s_{1,2} = \frac{-3 \pm \sqrt{9 + 4 \times 6}}{2 \times 1}$$

$$s_1 = -4.37$$

$$s_2 = +1.37$$

Since one of the roots lies on the right half of the S-plane, the system is unstable.

Example 10.33: A control system is described by the differential equation

$$\frac{d^3y(t)}{dt^3} = u(t)$$

where $y(t)$ is the observed output and $u(t)$ is the input. Express the system in the state-variable form

$$\dot{x} = Ax + Bu \qquad y = Cx + Du$$

Check the controllability and observability of the system.

Solution: The given system is of third order and, therefore, three state variables are required. Let the state variables be x_1, x_2 and x_3.

The state variables are defined as

$$x_1(t) = y(t)$$

$$\dot{x}_1(t) = x_2(t) = \frac{dy(t)}{dt}$$

$$\dot{x}_2(t) = x_3(t) = \frac{d^2y}{dt^2}$$

$$\dot{x}_3(t) = \frac{d^3y}{dt^3} = u(t)$$

The state equations can be written as

$$\dot{x}_1 = x_2 = 0 \cdot x_1 + 1 \cdot x_2 + 0 \cdot x_3$$

$$\dot{x}_2 = x_3 = 0 \cdot x_1 + 0 \cdot x_2 + 1 \cdot x_3$$

$$\dot{x}_3 = u(t) = 0 \cdot x_1 + 0 \cdot x_2 + 0 \cdot x_3 + 1 \cdot u(t)$$

The output equation can be written as:

$$y(t) = x_1(t) = 1 \cdot x_1 + 0 \cdot x_2 + 0 \cdot x_3 + 0 \cdot u$$

Therefore, the state model of the system can be written as

$$\begin{bmatrix} \dot{x}_1 \\ \dot{x}_2 \\ \dot{x}_3 \end{bmatrix} = \begin{bmatrix} 0 & 1 & 0 \\ 0 & 0 & 1 \\ 0 & 0 & 0 \end{bmatrix} \begin{bmatrix} x_1 \\ x_2 \\ x_3 \end{bmatrix} + \begin{bmatrix} 0 \\ 0 \\ 1 \end{bmatrix} u$$

$$y = [1 \quad 0 \quad 0] \begin{bmatrix} x_1 \\ x_2 \\ x_3 \end{bmatrix} + [0]u$$

This model is of the form

$$\dot{X} = Ax + Bu \qquad Y = Cx + Du$$

where
$$A = \begin{bmatrix} 0 & 1 & 0 \\ 0 & 0 & 1 \\ 0 & 0 & 0 \end{bmatrix}, \mathrm{B} = \begin{bmatrix} 0 \\ 0 \\ 1 \end{bmatrix}, C = [1 \ 0 \ 0], D = 0$$

10.20.7 Controllability Test

Since A is 3 × 3 matrix, the controllability matrix can be written as

$$Q_c = [B \ \vdots \ AB \ \vdots \ A^2B]$$

$$AB = \begin{bmatrix} 0 & 1 & 0 \\ 0 & 0 & 1 \\ 0 & 0 & 0 \end{bmatrix} \begin{bmatrix} 0 \\ 0 \\ 1 \end{bmatrix} = \begin{bmatrix} 0 \\ 1 \\ 0 \end{bmatrix}$$

$$A^2B = [A][AB]$$

$$= \begin{bmatrix} 0 & 1 & 0 \\ 0 & 0 & 1 \\ 0 & 0 & 0 \end{bmatrix} \begin{bmatrix} 0 \\ 0 \\ 1 \end{bmatrix} = \begin{bmatrix} 0 \\ 1 \\ 0 \end{bmatrix}$$

Therefore, $$Q_c = \begin{bmatrix} 0 & 0 & 1 \\ 0 & 1 & 0 \\ 1 & 0 & 0 \end{bmatrix}$$

The determinant of Q_c is

$$\begin{vmatrix} 0 & 0 & 1 \\ 0 & 1 & 0 \\ 1 & 0 & 0 \end{vmatrix} = -1$$

Since the determinant of Q_c is not equal to zero, the Q_c is 3 (order of A). Hence, the system is completely controllable.

10.20.8 Observability Test

Since A is 3 × 3 matrix, the observability matrix can be written as

$$Q_c = \begin{bmatrix} C \\ CA \\ CA^2 \end{bmatrix}$$

$$C = [1 \quad 0 \quad 0]$$

$$CA = [1 \quad 0 \quad 0] \begin{bmatrix} 0 & 1 & 0 \\ 0 & 0 & 1 \\ 0 & 0 & 0 \end{bmatrix} = [0 \quad 1 \quad 0]$$

$$CA^2 = [CA][A]$$

$$= [0 \quad 1 \quad 0] \begin{bmatrix} 0 & 1 & 0 \\ 0 & 0 & 1 \\ 0 & 0 & 0 \end{bmatrix} = [0 \quad 0 \quad 1]$$

Therefore, $$Q_0 = \begin{bmatrix} C \\ L \\ CA \\ L \\ CA^2 \end{bmatrix} = \begin{bmatrix} 1 & 0 & 0 \\ 0 & 1 & 0 \\ 0 & 0 & 1 \end{bmatrix}$$

The determinant of Q_c is

$$\begin{vmatrix} 0 & 1 & 0 \\ 0 & 0 & 1 \\ 0 & 0 & 0 \end{vmatrix} = 1$$

Since the determinant of Q_c is non-zero, the rank of Q_c is 3 (order of A). Hence, the system is completely observable.

EXERCISE

Fill in the Blanks

1. The system mode described by the state equation:

$$X = \begin{bmatrix} 0 & 1 \\ 2 & -3 \end{bmatrix} x + \begin{bmatrix} 0 \\ 1 \end{bmatrix} u,\ y = [1\ 1x] \text{ is } \_\_\_\_\_\_\_\_\_\_.$$

Ans: controllable and observable

2. The transfer function $Y(s)/U(s)$ of a system described by the state equations

$\dot{x}_1(t) = -2x(t) + 2u(t)\ y(t) = 0.5\ x(t)$ is __________.

Ans: $\dfrac{1}{(s + 2)}$

3. $\dot{X} = Ax + Bu$ and $\dot{W} = CW + Du$. The eigenvalues of the representations are computed as $[\lambda]$ and $[\mu]$. Which of the following statements is true?
   (a) $[\lambda] = [\mu]$ and $X = W$
   (b) $[\lambda] = [\mu]$ and $X \neq W$
   (c) $[\lambda] \neq [\mu]$ and $X = W$
   (d) $[\lambda] \neq [\mu]$ and $X \neq W$

Ans: $[\lambda] = [\mu]$ and $X = W$, option (a)

4 Consider the system $\dfrac{dx}{dt} = Ax + Bu$ with $A = \begin{bmatrix} 1 & 0 \\ 0 & 1 \end{bmatrix}$ and $B = \begin{bmatrix} p \\ q \end{bmatrix}$ where p and q are arbitrary real numbers. Which of the following statements about the controllability of the system is true?
   (a) The system is completely state controllable for any non zero values of p and q
   (b) only $p = 0$ and $q = 0$ result in controllability
   (c) The system is uncontrollable for all values of p and q
   (d) We cannot conclude about controllability from the given data

Ans: The system is uncontrollable for all values of p and q, option (c)

5. A linear second order single input continuous time system is described by the following set of differential equations:

$$\dot{x}_1(t) = -2x_1(t) + 4x_2(t)$$
$$\dot{x}_2(t) = 2x_1(t) - x_2(t) + u(t)$$

where $x_1(t)$ and $x_2(t)$ are the state variables and $u(t)$ is the control variable. The system is __________.

Ans: controllable but unstable

6. For the system described by the state equation $\dot{x} = \begin{bmatrix} 0 & 1 & 0 \\ 0 & 0 & 0 \\ 0.5 & 1 & 1 \end{bmatrix} x + \begin{bmatrix} 0 \\ 0 \\ 1 \end{bmatrix} u$ if the control signal u is given by $u(-0.5 - 3 - 5)x + v$ then the eigenvalues of the closed-loop system will be __________.

Ans: $0, -1, -2$

7. Given $A = \begin{bmatrix} 1 & 0 \\ 0 & 1 \end{bmatrix}$, the state transition matrix e^{At} is given by __________.

Ans: $\begin{bmatrix} e^t & 0 \\ 0 & e^t \end{bmatrix}$

8. A linear system is described by the following state equation $X(t) = Ax(t) + Bu(t), A = \begin{bmatrix} 0 & 1 \\ -1 & 0 \end{bmatrix}$. The state transition matrix of the system is __________.

Ans: $\begin{bmatrix} \cos t & \sin t \\ -\sin t & \cos t \end{bmatrix}$

Unsolved Questions

9. Define the following terms with respect to phase variable approach. State, state variables, state space, state vector, state model.
10. State the advantages of state space approach for analysis and designing of control systems.
11. What is meant by decomposition of a transfer function? Name the methods of decomposition. State their advantages.
12. What is meant by phase variables? What is the advantages and disadvantages of phase variables.
13. What is meant by state transition matrix (STM)? What is the significance of STM? List the properties of the STM.
14. Derive an expression for the solution of a homogeneous state equation.
15. Obtain the transfer function of a system from its state model
16. Obtain the state model of the mechanical system shown in Fig. 10.25.

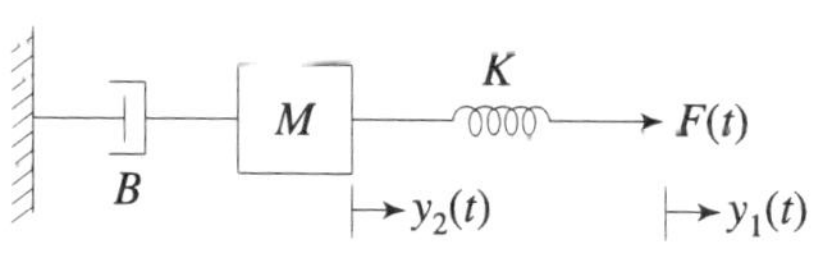

Fig. 10.25

Ans: $y = [1 \;\; 0]\begin{bmatrix} x_1 \\ x_2 \end{bmatrix} + \frac{1}{K} F(t)$

$$\begin{bmatrix} \&_1 \\ \& \end{bmatrix} \begin{bmatrix} 0 & 1 \\ 0 & \frac{-B}{M} \end{bmatrix} \begin{bmatrix} x_1 \\ x_2 \end{bmatrix} + \begin{bmatrix} 0 \\ \frac{1}{M} \end{bmatrix} F(t)$$

17. Obtain the state space representation of the mechanical system shown in Fig. 10.26.

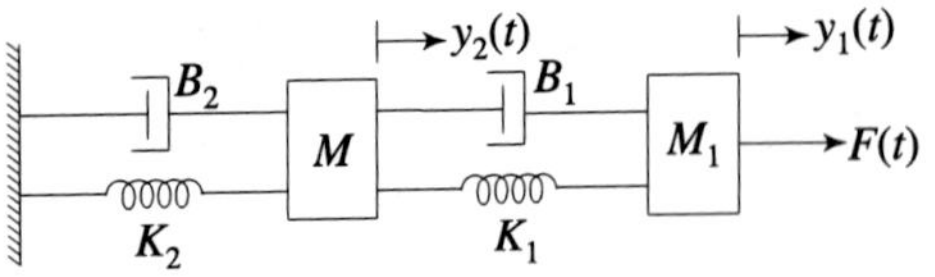

Fig. 10.26

Ans:
$$\begin{bmatrix} \dot{x}_1 \\ \dot{x}_2 \\ \dot{x}_3 \\ \dot{x}_4 \end{bmatrix} = \begin{bmatrix} 0 & 1 & 0 & 0 \\ -\dfrac{K_1}{M_1} & -\dfrac{B_1}{M_1} & \dfrac{K_1}{M_1} & \dfrac{B_1}{M_1} \\ 0 & 0 & 0 & 1 \\ \dfrac{K_1}{M_2} & \dfrac{B_1}{M_2} & \dfrac{-(K_1+K_2)}{M_2} & \dfrac{-(B_4+B_2)}{M_2} \end{bmatrix} \begin{bmatrix} x_1 \\ x_2 \\ x_3 \\ x_4 \end{bmatrix} \begin{bmatrix} 0 \\ \dfrac{1}{M_1} \\ 0 \\ 0 \end{bmatrix} u(t)$$

$$\begin{bmatrix} y_1 \\ y_2 \end{bmatrix} = \begin{bmatrix} 1 & 0 & 0 & 0 \\ 0 & 0 & 1 & 0 \end{bmatrix} \begin{bmatrix} x_1 \\ x_2 \\ x_3 \\ x_4 \end{bmatrix}$$

18. Obtain the phase variable state model for the system described by the differential equation

$$\frac{d^3y}{dt^3} + 2\frac{d^2y}{dt^2} + 3\frac{dy}{dt} + 4y = 5u(t)$$

Ans:
$$\begin{bmatrix} \dot{x}_1 \\ \dot{x}_2 \\ \dot{x}_3 \end{bmatrix} = \begin{bmatrix} 0 & 1 & 0 \\ 0 & 0 & 1 \\ -4 & -3 & -2 \end{bmatrix} \begin{bmatrix} x_1 \\ x_2 \\ x_3 \end{bmatrix} + \begin{bmatrix} 0 \\ 0 \\ 5 \end{bmatrix} u(t)$$

$$y = [1 \;\; 0 \;\; 0] \begin{bmatrix} x_1 \\ x_2 \\ x_3 \end{bmatrix}$$

19. Obtain the state model of the differential equation

$$4\frac{d^3y(t)}{dt^3} + 3\frac{d^2y(t)}{dt^2} + \frac{dy(t)}{dt} + 2y(t) = 5u(t)$$

Ans:
$$\begin{bmatrix} \dot{x}_1 \\ \dot{x}_2 \\ \dot{x}_3 \end{bmatrix} = \begin{bmatrix} 0 & 1 & 0 \\ 0 & 0 & 1 \\ -\dfrac{2}{4} & -\dfrac{1}{4} & -\dfrac{3}{4} \end{bmatrix} \begin{bmatrix} x_1 \\ x_2 \\ x_3 \end{bmatrix} + \begin{bmatrix} 0 \\ 0 \\ \dfrac{5}{4} \end{bmatrix} u(t)$$

$$y = [1 \;\; 0 \;\; 0] \begin{bmatrix} x_1 \\ x_2 \\ x_3 \end{bmatrix}$$

20. Draw the state block diagram for the transfer function $\dfrac{y(s)}{u(s)} = \dfrac{1}{s^2 + 5s + 6}$ and obtain the state model.

Ans:

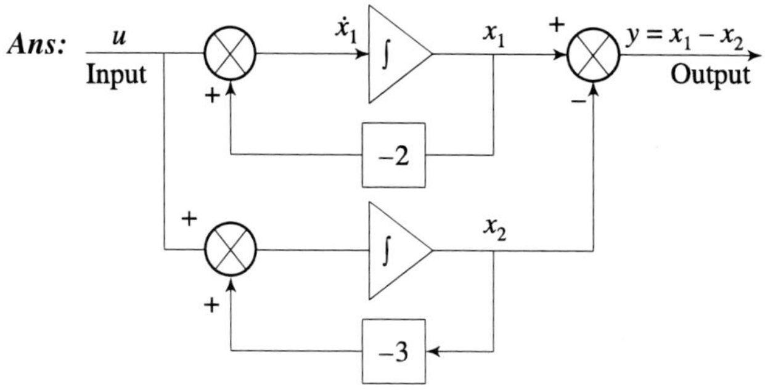

Fig. 10.27

$$\begin{bmatrix} \dot{x}_1 \\ \dot{x}_2 \end{bmatrix} = \begin{bmatrix} -2 & 0 \\ 0 & -3 \end{bmatrix} \begin{bmatrix} x_1 \\ x_2 \end{bmatrix} + \begin{bmatrix} 1 \\ 1 \end{bmatrix} u$$

$$y = [1 \ \ -1] \begin{bmatrix} x_1 \\ x_2 \end{bmatrix}$$

21. Obtain the state model of the system described by the transfer function
$$\frac{y(s)}{u(s)} = \frac{5}{s^3 + 6s + 7}.$$

Ans:
$$\begin{bmatrix} \dot{x}_1 \\ \dot{x}_2 \\ \dot{x}_3 \end{bmatrix} = \begin{bmatrix} 0 & 1 & 0 \\ 0 & 0 & 1 \\ -7 & -6 & 0 \end{bmatrix} \begin{bmatrix} x_1 \\ x_2 \\ x_3 \end{bmatrix} + \begin{bmatrix} 0 \\ 0 \\ 5 \end{bmatrix} u$$

$$y = [1 \ \ 0 \ \ 0] \begin{bmatrix} x_1 \\ x_2 \\ x_3 \end{bmatrix}$$

22. A system is described by transfer function
$$\frac{y(s)}{u(s)} = G(s) = \frac{20(10s + 1)}{s^3 + 3s^2 + 2s + 1}$$
Obtain a state model of the system

Ans:
$$\begin{bmatrix} \dot{x}_1 \\ \dot{x}_2 \\ \dot{x}_3 \end{bmatrix} = \begin{bmatrix} 0 & 1 & 0 \\ 0 & 0 & 1 \\ -1 & -2 & -3 \end{bmatrix} \begin{bmatrix} x_1 \\ x_2 \\ x_3 \end{bmatrix} + \begin{bmatrix} 0 \\ 0 \\ 20 \end{bmatrix} u$$

$$y = [1 \ \ 10 \ \ 0] \begin{bmatrix} x_1 \\ x_2 \\ x_3 \end{bmatrix}$$

23. Determine the STM for the scale model whose matrix A is given by

$$A = \begin{bmatrix} 0 & 1 \\ -6 & -5 \end{bmatrix}$$

Ans: $\phi(t) = \begin{bmatrix} 3e^{-2t} - 2e^{-3t} & e^{-2t} - e^{-3t} \\ -6(e^{-2t} - e^{-3t}) & -2e^{-2t} + 3e^{-3t} \end{bmatrix}$

24. Determine the STM for the state model whose matrix A is given by

$$A = \begin{bmatrix} -3 & 1 \\ -2 & 0 \end{bmatrix}$$

Ans: $\phi(t) = \begin{bmatrix} -e^{-t} + 2e^{-2t} & e^{-t} - e^{-2t} \\ -2e^{-t} + 2e^{-2t} & 2e^{-t} - e^{-2t} \end{bmatrix}$

25. List the properties of the state transition matrix

26. Consider the state equation

$$\begin{bmatrix} \dot{x}_1 \\ \dot{x}_2 \end{bmatrix} = \begin{bmatrix} 0 & 1 \\ -2 & -3 \end{bmatrix} \begin{bmatrix} x_1 \\ x_2 \end{bmatrix} + \begin{bmatrix} 0 \\ 1 \end{bmatrix} u(t)$$

Determine the state equation matrix $\phi(t)$ and the state vector $x(t)$ for $t \geq 0$ when the input is $u(t) = 1$ for $t > 0$

Ans: $\phi(t) = \begin{bmatrix} (2e^{-t} + e^{-2t}) & (e^{-t} - e^{-2t}) \\ (-2e^{-t} + 2e^{-2t}) & (-e^{-t} + 2e^{-2t}) \end{bmatrix}$

$$x(t) = \begin{bmatrix} (2e^{-t} - e^{-2t}) & (e^{-t} - e^{-2t}) \\ (-2e^{-t} + 2e^{-2t}) & (e^{-t} + 2e^{-2t}) \end{bmatrix} x(0) + \begin{bmatrix} 0.5 - e^{-t} + 0.5e^{-2t} \\ e^{-t} - e^{-2t} \end{bmatrix}, \ t \geq 0$$

27. A linear time invariant system is characterized by

$$\begin{bmatrix} \dot{x}_1 \\ \dot{x}_2 \end{bmatrix} = \begin{bmatrix} 1 & 4 \\ -2 & -5 \end{bmatrix} \begin{bmatrix} x_1 \\ x_2 \end{bmatrix} + \begin{bmatrix} 0 \\ 1 \end{bmatrix} u$$

$$y = [1 \ \ 0] \begin{bmatrix} x_1 \\ x_2 \end{bmatrix}$$

where u is a unit step function. The initial condition are $x_1(0) = 1$, $x_2(0) = 0$. Obtain

(i) zero input response (ii) zero state response

(iii) total response (iv) output response.

Ans: (i) $\begin{bmatrix} x_1 \\ x_2 \end{bmatrix} = \begin{bmatrix} 2e^{-t} & -e^{-3t} \\ -e^{-t} & 3e^{-3t} \end{bmatrix}$ (ii) $\begin{bmatrix} x_1 \\ x_2 \end{bmatrix} = \begin{bmatrix} \left(\frac{4}{3} - 2e^{-t}\right) & \left(+\frac{2}{3} - e^{-3t}\right) \\ \left(-\frac{1}{3} + e^{-t}\right) & \left(-\frac{2}{3} e^{-3t}\right) \end{bmatrix}$

(iii) $\begin{bmatrix} x_1 \\ x_2 \end{bmatrix} = \begin{bmatrix} \frac{4}{3} & -\frac{1}{3}e^{-3t} \\ -\frac{1}{3} & -\frac{1}{3}e^{-3t} \end{bmatrix}$ (iv) $y = [1 \;\; 0] \begin{bmatrix} x_1 \\ x_2 \end{bmatrix} = x_1 = \frac{4}{3} - \frac{1}{3}e^{-3t}$

28. The state space representation of a system is described by $\dot{x}_1 = -x_1 + u$

$$\dot{x}_2 = x_1 - 2x_2 + u$$

State whether the system is controllable or not

Ans: Not completely controllable?

29. A system is described by

$$\begin{bmatrix} \dot{x}_1 \\ \dot{x}_2 \end{bmatrix} = \begin{bmatrix} 1 & -1 \\ 1 & -1 \end{bmatrix} \begin{bmatrix} x_1 \\ x_2 \end{bmatrix} + \begin{bmatrix} 0 \\ 1 \end{bmatrix} u$$

$y = [1 \;\; 0] \begin{bmatrix} x_1 \\ x_2 \end{bmatrix}$ check the controllability and observability of the system.

Ans: Completely controllable and completely observation

30. Comment on the controllability of the system described by the state equation

$$\begin{bmatrix} \dot{x}_1 \\ \dot{x}_2 \end{bmatrix} = \begin{bmatrix} 3 & 0 \\ 2 & 4 \end{bmatrix} \begin{bmatrix} x_1 \\ x_2 \end{bmatrix} + \begin{bmatrix} 0 \\ 1 \end{bmatrix} u$$

Ans: Not controllable

31. Comment on the complete state observability of the system described by the state equation

$$\dot{X} = \begin{bmatrix} 0 & 1 & 0 \\ 0 & 0 & 1 \\ -6 & -11 & -6 \end{bmatrix} X$$

$$\dot{Y} = [1 \;\; -1 \;\; 1]X$$

Ans: Observable

32. Comment on the complete state observability of the system described by

$$X = \begin{bmatrix} 1 & 1 & 2 \\ 0 & 2 & 1 \\ 0 & 0 & 2 \end{bmatrix} X$$

$$Y = [1 \;\; -1 \;\; 0]\, X$$

Ans: Not observable.

Index